AF333698

# Sediment Provenance Studies in Hydrocarbon Exploration and Production

The Geological Society of London
**Books Editorial Committee**

**Chief Editor**
RICK LAW (USA)

**Society Books Editors**
JIM GRIFFITHS (UK)
DAVE HODGSON (UK)
HOWARD JOHNSON (UK)
PHIL LEAT (UK)
DANIELA SCHMIDT (UK)
RANDELL STEPHENSON (UK)
ROB STRACHAN (UK)
MARK WHITEMAN (UK)

**Society Books Advisors**
GHULAM BHAT (India)
MARIE-FRANÇOISE BRUNET (France)
MAARTEN DE WIT (South Africa)
JAMES GOFF (Australia)
MARIO PARISE (Italy)
SATISH-KUMAR (Japan)
MARCO VECOLI (Saudi Arabia)
GONZALO VEIGA (Argentina)

## Geological Society books refereeing procedures

The Society makes every effort to ensure that the scientific and production quality of its books matches that of its journals. Since 1997, all book proposals have been refereed by specialist reviewers as well as by the Society's Books Editorial Committee. If the referees identify weaknesses in the proposal, these must be addressed before the proposal is accepted.

Once the book is accepted, the Society Book Editors ensure that the volume editors follow strict guidelines on refereeing and quality control. We insist that individual papers can only be accepted after satisfactory review by two independent referees. The questions on the review forms are similar to those for *Journal of the Geological Society*. The referees' forms and comments must be available to the Society's Book Editors on request.

Although many of the books result from meetings, the editors are expected to commission papers that were not presented at the meeting to ensure that the book provides a balanced coverage of the subject. Being accepted for presentation at the meeting does not guarantee inclusion in the book.

More information about submitting a proposal and producing a book for the Society can be found on its website: www.geolsoc.org.uk.

It is recommended that reference to all or part of this book should be made in one of the following ways:

SCOTT, R. A., SMYTH, H. R., MORTON, A. C. & RICHARDSON, N. (eds) 2014. *Sediment Provenance Studies in Hydrocarbon Exploration and Production*. Geological Society, London, Special Publications, **386**.

ANDÒ, S., MORTON, A. & GARZANTI, E. 2014. Metamorphic grade of source rocks revealed by chemical fingerprints of detrital amphibole and garnet. *In:* SCOTT, R. A., SMYTH, H. R., MORTON, A. C. & RICHARDSON, N. (eds) *Sediment Provenance Studies in Hydrocarbon Exploration and Production*. Geological Society, London, Special Publications, **386**, 351–371. First published online June 20, 2013, http://dx.doi.org/10.1144/SP386.5

GEOLOGICAL SOCIETY SPECIAL PUBLICATION NO. 386

# Sediment Provenance Studies in Hydrocarbon Exploration and Production

EDITED BY

R. A. SCOTT
CASP, UK

H. R. SMYTH
HRS GeoLogic Ltd, UK

A. C. MORTON
HM Research Associates, UK

and

N. RICHARDSON
Dana Petroleum plc, UK

2014
Published by
The Geological Society
London

THE GEOLOGICAL SOCIETY

The Geological Society of London (GSL) was founded in 1807. It is the oldest national geological society in the world and the largest in Europe. It was incorporated under Royal Charter in 1825 and is Registered Charity 210161.

The Society is the UK national learned and professional society for geology with a worldwide Fellowship (FGS) of over 10 000. The Society has the power to confer Chartered status on suitably qualified Fellows, and about 2000 of the Fellowship carry the title (CGeol). Chartered Geologists may also obtain the equivalent European title, European Geologist (EurGeol). One fifth of the Society's fellowship resides outside the UK. To find out more about the Society, log on to www.geolsoc.org.uk.

**The Geological Society Publishing House** (Bath, UK) produces the Society's international journals and books, and acts as European distributor for selected publications of the American Association of Petroleum Geologists (AAPG), the Indonesian Petroleum Association (IPA), the Geological Society of America (GSA), the Society for Sedimentary Geology (SEPM) and the Geologists' Association (GA). Joint marketing agreements ensure that GSL Fellows may purchase these societies' publications at a discount. The Society's online bookshop (accessible from www.geolsoc. org.uk) offers secure book purchasing with your credit or debit card.

To find out about joining the Society and benefiting from substantial discounts on publications of GSL and other societies worldwide, consult www.geolsoc.org.uk, or contact the Fellowship Department at: The Geological Society, Burlington House, Piccadilly, London W1J 0BG: Tel. +44 (0)20 7434 9944; Fax +44 (0)20 7439 8975; E-mail: enquiries@geolsoc.org.uk.

For information about the Society's meetings, consult *Events* on www.geolsoc.org.uk. To find out more about the Society's Corporate Affiliates Scheme, write to enquiries@geolsoc.org.uk.

Published by The Geological Society from:
The Geological Society Publishing House, Unit 7, Brassmill Enterprise Centre, Brassmill Lane, Bath BA1 3JN, UK

The Lyell Collection: www.lyellcollection.org
Online bookshop: www.geolsoc.org.uk/bookshop
Orders: Tel. +44 (0)1225 445046, Fax +44 (0)1225 442836

The publishers make no representation, express or implied, with regard to the accuracy of the information contained in this book and cannot accept any legal responsibility for any errors or omissions that may be made.

© The Geological Society of London 2014. No reproduction, copy or transmission of all or part of this publication may be made without the prior written permission of the publisher. In the UK, users may clear copying permissions and make payment to The Copyright Licensing Agency Ltd, Saffron House, 6–10 Kirby Street, London EC1N 8TS UK, and in the USA to the Copyright Clearance Center, 222 Rosewood Drive, Danvers, MA 01923, USA. Other countries may have a local reproduction rights agency for such payments. Full information on the Society's permissions policy can be found at: www.geolsoc.org.uk/permissions

**British Library Cataloguing in Publication Data**

A catalogue record for this book is available from the British Library.
ISBN 978-1-86239-370-7
ISSN 0305-8719

**Distributors**
For details of international agents and distributors see:
www.geolsoc.org.uk/agentsdistributors

Typeset by Techset Composition India (P) Ltd, Bangalore and Chennai, India
Printed by Berforts Information Press Ltd, Oxford, UK

**This book is dedicated to the memory of Maria Mange (1935–2011) and Robert A. Scott (1960–2012)**

Maria Mange (photograph courtesy of Inga Sevastjanova)

Maria was a profoundly knowledgeable and skilled mineralogist and petrographer. As a testament to the value of her work it is a certainty that current and future generations of geologists working in the field of heavy mineral provenance studies will refer to and learn from her work.

Robert A. Scott (photograph courtesy of Xiaojing Zhang)

Robert, as part of his role on the Petroleum Group Committee of the Geological Society, proposed and co-convened the conference leading to this Special Publication. Robert was a passionate geologist who devoted more than 20 years to the study of Arctic geology through fieldwork, sediment provenance and plate reconstructions.

# Contents

## Looking forward: development of techniques and data handling

# Sediment provenance studies in hydrocarbon exploration and production: an introduction

HELEN R. SMYTH[1,2]*, ANDREW MORTON[2,3], NICHOLAS RICHARDSON[4] & ROBERT A. SCOTT[2†]

[1]*HRS GeoLogic Ltd, 1 Windmill Place, East Challow OX12 9RT, UK*

[2]*CASP, 181a Huntingdon Road, Cambridge CB3 0DH, UK*

[3]*HM Research Associates, 2 Clive Road, Balsall Common, West Midlands CV7 7DW, UK*

[4]*Dana Petroleum plc, King's Close, 62 Huntly Street, Aberdeen AB10 1RS, UK*

**Corresponding author (e-mail: hrsgeologic@gmail.com)*

**Abstract:** Sediment provenance studies concern the origin, composition, transportation and deposition of detritus, and are therefore an important part of understanding the links between basinal sedimentation, and hinterland tectonics and unroofing. Such studies can add value at many stages of hydrocarbon exploitation, from identifying regional-scale crustal affinities and sediment-dispersal patterns during the earliest stages of exploration to detailed correlation in producing reservoirs and understanding the impact of mineralogy on reservoir diagenesis.

This Special Publication records 20 of the papers given at the conference titled 'Sediment Provenance Studies in Hydrocarbon Exploration and Production' organized by the Petroleum Group of the Geological Society of London, and held in London from 5 to 7 December 2011. The observations drawn in this introductory section reflect the volume editors' experience, presentations at the conference and papers within this volume.

## Dedication

This book is dedicated to the memory of Maria Mange (1935–2011) and Robert Scott (1960–2012).

Maria Mange was a profoundly knowledgeable and skilled mineralogist and petrographer; one could not show her a mineral that she could not identify. In 2007, she and David Wright co-edited a monumental work, *Heavy Minerals in Use* (Mange & Wright 2007). As a testament to the value of her work, it is a certainty that current and future generations of geologists working in the field of heavy mineral provenance studies will refer to and learn from her work. Maria's death leaves a deep void in the quantitative study of heavy minerals. (https://www.geolsoc.org.uk/Geoscientist/Archive/May-2011/Maria-Mange-Rajetsky-1935–2011.)

The cover to this volume is a view looking north from south of Roonagh Quay, west of Louisburgh, County Mayo, Ireland. The rocks in the foreground are late Silurian fluviatile sandstones, part of the sedimentary fill of the Louisburgh Basin. The high ground on Clare Island, in the distance, lies near the northern margin of the basin. The sinistral–transtensional basin, some 12 km east–west and 4 km north–south, is part of the Ordovician–Silurian South Mayo Trough, an Ordovician forearc basin and Silurian arc basin, which was deformed by sinistral transpression in the late Silurian. Maria Mange conducted exhaustive and critical high-resolution heavy mineral analyses on this trough (Dewey & Mange 1999; Mange *et al.* 2003, 2010), in which she resolved many long-standing problems of provenance and correlation, and showed that detrital heavy minerals provide definitive solutions to such issues.

Robert Scott sadly did not see the completion of this Special Publication. Robert, as part of his role on the Petroleum Group Committee of the Geological Society, proposed and co-convened the conference leading to this Special Publication. Robert was a passionate geologist who devoted more than 20 years to the study of Arctic geology through fieldwork, sediment provenance and plate reconstructions. Robert's warm and engaging manner brought together academic and petroleum industry communities in the spirit of mutual collaboration. (https://www.geolsoc.org.uk/About/History/Obituaries-2001-onwards/Obituaries-2012/Robert-Andrew-Scott-1960-2012.)

## Sediment provenance studies in hydrocarbon exploration and production

This volume showcases the wide variety of techniques available, using examples and applications

---

†Deceased 26 September 2012

*From*: Scott, R. A., Smyth, H. R., Morton, A. C. & Richardson, N. (eds) 2014. *Sediment Provenance Studies in Hydrocarbon Exploration and Production*. Geological Society, London, Special Publications, **386**, 1–6.
First published online April 17, 2014, http://dx.doi.org/10.1144/SP386.21

from all aspects of sediment provenance research. The techniques utilized include petrography, heavy mineral analysis, mineral geochemistry, whole-rock geochemistry, geochronology and drainage capture analysis. The studies range from geographically diverse case studies that link source to sink (e.g. Bue & Andresen 2013; Lundmark *et al.* 2013; Morton *et al.* 2013) to discussions on how provenance controls diagenesis within reservoirs (Arribas *et al.* 2013; Caracciolo *et al.* 2013; Kettanah *et al.* 2013).

The integration of sediment provenance studies into geoscience workflows can provide critical insights at all stages of the exploration and production life cycle. From an exploration perspective, they can enhance the understanding of reservoir presence (distribution) and reservoir quality and, when applied to shales, evaluation of sealing capacities; this has a demonstrable impact on the exploration geoscientist's estimations of chance of success (risk) and resources at both the play and prospect level. Mineralogical and chemical techniques are most commonly applied as a tool for the establishment of reservoir zonation and stratigraphic architecture during static model development, thereby supporting development strategies at field level, and completion design at well level. However, these techniques also have important applications when considering dynamic reservoir behaviour during production; for example, when assessing the impact of permeability variations on compartmentalization, well deliverability/performance and decline rates resulting from spatial variability in diagenetic and facies effects related to sediment provenance. Provenance studies are therefore an important part of the industrial stratigrapher's toolkit and, in cases where alternative techniques yield ambiguous results, can ultimately have a major impact on project economics.

This volume is structured to present the papers into four sets around the following themes:

- Overview: applications of provenance information in hydrocarbon reservoir sandstones.
- Provenance, diagenesis and reservoir quality.
- Provenance studies linking sediment to source.
- Looking forward: development of techniques and data handling.

## Overview: applications of provenance information in hydrocarbon reservoir sandstones

The opening paper of this Special Publication is a review of sandstone mineral–chemical stratigraphy by **Hurst & Morton (2013)**. This paper considers the relationships between minerals, mineral chemistry and whole-rock composition, and discusses the important role that alluvial basins play in controlling mineral–chemical signatures. The application of mineral chemistry in the evaluation of sediment provenance and lithostratigraphic correlation is explored using examples from the North Sea region. Provenance data are widely used for correlation of hydrocarbon reservoir sandstones, especially in cases where conventional biostratigraphic resolution is poor. This topic is explored further in other volumes, notably those edited by Mange & Wright (2007) and Ratcliffe & Zaitlin (2010).

## Provenance, diagenesis and reservoir quality

Six of the papers focus on the influence that sediment provenance can have on reservoir quality and/or the relationship between sandstone composition and diagenetic evolution.

**Tobin & Schwarzer (2013)** model the impact of sediment provenance on reservoir quality preservation, drawing on the example of the deep-water turbidite sandstones reservoirs in the Gulf of Mexico. The modelling demonstrates the impact of mineralogy on both compaction and cementation. The authors predict that the windows for optimal reservoir quality preservation vary widely depending on the original character of the sediment.

A petrographic study exploring the relationship between sediment provenance, diagenesis and potential reservoir quality is presented by **Caracciolo *et al.* (2013)** for the Diligencia Basin, southern California. The traditional techniques of petrographic examination, point counting, and textural and compositional analysis by SEM (scanning electron microscope) reveal the composition and diagenetic evolution.

**Arribas *et al.* (2013)** also explore the relationship between sediment provenance and reservoir quality, using a case study of the Late Jurassic–Early Cretaceous succession of the Cameros Rift Basin, northern Spain. The authors demonstrate that of the two principal sandstone petrofacies in the region, quartzolithic and quartzofeldspathic arenites, it is the latter that have the greatest reservoir potential due to secondary porosity generation as a consequence of feldspar dissolution.

The **Kettanah *et al.* (2013)** study utilizes a multi-technique approach (petrography, heavy mineral analysis and microprobe data from garnet and tourmaline) to assess the reservoir characteristics of the Upper Triassic sandstones in the Bay of Fundy, Nova Scotia. The study revealed a significant intraformational provenance change within the Wolfville Formation, and the subsurface and surface sandstones are interpreted to have different sedimentary sources. The authors also highlight the value of the heavy mineral scheelite (calcium tungstate) as a provenance indicator in the Bay of Fundy.

The along-strike variation in sandstone composition derived from the uplift of the greater Western Caucasus, and supplied to the Black Sea, is documented in **Vincent et al. (2013)**. The authors predict that good quality, quartz-rich sandstone reservoirs, derived from the Russian Western Caucasus, are likely to be present in the Tuapse Trough and central Eastern Black Sea.

Detrital garnet data are utilized by **Kilhams et al. (2013)** to explore the provenance of the Paleocene–Eocene Sele Formation sandstone of the UK Central Graben. The garnet data have led to an interesting, and potentially important, reinterpretation of the sedimentary routing systems in the area and which may have a significant impact on exploration models.

## Provenance studies linking sediment to source

A set of 10 of the papers focuses on linkage of sedimentary rock and source region. The studies have a wide geographical diversity, from regions of frontier hydrocarbon exploration in the Arctic (Bue & Andresen 2013), North Africa (Meinhold et al. 2013) and SE Asia (Naing et al. 2013) to more established producing basins, such as the Norwegian Sea (Morton et al. 2013) and the North Sea (Lundmark et al. 2013).

Onshore–offshore provenance relationships between the Lofoten–Vesterålen area of northern Norway and the deep-water Vøring Basin are explored by **Morton et al. (2013)**. The paper demonstrates that sediment fed into the basin from Lofoten–Vesterålen was of minor importance, indicating that prospective hydrocarbon reservoir sandstones were mainly derived from either northern Norway or northern East Greenland.

A comprehensive example of source-to-sink analysis is provided by **Nicholson et al. (2013)**, who investigate the provenance of sediments along the Amur River and its delta in the Russian Far East. The implications of the dataset are wide reaching, as the authors demonstrate that caution must be taken when using stable mineral indices as regional correlation tools.

**Naing et al. (2013)** present a study of the Eocene–Miocene sandstones in the Indo-Burma ranges. The techniques used include framework grain, heavy mineral and detrital zircon analysis.

This study emphasizes the importance of sediment provenance studies in tectonic reconstructions.

In their study of the heavy mineral signatures of sediment from Trinidad and Barbados, **Vincent et al. (2014)** inform the exhumation history of the Andes. In this case study, the provenance history was determined by comparing heavy mineral assemblages from ancient sandstones to associations that are recognized in modern sands. The modern sands could reasonably be correlated with tectonic domains.

**Lundmark et al. (2013)** use a combination of U–Pb dating of zircon and rutile trace element data to investigate the provenance of late Devonian–early Permian sandstones of the Mid North Sea High. The data allow two sand types to be distinguished: one recycled from the Grampian High and Southern Uplands High–Mid North Sea High; and the other, derived from an Eastern Greenland source.

The value of U–Pb dating of detrital zircon data is demonstrated by the study of Mesozoic sandstones on Svalbard presented by **Bue & Andresen (2013)**. The data clearly demonstrate input from two distinct sedimentary sources, easterly Uralide- and westerly Greenland–Canada-derived detritus, therefore testing previous interpretations of regional sediment sources.

**Meinhold et al. (2013)** also use detrital zircons to explore sedimentary sources, in this case from the Devonian of Libya. The paper illustrates the added value of collecting different isotopic data on single minerals, in this case U–Pb, Lu–Hf and O from zircons.

In a break from the conventional U–Pb dating of zircon, **Pe-Piper et al. (2013)** explore the use of monazite geochronology as a sediment provenance tool. The study combines the monazite geochronological data with morphology, zoning, inclusions and major element geochemistry data to identify 20 discrete sedimentary sources.

In their paper **Amorosi & Sammartino (2013)** examine the value of bulk geochemistry in modern sediment of the Po Plain, Italy, and, in doing so, link heavy metal concentrations to the contribution from local ultramafic sources of sediment.

This section is concluded by **Bonne (2014)**, who describes a novel approach to sediment provenance analysis in the Niger River Delta. In a break from the traditional mineralogical techniques, this study relies on a multidisciplinary geographical information system (GIS)-based approach to study the drainage network and landscape evolution of the Niger River. These data are used to inform the palaeodrainage evolution of the Niger River, and so provide insights into drainage capture and, in turn, the character and reservoir potential of sediment derived from these source terranes since the Cretaceous.

## Looking forward: development of techniques and data handling

During the conference, several presentations focused on new developments or adaptations of existing techniques such as Qemscan, Raman spectroscopy, scanning electron microscopy–cathodoluminescence (SEM-CL), petrography or simply the way the age data from detrital zircons are plotted. These techniques build upon the seminal works of Maria Mange and others (e.g. Mange & Maurer 1992; Mange & Wright 2007) and advance the field of sediment provenance into the future. Three of the papers in this volume fall into this category.

**Andò *et al.* (2013)** present a paper that will be invaluable to those researching sedimentary successions derived from meta-igneous or metasedimentary source rocks. The paper explores the chemical composition of amphiboles and garnets derived from parent rocks of increasing metamorphic grade, using a case study of the Central Southern Alps. In addition, the article redefines a series of extremely valuable numerical indices that will help the end user assess the metamorphic grade of the sedimentary source rocks.

**Suggate & Hall (2013)** present an extensive garnet compositional database, compiled from published literature, and document a multistage methodology to enable the user to link garnet compositions to source rocks. The methodology is tested using a case study from the Neogene sandstone of northern Sabah, Borneo.

The final manuscript by **Andò & Garzanti (2013)** examines the potential of Raman spectroscopy. This is a fitting conclusion to the volume as the authors point to the value of this technique in future research in sediment provenance inasmuch as it is accurate, efficient and versatile.

## Conclusions

This volume demonstrates the value of sediment provenance studies in many areas of hydrocarbon exploration and production. The variety of techniques explored in the 20 papers show that insights can be gained from many different tools. What is apparent is that multi-technique, multi-facetted approaches are becoming much more prevalent, and that, in the future, sediment provenance studies will not rely simply on single mineral techniques, such as isotopic dating of zircon. Critically, it is the conveying of these data to the end user, the hydrocarbon geologist, who is often a non-specialist, that is becoming an important consideration to the community.

The authors look with interest to the future and hope that the new collaborations sparked during the conference will lead to the further advancement of sediment provenance studies.

We would like to thank Statoil and Maersk Oil for their generous sponsorship. The Petroleum Group, a specialist group of the Geological Society of London, supported this publication by providing funding towards colour figure production.

The editors of this Special Publication would like to extend our thanks to all of the manuscript reviewers and Publishing House staff for their efforts. Finally, we are grateful to every author who contributed to this publication and their patience for the delays resulting from the untimely death of Robert Scott.

## References

AMOROSI, A. & SAMMARTINO, I. 2013. Tracing provenance and pathways of late Holocene fluvio-deltaic sediments by heavy-metal spatial distribution (Po Plain–Northern Apennines system, Italy). *In*: SCOTT, R. A., SMYTH, H. R., MORTON, A. C. & RICHARDSON, N. (eds) *Sediment Provenance Studies in Hydrocarbon Exploration and Production*. Geological Society, London, Special Publications, **386**. First published online June 11, 2013, http://dx.doi.org/10.1144/SP386.6

ANDÒ, S. & GARZANTI, E. 2013. Raman spectroscopy in heavy-mineral studies. *In*: SCOTT, R. A., SMYTH, H. R., MORTON, A. C. & RICHARDSON, N. (eds) *Sediment Provenance Studies in Hydrocarbon Exploration and Production*. Geological Society, London, Special Publications, **386**. First published online June 20, 2013, http://dx.doi.org/10.1144/SP386.2

ANDÒ, S., MORTON, A. & GARZANTI, E. 2013. Metamorphic grade of source rocks revealed by chemical fingerprints of detrital amphibole and garnet. *In*: SCOTT, R. A., SMYTH, H. R., MORTON, A. C. & RICHARDSON, N. (eds) *Sediment Provenance Studies in Hydrocarbon Exploration and Production*. Geological Society, London, Special Publications, **386**. First published online June 20, 2013, http://dx.doi.org/10.1144/SP386.5

ARRIBAS, J., GONZÁLEZ-ACEBRÓN, L., OMODEO-SALÉ, S. & MAS, R. 2013. The influence of the provenance of arenite on its diagenesis in the Cameros Rift Basin (Spain). *In*: SCOTT, R. A., SMYTH, H. R., MORTON, A. C. & RICHARDSON, N. (eds) *Sediment Provenance Studies in Hydrocarbon Exploration and Production*. Geological Society, London, Special Publications, **386**. First published online June 10, 2013, http://dx.doi.org/10.1144/SP386.12

BONNE, K. P. M. 2014. Reconstruction of the evolution of the Niger River and implications for sediment supply to the equatorial Atlantic margin of Africa during the Cretaceous and Cenozoic. *In*: SCOTT, R. A., SMYTH, H. R., MORTON, A. C. & RICHARDSON, N. (eds) *Sediment Provenance Studies in Hydrocarbon Exploration and Production*. Geological Society, London, Special Publications, **386**. First published online April 9, 2014, http://dx.doi.org/10.1144/SP386.20

BUE, E. P. & ANDRESEN, A. 2013. Constraining depositional models in the Barents Sea region using detrital zircon U–Pb data from Mesozoic sediments in

Svalbard. *In*: SCOTT, R. A., SMYTH, H. R., MORTON, A. C. & RICHARDSON, N. (eds) *Sediment Provenance Studies in Hydrocarbon Exploration and Production*. Geological Society, London, Special Publications, **386**. First published online October 1, 2013, http://dx.doi.org/10.1144/SP386.14

CARACCIOLO, L., ARRIBAS, J., INGERSOLL, R. V. & CRITELLI, S. 2013. The diagenetic destruction of porosity in plutoniclastic petrofacies: the Miocene Diligencia and Eocene Maniobra formations, Orocopia Mountains, southern California, USA. *In*: SCOTT, R. A., SMYTH, H. R., MORTON, A. C. & RICHARDSON, N. (eds) *Sediment Provenance Studies in Hydrocarbon Exploration and Production*. Geological Society, London, Special Publications, **386**. First published online July 15, 2013, http://dx.doi.org/10.1144/SP386.9

DEWEY, J. F. & MANGE, M. A. 1999. Petrography of Ordovician and Silurian sediments in the western Irish Caledonides: tracers of a short-lived Ordovician continent–arc collision orogeny and the evolution of the Laurentian Appalachian–Caledonian margin. *In*: MAC NIOCAILL, C. & RYAN, P. D. (eds) *Continental Tectonics*. Geological Society, London, Special Publications, **164**, 55–107.

HURST, A. & MORTON, A. 2013. Provenance models. the role of sandstone mineral–chemical stratigraphy. *In*: SCOTT, R. A., SMYTH, H. R., MORTON, A. C. & RICHARDSON, N. (eds) *Sediment Provenance Studies in Hydrocarbon Exploration and Production*. Geological Society, London, Special Publications, **386**. First published online August 1, 2013, http://dx.doi.org/10.1144/SP386.11

KETTANAH, Y. A., KETTANAH, M. Y. & WACH, G. D. 2013. Provenance, diagenesis and reservoir quality of the Upper Triassic Wolfville Formation, Bay of Fundy, Nova Scotia, Canada. *In*: SCOTT, R. A., SMYTH, H. R., MORTON, A. C. & RICHARDSON, N. (eds) *Sediment Provenance Studies in Hydrocarbon Exploration and Production*. Geological Society, London, Special Publications, **386**. First published online October 24, 2013, http://dx.doi.org/10.1144/SP386.18

KILHAMS, B., MORTON, A., BORELLA, R., WILKINS, A. & HURST, A. 2013. Understanding the provenance and reservoir quality of the Sele Formation sandstones of the UK Central Graben utilizing detrital garnet suites. *In*: SCOTT, R. A., SMYTH, H. R., MORTON, A. C. & RICHARDSON, N. (eds) *Sediment Provenance Studies in Hydrocarbon Exploration and Production*. Geological Society, London, Special Publications, **386**. First published online October 3, 2013, http://dx.doi.org/10.1144/SP386.16

LUNDMARK, A. M., BUE, E. P., GABRIELSEN, R. H., FLAAT, K., STRAND, T. & OHM, S. E. 2013. Provenance of late Palaeozoic terrestrial sediments on the northern flank of the Mid North Sea High: detrital zircon geochronology and rutile geochemical constraints. *In*: SCOTT, R. A., SMYTH, H. R., MORTON, A. C. & RICHARDSON, N. (eds) *Sediment Provenance Studies in Hydrocarbon Exploration and Production*. Geological Society, London, Special Publications, **386**. First published online June 19, 2013, http://dx.doi.org/10.1144/SP386.4

MANGE, M. A. & MAURER, M. A. (eds) 1992. *Heavy Minerals in Colour*. Chapman & Hall, London.

MANGE, M. A. & WRIGHT, D. T. (eds) 2007. *Heavy Minerals in Use. Developments in Sedimentology*, **58**. Elsevier, Amsterdam.

MANGE, M. A., DEWEY, J. F. & WRIGHT, D. T. 2003. Heavy minerals solve structural and stratigraphic problems in Ordovician strata of the western Irish Caledonides. *Geological Magazine*, **140**, 25–30.

MANGE, M., IDLEMAN, B., YIN, Q.-Z., HIDAKA, H. & DEWEY, J. 2010. Detrital heavy minerals, white mica and zircon geochronology in the Ordovician South Mayo Trough, western Ireland: signatures of the Laurentian basement and the Grampian orogeny. *Journal of the Geological Society, London*, **167**, 1147–1160.

MEINHOLD, G., MORTON, A. C., FANNING, C. M., HOWARD, J. P., PHILLIPS, R. J., STROGEN, D. & WHITHAM, A. G. 2013. Insights into crust formation and recycling in North Africa from combined U–Pb, Lu–Hf and O isotope data of detrital zircons from Devonian sandstone of southern Libya. *In*: SCOTT, R. A., SMYTH, H. R., MORTON, A. C. & RICHARDSON, N. (eds) *Sediment Provenance Studies in Hydrocarbon Exploration and Production*. Geological Society, London, Special Publications, **386**. First published online June 11, 2013, http://dx.doi.org/10.1144/SP386.1

MORTON, A., FANNING, M. & BERRY, J. R. 2013. Heavy mineral provenance of the Mesozoic succession in Andøya, northern Norway: implications for sand transport in the Vøring Basin. *In*: SCOTT, R. A., SMYTH, H. R., MORTON, A. C. & RICHARDSON, N. (eds) *Sediment Provenance Studies in Hydrocarbon Exploration and Production*. Geological Society, London, Special Publications, **386**. First published online June 11, 2013, http://dx.doi.org/10.1144/SP386.3

NAING, T. T., BUSSIEN, D. A., WINKLER, W. H., NOLD, M. & VON QUADT, A. 2013. Provenance study on Eocene–Miocene sandstones of the Rakhine Coastal Belt, Indo-Burman Ranges of Myanmar: geodynamic implications. *In*: SCOTT, R. A., SMYTH, H. R., MORTON, A. C. & RICHARDSON, N. (eds) *Sediment Provenance Studies in Hydrocarbon Exploration and Production*. Geological Society, London, Special Publications, **386**. First published online July 22, 2013, http://dx.doi.org/10.1144/SP386.10

NICHOLSON, U., POYNTER, S., CLIFT, P. D. & MACDONALD, D. I. M. 2013. Tying catchment to basin in a giant sediment routing system: a source-to-sink study of the Neogene–Recent Amur River and its delta in the North Sakhalin Basin. *In*: SCOTT, R. A., SMYTH, H. R., MORTON, A. C. & RICHARDSON, N. (eds) *Sediment Provenance Studies in Hydrocarbon Exploration and Production*. Geological Society, London, Special Publications, **386**. First published online July 15, 2013, http://dx.doi.org/10.1144/SP386.7

PE-PIPER, G., PIPER, D. J. W. & TRIANTAFYLLIDIS, S. 2013. Detrital monazite geochronology, Upper Jurassic–Lower Cretaceous of the Scotian Basin: significance for tracking first-cycle sources. *In*: SCOTT, R. A., SMYTH, H. R., MORTON, A. C. & RICHARDSON, N. (eds) *Sediment Provenance Studies in Hydrocarbon Exploration and Production*. Geological Society,

London, Special Publications, **386**. First published online July 15, 2013, http://dx.doi.org/10.1144/SP386.13

RATCLIFFE, K. T. & ZAITLIN, B. A. (eds) 2010. *Application of Modern Stratigraphic Techniques: Theory and Case Histories*. SEPM Special Publications, Tulsa, **94**.

SUGGATE, S. M. & HALL, R. 2013. Using detrital garnet compositions to determine provenance: a new compositional database and procedure. *In*: SCOTT, R. A., SMYTH, H. R., MORTON, A. C. & RICHARDSON, N. (eds) *Sediment Provenance Studies in Hydrocarbon Exploration and Production*. Geological Society, London, Special Publications, **386**. First published online July 18, 2013, http://dx.doi.org/10.1144/SP386.8

TOBIN, R. C. & SCHWARZER, D. 2013. Effects of sandstone provenance on reservoir quality preservation in the deep subsurface: experimental modelling of deepwater sand in the Gulf of Mexico. *In*: SCOTT, R. A., SMYTH, H. R., MORTON, A. C. & RICHARDSON, N. (eds) *Sediment Provenance Studies in Hydrocarbon Exploration and Production*. Geological Society, London, Special Publications, **386**. First published

online September 12, 2013, http://dx.doi.org/10.1144/SP386.17

VINCENT, S. J., HYDEN, F. & BRAHAM, W. 2013. Along-strike variations in the composition of sandstones derived from the uplifting western Greater Caucasus: causes and implications for reservoir quality prediction in the Eastern Black Sea. *In*: SCOTT, R. A., SMYTH, H. R., MORTON, A. C. & RICHARDSON, N. (eds) *Sediment Provenance Studies in Hydrocarbon Exploration and Production*. Geological Society, London, Special Publications, **386**. First published online September 17, 2013, http://dx.doi.org/10.1144/SP386.15

VINCENT, H., WACH, G. & KETANNAH, Y. 2014. Heavy mineral record of Andean uplift and changing sediment sources across the NE margin of South America: a case study from Trinidad and Barbados. *In*: SCOTT, R. A., SMYTH, H. R., MORTON, A. C. & RICHARDSON, N. (eds) *Sediment Provenance Studies in Hydrocarbon Exploration and Production*. Geological Society, London, Special Publications, **386**. First published online February 6, 2014, http://dx.doi.org/10.1144/SP386.19

# Provenance models: the role of sandstone mineral–chemical stratigraphy

ANDREW HURST[1]* & ANDREW MORTON[2,3]

[1]*Department of Geology and Petroleum Geology, University of Aberdeen, King's College, Aberdeen AB24 3UE, UK*

[2]*HM Research Associates, 2 Clive Road, Balsall Common, CV7 7DW, UK*

[3]*CASP, University of Cambridge, Cambridge CB3 0DH, UK*

*Corresponding author (e-mail: ahurst@abdn.ac.uk)*

**Abstract:** Three distinct analytical approaches are embraced in mineral–chemical stratigraphy: mineralogy, whole-rock geochemistry and single-grain geochemical analysis. Mineralogical studies identify and quantify the clastic components of sandstone, even though any clast category may be geochemically diverse. Whole-rock geochemical studies (sometimes referred to as chemostratigraphy), by contrast, quantify the abundance of major and trace elements in sandstone, but provide no information on the distribution and location of the elements in minerals. These approaches are linked by single-grain geochemical analysis, which enables further characterization and subdivision of individual mineralogical components, and identifies sites where specific major and trace elements reside.

In this paper, we consider the relationships between minerals, mineral chemistry and whole-rock composition, before exploring the value of mineral–chemical stratigraphy for lithostratigraphic correlation and evaluation of sediment provenance, using published examples from the North Sea region, where the great majority of such studies have been undertaken. We conclude by discussing the important role that alluvial basins play in controlling mineral–chemical signatures.

## Mineral–chemical relationships in sandstones

The mineralogical components of sandstones are typically classified into four main groups, comprising framework grains (quartz, feldspar, rock fragments), heavy minerals, clay minerals, and authigenic phases. Of these, the framework components, the clay minerals, and the heavy minerals may all have some role to play in mineral–chemical stratigraphy. Although a wide variety of authigenic phases are identified in sandstones, their secondary nature precludes their use in reservoir correlation, except in unusual circumstances.

### Framework components

Despite their abundance, the framework components seldom provide a suitable basis for effective mineral–chemical stratigraphy. This is because changes in their relative proportions are rarely useful for defining lateral or vertical variations within sandstone units at a scale that corresponds to that of the sedimentary reservoir architecture. Quartz is usually the most abundant mineral in sandstone but, despite having characteristic cathodoluminescence properties (Boggs & Krinsley 2006) and internal texture (Basu *et al.* 1975), is generally not useful in mineral–chemical stratigraphy because of its perceived lack of chemical diversity. However, detailed analysis of size-fractionated quartz from sandstone reveals that inclusions, frequently of heavy minerals, can have significant influence on trace-element chemistry. This has particular relevance to Th concentration, since thoric micro-inclusions can dominate the Th budget in quartzose sandstone (Hurst & Milodowski 1996, 1998). Feldspar, although chemically diverse and of variable abundance, is sensitive to diagenetic dissolution and crystal growth (authigenesis) during weathering and burial. Hence, feldspar may have characteristics that reflect factors other than detrital variations. The rock fragment category includes mica and lithic clasts, both of which are mineralogically and geochemically diverse. Mica shows a range of mechanical and chemical stabilities, and its platy morphology makes it susceptible to hydraulic differentiation that distorts mineral provenance relationships. The rock frag ment category includes lithic clasts, which are common components of greywackes. Lithic clasts are commonly mechanically and chemically unstable and, when altered, are difficult to differentiate petrographically.

*From*: SCOTT, R. A., SMYTH, H. R., MORTON, A. C. & RICHARDSON, N. (eds) 2014. *Sediment Provenance Studies in Hydrocarbon Exploration and Production.* Geological Society, London, Special Publications, **386**, 7–26.
First published online August 1, 2013, http://dx.doi.org/10.1144/SP386.11

## Clay minerals

Detrital clay minerals are generally present in low abundances in sandstone, but show a variety of modes of occurrence that is in part caused by their physical and chemical instability relative to other grains present. They may be disseminated through the sandstone, but also occur as discrete clasts and as parts of lithic clasts. Although clay minerals have a wide variation in major and trace element composition, they are sometimes difficult to differentiate petrographically from, and are crystallographically similar to, authigenic clays. Complex micro-chemical and micro-textural relationships (Hurst 1999) and low abundance complicate their use in differentiating and correlating reservoir sandstones. Consequently, although clay mineralogy has been used to correlate within reservoir sandstone sequences (e.g. Jeans 1995; Pay *et al.* 1999), its application as a correlation tool is relatively limited.

## Heavy minerals

Detrital heavy minerals (HMs) are generally minor components of sandstones, rarely comprising more than a few percent by volume. Despite this, they comprise a diverse group with over 50 separate species, many diagnostic of specific provenance lithologies (Mange & Maurer 1992). Although their relative abundance is controlled both by provenance and by hydrodynamic, diagenetic and weathering processes, the effects of hydrodynamics, diagenesis and weathering are readily identifiable (Morton & Hallsworth 1999). Consequently, HM analysis is a proven and reliable method for correlating reservoir sandstone sequences. Morton & Hallsworth (1994) identify several HM ratios that are suitable for use in reservoir correlation (rutile:zircon, chrome spinel:zircon, monazite:zircon, garnet:zircon and apatite:tourmaline), since they are minimally affected by changes in hydraulic conditions and are stable under most burial diagenetic regimes.

Varietal HM data, which focus on the characteristics of a single mineral group, are useful, providing the mineral group is stable during burial diagenesis. Traditionally, varietal studies have focused on petrographic characteristics, such as colour, habit and zoning, which provide information on provenance and petrogenesis. Zircon (Poldervaart 1955; Lihou & Mange-Rajetzky 1996) and tourmaline (Krynine 1946; Mange-Rajetzky 1995) have proved especially useful in this context. The increasing availability of a variety of microbeam analytical methods has complemented and, in some cases, revolutionized varietal studies, since virtually every component of heavy mineral assemblages shows compositional variations in terms of either major or trace elements (Mange & Morton 2007).

## Heavy mineral–trace element relationships

The most reliable trace elements for mineral-chemical stratigraphic purposes are those that have low mobility, which include Ti, Zr, Nb, Ta, Hf, Yb, Cr and the rare-earth elements (REE) (Preston *et al.* 1998; Roser 2000). Although other trace elements may be useful, care is required when interpreting data from elements known to be mobile during burial diagenesis. Usefully, the low-mobility elements are preferentially hosted within heavy mineral phases. For example, zircon is the main host of Zr and Hf, rutile is the main host of Ti, Nb and Ta, and chrome spinel is the main host of Cr (Preston *et al.* 1998). It follows, therefore, that it may be possible to use some elemental ratios as proxies for provenance-sensitive heavy mineral ratios (Svendsen & Hartley 2002). For example, Cr:Zr and Ce:Zr are likely proxies for chrome spinel:zircon and monazite:zircon, respectively. Unfortunately the value of these relationships is undermined because they fail to take into account that elemental abundances derived from whole-rock determinations are partly controlled by hydraulic conditions as well as provenance. For example, the hydrodynamic control on elemental ratios is demonstrated by Zr abundances being consistently higher in sandstones compared with interbedded mudstones in the Triassic fluvio-lacustrine strata of the Beryl Field, North Sea (Fig. 1). Furthermore, Ti is also hosted by the diagenetic minerals anatase and brookite, and is also abundant in titanite, which is unstable during burial diagenesis.

The use of trace elements as proxies for mineral abundances is further complicated by the fact that virtually all mineral species have compositional ranges. For example, zircon ($ZrSiO_4$), one of the most persistent heavy minerals, is associated with radioelement (U and Th) enrichment, but zircon shows a very broad range of radioelement compositions (Table 1). Variations in zircon radioelement concentration are associated with textural variations, differences in resistance to physical and chemical processes, and the ultimate origin of the zircon. In particular, U and Th contents control the degree to which zircon becomes metamict, because the radiation dose zircons experience is a function of the U and Th content and the time since crystallization. Zircons with a low radiation dose at the time of sedimentary transport are among the most mechanically and chemically robust of detrital minerals, able to survive prolonged weathering and transport processes. By contrast, metamict zircons contain radiation-induced defects that render them liable to destruction both during

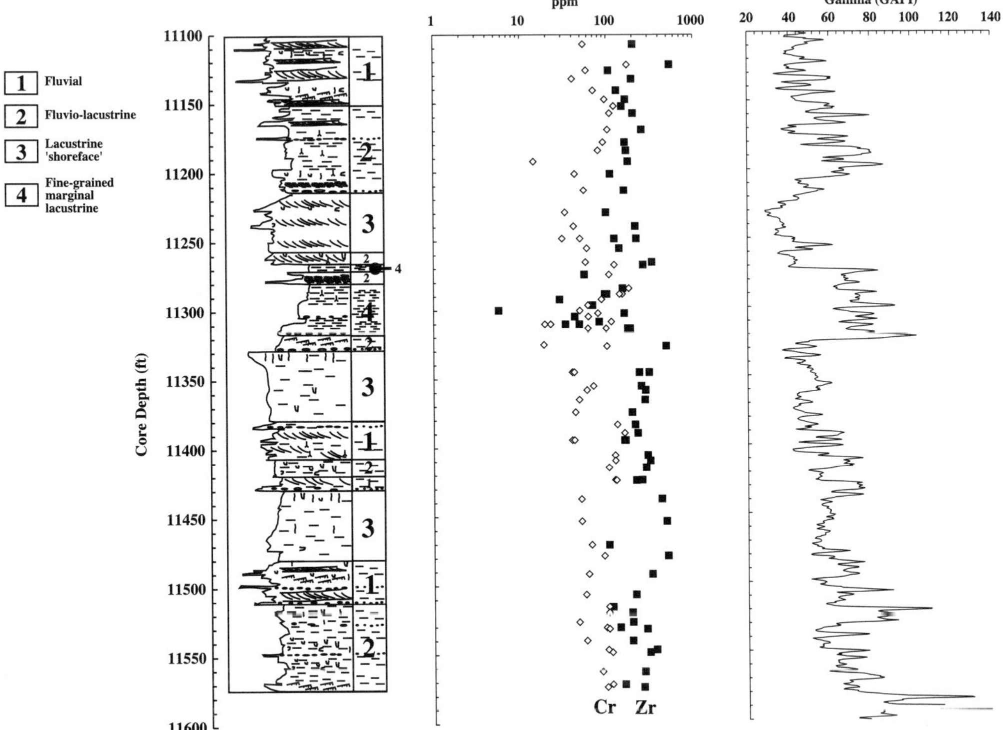


**Fig. 1.** Comparison of the Zr and Cr contents in interbedded sandstones and mudstones from the Lewis Formation (Triassic), well 9/13a-S48, North Sea (from Preston *et al.* 1998, fig. 12). The plot demonstrates that Zr is consistently higher in the sandstones than the associated mudstones, implying a gross hydraulic control on trace element abundance.

chemical weathering (Balan *et al.* 2001), and by mechanical abrasion during transportation. Mechanical degradation of metamict zircons tends to concentrate the most radioactive zircon in the finer grain-size fractions.

U and Th content in sedimentary zircons is thus a potential measure of sediment transport history and maturity. This is demonstrated by the comparison of U content and crystallization age of zircons between two Upper Carboniferous sandstones of contrasting provenance in the Pennine Basin, UK

(Fig. 2). The Halesowen Formation (Asturian) was sourced from the Variscan belt to the south. It is dominated by zircons of Variscan (323–357 Ma) and Cadomian (500–680 Ma) origin, but also contains a number of Proterozoic grains (Hallsworth *et al.* 2000). Some of the Variscan and Cadomian zircons have high U contents, whereas the earlier Proterozoic zircons are uniformly U-poor. This suggests that the Variscan and Cadomian grains reflect direct erosion of basement, but that the earlier Proterozoic zircons had a longer sedimentary

**Table 1.** *Radioelement (Th and U) concentrations in zircon, all values are in ppm*

| | DHZ | | FTR | | LAMP–ICP–MS | | SHRIMP | |
|---|---|---|---|---|---|---|---|---|
| | Mean value | Range of values | Mean value | Range of values | Mean value | Range of values | Mean value | Range of values |
| Th | 243 | 9–466 | | | 2062 | 47–24 597 | 206 | 0–4119 |
| U | 1431 | 23–4103 | 813 | 84–5000 | 243 | 131–12 038 | 457 | 3–13 696 |

DHZ are from Deer *et al.* (1997). FTR (fission track recognition) and LAMP–ICP–MS (laser-ablation microprobe–inductively coupled plasma–mass spectrometry) data are from Milodowski & Hurst (unpublished data). SHRIMP (sensitive high-resolution ion microprobe) data are from Carboniferous sandstones of the UK Pennine Basin and North Sea (Hallsworth *et al.* 2000; Morton *et al.* 2001).

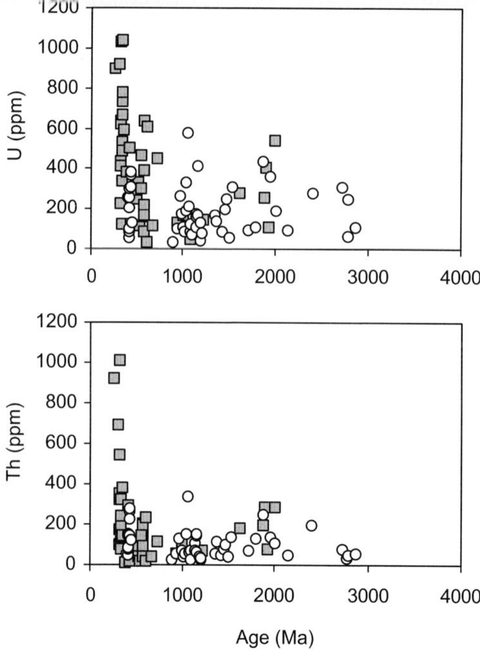

Fig. 2. Variation in U and Th contents and ages of zircons in the Clifton Rock (Langsettian) and Halesowen Formation (Asturian), Pennine Basin, UK. Data from Hallsworth *et al.* (2000).

history, and were probably recycled. By contrast, zircons in the westerly-derived Clifton Rock (Langsettian) are uniformly low in U (<400 ppm), despite showing a wide range in age from Early Palaeozoic (400–450 Ma), Proterozoic (mostly between 900–1200 Ma but extending as far back as 2150 Ma) and Archaean (2700–2900 Ma). This suggests elimination of metamict zircons as young as Devonian and suggests prolonged post-Caledonian sedimentary transport and/or recycling.

Most trace-element concentrations cannot be related directly to specific minerals. An example of the lack of simple relationships between trace elements and minerals is exemplified by U concentration in zircons and minor concentrations of U that are associated detrital ilmenite grains that are partly replaced by authigenic anatase (Fig. 3). Although U concentration is expected to be associated with zircon, fission-track recognition data show that some zircons are highly U-enriched whereas others are not, and that U is also associated with other minerals such as authigenic anatase. Thorium (Th), like U, is easily measured, including in boreholes with the almost ubiquitous gamma log, and is an attractive target for establishing mineral–chemical relationships. However, unless precise mineralogical data are available, spurious correlations may arise. In sandstones, a tiny concentration of the radioelement-enriched mineral monazite (a Th-bearing REE phosphate) may dominate the Th budget of sandstone, with 0.01% by volume

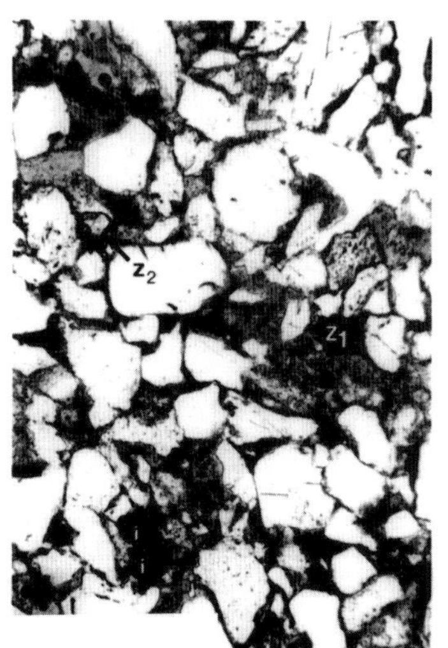


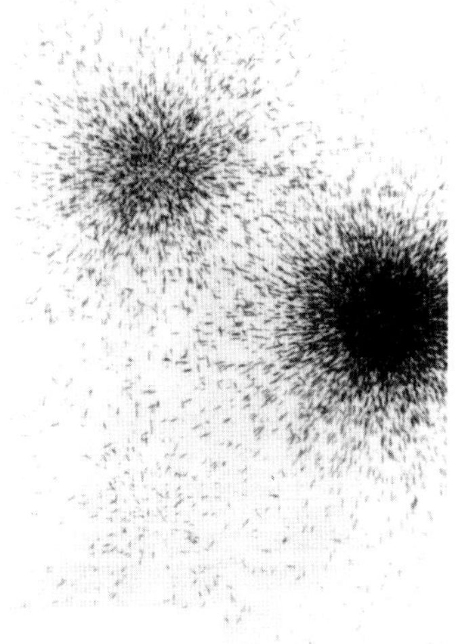

Fig. 3. Photomicrograph and fission track registration (FTR) image of a reservoir sandstone. The FTR image shows that U is associated with a highly radioactive (5000 ppm) metamict zircon ($z_1$), and small transparent zircon ($z_2$), and several small opaque ilmenite grains (i) that are partly replaced by anatase. Both diagrams are at the same scale. Scale bar (shown on photomicrograph) = 0.6 mm.

monazite accounting for an average concentration of approximately 4 ppm Th (Hurst & Milodowski 1996). A Th concentration of 4 ppm is typical for mineralogically-mature North Sea sandstone reservoirs.

Because heavy minerals are the most radioelement- and REE-enriched minerals in sandstone, intuitively the heavy mineral component is expected to exert strong control on the radioelement and REE budget (Fig. 4a). However, this is not always so, since although on an individual grain-by-grain basis the framework components (dominated by quartz) have low trace element concentrations, they may cumulatively control the elemental budget simply because of the presence of heavy mineral inclusions (Fig. 4b). Micro-inclusions of radioelement- and REE-enriched minerals confound the value of chemical data and establishing relationships between chemistry and mineralogy. In subsurface interpretation an association between Th-enrichment and kaolinite was established and used to differentiate kaolinite from other (less Th-rich) clay minerals (Hassan & Hossin 1975; Hassan *et al.* 1976) in sandstone. Kaolinite, however, does not contain structural Th, and any Th associated with kaolinite is *de facto* associated with thorium-rich micro-inclusions (Hurst 1990), which are not always present, and when present may be mixed with kaolinite without micro-inclusions. Thus the relationship between Th concentration and kaolinite abundance is spurious (Dombrowski & Murray 1984). Textural relationships between minerals and trace-element chemistry are summarized in Figure 5.

## Heavy mineral stability during weathering and burial diagenesis

The stability of HMs in sedimentary basins is governed largely by their susceptibility to dissolve in aqueous fluids at Earth-surface and near-Earth-surface conditions in weathering environments and at increasing temperatures during burial. All minerals are susceptible to dissolution in weathering environments, but several minerals, including some HMs, are remarkably stable, being at least as persistent as quartz and more persistent than feldspars. Zircon, rutile, tourmaline, kyanite, andalusite and sillimanite display high stability, whereas apatite, olivine and the pyroxenes are very unstable. Minerals with moderate stability (amphibole, epidote, garnet, staurolite, titanite, epidote) do not always have consistent stability relationships. This may be due partly to the wide compositional variation displayed by some of these minerals, but could also result from

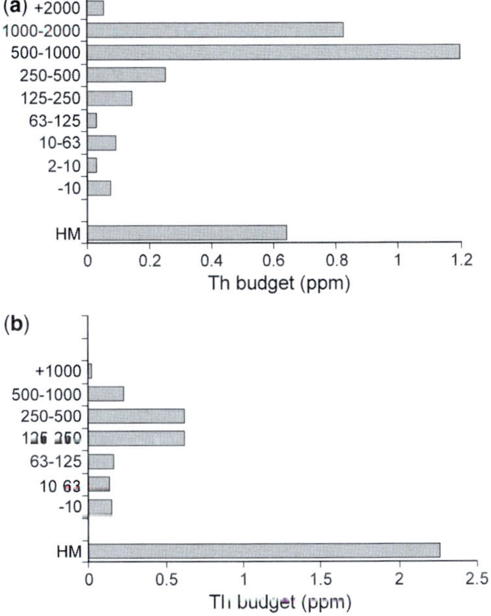

**Fig. 4.** Histograms of the Th budget in grain size ($Y$ axis, in microns) and heavy mineral (HM) fractions, adapted from Hurst & Milodowski (1996). Th budget = Th budget for fraction/Th budget total in ppm. (**a**) Sandstone with Th budget dominated by the heavy mineral fraction; (**b**) sandstone in which Th is associated with the sand-size quartz fraction.

| GRAIN TYPE | | EXAMPLES |
|---|---|---|
| DETRITAL | | monazite<br>zircon<br>anatase<br>apatite<br>etc. |
| DETRITAL | | *host*   *inclusion*<br>quartz  zircon<br>feldspar  monazite<br>mica  apatite<br>etc.  ilmenite<br>  magnetite |
| AUTHIGENIC | | monazite<br>zircon<br>anatase<br>illite<br>etc. |
| AUTHIGENIC | | *host*   *inclusion*<br>kaolinite  apatite<br>quartz  plumbogummite group<br>  anatase |

**Fig. 5.** Textural relationships of trace element-bearing minerals (after Hurst & Milodowski 1998). Note that trace element-bearing microinclusions can occur in both detrital and authigenic minerals.

variations in the composition and concentration of the groundwaters responsible for the leaching (Mitchell 1975). Furthermore, indigenous microorganisms significantly alter weathering patterns as a result of scavenging a limiting supply of nutrients in some environments (Bennett *et al.* 2001).

Burial diagenetic mineral stability is of direct interest to this paper because of the focus on sandstone hydrocarbon reservoirs. Temperature is the main driver of mineral reactions and stability during burial in sedimentary basins (Nadeau 2011) and HMs are known to undergo progressive dissolution in response to increased burial and temperature (Morton & Hallsworth 2007). HM stability patterns under burial diagenetic conditions follow a crudely inverse pattern to the Bowen reaction series of mineral growth during metamorphism. Although detrital minerals undergo mechanical compaction during early burial, only minor quantities of mineral components dissolve; for example, biogenic silica and carbonate. The major onset of temperature-driven silicate mineral reactions in mudstones and sandstones occurs at 60 °C (Nadeau *et al.* 2005; Bjørkum & Nadeau 2008) and the diversity and rates of detrital mineral dissolution and the growth of new minerals accelerate thereafter (Walderhaug 1996; Bjørkum *et al.* 1998; Walderhaug *et al.* 2001, 2004). Clay mineral transformations and growth in mudstone and sandstone have received particular attention (Hoffman & Hower 1979; Nadeau *et al.* 1984; Reynolds 1992) and they determine reservoir quality and seal character (Nadeau *et al.* 2002, 2005). However, some detrital heavy minerals, notably olivine, clinopyroxene and calcic amphibole, which are the least stable minerals during burial diagenesis (Morton & Hallsworth 2007), undergo dissolution at temperatures <60 °C. Walderhaug & Porten (2007) estimate that complete dissolution of calcic amphibole in sandstones from the Norwegian Sea has taken place in sandstones with maximum formation temperatures of 40 °C.

Textures formed by mineral dissolution during burial (Fig. 6) and the relative stabilities of HMs during burial are well known, having been documented in detail from many sedimentary basins around the world, including the North Sea, the Faroe–Shetland Basin, the Gulf of Mexico, the Bengal Basin, the Vienna Basin and the South Caspian Basin (see Morton & Hallsworth 2007 for further information). However, the calibration of HM stability with reference to formation temperature has, to our knowledge, been attempted only once (Walderhaug & Porten 2007), and the relationship between HM dissolution and formation temperatures is a subject that requires further research.

Dissolution of HMs during burial diagenesis typically forms well-defined dissolution surfaces that follow the crystallographically-defined planes of weakness in minerals. Consequently, HMs undergoing dissolution may have a euhedral form (Fig. 6b, c). In some cases, notably garnet, such forms have been attributed erroneously to crystal growth rather than dissolution (see Morton *et al.* 1989; Salvino & Velbel 1989, for further discussion).

The recognition that HM dissolution is widespread in the subsurface has two important implications for mineral–chemical stratigraphy. The first point is that correlations made on the basis of unstable mineral abundances or ratios are highly

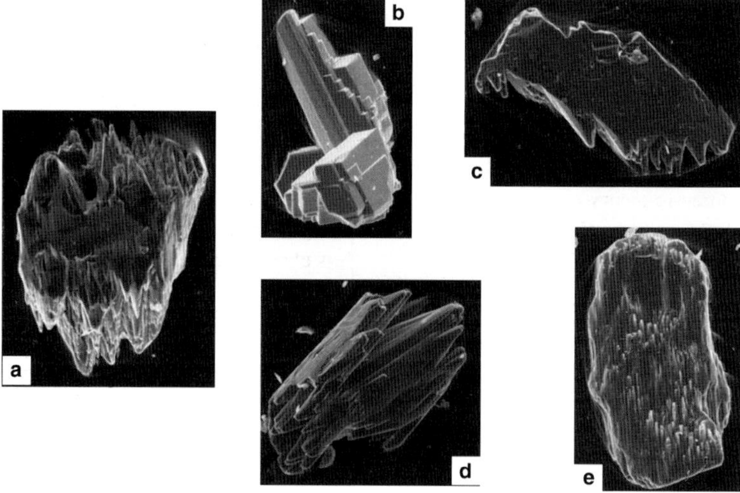

**Fig. 6.** Scanning-electron micrographs of diagenetically-etched heavy minerals extracted from sandstones: (**a**) staurolite; (**b**) garnet; (**c**) kyanite; (**d**) clinopyroxene; (**e**) amphibole.

suspect and it is, therefore, critical that the extent of diagenetic modification is fully appreciated. The second point concerns mobility of trace and rare earth elements. Partial or complete dissolution of minerals releases elements into the pore waters, thereby altering whole-rock chemistry to some degree. In many cases, mineral dissolution releases elements known to be highly mobile, such as calcium, which is hosted by clinopyroxene, amphibole, epidote and titanite, amongst others. However, mineral dissolution also releases elements considered to have low mobility. For example, titanite dissolution releases Ti. Likewise, garnets are heavy REE-enriched (Čopjaková *et al.* 2005) and, consequently, dissolution of garnet must lead to changes in REE element patterns in the whole rock at some scale (whether micro-, meso- or macro-). Garnet is one of the main hosts of REE in sandstones, along with apatite, monazite and zircon. The effects of burial diagenetic modification on whole-rock geochemical signals is another area that requires further research.

## Applications to lithostratigraphy

In this and the subsequent sections, we demonstrate the applications of conventional and geochemical heavy mineral studies to reservoir stratigraphy. We do not discuss whole-rock geochemistry (chemostratigraphy) to reservoir sandstones since, for the most part, successful applications of this techniques are based on mudstone-rich successions, such as the Westphalian of the Pennine Basin (Pearce *et al.* 1999) and the southern North Sea (Pearce *et al.* 2005). In cases where this technique has been applied to sandstones, it is generally calibrated using heavy mineral analysis (Ratcliffe *et al.* 2004, 2007). Neodymium isotope stratigraphy is another whole-rock technique that has been applied to clastic reservoir successions (Mearns *et al.* 1989; Dalland *et al.* 1995), but the technique appears to have become less popular since the discovery that post-depositional Sm–Nd fractionation is common in Jurassic reservoir sandstones on the Norwegian continental shelf (Ehrenberg & Nadeau 2002). Other techniques, such as Pb isotopic analysis of detrital feldspar (Tyrrell *et al.* 2012), appear useful for broad-scale provenance mapping, but at present do not appear to have sufficient resolution for correlation at the reservoir scale.

### Conventional heavy mineral studies

Fluvially-dominated sandstone reservoirs in the Lower Jurassic Statfjord Formation, northern North Sea, are laterally discontinuous and problematic to correlate between wells using subsurface data and sedimentological analysis. Although a regionally-robust biostratigraphic marker is present in marine mudstones above the Statfjord Formation, the paucity of biostratigraphic data within the Statfjord Formation led to the initiation of a HM study on four key exploration/appraisal wells, 34/7-3, -4, -6 and -10, in the Snorre Field. Significant variations in HM assemblages are present in all wells (Morton & Hurst 1995), with garnet and zircon in particular showing large differences in relative abundance (Fig. 7). These dramatically different HM assemblages in adjacent sandstones make it likely that if similar assemblages occur in other wells, they can be used to make field-wide correlations. The different assemblages may also be indicative of temporal variations in provenance.

When the HM data are cast as ratios, the differences between adjacent sandstones are accentuated, with GZi (garnet:zircon) being particularly sensitive (Fig. 8). The HM variations in sandstones across the Snorre Field show that progressively older sandstones are encountered at shallower levels in the section from right to left (Fig. 9). In immediate practical terms, this inter-well correlation reveals which sandstones are likely to have lateral connectivity and connected reservoir volume. Possibly uniquely, HMs allow direct correlation between reservoir volumes rather than non-reservoir to non-reservoir (e.g. interbedded mudstones) correlation. In more detail it is clear that the biostratigraphic correlation in the overlying mudstones is time-transgressive and that an erosive unconformity is present.

Of particular interest in well 34/7-6 is the sandstone between approximately 2526 and 2537 m, in which two very different HM assemblages are present, low GZi at the top and high GZi at the base. Viewed in continuous core the sandstone appears to be a single depositional unit but in the light of the HM data this is extremely unlikely. Comparison with adjacent wells shows that the upper sandstone thickens into well 34/7-6 and has approximately 9 m of mudstone between it and the lower sandstone in well 34/7/4 (Fig. 9). A significant erosional surface is inferred that has removed at least *c.* 9 m mudstone (well 34/7-6) and variably less elsewhere.

### Varietal studies

The first mineral–chemical stratigraphic study to use single-grain chemistry on sandstone reservoirs was made during the mid-1980s on the deltaic Brent Group (Middle Jurassic) reservoirs in the Oseberg Field, northern North Sea (Hurst & Morton 1988). Although the main lithostratigraphic units (Oseberg/Broom, Rannoch, Etive, Ness and Tarbert formations) were readily identifiable, the

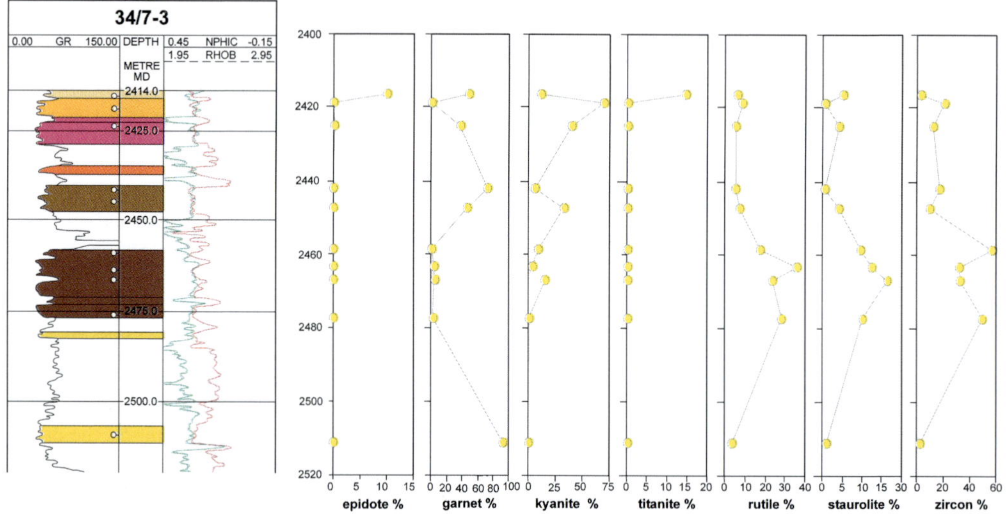

**Fig. 7.** Variations in key mineral–chemical parameters from Statfjord Formation (Triassic–Lower Jurassic) sandstones in Norwegian well 34/7-6, Snorre Field, North Sea, modified from Morton & Hurst (1995). GR, gamma-ray; NPHIC, neutron porosity; RHOB, formation density.

Etive–Ness and Ness–Tarbert boundaries were problematic to constrain, because sedimentologically-similar channelized units are present in all formations. Typical problems were: had the fluvial Ness Formation eroded into the underlying Etive shoreface, and were the uppermost sandstones part

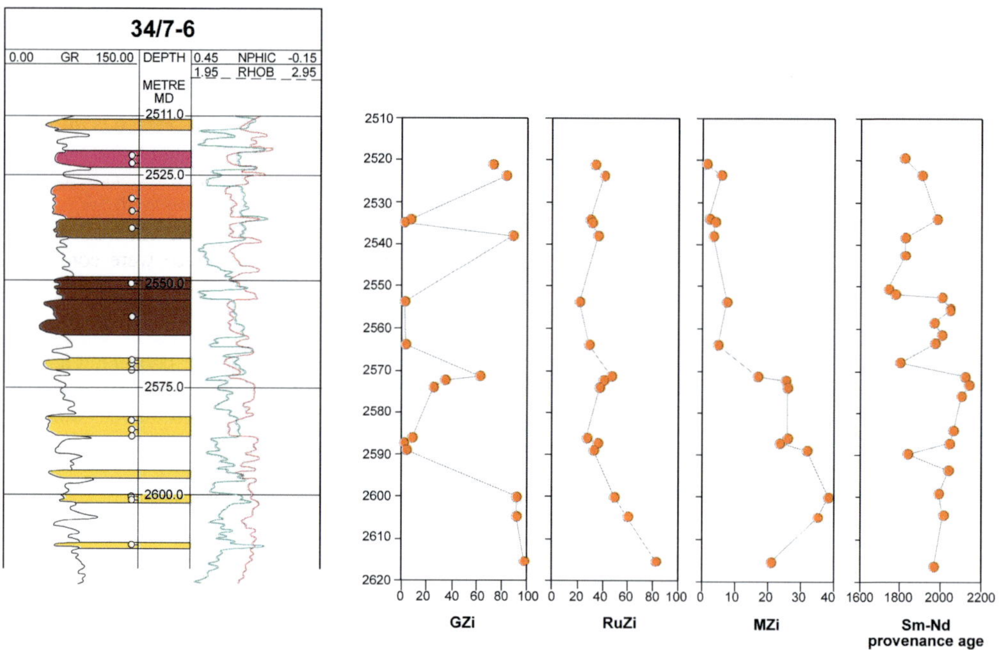

**Fig. 8.** Heavy-mineral ratios: GZi, garnet–zircon; RuZi, rutile–zircon; MZi, monazite–zircon. From well 34/7-6, Snorre Field, North Sea, modified from Morton & Hurst (1995). Sm–Nd provenance ages are from Mearns *et al.* (1989). Highlighted interval shows a sandstone bed that juxtaposes two distinctly different mineralogical zones, the lower with high GZi and the upper with low GZi.

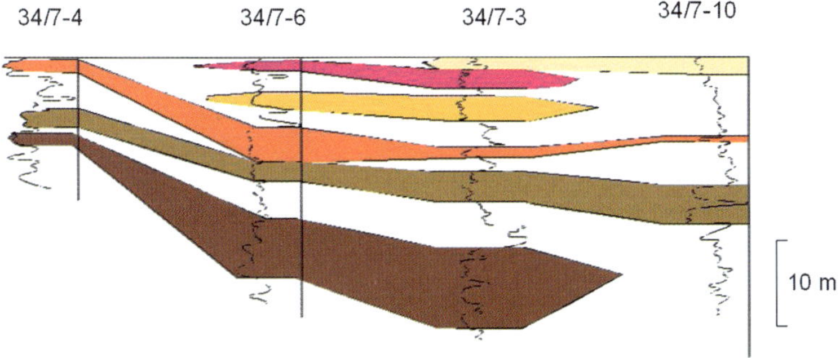



**Fig. 9.** Inter-well correlation using heavy-mineral zonation in the Statfjord Formation between four key appraisal wells from Snorre Field, North Sea (after Morton & Hurst 1995). Where sandstone units touched, for example in well 34/7-6, erosion at the base of the upper sandstone has removed mudstone that previously separated sandstones and a 'single' sandstone unit is preserved that in reality is a composite unit of two genetically-different sandstones.

of the Ness or Tarbert? To address these problems, HM data were acquired but failed to reveal systematic differences between stratigraphic units. The assemblages proved to comprise diagenetically-modified assemblages dominated by garnets with varying degrees of dissolution, together with apatite, rutile, tourmaline, zircon and minor amounts of other HMs (Fig. 10). When garnet chemistry was obtained, clear differences between sandstones in each formation became apparent (Fig. 11a). Subsequent work on the same stratigraphic units in other locations (Morton 1992) showed similar discrimination between the Etive, Ness and Tarbert formations was possible (Fig. 11b).

It is also possible to use HM grain morphology for stratigraphic subdivision and correlation. For example, Mange-Rajetzky (1995) identified stratigraphic changes in tourmaline, zircon and apatite morphologies in the Triassic of the UK central North Sea (Fig. 12). By contrast, conventional HM data do not reveal a clear stratigraphic pattern, and do not resolve the changes in provenance evident from the grain morphology data. The grain morphology data were used to develop a semi-regional correlation framework, and indicate that provenance evolved substantially during deposition. The variations in heavy mineral provenance have subsequently been augmented with palaeocurrent data by McKie *et al.* (2007), who interpreted the changes as a reflection of interplay between Scottish and Scandinavian sourcing.

## Applications to sequence stratigraphy

When HM data from the Middle Jurassic Brent Group of the Oseberg Field were acquired (Hurst & Morton 1988), sequence stratigraphic concepts and applications in the oil industry were in their infancy. Consequently the implications of the erosion of Ness Formation fluvial channels into the Etive Formation upper shoreface were considered

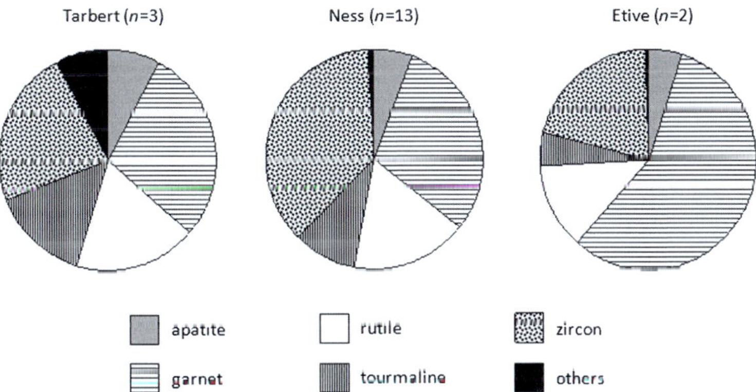



**Fig. 10.** Heavy-mineral assemblages from the Tarbert, Ness and Etive formations of the Brent Group (Middle Jurassic), North Sea. *n*, number of samples analysed.

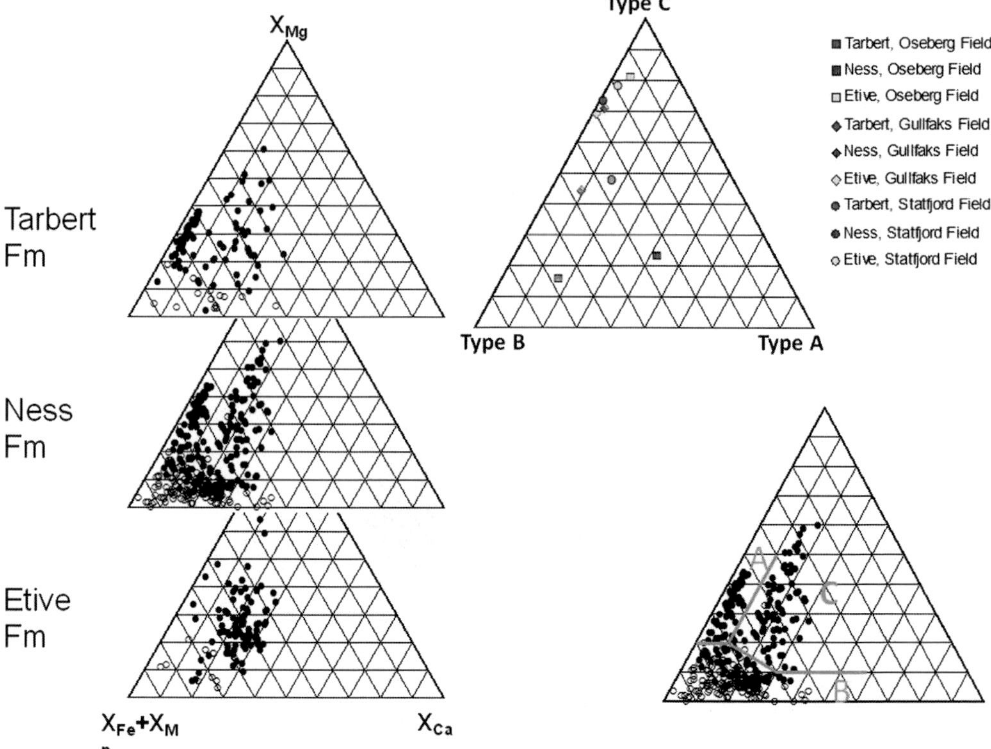



**Fig. 11.** Variations in garnet geochemistry of Etive, Ness and Tarbert sandstones (Middle Jurassic Brent Group) in the Oseberg, Statfjord and Gullfaks fields, northern North Sea. Garnet compositions are expressed in terms of $X_{Fe}$, $X_{Mn}$, $X_{Ca}$ and $X_{Mg}$, which represent the abundances of Fe, Mn, Ca and Mg in the garnet formula. Filled and open circles represent garnets with <5% and >5% $X_{Mn}$, respectively. Data are from Hurst & Morton (1988), Morton *et al.* (1989) and Morton (1992). Note that the greatest contrast between formations is within the Oseberg Field, which is in the most proximal part of the depositional system.

significant only in terms of determining reservoir-volume connectedness and geometry. By comparing the thickness of the Etive and Ness formations in adjacent wells in the Oseberg Field it seems likely that approximately 10 m of stratigraphy is missing due to the erosion. The lateral extent of the erosion surface is not known but the scale of down-cutting is commensurate with a 6th- or 7th-order bounding surface, a channel-belt or nested-valley boundary, respectively (Holbrook *et al.* 2006). Within the general setting of the evolution of the Brent Group in which deposition of the Ness Formation constitutes a significant seaward shift in deposition, the erosion surface is inferred to represent a response to a relative fall of sea level and a consequent ravinement. The absence of data to enable identification of the erosive surface in adjacent wells may reflect a more local control on the erosion and change in relative sea-level such as synsedimentary faulting.

In the Snorre Field, the terrestrial strata of the Lower Jurassic Statfjord Formation were deposited during a period of increasing humidity and ultimately relative sea-level rise. An erosive surface is inferred from the HM data (Fig. 8) and can be correlated over several kilometres (Fig. 9). Evidence for the regional extent of the erosion (from borehole data) reveals that in the order of 10 m (compacted) of stratigraphy is missing from well 34/7-6 and an erosive surface was formed that covered >50 km$^2$. Dramatically different GZi signatures are present in a single sandstone unit in well 34/7-6, with low GZi in the upper part and high in the lower part (Fig. 8). In addition the Sm–Nd provenance ages in the lower part are *c.* 1800 Ma but increase to *c.* 2000 Ma in the upper part. In combination, the GZi and Sm–Nd data make a strong case, not only for ravinement prior to deposition of the upper sandstones, but also for a significant shift in sediment sourcing that introduced detritus with significantly older model ages into the alluvial basin (Fig. 8). Ravinement is assumed to record a period of rapid base-level change.

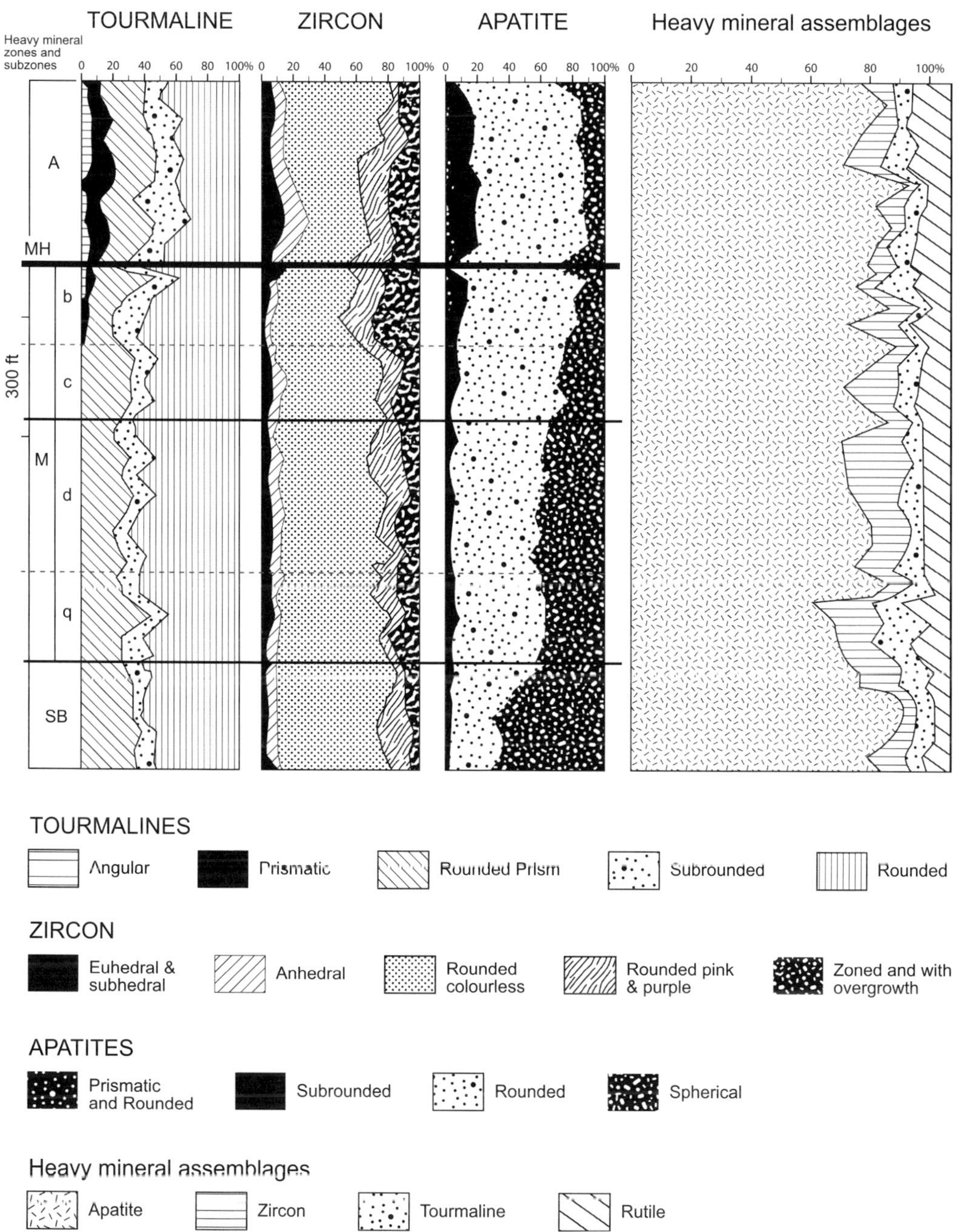



**Fig. 12.** Heavy mineral assemblages in the Triassic of well 22/24a-5, UK central North Sea. Although conventional heavy mineral data reveal little systematic variation, analysis of tourmaline, zircon and apatite morphologies shows distinct variations in provenance and transport history that can be used for correlative purposes. Adapted from Mange-Rajetzky (1995).

Because HM data tend to be acquired when conventional stratigraphic tools fail to deliver, there are few examples of HM studies that are constrained within an established sequence stratigraphic framework. A notable exception is from the Upper Jurassic succession in the Piper Field, Outer Moray Firth

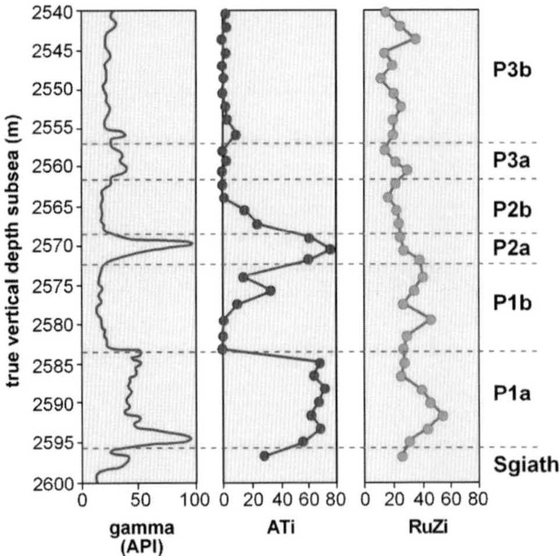



**Fig. 13.** Variations in apatite:tourmaline index (ATi) and rutile:zircon index (RuZi) in Upper Jurassic sandstones from the Piper Formation in well 15/17-B7 (Piper Field, UK central North Sea), showing the contrast in ATi between highstand (P1a, P2a) and lowstand (P1b, P2b) sandstones (adapted from Morton *et al.* 2012).

North Sea. In this field, zones P1a and P1b (basal Piper Formation), mark the change from highstand to lowstand conditions that coincides with a sudden decrease in ATi (apatite:tourmaline ratio) (Fig. 13). Because there is no other change in HM (e.g. rutile:zircon and garnet geochemistry) that may be indicative of a shift in provenance, the ATi changes are interpreted as resulting from the different relative stability of apatite and tourmaline in meteoric water, apatite being unstable and tourmaline stable. Assuming a similar area of sediment sourcing, the lower ATi in the lowstand deposits compared with the highstand is explained by the

lowstand exposing sediment in the drainage basin to more prolonged and intense weathering and thereby removing apatite (Morton *et al.* 2012).

## Provenance

Consideration of data acquired for mineral–chemical stratigraphy enables significant insight into provenance. HM assemblages have a relationship with the composition of source terrain but they can also provide insight into more subtle aspects of provenance that are related to processes in alluvial and marine sedimentary basins. In most depositional environments, the bulk of coarse-clastic sediment is derived from the erosion of weathered and/or mechanically-modified sediment in drainage basins. Sediment derived from drainage basins is transported down-dip into alluvial basins where sediment composition is modified by weathering, the sediment-storage capacity of the basin and time, collectively termed residence (Fig. 14; Hurst & Morton 2001). Many heavy minerals, especially apatite and, to a lesser extent, garnet, are prone to dissolution during residence. Developing the idea that a single source terrain can generate detritus with different mineral–chemical signatures is a critical factor when interpreting provenance. For example, data from the Statfjord Formation in the Brent Field indicate that despite dramatic changes in GZi between sandstones (Fig. 15), detrital zircon ages are remarkably similar (Morton *et al.* 1996). The zircon age data indicate that while provenance remained uniform, changes in GZi reflect variations in residence in alluvial basins. Likewise, the variations in ATi in the Piper Formation (described above) are related to differences in residence.

On a regional scale, data from Brent Group fields show that for each field, Etive, Ness and Tarbert

## RESIDENCE

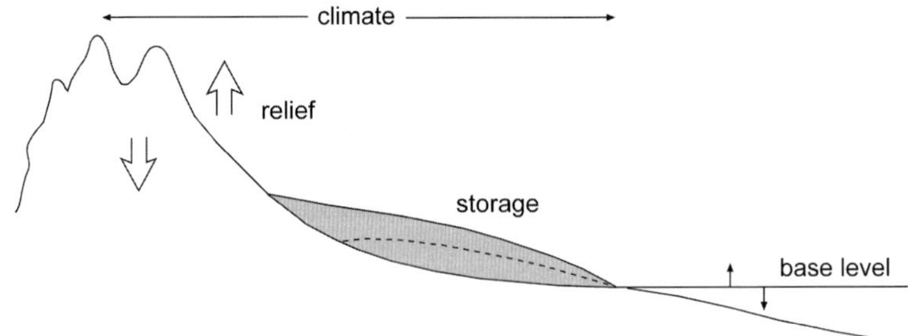



**Fig. 14.** A modified hypsometric curve to demonstrate the residence concept, R. Residence is a collective function of climate and sediment storage and controls the preservation of minerals; adapted from Hurst & Morton (2001).

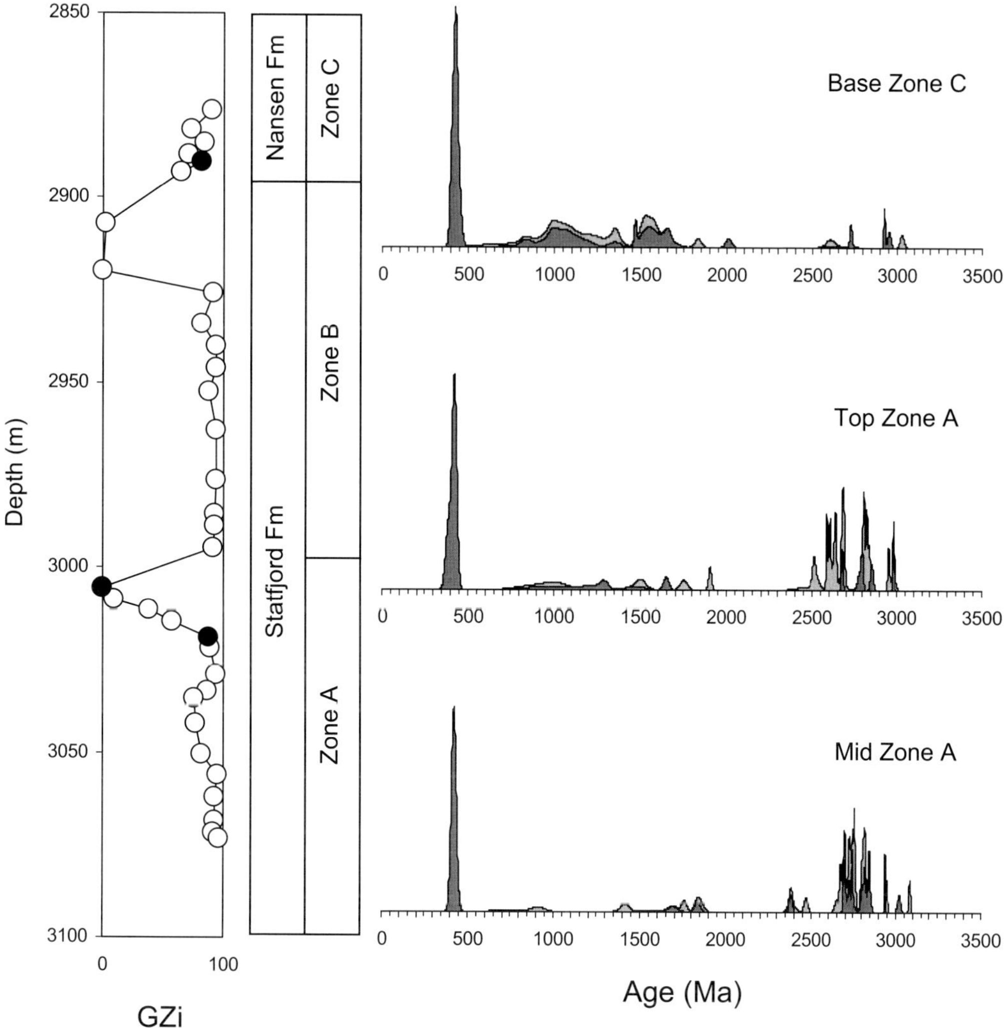



**Fig. 15.** Stratigraphic variations in garnet–zircon (GZi) in the Triassic–Lower Jurassic Statfjord Formation from well 211/29-6, Brent Field, UK northern North Sea. Zircon age spectra in the garnet-rich and the garnet-poor parts of Zone A are virtually indistinguishable, comprising two main components (Early Palaeozoic, *c.* 400 Ma and Archaean, *c.* 2.5–3.0 Ga) in similar proportions. Variations in the mineral–chemical signature are therefore attributed to differences in residence, rather than provenance; adapted from Morton *et al.* (1996).

formations are readily differentiated using garnet chemistry (Fig. 16), which records different provenance for each formation. Therefore, the three formations are not genetically related and because their garnet chemistry is fundamentally different, the Ness fluvial system could not have fed the Etive shoreface (Fig. 11). This early view of the lack of genetic relationships between formations in the Brent Group was later confirmed by the interpretation of seismic data (Mitchener *et al.* 1992). The differences in garnet provenance reflect changes in the character of the hinterland source area. The Etive Formation in all three fields is dominated by high-Ca, high-Mg garnets (Type C in the nomenclature of Mange & Morton 2007 and Fig. 11b), which are predominantly derived from high-grade metamafic rocks similar to those found in western Norway, including the Western Gneiss Region (Morton *et al.* 2004). The Type B (low-Mg, variable Ca and Mn) garnets that are abundant in the Ness Formation, especially in the Oseberg Field, have a variety of sources, but are derived

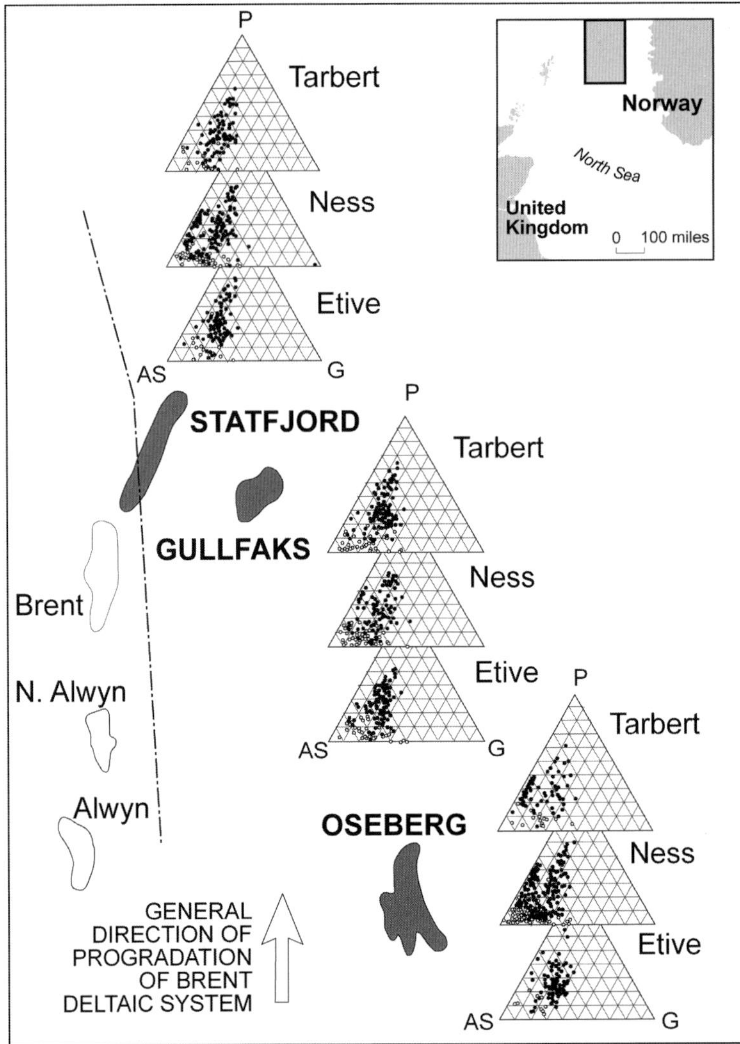



**Fig. 16.** Spatial variations in the garnet geochemistry of Etive, Ness and Tarbert formation sandstones (Middle Jurassic Brent Group) in the northern North Sea using data from the Oseberg, Statfjord and Gullfaks fields (see Fig. 12b). A northward progradation of the Brent delta occurred during this period. Data are from Hurst & Morton (1988), Morton *et al.* (1989) and Morton (1992). Note that the greatest contrast between formations is within the Oseberg field, which is in the most proximal part of the depositional system.

mainly from lower-grade (amphibolite-facies) metasediments and granitoids (Mange & Morton 2007). The high-Mg, low-Ca garnets (Type A) found in the Tarbert Fm. in Oseberg are usually, but not exclusively, associated with high-grade (granulite-facies) metasediments and charnockites (Mange & Morton 2007).

In detail there are significant spatial intra-formational differences within the Etive, Ness and Tarbert formations (Fig. 16) that record a secondary level of provenance variation. These variations are attributed to the time-transgressive nature of each formation. As similar facies prograded northward, different parts of the alluvial basin (with associated different levels of sediment maturity) were supplying sand. Consequently, the Etive, Ness and Tarbert formations preserved in the Statfjord Field are substantially younger than the same formations in the Oseberg Field, although the time differences are not quantifiable because they are below the level of resolution of biostratigraphic and chronostratigraphic analysis.

The Etive Formation displays little difference in garnet composition between the three fields, indicating that provenance did not vary during progradation of this unit. However, the Ness Formation is richer in Type B garnets in the south (Oseberg) than further northwards, where Type C is more common (Gullkaks and Statfjord). Hence, provenance evolved to a significant extent during Ness progradation. A similar south-to-north change is seen in the Tarbert Formation, although in this case it is manifested by a change from Type A dominance in Oseberg to Type C dominance in Gullfaks and Statfjord.

A simple end-member scenario to demonstrate how a single source terrain can create different mineral–chemical signatures is where an imaginary drainage basin with a granitic source terrain is supplying sediment to separate alluvial basins, one with minimal residence and the other with significantly higher residence. The implication is that the minimal-residence alluvial basin has low storage capacity, is little affected by weathering and that sediment is transported quickly through the basin into a marine environment. One would expect some key mineral–chemical differentiations between these basins, despite the common source terrain (Table 2). Accompanying the mineral–chemical variations one will, of course, expect variations in sediment grade and textural maturity related to changes in residence. In the context of mineralogy and, in particular, mineral–chemical stratigraphy, we introduce the term *mineral maturation zone* for depositionally-dominated alluvial basins (Fig. 17). Although sand and sandstone are primarily considered as allochthonous, modification of their mineralogy in the mineral maturation zone is an autochthonous process. One can argue that the heavy mineral data compel the realization that alluvial basins (at least those with lengthy records of sediment supply, storage and weathering) are likely to create spatially-variable heavy-mineral assemblages. Equally, it follows that erosion within alluvial basins will yield different heavy-mineral assemblages, depending on which areas of basins are eroded. These relationships may be complex, and we are unaware of quantitative studies that elucidate them.

When sediment exits from alluvial basins into the marine environment, chemical weathering stops, and the mineral–chemical characteristics of sand-grade detritus become essentially stable. However, it is possible that mechanical modifications in the marine environment may complicate the relationships between depositional mineralogy, source terrain and provenance. Most significant are hydraulic processes on shallow-marine shelves, where any differentiation in mineral–chemical stratigraphy created in the *mineral maturation zone* tends to be homogenized concomitant with textural homogenization (sorting). Hence, we term shallow-marine shelves *mineral mixing zones* (Fig. 17). Palaeogene sandstones of the central North Sea Basin (Forties Formation) and the Faroe–Shetland Basin (Vaila Formation) illustrate the differences between deep-water clastic systems that were derived from a common source terrain (Orkney–Shetland Platform) but had contrasting sediment-routing systems (Fig. 18a). The Forties Formation sandstones display high-frequency variations in ATi and GZi, suggesting little modification in the mineral mixing zone on the shallow-marine shelf with a direct physical linkage to an alluvial basin

**Table 2.** *Effects of low- and high-residence on selected mineral–chemical characteristics in sands or sandstones*

| Residence | Parameter | Preservation potential |
|---|---|---|
| Low | Unstable minerals | High |
| 1st cycle | Garnet:zircon | High |
| | Chrome spinel:zircon | High |
| | Apatite:tourmaline | High |
| | Cr/Zr | High |
| | Th/U | High |
| High | Unstable minerals | Low |
| 1st cycle | Garnet:zircon | Moderate |
| | Chrome spinel:zircon | High |
| | Apatite:tourmaline | Low |
| | Cr/Zr | High |
| | Th/U | Low |

Unstable minerals include micas, feldspars and most rock fragments, together with the heavy minerals epidote, amphibole, pyroxene, titanite and olivine. In both high- and low-residence cases the ages determined from isotopic data will be similar.

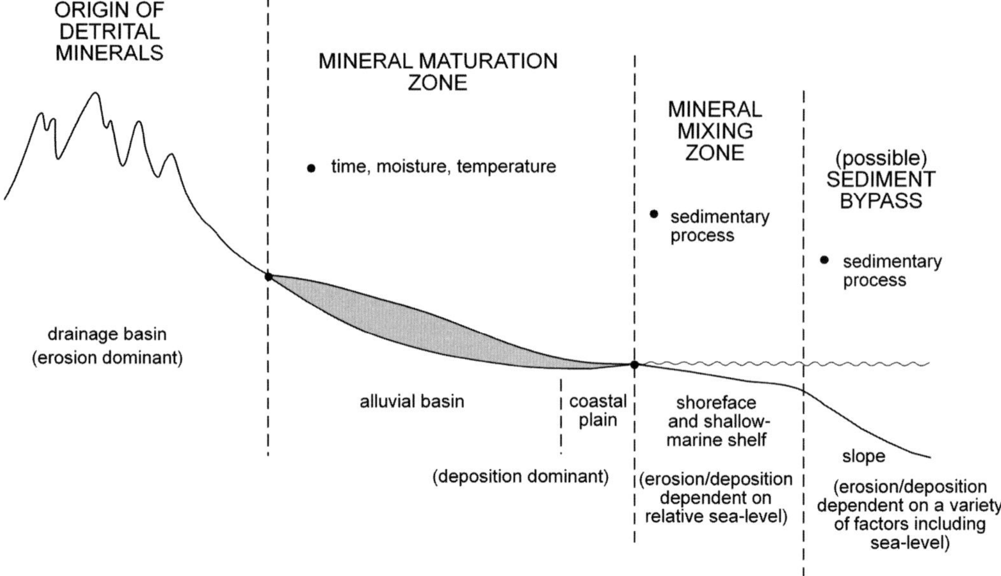



**Fig. 17.** An idealized cross-section (hypsometric curve) of a sedimentary basin that illustrates the distinctive zonation present from an elevated drainage basin to an eventual point of deposition. Relationships between heavy mineral stability and sedimentary processes are summarized. The relative stability of minerals in alluvial basins, termed the mineral maturation zone, is particularly significant as residence in alluvial basins determines the eventual mineralogy deposited in sedimentary environments.

(Fig. 18b), whereas the Vaila Formation has uniform mineralogy consistent with extensive homogenization on a shallow-marine shelf (Fig. 18c).

## Discussion

The main aim of mineral–chemical stratigraphy is to provide a correlation framework for reservoir sandstones, either to complement more conventional stratigraphic approaches, or as a stand-alone method for sandstones that have poor biostratigraphic control. Elucidation of sediment provenance is of less immediate importance, although development of reliable mineral–chemical stratigraphic frameworks does require an understanding of the processes that give rise to the correlatable events, of which provenance is the key. However, because the majority of coarse clastic minerals in sand and sandstone are ultimately derived from drainage basins that derive sediment from weathering of source terrains, mineral–chemical stratigraphic data make an important contribution to an understanding of provenance. Most mineral–chemical stratigraphic studies are made using subsurface (borehole) data, leading to generation of very detailed near-vertical stratigraphic profiles (Figs 7, 8 & 14) usually within a restricted

geographical area at oilfield scale (Fig. 9). This may, in turn, belong to a group of geologically-similar oilfields that occur over a larger area (Fig. 16). Ultimately, therefore, mineral–chemical stratigraphic studies can generate large amounts of data that give detailed stratigraphic and spatial insight, complementary to more conventional provenance data.

The obvious contribution that data from mineral–chemical stratigraphy make to investigation of provenance is the modification of mineral assemblages that occur in the *mineral maturation zone*. In this context, we have presented evidence of correlation between mineral–chemical data and (i) established sequence stratigraphy (Fig. 13), (ii) intra-formational compositional variations that are a response to variations in residence in alluvial basins (Fig. 18), and (iii) identification of probable ravinement surfaces (Figs 8 & 9). Recognition of the variability of residence in the *mineral maturation zone* is critical to understanding variability in heavy-mineral assemblages and mineral–chemical relationships when a single source terrain is providing clastic input.

Although a huge literature exists on the mineralogy of modern weathering profiles and soils, we are unaware of studies of modern alluvial basins in which spatial and temporal mapping of mineral chemistry of sand-sized minerals is undertaken.

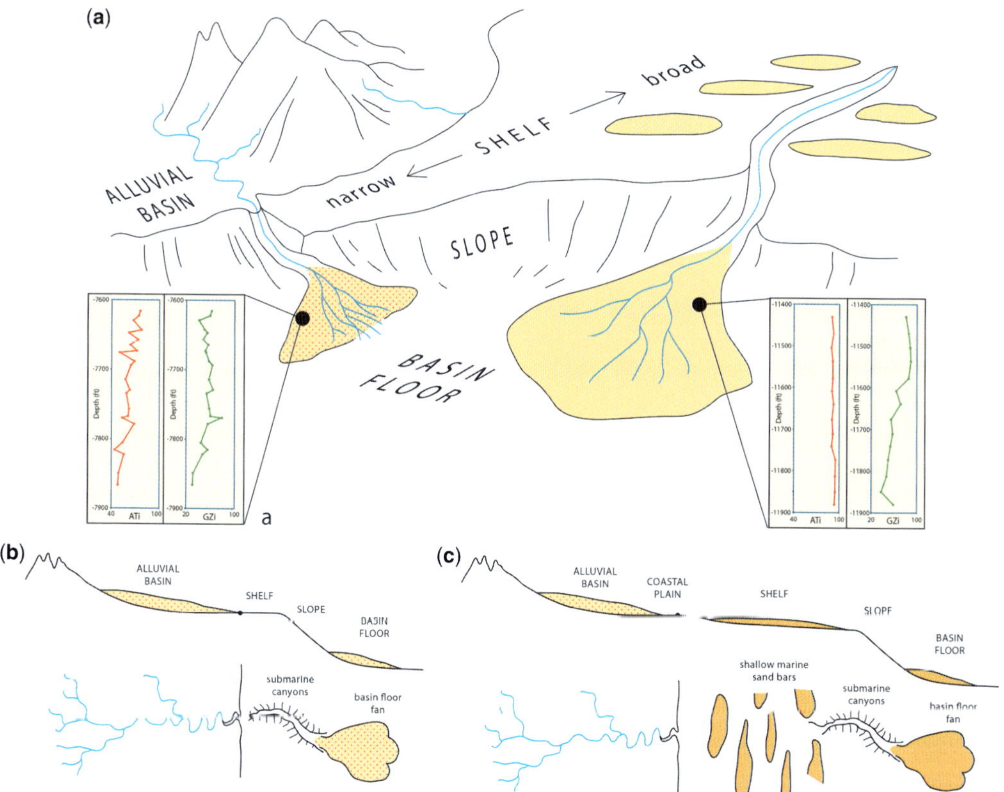



**Fig. 18.** (**a**) Drainage basins from a common source terrain feeding independent deep-water clastic systems: 1 – a submarine canyon that connects directly to a contemporaneous alluvial basin; and 2 – a submarine canyon that cuts back into a contemporaneous shallow-marine shelf. A more heterogeneous heavy-mineral signature, characterized by apatite– tourmaline (ATi) and garnet–zircon (GZi) ratios, is present in the former (Forties Formation) than in the latter (Vaila Formation); after Hurst & & Morton (2001). (**b**) A simplified cross-section and plan view of the Forties Formation scenario and (**c**) a similar representation for the Vaila Formation.

Equally, detailed considerations of residence that examine coarse clastic sediment storage and transport are not common in the geological literature, with the exception of studies of river systems in the equatorial zone of South America (Johnsson *et al.* 1988, 1991; Johnsson & Meade 1990; Savage & Potter 1991). In practical terms, however, mineral–chemical variations provide evidence of changes either of source area or of residence in the mineral maturation zone. By use of specific mineral–chemical parameters, distinction between these possibilities is demonstrably possible.

## Conclusion

Data acquired for mineral–chemical stratigraphy are complementary to conventional provenance studies that tend to focus on the lithological or geochemical character of source terrains, and do not account for modification of mineralogy during weathering and transport of sediment. Although many heavy minerals have diagnostic value in lithostratigraphy and are demonstrated to be sensitive to changes in source terrain, they also record changes during sediment weathering and transport that are attributable to residence in the *mineral maturation zone* of alluvial basins.

The common focus of many subsurface oil-industry applications of mineral–chemical stratigraphy on interwell-correlation within specific reservoir intervals of oil fields or other restricted areas of sedimentary basins tends to produce large sets of data. When similar studies are undertaken on closely adjacent oilfields, as is the case in the North Sea subsurface, substantial regional databases are generated. These databases provide an excellent foundation for consideration of provenance, since they identify intra-formational and intra-reservoir variations in mineral chemistry.

Mineral–chemical data acquisition should not depend on chemical data in isolation, because mineralogical and petrographic data demonstrate that key low-mobility trace elements vary in abundance both within individual mineral species and as components of micro-inclusions within single minerals. Failure to recognize the importance of mineralogical controls on trace-element distribution is likely to provide spurious correlations.

Because mineral–chemical stratigraphy examines sandstone reservoirs, it gives direct insight into correlation of reservoir volumes. This is unlike many other lithostratigraphic and biostratigraphic correlation methods, which use the characteristics of interbedded non-reservoir units for subsurface correlation.

Numerous colleagues, former colleagues and organizations have contributed to our understanding and application of mineral-chemical stratigraphy. Specific thanks are due to Claire Hallsworth, Malcolm Hole, John Still and, not in the least, to the late Maria Mange and Robert Knox. The British Geological Survey and Statoil both supported our early work, without which progress would undoubtedly have been much slower. Review comments from two anonymous reviewers were encouraging and helpful.

# References


BALAN, E., NEUVILLE, D. R., TROCELLIER, P., FRITSCH, E., MULLER, J. P. & CALAS, G. 2001. Metamictization and chemical durability of detrital zircon. *American Mineralogist*, **86**, 1025–1033.

BASU, A., YOUNG, S. W., SUTTNER, L. J., JAMES, W. C. & MACK, G. H. 1975. Re-evaluation of the use of undulatory extinction and polycrystallinity in detrital quartz for provenance interpretation. *Journal of Sedimentary Petrology*, **45**, 872–882.

BENNETT, P. C., ROGERS, J. R., CHOI, W. J. & HIEBERT, F. K. 2001. Silicates, silicate weathering, and microbial ecology. *Geomicrobiology Journal*, **18**, 3–19.

BJØRKUM, P. A. & NADEAU, P. H. 2008. Temperature controlled porosity/permeability reduction, fluid migration, and petroleum exploration in sedimentary basins. *Australian Petroleum Production and Exploration Association Journal*, **38**, 453–464.

BJØRKUM, P. A., OELKERS, E. H., NADEAU, P. H., WALDERHAUG, O. & MURPHY, W. M. 1998. Porosity prediction in sandstones as a function of time, temperature, depth, stylolite frequency and hydrocarbon saturation. *American Association of Petroleum Geologists Bulletin*, **82**, 637–648.

BOGGS, S. & KRINSLEY, D. 2006. *Application of Cathodoluminescence Imaging to the Study of Sedimentary Rocks*. Cambridge University Press, Cambridge, UK.

ČOPJAKOVÁ, R., SULOVSKÝ, P. & PATERSON, B. A. 2005. Major and trace elements in garnets from Variscan Moldanubian felsic granulites as sediment provenance indicators. *Lithos*, **82**, 51–70.

DALLAND, A., MEARNS, E. W. & MCBRIDE, J. J. 1995. The application of samarium-neodymium (Sm–Nd) provenance ages to correlation of biostratigraphically barren strata: a case study of the Statfjord Formation in the Gullfaks Oilfield, Norwegian North Sea. *In*: DUNAY, R. E. & HAILWOOD, E. (eds) *Non-Biostratigraphical Methods of Dating and Correlation.* Geological Society, London, Special Publications, **89**, 201–222.

DEER, W. A., HOWIE, R. A. & ZUSSMAN, J. 1997. *Rock-Forming Minerals, Volume 1A*. Orthosilicates. 2nd edn. Geological Society, London.

DOMBROWSKI, T. & MURRAY, H. H. 1984. Thorium – a key element in differentiating Cretaceous and Tertiary kaolins in Georgia and South Carolina. *In*: *Proceedings of the 27th International Geological Congress*, Moscow, 4–14 August. Taylor & Francis, **15**, 305–317.

EHRENBERG, S. H. & NADEAU, P. 2002. Postdepositional Sm/Nd fractionation in sandstones: implications for neodymium isotope stratigraphy. *Journal of Sedimentary Research*, **72**, 304–315.

HALLSWORTH, C. R., MORTON, A. C., CLAOUÉ-LONG, J. C. & FANNING, C. M. 2000. Carboniferous sand provenance in the Pennine Basin, UK: constraints from heavy mineral and SHRIMP zircon age data. *Sedimentary Geology*, **137**, 147–185.

HASSAN, M. & HOSSIN, A. 1975. Contribution a l'etudes comportements du thorium et du potassium dans les roches sedimentaires. *Comptes Rendue Academie Science, Paris*, **280**, 533–535.

HASSAN, M., HOSSIN, A. & COMBAZ, A. 1976. Fundamentals of the differential gamma log-interpretation technique. *In*: *Transactions of the SPWLA 17th Annual Logging Symposium*, Denver, June 9–12. Society of Petroleum Well Log Analysts (SPWLA), Houston, Paper H.

HOLBROOK, J., SCOTT, R. W. & OBOH-IKUENOBE, F. E. 2006. Base-level buffers and buttresses: a model for upstream versus downstream control on fluvial geometry and architecture within sequences. *Journal of Sedimentary Research*, **76**, 162–174.

HOFFMAN, J. & HOWER, J. 1979. Clay mineral assemblages as low grade metamorphic geothermometers: application to the thrust faulted disturbed belt of Montana, USA. *In*: SCHOLLE, P. & SCHLUGER, P. R. (eds) *Aspects of Diagenesis*. Society for Sedimentary Geology (SEPM), Special Publications, Tulsa, **28**, 55–79.

HURST, A. 1990. Natural gamma-ray spectrometry in hydrocarbon-bearing sandstone from the Norwegian Continental Shelf. *In*: HURST, A., LOVELL, M. & MORTON, A. (eds) *Geological Applications of Wireline Logs*. Geological Society, London, Special Publications, **48**, 211–222.

HURST, A. 1999. Textural and geochemical micro-analysis in the interpretation of clay mineral characteristics: lessons from sandstone hydrocarbon reservoirs. *Clay Minerals*, **34**, 137–150.

HURST, A. & MILODOWSKI, A. 1996. Thorium distribution in some North Sea sandstones: implications for petrophysical evaluation. *Petroleum Geoscience*, **2**, 59–68.

HURST, A. & MILODOWSKI, A. 1998. Natural gamma radiation in sandstone: causes and caution. *Dialog*, **6**, 6–8.

HURST, A. & MORTON, A. C. 1988. An application of heavy-mineral analysis to lithostratigraphy and reservoir modelling in the Oseberg Field, northern North Sea. *Marine and Petroleum Geology*, **5**, 157–170.

HURST, A. & MORTON, A. C. 2001. Generic relationships in the mineral–chemical stratigraphy of turbidite sandstones. *Journal of the Geological Society, London*, **158**, 401–404.

JEANS, C. V. 1995. Clay mineral stratigraphy in Palaeozoic and Mesozoic red-bed facies onshore and offshore UK. *In*: DUNAY, R. E. & HAILWOOD, E. A. (eds) *Non-Biostratigraphical Methods of Dating and Correlation*. Geological Society, London, Special Publications, **89**, 31–55.

JOHNSSON, M. J. & MEADE, R. H. 1990. Chemical weathering of fluvial sediments during alluvial storage: the Macuapanim Island point bar, Solimões River. *Journal of Sedimentary Petrology*, **59**, 768–781.

JOHNSSON, M. J., STALLARD, R. F. & MEADE, R. H. 1988. First-cycle quartz arenites in the Orinoco River basin, Venezuela and Colombia. *Journal of Geology*, **96**, 103–277.

JOHNSSON, M. J., STALLARD, R. F. & LUNDBERG, N. 1991. Controls on the composition of fluvial sands from a tropical weathering environment. Sands of the Orinoco drainage basin, Venezuela and Colombia. *Geological Society of America Bulletin*, **103**, 1622–1647.

KRYNINE, P. D. 1946. The tourmaline group in sediments. *Journal of Geology*, **54**, 65–87.

LIHOU, J. C. & MANGE-RAJETZKY, M. A. 1996. Provenance of the Sardona Flysch, eastern Swiss Alps: example of high resolution heavy mineral analysis applied to an ultrastable assemblage. *Sedimentary Geology*, **105**, 141–157.

MANGE, M. A. & MAURER, H. F. W. 1992. *Heavy Minerals in Colour*. Chapman & Hall, London.

MANGE, M. A. & MORTON, A. C. 2007. Geochemistry of heavy minerals. *In*: MANGE, M. & WRIGHT, D. T. (eds) *Heavy Minerals In Use. Developments in Sedimentology* **58**. Elsevier, Amsterdam, 345–391.

MANGE-RAJETZKY, M. A. 1995. Subdivision and correlation of monotonous sandstone sequences using high resolution heavy mineral analysis, a case study: the Triassic of the Central Graben. *In*: DUNAY, R. E. & HAILWOOD, E. A. (eds) *Non-Biostratigraphical Methods of Dating and Correlation*. Geological Society, London, Special Publications, **89**, 23–30.

MCKIE, T.,, JOLLEY, S. J. & KRISTENSEN, M. B. 2007. Stratigraphic and structural compartmentalization of dryland fluvial reservoirs: Triassic Heron Cluster, Central North Sea. *In*: JOLLEY, S. J., FISHER, Q. J., AINSWORTH, R. B., VROLIJK, P. J. & DELISLE, S. (eds) *Reservoir Compartmentalization*. Geological Society, London, Special Publications, **347**, 165–198.

MEARNS, E. W., KNARUD, R., RÆSTAD, N., STANLEY, K. O. & STOCKBRIDGE, C. P. 1989. Samarium-neodymium isotope stratigraphy of the Lunde and Statfjord Formations of Snorre oil field, northern North Sea. *Journal of the Geological Society, London*, **146**, 217–228.

MITCHELL, W. A. 1975. Heavy minerals. *In*: GIESEKING, J. E. (ed.) *Soil Components II. Inorganic*. Springer-Verlag, New York, 449–480.

MITCHENER, B. C., LAWRENCE, D. A., PARTINGTON, M. A., BOWMAN, M. B. J. & GLUYAS, J. 1992. Brent Group: sequence stratigraphy and regional implications. *In*: MORTON, A. C., HASZELDINE, R. S., GILES, M. R. & BROWN, S. (eds) *Geology of the Brent Group*. Geological Society, London, Special Publications, **61**, 45–80.

MORTON, A. C. 1992. Provenance of Brent Group sandstones: heavy mineral constraints. *In*: MORTON, A. C., HASZELDINE, R. S., GILES, M. R. & BROWN, S. (eds) *Geology of the Brent Group*. Geological Society, London, Special Publications, **61**, 227–244.

MORTON, A. C. & HALLSWORTH, C. R. 1994. Identifying provenance-specific features of detrital heavy mineral assemblages in sandstones. *Sedimentary Geology*, **90**, 241–256.

MORTON, A. C. & HALLSWORTH, C. R. 1999. Processes controlling the composition of heavy mineral assemblages in sandstones. *Sedimentary Geology*, **124**, 3–29.

MORTON, A. C. & HALLSWORTH, C. R. 2007. Stability of detrital heavy minerals during burial diagenesis. *In*: MANGE, M. & WRIGHT, D. T. (eds) *Heavy Minerals In Use. Developments in Sedimentology* **58**. Elsevier, Amsterdam, 215–245.

MORTON, A. C. & HURST, A. 1995. Correlation of sandstones using heavy minerals: an example from the Statfjord Formation of the Snorre Field, northern North Sea. *In*: DUNAY, R. E. & HAILWOOD, E. (eds) *Non-Biostratigraphical Methods of Dating and Correlation*. Geological Society, London, Special Publications, **89**, 3–22.

MORTON, A. C., STIBERG, J. P., HURST, A. & QVALE, H. 1989. Use of heavy minerals in lithostratigraphic correlation, with examples from the Brent sandstones of the northern North Sea. *In*: COLLINSON, J. (ed.) *Correlation in Hydrocarbon Exploration*. Graham and Trotman, London, 217–230.

MORTON, A. C., CLAOUÉ-LONG, J. & BERGE, C. 1996. Factors influencing heavy mineral suites in the Statfjord Formation, Brent Field, North Sea: constraints provided by SHRIMP U-Pb dating of detrital zircons. *Journal of the Geological Society, London*, **153**, 911–929.

MORTON, A. C., HALLSWORTH, C. R. & CLAOUÉ-LONG, J. C. 2001. Zircon age and heavy mineral constraints on provenance of North Sea Carboniferous sandstones. *Marine and Petroleum Geology*, **18**, 319–337.

MORTON, A. C., HALLSWORTH, C. R. & CHALTON, B. 2004. Garnet compositions in Scottish and Norwegian basement terrains: a framework for interpretation of North Sea sandstone provenance. *Marine and Petroleum Geology*, **21**, 393–410.

MORTON, A. C., MUNDY, D. J. C. & BINGHAM, G. 2012. High-frequency fluctuations in heavy mineral assemblages from Upper Jurassic sandstones of the Piper Formation, UK North Sea. relationships with sea level change and floodplain residence. *In*: RASBURY, E. T., HEMMING, S. R. & RIGGS, N. R. (eds) *Mineralogical and Geochemical Approaches to Provenance*. Geological Society of America Special Papers, Boulder, **487**, 163–176.

NADEAU, P. H. 2011. Earth's energy 'Golden Zone': a synthesis from mineralogical research. *Clay Minerals*, **46**, 1–24.

NADEAU, P. H., WILSON, M. J., McHARDY, W. J. & TAIT, J. M. 1984. Interstratified clay as fundamental particles. *Science*, **225**, 923–925.

NADEAU, P. H., WALDERHAUG, O., BJØRKUM, P. A. & HAY, S. 2002. Clay diagenesis, shale permeability and implications for petroleum systems analysis. Abstract G003, presented at the EAGE Annual Conference, Florence.

NADEAU, P. H., BJØRKUM, P. A. & WALDERHAUG, O. 2005. Petroleum system analysis: impact of shale diagenesis on reservoir fluid pressure, hydrocarbon migration and biodegradation risks. *In*: DORÉ, A. G. & VINING, B. (eds) *Petroleum Geology: North-West Europe and Global Perspectives – Proceedings of the 6th Petroleum Geology Conference*. Geological Society, London, 1267–1274.

PAY, M. D., ASTIN, T. R. & PARKER, A. 1999. Clay mineral distribution in the Devonian–Carboniferous sandstones of the Clair Field, west of Shetland, and its significance for reservoir quality. *Clay Minerals*, **35**, 151–162.

PEARCE, T. J., BESLY, B. M., WRAY, D. & WRIGHT, D. K. 1999. Chemostratigraphy: a method to improve interwell correlation in barren sequences – a case study using onshore Duckmantian/Stephanian sequences (West Midlands, UK). *Sedimentary Geology*, **124**, 197–220.

PEARCE, T. J., WRAY, D., RATCLIFFE, K., WRIGHT, D. K. & MOSCARIELLO, A. 2005. Chemostratigraphy of the Upper Carboniferous Schooner Formation, southern North Sea. *In*: COLLINSON, J. D., EVANS, D. J., HOLLIDAY, D. S. & JONES, N. S. (eds) *Carboniferous Hydrocarbon Geology: The Southern North Sea and Surrounding Onshore Areas*. Yorkshire Geological Society Occasional Publication, W. & G. Baird, Antrim, Northern Ireland, **7**, 147–163.

POLDERVAART, A. 1955. Zircon in rocks, 1: sedimentary rocks. *American Journal of Science*, **253**, 433–461.

PRESTON, J., HARTLEY, A., HOLE, M., BUCK, S., BOND, J., MANGE, M. & STILL, J. 1998. Integrated whole-rock trace element geochemistry and heavy mineral chemistry studies: aids to correlation of continental red-bed reservoirs in the Beryl Field, UK North Sea. *Petroleum Geoscience*, **4**, 7–16.

RATCLIFFE, K. T., WRIGHT, A. M., HALLSWORTH, C., MORTON, A., ZAITLIN, B. A., POTOCKI, D. & WRAY, D. S. 2004. An example of alternative correlation techniques in a low accommodation setting, non-marine hydrocarbon system: the (Lower Cretaceous) Mannville Basal Quartz succession of southern Alberta. *American Association of Petroleum Geologists Bulletin*, **88**, 1419–1432.

RATCLIFFE, K. T., MORTON, A. C. & RITCEY, D. H. 2007. Whole-rock geochemistry and heavy mineral analysis as petroleum exploration tools in the Bowser and Sustut basins, British Columbia, Canada. *Bulletin of Canadian Petroleum Geology*, **55**, 320–333.

REYNOLDS, R. C. 1992. X-ray diffraction studies of illite/smectite from rocks, <1 mm randomly oriented powders, and <1 mm randomly oriented aggregates: the absence of laboratory induced artifacts. *Clays and Clay Minerals*, **40**, 387–396.

ROSER, B. P. 2000. Whole-rock geochemical studies of clastic sedimentary suites. *Memoir of the Geological Society of Japan*, **57**, 73–89.

SALVINO, J. F. & VELBEL, M. A. 1989. Faceted garnets from sandstones of the Munising Formation (Cambrian), northern Michigan: petrographic evidence for origin by intrastratal dissolution. *Sedimentology*, **36**, 371–379.

SAVAGE, K. M. & POTTER, P. E. 1991. Petrology of modern sands of the Rios Guavare and Inírada, southern Colombia. *Journal of Geology*, **99**, 289–298.

SVENDSEN, J. B. & HARTLEY, N. R. 2002. Synthetic heavy mineral stratigraphy: applications and limitations. *Marine and Petroleum Geology*, **19**, 389–405.

TYRRELL, S., HAUGHTON, P. D. W., SOUDERS, A. K., DALY, J. S. & SHANNON, P. M. 2012. Large-scale, linked drainage systems in the NW European Triassic: insights from the Pb isotopic composition of detrital K-feldspar. *Journal of the Geological Society, London*, **169**, 279–295.

WALDERHAUG, O. 1996. Kinetic modeling of quartz cementation and porosity loss in deeply buried sandstone reservoirs. *American Association of Petroleum Geologists Bulletin*, **80**, 731–745.

WALDERHAUG, O. & PORTEN, K. W. 2007. Stability of detrital heavy minerals on the Norwegian continental shelf as a function of depth and temperature. *Journal of Sedimentary Research*, **77**, 992–1002.

WALDERHAUG, O., BJØRKUM, P. A., NADEAU, P. H. & LANGNES, O. 2001. Quantitative modeling of basin subsidence caused by temperature-driven silica dissolution and reprecipitation. *Petroleum Geoscience*, **7**, 107–113.

WALDERHAUG, O., OELKERS, E. H. & BJØRKUM, P. A. 2004. An analysis of the role of stress, temperature and pH in chemical compaction of sandstones: discussion. *Journal of Sedimentary Research*, **74**, 447–449.

# Effects of sandstone provenance on reservoir quality preservation in the deep subsurface: experimental modelling of deep-water sand in the Gulf of Mexico

RICK C. TOBIN* & D. SCHWARZER

*Maersk Oil, Inc., 2500 CityWest Boulevard, Houston, Texas 77042, USA*

**Corresponding author (e-mail: rick.tobin@maerskoil.com)*


**Abstract:** Deep-water turbidite sandstone reservoirs in the Gulf of Mexico have been sourced from a variety of provenance terrains. As a result, the framework composition of each reservoir in the basin varies widely, including volcanic-rich litharenite (Oligocene Vicksburg), feldspathic-rich lithic arkose (Oligocene Frio), metamorphic-rich feldspathic litharenite (Palaeogene Wilcox), lithic-poor quartzarenite (Miocene), quartz-rich sublitharenite (Cretaceous) and quartz- and feldspar-rich subarkose (Norphlet).

Provenance-driven differences in composition have a complex but critical influence on how each of these reservoirs responds to burial-induced changes in depth, fluid pressure, effective stress and temperature. A combination of Petromod® and Touchstone™ modelling programs are used in this study to simulate the influence of provenance on compaction and cementation of the main reservoir types in the Gulf of Mexico. For example, modelling results predict that at higher levels of thermal exposure, some lithic-rich sands, although more ductile and highly compacted, will experience less quartz cementation than less ductile, quartz-rich sands, thereby preserving a higher range of porosity and permeability. Furthermore, modelling results predict that temperature/effective stress/depth windows for optimal reservoir quality preservation vary widely depending on sandstone provenance.


The evaluation of sandstone grain composition and subsequent interpretation of sand provenance has proven to be a useful methodology for oil and gas exploration for many years. Specific applications of provenance studies are varied and have included the following: assistance in the interpretation of basin evolution and tectonic setting (Dickinson 1985), help in discriminating sandstone petrofacies and source areas (Ingersoll 1990), interpretation of relief and climate of source areas (Folk 1980), help with the interpretation of palaeo-drainage and sediment dispersal patterns (Lowe *et al.* 2011), and help with improving stratigraphic correlation (Morton & Hurst 1995). Recent workers have used provenance as a means of discriminating potential areas of basins with higher reservoir quality potential (Vincent *et al.* 2010; 2013). Provenance reconstruction has also become a critical step in selecting the most appropriate analogues for predictive reservoir quality modelling using diagenetic modelling software such as Exemplar® (Lander & Walderhaug 1999) and Touchstone™ (Lander *et al.* 2008). What has not been reported as yet is the degree to which provenance affects predictive modelling results. With this in mind, we pose a question: Could provenance related differences in sandstone composition within a single basin be large enough to significantly affect predictive modelling results?

This study is intended to address that question by comparing a variety of sandstone compositions across a wide range of simulated burial histories. Specifically, sandstone compositions representing the Mesozoic through Tertiary Gulf of Mexico reservoirs of the deep-water basin are used in a series of test cases. This basin provides a natural laboratory for testing the effects of provenance on reservoir quality prediction, and the findings from this study are intended to be used for global application.

## Provenance and reservoir quality modelling

Provenance, transport distance, climate and depositional setting all affect the initial detrital composition, and to some extent the textural attributes of sandstone reservoirs at the onset of burial. Depositional texture and composition greatly control the rate and style of compaction and other diagenetic processes that occur during burial. Specific subsurface processes that are affected by initial sand composition include the following:

(1)  mechanical compaction;
(2)  chemical compaction;
(3)  dissolution of unstable minerals;
(4)  precipitation of quartz cement;
(5)  precipitation of other cements (e.g. carbonates, zeolites, sulphates);

*From*: SCOTT, R. A., SMYTH, H. R., MORTON, A. C. & RICHARDSON, N. (eds) 2014. *Sediment Provenance Studies in Hydrocarbon Exploration and Production*. Geological Society, London, Special Publications, **386**, 27–47. First published online September 12, 2013, http://dx.doi.org/10.1144/SP386.17

(6)    pore lining, pore-bridging and pore-filling clay authigenesis;
(7)    mineralogical replacement of unstable grains.

These predominantly subsurface processes are varied and relate in complex ways to each other. Understanding and quantitatively predicting subsurface processes and resulting reservoir quality attributes requires use of modelling software that is capable of simulating a variety of diagenetic effects. For this study, Touchstone was chosen as the preferred model prediction software because it can simultaneously simulate the effects of compaction and a wide variety of other diagenetic processes. It also facilitates the prediction and comparison of dissimilar reservoir petrofacies. Touchstone modelling software is based on the conceptual foundation established by Walderhaug (1996) and Lander & Walderhaug (1999). It incorporates forward numerical models that simulate sandstone diagenesis through geological time based on burial history inputs and measured rock properties.

## Gulf of Mexico background

The geological framework, palaeogeography and depositional history of the Gulf of Mexico have been reviewed recently by Galloway (2008). The Gulf of Mexico basin is filled with a thick succession of reservoir and non-reservoir strata ranging in age from Jurassic to Holocene. In the more distal deep-water portions of the basin, siliciclastic deposits form petroleum reservoirs that are the subject of exploration for various oil companies. Reservoir targets of interest have included Jurassic (Norphlet), Cretaceous (Woodbine/Tuscaloosa), Lower Palaeogene (Wilcox), Oligocene (Frio) and Plio-Miocene.

From Mesozoic (Cretaceous) through Tertiary times, sand has been supplied to the basin from a variety of continental fluvial systems that have fed marginal marine deltas along the northern Gulf of Mexico coastline (Fig. 1). These coastal deltaic systems have provided a variety of sediment sources that have switched positions along the coastline during basin evolution. Sand has been

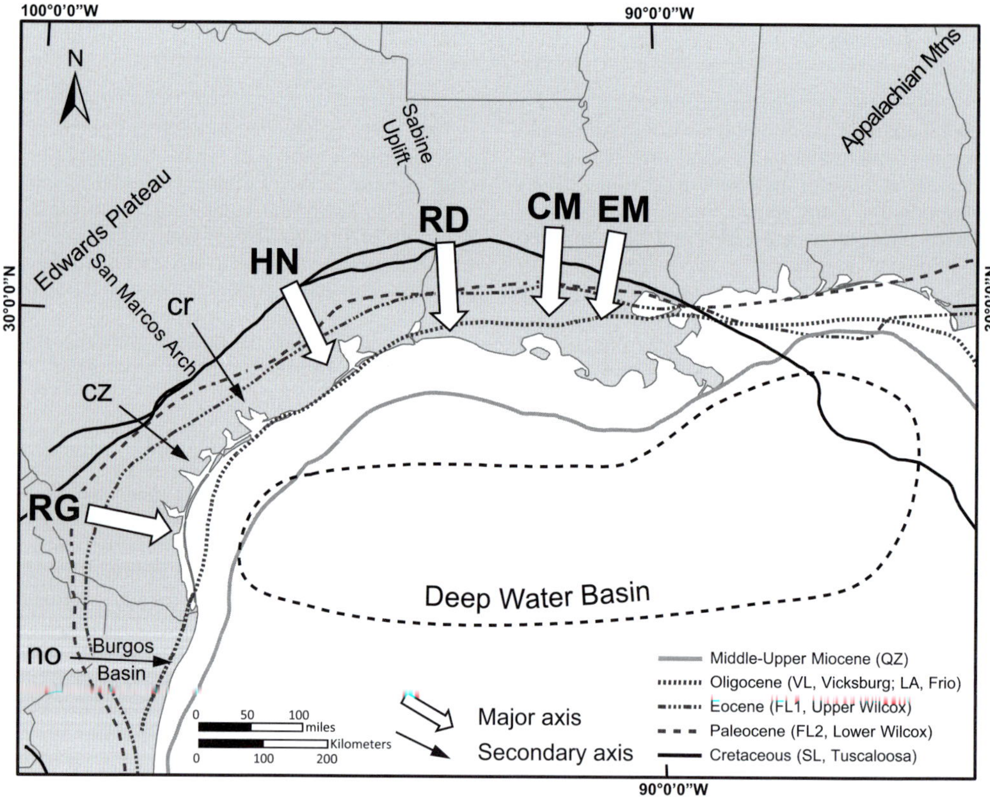


**Fig. 1.** Location map for the deep-water basin in the Gulf of Mexico, showing major and minor sediment dispersal axes (modified after Galloway 2008). Sediment sources (from east to west): EM, East Mississippi; CM, Central Mississippi; RD, Red River; HN, Houston; cr, Corsair; cz, Carrizo; RG, Rio Grande; no, Norias.

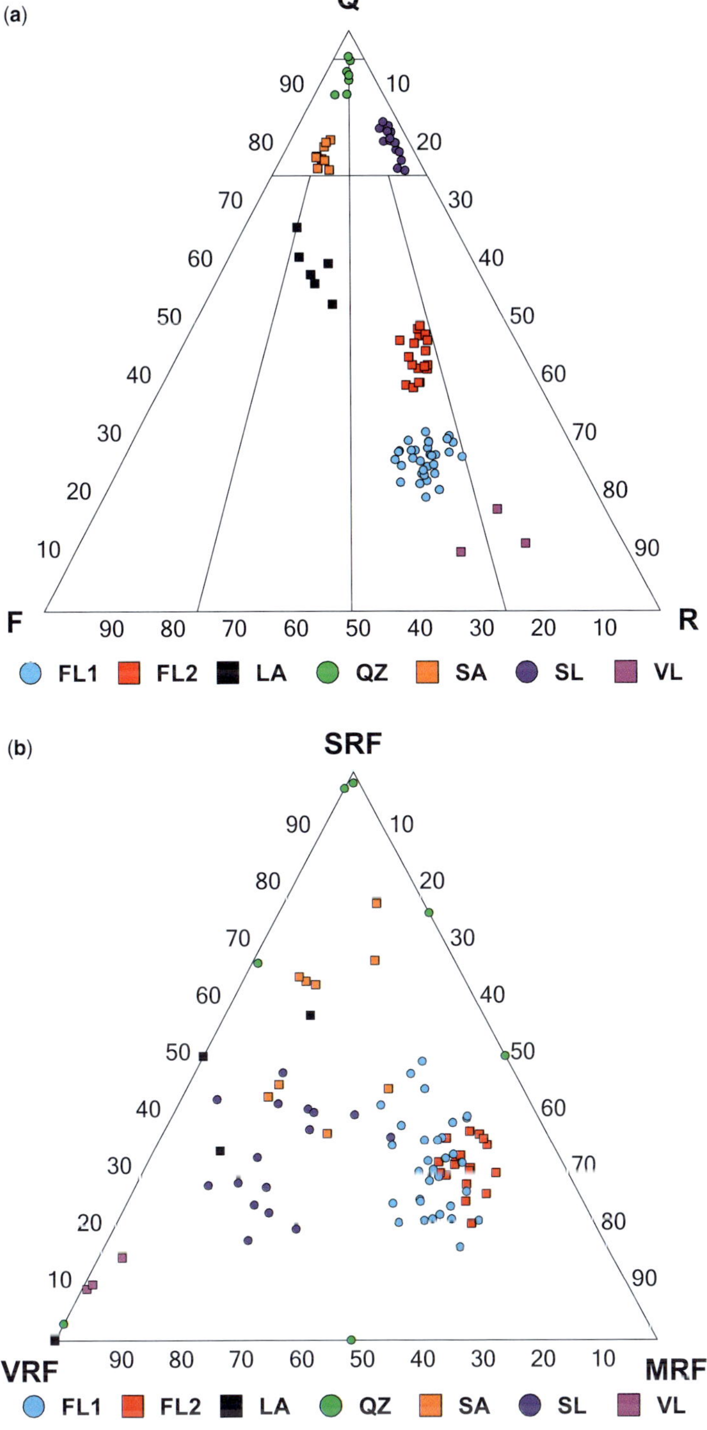



**Fig. 2.** Compositional ternary plots for Gulf of Mexico provenance groupings modelled in this study: (**a**) QFR (quartz–feldspar–rock fragments) distribution plot; (**b**) lithics distribution plot, showing SRF (sedimentary rock fragments), VRF (volcanic rock fragments) and MRF (metamorphic rock fragments).

transported from these local nearshore entry points across the shelf and slope via canyon systems into deep-marine abyssal settings as far as 300 miles away from the palaeo-coastline. Hence, when we consider the provenance of distal, deep-water turbidite sands in this basin, we are referring to the deltaic system that sourced the sand, not the ultimate continental sand source (e.g. Appalachians, Rocky Mountains).

## Methods used

The purpose of this study was to test the importance of initial sand composition on the prediction of reservoir quality within a single basin. To address that question, a limited set of sandstone compositional data from the Gulf of Mexico was evaluated from a wide variety of ages, provenance sources and compositions. The sample set includes seven distinctly different compositional groupings (petrofacies) with a wide variety of quartz, feldspar and rock fragment contents (Fig. 2). Within the rock fragment component, the sample set includes a variety of lithic types, including sedimentary, volcanic, metamorphic and plutonic rock fragments.

Petrographic data (91 samples in total) were collected from all seven petrofacies from the Gulf of Mexico (Table 1). Each sample was point counted (minimum 300 points each), and results were formatted for Touchstone modelling purposes. For each compositional grouping (petrofacies), a preliminary Touchstone model was built.

For most normal Touchstone modelling applications, model simulations utilize run parameters that are carefully calibrated by comparing actual rock data measurements with model predictions at analogue well locations. However, for this study, sandstone data were not evaluated using this traditional calibration modelling approach. Instead, a single set of generic model run parameters was used to drive each simulation. In this way, each petrofacies represents a starting composition for which provenance is the dominant variable. Local depositional or early diagenetic effects that might obscure the effects of provenance on model results are thereby nullified. The run parameters used are averages of parameters used successfully in previous Touchstone simulations in the Gulf of Mexico (Table 2).

In order to isolate the effects of provenance-controlled composition, some rock attributes were held constant, as follows. (1) Whenever possible, samples were collected from the same or similar depositional setting (proximal/axial turbidite fan system, such as inner lobe sheet sands and confined channel sands, with the exception of the Norphlet samples, which are aeolian). (2) Mean grain size and sorting for each sample were held constant (0.160 mm and 1.85 Trask). (3) Samples with high detrital clay matrix (greater than 5%) were eliminated from the study. (4) The degree of grain coating was held constant (15% coating used for all samples). Grain coatings, especially authigenic chlorite coats, could be related to provenance provided that the coatings originated from the alteration of volcanic rock fragments during burial. However, for the sample set used in this study, the only petrofacies with a significant volcanic grain content that could serve as seed material for chlorite coatings is the VL petrofacies (Table 1). For comparison, this petrofacies was modelled using both constant 15% grain coatings and actual grain coatings (78–83%). Therefore, any variability in predicted reservoir quality attributes were designed to be a result of compositional differences, and not in any way related to textural properties inherited from the local depositional setting.

Model simulations were run using a range of burial histories that would be expected to occur within the Gulf of Mexico. This enabled the simulation of a wide range of exploration opportunities and risk factors that might be present within the basin (from shallow and cool burial to deep and hot burial). Depth, vertical effective stress and maximum burial temperature ranges are given in Table 3. Burial histories are shown in Figure 3. To offset any temporal masking of provenance effects, all seven provenance groups were treated as having the same depositional age (65 Ma), despite the fact that their actual depositional ages differ.

The model parameters described above were designed specifically to facilitate the comparison of provenance-derived compositional effects on reservoir quality. Given the artificial run parameters used, the modelled results shown in the remainder of this publication should not be used for direct exploration applications. However, one set of predictive simulations was run for a local exploration prospect under evaluation (Wilcox subsalt reservoir). For that model, local well analogues were used. Each of those analogue wells was calibrated using Touchstone, and the calibrated models were then used for the local prospect prediction modelling.

## Qualitative process expectations

To some extent, burial processes can be qualitatively predicted without detailed modelling. The most quartzose of the sample groupings in this test are the QZ, SA and SL petrofacies (Table 1). These groupings have the highest detrital quartz content (80–90%) of the seven groups and are therefore the most mechanically rigid (with the greatest

**Table 1.** *Gulf of Mexico petrofacies modelled in Touchstone simulations*

| Petrofacies | Rock type* | Formation name | Age | Environment | Samples |
|---|---|---|---|---|---|
| QZ | Quartzarenite | Miocene | Miocene | Deep-water turbidite | 7 |
| LA | Lithic arkose | Frio | Oligocene | Deltaic | 6 |
| VL | Volcanic litharenite | Vicksburg | Oligocene | Deltaic | 3 |
| FL1 | Feldspathic litharenite | Upper Wilcox | Eocene | Deep-water turbidite | 32 |
| FL2 | Feldspathic litharenite | Lower Wilcox | Paleocene | Deep-water turbidite | 18 |
| SL | Sublitharenite | Tuscaloosa | Cretaceous | Deep-water turbidite | 16 |
| SA | Subarkose | Norphlet | Jurassic | Aeolian | 9 |

*Dominant rock type after Folk (1980) classification.

mechanical strength and therefore highest resistance to compaction). However, because of the high detrital quartz content, they also have the greatest depositional nucleation surface area (DNSA) available for precipitation of quartz cement during burial (Pittman *et al.* 1992; Walderhaug 1996; Makowitz & Sibley 2001; Lander *et al.* 2008). Therefore, those petrofacies are expected to have little resistance to quartz cementation under conditions of high thermal stress. Because of the limited amount of unstable feldspars and lithics, they are also expected to develop only minor amounts of microporosity and secondary porosity.

In contrast, the LA petrofacies is a lithic arkose with only about 53% detrital quartz. It also has the most abundant feldspar content (27%). Like the previous three groups, this petrofacies also has high mechanical strength and, because of the lower quartz content (and therefore lower DNSA), this facies is expected to be less susceptible to quartz cementation. Because of the higher feldspar and lithic content, it is expected to develop more secondary porosity and microporosity than the more quartzose facies previously described.

The FL2 petrofacies consists of feldspathic litharenites with even lower quartz content (40%), and more abundant ductile lithic grains, such as shale, schist, cherts and volcanic rock fragments. This facies should be less resistant to mechanical compaction than the others, but because of the lower quartz content and low DNSA, quartz cementation should be less common. Moderate amounts of

**Table 2.** *Touchstone model run parameters used in study*

| | |
|---|---|
| Compaction | (a) Used common grain ductility ($IGV_f$) values for each sand grain type |
| | (b) $IGV_f$ range used from 2.5% (biotite) to 25.7% (quartz) |
| | (c) Used common stiffness (gamma) values for pore-filling material |
| | (d) Gamma range from 0.29 (sericite) to 2.46 (quartz) |
| | (e) Beta (1/MPa) used in model = 0.06 |
| | (f) Depositional IGVs from Reservoir Quality Consortium experiments |
| Quartz cement | (a) Kinetic model dependent on grain size used for all simulations |
| | (b) Used common activation energy ($E_a = 58$ kJ mol$^{-1}$) and slope = $-2.02$ |
| Paragenesis | (a) Used common timing based on multiwell analogues (five categories): |
| |     20–40 °C, early diagenesis |
| |     45–100 °C, intermediate diagenesis |
| |     75–110 °C, late diagenesis |
| |     75–110 °C, secondary framework grain dissolution |
| | (b) Used illite kinetics model, from Lander & Bonnell (2010) |
| | (c) Used default crystal growth, nucleation and mass balance parameters |
| | (d) All other replacement reactions based on analogues |
| Core porosity | (a) Microporous fraction matched using 23 microporosity groupings |
| | (b) Micropore fraction range 6.2% (sandstone rock fragments) to 56.9% (kaolinite) |
| Permeability | (a) Utilized 26 surface area classes |
| | (b) Range from 29.5 (sulphate cement) to 684.4 mm$^2$ mm$^{-3}$ (chlorite) |
| | (c) Secondary porosity tortuosity = 5.0 |
| | (d) Maximum tortuosity = 5000 |
| | (e) Microporosity effectiveness = 0.55 |
| | (f) Correction factor = 1.00 |

**Table 3.** *Burial history ranges used in Touchstone simulations*

| History | $T_{max}$ (°C) | Depth | OP (MPa) | VES (MPa) |
|---|---|---|---|---|
| 1 | 60 | 7 961 | 12.6 | 14.4 |
| 2 | 80 | 11 446 | 27.6 | 15.9 |
| 3 | 100 | 14 932 | 42.5 | 16.5 |
| 4 | 120 | 18 418 | 54.7 | 18.4 |
| 5 | 140 | 21 904 | 66.8 | 20.4 |
| 6 | 160 | 25 390 | 77.4 | 23.1 |
| 7 | 180 | 28 875 | 95.5 | 29.0 |
| 8 | 200 | 32 361 | 113.4 | 34.6 |
| 9 | 220 | 35 847 | 128.7 | 41.8 |
| 10 | 240 | 39 333 | 142.7 | 49.5 |
| 11 | 260 | 42 819 | 150.4 | 52.2 |

Depth denotes true vertical depth in feet below mudline; OP, overpressure, VES, vertical effective stress.

secondary porosity and microporosity are expected to be present.

The most lithic of the provenance groupings are the FL1 and VL petrofacies. Both have very little detrital quartz, very low DNSA and are therefore expected to be somewhat more resistant to quartz cementation, except perhaps at very high levels of thermal stress. Both should be highly susceptible to mechanical compaction, especially the volcanic rock fragment-rich VL petrofacies. Moderate amounts of secondary porosity are expected, and for the VL petrofacies, high amounts of microporosity are expected to be associated with microporous volcanics and associated chlorite.

## Results of model simulations

### Compaction effects

For most sandstones, the dominant porosity reduction mechanism in the subsurface is mechanical compaction. The state of compaction is determined by the interaction between vertical effective stress (VES) and the mechanical properties of the sand (Houseknecht 1987; Pittman & Larese 1991; Paxton *et al.* 2002). The more ductile the sand grains, the less VES is required for compaction to reduce porosity. For this study, the different provenance petrofacies have been subjected to a simulated range of maximum burial depths (11 000–32 000 ft), and maximum VES (up to 52 MPa). Only mechanical compaction was simulated in these models. The predicted compaction profiles described in this study could vary in natural systems if chemical compaction (i.e. pressure solution) were to occur (Bjorkum 1996), but this process was not explicitly simulated in the Touchstone models.

Partially offsetting the effect of grain ductility is the presence of quartz and other rigid intergranular cements, which act to stiffen the sand pack and reduce the overall amount of compaction (Bernabe *et al.* 1992; Makowitz *et al.* 2006). For the burial histories used in this study, reservoir temperatures sufficient to begin generating significant quartz cement (>100 °C) are not predicted to occur until an average palaeo-burial depth of *c.* 15 000 ft. The reservoir VES at that palaeo-burial depth ranges from 15.4 to 16.3 MPa. For this range of VES, the starting depositional intergranular volume (IGV) has been reduced from 39% to a range of 24–31%. The final IGV range for the model simulations is 22–30%. Therefore, almost 90% of the modelled compaction is simulated to have occurred prior to any significant precipitation of quartz cement. Hence, any compensating effect of quartz cement rigidity affected the burial simulations primarily after significant compactional porosity loss had already occurred.

Model results were used to compare intergranular volume (IGV) v. geological time for the range of burial histories, as shown in Figure 3. One example of these is described in Figure 4, which plots predicted compaction behaviour for a moderate temperature/VES burial history (18 MPa VES and 120 °C). IGV is an indicator of the degree of compaction that the sand has experienced during burial (Houseknecht 1987). The higher the IGV value, the less compaction has occurred. Because grain size and sorting were held constant, all seven provenance petrofacies started out with a depositional IGV of 39%. The most mechanically stable, quartz-rich of these is the QZ petrofacies, which has been compacted to a final average IGV of *c.* 30%, meaning that the sand has lost about nine porosity units from mechanical compaction alone. This represents a 24% loss of the original pore volume from compaction. In contrast, the most ductile of the seven petrofacies (VL) has been compacted to a final IGV of 24%, meaning that the sand has lost about 15 porosity units from compaction alone. This represents a 51% loss of the original porosity from compaction, about twice that of the QZ petrofacies. Between these two end members, there is a good correlation between the overall ductility of the sand and the resulting final IGV. The more ductile the sand composition, the more compacted the sand and the lower the IGV.

Based on the model results shown in Figure 4, most of the compactional loss of porosity occurred during the first 13 myr of burial. The resulting compaction rate of the more rigid QZ petrofacies (0.44 pu per million years) is about half that of the more ductile VL petrofacies (0.83 pu per million years). From *c.* 52 Ma to the present day, compaction rate has stabilized for all provenance petrofacies, and is more than an order of magnitude less than it was during the first 13 myr. This rate reduction is

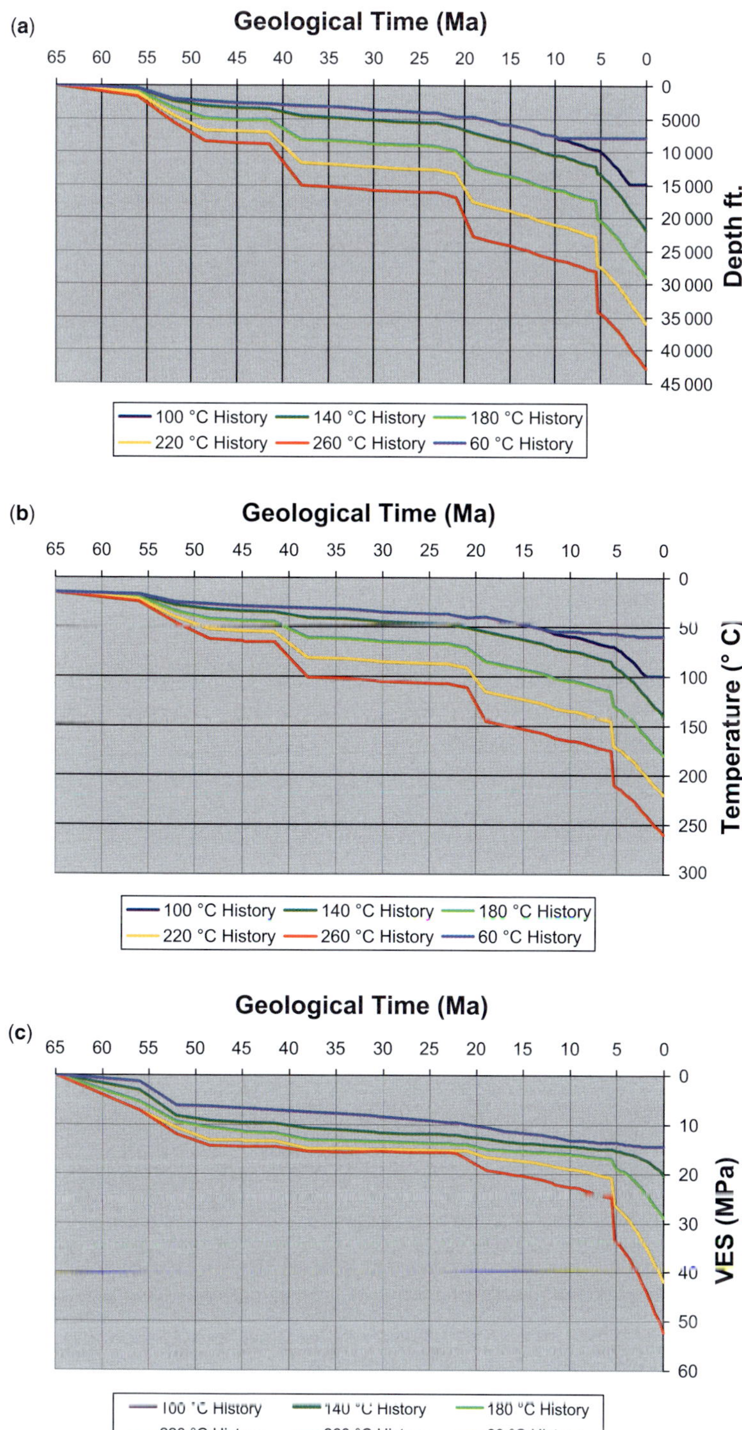


**Fig. 3.** Range of burial history curves used for Touchstone modelling in this study: (**a**) geological time v. burial depth (true vertical depth below mudline); (**b**) geological time v. maximum reservoir temperature; (**c**) geological time v. VES (vertical effective stress).

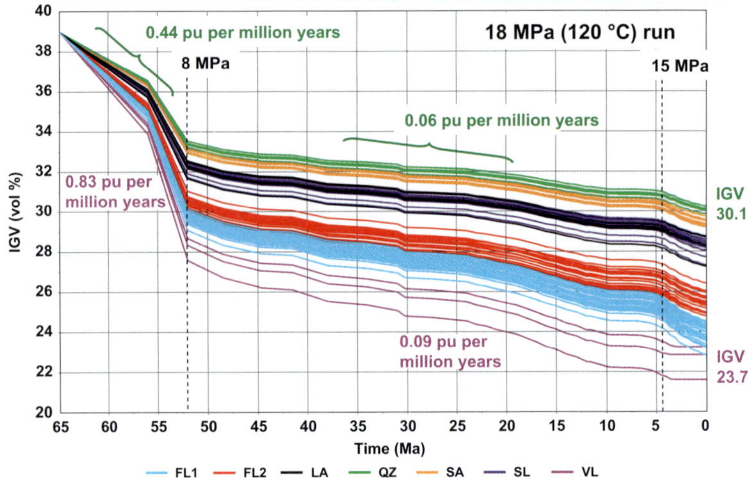



**Fig. 4.** Plot of simulated intergranular volume (IGV) v. geological time for all provenance groupings using the 120 °C/ 18 MPa burial history as an example. All 91 samples are shown as individual curves. VES at various time steps is shown with black vertical lines. Intergranular porosity decline rates (in porosity units (pu) per million years) are shown for two time segments for the QZ (green) and VL (pink) groups. Final IGV values for each group are also given for the final (0 Ma) time step.

a result of several factors. (1) Sedimentation rate was slower during the first 13 myr of burial (177 ft per million years), so there was little overpressure development and consequent rapid increase in VES (from 0 to 8 MPa, or 0.6 MPa per million years). During the last 52 myr, sedimentation rate was substantially higher (310 ft per million years), and overpressure increased more rapidly, resulting in a slower increase in VES (from 8 to 18.4 MPa, or 0.2 MPa per million years). (2) For all petrofacies, especially those with higher ductile grain compositions, mechanical compaction rates naturally slow down at an exponential rate with increasing VES (Pittman & Larese 1991). Therefore, increasing VES during the last 52 myr of burial had a diminishing impact on IGV change. (3) During the first 13 myr of burial, reservoir temperatures were low (<30 °C) and no quartz cement was precipitated, so there was no stiffening effect available to reduce mechanical compaction. During the last 52 myr of burial, temperatures gradually increased to a maximum of 120 °C, resulting in quartz cement stiffening of the sand. However, this effect was minimal except in the most detrital quartz-rich petrofacies (QZ and SL), and only during the last 5 myr of burial when temperatures were in excess of 80 °C. However, for higher-temperature simulations, the effect of quartz cement stiffening is more significant (e.g. the modelled 220 °C burial history simulates temperatures above 80 °C for the last 40 myr).

Similarly, model results were used to compare compaction state (IGV) v. maximum effective stress for the range of burial histories shown in Figure 3. Figure 5 summarizes the average IGV for each provenance petrofacies v. increasing VES for the burial scenarios given in Table 3. This plot demonstrates that the more mechanically ductile the sand, the greater the effect of VES on mechanical compaction. The most mechanically ductile petrofacies (VL) has lost 17.5 pu from mechanical compaction for all of the burial histories with VES greater than 23 MPa (0.8 pu MPa$^{-1}$). In contrast, the most rigid petrofacies (QZ) has lost 9.3 pu from mechanical compaction for all of the burial histories with VES greater than 23 MPa (0.4 pu MPa$^{-1}$).

Compaction behaviour was also evaluated by comparing IGV loss against palaeo-burial depth. Figure 6 presents a comparison of predicted compaction curves for each provenance petrofacies for three different burial histories (16 MPa/80 °C run, 18 MPa/120 °C run and 35 MPa/200 °C run). All three simulations show a more rapid decline of IGV within the first few thousand feet of burial followed by stable, minor compaction with continued burial. For the most mechanically rigid petrofacies (QZ), more than half of the total intergranular pore volume loss from mechanical compaction has occurred within the first 2000 ft of burial (82–86% of total pore volume loss by 6000 ft of burial). Although the total pore volume loss from compaction is higher for the most ductile facies (VL), a similar trend with depth is observed (53–68% loss by 2000 ft of burial, and 88–94% loss by 6000 ft of burial). Mechanical

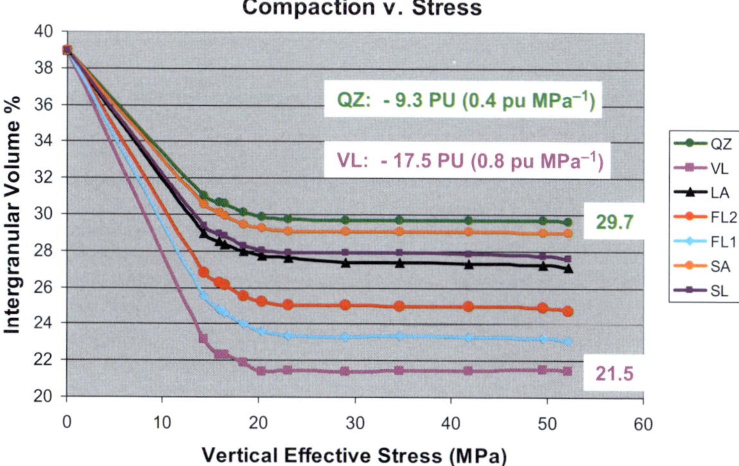



**Fig. 5.** Average IGV for each provenance group v. maximum VES. Each point shown on the curves represents the average IGV for the individual petrofacies v. the maximum VES for separate burial histories given in Table 3. These curves do not represent present-day VES v. IGV trends. Total intergranular porosity loss (in porosity units) and porosity loss per MPa are given for the QZ (green) and VL (pink) petrofacies. For all petrofacies, IGV loss levels off for all of the burial history scenarios with VES above 23MPa.

stiffening of the sand and consequent reduction in compaction for all petrofacies was effective only for those simulations below a palaeo-burial depth of *c.* 15 000 ft at reservoir temperatures greater than 100 °C.

## Quartz cementation effects

The volume of quartz cement precipitated in sand is dependent on three main factors: (1) the thermal history of the reservoir (i.e. temperature and

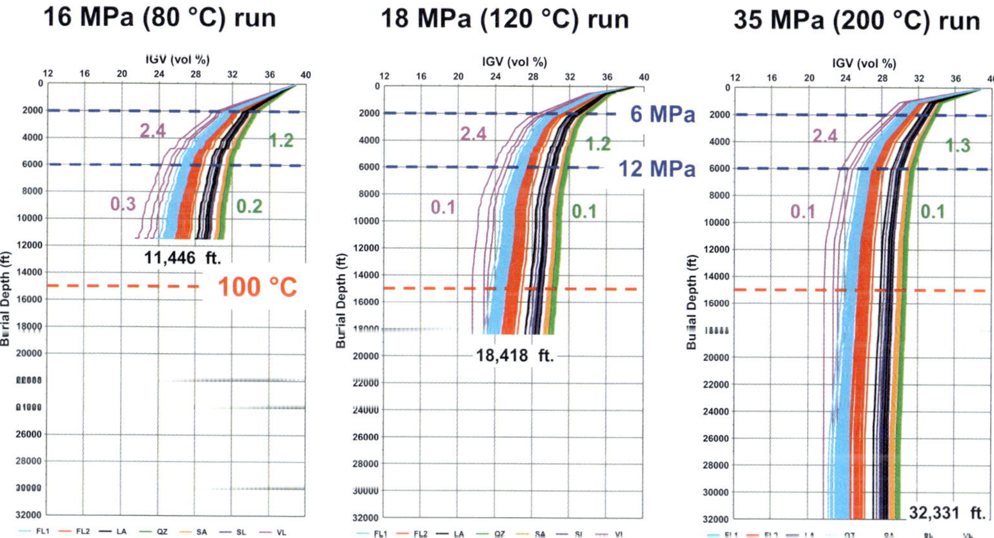



**Fig. 6.** Compaction curves for three representative burial histories. Each plot shows the decline of IGV with increasing palaeo-burial depth for all samples. Equivalent VES values are shown at 2000 and 6000 ft depths. IGV decline rates (roughly equivalent to intergranular porosity decline rates, and given in pu per 1000 ft) are shown for the QZ (green) and VL (pink) petrofacies above and below 6000 ft. Note that IGV decline rates are substantially higher above 6000 ft palaeo-burial depth in all cases.

heating duration), (2) the amount of pore space available for quartz to fill ('quartz accommodation space') and (3) quartz nucleation surface area (NSA). For this study, a variety of burial histories have been simulated to capture a wide range of maximum temperature (60–260 °C). Quartz accommodation space is generally reduced during burial by compaction that occurs prior to quartz cementation combined with any competition for pore space from other cements and authigenic clay minerals. For this sample set, compaction is the dominant control on reduction of quartz accommodation space during burial. Other non-quartz cements and authigenic clays combined range from an average of only 3.5% in the lowest temperature scenario to 5.5% in the highest temperature scenario. Quartz NSA changes throughout burial. The initial DNSA is controlled by the amount of detrital quartz sand grains, grain size and detrital grain coatings (Pittman *et al.* 1992; Walderhaug 1996; Makowitz & Sibley 2001; Lander *et al.* 2008). NSA is generally reduced during burial as authigenic grain coatings form, grain-to-grain contact surface area increases from compaction, and quartz cement and other cements precipitate. For this study, detrital clay grain coatings are minimal (0–1.4%, averaging 0.3%). Grain size and authigenic grain coatings were held artificially constant, but detrital quartz sand content was not adjusted. Therefore, for the conditions of this study, quartz cementation was strongly controlled by compositional differences related to provenance. Differences in

detrital sand grain composition greatly affected the degree of pre-quartz cement compaction (and therefore quartz accommodation space), as well as the evolving NSA during burial.

Some of these effects are illustrated in Figure 7 for one of the high thermal stress burial histories (200 °C, 35 MPa VES). In this simulation, quartz cementation starts at *c.* 48 Ma, but the volume precipitated for all provenance petrofacies is low until each petrofacies reaches a temperature greater than *c.* 100 °C. As the temperature increases, three effects are observed. (1) The higher the temperature the greater the volume of quartz cement precipitated. (2) Those petrofacies with the greatest detrital quartz sand content (and consequently the highest DNSA) are predicted to have the highest volume of quartz cement. Petrofacies with low detrital quartz content have smaller amounts of quartz cement. (3) At low burial temperatures (100 °C or slightly higher) the difference in modelled quartz cement is minimal, and provenance therefore has a minor impact on quartz cement-related porosity loss. However, at the highest maximum temperatures experienced, the provenance petrofacies display the greatest amount of separation in modelled quartz cement. Under present-day conditions (200 °C), a wide range of average quartz cement values are simulated, ranging from 1.5% (VL petrofacies with only 13% detrital quartz grains and 5 mm$^2$ DNSA) to 28.1% (QZ petrofacies with 92% detrital quartz grains and 158 mm$^2$ DNSA). Figure 8 illustrates the correlations between both

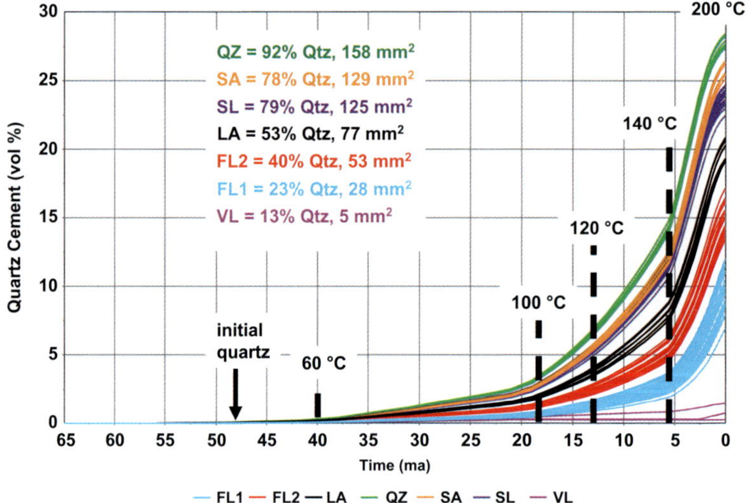


**Fig. 7.** Plot of simulated quartz cement volume v. geological time for all provenance petrofacies using the 200 °C/35 MPa burial history as an example. All 91 samples are shown as individual curves. Burial temperature at various time steps is shown with black vertical lines. Detrital quartz sand content (%) and corresponding DNSA (in mm$^2$) are shown for each provenance group.

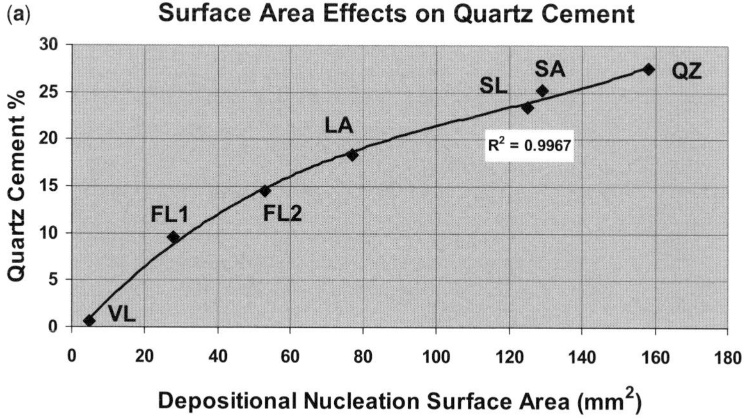



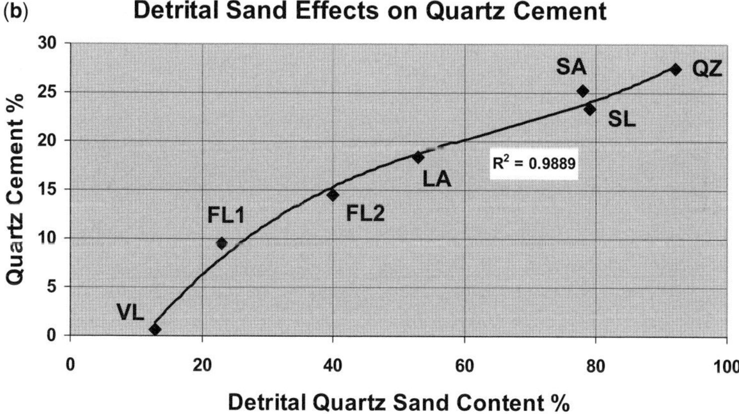



**Fig. 8.** Comparison of quartz cement (%) v. depositional nucleation controls: (**a**) average quartz cement v. average DNSA (in mm$^2$), and (**b**) average quartz cement v. average volume % detrital quartz sand. Both curves indicate a positive correlation between average quartz cement and the amount of surface area available for quartz nucleation at the time of deposition. Provenance petrofacies designations are referenced in Table 1.

DNSA and detrital quartz content v. precipitated quartz cement.

Another subtle observation is the interplay between quartz cementation and compaction during burial history. For example, the average IGV of the QZ petrofacies has been reduced by mechanical compaction from 39% at the time of deposition to a final value of 29.7% (at 200 °C and 35 MPa VES). By this point, an average of 28.1% quartz has been precipitated as pore filling cement (Fig. 7). Therefore, 28.1% of the available 29.7% intergranular accommodation space has been filled with quartz cement, yielding a cementation efficiency of 95% (i.e. 95% of the available space has been filled). In contrast, the FL2 petrofacies has been compacted to a final IGV of 25.0% at the end of its burial history. By this point, an average of 15.2% quartz has been precipitated as pore filling cement. Therefore, 15.2% of the available 25.0% intergranular accommodation space has been filled with quartz cement, yielding a cementation efficiency of only 61%. Therefore, quartz cementation in the FL2 petrofacies is slightly less than two-thirds as efficient as the QZ petrofacies given the same time and temperature conditions. The least quartzose petrofacies (VL) has been compacted to a final IGV of 21.5% at the end of its burial history. By this point, an average of only 1.5% quartz has been precipitated as pore filling cement, yielding a cementation efficiency of only 7%. The VL petrofacies is the only provenance grouping that contains sufficient detrital volcanic rock fragments that potentially could be altered to authigenic chlorite in the subsurface. Hence, this petrofacies could have more than the constant 15% grain coating used for all samples in this study. However, a repeat simulation of the 200 °C/35 MPa model using actual VL grain coatings (78–83%) yielded only a minor reduction in quartz cement for the VL petrofacies (0.1–0.9% quartz cement with high grain coatings v. 0.4–2.8% quartz cement using 15% grain coatings).

Quartz cementation is also dependent on thermal history. Figure 9 compares average quartz cement v. maximum burial temperature (up to 260 °C) for each provenance petrofacies for the burial scenarios given in Table 3. This plot also illustrates the strong dependency of quartz cement volume on detrital quartz content. The more quartzose petrofacies (QZ, SA) have higher modelled quartz cement volumes for all temperature histories than the more lithic petrofacies (FL1, VL).

A final observation is that quartz cementation does not correlate directly with palaeo-burial depth for all facies. Figure 10 compares quartz cement v. palaeo-burial depth for the 200 °C and 35 MPa VES model simulation. Quartz cement is predicted to be less than 1% at depths shallower than *c.* 10 000 ft. For this burial history, quartz cement volumes increase substantially at palaeo-burial depths between 15 000 and 20 000 ft. There is a dramatic decrease in additional quartz cementation predicted from *c.* 20 000 to 24 000 ft. This decrease in quartz cementation rate with depth is explained by the Arrhenius-based algorithm used in Touchstone, where both time and temperature are important components. In this example, burial from 20 000 to 24 000 ft is accompanied by an increase of 23 °C, but the burial is quite rapid, occurring in just 1 Ma, so a short-lived deviation

in the quartz cement curve is observed. At depths greater than 24 000 ft, quartz cementation is more significant because of the additional time (6 myr) and temperature (+48 °C) effects.

## Dissolution effects

Another reservoir quality consideration is the effect of provenance on the development of secondary porosity, which has been described previously as a mechanism that can offset the destructive effects of compaction and cementation (Schmidt & McDonald 1979). Secondary dissolution of labile grains, such as plagioclase, potassium feldspar and various lithics, adds additional porosity, although the net effect on reservoir quality is debatable. Although additional intragranular macroporosity is added to the pore system, secondary pores are generally more tortuous than primary intergranular pores, and are associated with the precipitation of clay cements (illite, kaolinite, chlorite) and other cements (albite, potassium feldspar, zeolites), thereby reducing permeability (Bjørlykke 1984; Giles & Marshall 1986). Recent work has demonstrated that in deep burial settings lacking hydrothermal fluid flow, diagenetic reactions occur within closed or nearly closed systems, and large-scale addition of porosity from mineral dissolution is unlikely (Bjørlykke & Jahren 2012).

For this study, most of the provenance groupings are modelled to have minor amounts of secondary porosity (Fig. 11). Those petrofacies with higher amounts of labile grains such as potassium feldspar, plagioclase, volcanic rock fragments, chert and

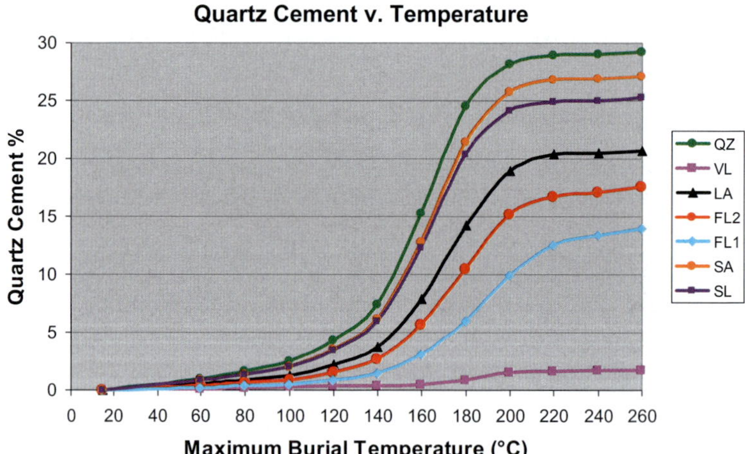


**Fig. 9.** Comparison of average quartz cement volume % v. increasing maximum burial temperature (°C) for all provenance petrofacies. Each point shown on the curves represents the average volume % quartz cement for the individual petrofacies v. the maximum temperature for separate burial histories given in Table 3. These curves do not represent present-day temperature v. quartz cement trends.

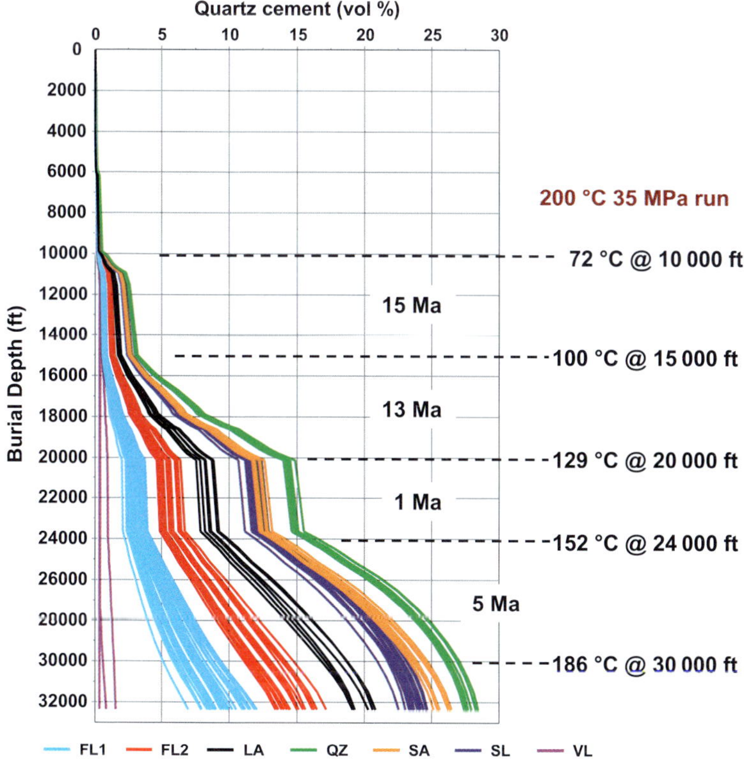


**Fig. 10.** Prediction of quartz cement v. palaeo-burial depth for all samples and provenance groupings for the 200 °C/35 MPa burial history. Time and palaeo-temperature steps are shown for various palaeo-burial depths.

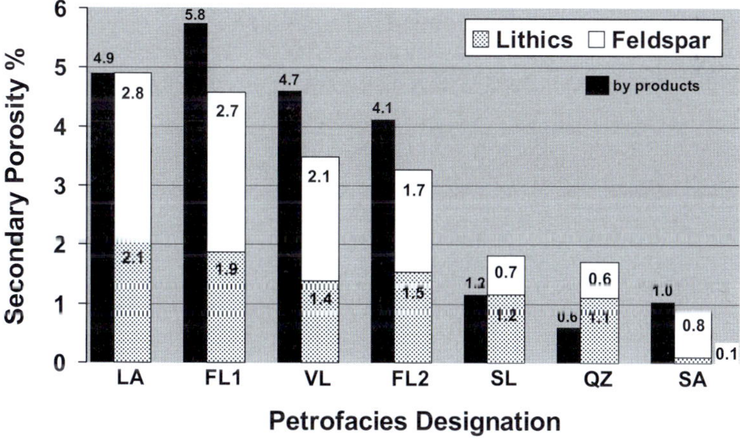


**Fig. 11.** Average secondary porosity for all seven provenance petrofacies for the 260 °C Touchstone simulation. Stacked bars represent the average volume of secondary porosity, subdivided by parent grain type (K-feldspar + plagioclase combined, and total lithic grains combined). Lithic parent grains include altered volcanic rock fragments and chert. Black bars represent the combined solid dissolution by-products for each petrofacies (illite, kaolinite, chlorite, albite, potassium feldspar, zeolites).

other unstable lithics (LA and FL1 petrofacies) are predicted to have slightly higher amounts of secondary porosity (5% v. other petrofacies with 1–3%). The simulated secondary porosity shown in Figure 11 is differentiated by parent grain type (feldspars and lithics). In all cases, the Touchstone results are based on the use of the fibrous illite model, which predicts secondary porosity formation from feldspar and kaolinite dissolution. Secondary porosity from the dissolution of lithics is based on petrographically observed secondary pores that have been attributed to dissolved lithic grains. For comparison, dissolution by-products are also shown in Figure 11. The average amount of combined (solid) dissolution by-products (illite, kaolinite, chlorite, albite, potassium feldspar, zeolites) equals or slightly exceeds the amount of secondary porosity observed for most petrofacies. Hence, the net effect on reservoir quality for this sample set is minor to no increase in porosity, and a general reduction of permeability.

*Authigenic clay effects*

Similarly, the effect of authigenic clay on reservoir quality must be considered in the model simulations. Insignificant amounts of authigenic clay are predicted at lower burial temperatures. At temperatures below *c.* 90 °C, all petrofacies have less than 4% authigenic clay. At higher temperatures (up to 260 °C), all of the provenance petrofacies are still predicted to have minor amounts of authigenic clay (<7%), except for the volcanic lithic-rich VL petrofacies. This petrofacies is predicted to have over 17% authigenic clay, mostly from the alteration of volcanic rock fragments to chlorite.

Authigenic clay content for all petrofacies is a mixture of illite, kaolinite and chlorite. Illite predictions are based on the Touchstone fibrous illite model (Lander & Bonnell 2010). Kaolinite abundance is based on petrographically observed kaolinite in the sample set. Chlorite abundance is based partially on petrographic observations from the sample set. For Touchstone simulations with burial temperatures exceeding actual well temperatures from the sample set, a proprietary Maersk empirical relationship between chlorite volume abundance and temperature was used to estimate chlorite volumes.

*Other cementation effects*

Other cements that are included in the models are calcite, ankerite, pyrite, zeolites, albite, potassium feldspar, and both iron and titanium oxides. The abundance of these cements (<1–4%) is not simulated by Touchstone but is based on petrographic point-count data. In the Touchstone model, the timing of each cement is based on the paragenetic constraints previously outlined. The VL petrofacies has the greatest abundance of these other cement types (4% total). Most of this cement (3%) is zeolite that formed from the alteration of volcanic rock fragments.

*Overall effects on porosity and permeability*

As stated previously, sandstone grain size, sorting and grain coating have been held constant in our models to remove their effects on porosity and permeability simulation. Therefore, the simulated values are mainly the end result of provenance-controlled differences in compaction, quartz cementation, other cements, authigenic clays and sand grain dissolution (although formation water chemistry could have a local control over some pore-filling cements such as carbonate cements). Figure 12 illustrates the combined effect of these processes on total porosity for a low-temperature, low-stress burial history simulation (120 °C and 18 MPa). In this case, the model simulates rapid porosity loss from 65 Ma to *c.* 52 Ma, largely in response to rapid compaction for all provenance petrofacies. From 52 Ma to *c.* 5 Ma, porosity loss is minor and very gradual. From 5 Ma to the present day, there is another slight increase in the rate of porosity loss, mainly due to the onset of quartz cementation and the precipitation of other cements. At this level of thermal maturity and effective stress, petrofacies with the highest porosities are those with high resistance to mechanical compaction (QZ, SA, SL). More ductile provenance petrofacies like FL1 and FL2 are more compacted, and therefore have the lowest porosity. One exception to this is the lithic-rich, highly ductile VL petrofacies, which has the highest porosity of all the provenance petrofacies. However, most of the VL pore system consists of microporosity within authigenic clays and altered volcanic rock fragments. Even though the total porosity is high, the permeability is expected to be quite low, so the overall reservoir quality is low.

In comparison, Figure 13 illustrates a similar porosity v. time profile for a high-temperature, high VES burial history (200 °C and 35 MPa). In this case, the porosity v. time profile is similar to the previous example for the early stages of burial history, with rapid porosity loss from 65 Ma to *c.* 52 Ma, largely in response to rapid compaction. From 52 Ma to *c.* 19 Ma, porosity loss is minor and very gradual. From 19 Ma to 0 Ma, there is another more rapid increase in the rate of porosity loss, mainly due to quartz and other cements. At this higher level of thermal maturity and VES, the petrofacies with the best porosities are those with higher resistance to quartz cementation (such as LA and FL1). The more quartzose petrofacies like

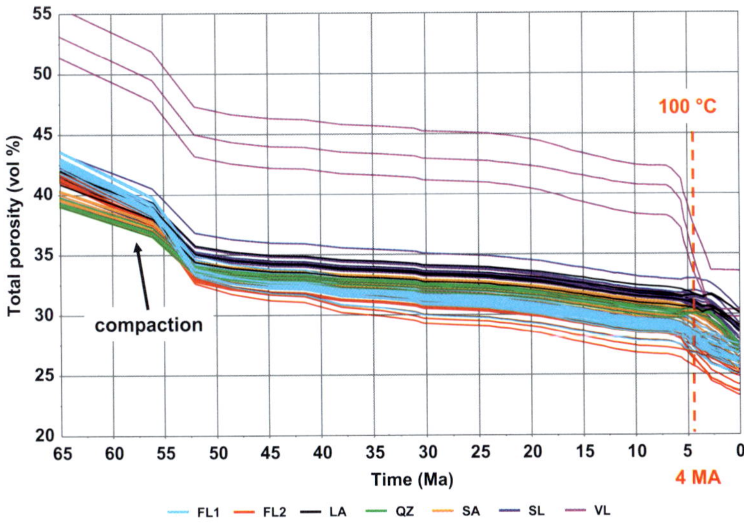



**Fig. 12.** Total porosity v. geological time for all provenance petrofacies using the 120 °C/18 MPa burial history as an example. All 91 samples are shown as individual curves. Significant burial processes are indicated, as well as the 100 °C isotherm.

QZ, SA and SL are more highly cemented, and therefore have the lowest porosity predictions. The cross-over from better to poorer reservoir quality for those three facies occurs between 10 and 17 Ma (105–120 °C). At temperatures greater than c. 140–150 °C, these three quartzose petrofacies have demonstrably poorer reservoir quality than all other petrofacies.

Figure 14 presents predicted permeability v. time for a low-temperature, low-VES burial history simulation (120 °C and 18 MPa burial history). In this case, rapid permeability loss occurs from 65 Ma to c. 52 Ma, largely in response to rapid compaction. From 52 Ma to c. 4 Ma, permeability loss is minor and very gradual. From 4 Ma to 0 Ma there is another slight increase in the rate of permeability

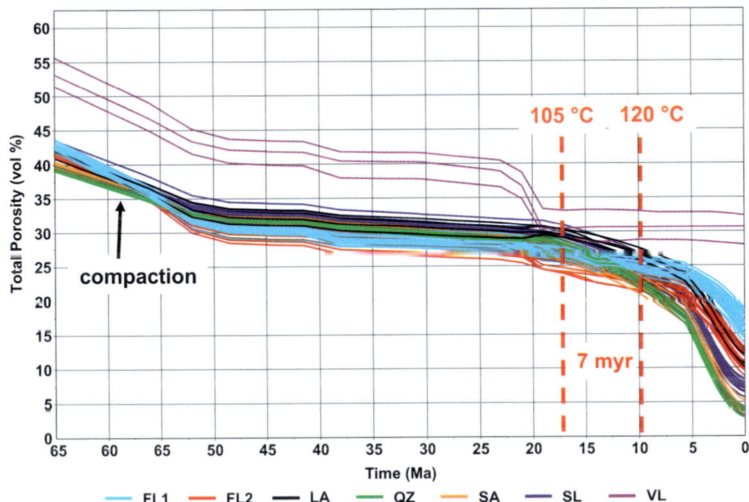



**Fig. 13.** Total porosity v. geological time for all provenance petrofacies using the 200 °C/35 MPa burial history as an example. All 91 samples are shown as individual curves. Significant burial processes are indicated, as well as the 105 °C and 120 °C isotherms.

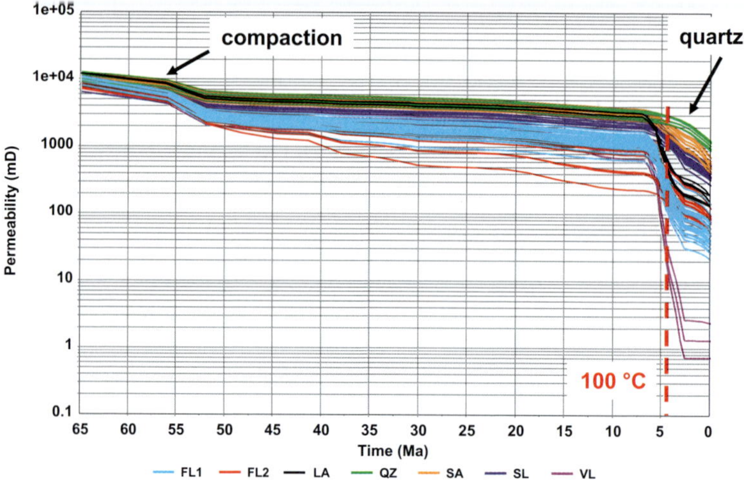



**Fig. 14.** Air permeability v. geological time for all provenance petrofacies using the 120 °C/18 MPa burial history as an example. All 91 samples are shown as individual curves. Significant burial processes are indicated, as well as the 100 °C isotherm.

loss, mainly caused by the onset of quartz and other cements. At this level of thermal maturity and VES, the petrofacies with the best permeability are clearly those with high resistance to mechanical compaction (QZ, SA and SL). The more ductile petrofacies like FL1 and FL2 are more compacted, and therefore have the lowest permeability. As predicted previously, the VL petrofacies has high porosity, but most of the pore system consists of microporosity, so even though the total porosity is high, the permeability and overall reservoir quality are by far the lowest of the sample database.

In comparison, Figure 15 illustrates a similar permeability v. time profile for a high-temperature, high-VES burial history (200 °C and 35 MPa). In this case, rapid permeability loss is predicted from 65 Ma to c. 52 Ma, largely in response to rapid compaction. From 52 Ma to c. 19 Ma, permeability loss is minor and very gradual. From 19 Ma to 0 Ma, there is another more rapid increase in the rate of permeability loss, mainly due to quartz and other

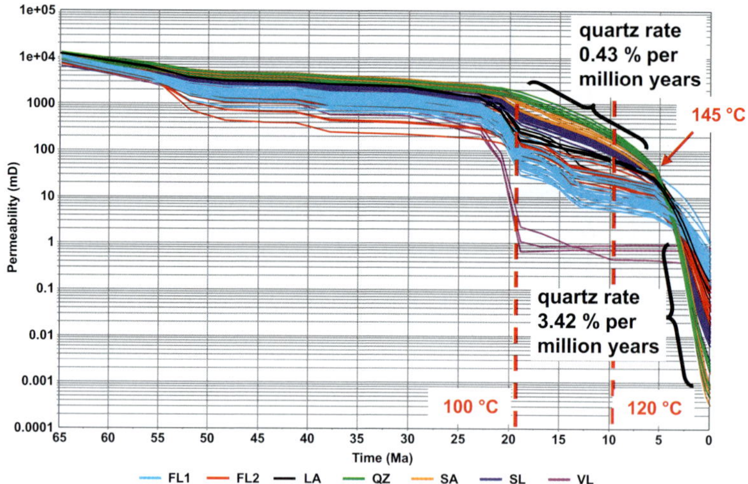



**Fig. 15.** Air permeability v. geological time for all provenance petrofacies using the 200 °C/35 MPa burial history as an example. All 91 samples are shown as individual curves. Significant burial processes are indicated, as well as various temperature isotherms.

cements. At this higher level of thermal maturity and VES, the petrofacies with the best final permeability values are those with higher resistance to quartz cementation (LA and FL1). The more quartzose petrofacies like QZ, SA and SL are more cemented, and therefore have the lowest permeability. The cross-over from better to poorer reservoir quality for those three petrofacies occurs between 4 Ma and 7 Ma (by 145 °C). At higher temperatures, those three quartzose petrofacies have considerably poorer reservoir quality than all other facies. Average quartz cement rates for those petrofacies are considerably higher above 145 °C (average of 3.42% per million years above 145 °C, compared with 0.43% per million years between 100 °C and 145 °C).

Porosity and permeability decline v. maximum burial temperature (up to 260 °C) are also illustrated in Figure 16. Given the absence of high grain coatings (previously discussed), the model simulations indicate that the poorest petrofacies (lowest porosity and permeability) at very high levels of thermal stress (above 150 °C) are those with high detrital quartz sand content and associated advanced quartz cementation. Better reservoir quality (higher porosity and permeability) is predicted for those petrofacies with lower detrital quartz sand content and lower quartz cement. The VL petrofacies is predicted to have the highest porosity of all the provenance petrofacies at very high temperatures, with essentially no change in total porosity from

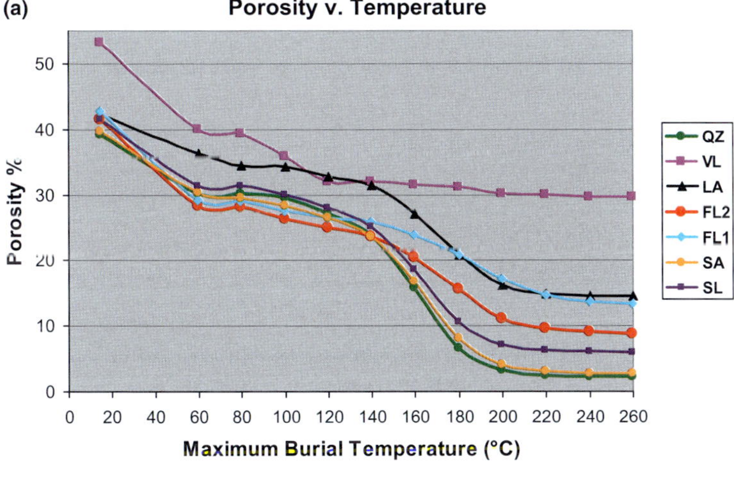


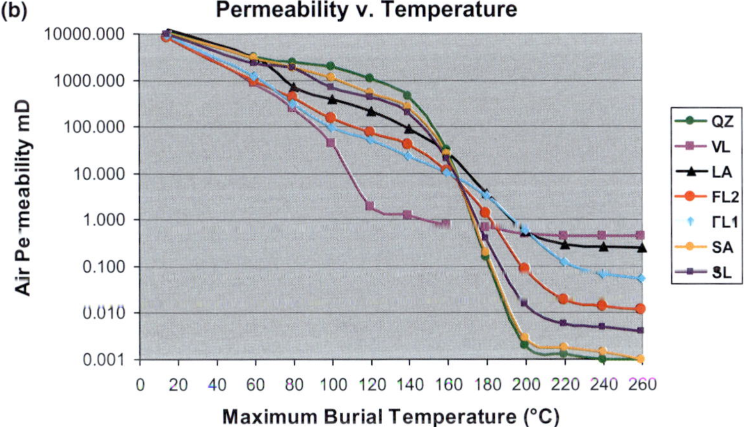


**Fig. 16.** Reservoir quality attributes v. maximum burial temperature: (**a**) average total porosity v. maximum burial temperature for all seven provenance petrofacies; (**b**) average air permeability v. maximum burial temperature for all seven provenance petrofacies. Each point shown on the curves represents the average porosity (or permeability) for the individual petrofacies v. the maximum temperature for separate burial histories given in Table 3. These curves do not represent present-day temperature v. porosity or permeability trends.

*c.* 120 °C to 260 °C. However, as discussed previously, most of the VL pore system consists of microporosity, which is accompanied by low permeability for the same thermal range.

### Local prospect evaluation model

Clearly, provenance-controlled differences in sand composition across stratigraphic boundaries greatly impact reservoir quality. The next part of the study is focused on the impact that less pronounced differences in provenance within a single formation might have on reservoir quality. Figure 17 is a seismic cross-section showing two exploration drilling leads in the deep-water Gulf of Mexico. Potential traps are located for a single formation of interest, in this case the Palaeogene Wilcox reservoir interval. Because of seismic image quality and well tie-in limitations, only a single seismic reflector can be identified that is probably the Wilcox Formation. In this case, Upper and Lower Wilcox divisions cannot be picked with certainty. These two leads are located in a portion of the basin where two different provenance interpretations are possible (i.e. the sand could have been sourced by two different deltaic complexes onshore). For this scenario, petrographic studies from analogue wells suggest that any sandstone reservoir strata encountered at both locations (*c.* 70 miles apart) were likely derived from the same provenance (a northwestern source).

The burial depth of the Wilcox Formation reflector at both lead locations is roughly the same (*c.* 32 000 ft total vertical depth subsea, TVDss). However, location A is positioned under a thick salt canopy (18 000 ft), whereas location B is positioned under a much thinner salt body (4000 ft). In both cases, the salt cover is widespread laterally to the extent that it is expected to affect the local temperature and VES history, and is interpreted to have been present since Early Miocene time (last 15–20 Ma). Because of sediment density and thermal conductivity differences, basin modelling results predict a slightly higher maximum temperature at location B (6 °C higher) and a higher vertical effective stress (4 MPa higher). These observations are consistent with similar subsalt observations and Touchstone modelling by Taylor *et al.* (2010) for the deep-water Gulf of Mexico. Using these two burial simulations, the Touchstone model predicts slightly higher quartz and other cements and higher compaction at location B. Therefore, the predicted porosity and permeability at location B (16.5% porosity and 3.5 mD air permeability) are lower than at location A (20.6% porosity and 19.6 mD air permeability). This decrease in porosity and permeability is largely caused by increased mechanical compaction and a minor increase in predicted quartz cementation. With this scenario, an economically viable outcome cannot be modelled when trying to simulate flow from the Wilcox at location B, so the decision is to drop this lead in favour of location A.

Figure 18 describes an alternative hypothesis for provenance in this area. In this case the two leads are interpreted to have been sourced from two

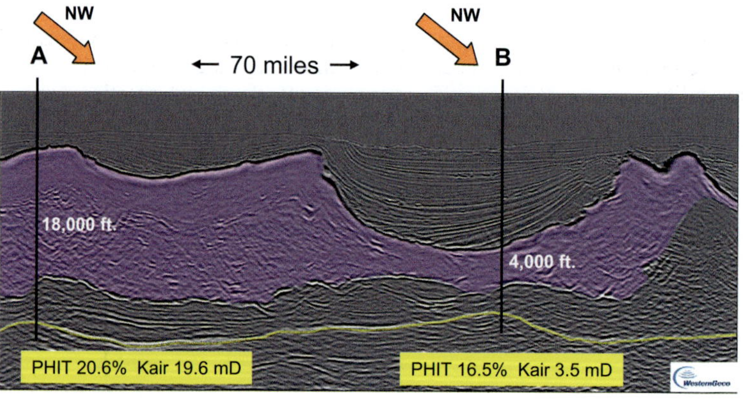



**Fig. 17.** Seismic example of two exploration drilling leads (A and B) in the deep-water Gulf of Mexico. The reservoir target is the Palaeogene Wilcox horizon (yellow line). Note the difference in the overlying salt layer thickness (18 000 ft at A and 4000 ft at B) and resulting temperature and VES conditions. Both locations are interpreted to have been sourced from the same provenance. Resulting modelled Touchstone predictions for porosity and permeability are given. PHIT is total porosity; Kair is air permeability.

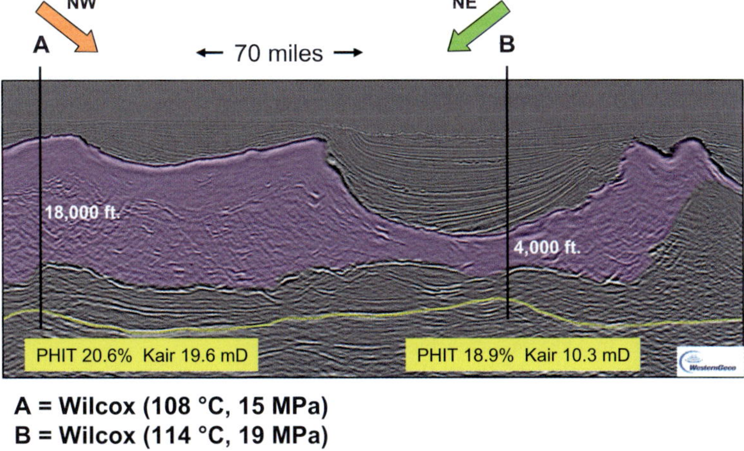


**Fig. 18.** Seismic example of two exploration drilling leads (A and B) in the deep-water Gulf of Mexico. The reservoir target is the Palaeogene Wilcox horizon (yellow line). Note the difference in the overlying salt layer thickness (18 000 ft at A and 4000 ft at B) and resulting temperature and VES conditions. In this scenario, location B is interpreted to have been sourced from a different provenance, resulting in more favourable modelled Touchstone predictions for porosity and permeability. Abbreviations as Figure 17.

different sand provenances (different delta systems). The Wilcox sand at location A is interpreted to have been sourced from a northwestern provenance (with more ductile metamorphic and sedimentary rock fragments, especially schist and shale, and with slightly lower quartz and feldspar content). Location B is interpreted to have been sourced from a different delta system to the NE. This source provides sand with fewer ductile plutonic and metamorphic rock fragments (especially gneiss and quartzite), and slightly more detrital quartz and feldspar. In this case, Touchstone predicts less compaction, and therefore a better porosity and permeability outcome at location B, even though the thermal and effective stress histories are less favourable (18.9% total porosity and 10.3 mD air permeability). With this outcome, economic flow rates can be simulated at both locations. Consequently, the business decision would be to include both leads in the exploration portfolio. This revised interpretation clearly demonstrates that even within a single formation of interest, provenance can impact predictive results. In this example, the interpretation of sediment dispersal (and therefore provenance) can easily impact economic drill or drop decisions like this.

## Discussion

Provenance commonly exerts major controls on sand composition, which affects compaction, diagenesis and ultimately reservoir quality. In the deep-water Gulf of Mexico, provenance reconstruction is critical for interpreting sediment transport fairways and predicting both the presence and quality of potential sandstone reservoirs. Touchstone modelling has demonstrated that provenance-controlled differences in sand composition greatly affect predicted porosity and permeability under a wide range of thermal and effective stress histories.

For all provenance petrofacies examined in this study and the burial histories modelled, compaction rates are highest during the first 13 myr of burial (first 2000–3000 ft). Vertical effective stress at those depths are less than 14 MPa (2 000 psi). At greater depth and effective stress, sand compaction is stabilized. Quartz and feldspar-rich petrofacies are more rigid and are less affected by mechanical compaction than more ductile, lithic-rich petrofacies (especially those containing schist, shale and some volcanic lithics). Ductile petrofacies compact at twice the rate of rigid facies.

Quartz cementation is highest for detrital quartz-rich facies, lower for feldspathic and lithic-rich facies, and lowest for ultra lithic facies. At any level of thermal stress, quartz-rich facies are predicted to have the highest quartz nucleation surface area, and subsequently the highest predicted quartz cement. The higher the level of thermal stress, the greater the effect of provenance that is observed.

The dissolution of feldspars and some lithics creates secondary intragranular macroporosity. The abundance of secondary porosity is provenance-controlled. Those facies with the highest plagioclase, potassium feldspar, altered volcanic rock fragments and chert have the highest amounts of observed and modelled secondary porosity. However, model simulations indicate that those petrofacies with higher amounts of secondary porosity also have higher volumes of dissolution by-products (authigenic clay and other cements), so there does not appear to be an overall improvement in reservoir quality resulting from dissolution. Similarly, authigenic clays are provenance-controlled. Feldspathic and lithic facies develop more clay during diagenesis, mainly from the alteration of feldspars to illite and kaolinite, and the alteration of volcanic rock fragments to chlorite. In addition, other minor cements (carbonates, pyrite, zeolite, feldspars, FeO and TiO) are slightly more common in facies with more lithic and feldspathic content.

Under low temperature and VES conditions (less than 100–120 °C and 16–20 MPa VES) the best reservoirs are those derived from sources with high feldspar and quartz content. For those facies even high thermal stress does not induce abundant cements, and the high associated mechanical strength inhibits mechanical compaction. At higher temperature and VES conditions (greater than 140 °C and 22 MPa VES) the best reservoirs are those derived from sources with more moderate amounts of lithics and feldspar. At higher temperatures, the thermal drive to precipitate quartz cement destroys porosity and permeability in more quartzose facies.

At the local (lead, prospect) level, sand provenance for a single horizon can have a moderate effect on predicted reservoir quality, and can make enough of a difference to alter drill or drop decisions.

The authors would like to express their appreciation to Maersk Oil for funding support and for permission to publish this study. WesternGeco kindly provided permission to publish the seismic line shown in Figures 17 and 18. This article benefited considerably from comments and suggestions by reviewers Robert Lander and Thomas Taylor.

# References

BERNABE, Y., FRYER, D. & HAYES, J. 1992. The effect of cement on the strength of granular rocks. *Geophysical Research Letters*, **19**, 1511–1514.

BJØRKUM, P. A. 1996. How important is pressure in causing dissolution of quartz in sandstones? *Journal of Sedimentary Research*, **66**, 147–154.

BJØRLYKKE, K. 1984. Formation of secondary porosity: how important is it? *In*: MACDONALD, D. A. & SURDAM, R. C. (eds) *Clastic Diagenesis*. American Association of Petroleum Geologists, Memoirs, **37**, 277–286.

BJØRLYKKE, K. & JAHREN, J. 2012. Open or closed geochemical systems during diagenesis in sedimentary basins: constraints on mass transfer during diagenesis and the prediction of porosity in sandstone and carbonate reservoirs. *American Association of Petroleum Geologists Bulletin*, **96**, 2193–2214.

DICKINSON, W. R. 1985. Interpreting provenance relations from detrital modes of sandstones. *In*: ZUFFA, G. G. (ed.) *Provenance of Arenites*. Reidel Publishing Company, Dordrecht, 331–361.

FOLK, R. L. 1980. *Petrology of Sedimentary Rocks*. Hemphill Publication Company, Austin.

GALLOWAY, W. E. 2008. Depositional evolution of the Gulf of Mexico sedimentary basin. *In*: MIALL, A. D. (ed.) *The Sedimentary Basins of the United States and Canada*. Elsevier, Amsterdam, 505–549.

GILES, M. R. & MARSHALL, J. D. 1986. Constraints on the development of secondary porosity in the subsurface: re-evaluation of processes. *Marine and Petroleum Geology*, **3**, 243–255.

HOUSEKNECHT, D. W. 1987. Assessing the relative importance of compaction processes and cementation to reduction of porosity in sandstones. *American Association of Petroleum Geologists Bulletin*, **71**, 633–642.

INGERSOLL, R. V. 1990. Actualistic sandstone petrofacies: discriminating modern and ancient source rocks. *Geology*, **18**, 733–736.

LANDER, R. H. & BONNELL, L. M. 2010. A model for fibrous illite nucleation and growth in sandstones. *American Association of Petroleum Geologists Bulletin*, **94**, 1161–1187.

LANDER, R. H. & WALDERHAUG, O. 1999. Porosity prediction through simulation of sandstone compaction and quartz cementation. *American Association of Petroleum Geologists Bulletin*, **83**, 433–449.

LANDER, R. H., LARESE, R. E. & BONNELL, L. M. 2008. Toward more accurate quartz cement models: the importance of euhedral versus noneuhedral growth rates. *American Association of Petroleum Geologists Bulletin*, **92**, 1537–1563.

LOWE, D. G., SYLVESTER, P. J. & ENACHESCU, M. E. 2011. Provenance and paleodrainage patterns of Upper Jurassic and Lower Cretaceous synrift sandstones in the Flemish Pass Basin, offshore Newfoundland, east coast of Canada. *American Association of Petroleum Geologists Bulletin*, **95**, 1295–1320.

MAKOWITZ, A. & SIBLEY, D. F. 2001. Crystal growth mechanisms of quartz overgrowths in a Cambrian quartz arenite. *Journal of Sedimentary Research*, **71**, 809–816.

MAKOWITZ, A., LANDER, R. H. & MILLIKEN, K. L. 2006. Diagenetic modeling to assess the relative timing of quartz cementation and brittle grain processes during compaction. *American Association of Petroleum Geologists Bulletin*, **90**, 873–885.

MORTON, A. & HURST, A. 1995. Correlation of sandstones using heavy minerals: an example from the Statfjord Formation of the Snorre Field, Northern North Sea. *In*: DUNAY, R. E. (ed.) *Non-biostratigraphical Methods of Dating and Correlation*. Geological Society, London, Special Publications, **89**, 3–22.

PAXTON, S. T., SZABO, J. O., AJDUKIEWICZ, J. M. & KLIMENTIDES, R. E. 2002. Construction of an intergranular compaction curve for evaluating and predicting compaction and porosity loss in rigid grained sandstone reservoirs. *American Association of Petroleum Geologists Bulletin*, **86**, 2047–2067.

PITTMAN, E. D. & LARESE, R. E. 1991. Compaction of lithic sands: experimental results and applications. *American Association of Petroleum Geologists Bulletin*, **75**, 1279–1299.

PITTMAN, E. D., LARESE, R. E. & HEALD, M. T. 1992. Clay coats: occurrence and relevance to preservation of porosity in sandstones. *In*: HOUSEKNECHT, D. W. & PITTMAN, E. D. (eds) *Origin, Diagenesis, and Petrophysics of Clay Minerals in Sandstones*. SEPM Special Publications, Tulsa, **47**, 241–255.

SCHMIDT, V. & McDONALD, D. A. 1979. The role of secondary porosity in the course of sandstone diagenesis. *In*: SCHOLLE, P. & SCHLUGER, P. (eds) *Aspects of Diagenesis*. SEPM Special Publications, Tulsa, **26**, 175–207.7

TAYLOR, T. R., GILES, M. R. *ET AL.* 2010. Sandstone diagenesis and reservoir quality prediction: models, myths, and reality. *American Association of Petroleum Geologists Bulletin*, **94**, 1093–1132.

VINCENT, S. J., MORTON, A. C., HYDEN, F., GARZANTI, E. & VEZZOLI, G. 2010. Sandstone provenance and reservoir quality prediction in the Eastern Black Sea. Abstract presented at the 2010 AAPG European Region Annual Conference, Kiev, Ukraine, 17–19 October 2010.

VINCENT, S. J., HYDEN, F. & BRAHAM, W. 2013. Along-strike variations in the composition of sandstones derived from the uplifting western Greater Caucasus – causes and implications for reservoir quality prediction in the Eastern Black Sea. *In*: SCOTT, R. A., SMYTH, H. R., MORTON, A. C. & RICHARDSON, N. (eds) *Sediment Provenance Studies in Hydrocarbon Exploration and Production*. Geological Society, London, Special Publications, **386**. First published online September 17, 2013, http://dx.doi.org/10.1144/SP386.15.

WALDERHAUG, O. 1996. Kinetic modeling of quartz cementation and porosity loss in deeply buried sandstone reservoirs. *American Association of Petroleum Geologists Bulletin*, **80**, 731–745.

# The diagenetic destruction of porosity in plutoniclastic petrofacies: The Miocene Diligencia and Eocene Maniobra formations, Orocopia Mountains, southern California, USA


L. CARACCIOLO[1]*, J. ARRIBAS[2], R. V. INGERSOLL[3] & S. CRITELLI[1]

[1]*Dipartimento di Scienze della Terra, Università della Calabria, via P.Bucci, cubo 15b, 87036, Rende, Italy*

[2]*Departamento de Petrología y Geoquímica, UCM – IGEO (UCM-CSIC), Madrid, Spain*

[3]*Department of Earth and Space Sciences, University of California, Los Angeles, California 90095-1567, USA*

*Corresponding author (e-mail: luca.caracciolo@unical.it)*



**Abstract:** The Miocene Diligencia Formation of southern California was deposited in an alluvial/ fluvial depositional system developed within the Diligencia basin. This basin formed as an extensional half-graben during latest Oligocene–Early Miocene crustal extension. This continental half-graben was superposed on an Eocene fault-controlled submarine canyon, in which the Maniobra Formation was deposited.

The arkosic composition of the sandstone framework would predict a high-quality reservoir with preserved primary porosity or at least important intergranular volume (IGV). However, inspection of sandstone samples indicates that diagenesis was intense, as several diagenetic processes drastically reduced the reservoir quality. Dominant diagenetic processes were compaction and cementation. Mechanical compaction, expressed as intense deformation of ductile grains and chemical compaction in the form of pressure solution, reduced the IGV of sandstone at the bases of the Maniobra and Diligencia formations to low levels (10–20%). In the upper parts of both units, early diagenetic cements partially inhibited compaction, maintaining IGV values close to 35%.

Several mineral phases constitute cements responsible for occlusion of primary porosity. Many of these phases also replaced framework components and early cements. The more important cements are: (1) quartz and K-feldspar, which appear mainly as grain overgrowths, although quartz mosaics have been observed at the base of the Maniobra Formation; (2) carbonates, such as ankerite, dolomite and calcite in diverse textures and with significant occluding character; (3) phyllosilicates, such as kaolin, that developed mainly in Diligencia sandstone as early pore fillings, and locally as illite coats around detrital grains; and (4) other mineral phases, such as Fe-oxides and fluorite, which occur exclusively at the base of the Maniobra Formation, exhibiting aggressive textures against framework grains and older cements.

The chronology of diagenetic processes includes marine early diagenesis (eodiagenesis) for Maniobra sandstone, characterized by K-feldspar, ankerite and dolomite cements. Continental early diagenesis is identified in Diligencia sandstone, manifested by the presence of smectite (replaced by illite), kaolin and Fe-oxide cements. Mesodiagenetic processes are similar in both formations. In addition, hydrothermal phases (fluorite and quartz mosaics) are identified at the base of the Maniobra Formation. Diagenetic mineral associations suggest palaeotemperatures above 130 °C.

The geotectonic scenario in which diagenesis occurred explains these post-depositional processes. During latest Oligocene–Early Miocene crustal extension, the Diligencia basin developed in an area of high heat flow, as expressed in the eruption of interbedded basaltic–andesitic lavas. High heat flow favoured compaction and cementation, accelerating these diagenetic processes during a relative short time interval. Hydrothermal fluxes produced mineral phases that contributed to the destruction of a potentially good reservoir by intense diagenesis.



**Supplementary material:** Optical analysis: point count and recalculated parameters are available at http://www.geolsoc.org.uk/SUP18650


Plutoniclastic deposits are generated in diverse tectonic settings related to erosion of felsic crustal terranes, such as basement uplifts associated with rift basins and transcurrent faults (Dickinson 1985). These deposits generally form thick strata of continental alluvial/fluvial or marine origin. Plutoniclastic sand(stone) has framework compositions consisting dominantly of monomineralic quartz


*From*: SCOTT, R. A., SMYTH, H. R., MORTON, A. C. & RICHARDSON, N. (eds) 2014. *Sediment Provenance Studies in Hydrocarbon Exploration and Production*. Geological Society, London, Special Publications, **386**, 49–62. First published online July 15, 2013, http://dx.doi.org/10.1144/SP386.9

and feldspars, and negligible aphanitic lithic fragments. This monomineralic composition is due to the coarse crystal size of plutonic sources (Pettijohn *et al.* 1972). Quartz and feldspar provide consistent framework rigidity, which helps maintain intergranular space during burial diagenesis (Bloch 1991). In addition, secondary porosity generated by feldspar dissolution during diagenesis may enhance reservoir quality (Schmidt & McDonald 1979). These features make plutoniclastic sandstone attractive as potential reservoirs. Good examples of high-quality quartzofeldspathic reservoirs have been widely documented (e.g. Frio Formation in the Gulf of Mexico, Land *et al.* 1987; Wilson 1994).

First-cycle basins are generally characterized by porosity reduction in sandstone with increasing depth, providing useful tools for hydrocarbon exploration, such as regional porosity–depth and porosity–temperature curves (e.g. Schmoker & Gautier 1988; Schmoker & Schenk 1994; Ehrenberg *et al.* 2008; Taylor *et al.* 2010). As a consequence, processes or conditions limiting compaction and cementation, as well as those enhancing porosity through dissolution, determine porosity–depth trends in best-quality reservoir sandstone. Moreover, the interplay of controlling factors, such as sand(stone) composition, texture and fluid types, as well as temperature variation and tectonic conditions, is crucial in determining reservoir porosity and permeability.

All these processes and conditions are likely to change, with early compaction and shallow groundwater systems influencing the type and abundance of early diagenetic attributes that determine the dissolution of unstable grains and consequent formation of secondary porosity, as well as the precipitation of early cements (Bjørlykke 1983; Morad *et al.* 2010).

The aim of this paper is to document how a potentially good reservoir, based on its original composition, has been degraded by diagenesis in a tectonic setting with high heat flow.

## Geology and stratigraphy

The Diligencia basin of the eastern Orocopia Mountains in southern California (Fig. 1) is characterized by 2500–3000 m of Tertiary siliciclastic, and subordinate limestone and evaporitic strata, locally intercalated with basaltic lavas (Crowell 1975; Spittler & Arthur 1982; Law *et al.* 2001; Ingersoll 2009). This basin developed under a complex tectonic regime, synchronous with the transition from a convergent to a transform plate margin (e.g. Atwater 1970; Crowell 1987; Graham *et al.* 1989; Nicholson *et al.* 1994; Bohannon & Parsons 1995; Dickinson 1997; Atwater & Stock

1998). Sedimentological, stratigraphic and structural analyses indicate Late Oligocene to Miocene extension as a half-graben, followed by Late Miocene to Pliocene inversion, in a transpressional regime (Law *et al.* 2001; Ingersoll 2009).

The Orocopia and Hayfield Mountains (Fig. 1) consist of crystalline Proterozoic gneiss and plutonic rocks, and Upper Cretaceous–Palaeogene Orocopia Schist, as well as the sedimentary marine Eocene Maniobra Formation, nonmarine Oligo-Miocene Diligencia Formation and diverse nonmarine Pliocene–Quaternary deposits (Crowell 1975; Spittler & Arthur 1982; Law *et al.* 2001; Jacobson *et al.* 2007). The crystalline basement consists of a medium-grade lower plate of Orocopia Schist, which is tectonically overlain by Proterozoic gneiss, Proterozoic mafic to ultramafic plutonic rocks and Late Cretaceous granite. Multiple extensional faults cut these crystalline rocks along the southwestern edge of the Diligencia basin (Jacobson *et al.* 2007). Only Mesozoic granitic plutons and related gneiss north of the basin contributed detritus to the Maniobra and basal Diligencia strata examined in the present study (Ingersoll 2009).

The Maniobra Formation nonconformably overlies the Mesozoic granitoids, and locally is in tectonic contact with these granitoids. The Diligencia Formation nonconformably overlies Proterozoic gneiss on the south and Mesozoic granitoids in the west, and is in angular unconformity on the Maniobra Formation on the north, where the present study was conducted. At a few locations in the west, the Diligencia Formation is in fault contact with Proterozoic and Mesozoic crystalline rocks (Fig. 1).

### Basement complex

The Orocopia Schist is part of the regionally extensive Pelona–Orocopia–Rand (POR) Schist terrane, which is part of a subduction complex related to low-angle subduction of the latest Cretaceous–Palaeogene Laramide orogeny (Dickinson & Snyder 1978; Livaccari *et al.* 1981; Bird 1988; Ingersoll 1997; Grove *et al.* 2003; Jacobson *et al.* 2007, 2011). The Orocopia Schist consists dominantly of arkosic metasandstone and subordinate metabasalt, metachert, marble, serpentinite and talc-actinolite rock (Jacobson *et al.* 2007).

Tectonically above the Orocopia Schist are Proterozoic augen gneiss underlying the Diligencia Formation, and blue/purple-quartz gneiss intruded by an anorthosite–syenite complex and Cretaceous granite (Jacobson *et al.* 2007). The north side of the basin is dominated by Late Cretaceous granite, granodiorite and quartz monzonite, and local small pods of migmatite and gneiss.

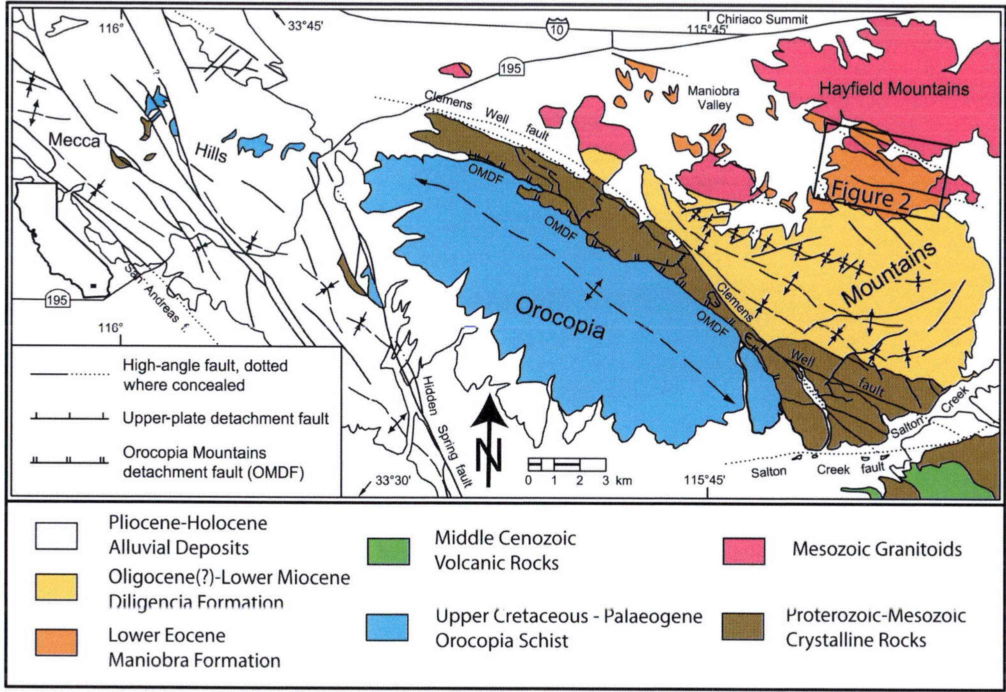


**Fig. 1.** Geological map of the Orocopia Mountains area (after Crowell 1975; Jacobson *et al.* 2007). The Maniobra Formation is unconformable on Mesozoic granitoids; locally they are in fault contact. The Diligencia Formation is in angular unconformity on the Maniobra Formation to the north and is unconformable on Proterozoic–Mesozoic crystalline rocks to the south. The Clemens Well fault locally forms this boundary to the SW. Location map for this figure is shown on map of California on left side. Location of Figure 2 is shown by box on right.

## Stratigraphy

*The Maniobra Formation.* The Maniobra Formation consists of 1500 m of Lower Eocene marine strata of mudrock, sandstone and conglomerate, and local sandy limestone and conglomerate (Fig. 2) (Crowell & Susuki 1959; Advocate *et al.* 1988). Nonmarine deposits at the base of the Maniobra Formation are derived directly from nonconformably underlying Late Cretaceous granite (Crowell & Susuki 1959), and locally contain alteration profiles consisting of fresh granite transitional upward to weathered granite to feldspathic grus to feldspathic conglomerate and sandstone, and local palaeosols (Advocate *et al.* 1988). The remainder of the Maniobra Formation consists of conglomerate, breccia, sandstone, mudrock and limestone, including numerous gravity-flow deposits, convoluted beds and slump features (Crowell & Susuki 1959; Advocate *et al.* 1988).

The thin nonmarine interval is directly overlain by marine deposits, testifying to the onset of rapid subsidence following initiation of sedimentation (Advocate *et al.* 1988). Granitic boulders up to 10 m in diameter occur at different positions along the exposed margins, indicating rugged topography and fault-controlled deposition in marine environments (Crowell & Susuki 1959; Advocate *et al.* 1988). The general westward decrease in grain size suggests eastern source areas (Crowell & Susuki 1959). Local southward- and northward-derived coarse debris indicates an east–west trough surrounded by rugged terrain (Advocate *et al.* 1988).

Advocate *et al.* (1988) recognized the following facies: (1) basal transgressive unit of nonmarine grus and shallow-marine conglomerate and sandstone; (2) overlying slope mudstone; and (3) submarine-canyon deposits of mudrock, sandstone and conglomerate. The submarine-canyon deposits are dominantly mudrock, with abundant channel deposits, slump deposits and truncated beds; most of the southern margin of the submarine canyon is covered by the Diligencia Formation (Advocate *et al.* 1988).

*The Diligencia Formation.* The Lower Miocene Diligencia Formation (Fig. 2) is a nonmarine dominantly siliciclastic succession up to 2000 m in thickness, and including in its lower part, basaltic

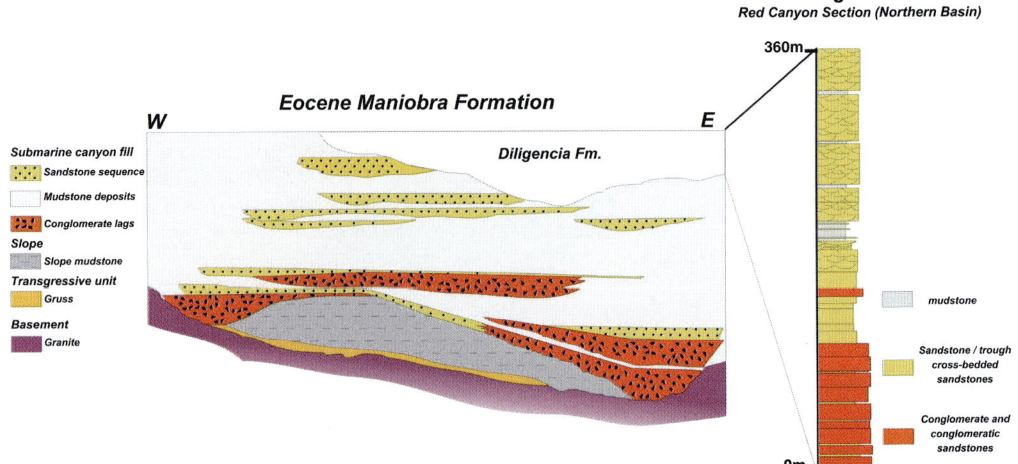



**Fig. 2.** West-east cross-section of Maniobra Formation (left side) and generalized stratigraphic section of Diligencia Formation (right side) from northern (bottom) and southern (upper) part of the basin. Location shown in Figure 1. Modified from Law *et al.* (2001).

lavas and andesitic intrusions emplaced between 23.6 and 21.3 Ma (K–Ar; Spittler & Arthur 1982; Frizzell & Weigand 1993; Law *et al.* 2001). The northern preserved boundary of the Diligencia Formation unconformably overlies the marine Maniobra Formation (Crowell & Susuki 1959; Law *et al.* 2001), whereas on the southeastern margin it nonconformably overlies Proterozoic crystalline basement.

The Diligencia Formation has been divided into four sedimentary members (Td1–Td4), plus a volcanic member (Tdv, between Td1 and Td2) (Spittler & Arthur 1982). Td1 consists of coarse-grained alluvial siliciclastic strata. The basal conglomerate contains primarily granitic clasts in the northern part of the basin, with minor Maniobra clasts, whereas in the southern part of the basin, conglomerate clasts consist mainly of augen gneiss (Spittler & Arthur 1982; Crowell 1993; Law *et al.* 2001). No Orocopia Schist clasts have been found anywhere in either the Maniobra or the Diligencia formations (Crowell 1993).

The basal conglomerate (more than 100 m thick) is overlain by more than 300 m of mostly red sandstone, conglomerate, mudrock and tuff (Law *et al.* 2001). Law *et al.* (2001) suggested that Td1 was deposited on high-gradient alluvial fans derived from fault escarpments. Td1 fines upward into lacustrine deposits, and is capped by basaltic flow rocks and andesitic sills and dykes (Spittler & Arthur 1982).

Most of the middle and upper Diligencia Formation (Td2, Td3 and Td4) consists of arkosic sandstone and mudrock, with local conglomerate, tuffaceous beds, limestone and evaporitic deposits (Spittler & Arthur 1982). As argued by Spittler & Arthur (1982), the dominant source of most of the middle and upper Diligencia Formation was the Hayfield granite and associated units north, NE and NW of the Diligencia basin, based on palaeocurrent measurements, clast size and clast composition.

## Methods

Thirty-seven medium to coarse sandstone samples were thin-sectioned for point counting, 12 from the Maniobra Formation and 23 from the northern part of the lower Diligencia Formation (Td1). An average of 500 points was counted for each thin section (etched and stained for plagioclase and potassium feldspars) according to the Gazzi–Dickinson method (Ingersoll *et al.* 1984; Zuffa 1985, 1987). Point-count results can be found in the Supplementary material. Grain parameters and recalculated parameters are those of Dickinson (1970, 1985), Zuffa (1985), Critelli & Le Pera (1994) and Arribas *et al.* (2003). Volcanic particles were subdivided using textural (Marsaglia & Ingersoll 1992) and temporal (Zuffa 1987) criteria. Volcanic grains were characterized as palaeovolcanic or neovolcanic, following criteria of Critelli & Ingersoll (1995) and Critelli *et al.* (2002). Confidence regions of detrital-mode means have been calculated following procedures described by Aitchison (1997).

Confidence regions (90 and 95%) were computed using log-ratio transformation of compositional data using 'R' and its software package called 'Composition' (Boogaart & Tolosana-Delgado 2008). Geometric means were used and confidence regions were plotted according to Weltje (2002).

Selected samples were analysed by SEM for detailed textural and compositional features, using a Jeol-6400 with an EDAX unit. Carbonate cements were analysed using a microprobe analyser Jeol Superprobe JXA-8900M. Both types of analysis were performed in the CNME (Centro Nacional de Microscopía Electrónica, Madrid, Spain).

## Petrology

Analysed sandstone from the Maniobra and Diligencia formations is arkose, as shown by average $Q_{31}F_{31}R_{38}-Qm_{47}F_{50}Lt_3$ compositions (e.g. Pettijohn *et al.* 1972; Dickinson 1985, respectively, Fig. 3). All analysed samples are medium to coarse-grained and poorly sorted, with general sub-angular to angular texture; most Maniobra and Diligencia sandstone is well compacted.

Quartz and feldspars grains represent the bulk of the sandstone framework, with K-feldspar dominating over plagioclase, $Qm_{49}Fk_{35}Fp_{16}$. Plagioclase is commonly altered to sericite, whereas K-feldspar is commonly replaced by kaolin. Phaneritic rock fragments – $Rg_{64}Rv_1Rm_{35}$ – are dominated by plutonic (predominantly granite) and subordinate metamorphic lithotypes (predominantly gneiss). Minor sandstone and volcanic rock fragments (lathwork and felsitic textures) have been documented. Nonvolcanic aphanitic lithic fragments are uncommon and consist of fine-grained micaschist, phyllite, slate, chert and siltstone grains, commonly deformed into pseudomatrix (e.g. Dickinson 1970). Accessory minerals are generally rare and are represented primarily by zircon and hornblende. As expressed in Figure 3, Maniobra and Diligencia detrital modes are quite similar, although differences exist. Ternary plots ($Qm_{Man54;Dil43}F_{Man43;Dil54}Lt_{Man3;Dil3}$; $Q_{Man35;Dil28}F_{Man36;Dil40}R_{Man29;Dil32}$) indicate that Maniobra sandstone has higher monocrystalline quartz, lower feldspar and comparable amounts of phaneritic and aphanitic rock fragments

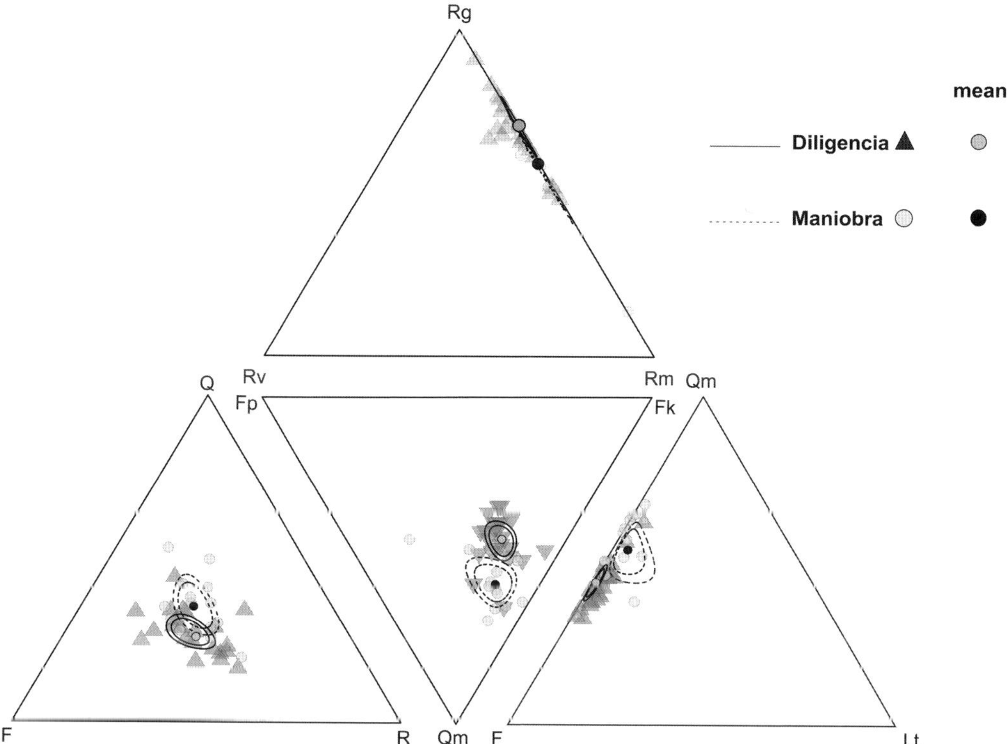


**Fig. 3.** Recalculated compositional parameters for the Maniobra and Diligencia Fms. Q, total quartz; Qm, monocrystallyne quartz; F, total feldspars; Fk, K-feldspar; Fp, plagioclase; R, phaneritic rock fragments; Rg, phaneritic granitoid rock fragments; Rm, phaneritic metamorphic rock fragments; Rv, phaneritic volcanic rock fragments; Lt, aphanitic lithic fragments.

($Rg_{Man58;Dil69}Rv_{Man0;Dil1}Rm_{Man42;Dil30}$). The difference between the quartz and feldspar content is accompanied by higher contents of plutonic rock fragments within the Diligencia Formation. Sandstone and siltstone rock fragments are common in samples from the basal conglomerate of the Diligencia Formation; the ratio between metamorphic and plutonic rock fragments is similar to those of the Maniobra Formation.

## Diagenesis

### Compaction

Mechanical compaction in Maniobra and Diligencia sandstone is manifested as deformation of ductile grains (micas and silty-clay intraclasts). Mechanical compaction is favoured by the presence of these grains, but is inhibited by early diagenetic cements. Coarse-grained deposits at the base of the Maniobra Formation are characterized by loss of original porosity by compaction that reduced intergranular volume (IGV) to levels lower than 10% (Fig. 4). In contrast, sandy deposits at the top of the formation maintain levels of IGV close to 35% due to the presence of early carbonate cementation (see below). Similar mechanical compaction is observed in Diligencia sandstone: coarse-grained deposits from the base and top of the studied succession are those with IGV lower than 20%, whereas intermediate sandy deposits have

IGV from 20 to 30% (Fig. 4), preserved by early ankerite and subsequent carbonate cementation.

Chemical compaction is evidenced by pressure solution, dominantly between quartz grains, generating long and concave–convex contacts.

### Cements and replacements

*Quartz cement.* Quartz cement is not well developed and occurs only locally. Percentages are generally lower than 1%, but at the base of the Maniobra Formation, it is up to 18% of rock bulk volume. Two different textures can be recognized: (1) grain overgrowth (Fig. 5b); and (2) equant mosaic (Fig. 5c). Overgrowths form discontinuous and thin coatings around quartz grains. This texture developed in samples where chemical compaction is manifested. Thus, the origin of this cement can be related to an internal source of silica provided by pressure solution (Fig. 5d) (i.e. Houseknecht 1988; Bjørlykke & Egeberg 1993). Quartz cement consisting of microcrystalline prismatic quartz occluded intergranular porosity. Textural relationships with other cements and grains suggest a late, postcompactional, diagenetic origin during burial. Prismatic quartz has been associated with supersaturated deep basinal brines (e.g. Hendry & Trewin 1995).

*K-feldspar cement.* K-feldspar cement at the base of the Maniobra Formation is represented by thin and discontinuous syntaxial overgrowths, forming

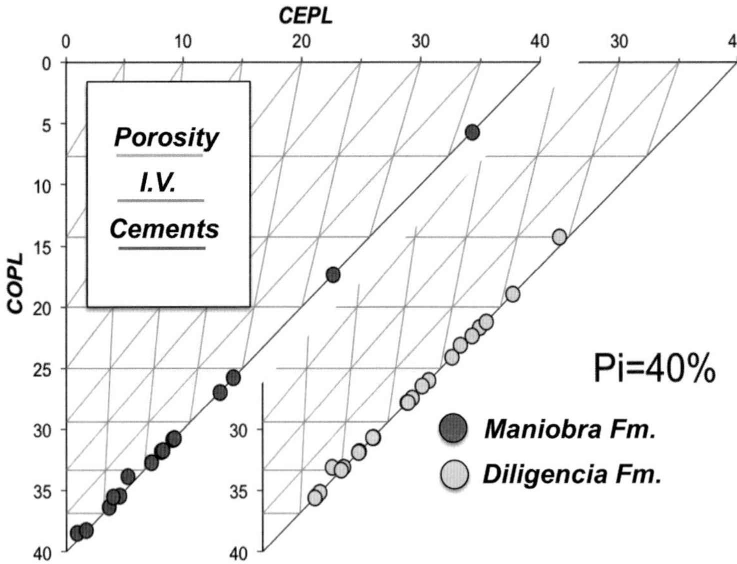


**Fig. 4.** Compactional (COPL) v. cementational (CEPL) porosity-loss diagram (modified from Lundegard, 1992) in sandstone from Maniobra and Diligencia formations.

**Fig. 5.** Petrographic microphotographs of sandstone from Maniobra and Diligencia formations showing diagenetic features. (**a**) Silty-clay intraclast deformed by mechanical compaction. Maniobra Fm., crossed nichols. (**b**) Quartz overgrowth cement. Maniobra Fm., crossed nichols. (**c**) Prismatic microcrystalline quartz mosaic. Maniobra Fm. SEM image. (**d**) Long and concave–convex contact between quartz grains evidencing pressure solution. Note microcrystalline quartz cement between detrital grains (red arrow), crossed nichols. (**e**) Euhedral K-feldspar overgrowth (red arrow) predating quartz cementation, which encloses it, crossed nichols. (**f**) Oxidized ankerite crystals showing euhedral rhombohedral shape bordered by a late dolomitic phase (low content in Mn and Fe). Diligencia Fm., crossed nichols (**g**) Fe-oxide spheroidal crystals generated by calcitization from Fe–Mn carbonate cement. Diligencia Fm. (**h**) Dolomitic crystals generated in veins and partially calcitized (red stained). Note high proportion of opaque residues. Diligencia Fm., parallel nicols. White bars are for scale and include magnification.

small euhedral prisms around detrital K-feldspar grains. Growth of this cement preceded quartz overgrowth, which includes it (Fig. 5e); thus, it is considered an early phase of cementation.

*Carbonate cements*

*Ankerite.* Ankerite represents an early phase of cementation, appearing as euhedral and sub-euhedral semi-opaque rhombs of $50–100\ \mu m$.

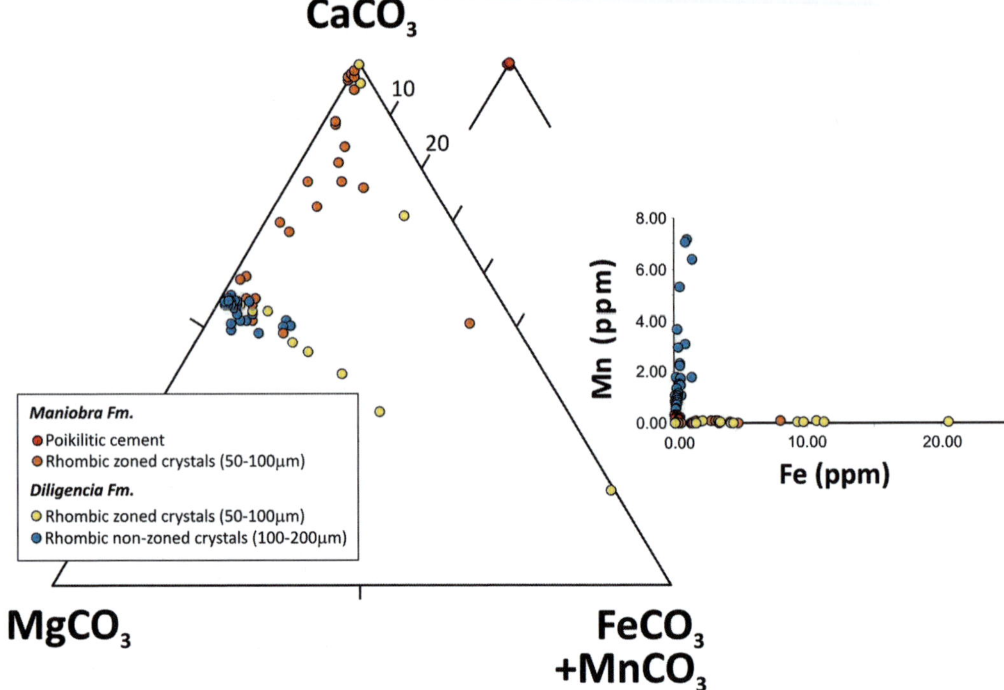



**Fig. 6.** Chemical composition of carbonate cements from Maniobra and Diligencia sandstone.

Ankerite crystals grew on detrital grain surfaces, suggesting early precipitation. In addition, these crystals appear included in carbonate mosaics and poikilotopic cements. Ankerite is common in Diligencia sandstone, where crystals are zoned, showing cloudy opaque iron-rich cores bordered by thin and clean dolomitic phases (Fig. 5f). The content of $FeCO_3$ in ankerite crystals varies from 0.5 (close to a stoichiometric dolomite) to 45% (Fig. 6). Calcitization of this cement produces Fe-oxides as opaque crystal coatings or as spheroidal remains (Fig. 5g).

*Dolomite.* Dolomite occurs as rhombohedral crystals, mosaics and poikilotopic cement in both the Maniobra and Diligencia formations. Composition of this cement in the Maniobra Formation varies from pure dolomite (mainly zoned rhombohedral crystals) to pure calcite (mainly poikilotopic textures with opaque inclusions) with low Fe content (less than 15% in $FeCO_3$, Fig. 6). This cement is common in sandstone with foraminifera and unidentified bioclasts. Sandstone with dolomite has IGV from 15 to 30% of total volume, suggesting early pre-compactional cementation that occluded original pores. As a consequence, the presence of these mosaics inhibited later quartz cementation. Intergranular and vein-filling dolomite is locally replaced by calcite and Fe-oxides, concentrated according to its original iron zonation (Fig. 5h). Diligencia sandstone contains pure dolomite with lower than 10% Fe, but with Mn content as high as 8% (Fig. 6).

*Patchy calcite.* This cement occurs mainly in Maniobra sandstone, where it occurs as small mosaics or single crystals, which formed as corrosion and grain replacement of siliciclastic grains (mainly feldspar). In Diligencia sandstone, this cement commonly occurs as iron-rich calcite formed by calcitization of older carbonate cement (i.e. ankerite). Growth of this cement was probably late, based on its occurrence dominantly in well-compacted sandstone. Generally, this latest cement is found in sandstone with IGV lower than 15%.

*Phyllosilicate cements*
*Kaolin.* Kaolin cements are well developed as pore filling, mainly in the younger part of studied Diligencia sandstone (Fig. 7a). Kaolin plates are less than 10 µm and normally occur close to detrital framework grains. Kaolin sheets commonly appear included in younger carbonate cement (Fig. 7b). Some kaolin has been transformed into illite, forming face-to-face aggregates (Fig. 7c) and kaolin

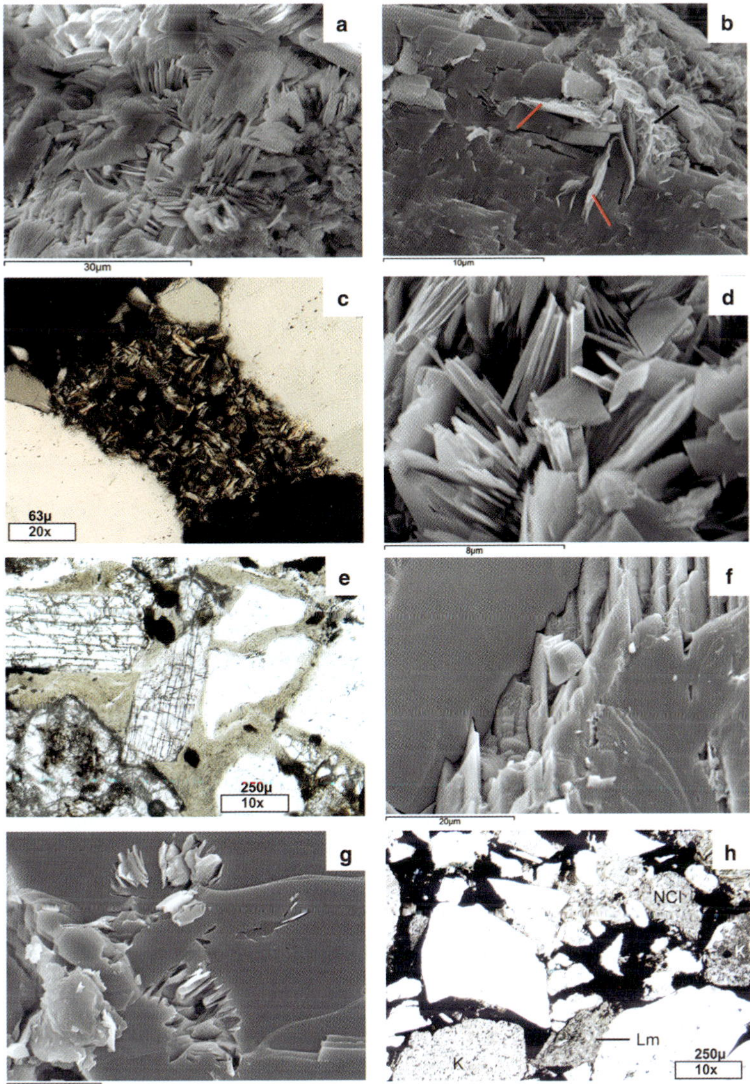

**Fig. 7.** Petrographic microphotographs of Maniobra and Diligencia sandstone showing diagenetic features. (**a**) SEM image of booklets of kaolin pore-filling. (**b**) Kaolin platelets included in a dolomite crystal (red arrows). In turn, kaolinite is replaced by an illite phase (black arrow). (**c**) Illite epimatrix after feldspar. Note face-to-face arrangement of illite crystals that suggests a kaolinite precursor. Crossed nicols (**d**) Kaolin plates interlayered in thick blocky dickite crystals. (**e**) Illite clay coats arround detrital grains. Parallel nichols. (**f**) Fluorite crystal (upper left) showing replacing and aggressive textures against a K-feldspar grain. (**g**) Kaolin crystal relics included and partially replaced by fluorite cement. (**h**) Fe-oxide cement around detrital grains: K-feldspars (K), low-grade metamorphic grains (Lm) and non-carbonate intrabasinal grains (NCI), parallel nicols. White bars are for scale and include magnification.

commonly is associated with dickite crystals (Fig. 7d).

All these features suggest an early eogenetic origin for kaolin pore filling related to fluxes of meteoric fresh waters. During burial, this mineral may transform into illite or dickite at temperatures greater than 130 °C, with or without potassium input, respectively (Bjørlykke 1983; Morad *et al.* 1994; Beaufort *et al.* 1998; Worden & Morad 2003).

*Illite pore lining*. This cement at the base of the Diligencia Formation consists of illite clay coats characterized mainly by discontinuous coatings tangentially oriented to grain surfaces (Fig. 7e).

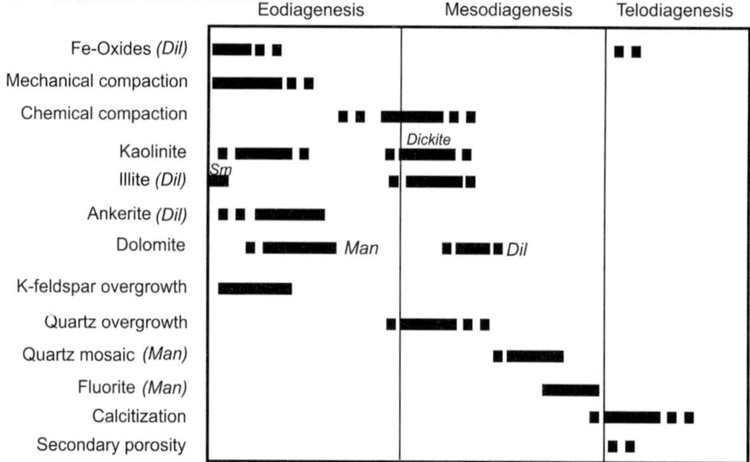


**Fig. 8.** Diagenetic sequence of main diagenetic processes in Maniobra and Diligencia sandstone, as interpreted from petrographic analysis.

These features suggest a geopetal distribution of coatings. Illite ridges and bridges between detrital grains have also been observed. All these textures suggest very early diagenesis related to mechanical infiltration of clay, probably a smectitic precursor, in a vadose environment (e.g. Moraes & De Ros 1990, 1992, and references therein).

*Other cements*

*Fluorite.* Fluorite cement occurs exclusively at the base of the Maniobra Formation, where it appears as irregular violet or colourless coarse crystals. This cement shows aggressive textures against framework, corroding and replacing mainly K-feldspar grains, and older cements (Fig. 7f), incorporating kaolin relics (Fig. 7g). These textures suggest a very late genesis of this cement, postdating compaction and all diagenetic mineral phases, indicating late-burial growth related to hydrothermal fluids (Deer *et al.* 1966).

*Fe-oxides.* Early-precipitated Fe-oxides occur in some lower Diligencia sandstone as continuous opaque coatings around detrital grains. These coatings are locally thick and completely occlude original pore space (Fig. 7h). In addition, Fe-oxides occur in association with calcitization of Fe-bearing carbonate cements as ankerite or ferroan dolomite. This process can be related to a telodiagenetic phase during exhumation (i.e. Morad *et al.* 1994).

*Diagenetic matrix.* The clay-mineral fraction of less than 30 μm ('matrix' after Dott 1964) formed during diagenesis due to several processes. Infiltration of clay minerals (smectites) by vadose water

circulation is deduced at the base of the Diligencia Formation by the presence of illite clay coats, which implies mineral transformation/recrystallization of a potential dioctahedral smectitic precursor (e.g. Worden & Morad 2003). In addition, and included as diagenetic matrix, kaolin cement is widely distributed, mainly in pore space of Diligencia sandstone. Associated in time with these pore fillings, feldspar transformed into kaolin (epimatrix), as a result of early kaolinitization by meteoric water. In turn, these minerals transformed during burial into more stable mineral phases (i.e. dickite) or other clay minerals (i.e. illite). Pseudomatrix is also present, mainly at the base of the Diligencia Formation, as a product of deformation and disaggregation of silt–clay intrabasinal clasts by the action of mechanical compaction (Fig. 5a).

*Evolution of original porosity*

Original porosity can be estimated as between 30.7 and 39%, using framework sorting values (from poorly to well sorted, respectively; Beard & Weyl 1973). However, to analyse the evolution of porosity loss of all the sandstone during diagenesis, an initial porosity of 40% (Fig. 4) has been assumed (e.g. Lundegard 1992). Both mechanical compaction and cementation reduced initial porosity to less than 2%. Compaction was the most important process controlling porosity loss in both the Maniobra and Diligencia formations, as suggested by ICOMPACT (Lundegard 1992) average values (0.74 and 0.69, respectively). Secondary porosity is negligible and consists of local dissolution of feldspar and carbonate cement.

## Discussion

### Composition and provenance

Homogeneous plutoniclastic detrital modes of analysed Diligencia and Maniobra sandstone indicate that both were derived primarily from the same granitic basement. The slightly higher quartz content of Maniobra sandstone may result from more intense weathering of source rocks during the warmer and more humid conditions of the Eocene (e.g. Hall 2007) or slightly higher content of gneissic rock fragments (Fig. 3), or both. More intense weathering would favour the preservation of gneiss fragments v. granitic fragments, and quartz over feldspar.

### Chronology of diagenetic processes

Textural relationships of cements and replacements suggest the chronology of eo-, meso- and telodiagenetic stages represented in Figure 8. During early diagenesis (eodiagenesis), Maniobra sandstone was affected mainly by mechanical compaction, and K-feldspar and ferroan dolomite cementation, generated in an alkaline and reduced system related to marine environments (i.e. Füchtbauer 1967; Almon et al. 1976; Hawkins 1978; Hurst & Irwin 1982; Morad 1998). Diligencia sandstone is characterized by early mechanical compaction and illite after smectite (clay coats), kaolin and Fe-oxides cementation. These processes can be related to meteoric freshwater fluxes in an oxidizing continental environment (i.e. Bjørlykke & Aagaard 1992; Bjørlykke 1994; Worden & Morad 2003). In addition, ankerite cement developed as a late eodiagenetic phase under reducing conditions (i.e. Morad 1998). The mesodiagenetic stage in Maniobra sandstone is characterized by intense chemical compaction, especially at the base, which culminated with development of quartz overgrowth and mosaic cements, and subsequently, by fluorite precipitation. The occurrence of prismatic quartz and fluorite is strongly related to oversaturated hydrothermal fluxes (Deer et al. 1966; Ahmed 2002; Ohba & Kitade 2005). Burial diagenesis is primarily represented in Diligencia sandstone by dolomite cement (non-ferroan dolomite, Fig. 6) that shows aggressive textures on framework grains, particularly replacing feldspar. This dolomite cement contains significant Mn, suggesting an origin related to increased temperature, related to hydrothermal activity (e.g. Chow et al. 1996; Morad et al. 2000). Kaolin transformation into illite or dickite indicates temperatures greater than 130 °C, with or without potassium input, respectively (Bjorlykke 1983; Morad et al. 1994; Beaufort et al. 1998; Worden & Morad 2003). Diagenetic processes related to exhumation (telodiagenetic stage) consist, in both formations, of calcitization of Fe-carbonate cement (Morad et al. 1994). This process produced Fe-oxide rims and cores in rhombohedral calcite crystals. Scarce secondary porosity can be related to this uplift stage as a consequence of carbonate dissolution caused by influx of fresh meteoric water.

### Diagenesis, framework composition and sedimentary environment

Provenance exerted an important control on diagenesis of Maniobra and Diligencia sandstone. The quartzofeldspathic composition was determined by the erosion and deposition of Cretaceous granitoids, with minor metamorphic input. Original composition favoured development of diagenetic phyllosilicates (kaolin, illite and dickite), as well as quartz and K-feldspar cements. Moreover, the presence of intrabasinal silt–clay clasts in the lower Diligencia Formation (primarily recycled from the Maniobra Formation) favoured mechanical compaction, thereby reducing original porosity. In addition, dissolution of carbonate intrabasinal clasts (bioclasts) in Maniobra sandstone could have assisted early carbonate cementation.

Early diagenesis was governed mainly by geochemical properties of environmental waters. The chronology of eodiagenetic processes (Fig. 8) differs between Maniobra and Diligencia sandstone, resulting from marine early diagenesis in Maniobra sandstone, and meteoric freshwater diagenesis in Diligencia sandstone. However, mesodiagenetic processes (Fig. 8) were similar in both formations, characterized by generally closed-system diagenesis, without significant mass transfer. At the base of the Maniobra Formation, diagenesis controlled by hydrothermal waters is recorded (i.e. microcrystalline quartz and fluorite cements), related to high heat flow related to Early Miocene extensional tectonics that resulted in eruption of primarily basaltic flows (Spittler & Arthur 1982). In addition, the presence of Mn in late dolomite cements from Diligencia sandstone suggests precipitation of this element by hydrothermal fluxes (e.g. Morad et al. 2000). Telodiagenetic processes (Fig. 8) are similar in both formations, associated with meteoric water related to exhumation.

## Conclusions

This petrographic study of sandstone from the Eocene Maniobra and Miocene Diligencia formations illustrates relationships among provenance, diagenesis and potential reservoir quality. Similar granitic provenance has been deduced for both formations. A more mature composition of Maniobra sandstones can be related to warmer and more humid climatic conditions during the Eocene. The original quartzofeldspathic composition of both formations suggests potential good porosity, based on their rigid frameworks of monocrystalline grains.

Analysis of diagenetic processes reveals a succession of early diagenetic mineral phases (cements) that is consistent with the geochemistry of waters from the Eocene marine Maniobra Formation and the Miocene continental Diligencia Formation. Burial diagenesis is manifested by homogeneity of processes in both formations. Highly replacive mineral phases at the base of the Maniobra Formation can be related to hydrothermal fluxes. Compaction that reduced intergranular volume was intense, but this process was inhibited by early carbonate cement.

Ingersoll thanks participants in the graduate field seminar (GIFE) for help in collecting samples and providing the palaeotectonic context for this study. The authors acknowledge the two reviewers, L. F. De Ros and O. Walderhaug for their criticism and useful suggestions.

# References

ADVOCATE, D. M., LINK, M. H. & SQUIRES, R. L. 1988. Anatomy and history of an Eocene submarine canyon: the Maniobra Formation, southern California. *In*: FILEWICZ, M. V. & SQUIRES, R. L. (eds) *Paleogene Stratigraphy, West Coast of North America*. Pacific Section, Society of Economic Paleontologists and Mineralogists, Los Angeles, **58**, 45–58.

AHMED, W. 2002. Effects of heat-flow and hydrothermal fluids from volcanic intrusions on authigenic mineralization in sandstone formations. *Bulletin of Chemical Society of Ethiopia*, **16**, 37–52.

AITCHISON, J. 1997. The one hour course in compositional data analysis of compositional data analysis is simple. *In*: PAWLOWSKY GLAHN, V. (ed.) *Proceedings of IAMG '97 – The Third Annual Conference of the International Association for Mathematical Geology*. International Center for Numerical Methods in Engineering (CIMNE), Barcelona (E), 3–35.

ALMON, W. R., FULLERTON, L. B. & DAVIES, D. K. 1976. Pore space reduction in Cretaceous sandstones through chemical precipitation of clay minerals. *Journal of Sedimentary Petrology*, **46**, 89–96.

ARRIBAS, J., ALONSO, A., MAS, R., RODAS, M., BARRENECHEA, J. F., ALONSO-AZCÁRATE, J. & ARTIGAS, R. 2003. Sandstone petrography of continental depositional sequences of an intraplate rift basin: western Cameros Basin (North Spain). *Journal of Sedimentary Research*, **73**, 309–327.

ATWATER, T. 1970. Implications of plate tectonics for the Cenozoic tectonic evolution of western North America. *Geological Society of America Bulletin*, **81**, 3513–3536.

ATWATER, T. & STOCK, J. 1998. Pacific-North America plate tectonics of the Neogene southwestern United States: an update. *International Geology Review*, **40**, 375–402.

BEARD, D. C. & WEYL, P. K. 1973. Influence of texture on porosity and permeability of unconsolidated sand. *American Association of Petroleum Geologists Bulletin*, **57**, 349–369.

BEAUFORT, D., CASSAGNABÈRE, A., PETIT, S., LANSON, B., BERGER, G., LACHARPAGNE, J. C. & JOHANSEN, H. 1998. Kaolinite-to-dickite reaction in sandstone reservoirs. *Clay Minerals*, **33**, 297–316.

BIRD, P. 1988. Formation of the Rocky Mountains, western United States: a continuum computer model. *Science*, **239**, 1501–1507.

BJØRLYKKE, K. 1983. Diagenetic reactions in sandstones. *In*: PARKER, A. & SELLWOOD, B. W. (eds) *Sediment Diagenesis*. NATO ASI series 115. Reidel, Dordrecht, 169–213.

BJØRLYKKE, K. 1994. Fluid-flow processes and diagenesis in sedimentary basins. *In*: PARNELL, J. (ed.) *Geofluids: Origin, Migration and Evolution of Fluids in Sedimentary Basins*. Geological Society, London, Special Publications, **78**, 127–140.

BJØRLYKKE, K. & AAGAARD, P. 1992. Clay minerals in North Sea sandstones. *In*: HOUSEKNECHT, D. W. & PITTMAN, E. P. (eds) *Origin, Diagenesis and Petrophysics of Clay Minerals in Sandstones*. Society of Economic Paleontologists and Mineralogists, Tulsa, Special Publications, **47**, 65–80.

BJØRLYKKE, K. & EGEBERG, P. K. 1993. Quartz cementation in sedimentary basins. *American Association of Petroleum Geologists Bulletin*, **77**, 1538–1548.

BLOCH, S. 1991. Empirical prediction of porosity and permeability in sandstones. *American Association of Petroleum Geologists Bulletin*, **75**, 1145–1160.

BOHANNON, R. G. & PARSONS, T. 1995. Tectonic implications of post-30 Ma Pacific and North American relative plate motions. *Geological Society of America Bulletin*, **107**, 937–959.

BOOGAART, K. G.v.D. & TOLOSANA-DELGADO, R. 2008. 'Compositions': a unified R package to analyze compositional data. *Computer Geoscience*, **34**, 320–338.

CHOW, N., MORAD, S. & AL-AASM, I. S. 1996. Origin of authigenic carbonates in Eocene to Quaternary sediments from the Arctic and Norwegian–Greenland Sea. *In*: MYHRE, A., THIEDE, J., FIRTH, J., RUDDIMAN, W. F. & JOHNSSON, L. (eds) *Proceedings of the Ocean Drilling Program, Scientific Results*, **151**, 415–434.

CRITELLI, S. & INGERSOLL, R. V. 1995. Interpretation of neovolcanic versus palaeovolcanic sand grains: an example from Miocene deep-marine sandstone of the Topanga Group (Southern California). *Sedimentology*, **42**, 783–804.

CRITELLI, S. & LE PERA, E. 1994. Detrital modes and provenance of Miocene sandstones and modern sands of the Southern Apennines thrust-top basins (Italy). *Journal of Sedimentary Research*, **64**, 824–835.

CRITELLI, S., MARSAGLIA, K. M. & BUSBY, C. J. 2002. Tectonic history of a Jurassic backarc-basin sequence (the Gran Canon Formation, Cedros Island, Mexico), based on compositional modes of tuffaceous deposits. *Geological Society of America Bulletin*, **114**, 515–527.

CROWELL, J. C. 1975. *Geologic sketch of the Orocopia Mountains, southeastern California*. California

Division of Mines and Geology Special Report **118**, 99–110.

CROWELL, J. C. 1987. Late Cenozoic basins of onshore southern California: complexity is the hallmark of their tectonic history. *In*: INGERSOLL, R. V. & ERNST, W. G. (eds) *Cenozoic Basin Development of Coastal California [Rubey Volume VI]*. Prentice-Hall, Englewood Cliffs, 207–241.

CROWELL, J. C. 1993. The Diligencia Formation, Orocopia Mountains, southeastern California. *United States Geological Survey Bulletin*, **2053**, 239–242.

CROWELL, J. C. & SUSUKI, T. 1959. Eocene stratigraphy and paleontology, Orocopia Mountains, southeastern California. *Geological Society of America Bulletin*, **70**, 581–592.

DEER, W. A., HOWIE, R. A. & ZUSSMAN, J. 1966. *An Introduction to the Rock Forming Minerals*. Longman, Harlow.

DICKINSON, W. R. 1970. Interpreting detrital modes of graywacke and arkose. *Journal of Sedimentary Petrology*, **40**, 695–707.

DICKINSON, W. R. 1985. Interpreting provenance relations from detrital modes of sandstones. *In*: ZUFFA, G. G. (ed.) *Provenance of Arenites*. Reidel, Dordrecht, 333–361.

DICKINSON, W. R. 1997. Tectonic implications of Cenozoic volcanism in coastal California. *Geological Society of America Bulletin*, **109**, 936–954.

DICKINSON, W. R. & SNYDER, W. S. 1978. Plate tectonics of the Laramide orogeny. *In*: MATTHEWS, V. III. (ed.) *Laramide Folding Associated with Basement Block Faulting in the Western United States*. Geological Society of America, Memoir, **151**, 355–366.

DOTT, R. H. 1964. Wacke. greywacke and matrix, what approach to immature sandstone classification? *Journal of Sedimentary Petrology*, **34**, 625–632.

EHRENBERG, S. N., NADEAU, P. H. & STEEN, Ø. 2008. A megascale view of reservoir quality in producing sandstones from the offshore Gulf of Mexico. *American Association of Petroleum Geologists Bulletin*, **92**, 145–164, http://dx.doi.org/10.1306/09280707062

FRIZZELL, V. A. JR. & WEIGAND, P. W. 1993. Wholerock K–Ar ages and geochemical data from middle Cenozoic volcanic rocks, southern California: a test of correlations across the San Andreas fault. *In*: POWELL, R. E., WELDON, R. J & MATTI, J. C. (eds) *The San Andreas fault system: Displacement, palinspastic reconstruction, and geologic evolution*. Geological Society of America, Memoir, **178**, 213–287.

FÜCHTBAUER, H. 1967. Influence of different types of diagenesis on sandstone porosity. *In*: *VII World Petroleum Congress Proceedings*. Elsevier, Mexico, **2**, 353–369.

GRAHAM, S. A., STANLEY, R. G., BENT, J. V. & CARTER, J. B. 1989. Oligocene and Miocene paleogeography of central California and displacement along the San Andreas fault. *Geological Society of America Bulletin*, **101**, 711–730.

GROVE, M., JACOBSON, C. E., BARTH, A. P. & VUCIC, A. 2003. Temporal and spatial trends of Late Cretaceous-early Tertiary underplating of Pelona and related schist beneath southern California and southwestern Arizona. *In*: JOHNSON, S. E., PATERSON, S.

R., FLETCHER, J. M., GIRTY, G. H., KIMBROUGH, D. L. & MARTIN-BARAJAS, A. (eds) *Tectonic Evolution of Northwestern Mexico and the Southwestern U.S.A.* Geological Society of America Special Papers, **374**, 381–406.

HALL, C. A. JR. 2007. *Introduction to the Geology of Southern California and its Native Plants*. University of California Press, Los Angeles.

HAWKINS, P. J. 1978. Relationship between diagenesis, porosity reduction, and oil emplacement in late Carboniferous sandstones reservoirs. Bothamsall Oilfield, East Midlands. *Journal of the Geological Society, London*, **135**, 7–24.

HENDRY, J. P. & TREWIN, N. H. 1995. Authigenic quartz microfabrics in Cretaceous turbidites: evidence for silica transformation processes in sandstones. *Journal of Sedimentary Research*, **A65**, 380–392.

HOUSEKNECHT, D. W. 1988. Intergranular pressure solution in four quartzose sandstones. *Journal of Sedimentary Petrology*, **58**, 228–246.

HURST, A. & IRWIN, H. 1982. Geological modelling of clay diagenesis in sandstones. *Clays and Clay Minerals*, **17**, 5–22.

INGERSOLL, R. V. 1997. Phanerozoic tectonic evolution of central California and environs. *International Geology Review*, **39**, 957–972.

INGERSOLL, R. V. 2009. Diligencia basin (southern California) revisited: sedimentation in half graben bounded on the northeast by normal fault. *Geological Society of America Abstracts with Programs*, **41**, 569.

INGERSOLL, R. V., BULLARD, T. F., FORD, R. L., GRIMM, J. P., PICKLE, J. D. & SARES, S. W. 1984. The effect of grain size on detrital modes: a test of the Gazzi–Dickinson point–counting method. *Journal of Sedimentary Petrology*, **54**, 103–116.

JACOBSON, C. E., GROVE, M., VUCIC, A., PEDRICK, J. N. & EBERT, K. A. 2007. Exhumation of the Orocopia Schist and associated rocks of southeastern California: relative rates of erosion, synsubduction tectonic denudation, and middle Cenozoic extension. *In*: CLOOS, M., CARLSON, W. D., GILBERT, M. C., LIOU, J. G. & SORENSEN, S. S. (eds) *Convergent Margin Terranes and Associated Regions (A tribute to W.G. Ernst)*. Geological Society of America Special Papers, **419**, 1–37.

JACOBSON, C. E., GROVE, M., PEDRICK, J. N., BARTH, A. P., MARSAGLIA, K. M., GEHRELS, G. E. & NOURSE, J. A. 2011. Late Cretaceous–early Cenozoic tectonic evolution of the southern California margin inferred from provenance of trench and forearc sediments. *Geological Society of America Bulletin*, **123**, 485–506.

LAND, L. S., MILLIKEN, K. L. & MCBRIDE, E. F. 1987. Diagenetic evolution of Cenozoic sandstones, Gulf of Mexico sedimentary basin. *Sedimentary Geology*, **50**, 195–225.

LAW, R. D., ERIKSSON, K. & DAVISSON, C. 2001. Formation, evolution, and inversion of the middle Tertiary Diligencia basin, Orocopia Mountains, southern California. *Geological Society of America Bulletin*, **113**, 196–221.

LIVACCARI, R. F., BURKE, K. & SENGOR, A. M. C. 1981. Was the Laramide orogeny related to subduction of an oceanic plateau? *Nature*, **289**, 276–278.

LUNDEGARD, P. D. 1992. Sandstone porosity loss: a 'big picture' view of the importance of compaction. *Journal of Sedimentary Petrology*, **62**, 250–260.

MARSAGLIA, K. M. & INGERSOLL, R. V. 1992. Compositional trends in arc-related, deep-marine sand and sandstone: a reassessment of magmatic-arc provenance. *Geological Society of America Bulletin*, **104**, 1637–1649.

MORAD, S. 1998. Carbonate cementation in sandstones: distribution patterns and geochemical evolution. *In*: MORAD, S. (ed.) *Carbonate Cementation in Sandstones*, International Association of Sedimentologists Special Publications, **26**, 1–26.

MORAD, S., BEN ISMAIL, H., DE ROS, L. F., AL-AASM, I. S. & SERRHINI, N. E. 1994. Diagenesis and formation water chemistry of Triassic reservoir sandstones from southern Tunisia. *Sedimentology*, **41**, 1253–1272.

MORAD, S., KETZER, J. M. & DE ROS, F. 2000. Spatial and temporal distribution of diagenetic alterations in siliciclastic rocks: implications for mass transfer in sedimentary basins. *Sedimentology*, **47**, 95–120.

MORAD, S., AL-RAMADAN, K., KETZER, J. M. & DE ROS, L. F. 2010. The impact of diagenesis on the heterogeneity of sandstone reservoirs: a review of the role of depositional facies and sequence stratigraphy. *American Association of Petroleum Geologists Bulletin*, **94**, 1267–1309, http://dx.doi.org/10.1306/04211009178

MORAES, M. A. S. & DE ROS, L. F. 1990. Infiltrated clays in fluvial Jurassic sandstones of Recôncavo Basin, northeastern Brazil. *Journal of Sedimentary Petrology*, **60**, 809–819.

MORAES, M. A. S. & DE ROS, L. F. 1992. Depositional, infiltrated and authigenic clays in fluvial sandstones of the Jurassic Sergi formation, Recôncavo Basin, northeastern Brazil. *In*: HOUSEKNECHT, D. W. & PITTMAN, E. P. (eds) *Origin, Diagenesis and Petrophysics of Clay Minerals in Sandstone*. Society for Sedimentary Geology (SEPM) Special Publications, Tulsa, **47**, 197–208.

NICHOLSON, C., SORLIEN, C. C., ATWATER, T., CROWELL, J. C. & LUYENDYK, B. P. 1994. Microplate capture, rotation of the western Transverse Ranges, and initiation of the San Andreas transform as a low-angle fault system. *Geology*, **22**, 491–495.

OHBA, T. & KITADE, Y. 2005. Subvolcanic hydrothermal systems: implications from hydrothermal minerals in hydrovolcanic ash. *Journal of Volcanology and Geothermal Research*, **145**, 249–262.

PETTIJOHN, F. J., POTTER, P. E. & SIEVER, R. 1972. *Sand and Sandstone*. Springer-Verlag, New York.

SCHMIDT, V. & MCDONALD, D. A. 1979. The role of secondary porosity in the course of sandstone diagenesis. *In*: SCHOLLE, P. A. & SCHLUGER, P. R. (eds) *Aspects of Diagenesis*. Society for Sedimentary Geology (SEPM) Special Publications, Tulsa, **26**, 175–207.

SCHMOKER, J. W. & GAUTIER, D. L. 1988. Sandstone porosity as a function of thermal maturity. *Geology*, **16**, 1007–1010.

SCHMOKER, J. W. & SCHENK, C. J. 1994. Regional trends of the upper Jurassic Norphlet Formation in southwestern Alabama and vicinity, with comparisons to formations of other basins. *American Association of Petroleum Geologists Bulletin*, **78**, 166–180.

SPITTLER, T. E. & ARTHUR, M. A. 1982. The Lower Miocene Diligencia Formation of the Orocopia Mts., southern California: stratigraphy, petrology, sedimentology and structure. *In*: INGERSOLL, R. V. & WOODBURNE, M. O. (eds) *Cenozoic Nonmarine Deposits of California and Arizona*. Pacific Section, Society of Economic Paleontologists and Mineralogists, Los Angeles, 83–99.

TAYLOR, T. R., GILES, M. R. *ET AL.* 2010. Sandstone diagenesis and reservoir quality prediction: models, myths, and reality. *American Association of Petroleum Geologists Bulletin*, **94**, 1093–1132.

WELTJE, G. J. 2002. Quantitative analysis of detrital modes: statistically rigorous confidence regions in ternary diagrams and their use in sedimentary petrology. *Earth-Science Reviews*, **57**, 211–253.

WILSON, M. D. 1994. Case history. Jurassic sandstones, Viking Graben, North Sea. *In*: WILDON, M. D. (ed.) *Reservoir Quality Assessment and Prediction in Clastic Rocks*. Society for Sedimentary Geology (SEPM) Short Course, Tulsa, **30**, 367–384.

WORDEN, R. H. & MORAD, S. 2003. Clay minerals in sandstones: controls on formation, distribution and evolution. *In*: WORDEN, R. H. & MORAD, S. (eds) *Clay Mineral Cements in Sandstones*. International Association of Sedimentologists Special Publications, **34**, 3–41.

ZUFFA, G. G. 1985. Optical analyses of arenites: influence of methodology on compositional results. *In*: ZUFFA, G. G. (ed.) *Provenance of Arenites*. Reidel, Dordrecht, 165–189.

ZUFFA, G. G. 1987. Unravelling hinterland and offshore palaeogeography from deep-water arenites. *In*: LEGGETT, J. K. & ZUFFA, G. G. (eds) *Marine Clastic Sedimentology*. Graham and Trotman, London, 39–61.

# The influence of the provenance of arenite on its diagenesis in the Cameros Rift Basin (Spain)

J. ARRIBAS[1]*, L. GONZÁLEZ-ACEBRÓN[2], S. OMODEO-SALÉ[1] & R. MAS[2]

[1]*Dpto. de Petrología y Geoquímica, Facultad de Ciencias Geológicas, UCM-IGEO (CSIC), José Antonio Novais 12, 28040 Madrid, Spain*

[2]*Dpto. de Estratigrafía, Facultad de Ciencias Geológicas, UCM-IGEO (CSIC), José Antonio Novais 12, 28040 Madrid, Spain*

**Corresponding author (e-mail: arribas@geo.ucm.es)*

**Abstract:** The intraplate Cameros Rift Basin in northern Spain, which has sediments some 6500 m thick, developed between the Late Jurassic and Early Albian. Its facies and their distribution in the sedimentary record suggest the basin may contain hydrocarbon systems. The arenite composition of the basin reveals two main petrofacies: (1) a quartzolithic petrofacies, the provenance of which is related to recycling processes that took place in the pre-rift sedimentary cover; and (2) a quartzofeldspathic petrofacies mainly related to the erosion of a plutonic and metamorphic source of arenite. The succession of these petrofacies reflects two main cycles representing the progressive erosion of their sources, one of 10 Ma, the other of 30 Ma. Such succession is typical of a non-volcanic rift basin. The quartzolithic petrofacies shows early carbonate cements that inhibited compaction and later quartz, feldspar and clay mineral diagenetic phases. The quartzofeldspathic petrofacies has a rigid framework that maintained the original pores of the arenite during burial diagenesis. Quartz and K-feldspar overgrowths are common, with secondary porosity occurring as a product of feldspar dissolution. The quartzofeldspathic petrofacies has a greater potential to act as a hydrocarbon reservoir. This study corroborates the close relationship between the provenance of arenite and its reservoir potential in continental rift basins.

Recent decades have seen much effort invested in trying to better understand the development of petroleum systems and plays in sedimentary basins. The literature now contains a great deal of information on the configuration of sedimentary basins (e.g. Harff *et al.* 1993; Busby & Ingersoll 1995; Einsele 2000; Allen & Allen 2005), the stratigraphy of their deposits (e.g. Doust & Sumner 2007; Gregory *et al.* 2007; Veeken 2007), their depositional sequences (e.g. Ketzer 2002; Strohmenger *et al.* 2006), and their thermal evolution (e.g. Makhous & Galushkin 2005). Certainly, the widespread use of thermal evolution modelling software (e.g. BasinMod©, PetroMod©) has allowed a more accurate assessment of the maturity of the organic matter and the oil-bearing potential of sedimentary basins (Welte & Yukler 1981; Higley *et al.* 2006). The configuration of reservoirs, their porosity and permeability characteristics and their potential to hold oil and gas have also been the focus of attention (Barwis *et al.* 1990; Ashton 1993). Many authors have shown that the reservoir properties of sandstone bodies depend strongly on the composition – and therefore provenance – of that sandstone (Barwis *et al.* 1990; Bloch 1991; De Ros *et al.* 1994; Willson 1994; Dias Lima & De Ros 2002).

The Cameros Rift Basin (Mas *et al.* 2003) in northern Spain has been interpreted as a Mesozoic intracratonic rift basin that is non-productive in terms of oil and gas (Fig. 1). Several authors have reported the presence of abundant black shale deposits (mainly of continental/lacustrine origin) interbedded with sandstones. The generation and accumulation of hydrocarbons could occur in such deposits (Mas *et al.* 2003, 2008, 2011). Thus, potential petroleum systems may have been active during the diagenetic development of the basin (Mas *et al.* 2003, 2008). As a result, the basin offers a natural laboratory in which the relationship between sandstone composition (provenance) and the reservoir-favouring properties (quality) acquired during the course of diagenesis can be examined.

This paper provides a synthesis of the different studies (Arribas *et al.* 2002, 2003a, b, 2007; Ochoa *et al.* 2004; González-Acebrón *et al.* 2007, 2010a, Omodeo-Salé *et al.* 2011) performed on the arenites belonging to the sedimentary record of the Cameros Rift Basin, with the aim of finding relationships between arenite provenance

*From:* Scott, R. A., Smyth, H. R., Morton, A. C. & Richardson, N. (eds) 2014. *Sediment Provenance Studies in Hydrocarbon Exploration and Production.* Geological Society, London, Special Publications, **386**, 63–73.
First published online July 10, 2013, http://dx.doi.org/10.1144/SP386.12

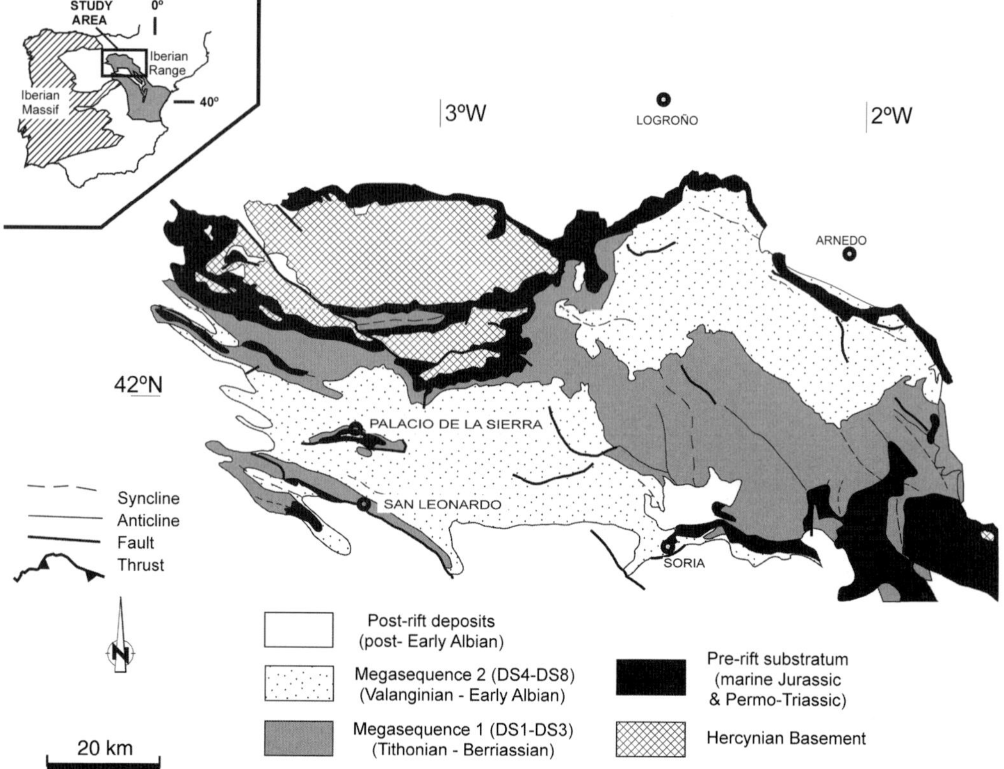



**Fig. 1.** Geological map of the Cameros Rift Basin showing the sedimentary infill divided into two megasequences (MS-1 and MS-2). MS1 includes depositional sequences DS1–DS3. MS-2 includes DS4–DS8.

and its characteristics as a potential hydrocarbon reservoir.

## Geological setting

The Cameros Rift Basin is a continental basin overlying a south-dipping ramp linking two flats (7000 and 11 000 m deep) in a crustal-scale horizontal extensional fault. It has, therefore, been defined as an 'extensional ramp basin' (Mas *et al.* 1993; 2002, 2004; Guimerà *et al.* 1995). The basin was configured during the Tithonian, the different depositional sequences that filled it laying unconformably over a marine Jurassic substratum. The sedimentary record is organized as eight depositional sequences (DS) (Fig. 2) separated by unconformities. Together they show a particular architecture consisting of progressive on-lapping of DS over the substratum towards the north. The progressive migration of depocentres therefore occurred in this direction (Mas *et al.* 2003, 2011; Omodeo-Salè *et al.* 2011). This resulted in a stratigraphic record of more than 9000 m of sediments dating from Tithonian to Early Albian times. The

maximum vertical thickness at the depocentre of the basin reaches some 6500 m (Mas *et al.* 2003, 2011).

Although some marine influence is manifested at definite intervals, the DS usually consist of continental deposits. These change from alluvial/fluvial at the base of the sequence to lacustrine clastic and carbonate deposits at the top. The same change occurs distally.

A large unconformity is seen between DS3 and DS4, manifested by an obvious change in the progression of the depocentre areas and the sources of sediment (Mas *et al.* 2003; Arribas *et al.* 2003a, 2007; Omodeo-Salè *et al.* 2011). This unconformity defines two main megasequences (MS): MS-1, made up of DS1–DS3, and MS-2 made up of DS4–DS8 (Arribas *et al.* 2003a, b, 2007; Mas *et al.* 2003) (Fig. 2).

Two hydrothermal processes affected the deposits of the eastern sector of the basin during the Late Cretaceous and Eocene (Casquet *et al.* 1992; Alonso-Azcárate *et al.* 1995, 1999; Barrenechea *et al.* 1995; Mantilla-Figueroa *et al.* 1998, 2002; González-Acebrón *et al.* 2011). This hydrothermal

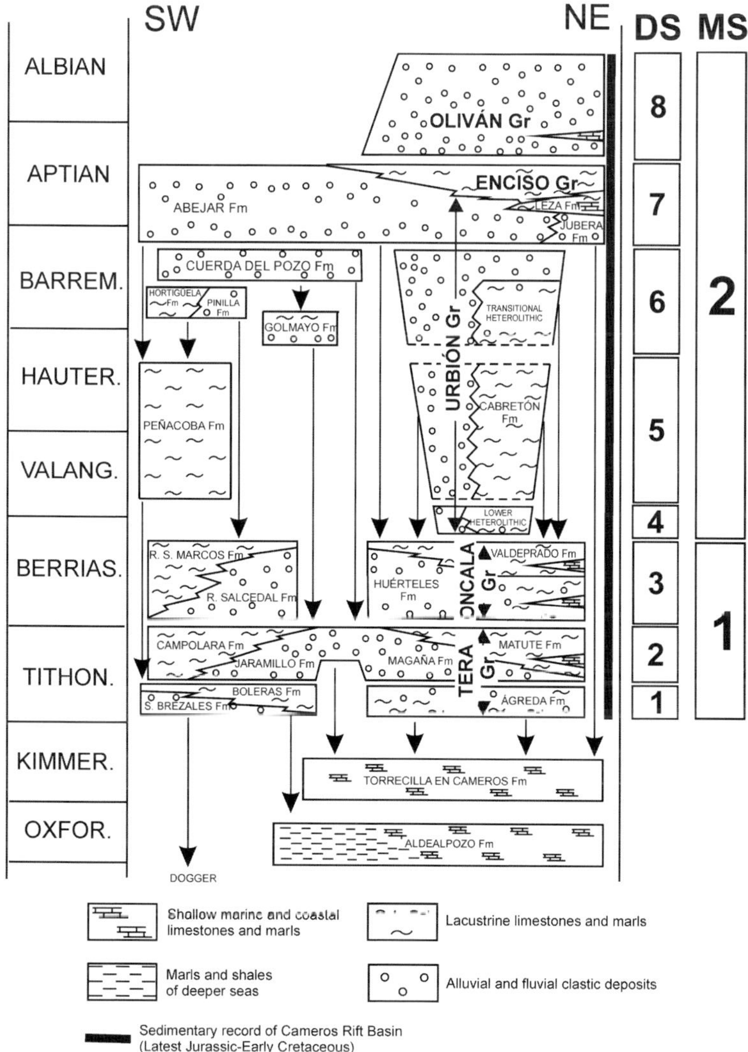


**Fig. 2.** Chronolithostratigraphic chart of the Cameros Rift Basin showing depositional sequences (DS), megasequences (MS) and lithostratigraphic units (formations [Fm] and groups [Gr]). Arrows refer to the stratigraphic lying relationships. Modified from Mas *et al.* (2011).

metamorphism destroyed the porosity of the sediments along with any potential hydrocarbons in the possible reservoirs provided by them (Mas *et al.* 2002; Ochoa *et al.* 2007).

## Petrographic methods providing the data for this review

Classic petrographic analyses have been performed (Arribas *et al.* 2002, 2003*a*, 2007; González-Acebrón *et al.* 2007, 2010*a*) on 420 medium- to fine-grained arenite samples from the Cameros Rift Basin. These involved point counting of framework grains and intergranular space on double-polished thin sections, stained for carbonate and K-feldspar identification. More than 600 point analyses were performed on each sample, following the Gazzi (1966) Dickinson (1970) (G–D) point count method, and using the petrographic classes of Zuffa (1980, 1987). Porosity analyses were performed by impregnating arenite samples with coloured epoxy resin. These compositional data were used to define the different petrofacies of the

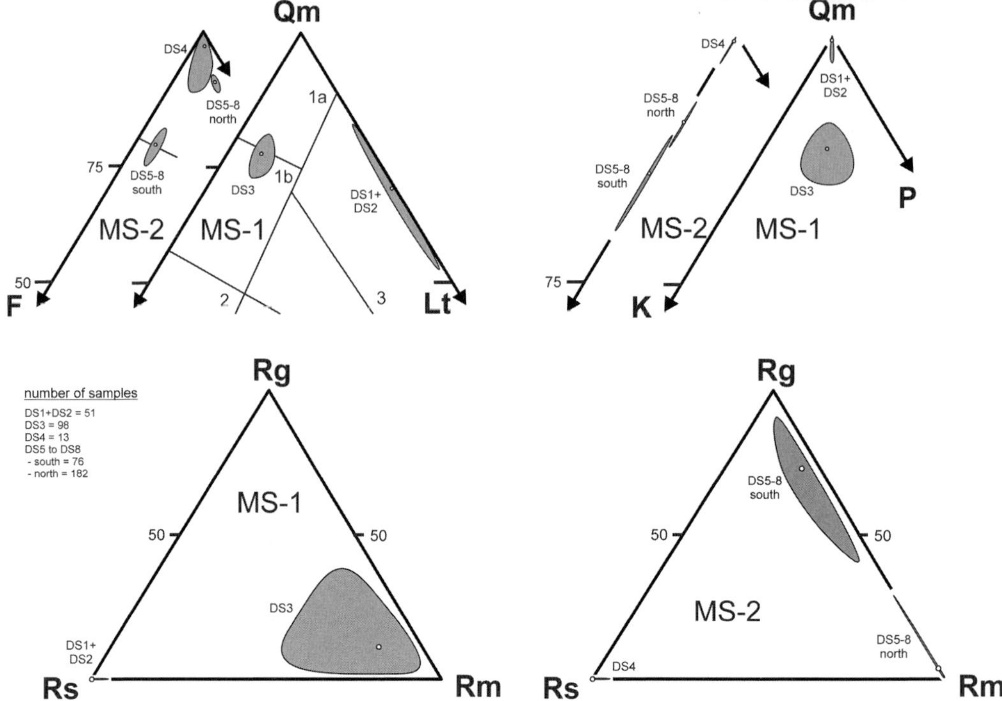


**Fig. 3.** Compositional characteristics of the studied petrofacies, as determined from QmFLt, QmKP (Dickinson 1970) and RsRgRm (Arribas *et al.* 1990) ternary diagrams. Grey fields represent the 95% confidence regions of the population mean. Qm, monocrystalline quartz; F, total feldspars; Lt, aphanitic rock fragments; K, K-feldspar; P, plagioclase; Rs, sedimentary rock fragments (aphanitic plus phaneritic); Rg, coarse-grained plutonic rock fragments; Rm, metamorphic rock fragments (aphanitic plus phaneritic). Provenance fields from Dickinson *et al.* (1983): 1a, 'stable craton'; 1b, 'continental transition'; 2, 'basement uplift'; 3, 'recycled orogen'. Data for Figure 3 are from Arribas *et al.* (2002, 2003*a*, 2007) and González-Acebrón *et al.* (2007, 2010*a*).

stratigraphic record. Framework petrographic variables were plotted on provenance diagrams (see Dickinson *et al.* 1983) (Fig. 3).

Intergranular diagenetic species and porosity values and types have been used in the examination of the diagenesis of the basin's arenites (Arribas *et al.* 2002, 2003*a*, 2007; González-Acebrón *et al.* 2007, 2010*a*). SEM analysis was also used to describe the textural relationships between the detrital grains and the diagenetic components. Electron microprobe analysis of the feldspars allowed the discrimination between detrital and albitized feldspars (González-Acebrón *et al.* 2010*b*). Fluid inclusion petrography and microthermometry analysis of the quartz cement phases (Ochoa 2004; González-Acebrón *et al.* 2011) provided the temperature conditions of the overgrowths. As a result, the chronology of diagenetic processes was established for each group of sandstone petrofacies (Arribas *et al.* 2002, 2003*b*; González-Acebrón *et al.* 2011). The data from the above papers are here used to assess the reservoir quality of the

different arenites of the basin via the analysis of original porosity loss during diagenesis.

## Arenite composition and provenance

Petrographic analysis of the basin's framework arenites suggests the presence of two main petrofacies (Arribas *et al.* 2003*a*, *b*, 2007): (1) a quartzolithic petrofacies, the provenance of which is related to recycling of the eroded, pre-rift sedimentary cover ('sedimenticlastic' according to Arribas & Tortosa (2003)); and (2) a quartzofeldspathic petrofacies mainly related to the erosion of plutonic and metamorphic source rocks (Fig. 3).

Sedimenticlastic petrofacies consist of quartzarenites and sedarenites (Folk 1968) with variable amounts of carbonate rock fragments (Fig. 4a, b). Rounded, monocrystalline (Fig. 4b) quartz grains dominate, with some showing abraded overgrowths (González-Acebrón *et al.* 2007, 2010*a*). Labile grains such as K-feldspars, plagioclase and

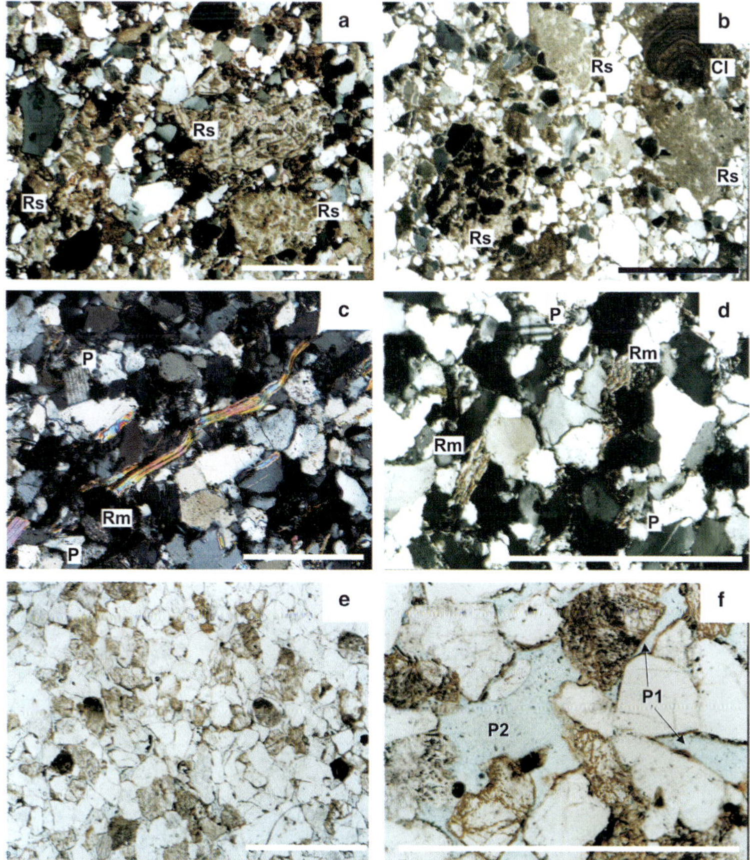

**Fig. 4.** Framework petrographic features of the studied petrofacies. (**a**) Quartzolithic petrofacies from DS1 showing abundant carbonate rock fragments (Rs). Cross-polarized light. (**b**) Quartzolithic petrofacies (DS4) with micritic carbonate rock fragments (Rs) and carbonate intrabasinal (oncoid) grains (CI). The quartz grains show rounded borders, suggesting a recycled origin. Cross-polarized light. (**c**) Quartzofeldspatholithic petrofacies (DS3) with monocrystalline quartz, mica and plagioclase grains (P). Fine grained metamorphic rock fragments (Rm) are also visible. Cross-polarized light. (**d**) Quartzofeldspatholithic petrofacies (DS3) showing plagioclase (P), fine-grained metamorphic rock fragments (Rm) and abundant monocrystalline quartz grains. Cross-polarized light. (**e**) Quartzofeldspathic petrofacies (DS7) with abundant K-feldspar (yellow stained grains). Primary porosity (blue) is also visible. Plane-polarized light. (**f**) Quartzofeldspathic petrofacies (DS7): close-up showing stained K-feldspars grains. The porosity (blue) is both primary (P1) and secondary (P2). Plane-polarized light. Scale bar in all photomicrographs = 1 mm.

metamorphic lithic grains appear in low proportions. Several mineralogical and textural types of carbonate rock fragment (micritic and sparitic limestones and dolostones) dominate the lithic population. The present quartzolithic petrofacies represents the recycling of the sedimentary cover made up mainly of Jurassic marine carbonate deposits. As a consequence, its arenite plots near the QmLt edge, and has a 'recycled orogen' type of provenance (Dickinson *et al.* 1983). This confirms the sedimenticlastic origin of this petrofacies (Fig. 2). At the top of the sequence, and in areas of maximum subsidence, additional supplies from

fine-grained schists and slates are deduced (Arribas *et al.* 2003a, 2007).

The quartzolithic petrofacies developed during the initial stages of rifting (DS1 and 2 at the base of MS-1), and at the base of the MS-2 (DS4), associated with the beginning of active tectonic phases (Arribas *et al.* 2003a). This petrofacies is recorded in a relatively thin succession of sediments (<100 m) showing little extension, mainly in the central southern part of the basin. This petrofacies is equivalent to the undissected–transitional stage of the non-volcanic rifted basin in the Gulf of Aden (Yemen) (Garzanti *et al.* 2001, 2003) and other, similar, geological settings (Zuffa *et al.* 1980; Evans 1990).

A metasedimenticlastic petrofacies (quartzofeldspatholithic sandstones with P > K) (Fig. 4c, d), intermediate between the quartzolithic and quartzofeldspathic petrofacies, appears as a product of the erosion of metasediments of the low grade metamorphic substratum of the basin; this occurred after the sedimentary cover was eroded. This petrofacies developed in the upper part of the MS-1 (DS2 and 3). It characterizes a sedimentary record more than 400 m thick (Arribas *et al.* 2007) in the western part of the basin, and up to 3500 m thick

in the eastern part (Mas *et al.* 2011). An equivalent petrofacies in a similar tectonic setting in the Gulf of Aden has been described as a product of an evolved phase of erosion in a rift-shoulder (Garzanti *et al.* 2001).

The quartzofeldspathic (plutoniclastic) petrofacies (Fig. 4e, f) contains medium- to coarse-grained arkoses that evolve spatially to quartzarenites, the consequence of maturation during transport in a humid climate (Ochoa *et al.* 2007). Quartz is the main framework component, dominating the monocrystalline types. K-feldspar dominates over plagioclase grains, its content progressively

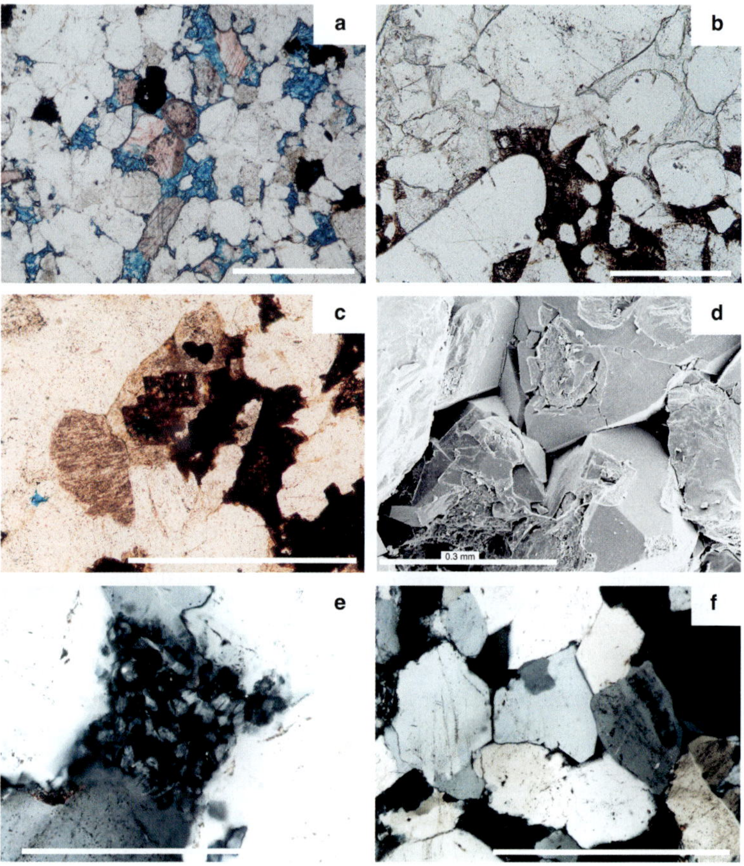

**Fig. 5.** Diagenetic features of the studied petrofacies. (**a**) Quartzolithic petrofacies (DS1) showing calcite grains (red stained grains) and ferroan dolomite cement (intense blue stain) occluding the intergranular spaces. Plane-polarized light. (**b**) Close-up of the quartzolithic petrofacies (DS1) in which Fe-oxides (ankerite baroque ghost) replace an early carbonate cement that occluded the intergranular space. Plane-polarized light. (**c**) Quartzolithic petrofacies (DS1) showing intense ankerite replacement over early intergranular cements (middle-right) and feldspar grains (centre). Plane-polarized light. (**d**) SEM image of the quartzofeldspathic petrofacies showing intense quartz overgrowth. Note the preservation of low surface area primary pores. Scale bar = 0.5 mm. (**e**) Kaolinite pore-filling postdating the quartz cement in the quartzofeldspathic petrofacies (DS6). Cross-polarized light. (**f**) Quartzofeldspathic petrofacies (DS6) showing intense quartz overgrowth and the remains of primary porosity. Cross-polarized light. Scale bar in all photomicrographs = 1 mm (except for image (d).

increasing from older to younger deposits. Lithic grains are scarce and consist mainly of aphanitic low-grade metamorphic material, along with coarse-grained phaneritic crystalline quartz–feldspar rock fragments.

This extensive petrofacies developed during the most active phase of rifting and characterizes a very thick succession up to 6000 m thick in the eastern part of the basin (Mas *et al.* 2011), and up to 1600 m thick in the west (Arribas *et al.* 2007). The quartzofeldspathic petrofacies is clear between DS5 and DS8 (the mid-upper part of MS-2). It appears in association with low grade metamorphic supplies that gradually became diluted during the sedimentation of the successive DS. A progressive reduction in metamorphic rock fragments (Lm content) and an increase in K-feldspar grains can be seen. Coarse crystalline rocks (granites and gneisses) from the Variscan substratum (Central Iberian Zone) have been proposed as the main source for this petrofacies (Arribas *et al.* 2003*a*, 2007). The provenance of this petrofacies type has been related to the mature stages of rift-shoulder erosion (Garzanti *et al.* 2001, 2003). Finally, the quartzarenitic composition of the sandy detritus in the quartzofeldspathic petrofacies is interpreted as the result of the in-transport maturation of the quartzofeldspathic sediment in a transport-limited denudation regime (Ochoa *et al.* 2004).

The locations of the quartzolithic and quartzofeldspathic petrofacies in the sedimentary record of the basin agree well with the hierarchy of the main bounding surfaces between the DS and MS. Thus, a direct relationship exists between these petrofacies and the main rifting stages (Arribas *et al.* 2003*a*, 2007). As a consequence, two main 'provenance cycles' (Arribas *et al.* 2007) can be identified involving the succession of two pairs of these petrofacies. These provenance cycles coincide with the defined stratigraphic MS and show the original composition of sediments in the stratigraphic record.

## Diagenetic alterations and the evolution of porosity

The quartzolithic petrofacies shows carbonate cements from both the early (eogenetic) and late (mesogenetic) stages of burial. Early microcrystalline mosaic and poikilotopic textures of calcite are common. Replacements of ferroan dolomite and ankerite (ghosts) have been interpreted as belonging to a late diagenetic phase (Fig. 5a, b). Carbonate cements (in the main) help retain the original intergranular volume, but drastically occlude the original porosity. The ankerite also replaces the

framework grains, especially those of K-feldspar (Fig. 5c). The origin of the early carbonate cements might be related to the dissolution of the carbonate clasts (both intrabasinal and extrabasinal), common in the framework of the quartzolithic petrofacies. Quartz and K-feldspars cements are scarce. Also rare are diagenetic clay minerals (e.g. kaolinite, illite); these only appear very locally. Mechanical compaction is in part inhibited by early carbonate cementation, helping to maintain high intergranular volume values (21–24%) (Fig. 6). Porosity, if present at all, is very low ($<1\%$) and corresponds to secondary porosity generated by local carbonate dissolution during the inversion of the basin.

The diagenesis in the quartzofeldspathic petrofacies is characterized by an intense reduction of the intergranular volume (which shows values of 11–21%), the result of mechanical and chemical compaction (Fig. 6). Quartz and K-feldspar overgrowths are the most abundant cements (Fig. 5d). Kaolinite also appears as a well-developed diagenetic phase in the form of early pore fillings and

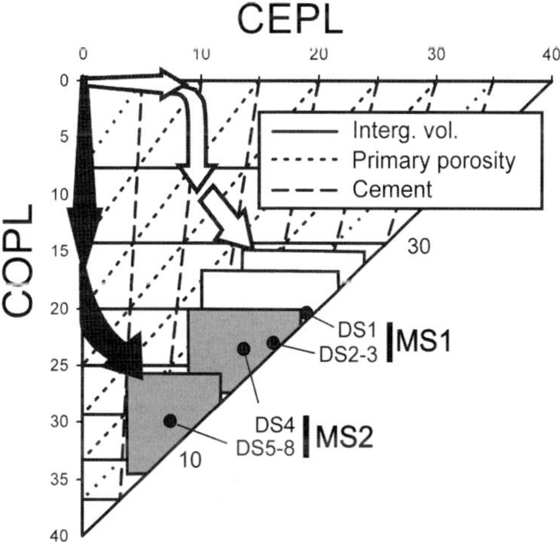



**Fig. 6.** Compactional (COPL) v. cementational (CEPL) loss of original porosity of the studied petrofacies during burial diagenesis (modified from Lundegard [1992]). The arenite samples were grouped into quartzolithic and quartzofeldspathic petrofacies from both megasequences (MS-1 and MS-2). Fields (white for MS-1 and grey for MS-2) were constructed using standard deviation values around the mean (black circles) of the corresponding population. Arrows show tentative pathways of porosity loss during burial. The white arrow represents the evolution of the quartzolithic petrofacies under early carbonate cementation. The black arrow represents the evolution of the quartzofeldspathic petrofacies under intense compaction. Notice that this last petrofacies may retain some of its primary porosity.

K-feldspar replacements (epimatrix) (Fig. 5e). Illite locally appears as the replacement of an early pore lining coating the grains, or as a mesodiagenetic replacement of kaolinite pore filling and epimatrix. In the southern DS5–DS7 arenites some 5–7% of the primary porosity is preserved (Fig. 5d). Secondary porosity is also present, generated by the dissolution of K-feldspar grains producing oversized intergranular pores (Fig. 4f). The absence of early phases of carbonate cementation enhanced compaction and the reduction of the intergranular volume. However, in the southwestern part of the basin, primary pores were maintained owing to the rigid quartzofeldspathic framework of the arenites.

In the central and northern parts of the basin, the strong development of quartz overgrowth, spurred by hydrothermal processes (Mas *et al.* 2011), led to drastically occluded porosity in both the studied petrofacies (Ochoa *et al.* 2007; González-Acebrón *et al.* 2011). Different stages of quartz cementation and recrystallization have been established using SEM-CL (Fig. 5f), with homogenization temperatures affecting the primary fluid inclusions ranging from 114 °C to 141 °C in DS2 (González-Acebrón *et al.* 2011), and reaching 226 °C in the northern part of DS7 (Ochoa *et al.* 2007). These high temperatures are related to hydrothermalism.

The quartzofeldspathic petrofacies shows signs of an albitization process, at least in DS2 (González-Acebrón *et al.* 2010b). Both the plagioclases and K-feldspars are albitized and have a very Na-rich composition (mean composition: $Al_{94.0}$ $An_{4.5}$ $Or_{1.5}$). Chemically pure albite (>99%) is relatively common. The homogenization temperatures affecting the primary fluid inclusions in the albitized plagioclases were between 83 °C and 115 °C. As a result of the albitization of the Ca-plagioclases an associated phase of calcite cement was generated, causing a slight reduction in porosity (González-Acebrón *et al.* 2010b).

The potential of the arenitic petrofacies to act as a reservoir can be determined by analysing the original loss of porosity via compactional and cementational processes (Lundegard 1992) over the course of diagenesis (Fig. 6). The quartzofeldspathic petrofacies suffered a very intense loss of porosity via compaction, which drastically reduced the intergranular space. In contrast, the quartzolithic petrofacies show high intergranular space values, mainly due to early carbonate cementation. No

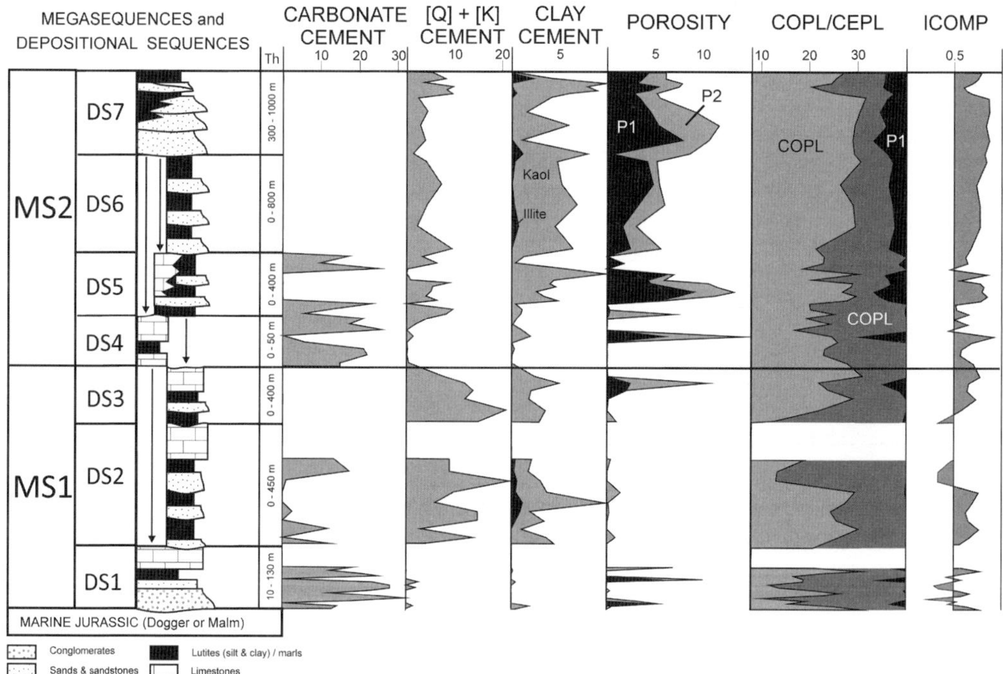


**Fig. 7.** Quantification of the main diagenetic processes affecting the studied petrofacies in the SW of the Cameros Basin (modified from Arribas *et al.* 2002, 2003b). Idealized stratigraphic log shows dramatic thickness variations (Th) of DS in this area, and the stratigraphic lying relationships (arrows). Note that DS8 does not appear in this part of the basin. [Q], Quartz cement; [K], K-feldspar cement. Kaol, Kaolinite; P1, primary porosity; P2, secondary porosity; COPL, compactional porosity loss; CEPL, cementational porosity loss; ICOMP, compactional index (after Lundegard 1992).

primary porosity is, however, preserved. The quartzofeldspathic petrofacies maintained its primary pores because the framework was sufficiently rigid. These pores commonly have crystal overgrowth faces showing borders with a small surface area (Wilson 1984). Secondary porosity, created by K-feldspar dissolution/replacement, is also common.

## Discussion

The Cameros Rift Basin has been interpreted as an inverted basin non-productive in terms of oil and gas (Mas *et al.* 2003, 2011). However, the petrographic and geochemical evidence suggests that hydrocarbons were generated, but migrated away during the basin's evolution (Mantilla Figueroa *et al.* 1998; Ochoa *et al.* 2007; Mas *et al.* 2011).

Several authors (e.g. Mas *et al.* 2003) have reported the presence of abundant black shale deposits (mainly of continental/lacustrine origin) in the sedimentary record, interbedded with the described petrofacies. These black shales are the most likely potential source rocks of the basin's petroleum system (Mas *et al.* 2003, 2008); the arenite petrofacies could have acted as potential reservoirs. The thermal model for the basin (Omodeo-Salé *et al.* 2012) shows that hydrocarbon generation could have taken place during diagenetic burial processes. The above-mentioned metamorphic hydrothermal event is probably responsible for the transformation and disappearance of the hydrocarbons (Ochoa *et al.* 2007; Mas *et al.* 2011).

### Cyclicity of arenite provenance

The arenites in the sedimentary record of the basin are characterized by two petrofacies: quartzolithic and quartzofeldspathic. The succession of pairs of these petrofacies constitutes a provenance cycle that represents the evolution of eroded lithologies (from the sedimentary cover to the crystalline basement) and an unroofing sequence during an active tectonic stage. Similar cyclicity has been observed by other authors in other rift basins (e.g. Garzanti *et al.* 2001, 2003). The Cameros Rift Basin has two main provenance cycles, one approximately 500–3500 m thick, the other 1600–6000 m thick, and covering some 10 Ma and 30 Ma intervals, respectively. Throughout the basin each cycle consists of a lower, thin and little-extended sedimenticlastic (quartzolithic) succession (<100 m) followed by a thick, well-developed plutoniclastic (quartzofeldspathic) succession. The great difference in the sediment volume between the petrofacies is related to the capacity of sources to produce sandy detritus. Plutonites generate high volumes of quartzofeldspathic sediment (Palomares & Arribas

1993), unlike sedimentary source rocks which generate little quartzolithic detritus (Arribas & Tortosa 2003; Arribas *et al.* 2007).

### Cyclicity of diagenetic processes

The course of diagenesis for the studied arenites can be interpreted as a function of their framework composition. Carbonate cementation during diagenesis is mainly seen in the quartzolithic petrofacies; this maintained high intergranular volume values, but led to the porosity becoming totally occluded (Fig. 7). Silicate diagenesis is well developed in the quartzofeldspathic petrofacies, with quartz and K-feldspar overgrowths, kaolinite and illite cements and replacements. However, the rigid framework of this petrofacies prevented the total compaction and occlusion of all porosity. The existence of primary pores and the development of secondary porosity make this petrofacies the most likely as a potential hydrocarbon reservoir (Fig. 7).

## Conclusions

The petrology of the sandstones deposited in the Cameros Rift Basin demonstrates the relationship between provenance and the potential to act as a reservoir. The quartzofeldspathic petrofacies has characteristics that favour it over the quartzolithic petrofacies as a hydrocarbon reservoir. Rigid sandstone frameworks made up of quartz and K-feldspar grains in the quartzofeldspathic petrofacies favour the preservation of the primary pores, even though the intergranular volume is strongly reduced by compaction. In addition, the silicate diagenetic phases (quartz, K-feldspar, kaolinite, illite) do not occlude the porosity drastically, unlike the carbonate in the quartzolithic petrofacies. In addition, the source rocks of the quartzofeldspathic petrofacies (coarse plutonites) generate thick and extensive sequences of potential clastic reservoirs.

The quartzolithic petrofacies show large intergranular spaces (carbonate cements) that could be transformed into secondary porosity if dissolution processes took place. This petrofacies is recorded in thin, little-extensive, sedimentary successions. Thus, its potential to act as a hydrocarbon reservoir is low.

The cycling of the arenite petrofacies as the extension of the basin proceeded led to the cyclicity of diagenetic processes and eventually the different potentials of the petrofacies to act as hydrocarbon reservoirs.

This paper is a contribution to the Spanish DGICYT research projects CGL2008 01648 and CGL2011-22709. The authors are grateful to E. Le Pera (UNICAL, Italy) for improving the original manuscript.

# References


ALONSO-AZCÁRATE, J., BARRENECHEA, J. F., RODAS, M. & MAS, J. R. 1995. Comparative study of the transition between very low-grade and low-grade metamorphism in siliciclastic and carbonate sediments: early Cretaceous, Cameros Basin (Northern Spain). *Clay Minerals*, **30**, 407–419.

ALONSO-AZCÁRATE, J., RODAS, M., BOTRELL, S. H., RAISWELL, R., VELASCO, F. & MAS, R. 1999. Pathways and distances of fluid flow during low-grade metamorphism: evidence from pyrite deposits of the Cameros Basin, Spain. *Journal of Metamorphic Geology*, **17**, 339–348.

ALLEN, P. A. & ALLEN, J. J. 2005. *Basin Analysis: Principles and Applications*. 2nd edn. Blackwell Publishing, Oxford.

ARRIBAS, J. & TORTOSA, A. 2003. Detrital modes in sedimenticlastic sand from low-order streams in the Iberian Range, Spain: the potential for sand generation by different sedimentary rocks. *Sedimentary Geology*, **159**, 275–303.

ARRIBAS, J., GÓMEZ-GRAS, D., ROSELL, J. & TORTOSA, A. 1990. Estudio comparativo entre las areniscas Paleozoicas y Triásicas de la Isla de Menorca: evidencias de procesos de reciclado. *Revista de la Sociedad Geológica de España*, **3**, 105–116.

ARRIBAS, A., MAS, R., OCHOA, M. & ALONSO, A. 2002. Composición y diagénesis del registro detrítico en el borde suroccidental de la cuenca de Cameros. *Zubía*, **14**, 99–119.

ARRIBAS, J., ALONSO, A. *ET AL.* 2003*a*. Sandstone petrography of continental depositional sequences of a intraplate rift basin: Western Cameros Basin (North Spain). *Journal of Sedimentary Research*, **73**, 309–327.

ARRIBAS, J., MAS, R., ALONSO, A. & OCHOA, M. 2003*b*. Diagenetic processes controlling quality of potential clastic reservoirs in a continental Rift Basin: Western Cameros Basin, Spain. *AAPG International Conference and Exhibition*, 21–24 September, Barcelona. Abstract book, A5.

ARRIBAS, A., MAS, R., ARRIBAS, M. E., OCHOA, M. & GONZÁLEZ, L. 2007. Sandstone petrofacies in the Northwestern sector of the Iberian Basin. *Journal of Iberian Geology*, **33**, 191–206.

ASHTON, M. (ed.) 1993. *Advances in Reservoir Geology*. Geological Society of London, London, Special Publications, **69**.

BARWIS, J. H., MCPHERSON, J. G. & STUDLICK, J. R. J. (eds) 1990. *Sandstone Petroleum Reservoirs*. Springer-Verlag, Berlin.

BARRENECHEA, J. F., RODAS, M. & MAS, J. R. 1995. Clay mineral variations associated with diagenesis and low-grade metamorphism of early Cretaceous sediments in the Cameros Basin, Spain. *Clay Minerals*, **30**, 119–133.

BUSBY, C. J. & INGERSOLL, R. V. 1995. *Tectonic of Sedimentary Basins*. Blackwell Science, Oxford.

BLOCH, S. 1991. Empirical prediction of porosity and permeability in sandstones. *American Association of Petroleum Geologists Bulletin*, **75**, 1145–1160.

CASQUET, C., GALINDO, C. *ET AL.* 1992. El metamorfismo en la cuenca de los Cameros. Geocronología e implicaciones tectónicas. *Geogaceta*, **11**, 22–25.

DE ROS, L. F. S., MORAD, S. & PAIM, P. S. G. 1994. The role of detrital composition and climate on the diagenetic evolution of continental molasses: evidence from the Cambro-Ordovician Guaritas Sequence. Southern Brazil. *Sedimentary Geology*, **92**, 197–228.

DIAS LIMA, R. D. & DE ROS, L. F. 2002. The role of depositional setting and diagenesis on the reservoir quality of Devonian sandstones from the Solimoes Basin, Brazilian Amazonia. *Marine and Petroleum Geology*, **19**, 1047–1071.

DICKINSON, W. R. 1970. Interpreting detrital modes of graywacke and arkose. *Journal of Sedimentary Petrology*, **40**, 695–707.

DICKINSON, W. R., BEARD, L. S. *ET AL.* 1983. Provenance of North American Phanerozoic sandstones in relation to tectonic setting. *Geological Society of America Bulletin*, **94**, 222–235.

DOUST, H. & SUMNER, H. S. 2007. Petroleum systems in rift basins – a collective approach in Southeast Asian basins. *Petroleum Geoscience*, **13**, 127–144.

EINSELE, G. 2000. *Sedimentary Basins – Evolution, Facies, and Sediment Budget*. 2nd edn. Springer, Berlin.

EVANS, A. L. 1990. Miocene sandstone provenance relations in the Gulf of Suez: insights into synrift unroofing and uplift history. *American Association of Petroleum Geologists Bulletin*, **74**, 1386–1400.

FOLK, R. L. 1968. *Petrology of Sedimentary Rocks*. Hemphill's, Austin, TX.

GARZANTI, E., VEZZOLI, G., ANDÒ, S. & CASTIGLIONI, G. 2001. Petrology of rifted-margin sand (Red Sea and Gulf of Aden, Yemen). *Journal of Geology*, **109**, 277–297.

GARZANTI, E., ANDÒ, S., VEZZOLI, G. & DELL'ERA, D. 2003. From rifted margins to foreland basins: investigating provenance and sediment dispersal across desert Arabia (Oman, U.A.E.). *Journal of Sedimentary Research*, **73**, 572–588.

GAZZI, P. 1966. Le arenarie del flysch sopracretaceo dell'Appennino modenese; correlazioni con il Flysch di Monghidoro. *Mineralogia e Petrografia Acta*, **12**, 69–97.

GONZÁLEZ-ACEBRÓN, L., ARRIBAS, J. & MAS, R. 2007. Provenance of fluvial sandstones at the start of late Jurassic–early Cretaceous rifting in the Cameros Basin (N. Spain). *Sedimentary Geology*, **202**, 138–157.

GONZÁLEZ-ACEBRÓN, L., ARRIBAS, J. & MAS, R. 2010*a*. Sand provenance and implications for paleodrainage in a rift basin: the Tera Group. *Journal of Iberian Geology*, **36**, 179–184.

GONZÁLEZ-ACEBRÓN, L., ARRIBAS, J. & MAS, R. 2010*b*. The role of sandstone provenance in diagenetic albitization of feldspars. A case study in the Jurassic Tera Group sandstones (Cameros Basin, NE Spain). *Sedimentary Geology*, **229**, 53–63.

GONZÁLEZ-ACEBRÓN, L., GOLDSTEIN, R. H., MAS, R. & ARRIBAS, J. 2011. Criteria for recognition of localization and timing of multiple events of hydrothermal alteration in sandstones illustrated by petrographic, fluid inclusion, and isotopic analysis of the Tera Group, Northern Spain. *International Journal of Earth Sciences*, **100**, 1811–1826.

GREGORY, F. J., COPESTAKE, P. & PEARCE, J. M. S. 2007. Key issues in petroleum geology: stratigraphy. *Geological Society of London, London. Key Issues*, **3**, 399.

GUIMERÀ, J., ALONSO, A. & MAS, R. 1995. Inversion of an extensional-ramp basin by a newly formed thrust: the Cameros Basin (N Spain). *In*: BUCHANAN, J. G. & BUCHANAN, P. G. (eds) *Basin Inversion*. Geological Society, London, Special Publications, **88**, 433–453.

HARFF, J. E., DAVIS JOHN, C. & EISERBECK, W. 1993. Prediction of hydrocarbons in sedimentary basins. *Mathematical Geology*, **25**, 925–936.

HIGLEY, D. K., LEWAN, M., ROBERTS, R. N. L. & HENRY, M. E. 2006. *Petroleum system modeling capabilities for use in oil and gas resource assessment*. United States Geological Survey Open-file report 2006-1024.

KETZER, J. 2002. *Diagenesis and Sequence Stratigraphy. an integrated approach to constrain evolution of reservoir quality in sandstones*. Acta Universitatis Upsaliensis, Uppsala, **762**.

LUNDEGARD, P. D. 1992. Sandstone porosity loss. A 'big picture' view of the importance of Compactation. *Journal of Sedimentary Petrology*, **62**, 250–260.

MAKHOUS, M. & GALUSHKIN, Y. I. 2005. *Basin Analysis and Modeling of the Burial, Thermal and Maturation Histories in Sedimentary Basins*. Editions Technip, Paris.

MAS, R., ALONSO, A. & GUIMERÀ, J. 1993. Evolución tectonosedimentaria de una cuenca extensional intraplaca: la cuenca finijurásica- eocretácica de Los Cameros (La Rioja- Soria). *Revista Sociedad Geológica de España*, **6**, 129–144.

MAS, R., BENITO, M. I., ARRIBAS, J., SERRANO, A., GUIMERÀ, J., ALONSO, A. & ALONSO-AZCARATE, J. 2002. La Cuenca de Cameros: desde la extensión finijurásica-eocretácica a la inversión terciaria – implicaciones en la exploración de hidrocarburos. *Zubía*, **14**, 9–64.

MAS, R., BENITO, M. I., ARRIBAS, J., SERRANO, A., ALONSO, A. & ALONSO-AZCARATE, J. 2003. The Cameros Basin: from Late Jurassic–Early Cretaceous extension to Tertiary contractional inversion – Implications of hydrocarbon exploration. *In*: *AAPG International Conference and Exhibition*, 21–24 September, Barcelona, Spain. Geological Field Trip 11.

MAS, R. & GARCÍA, A. (coord.) *ET AL*. 2004. 5.3.3. Segunda fase de rifting: Jurásico Superior-Cretácico Inferior. *In*:VERA, J. (ed.) *Geología de España*. Sociedad Geológica de España – Instituto Geológico y Minero de España, Madrid, 503–509.

MAS, R., BENITO, M. I. Y., & ARRIBAS, J. 2008. *The Cameros Basin (Northwest Iberian Chain, North Spain): From Late Jurassic-Early Cretaceous Extensions to Tertiary Contractional Inversion - Implications of Hydrocarbon Exploration*. Field trip guide, Repsol-YPF Brasil and Petrobras.

MAS, R., BENITO, M. I. Y. *ET AL*. 2011. Evolution of an intra-plate rift basin: the Latest Jurassic–Early Cretaceous Cameros Basin (Northwest Iberian Ranges, North Spain). *In*: ARENAS, C., POMAR, L. & COLOMBO, F. (eds) *Post-Meeting field trips 28th IAS Meeting. Geoguías*, **8**, 117–154.

MANTILLA, L. C., CASQUET, C. & MAS, R. 1998. Los paleofluidos en el Grupo Oncala, Cuenca de Cameros (La Rioja, España): datos de inclusiones fluidas, isótopos de oxígeno y SEM. *Geogaceta*, **24**, 207–210.

MANTILLA, L. C., CASQUET, C., GALINDO, C. & MAS, R. 2002. El metamorfismo hidrotermal cretácico y paleógeno en la cuenca de Cameros (Cordillera Ibérica, España). *Zubía*, **14**, 143–154.

OCHOA, M., ARRIBAS, J. & MAS, R. 2004. Changes in sandstone composition during Lower Cretaceous syn-rift fluvial sedimentation (Cameros Basin, Spain). *32nd International Geological Congress*, 20–28 August, Florence (Italy). Abstract CD, Session 242–34.

OCHOA, M., ARRIBAS, J., MAS, R. & GOLDSTEIN, R. H. 2007. Destruction of a fluvial reservoir by hydrothermal activity. *Sedimentary Geology*, **202**, 158–173.

OMODEO-SALÈ, S., ARRIBAS, J., GUIMERÀ, J., MAS, R., LÓPEZ, V., QUIJADA, E. & SUAREZ, P. 2011. The architecture of depositional sequences in an inverted rift basin: tectonic controls and genetic insight (the Late Jurassic–Early Cretaceous Cameros Basin, N Spain). *28th IAS meeting of Sedimentology*, 5–8 July, Zaragoza, Spain. Abstract book, 445.

OMODEO-SALÉ, S., ARRIBAS, J., GUIMERÀ, J., MARTINEZ, L., MAS, R., SALAS, R. & SUÁREZ-RUIZ, I. 2012. 1D thermal modeling of an extensional basin: the Cameros Basin (NE Spain). *29th IAS meeting of Sedimentology*, 10–13 September, Schladming, Austria. Abstract book.

PALOMARES, M. & ARRIBAS, J. 1993. Modern stream sands from compound crystalline sources: Composition and sand generation index. *In*: JOHNSSON, M. J. & BASU, A. (eds) *Processes Controlling the Composition of Clastic Sediments*. Geological Society of America Special Papers, **284**, 313–322.

STROHMENGER, CH. J., WEBER, L. J. *ET AL*. 2006. High-resolution sequence stratigraphy and reservoir characterization of Upper Thamama (Lower Cretaceous) reservoirs of a Giant Abu Dhabi Oil Field, United Arab Emirates. *In*: HARRIS, P. M. & WEBER, L. J. (eds) *Giant Hydrocarbon Reservoirs of the World: From Rocks to Reservoir Characterization and Modeling*. American Association of Petroleum Geologists, Memoirs, **88**, 469.

VEEKEN, P. C. H. 2007. Seismic stratigraphy, basin analysis and reservoir characterisation. *In*: *Handbook of Geophysical Exploration*, Elsevier, Amsterdam, **37**, 509.

WELTE, D. H. & YUKLER, M. A. 1981. Petroleum origin and accumulation in Basin Evolution – a quantitative model. *American Association of Petroleum Geologists Bulletin*, **65**, 1387–1396.

WILSON, M. D. 1984. Clastic diagenesis. *In*: *Geology for Engineers*. Short Course Notes: The Petroleum Society of CIM, Montreal, Calgary Section.

WILSON, M. D. 1994. *Reservoir quality assessment and prediction in clastic rocks*. Society for Sedimentary Geology (SEPM) *Short Course, Tulsa*, **30**, 432.

ZUFFA, G. G. 1980. Hybrid arenites: their composition and classification. *Journal of Sedimentary Petrology*, **50**, 21–29.

ZUFFA, G. G. 1987. Unravelling hinterland and offshore palaeogeography from deep-water arenites. *In*: LEGGETT, J. K. & ZUFFA, G. G. (eds) *Marine Clastic Sedimentology: Concepts and Case Studies*. Graham & Trotman, California, 39–61.

ZUFFA, G. G., CLAUDIO, W. & ROVITO, S. 1980. Detrital mode evolution of the rifted continental margin Longobucco sequence (Jurassic), Calabrian Arc, Italy. *Journal of Sedimentary Petrology*, **50**, 51–61.

# Provenance, diagenesis and reservoir quality of the Upper Triassic Wolfville Formation, Bay of Fundy, Nova Scotia, Canada


YAWOOZ A. KETTANAH[1]*, MUHAMMAD Y. KETTANAH & GRANT D. WACH

*Department of Earth Sciences, Dalhousie University, Halifax, Nova Scotia, Canada*

[1]*Present address: Department of Geology, Salahaddin University, Erbil, Iraq*

**Corresponding author (e-mail: kettanah@dal.ca)*



**Abstract:** The reservoir characterization and provenance of the Wolfville Formation were investigated using petrography, heavy minerals and microprobe analysis of tourmaline and garnet. Sandstone samples were taken from exposures at Rainy Cove and Cambridge Cove, and from the subsurface at Chinampas N-37 well beneath the Bay of Fundy. The surface and subsurface rocks have differences in their relative content and/or type of detritus, cement and heavy minerals (opaques, garnet, scheelite, tourmaline, rutile, apatite and others). These sandstones have continental block provenance in the subsurface rocks and recycled orogen provenance in the exposures. The main sources of the exposed Wolfville Formation sediments were the Palaeozoic rocks of the Meguma Supergroup, South Mountain Batholith, Horton and Windsor groups; meanwhile the subsurface sandstones might have been derived from the same sources or the Avalon Terrane and/or Gondwana. The sandstones were deposited during early stages of rifting post-dating earlier Palaeozoic collision orogenies that culminated with the Appalachian orogeny. The exposures of Wolfville Formation have low porosity (*c.* 6%) that diminishes to negligible in the subsurface. The Wolfville Formation has a considerable thickness beneath the Bay of Fundy where it overlies the Horton Bluff Formation, Meguma and/or Avalon terranes. However, its reservoir potential is not encouraging.


The Fundy Basin has long been recognized as a potential basin offshore eastern Canada. Petroleum companies became interested in the Triassic synrift sediments and started exploration in the Bay of Fundy during the period 1968–1983. As a result, detailed seismic exploration was conducted and two deep exploration wells were drilled in the southwestern part of the Bay of Fundy (Wade *et al.* 1996), Chinampas N-37 (3661.6 m deep) and Cape Spencer-1 (2587 m deep). Both wells drilled through the Wolfville Formation, which was the reservoir target, but were dry (Fig. 1).

The Late Triassic Wolfville Formation was deposited under fluvial, alluvial and aeolian conditions (Klein 1962; Hubert & Forlenza 1988; Tanner 2000; Fensome & Williams 2001). It has been studied as an outcrop example of subsurface fluvial reservoirs (Leleu *et al.* 2009; Leleu & Hartley 2010). The Wolfville Formation overlies the Tournaisian Horton Bluff Formation, which is made up of dark-coloured lacustrine shales, siltstones and sandstones (Figs 1 & 2). Elsewhere in Atlantic Canada, such as in the Stoney Creek field (St Peter 1993; Mukhopadhyay *et al.* 2000, Mukhopadhyay 2008), Tournaisian organic-rich shales are an important source rock and the interbedded sandstones were an exploration target on land. Seismic data show that both the Horton Bluff Formation and the Wolfville Formation extend

below the Bay of Fundy to a depth of *c.* 4 km (Fig. 1). If the overlying shales of the Blomidon Formation form an effective seal, the system could be an attractive hydrocarbon future target.

This study examines the reservoir characteristics of the Wolfville Formation including petrography, provenance as indicated by heavy minerals, and diagenesis and its influence on porosity. Samples were taken from the northern margin of the basin, in the Chinampas N-37 exploratory well and from the southern part of the basin from surface exposures at Rainy Cove and Cambridge Cove in the Minas sub-basin (Fig. 1).

## Geological setting

The Bay of Fundy was formed during the early rifting stages of Pangaea in the Early Mesozoic. It was formed along the suture line of the Appalachian Avalon and Meguma terranes, marked by the extension of the Cobequid Chedabucto (Minas) fault system (Fig. 1). Upper Triassic and Lower Jurassic sediments were deposited during various stages of rifting and the early development of the Fundy Basin. The exposed portion of Mesozoic strata along the beaches and cliffs of the Minas subbasin is small in comparison to its subsurface extent beneath the Bay of Fundy. The Upper Triassic


*From*: SCOTT, R. A., SMYTH, H. R., MORTON, A. C. & RICHARDSON, N. (eds) 2014. *Sediment Provenance Studies in Hydrocarbon Exploration and Production.* Geological Society, London, Special Publications, **386**, 75–110.
First published online October 24, 2013, http://dx.doi.org/10.1144/SP386.18

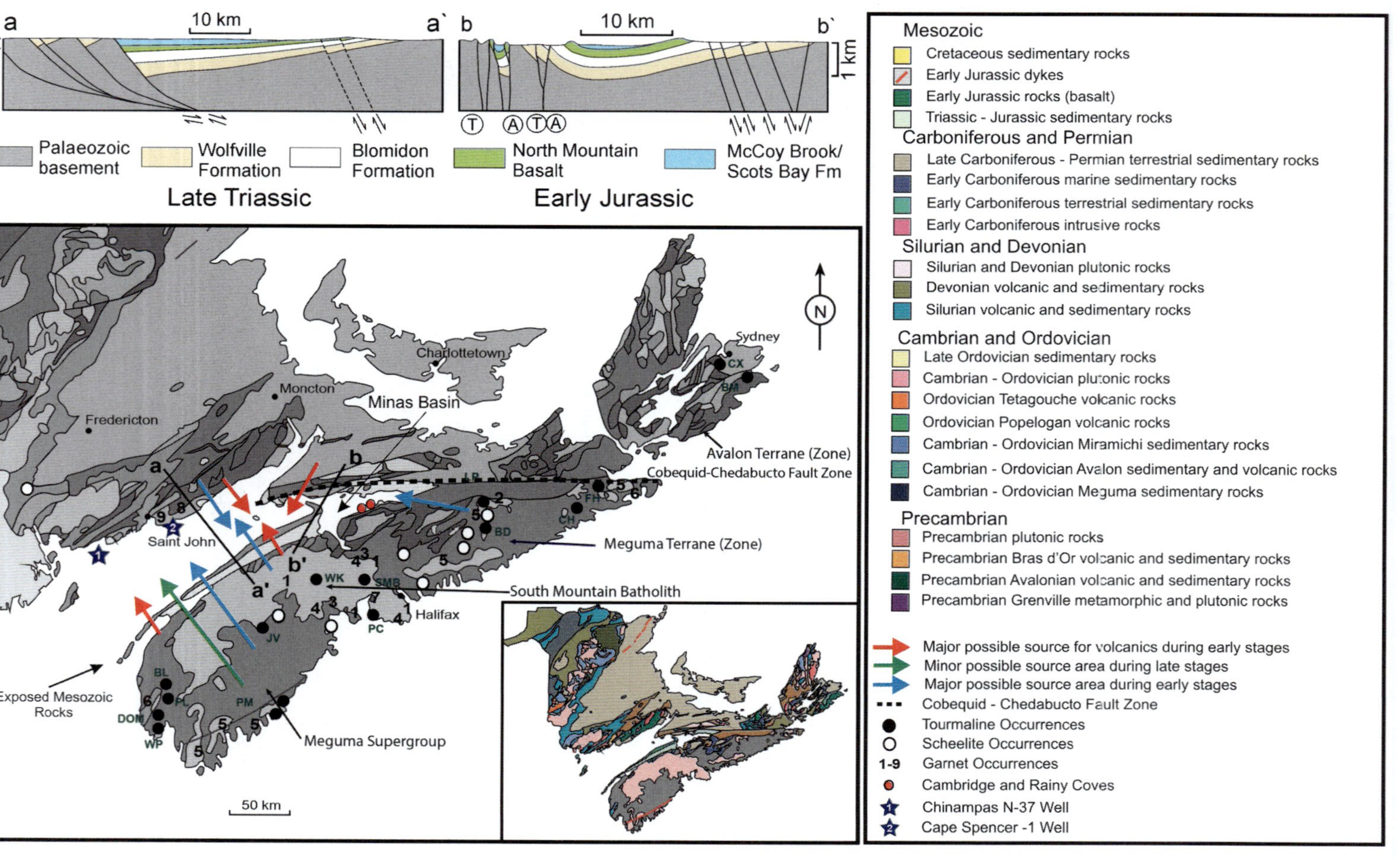


**Fig. 1.** Geological map of the Maritimes, showing the location of the studied areas (from Fensome & Williams 2001). The two cross-sections through the Minas Basin showing the stratigraphy of Mesozoic formations beneath the basin and onshore are from Olsen & Schlische (1990) (redrafted; permission granted by the publishers). The letter and number symbols within the map are the locations of tourmalines and garnets, respectively; their meaning and details are given in Table 1 and Figures 8 and 9.

succession of the Minas subbasin can be divided into the lower and middle Wolfville Formation (fluvial–alluvial fan-dominated unit), the upper Wolfville Formation (240 m-thick transitional fluvial–playa unit) and the Blomidon Formation (playa-dominated package) (Leleu & Hartley 2010).

Fundy Basin forms a half-graben at the boundary of the Meguma and Avalon terranes of the Appalachian Orogen (Schlische & Olsen 1990; Wade *et al.* 1996). The Meguma and Avalon terranes are separated by the Cobequid–Chedabucto fault system (Fig. 1). Both areas are part of the Appalachians and their rocks are dominated by igneous rocks (mostly granitoids) and metamorphic suites. The southern margin of the Fundy Basin onlaps the Meguma Terrane, which forms the southern part of Nova Scotia and the adjacent continental shelf. This terrane consists of thick turbiditic sediments of the Cambrian–Ordovician Meguma Supergroup (MSG), Silurian metasediments and volcanics of the White Rock, Kentville and New Canaan formations, and the Early Devonian Torbrook Formation. The MSG consists of sand-dominated, metamorphosed turbidites of the Goldenville Formation and overlying silt/shale-dominated metamorphosed turbidites of the Halifax Formation. The Late Devonian South Mountain Batholith (SMB) has intruded the older formations of the metapelite-dominated Halifax Group and the metapsammite-dominated Goldenville Group of the MSG (Fig. 1) (McKenzie & Clarke 1975; Schenk 1997; Hicks *et al.* 1999; MacDonald 2001; White 2008).

The Avalon Terrane occupies southern New Brunswick and northern Nova Scotia and consists of late Precambrian and early Palaeozoic mafic volcanics and continental metasediments, weakly metamorphosed and intruded by granites (Rast *et al.* 1976; Keppie 1979). Both terranes were affected by many tectonic events including the Acadian Orogeny during the mid-Palaeozoic, which resulted in the deformation and metamorphism of these rocks. The Mesozoic formations are underlain beneath the Bay of Fundy by Carboniferous formations in the northeastern parts of the bay and by Meguma and/or Avalon zone rocks towards the southwestern parts, as indicated from the seismic profiles and sections studied by Wade *et al.* (1996) (Fig. 1). The Horton Group in the Minas Basin covers a large area, unconformably overlies the lower Palaeozoic rocks of the MSG and is conformably to unconformably overlain by the Windsor Group of Early Carboniferous age (Fig. 1). The Mesozoic formations underlying the Fundy subbasin (Fundy Group) in these sections are, from older to younger, the Wolfville, Blomidon, North Mountain Basalt and Scots Bay/McCoy Brook formations (Fig. 1).

## The Wolfville Formation

The Wolfville Formation comprises the first thick synrift sediments deposited in the Fundy Basin, mostly by braided rivers and also in parts as alluvial fans and aeolian sands under continental semi-arid oxidizing conditions (Olsen 1997; Hamblin 2004). Its thickness varies in outcrop from 60 to 833 m (Williams *et al.* 1985). It is exposed on both sides of the Minas Basin and forms much of its floor

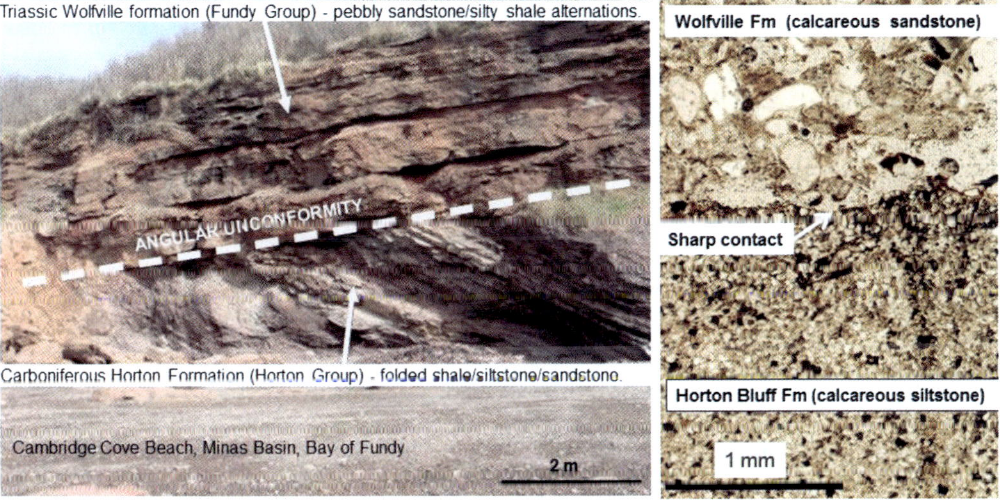


**Fig. 2.** Field view of the studied Wolfville Formation at Cambridge Cove unconformably overlying Horton Bluff Formation. The image to the right is a photomicrograph for a sample taken from the contact of the two formations.

beneath the water of the Bay of Fundy. The thickness of the Wolfville Formation beneath the Bay of Fundy increases toward the SW from 1308 m in the Irving Cape Spencer no. 1 well to >1718 m in the Mobil–Gulf Chinampas N-37 well, and is believed to increase to >3000 m east of Grand Manan Island (Wade *et al.* 1996).

Klein (1962) described the Wolfville Formation as stratified coarse- to medium-grained clastic rocks, which include conglomerates containing boulder to pebble-size clasts from lower and middle Palaeozoic metamorphic rocks, vein quartz, metamorphic quartzite, Mississippian sedimentary rocks, and intraformational siltstone and claystone. The sizes of clasts vary between 2 and 35 cm and consist of lower Palaeozoic metamorphic rocks, granite and rhyolite, which were derived from the Cobequid Mountains and sandstone derived from the Pennsylvanian strata. The Wolfville sandstones are mostly texturally submature and range in grain size from fine to coarse, falling in the fields of low-rank greywacke, orthoquartzite, high-rank greywacke and lesser arkose based on Krynine's (1948) classification on the quartz–mica–feldspar (QMF) diagram.

The Wolfville Formation overlies various rock units depending on the location, including the Avalon Zone, the Meguma Zone (Meguma Group, Cambrian to Ordovician), the Kentville (Silurian) and Torbrook (Early Devonian) formations, the SMB (Lower Devonian), the Horton and Windsor groups (Lower Carboniferous) and the Canso (Mabou) Group (Upper Carboniferous). It is conformably overlain by the Upper Triassic Blomidon Formation or the North Mountain Basalt (Williams *et al.* 1985).

The studied areas of Rainy Cove and Cambridge Cove are characterized by a sharp angular unconformity between the oppositely dipping deformed beds of the Horton Bluff Formation and the gently dipping strata of the Wolfville Formation (Fig. 2). The beds of the Wolfville Formation are parallel to the angular unconformity surface trending nearly east–west and gently dipping *c.* 20°N–NNE. Both formations are dissected by a number of steeply dipping normal faults. The coastal exposures exist as vertical cliffs up to *c.* 17 m in thickness, produced by wave erosion. The Wolfville Formation consists of thickly bedded sandstone, pebbly sandstone and conglomerate separated by thin reddish brown softer clayey/silty layers (Fig. 2). They are red to brown, poorly sorted sediments, and show large-scale sedimentary structures, such as trough cross-bedding, and smaller-scale structures, such as current ripples and imbricated clasts, indicating their deposition by traction currents in rivers under continental oxidizing conditions.

## *The Horton Bluff Formation*

The Early Carboniferous Horton Bluff Formation is widely exposed in the Windsor Basin, south of the Bay of Fundy, and is disconformably overlain by the Cheverie Formation (Bell 1958). The Horton Bluff Formation is >525 m thick in the type area and 1015 m thick in Soquip Noel #1 well at Kennetcook (Martel & Gibling 1994, 1996). The Horton Group has been deposited in fluvial–lacustrine environments as alternating grey to black, very fine- to very coarse-grained clastic sediments ranging from shale to conglomerate.

## Materials and methods

Thirty-four samples of sandstones and pebbly sandstones/conglomerates were collected from the Wolfville Formation for detailed study (Appendices Ia–d & II). They included 25 samples from exposures along the Minas Basin at Rainy and Cambridge coves and nine core samples from the subsurface at 3600.6 to 3603.4 m depth in the lower part of the Wolfville Formation in the Chinampas N-37 well on the northern side of the Bay of Fundy (Fig. 1). Thin sections were made from these samples, and part of each sample was disaggregated using a porcelain mortar and pestle. The disaggregated samples were sieved using standard sieves with sizes of >500, 250, 125, 62.5 and <62.5 $\mu$m. The 62.5 to 250 $\mu$m sizes were used for heavy mineral separation, using Na-polytungstate heavy liquid diluted to a specific gravity of 2.89. Thin and polished thick sections were prepared from the separated heavy minerals for microscopic study and electron probe microanalysis (EPMA). An Olympus BX51 polarizing research microscope with an advanced Olympus DP71 digital camera and Image ProPlus-7 software capable of measuring the area of pore spaces was used to study the rock and heavy mineral thin sections. An X-ray diffraction apparatus was used to analyse the clay fraction separated from one of the sandstones of the Cambridge Cove. A point counter attached to the microscope was used for grain counting. The garnet and tourmaline grains were analysed with a fully automated JEOL 8200 electron microprobe equipped with five wavelength spectrometers, operating under the following conditions in wavelength-dispersion mode: accelerating voltage of 15 kV, beam current of 20 nA, beam diameter of 1 $\mu$m, peak count-time of 20 s and background count-time of 10 s. The standards used were PETJ sanidine (K), PETJ kanganui kaersutite (Ca), PETJ kanganui kaersutite (Ti), PETJ Cr-metal (Cr), TAPH jadeite (Na), TAPH almandine and TAPH kanganui kaersutite (Mg), TAP almandine and TAP sanidine (Al), TAP almandine and TAP sanidine (Si), LIFH pyrolusite (Mn), LIFH almandine (Fe) and TAPH Durango fluorapatite (F). B was calculated by stoichiometric methods.

## Results

### Petrography of Wolfville Formation

Thin section and sieve analysis studies of the Wolfville Formation sandstones showed that their grain size distributions are variable and mostly bimodal. They dominantly range between fine- and coarse-grained sand (medium on average), although sizes as fine as silt and as large as pebbles exist in some samples. The shape of detrital grains varies from subangular to subrounded. The slates, shales and siltstones, which are the dominant rock fragments in the studied sandstones, are mostly disc-shaped. The feldspars are mostly bladed in shape, while the more resistant quartz, granites and quartzite grains are mostly equant to prolate in shape

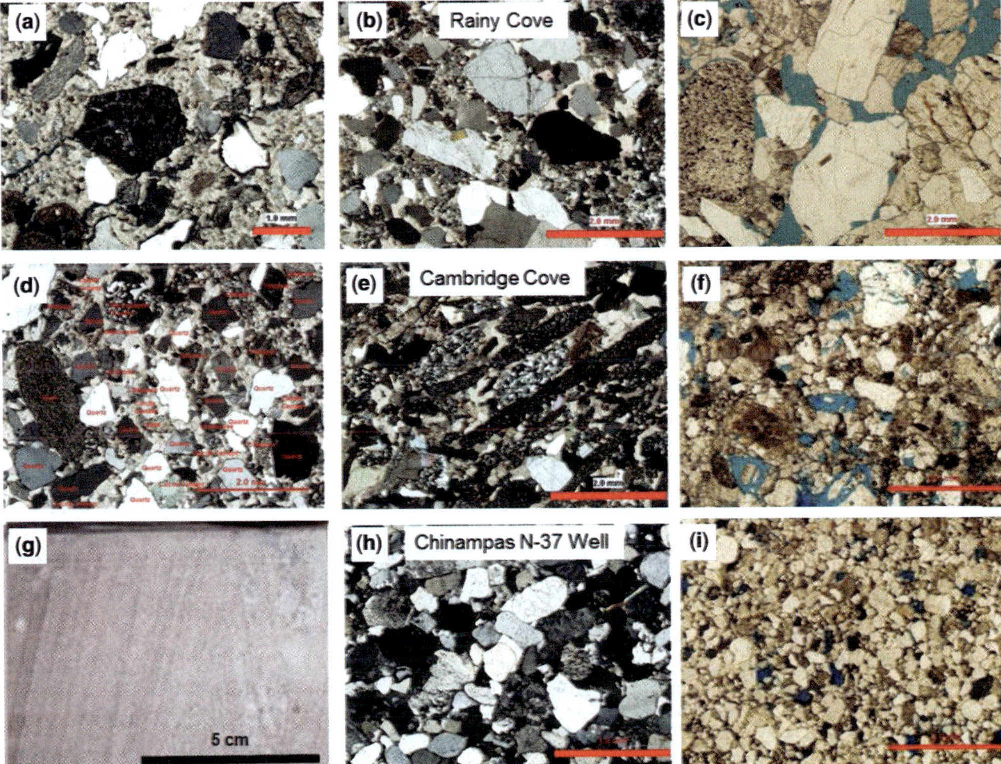


**Fig. 3.** Photomicrographs showing the composition, texture and porosity of the studied sandstones of Wolfville Formation from (a–c) Rainy Cove, (d–f) Cambridge Cove and (g–i) Chinampas N-37 well. (a, b) Cement-supported sandstone showing quartz grains (grey and white with microfractures) and lithics of granitoid (black) and slates (dark brown, elongate) surrounded by sparry calcite cement; (c) sandstone with good porosity–permeability (blue) containing microfractured quartz grains (white) and lithics (dark grey to brown), cemented by sparry calcite (light brown); (d) cement-supported sandstone showing quartz grains (grey and white with microfractures) and lithics of granitoid with micrographic texture (black and black-white) and slates (dark brown, elongate) surrounded by sparry calcite cement; (e) sandstone with imbricated texture that consists of elongate, orientated and parallel grains of slate (dark grey to brown) and quartzite (white-grey), as well as subrounded quartz grains (white and grey) embedded in sparry calcite cement (light brown); (f) sandstone with good porosity–permeability (blue) containing microfractured quartz grains (white) and lithics (different shades of brown), cemented by sparry calcite (light brown); (g) a core image of reddish-brown sandstone from the subsurface Chinampas N-37 well shows cross-bedding and in parts a pebbly structure; (h) grain-supported sandstone showing very well compacted sand grains dominated with quartz and lesser amounts of feldspars and lithics, cemented by quartz as overgrowths; (i) grain-supported sandstone with negligible porosity (blue isolated spots) showing very well compacted sand grains dominated with quartz (white) and lesser amounts of feldspars (light brown) and lithics (dark brown), cemented by quartz as overgrowths.

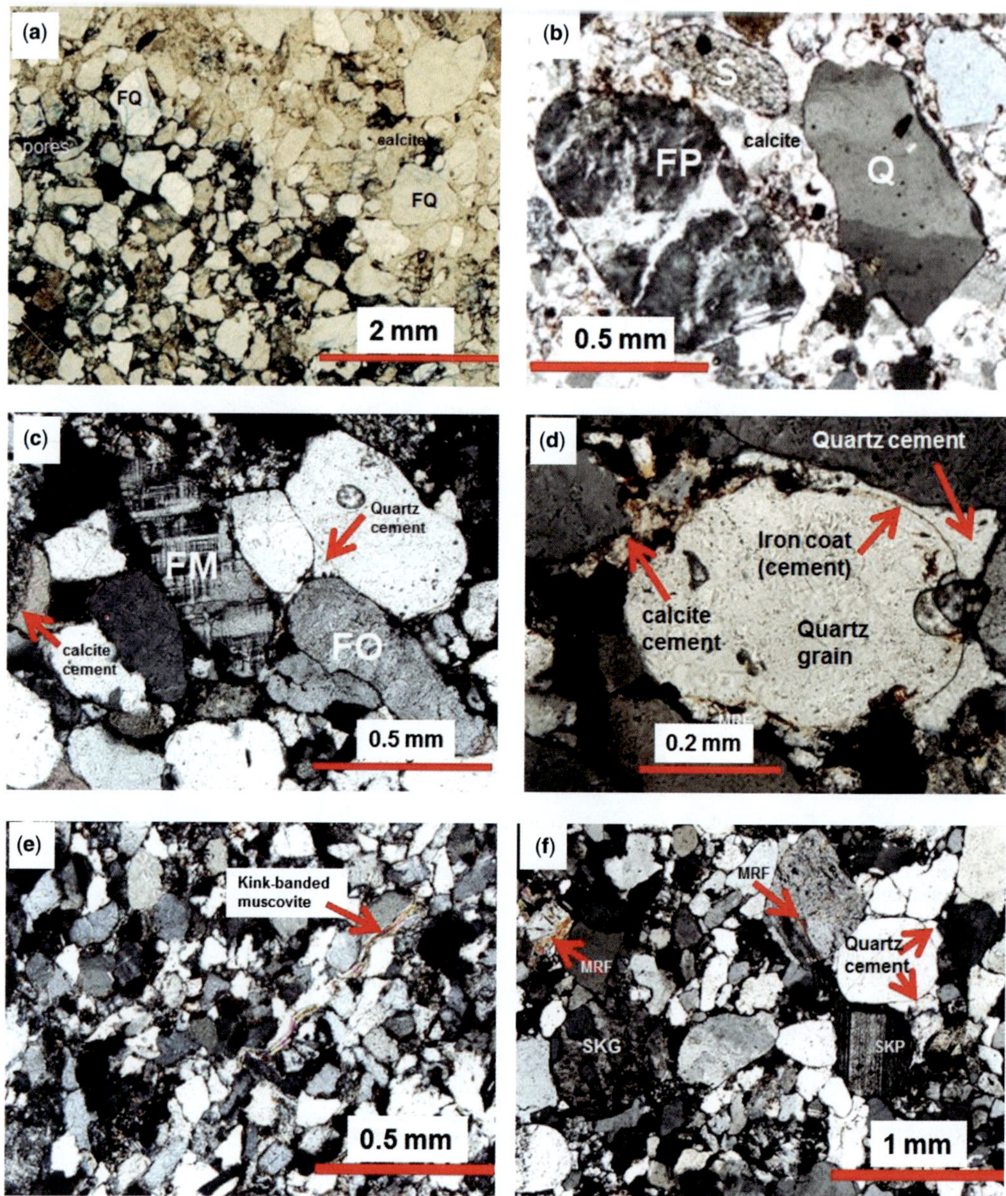



**Fig. 4.** Photomicrographs showing the important diagenetic type in the Wolfville Formation sandstones: (**a**) dissolution of calcite cement in the lower left corner (the upper right corner was not affected in the surface sandstone at Cambridge Cove); (**b**) calcite-cement supported sandstone of Rainy Cove showing fresh lithics of perthite (FP), slate (S) and quartz (Q) separated by sparry calcite cement; (**c**) well-compacted, grain-supported lithics including relatively fresh microcline (FM) and orthoclase (FO) in the subsurface Chinapmas sandstones with quartz cement as authigenic overgrowths and minor sparry calcite; (**d**) a thin coating of iron oxide cement enveloping a rounded quartz grain in the sandstone of Chinampas well, which is overgrown on three sides by authigenic quartz cement and by sparry calcite on the left side; (**e**) a grain-supported sandstone of Chinampas well showing a kink-banded detrital muscovite flaky grain; (**f**) a grain-supported sandstone of Chinampas well showing muscovite partially replacing feldspars (MRF) and slightly kaolinized plagioclase (SKP) and granite lithics (SKG) as well as authigenic quartz cement overgrowing quartz grains and tightly filling intergranular spaces.

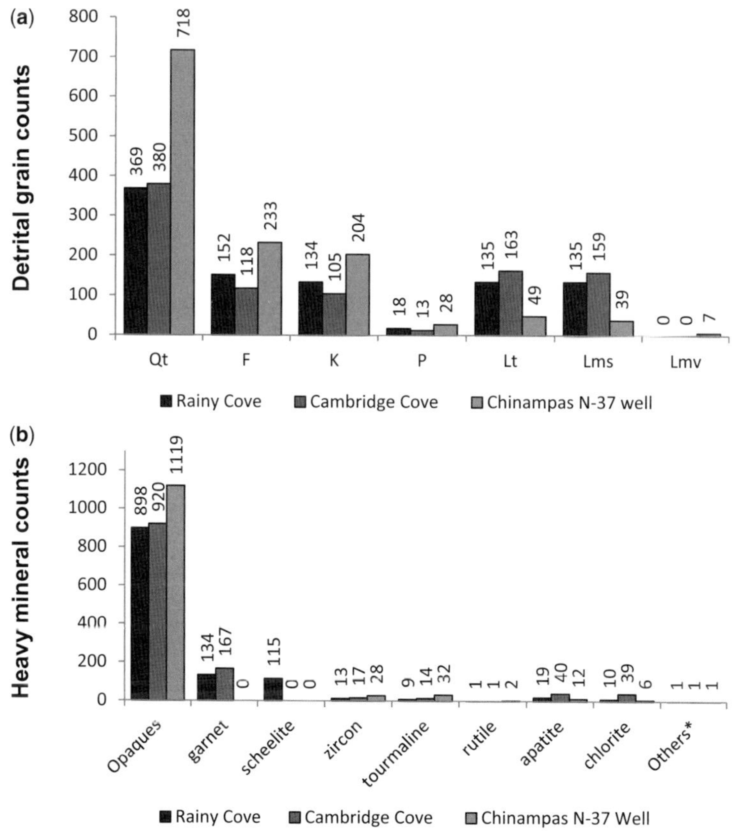



**Fig. 5.** Relative distribution of (**a**) detrital grains and (**b**) heavy minerals in the sandstones of Wolfville Formation. *Others include hornblende, pyroxene, staurolite, epidote and zoisite.

according to Zingg diagram terminology (Folk 1968) (Figs 3 & 4).

Point counting of 1200 points was conducted on each studied thin section (Appendix Ia–c), using the Gazzi–Dickinson method (Dickinson 1985; Ingersoll & Suczek 1979). Framework grains consist dominantly of quartz, rock fragments and feldspars, with minor amounts of mica, chlorite and a few heavy minerals (Appendix Ia–d; Figs 3 & 4). The relative proportions of the major three groups of grains in the studied localities are shown in Figure 5a and Appendix Id, and the relative abundances of heavy minerals are shown in Figure 5b. The detrital grains in the surface sections are well or poorly cemented by sparry calcite and minor amounts of iron oxide and clay (Figs 3a–f & 4a, b), and dominantly by quartz in the case of subsurface Chinampas N-37 core samples (Figs 3h, i & 4c–f) (Appendix Ia–c).

Quartz is the dominant grain and is more abundant relative to other framework constituents at Chinampas compared with the exposures of Rainy and Cambridge coves (Fig. 5a). Many quartz grains in the Chinampas N-37 core have authigenic quartz overgrowths, but such features are not observed in the outcrop sections. Many of the quartz grains also show microfractures, which form an effective part of the secondary porosity in the sandstones of Rainy and Cambridge coves (Fig. 3c, f).

Rock fragments (lithics) are the next dominant grain types in the studied samples (Appendix Ia–c; Fig. 5a). The most important types of lithics in the studied sandstones are metamorphic (slate, phyllite, schist, quartzite), sedimentary (siltstone, sandstone, limestone, chert) and magmatic (granitoid) rocks. Volcanic rock fragments exist in a few sandstones of the Chinampas N-37 core (basalt, rhyolite and possibly altered trachyte), but are extremely rare in the outcrop sections. Only one sample taken from the contact zone between the Wolfville Formation and the Horton Bluff Formation at Cambridge Cove contained two grains of felsic volcanics (rhyolite/rhyodacite and possibly altered trachyte). Sandstones of the Wolfville Formation near the contact with the underlying Horton Bluff Formation are richer in lithics and are dominated by lithologies similar to those of the Horton Bluff Formation such as phyllites, slate, shale, siltstone, microquartzite and limestone. This indicates that these metasedimentary lithics were mostly derived from the exposures of the Horton Group. These sandstones also contain minor granite lithics.

Feldspars, which form the third important framework grain type, are present as perthite, orthoclase and plagioclase, and are present as part of the granitic rock fragments, together making up 5–27% of the studied rocks. Feldspars are much more abundant at Chinampas relative to the exposed sections (Fig. 5a; Appendix Id). Most of the feldspars have undergone various degrees of alteration by sericitization, kaolinization and calcitization. Many empty spaces in the rock samples are the result of dissolution of feldspars, limestone lithics and calcareous cementing material. Detrital mica and chlorite occur in minor amounts (<2%) in the studied rocks. Chlorite is usually an alteration product of biotite.

The QtFL triangle of Folk showed that the sandstones of the surface sections at Cambridge and Rainy coves are divided between feldspathic litharenites and lithic feldsarenites, while those of the subsurface Chinampas well are divided between feldsarenites (arkoses) and subarkoses. They are dominantly feldspathic litharenites at Cambridge Cove, lithic feldsarenites at Rainy Cove and feldsarenites at Chinampas (Fig. 6).

*Heavy minerals*

Heavy minerals were counted in thin sections with other constituents as part of the petrographic studies of the sandstones, with abundance ranging from 0 to 4.8% (Appendix Ia–c), and were also studied in detail in polished thin sections of the heavy mineral concentrates. The dominant opaque minerals are mostly iron and iron–titanium oxides (hematite, magnetite and ilmenite, as well as goethite and leucoxene as their alteration products). The transparent group, in the order of abundance, comprises garnet, scheelite, apatite, chlorite, zircon, tourmaline, rutile and rarely others (hornblende, pyroxene, staurolite, epidote and zoisite) (Appendix II; Fig. 5b). Biotite and muscovite micas, which exist in most samples in variable amounts, were excluded from counting because of their incomplete separation by heavy liquid, as some of the flaky grains float rather than sink during the separation process. Opaques, zircon, tourmaline and rutile are more abundant in the subsurface samples of Chinampas N-37 well relative to the surface samples of Cambridge and Rainy coves, and apatite and chlorite are less abundant (Fig. 5b). Garnet is the

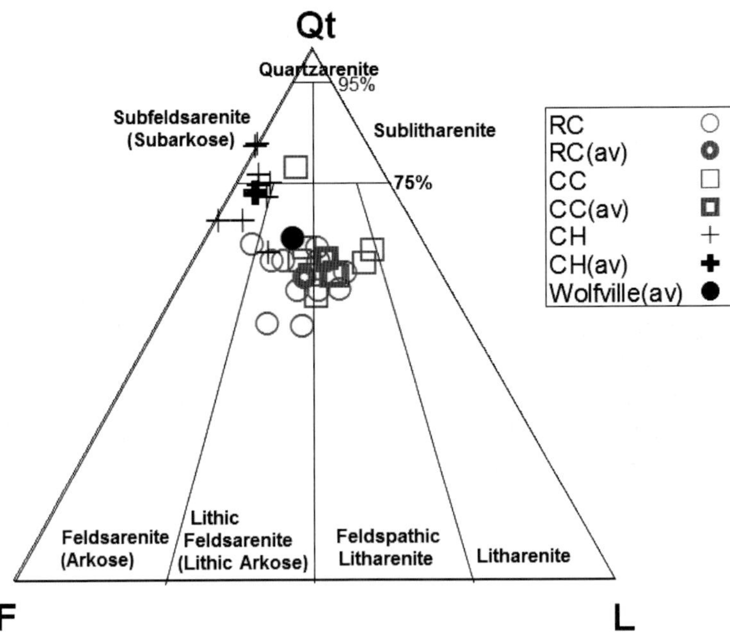


**Fig. 6.** QtFL classification of Folk (1968) for the Wolfville Formation sandstones. RC, Rainy Cove; CC, Cambridge Cove; CH, Chinampas N-37 well; av, average.

dominant non-opaque mineral among the heavy minerals from the surface outcrops; however, the subsurface Chinampas well samples do not contain any garnet (Appendix II; Figs 5b & 7). Scheelite was found as the dominant transparent mineral in some surface samples at Rainy Cove section where it exists as small, perfect rhombic crystals with high relief, moderate birefringence and vivid interference colours. Garnet and tourmaline have been studied in detail and chemically analysed by EPMA because of their importance as provenance indicators (Appendices III and IV, respectively). The importance of these two mineral groups as provenance indicators comes from the fact that they have diverse composition and origin and can be found as heavy minerals in many types of sandstone (Wright 1938; Sobolev 1964; Rub *et al.* 1977; Sabeen *et al.* 2002; Morton & Hallsworth 2007;

Hallsworth & Chisholm 2008; Kettanah *et al.* 2008). The analysis results were plotted on the triangles of Wright (1938) for garnets (Fig. 8) and Henry & Guidotti (1985) for tourmaline (Fig. 9). EPMA results for 56 garnets in five sandstones from Cambridge and Rainy coves showed that they are of two types, dominant almandine-rich spessartine and few spessartine-rich almandine garnets (Appendix III; Fig. 8). The almandine-rich spessartines are usually cloudy grey in colour, euhedral (equant) in shape, nearly of the same size and have rough-looking surfaces under the microscope, while the spessartine-rich almandines are clear pinkish to colourless in colour, anhedral in shape, fractured, variable in size and generally larger than those of the other type. EPMA results for 41 tourmalines in six sandstones (four from the surface sections of Rainy and Cambridge coves and two from

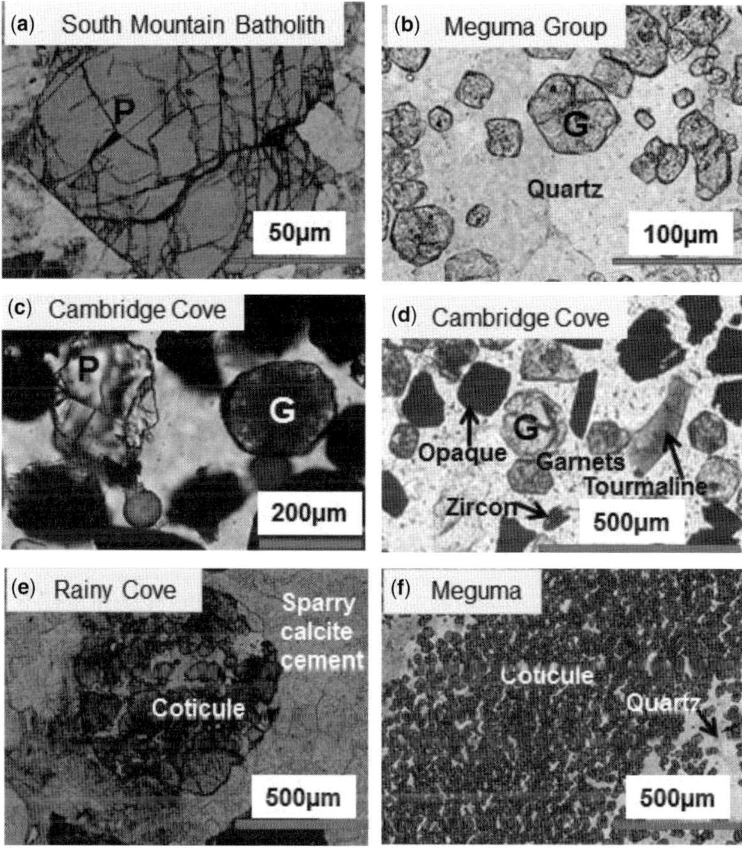



**Fig. 7.** Photomicrographs showing comparison between garnets from the studied Wolfville Formation with those from the Meguma Group and SMB in Nova Scotia: (**a**) Garnet from SMB, clear pinkish, fractured; (**b**) garnets from Meguma Group, grey, cloudy, euhedral surrounded by quartz; (**c**) the two types of garnet in the sandstones of Cambridge Cove; (**d**) the abundant heavy minerals in the sandstones of Cambridge Cove including garnet, opaques, tourmaline, zircon; (**e**) a lithic of coticule (garnet-quartz rock) in the sandstones of Rainy Cove; (**f**) a coticule from MSG for comparison with the coticule in (**e**).

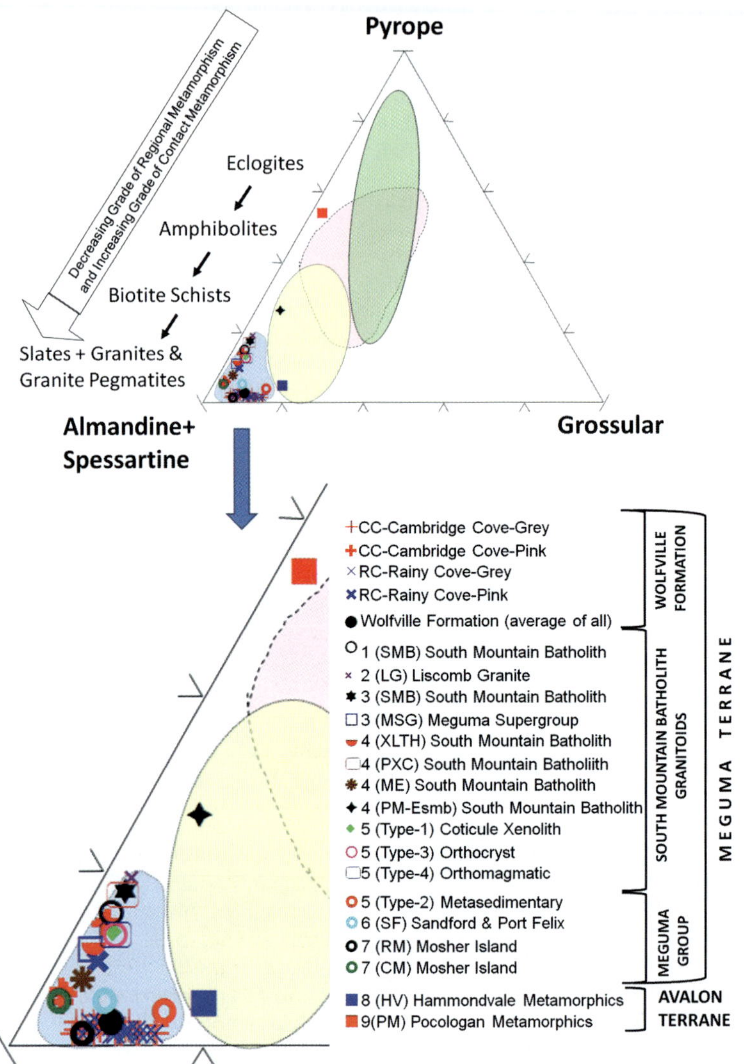



**Fig. 8.** Composition diagram modified from Wright (1938) for garnets of the Wolfville Formation sandstones and those in Meguma and Avalon terranes taken from many references. Details of the symbols and references used for comparative reasons are given in Table 1, and their locations are illustrated in Figure 1.

the subsurface Chinampas well) showed that the majority are plotted closer to the dravite end member and few near the schorl and buergerite end members (Appendix IV; Fig. 9). The tourmaline grains are rounded to subrounded and rectangular in shape and showed various colours under the microscope.

### Cementing material

There are distinct differences in the type and relative abundances of the cementing material between the studied sections (Appendix Ia–c; Figs 3 & 4). Sparry calcite is the dominant cement with minor amounts of iron oxide and clays in the surface sections of Rainy and Cambridge coves (11–53%) (Appendix Ia–d; Figs 3a–f & 4a, b), but is relatively much less abundant in the subsurface Chinampas N-37 well (0.3–8%) (Appendix Ic, d; Figs 3g–i & 4c, d). Quartz is the dominant cement in the subsurface sandstones (Figs 3h & 4c, d). Ferruginous cement is the earliest type of cement, indicated by a coating of iron oxide around most of the grains, which are covered or surrounded by calcite and/or

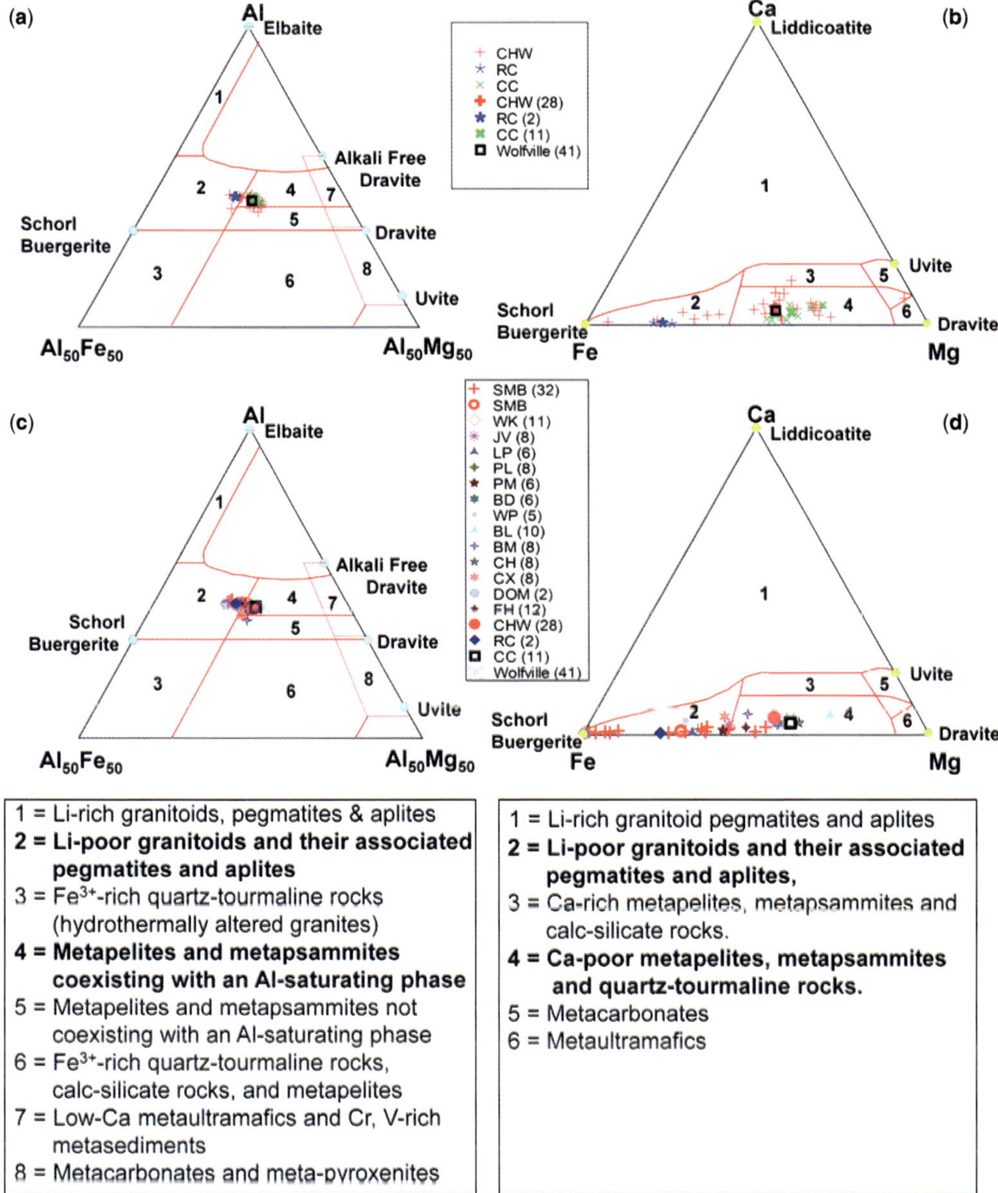



**Fig. 9.** Composition diagram of Henry & Guidotti (1985) for tourmalines of the Wolfville Formation sandstones (a, b) and those in various rocks of the Meguma and Avalon terranes taken from many references (c, d). CHW, Chinampas Well; RC, Rainy Cove; CC, Cambridge Cove; Wolfville, Wolfville Formation. The details of these symbols and the others, as well as the used references, are given in Table 1. The numbers in brackets indicate the average values for those analysis numbers. Their locations are illustrated in Figure 1.

quartz (Fig. 4d). The relationship between the ferruginous and clay cements is not obvious. Iron oxide cementation was followed by the deposition of calcite as coarse crystalline infill and in many cases as radiating crystals surrounding framework grains, especially around the rounded limestone lithics. X-ray diffraction analysis of a sandstone from Cambridge Cove indicated that the clay minerals consist of kaolinite as booklets, and illite, montmorillonite and chlorite as spherulitic aggregates

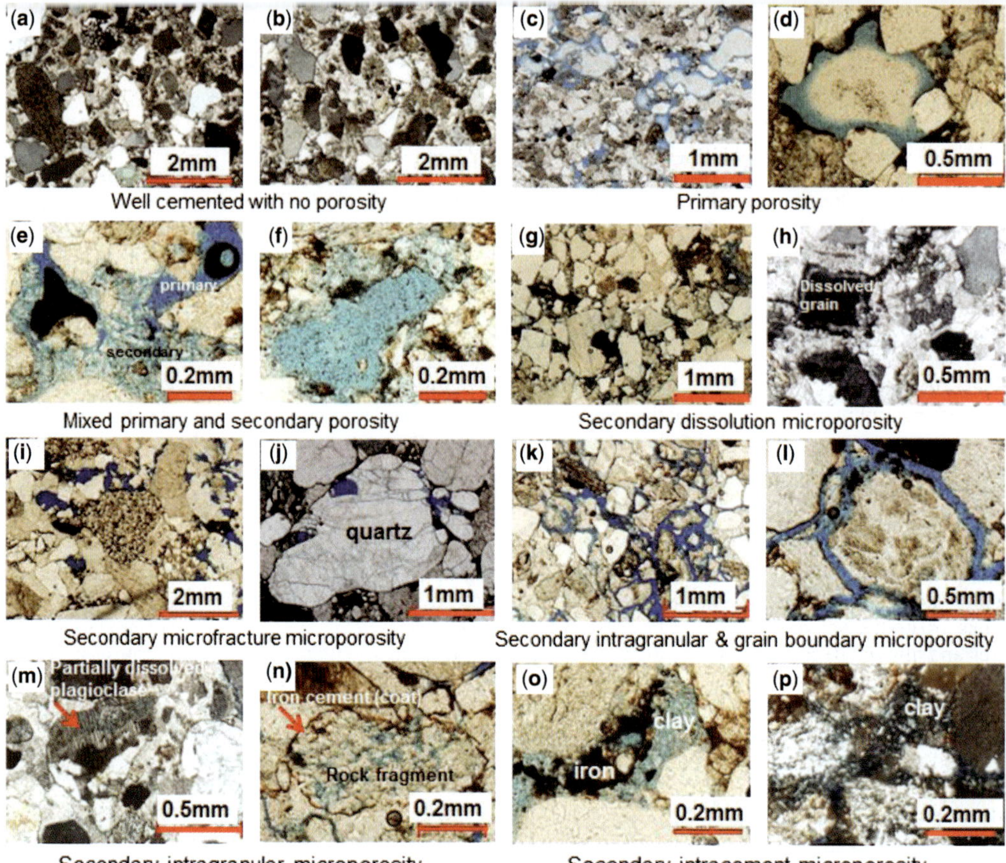



**Fig. 10.** Porosity types in the sandstones of the outcrop sections of the Wolfville Formation.

filling some of the pore spaces. The order of paragenesis of cementation in the subsurface core samples is from iron oxide, quartz, calcite and finally clays.

## Porosity

Many pore spaces can be observed within the cement or within and between the grains (Fig. 10). The sandstones of the Wolfville Formation are cement-supported in the surface sections of Rainy and Cambridge coves (Fig. 3c, f, respectively), where porosity ranges from 1 to 17%, and grain-supported in the subsurface at Chinampas (0–2%) (Figs 3–9) (Appendix Ia–c). The considerable decrease in porosity values of the subsurface section is due to compaction and the growth of quartz cement blocking many intergranular spaces. Detailed microscopic study showed both primary and secondary porosity, especially in the surface sections of Rainy and Cambridge coves, with dissolution, microfracture, grain boundary, intragranular and intra-cement secondary porosity types (Fig. 10). Other than the intra-cement term, the other types are named according to the porosity classification described by Scholle (1979).

## Discussion

### Heavy mineral provenance

The heavy minerals in the studied sandstones can be grouped into three groups based on Folk's (1968) classification (Fig. 5b; Appendix II). The first and predominant group comprises the opaques (hematite, magnetite, ilmenite, as well as goethite and leucoxene as their alteration products). The second group comprises the metastable minerals (garnet, apatite, zoisite, epidote, staurolite, hornblende and pyroxene), and the third group the ultrastable minerals (zircon, tourmaline and rutile). Scheelite, which has a moderate hardness of 4.5–5 with a specific gravity of 6.1 (Anthony *et al.* 2005) and decomposes in HCl (Mange & Maurer 1992), is not reported in the heavy minerals list of Folk (1968) and Pettijohn *et al.* (1973), possibly because it is not a common heavy mineral in sediments. Based on its physical and chemical characteristics, scheelite can be included with the metastable group of Folk (1968).

The ultrastable zircon–tourmaline–rutile group (ZTR) forms 2.7% of the heavy minerals. The opaques are not as diagnostic for provenance as some transparent heavy minerals. Based on Folk (1968), the ZTR group is diagnostic of source rocks such as recycled older sediments, granites, pegmatities and schists; meanwhile, garnet may be derived from plutonic rocks, pegmatites and metamorphic rocks. Apatite may be derived from volcanic and/or basic to acidic igneous rocks. Epidote and zoisite are mostly of metamorphic or hydrothermal origin. Staurolite is highly diagnostic of metamorphic source rocks. The source of hornblende and pyroxene is either plutonic or metamorphic rocks. Scheelite occurs in contact metamorphic rocks, hydrothermal veins, greisens, granite pegma tites and alluvial deposits (Anthony *et al.* 2005). In addition to these source rocks, it can also be found in amphibolites and regionally metamorphosed impure limestones (Mange & Maurer 1992). Granitoid and metamorphic rocks are present in both the Meguma Terrane, south of the Bay of Fundy, and in the Avalon Terrane to the north. Both terranes could have acted as the source rocks for the Wolfville Formation.

The studied garnets are almandine-rich spessartines (cloudy grey under a polarized microscope) or spessartine-rich almandines (clear pink to colourless) (Figs 7 & 8). The range and average end member composition of these two garnet groups are $(Sp_{35-71}Alm_{18-54}Gro_{6-15}Py_{1-3})$ $(Sp_{57}Alm_{31}Gro_{10}Py_2)$ and $(Sp_{8-23}Alm_{67-75}Gro_{3-5}Py_{5-14})$ $(Sp_{17}Alm_{71}Gro_4Py_9)$, respectively (Appendix III). When plotted on the modified, pyrope–(almandine + spessartine)–grossular (P–AS–G) triangular composition diagram of Wright (1938), all of the studied garnets fall in the field of slates + granites and granite pegmatites (Fig. 8) and are similar in chemical composition and optical properties to garnets from the Meguma Group and SMB (Figs 7 & 8; Table 1). The grey garnets cluster together with the coticule garnets from the Meguma Group, which were studied by Hick (1996) and Schiller & Taylor (1965), and metasediment-hosted, type-2 garnets of Scallion *et al.* (2011) (Figs 1 & 8; Table 1). Similarly, the pink to colourless garnets were clustered with those from the SMB near the contacts with the Meguma Group studied by Allen & Clarke (1981), Cameron & Zentilli (1997), Erdmann (2006), Erdmann *et al.* (2009) and

SMB-hosted types 1 (coticule xenolith), 3 (orthocryst) and 4 (orthomagmatic) garnets of Scallion *et al.* (2011) (Figs 1 & 8; Table 1). A comparison of garnets hosted by metamorphic rocks in the Avalon Terrane of southern New Brunswick studied by White *et al.* (2001, 2006) (Figs 1 & 8; Table 1) shows that they are different from the garnets of the Wolfville Formation as they are generally richer in Fe, Ca and Mg and poorer in Mn. The average end member composition of these Avalon Terrane garnets is either spessartine-rich pyropes $(Py_{54}Sp_{43}Gr_3Al_{0.1})$ in the Pocologan metamorphic suites studied by White *et al.* (2006), or grossular-rich almandines $(Al_{66}Gr_{18}Sp_{12}Py_4)$ in the Hammondvale metamorphic rocks studied by White *et al.* (2001). When plotted on the P–AS–G triangle, they fall in different fields away from those of the Wolfville Formation (Figs 1 & 8; Table 1). There are garnets as an accessory mineral in many Avalonian rocks (e.g. Miller *et al.* 1996; Barr *et al.* 1998). However, the lack of chemical analysis of these garnets does not allow comparison to be made with the Wolfville Formation garnets. Garnet aggregates as coticules were found as detrital sand grains within the conglomerates of Rainy Cove, similar to the coticules commonly found in the Meguma Group near the contact with the SMB (Fig. 7f). Coticules were reported in many such contact zones in Nova Scotia, including localities near Port Felix south of Lundy and near Sandford (Schiller & Taylor 1965), Whitehead Harbour (Poole 1984; Raeside *et al.* 1988), Spears Lake and south of Mt Uniacke (Romer *et al.* 2011). Similar coticules sampled by the authors and taken exactly from the sharp contact line between Meguma Group and the SMB in Purcell's Cove on the southern coast of the NW Arm of Halifax showed that the garnet aggregates were concentrated within the metapsammite laminae rather than in the adjacent alternating metapelite laminae. Garnet coticules exist in few other localities within the Avalon Terrane of New Brunswick (Romer *et al.* 2011). Further comparison of the petrographic properties of the garnets from Wolfville Formation and those taken from the Meguma Group and SMB also revealed a close similarity in shape and optical properties (Fig. 7a–f). The grey coloured, almandine-rich spessartine garnets are very similar to those of coticules within the MSG, while the pink to colourless, spessartine-rich almandine garnets are similar to those which exist within the granitoids of SMB. All these comparisons and indications suggest that the metasedimentary rocks and granitoids of the SMB of the Meguma Terrane were the major sources for the studied garnets in the exposed parts of the Wolfville Formation with a possible minor contribution from the rocks of the Avalon Terrane.

**Table 1.** *Brief description of garnets and tourmalines used in Figures 1, 8 and 9 including their symbols, number of averaged chemical analysis, locations, host rocks and the references from which the data were derived*

| | | Symbol (no. of analysis) | Location | Host rocks | References |
|---|---|---|---|---|---|
| Garnet | Wolfville | WF (56)<br>CCg (34)<br>CCp (5)<br>RCg (16)<br>RCp (1) | Southern Minas Basin, Bay of Fundy, NS<br>Cambridge Cove (grey garnets), NS<br>Cambridge Cove (pink garnets), NS<br>Rainy Cove (grey garnets), NS<br>Rainy Cove (pink garnets), NS | Triassic sandstones and conglomerates | Current study |
| | Miscellaneous | 1 (8) SMB | Chebucto Head, East Dover, East Chester, Aspotogan, New Germany, Kearney Lake, Mount Uniacke | Granodiorites and monzogranites of SMB | Allen & Clarke (1981) |
| | | 2 (10) LG | Eastville, NS | Liscomb Granite | Cameron & Zentilli (1997) |
| | | 3 (11) SMB | SMB, NS | SMB | Erdmann *et al.* (2009) |
| | | 3 (7) MSG | MSG, NS | MSG | |
| | | 4 (4) XLTH, PXC, ME, PM(Esmb) | SMB, NS | SMB | Erdmann (2006) |
| | | 5 (14) Types 1, 2, 3, & 4 | SMB and MSG, NS | Type-1 (coticule xenolith); Type-2 (metasedimentary); Type-3 (orthocryst); Type-4 (orthomagmatic) | Scallion *et al.* (2011) |
| | | 6 (1) SF | Port Felix and Sandford, NS | Coticule in Halifax Formation, Meguma Group | Schiller & Taylor (1965) |
| | | 7 (RM) | Mosher Island, NS | Coticule in Halifax Formation, Meguma Group | Hick (1996) |
| | | 7 (CM ) | Mosher Island, NS | Coticule in Halifax Formation, Meguma Group | Hick (1996) |
| | | 8 (18) | Hammondvale, NB | Metamorphic rocks of Avalone Terrane | White *et al.* (2001) |
| | | 9 (16) PMS | Southern NB, near Saint John | Pocologan metamorphic suite, Avalon Terrane | White *et al.* (2006) |

| | | | | Wolfville Formation sandstones | Current study |
|---|---|---|---|---|---|
| Tourmaline | Wolfville | CHW (28) | Chinampas well, Bay of Fundy | | |
| | | RC (2) | Rainy Cove, NS | | |
| | | CC (11) | Cambridge Cove NS | | |
| | | Wolfville (41) | Wolfville Formation (all above) | | |
| | Miscellaneous | SMB (32) | SMB, NS | Aplites & pegmatites in SMB | MacDonald *et al.* (1987) in Clarke *et al.* (1989) |
| | | WK (11) | Walker deposit | Aplites & pegmatites in SMB | O'Reilly *et al.* (1982); Charest *et al.* (1985) in Clarke *et al.* (1989) |
| | | JV (8) | Jumbo vein, NS | Quartz veins in Meguma Group | O'Reilly (1986) in Clarke *et al.* (1989) |
| | | LP (6) | Liscomb Pluton, NS | Gneisses and granitoids | Giles & Chatterjee (1986) in Clarke *et al.* (1989) |
| | | PL (8) | Pearl Lake, NS | Metallic quartz veins in Meguma Group | Anonymous (1983) in Clarke *et al.* (1989) |
| | | PM (6) | Port Mouton Pluton, NS | Pegmatites & dykes in Granitoids | Douma (1988) in Clarke *et al.* (1989) |
| | | BD (6) | Beaver Dam, NS | Au-quartz veins in Meguma Group | Smith & Kontak (1987) in Clarke *et al.* (1989) |
| | | WP (5) | Wedgeport pluton, NS | Quartz veins in granitoids | Wolfson (1983) in Clarke *et al.* (1989) |
| | | BL (10) | Brazil Lake, NS | Pegmatite in White Rock Formation | Hutchinson (1982) in Clarke *et al.* (1989) |
| | | BM (8) | Blue Mountain, NS | Veins in granitoids of Avalone Terrane | Chatterjee (1977); MacDonald & Scott (1980) in Clarke *et al.* (1989) |
| | | CH (8) | Cochrane Hill, NS | Quartz veins in granitoids of SMB | Smith (1983) in Clarke *et al.* (1989) |
| | | DOM (2) | Dominique, NS | Metallic quartz veins in Meguma Group | Anonymous (1983) in Clarke *et al.* (1989) |
| | | PC (13) | Peggy's Cove, NS | Aplites & pegmatites in SMB | Kontak *et al.* (2002) |
| | | FH (12) | Forest Hill, NS | Au-quartz veins in Meguma Group | Haynes (1983); Smith & Kontak (1988) in Clarke *et al.* (1989) |
| | | CX (8) | Coxheath, NS | Granitoids of Avalone Terrane | Oldale (1966) in Clarke *et al.* (1989) |
| | | CX (36) | Coxheath, NS | Veinlets in porphyry Au–Cu–Mo, Avalone Terrane | Lynch & Ortega (1997) |

WF, Wolfville Formation; CCg, Cambridge Cove (grey garnet); CCp, Cambridge Cove (pink garnet); RCg, Rainy Cove (grey garnet); RCp, Rainy Cove (pink garnet); SMB, South Mountain Batholith; LG, Liscomb Granite; MSG, Meguma Supergroup (XLTH, xenolith, PXC, ME, PM (Esmb) are garnets from South Mountain Batholith and contact zone with Meguma Group; Type-1, coticule xenolith; Type-2, metasedimentary; Type-3, orthocryst; Type-4 orthomagmatic; SF, Port Felix and Sandford; RM, low-grade regionally metamorphosed Meguma slates and sandstones of Mosher's Island Member, Halifax Formation; CM, contact-metamorphosed Meguma slates of Mosher's Island Member, Halifax Formation; PMS, Pocologan Metamorphic Suite, southern New Brunswick; CHW, Chinampas Well; RC, Rainy Cove; CC, Cambridge Cove; WK, Walker Deposit; JV, Jumbo Vein; LC, Liscomb Pluton; PL, Pearl Lake; PM, Port Mouton Pluton; BD, Beaver Dam; WP, Wedgeport Pluton; BL, Brazil Lake; BM, Blue Mountain; CH, Cochrane Hill; DOM, Dominique; PC, Peggy's Cove; FH, Forest Hill; CX, Coxheath.

The perfect suitability of the P–AS–G ternary diagram of Wright (1938) for the studied garnets from the Fundy Basin has encouraged us to revise this diagram to reflect the geology of Nova Scotia where contact metamorphism between the SMB with the older Meguma Group slates resulted in the formation of garnets at and near their contact zones. Garnet is abundant at contact zones between the SMB and the Meguma Group and becomes less abundant away from the contacts (Hale 2007). For this reason, we added slates to the field of granites and granite pegmatites, and also 'increasing grade of contact metamorphism' to the 'decreasing grade of regional metamorphism' trend along the (pyrope)–(almandine + spessartine) line of the triangle because both types of metamorphism have affected the area (Fig. 8).

The total absence of garnet from the studied subsurface section at Chinampas N-37 well is puzzling. This absence of garnet can theoretically be attributed to different factors such as different provenances, or dissolution during deposition or under burial diagenesis conditions. Regarding the provenance, it is possible to assume that the rocks of the Avalon Terrane to the north of the Bay of Fundy in New Brunswick were the major source for the older sediments at the lower part of the Wolfville Formation at Chinampas. However, this assumption is not very conclusive because there are many occurrences of garnet as coticules and within granitoids of the Avalon Terrane in New Brunswick (e.g. White *et al.* 2001, 2006) that could have acted as the source (Fig. 1; Table 1). The other possible source is the Argana Basin of Morocco (northwestern Africa), which is believed to have been adjacent to the Fundy Basin of Nova Scotia and New Brunswick (eastern Canada) during the Late Triassic (Olsen *et al.* 2000). The stratigraphy and sedimentary facies of the well-exposed, thick sequence of Triassic–Liassic deposits in the Argana Valley of Morocco, northwestern Africa, are remarkably similar to those of the Bay of Fundy Basin of Nova Scotia (Smoot & Olsen 1988; Smoot & Castens-Seidell 1994), suggesting a predrift proximity of both areas (Hofmann *et al.* 2000). Coticule-bearing sedimentary layers occur in Ordovician siliciclastic sedimentary rocks that were deposited on the Avalonian shelf as part of the Acadian and Variscan orogenic belts and subsequently metamorphosed between 340 and 380 Ma (Romer *et al.* 2011). Waldron & White (2012) argued that the Nova Scotian coticule samples studied by Romer *et al.* (2011) are of Cambrian age and that they were deposited in a deep-water turbidite basin of Megumia, not on the Avalonian shelf. However, Romer & Kroner (2012) insisted on the Early Ordovician age of these coticlues and considered Megumia as part of the

Avalonian plate, and that the Mn-rich sediments were deposited on its shelf. The isotopic composition and geochemical fingerprints of the coticules and their Mn-poor sedimentary hosts suggest that the clastic input was dominated by material derived from Gondwana rather than from Avalonia, although the occurrence of coticules seemed to be restricted to the Avalonian shelf. Based on these ideas, it is possible that the early sediments of the Wolfville Formation at Chinampas well that do not contain any garnets were derived from Gondwana (northwestern Africa) prior to drifting. Meanwhile the garnet-rich, late deposits at the surface exposures of Rainy and Cambridge coves were dominantly derived from the Meguma Terrane and possibly partly from the Avalonian Terrane.

The other possible reason for the absence of garnet in the subsurface Chinampas sandstones is dissolution under weathering and/or burial diagenesis conditions. Garnet has intermediate persistence among heavy minerals in weathering and the sedimentary cycle (Velbel 1999; Velbel *et al.* 2007), and also in sediments and sedimentary rocks (e.g. Morton & Hallsworth 1999; Velbel *et al.* 2007). Experimental work has indicated that the dissolution rate of garnet increases in acidic solutions (Nickel 1973) and disappears at a geologically rapid rate in acidic soils during surficial weathering environments (e.g. Velbel 1984). Garnet surface textures have been previously studied from several parts of the sedimentary cycle including weathered garnets, garnets eroded, transported and deposited by moving water or ice, and diagenetically modified garnets in sandstones (Velbel *et al.* 2007). Low-temperature mineral–solution reactions involving garnet during weathering and burial diagenesis have suggested that grain-surface textures of garnets show more variation after alteration under different environments than other silicate minerals, possibly indicating its greater sensitivity to variations in the geochemical environment of alteration than other silicates (Velbel 1984, 1993*a*; Salvino & Velbel 1989). Garnet dissolution has been recognized in many deep sedimentary basins, including the Cleveland Basin of Yorkshire (Smithson 1941), the central North Sea (Morton 1984), offshore New Zealand (Smale & Morton 1987) and the Voring Basin, offshore Norway (Morton *et al.* 2005*a*, *b*). Two studies can be given as examples concerning the stability and dissolution of garnet under burial conditions. These studies were conducted in two deep sedimentary basins and showed contrasting results regarding the dissolution of garnet due to burial diagenesis. Dissolution of unstable heavy minerals under burial depth conditions in Paleocene–Eocene sandstones of the central North Sea showed a decrease in mineral diversity with increasing depth, and many of these minerals disappeared at certain depths. For example, calcic-amphibole, epidote, titanite, kyanite and staurolite disappeared successively at 600, 1100, 1400, 1800 and 2400 m depths, respectively (Morton 1984; Morton & Hallsworth 1999, 2007). Garnet declined markedly in abundance relative to the stable accompanying zircon, but did not disappear entirely. However, it was rarely present at burial depths >3500 m in the underlying Jurassic rocks of the central North Sea. Dissolution of heavy minerals under burial conditions in Paleocene sandstones from the Foinaven Subbasin in Shetland (Scotland) showed the disappearance of amphibole, epidote, titanite and kyanite at depths of 1500, 1620, 1700 and 1760 m, respectively, but staurolite and garnet were persistent throughout (Morton *et al.* 2002, 2007). Other stable minerals such as apatite, chrome spinel, monazite, rutile, tourmaline and zircon showed no evidence for depletion with increasing burial depth. Garnet did not disappear with depth; in fact, its relative abundance increased with increasing burial between 1400 m and 1800 m depth, compensating for the disappearance of the unstable minerals mentioned above. Some samples below 2200 m depth showed relatively low garnet abundances, but this is considered to be a provenance effect related to deep weathering of the Cretaceous land surface (Morton *et al.* 2002), rather than depletion during burial diagenesis (Morton & Hallsworth 2007). In the currently studied sections, garnet is associated with apatite and other very rarely found heavy minerals such as epidote, zoisite, staurolite, hornblende and/or pyroxene. Garnet is more stable than these minerals in surficial weathering environments, which explains why it is shown above these minerals in the 'order of persistence of heavy minerals' of Bateman & Catt (1985). Therefore, the absence of garnet in the subsurface Chinampas section does not necessarily mean that it has been removed by weathering during deposition under surficial conditions. Although garnet is less stable than the ultrastable heavy minerals (zircon, tourmaline, rutile) and apatite under burial diagenesis conditions, it is more stable than many other minerals such as staurolite, epidote, amphiboles and pyroxenes (Van Loon & Mange 2007). The presence of minerals less stable than garnet, especially pyroxene, even very rarely in the subsurface Chinampas sandstones, throws doubt on the possibility of complete dissolution and removal of garnet. However, there remains a possibility that garnet disappeared in subsurface sandstones due to burial diagenesis because the less stable minerals such as pyroxene were only rarely found and only in a few samples. It is also not logical to imagine that all garnets in the subsurface Chinampas sandstones have been dissolved and eliminated during or after deposition under

chemical weathering due to acid leaching, or under burial conditions. This is because there is no obvious difference in depositional conditions between the Wolfville Formation at outcrop compared with the subsurface. Both were mostly deposited by braided rivers and also partly as alluvial fans and aeolian sands under continental semi-arid oxidizing conditions (Olsen 1997; Hamblin 2004). The dominant red coloration of these sandstones due to ferruginous cementation indicates their deposition under oxidizing and acidic conditions. Although garnets are very sensitive to acid leaching, they can survive recycling conditions and to a certain degree burial diagenesis, as faceted garnets are often found together with highly resistant zircon, tourmaline, rutile and, not uncommonly, apatite in deeply buried rocks (Mange & Maurer 1992).

Tourmaline, like garnet, is another important heavy mineral in provenance studies of sandstones. It consists of many groups and subgroups (Henry *et al.* 2011). The variation and wide range of potential compositions as well as its stability under weathering and diagenetic conditions make tourmaline an important and ideal mineral to be used as a provenance indicator and geochemical discriminator of sediments (Morton & Hallsworth 1999, 2007; Mange & Morton 2007). It has also been used as a guide to mineral exploration (Clarke *et al.* 1989; Yu & Jiang 2003; Galbraith *et al.* 2009). The Al–$Fe_t$–Mg and Ca–$Fe_t$–Mg provenance-discriminant ternary diagrams, established by Henry & Guidotti (1985) and Henry & Dutrow (1992) and later adapted by Preston *et al.* (2002), have effectively enhanced the use of tourmaline by many researchers as a discriminator for many types of source rocks. The analysed tourmalines from all three studied sections of the Wolfville Formation are generally of similar chemistry (Appendix IV). They are distributed in two fields when plotted on the $Al_{50}Fe_{50}$–Al–$Al_{50}Mg_{50}$ and Fe–Ca–Mg triangles of Henry & Guidotti (1985) (Fig. 9a, b). All of the Cambridge Cove and most Chinampas tourmalines are clustered in the field of metapelites and metapsammites coexisting with an Al-saturated phase and Ca-poor metapelites, metapsammites and quartz-tourmaline rock. Meanwhile the samples of Rainy Cove and a few of the Chinampas samples fall within the field of Li-poor granitoids and their associated pegmatites and aplites. Tourmalines within various rocks of the Meguma and Avalon terranes of Nova Scotia studied by many researchers were used and plotted for comparative reasons (Figs 1 & 9c, d; Table 1). These tourmalines, which mostly exist within the SMB and the Meguma Group fall exactly within the same fields and match those of the studied Wolfville Formation (Fig. 9a–d). They mostly occur in granites, granodiorite, monzogranites, aplites, pegmatites, metalliferous porphyries and veins hosted by the granitoids of the SMB and metasediments of the Meguma Group (Fig. 1; Table 1) (Oldale 1966; Chatterjee 1977; MacDonald & Scott 1980; Hutchinson 1982; O'Reilly *et al.* 1982; Anonymous 1983; Haynes 1983; Smith 1983; Wolfson 1983; Charest *et al.* 1985; Giles & Chatterjee 1986; MacDonald *et al.* 1987; Smith & Kontak 1987; Douma 1988; Clarke *et al.* 1989; Lynch & Ortega 1997; Kontak *et al.* 2002*b*). Tourmaline also exists in many other parts of Nova Scotia, such as the pegmatites and adjacent siliceous sedimentary rocks in Yarmouth County (Kontak & Kyser 2009), the Beaver Dam gold veins located within the metaturbidites of the Meguma Terrane of southern Nova Scotia (Kontak & Kerrich 1995), the Seal Island pluton and the pegmatite and aplite dykes associated with Shelburne pluton (White 2003). These comparisons for tourmalines support the results obtained from the analysed garnets separated from the same samples, which fell in the field of slates + granites and granite pegmatites. Thus, both garnet and tourmaline geochemistry point to derivation from metapelites, metapsammites and quartz-tourmaline rocks such as the slates and metaquartzites of the Meguma Group and the granites and granodiorites of the SMB.

Scheelite, which was found in many of the Rainy Cove sandstones (Appendix Ia), again seems to be a very diagnostic and useful provenance indicator mineral for the studied area. Scheelite occurrences from seven localities in several metaturbidite-hosted gold-bearing quartz veins of the Meguma Terrane in Nova Scotia were studied by Dostal *et al.* (2009) (Fig. 1). These scheelite-bearing veins are mostly concentrated in the southern central part of Nova Scotia. Scheelite also exists in Lake George as part of scheelite–molybdenite stockwork mineralization within the polymetallic hydrothermal deposits of Late Silurian age in the Fredericton trough of the northern Appalachians in New Brunswick (Seal *et al.* 1987) (Fig. 1). Because scheelite was found in the exposed part of Wolfville Formation that extends mostly along the southern part of Nova Scotia and around the Minas Basin (Fig. 1), it is logical to expect that its source was the scheelite-bearing veins that are widely distributed in the Meguma Terrane.

In addition to the three diagnostic provenance indicator minerals – garnet, tourmaline and scheelite – the other heavy minerals that were observed within the Wolfville sandstones in various amounts are also helpful for provenance interpretations. These minerals are opaques (hematite, magnetite, ilmenite and their alteration products), apatite, zircon, chlorite, rutile, muscovite, biotite, hornblende, pyroxene, staurolite, epidote and zoisite (Fig. 5b; Appendix II). These minerals also exist within the rocks of the Meguma Group, the SMB and the other rock units exposed in the Meguma Terrane of Nova Scotia and also in various rocks of the Avalon Terrane of Nova Scotia and New Brunswick. Many researchers have studied these minerals, including Clarke & Carruzzo (2007) (ilmenite and rutile); Clarke & Bogutyn (2003), Hicks *et al.* (1999) and Pattison *et al.* (1999) (muscovite, biotite); Krogh & Keppie (1990) (zircon); Lackey *et al.* (2011) (garnet, zircon); McKenzie & Clarke (1975) (apatite, zircon, opaques, biotite, muscovite, garnet, tourmaline, andalusite, topaz); Smith & Kontak (1986), Kontak & Smith (1993*a*, *b*) and Dostal *et al.* (2009) (gold, scheelite, rutile); White & Bar (2012) (chlorite, staurolite, andalusite, epidote, garnet, cordierite, chloritoid); Kontak *et al.* (2002*a*) (pyroxene); and Owen *et al.* (2010) and Pe-Piper (1988, 2007) (amphibole).

The Meguma Terrane, which was accreted to North America (Laurentia) during the Acadian Orogeny, again forms the most easterly part of the northern Appalachians (Dostal *et al.* 2009). This terrane, which extends across mainland Nova Scotia for *c.* 750 km, consists mainly of 10 km-thick metasedimentary rocks of the Meguma Group and was intruded by post-tectonic granitoids dominated by the Late Devonian SMB, which is exposed over an area of 7300 km$^2$ (MacDonald 2001; Dostal *et al.* 2009). This indicates that the rocks of the Meguma Group and the SMB, which abundantly contains the three minerals used in this study as provenance indicators (garnet, tourmaline and scheelite) and all other heavy minerals detected within the Wolfville Formation sandstones and conglomerates, have acted as the predominant source of these minerals (Fig. 1). The Avalon Terrane to the north of Meguma Terrane and the Bay of Fundy might have contributed together with the Meguma Terrane as the source for the deeper-part clastics of the Wolfville Formation in the early stages, but that role diminished considerably during deposition of the upper part of the exposed sediments. This is indicated by the lack of garnet and scheelite and the dominance of tourmaline in the sandstones of Chinampas well in the northern part of the Bay of Fundy. Meanwhile, garnet and scheelite became dominant, accompanied by lesser amounts of tourmaline in the surface sections at Cambridge and Rainy coves of the southern part of the Bay of Fundy.

## Tectonic setting and provenance of the sandstones

The results for all detrital grain counts (minerals and lithics) (Appendix Ia–d) were plotted on the triangular diagrams of Folk (1968) (Fig. 6) and the different triangular diagrams of Dickinson &

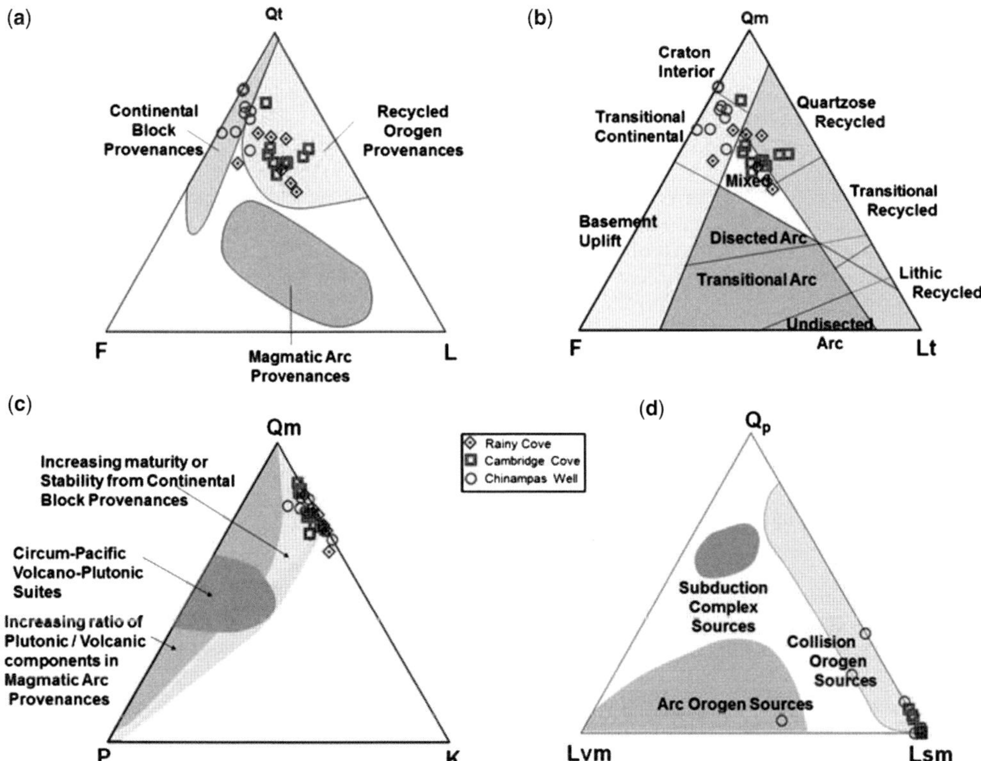



**Fig. 11.** Distribution of major framework groups in the sandstones of Wolfville Formation plotted on provenance indicator triads of Dickinson & Suczek (1979), Dickinson *et al.* (1982, 1983) and Dickinson (1985). (**a, b**) FQtL & FQmLt: Cambridge Cove and Rainy Cove sandstones fall in the field of 'recycled orogenic provenance (quartzose recycled and mixed)', while Chinampas N-37 sandstones are in the 'continental block provenance' field. (**c**) PQmK: all sandstones fall in the field of 'increasing maturity' or stability from 'continental block provenances'. (**d**) LvmQpLsm: all sandstones except one sample fall in the field of 'collision orogen sources'. The exceptional sample is a conglomerate from Chinampas N-37 well, which falls in the field of 'arc orogen sources'.

Suczek (1979), Dickinson *et al.* (1982, 1983) and Dickinson (1985) (Fig. 11). The QtFL triangle of Folk showed that the sandstones of the surface sections at Cambridge and Rainy coves are divided between feldspathic litharenites and lithic feldsarenites, while those of the subsurface Chinampas well are divided between feldsarenites (arkoses) and subarkoses. They are dominantly feldspathic litharenites at Cambridge Cove, lithic feldsarenites at Rainy Cove, and feldsarenites at Chinampas (Fig. 6). This indicates some differences between the exposed and subsurface sandstones of the Wolfville Formation and ultimately differences in their provenance. The surface sandstones are richer in lithics and poorer in quartz and feldspars relative to subsurface ones (Figs 5a, 6 & 11). The noticeable difference in their lithics content is that the subsurface sandstones contain different types of volcanics that are almost absent in surface rocks.

The QtFL and QmFLt plots of Dickinson show that the Cambridge and Rainy coves sandstones fall in the field of 'recycled orogen provenance' (quartzose recycled & mixed). while the Chinampas N-37 well sandstones fall in the 'continental block provenance' (transitional continental) field (Fig. 11). The QmPK triangle indicates that all sandstones fall in the field of 'increasing maturity or stability from continental block provenances'. The LvmQpLsm triangle shows that all sandstones except one fall in the field of 'collision orogen sources'. These results indicate that the surface samples have different tectonic settings from the sub surface ones, but their accumulation and formation are initially related to and post dated the collision orogen sources that can be interpreted as the collision between the Meguma and Avalon terrains. However, they show different fields in the FQtL and FQmLt plots, which can be explained by different sources or diagenetic history. The differences in abundance and composition of volcanic clasts at Chinampas compared with surface outcrops suggests a source from the Avalon Terrane to the north. The geology of the Avalon Terrane is dominated by metamorphic complexes and volcanics interbedded with both marine and non-marine sedimentary units (Lackey *et al.* 2007), in contrast to the very limited extent of volcanic rocks in the Meguma Terrane, which was once part of northern Gondwana (Murphy & Keppie 2005). The geology of the Meguma Terrane is dominated by clastic sedimentary and metasedimentary rocks of Precambrian to Cambrian, intruded by Devonian granitoids of the SMB (Keppie 2000). According to Hamblin (2004), the Fundy Rift System evolved through four major geological phases: (1) the Acadian/Alleghanian thrusting/collision during Late Devonian–Late Carboniferous, which assembled the Avalon and Meguma terranes, and subsequent deposition of the marine/continental deposits of Horton, Windsor and Mabou groups in an Early Carboniferous rift; (2) deposition of thick red fluvial clastics of the Pictou Group, followed by local erosion during the Late Carboniferous–Middle Triassic; (3) development of the Fundy Rift during the Middle Triassic–Early Jurassic as a result of tectonic relaxation and subsidence of the Meguma Terrane down the Cobequid–Jedabucto Fault in a deep fault-bounded basin that was filled with synrift continental deposits and basalts of the Fundy Group; and (4) further extension, subsidence and accumulation of younger deposits, followed by erosion during the Early Jurassic to Recent. Triassic sediments in the Argana Basin of Morocco have many similarities with the Triassic sediments of the Bay of Fundy and it is believed that they were lying side by side before drifting (Smoot & Olsen 1988; Mertz & Hubert 1989; Smoot & Castens-Seidell 1994; Hoffman *et al.* 2000). These indicate the active and complicated nature of the major tectonic/geological events that affected the surroundings of the Bay of Fundy from the Late Devonian. These events might help explain the differences between the studied subsurface and exposed clastics of the Wolfville Formation. Lithological differences between the studied subsurface and surface sandstones, such as the existence of different volcanic lithics in the Chinampas N-37 core and their absence or extremely rare presence in the surface Rainy Cove and Cambridge Cove sandstones, the total absence of garnet and abundance of tourmaline and zircon in the heavy mineral fractions, as well as the much lower lithic fragments relative to the quartz and feldspars in the subsurface section, suggests effective changes in the source area from which the sediments were derived as well as effective changes from tectonism. Tectonism (rifting and transform faulting) might have changed the paths of rivers carrying sediments from the source area towards the Bay of Fundy. In the early stages where the deeper parts of the Wolfville sandstones were depositing (represented by the studied Chinampas N-37 core), the Bay of Fundy area was probably a 'transitional continental' environment undergoing rifting, but with time the tectonic conditions were changing towards a 'recycled orogenic environment' where the sediments of the younger parts of the Wolfville Formation were forming (represented by the studied surface sections of Rainy Cove and Cambridge Cove). The early subsurface sandstones of the Wolfville Formation were probably derived from the Gondwana sources before and during drifting, representing the 'transitional continental/continental block provenance' environments in Figure 11. Meanwhile, the younger upper exposed parts were dominantly derived from the Meguma Terrane, with a possible minor contribution from Avalonia representing the 'mixed/recycled orogenic provenance' environment.

## Diagenesis

The mineralogy and petrophysical properties of the sandstones of the Wolfville Formation have undergone many diagenetic modifications (Fig. 12). The main diagenetic processes that have affected these sandstones include mechanical and chemical compaction, the precipitation of iron oxides, carbonates, clays and quartz, partial dissolution of unstable minerals such as K-feldspars, plagioclase, granitoid lithics and calcite cement, as well as formation of various types of secondary porosities (Figs 3, 4 & 10).

Mechanical compaction is indicated by the formation of grain-supported texture in the deeply buried sandstones at Chinampas well (Figs 3h & 4c–f), parallel orientation of the elongate rock fragments (Figs 3–5), the kink-banding of elongate flaky minerals such as muscovite (Figs 3h & 4e), abundant microfracturing of brittle minerals such as quartz (Fig. 3b, c, f) and occasionally feldspars. The mechanical compaction effects have taken place in many stages during burial diagenesis, and this effect is much more pronounced in the sandstones of the subsurface Chinampas section. The sandstones of the surface sections at Rainy and Cambridge coves have cement-supported textures with distinct porosity (Figs 3a–f & 4a, b), in contrast to the grain-supported texture with negligible porosity of the subsurface Chinapmas section (Figs 3f–i & 4c–f), which is a reflection of the differential mechanical compaction in the two locations.

The chemical compaction and diagenesis are indicated as cementation, alteration, replacement and dissolution. Both surface and subsurface

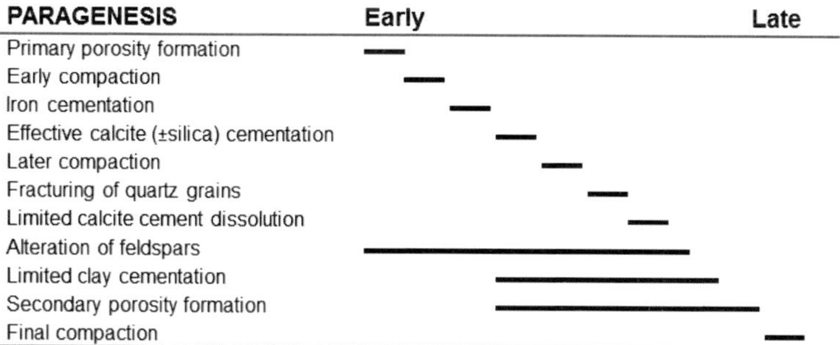



**Fig. 12.** Paragenesis of the diagenetic events of the Wolfville Formation sediments.

sandstones have iron cement, which is earlier in origin than the other cement types. The iron cement was deposited on the surface of detrital grains as thin coats enveloped by other cements such as calcite and quartz in many cases (Fig. 4d). The iron cement, as goethite, originated from the hydration of detrital iron oxides (magnetite and hematite), which are abundant heavy minerals in these sandstones (Fig. 5b). The calcite, iron and clay cementation has dominated the surface rocks, while quartz and iron are effective cementing material in the subsurface. Clays (kaolinite booklets and spherulitic aggregates of illite, montmorillonite and chlorite), which partially or completely fill pore spaces (Fig. 10j, p), were derived from the weathering and alteration of feldspar and granite lithics (Fig. 4f). Timing of clay cement formation relative to calcite cementation is not clear, as it sometimes seems to be earlier and in other cases later. It was possibly formed during more than one stage due to the alterations of feldspars. Carbonate cement, deposited from the interstitial pore fluids and partly derived from the dissolution of abundant carbonate detritus, was deposited and recrystallized during burial, filling most of the open spaces between detrital grains and resulted in sparry calcite cement-supported sandstones of the surface sections at Rainy and Cambridge coves (Figs 3a–f & 4a, b). The stronger mechanical compaction in the deeper subsurface sandstones has promoted chemical compaction and diagenesis, which is seen in the partial dissolution of unstable minerals such as calcite and clay cement, feldspars and even quartz grains at grain-to-grain contacts. This partial dissolution has liberated silica from quartz, feldspars and clays, which were deposited mostly on the surface of quartz grains as authigenic quartz cement (Figs 3h & 4c–f). Pressure solution is thought by many workers to be a major source of silica for cementation in sandstones. This process has a significant effect on reservoir quality by removing intergranular porosity due to greater compaction and increased quartz cementation, which is the

main cause of porosity loss in sandstones buried in excess of 2–3 km (Giles *et al.* 2000). In spite of a burial depth of *c.* 4 km of the Wolfville Formation sandstones at Chinampas well, stylolites microstructures, one of the manifestations of pressure solution that take place in a minimum temperature range of 86–136 °C at depths >2.5 km (Baron & Parnell 2007), were not found. The minimum burial depth required for the formation of stylolite in sandstones is uncertain (Baron & Parnell 2007), although most sandstones that have experienced temperatures of >250 °C have developed some evidence of pressure solution (Dewer & Hajash 1995). Many factors are believed to influence the development of stylolites in sandstones, including temperature, pressure and the mineralogy of the sandstone and of the stylolite. The importance of pressure is still a matter of much debate between those who believe that dissolution is dependent on the stress at grain contacts (e.g. Sheldon *et al.* 2003) and those who believe that pressure has only a limited effect on stylolite formation in sandstones (e.g. Oelkers *et al.* 2000). Many studies have also indicated that the presence of mica and/or illite clays promotes pressure solution in quartz (e.g. Pittman *et al.* 1992; Bjørkum 1996; Oelkers *et al.* 1996). These studies have shown that pressure solution in some quartzose sandstones occurred only at interfaces between quartz and mica/illite clays, with no pressure solution occurring at quartz/quartz interfaces. This might suggest that the amount of clays (detrital and/or diagenetic) in the subsurface Wolfville Formation sandstones was very low, which is reflected in the absence of stylolites in these sandstones, although muscovite mica exists in minor amounts (Fig. 4e, f).

The partial dissolution of the unstable calcareous cement and grains, such as feldspars and rock fragments (carbonates, granites, etc.), in the surface sandstones is reflected in the formation of various types of secondary porosities (Figs 4a & 10). These secondary microporosities are of dissolution (Fig. 10g, h), microfracture (Fig. 10i, j), intergranular and grain boundary (Fig. 10k, l), intra-granular (Fig. 10m, n) and intra-cement (Figs 4a & 10o, p) types. In addition to these types of secondary porosities that were formed by diagenesis, there are what seem to be primary porosities indicated by empty pore spaces between grains (Fig. 10c, d), which can sometimes be seen side by side with secondary porosity (Fig. 10e, f). The dissolution of these minerals in the subsurface sandstones has not resulted in increasing porosity because of the counter action of effective burial compaction. Feldspars were altered only partially to muscovite (sericite) and clays and/or partially dissolved (Fig. 4f). Relatively fresh K-feldspars (orthoclase and microcline) and plagioclase are also common in both surface and subsurface sandstones (Figs 3a & 4b, c). The absence of carbonate lithics in the subsurface sandstones, which is in contrast with surface samples, either indicates they have been removed by dissolution or that they were never present.

The studied subsurface sandstones of Chinampas N-37 well are different from those of surface sections in some characteristics (Figs 3–6). The subsurface sandstones consist of well-compacted sand grains, cemented by quartz, iron and minor calcite, and have negligible porosity. It is evident that the deep burial conditions under a rock column *c.* 4 km beneath the Bay of Fundy resulted in reduction of most of the original cementing material (calcite and clays) by dissolution, bringing grains tightly together and considerably reducing porosity. The amount of calcite and clay cement is relatively much lower than that of the surface sandstones; this could be due to the relatively low amount of these minerals in the source rocks from which these sandstones were derived, and/or their dissolution during burial. The quartz cement, which was deposited as authigenic overgrowths on quartz grains, sealed the remaining pore spaces, further reducing the porosity (Figs 3h, i & 4c–f). This silica, which can be derived from dissolution at grain-to-grain contacts under the effect of solution pressure (Greensmith 1978; Hutcheon 1990) and also from alteration of feldspars and minor clays, was the main source of quartz cement. In general, the extent of diagenesis is considered to be relatively mild in both surface and subsurface samples, because unstable minerals such as feldspar are mostly fresh or only slightly altered throughout.

## *Hydrocarbon potential of the Wolfville Formation*

Hydrocarbon source rocks have been reported within the late Palaeozoic Maritime Basin in Nova Scotia (Calder *et al.* 1998). These source rocks include organic-rich shales of the Horton Group, organic-rich carbonates of the Windsor Group and sapropelic shales, coals and organic-rich limestones and shales of the Cumberland Group (Bell 1958; Calder *et al.* 1998). The strata at the surface of the Maritime Basin are mostly within the oil or gas windows (Mukhophadyay *et al.* 1991) and oil seeps have been reported (Bell 1958).

Based on the occurrence of oil in the Windsor and Horton groups (Bell 1958), the Cheverie Formation has been considered a potential hydrocarbon exploration target. Two wells (Kennetcook #1 and #2) were drilled in 2007 in the Windsor Basin of Nova Scotia to investigate the hydrocarbon potential of the Horton Bluff Formation shales (Anonymous 2008). A related news release announced the discovery of substantial amounts of gas with an estimated resource of $89–109 \times 10^9$ SCF of oil and gas in place per square mile (http://www.wallstreet-online.de/diskussion/1137 920-neustebeitraege/news-news-triangle-pet). The total gas content of the shales ranges from 7.9 to 190 ft$^3$/ton, with an average total organic carbon (TOC) of 10% for all shales containing organic matter of type II/III to III. Such shales have a maturity of 1.53–2.07%, placing them within the peak window for natural gas generation.

Triassic sandstones of the Wolfville Formation overlie the Horton Group, and the overlying Blomidon Formation shales form a potential seal. A petroleum system has been demonstrated in the underlying Horton Group. This study has shown only 1–17% porosity within the upper part of the Wolfville Formation and low porosity in the deeply buried strata at Chinampas. The two discovery wells drilled in the northern part of the Bay of Fundy, Chinampas N-37 and Cape Spencer-1, which have targeted the deeper parts of the Wolfville Formation, were not successful because they did not encounter source rocks within the formation. Furthermore, the porosity of the sandstones is negligible. The studied subsurface sandstones and conglomerates do not have reservoir characteristics because of their tight compaction and quartz cementation, which resulted in negligible porosity and permeability. The exposed part of the Wolfville Formation has better porosity. Therefore, it is possible that the upper parts of the formation could act as a hydrocarbon reservoir, and interbedded thin shales could form local seals. Further investigation is warranted to better understand the factors leading to localized good porosity within the Wolfville Formation.

## Conclusions

An outcrop on the southern margin of the Fundy Basin, the fluvial Wolfville Formation, comprises calcite cement-supported sandstones dominated by quartz > lithics > feldspars, with almost no volcanics and abundant opaques and garnet as heavy mineral. The subsurface rocks in the Chinampas well, on the northern margin of the basin, are well compacted, grain-supported with very little cement, dominated by quartz > feldspars > lithics, with minor volcanics and no garnet. The porosity of exposed rocks is low (*c.* 6% on average), but much higher than that of the subsurface rocks (*c.* 0.5% on average).

The exposed Wolfville Formation sediments have 'recycled orogenic provenance (mixed and quartzose recycled zones)' of 'collision tectonic setting (subduction complex or fold–thrust belt)'. In the subsurface, the Wolfville Formation falls in the field of 'continental block provenance (passive platform or rift shoulder setting)'. This means that either the two areas had different source rocks or there was a change in the provenance during deposition of the Wolfville Formation between collision and rifting, These active tectonic events were the collision between Meguma and Avalon terrains during the late Palaeozoic (Permo-Carboniferous) and the active transform faulting of the Cobequid–Minas fault system along the suture zone dividing the Meguma and Avalon terrains, and the subsequent active rifting of Pangaea in the Bay of Fundy Basin during the Late Triassic. The active movement during the Late Triassic along the Cobequid–Minas fault zone resulted in shifting of the relative positions of the Avalon and Meguma terranes, which might have had an important role in this change in provenance between the early and late stages of the deposition of the Wolfville Formation.

The varietal compositions of garnet and tourmaline indicate that the exposed Wolfville Formation sediments were derived mostly from metapelites, metapsammites and granitoids. The mineral chemistry from the southern margin of the basin corresponds to many known mineral occurrences in the Meguma Terrane. Lithic clasts confirm that the main source rocks for the southern part of the basin were the MSG and the SMB, with contributions from the Horton Group and the carbonate-bearing formations such as the Windsor Group (Fig. 1). As a result, the sediments have 'recycled orogenic provenance (mixed and quartzose recycled zones)' of 'collision tectonic setting (subduction complex or fold–thrust belt)'. The existence of scheelite supports the conclusions from garnet and tourmaline geochemistry and proves the mineral is a good provenance indicator. The tourmalines of both surface and subsurface sandstones are similar and match very well with those reported in many occurrences of the Meguma Terrane, suggesting their common origin. However, the absence of garnet and scheelite in the subsurface rocks might suggest alternative source rocks. This alternative source could be Gondwana (currently located in the northwestern corner of Africa) or the Avalon Terrane of New Brunswick. Another possibility is that the absence of these two heavy minerals might be due to their dissolution under burial conditions, although there is no conclusive evidence to support this assumption.

The occurrence of volcanic rock fragments in the studied subsurface section and their very rare occurrence in the surface sections imply that the volcanic rock exposures, which are sporadically distributed in the Meguma (Palaeozoic) and Avalon (Palaeozoic and older) terranes of Nova Scotia and possibly New Brunswick, contributed as the source for the lithics in the early stages of deposition of the Wolfville Formation. The rare occurrence of these volcanic lithics in the surface exposures suggests that the supply from volcanic rocks stopped for some reason, due to active tectonic events in the area, changes in river courses or erosion of the volcanic bodies. The lack of heavy minerals attributable to volcanics and high-grade metamorphic rocks in the exposed studied rocks exclude the Avalon Terrane and other rock units of the Appalachian Mountains in New Brunswick as contributing sources, at least in the later stages of deposition of the Wolfville Formation sediments.

The reservoir quality of Wolfville Formation sandstones is not encouraging, especially in the deeper parts of the subsurface, due to low porosity/permeability. Nevertheless, there is a petroleum system with considerable reserves of gas in the onshore carbon-rich black shales within the Carboniferous Horton Bluff Group that unconformably underlie the Wolfville Formation.

We acknowledge the help of D. Brown, M. J. Verrall and N. White from the Canada–Nova Scotia Offshore Petroleum Board–Geoscience Research Centre in providing samples from Chinampas N-37 well and for their help during sampling. From the Department of Earth Sciences at Dalhousie University, we thank G. Brown for preparing thin and polished thick sections of the studied samples, and D. MacDonald for his help during microprobe analysis. We also thank K. McVicar and D. Tobey for redrafting the map used in Figure 1.

**Appendix Ia.** *Statistics of the point-counting results of the Wolfville Formation sandstones and conglomerates at Rainy Cove*

| Grain groups | | | Grain type | Rainy Cove – sandstones | | | | | | | Rainy Cove – conglomerates | | | | | Av | Min | Max |
|---|---|---|---|---|---|---|---|---|---|---|---|---|---|---|---|---|---|---|
| | | | | R3 | R4 | R5 | R6 | R7 | R8 | R9 | RC1 | RC2 | RC3 | RC4 | RC5 | | | |
| Qt | Qm | | Quartz | 367 | 334 | 305 | 299 | 401 | 393 | 402 | 395 | 353 | 279 | 476 | 357 | 363 | 279 | 476 |
| | | | Quartzite | 0 | 4 | 3 | 7 | 3 | 1 | 0 | 15 | 11 | 0 | 8 | 13 | 5 | 0 | 15 |
| | Qp | | Chert | 1 | 0 | 0 | 0 | 0 | 0 | 0 | 4 | 0 | 0 | 0 | 0 | 0 | 0 | 4 |
| F | | | K-feldspar | 128 | 225 | 159 | 101 | 132 | 113 | 120 | 162 | 160 | 87 | 123 | 96 | 134 | 87 | 225 |
| | | | Plagioclase | 27 | 13 | 23 | 37 | 7 | 5 | 12 | 28 | 13 | 26 | 21 | 7 | 18 | 5 | 37 |
| Lt = L + Qp | L | | Slate | 47 | 75 | 122 | 38 | 61 | 81 | 69 | 7 | 56 | 57 | 96 | 91 | 67 | 7 | 122 |
| | | | Siltstone | 50 | 59 | 40 | 34 | 145 | 69 | 76 | 52 | 85 | 67 | 60 | 83 | 68 | 34 | 145 |
| | | | Volcanics | 0 | 0 | 0 | 0 | 0 | 0 | 0 | 0 | 0 | 0 | 0 | 0 | 0 | 0 | 0 |
| Carbonates | | | Limestone | 30 | 24 | 30 | 28 | 0 | 11 | 17 | 4 | 7 | 26 | 6 | 0 | 15 | 0 | 30 |
| Cement | | | Calcite | 413 | 394 | 380 | 529 | 261 | 390 | 328 | 327 | 425 | 641 | 363 | 398 | 404 | 261 | 641 |
| | | | Iron stain | 33 | 21 | 37 | 4 | 33 | 50 | 51 | 11 | 0 | 0 | 0 | 59 | 25 | 0 | 59 |
| | | | Clay + sericite | 7 | 5 | 3 | 2 | 44 | 37 | 28 | 13 | 4 | 4 | 18 | 6 | 14 | 2 | 44 |
| Mica & chlorite | | | Mica | 5 | 14 | 11 | 0 | 3 | 7 | 3 | 4 | 2 | 0 | 10 | 13 | 6 | 0 | 14 |
| | | | Chlorite | 0 | 0 | 5 | 0 | 0 | 2 | 0 | 0 | 0 | 0 | 0 | 0 | 1 | 0 | 5 |
| Heavy minerals | | | Opaque | 9 | 5 | 20 | 4 | 5 | 15 | 4 | 0 | 0 | 0 | 6 | 11 | 7 | 0 | 20 |
| | | | Transparent | 2 | 1 | 1 | 0 | 0 | 1 | 0 | 0 | 0 | 0 | 0 | 0 | 0 | 0 | 2 |
| Others | | | Alterites | 0 | 0 | 16 | 57 | 5 | 6 | 3 | 0 | 0 | 0 | 0 | 6 | 8 | 0 | 57 |
| Pore spaces | | | Voids | 81 | 26 | 45 | 61 | 101 | 20 | 89 | 179 | 84 | 14 | 14 | 59 | 64 | 14 | 179 |
| Total counts | | | | 1200 | 1200 | 1200 | 1200 | 1200 | 1200 | 1200 | 1200 | 1200 | 1200 | 1200 | 1200 | 1200 | 1200 | 1200 |
| Porosity (%) | | | | 6.8 | 2.2 | 3.7 | 5.1 | 8.4 | 1.7 | 7.5 | 14.9 | 7 | 1.2 | 1.1 | 4.9 | 5.4 | 1.1 | 14.9 |
| Cement counts | | | | 454 | 420 | 420 | 535 | 338 | 477 | 406 | 351 | 429 | 644 | 380 | 463 | 443 | 338 | 644 |
| Total counts–pores | | | | 746 | 780 | 780 | 665 | 862 | 723 | 794 | 849 | 771 | 556 | 820 | 737 | 757 | 556 | 862 |
| Cement/grain counts (%) | | | | 61 | 54 | 54 | 81 | 39 | 66 | 51 | 29 | 36 | 54 | 32 | 39 | 49.5 | 29.2 | 80.5 |
| Quartz/total counts (%) | | | | 31 | 28 | 25 | 25 | 33 | 33 | 33 | 33 | 29 | 23 | 40 | 30 | 30.3 | 23.2 | 39.7 |
| Lithics/total count (%) | | | | 11 | 13 | 16 | 8 | 17 | 13 | 13 | 5 | 12 | 13 | 13 | 15 | 12.5 | 5.2 | 17.2 |
| Feldspars/total counts (%) | | | | 13 | 20 | 15 | 11 | 12 | 10 | 11 | 16 | 14 | 9 | 12 | 9 | 12.7 | 8.6 | 19.8 |
| Plagioclase/K-feldspars (%) | | | | 21 | 6 | 14 | 36 | 5 | 4 | 10 | 17 | 8 | 30 | 17 | 8 | 14.7 | 4.2 | 36.3 |
| Cement/total counts (%) | | | | 38 | 35 | 35 | 45 | 28 | 40 | 34 | 29 | 36 | 54 | 32 | 39 | 36.9 | 28.2 | 53.7 |
| Calcite/total counts (%) | | | | 34 | 33 | 32 | 44 | 22 | 32 | 27 | 27 | 35 | 53 | 30 | 33 | 33.7 | 21.7 | 53.4 |
| (Mica + chlorite)/total counts (%) | | | | 0.4 | 1.2 | 1.4 | 0 | 0.2 | 0.7 | 0.2 | 0.3 | 0.2 | 0 | 0.8 | 1.1 | 0.5 | 0 | 1.4 |
| Heavy minerals/total counts (%) | | | | 0.2 | 0.1 | 1.4 | 4.8 | 0.4 | 0.6 | 0.2 | 0 | 0 | 0 | 0.5 | 0.9 | 0.8 | 0 | 4.8 |

*Note*: In the case of conglomerates, counting was conducted on all grains including the pebbles, but only their sand size constituents were used in this table after recalculating them to 1200 counts to match the normal pebble-free sandstones. (Qm, moncrystalline quartz; Qp, polycrystalline quarts; Qt, total quartz; K, K-feldspars; P, plagioclase feldspar; Lms, metasedimentary lithics; Lmv, metavolcanic lithics; L, total lithics; Lt, total lithics & polycrystalline quartz).

**Appendix Ib.** *Statistics of the point-counting results of Wolfville Formation sandstones at Cambridge Cove*

| Grain groups | | | Grain type | Cambridge Cove – sandstones | | | | | | | | | | | Av | Min | Max |
|---|---|---|---|---|---|---|---|---|---|---|---|---|---|---|---|---|---|
| | | | | C1 | C2 | C3 | C4 | C5 | C6 | C7 | C8 | C9 | C10 | C11 | | | |
| Qt | Qm | | Quartz | 397 | 362 | 384 | 349 | 355 | 312 | 395 | 337 | 363 | 313 | 600 | 379 | 312 | 600 |
| | | | Quartzite | 25 | 14 | 37 | 12 | 31 | 22 | 12 | 13 | 10 | 9 | 35 | 20 | 9 | 37 |
| | Qp | | Chert | 3 | 2 | 18 | 7 | 0 | 10 | 1 | 1 | 1 | 0 | 1 | 4 | 0 | 18 |
| F | | | K-feldspar | 120 | 105 | 62 | 133 | 114 | 99 | 143 | 63 | 105 | 108 | 112 | 106 | 62 | 143 |
| | | | Plagioclase | 14 | 12 | 3 | 7 | 13 | 8 | 33 | 8 | 17 | 12 | 4 | 12 | 3 | 33 |
| Lt = L + Qp | L | | Slate | 146 | 127 | 65 | 88 | 49 | 96 | 73 | 95 | 59 | 66 | 35 | 82 | 35 | 146 |
| | | | Siltstone | 40 | 40 | 144 | 50 | 63 | 60 | 113 | 78 | 106 | 33 | 40 | 70 | 33 | 144 |
| | | | Volcanics | 0 | 0 | 0 | 0 | 0 | 0 | 0 | 0 | 0 | 0 | 0 | 0 | 0 | 0 |
| Carbonates | | | Limestone | 7 | 7 | 46 | 17 | 24 | 21 | 19 | 11 | 21 | 23 | 0 | 18 | 0 | 46 |
| Cement | | | Calcite | 223 | 244 | 311 | 444 | 467 | 462 | 255 | 474 | 461 | 510 | 128 | 362 | 128 | 510 |
| | | | Iron stain | 25 | 13 | 35 | 34 | 6 | 18 | 3 | 17 | 4 | 2 | 5 | 15 | 2 | 35 |
| | | | Clay + sericite | 126 | 104 | 22 | 21 | 2 | 10 | 98 | 0 | 1 | 2 | 17 | 37 | 0 | 126 |
| Mica & chlorite | | | Mica | 7 | 8 | 0 | 0 | 7 | 7 | 1 | 5 | 2 | 4 | 15 | 5 | 0 | 15 |
| | | | Chlorite | 1 | 2 | 0 | 0 | 1 | 0 | 1 | 0 | 0 | 0 | 1 | 1 | 0 | 2 |
| Heavy minerals | | | Opaque | 18 | 10 | 25 | 3 | 10 | 6 | 12 | 4 | 4 | 3 | 6 | 9 | 3 | 25 |
| | | | Transparent | 0 | 1 | 0 | 1 | 4 | 2 | 2 | 2 | 0 | 1 | 0 | 1 | 0 | 4 |
| Others | | | Alterites | 7 | 15 | 14 | 1 | 20 | 10 | 2 | 48 | 15 | 6 | 2 | 13 | 1 | 48 |
| Pore spaces | | | Voids | 41 | 134 | 34 | 33 | 34 | 57 | 37 | 44 | 31 | 108 | 199 | 68 | 31 | 199 |
| Total counts | | | | 1200 | 1200 | 1200 | 1200 | 1200 | 1200 | 1200 | 1200 | 1200 | 1200 | 1200 | 1200 | 1200 | 1200 |
| Porosity (%) | | | | 3.4 | 11.2 | 2.8 | 2.8 | 2.8 | 4.8 | 3.083 | 3.7 | 2.6 | 9 | 16.6 | 5.7 | 2.6 | 16.6 |
| Cement counts | | | | 374 | 361 | 368 | 499 | 475 | 490 | 356 | 491 | 466 | 514 | 150 | 413 | 150 | 514 |
| Total counts – pores | | | | 1159 | 1066 | 1166 | 1167 | 1166 | 1143 | 1163 | 1156 | 1169 | 1092 | 1001 | 1132 | 1001 | 1169 |
| Cement/grain counts (%) | | | | 32 | 34 | 32 | 43 | 41 | 43 | 31 | 42 | 40 | 47 | 15 | 36.4 | 15 | 47 |
| Quartz/total counts (%) | | | | 33 | 30 | 32 | 29 | 30 | 26 | 33 | 28 | 30 | 26 | 50 | 31.5 | 26 | 50 |
| Lithics/total count (%) | | | | 16 | 15 | 21 | 13 | 11 | 15 | 17 | 15 | 16 | 10 | 6 | 14.1 | 6.3 | 21.3 |
| Feldspars/total counts (%) | | | | 11 | 10 | 5 | 12 | 11 | 9 | 15 | 6 | 10 | 10 | 10 | 9.8 | 5.4 | 14.7 |
| Plagioclase/K-feldspars (%) | | | | 12 | 11 | 5 | 5 | 11 | 8 | 23 | 13 | 16 | 11 | 4 | 10.8 | 4 | 23 |
| Cement/total counts (%) | | | | 31 | 30 | 31 | 42 | 40 | 41 | 30 | 41 | 39 | 43 | 13 | 34.6 | 13 | 43 |
| Calcite/total counts (%) | | | | 19 | 20 | 26 | 37 | 39 | 39 | 21 | 40 | 38 | 43 | 11 | 30.3 | 11 | 43 |
| (Mica + chlorite)/total counts (%) | | | | 0.7 | 0.8 | 0 | 0 | 0.6 | 0.6 | 0.2 | 0.4 | 0.2 | 0.3 | 1.3 | 0.5 | 0 | 1.3 |
| Heavy minerals/total counts (%) | | | | 1.5 | 0.9 | 2.1 | 0.3 | 1.3 | 0.7 | 1.3 | 0.5 | 0.3 | 0.3 | 0.5 | 0.9 | 0.3 | 2.1 |

*Note*: For abbreviations, see footnote to Appendix Ia.

**Appendix Ic.** *Statistics of the point-counting results of Wolfville Formation sandstones at Chinampas N-37 well*

| Grain groups | | Grain type | Chinampas N-37 well – sandstones and pebbly sandstones | | | | | | | | | Av | Min | Max |
|---|---|---|---|---|---|---|---|---|---|---|---|---|---|---|
| | | | CH–12002 | CH–12003.1 | CH-12004.1 | CH-12006 | CH–12006.6 | CH–12007.9 | CH–12008.8 | CH–12010.4 | CH–12011.2 | | | |
| Qt | Qm | Quartz | 778 | 670 | 733 | 752 | 675 | 648 | 746 | 695 | 703 | 711 | 648 | 778 |
| | | Quartzite | 8 | 0 | 0 | 1 | 0 | 16 | 7 | 1 | 32 | 7 | 0 | 32 |
| | Qp | Chert | 9 | 0 | 1 | 0 | 0 | 6 | 5 | 0 | 7 | 3 | 0 | 9 |
| F | | K-feldspar | 177 | 319 | 145 | 193 | 156 | 240 | 200 | 279 | 129 | 204 | 129 | 319 |
| | | Plagioclase | 43 | 0 | 19 | 32 | 0 | 53 | 29 | 9 | 69 | 28 | 0 | 69 |
| Lt = L + Qp | L | Slate | 28 | 11 | 2 | 41 | 2 | 57 | 57 | 23 | 49 | 30 | 2 | 57 |
| | | Siltstone | 4 | 0 | 0 | 3 | 0 | 23 | 16 | 29 | 12 | 10 | 0 | 29 |
| | | Volcanics | 5 | 0 | 0 | 1 | 0 | 55 | 0 | 0 | 0 | 7 | 0 | 55 |
| Carbonates | | Limestone | 0 | 0 | 0 | 0 | 0 | 0 | 0 | 0 | 1 | 0 | 0 | 1 |
| Cement | | Calcite | 5 | 14 | 4 | 15 | 95 | 12 | 24 | 48 | 57 | 30 | 4 | 95 |
| | | Quartz + iron | 125 | 154 | 259 | 109 | 160 | 78 | 93 | 54 | 119 | 128 | 54 | 259 |
| | | Clay + sericite | 1 | 18 | 26 | 0 | 31 | 0 | 8 | 13 | 13 | 12 | 0 | 31 |
| Mica & chlorite | | Mica | 13 | 6 | 3 | 14 | 8 | 3 | 1 | 20 | 2 | 8 | 1 | 20 |
| | | Chlorite | 0 | 0 | 1 | 0 | 15 | 0 | 1 | 0 | 0 | 2 | 0 | 15 |
| Heavy minerals | | Opaque | 4 | 10 | 7 | 15 | 36 | 9 | 14 | 26 | 4 | 14 | 4 | 36 |
| | | Transparent | 0 | 0 | 0 | 0 | 0 | 0 | 0 | 0 | 0 | 0 | 0 | 0 |
| Others | | Alterites | 0 | 0 | 0 | 0 | 0 | 0 | 0 | 0 | 0 | 0 | 0 | 0 |
| Pore spaces | | Voids | 1 | 0 | 0 | 24 | 21 | 2 | 0 | 4 | 4 | 6 | 0 | 24 |
| Total counts | | | 1200 | 1200 | 1200 | 1200 | 1200 | 1200 | 1200 | 1200 | 1200 | 1200 | 1200 | 1200 |
| Porosity (%) | | | 0.1 | 0 | 0 | 2 | 1.8 | 0.2 | 0 | 0.3 | 0.3 | 0.5 | 0 | 2 |
| Cement counts | | | 131 | 185 | 289 | 124 | 287 | 89 | 125 | 115 | 189 | 170 | 89 | 289 |
| Total counts – pores | | | 1069 | 1015 | 911 | 1076 | 913 | 1111 | 1075 | 1085 | 1011 | 1030 | 911 | 1111 |
| Cement/grain counts (%) | | | 12 | 18 | 32 | 12 | 31 | 8 | 12 | 11 | 19 | 17.1 | 8.1 | 31.7 |
| Quartz/total counts (%) | | | 65 | 56 | 61 | 63 | 56 | 54 | 62 | 58 | 59 | 59.3 | 54 | 64.8 |
| Lithics/total count (%) | | | 3 | 1 | 0 | 4 | 0 | 11 | 6 | 4 | 5 | 3.9 | 0.2 | 11.2 |
| Feldspars/total counts (%) | | | 18 | 27 | 14 | 19 | 13 | 24 | 19 | 24 | 17 | 19.4 | 13 | 26.6 |
| Plagioclase/K-Feldspars (%) | | | 24 | 0 | 13 | 17 | 0 | 22 | 15 | 3 | 53 | 16.4 | 0 | 53.4 |
| Cement/total counts (%) | | | 11 | 15 | 24 | 10 | 24 | 7 | 10 | 10 | 16 | 14.2 | 7.5 | 24.1 |
| Calcite/total counts (%) | | | 0.4 | 1.1 | 0.3 | 1.2 | 7.9 | 1 | 2 | 4 | 4.8 | 2.5 | 0.3 | 7.9 |
| (Mica + chlorite)/total counts (%) | | | 1.1 | 0.5 | 0.3 | 1.1 | 1.9 | 0.2 | 0.2 | 1.7 | 0.2 | 0.8 | 0.2 | 1.9 |
| Heavy minerals/total counts (%) | | | 0.3 | 0.8 | 0.6 | 1.2 | 3 | 0.7 | 1.1 | 2.1 | 0.3 | 1.1 | 0.3 | 3 |

*Note*: For abbreviations, see footnote to Appendix Ia.

**Appendix Id.** *Summary statistics of the point-counting results of Wolfville Formation sandstones and conglomerates at the three studied locations*

| Location | Rock type | Sample no. | Qm | Qp | Qt = Qm + Qp | K | P | F = K + P | Lms | Lmv | L = Lms + Lmv | Lt = L + Qp |
|---|---|---|---|---|---|---|---|---|---|---|---|---|
| Rainy Cove | Sandstones | R3 | 367 | 1 | 368 | 128 | 27 | 155 | 97 | 0 | 97 | 98 |
| | | R4 | 338 | 0 | 338 | 225 | 13 | 238 | 134 | 0 | 134 | 134 |
| | | R5 | 308 | 0 | 308 | 159 | 23 | 182 | 162 | 0 | 162 | 162 |
| | | R6 | 306 | 0 | 306 | 101 | 37 | 138 | 71 | 0 | 71 | 71 |
| | | R7 | 404 | 0 | 404 | 132 | 7 | 138 | 206 | 0 | 206 | 206 |
| | | R8 | 394 | 0 | 394 | 113 | 5 | 118 | 150 | 0 | 150 | 150 |
| | | R9 | 402 | 0 | 402 | 120 | 12 | 131 | 145 | 0 | 145 | 145 |
| | Conglomerates | RC1 | 410 | 4 | 414 | 162 | 28 | 190 | 59 | 0 | 59 | 63 |
| | | RC2 | 364 | 0 | 364 | 160 | 13 | 173 | 141 | 0 | 141 | 141 |
| | | RC3 | 279 | 0 | 279 | 87 | 26 | 113 | 125 | 0 | 125 | 125 |
| | | RC4 | 484 | 0 | 484 | 123 | 21 | 144 | 156 | 0 | 156 | 156 |
| | | RC5 | 370 | 0 | 370 | 96 | 7 | 104 | 174 | 0 | 174 | 174 |
| | Average | | 369 | 0 | 369 | 134 | 18 | 152 | 135 | 0 | 135 | 135 |
| | Minimum | | 279 | 0 | 279 | 87 | 5 | 104 | 59 | 0 | 59 | 63 |
| | Maximum | | 484 | 4 | 484 | 225 | 37 | 238 | 206 | 0 | 206 | 206 |
| Cambridge Cove | Sandstones | C1 | 422 | 3 | 425 | 120 | 14 | 134 | 186 | 0 | 186 | 189 |
| | | C2 | 376 | 2 | 378 | 105 | 12 | 117 | 167 | 0 | 167 | 169 |
| | | C3 | 421 | 18 | 439 | 62 | 3 | 65 | 209 | 0 | 209 | 227 |
| | | C4 | 361 | 7 | 368 | 133 | 7 | 140 | 138 | 0 | 138 | 145 |
| | | C5 | 386 | 0 | 386 | 114 | 13 | 127 | 112 | 0 | 112 | 112 |
| | | C6 | 334 | 10 | 344 | 99 | 8 | 107 | 156 | 0 | 156 | 166 |
| | | C7 | 407 | 1 | 408 | 143 | 33 | 176 | 186 | 0 | 186 | 187 |
| | | C8 | 350 | 1 | 351 | 63 | 8 | 71 | 173 | 0 | 173 | 174 |
| | | C9 | 373 | 1 | 374 | 105 | 17 | 122 | 165 | 0 | 165 | 166 |
| | | C10 | 322 | 0 | 322 | 108 | 12 | 120 | 99 | 0 | 99 | 99 |
| | | C11 | 635 | 1 | 636 | 112 | 4 | 116 | 75 | 0 | 75 | 76 |
| | Average | | 399 | 4 | 403 | 106 | 12 | 118 | 151 | 0 | 151 | 155 |
| | Minimum | | 322 | 0 | 322 | 62 | 3 | 65 | 75 | 0 | 75 | 76 |
| | Maximum | | 635 | 18 | 636 | 143 | 33 | 176 | 209 | 0 | 209 | 227 |
| Chinampas N-37 well | Sandstones | CH12002.0 | 786 | 9 | 795 | 177 | 43 | 219 | 32 | 5 | 37 | 46 |
| | | CH12003.1 | 670 | 0 | 670 | 319 | 0 | 319 | 11 | 0 | 11 | 11 |
| | | CH12004.1 | 733 | 1 | 734 | 145 | 19 | 165 | 2 | 0 | 2 | 3 |
| | | CH12006.0 | 753 | 0 | 753 | 193 | 32 | 226 | 44 | 1 | 45 | 45 |
| | | CH12006.6 | 675 | 0 | 675 | 156 | 0 | 156 | 2 | 0 | 2 | 2 |
| | | CH12007.9 | 663 | 6 | 669 | 240 | 53 | 293 | 80 | 55 | 135 | 141 |
| | | CH12008.8 | 753 | 5 | 758 | 200 | 29 | 229 | 73 | 0 | 73 | 77 |
| | | CH12010.4 | 696 | 0 | 696 | 279 | 9 | 287 | 52 | 0 | 52 | 52 |
| | | CH12011.2 | 735 | 7 | 742 | 129 | 69 | 199 | 60 | 0 | 60 | 67 |
| | Average | | 718 | 3 | 721 | 204 | 28 | 233 | 39 | 7 | 46 | 49 |
| | Minimum | | 663 | 0 | 669 | 129 | 0 | 155 | 2 | 0 | 2 | 2 |
| | Maximum | | 786 | 9 | 795 | 319 | 69 | 319 | 80 | 55 | 135 | 141 |
| Average of all | | | 486 | 3 | 489 | 147 | 20 | 165 | 112 | 3 | 115 | 118 |

*Note* For abbreviations, see footnote to Appendix Ia.

**Appendix II.** *Heavy mineral counts in the sandstones of Wolfville Formation*

| Site | No. | Opaque heavy minerals | | | | | Transparent heavy minerals | | | | | | | | | Total of all counts |
|---|---|---|---|---|---|---|---|---|---|---|---|---|---|---|---|---|
| | | Hematite & magnetite | Ilmenite | Goethite | Leucoxene | Total opaques | Garnet | Zircon | Tourmaline | Rutile | Apatite | Scheelite | Chlorite | Others* | Total transparents | |
| Rainy Cove | R1 | 297 | 24 | 512 | 208 | 1041 | 90 | 12 | 12 | 3 | 16 | 12 | 13 | 1 | 159 | 1200 |
| | R2 | nd | nd | nd | nd | 1057 | 68 | 7 | 6 | 1 | 16 | 10 | 35 | 0 | 143 | 1200 |
| | R3 | nd | nd | nd | nd | 1042 | 109 | 12 | 10 | 0 | 15 | 0 | 12 | 0 | 158 | 1200 |
| | R4 | nd | nd | nd | nd | 1079 | 89 | 7 | 2 | 1 | 12 | 0 | 8 | 2 | 121 | 1200 |
| | R5 | 151 | 6 | 297 | 118 | 572 | 157 | 17 | 11 | 6 | 22 | 404 | 11 | 0 | 628 | 1200 |
| | R6 | nd | nd | nd | nd | 541 | 151 | 3 | 13 | 0 | 16 | 456 | 20 | 0 | 659 | 1200 |
| | R7 | nd | nd | nd | nd | 1057 | 116 | 6 | 5 | 1 | 11 | 0 | 2 | 2 | 143 | 1200 |
| | R8 | 77 | 0 | 168 | 65 | 310 | 65 | 26 | 13 | 0 | 39 | 722 | 13 | 13 | 890 | 1200 |
| | R9 | 246 | 41 | 439 | 234 | 959 | 189 | 8 | 8 | 3 | 26 | 0 | 7 | 0 | 241 | 1200 |
| | RC-base | nd | nd | nd | nd | 1079 | 84 | 7 | 13 | 0 | 15 | 0 | 1 | 1 | 121 | 1200 |
| | RC1 | 222 | 178 | 403 | 111 | 913 | 223 | 33 | 14 | 1 | 16 | 0 | 0 | 0 | 287 | 1200 |
| | RC2 | nd | nd | nd | nd | 1003 | 161 | 8 | 6 | 0 | 16 | 0 | 5 | 1 | 197 | 1200 |
| | RC3 | nd | nd | nd | nd | 1022 | 136 | 9 | 6 | 0 | 21 | 0 | 6 | 0 | 178 | 1200 |
| | RC4 | 306 | 28 | 294 | 272 | 899 | 235 | 21 | 6 | 3 | 23 | 0 | 13 | 0 | 301 | 1200 |
| | Average | 217 | 46 | 352 | 168 | 898 | 134 | 13 | 9 | 1 | 19 | 115 | 10 | 1 | 302 | 1200 |
| | Minimum | 77 | 0 | 168 | 65 | 310 | 65 | 3 | 2 | 0 | 11 | 0 | 0 | 0 | 121 | 1200 |
| | Maximum | 306 | 178 | 512 | 272 | 1079 | 235 | 33 | 14 | 6 | 39 | 722 | 35 | 13 | 890 | 1200 |
| Cambridge Cove | C1 | nd | nd | nd | nd | 803 | 181 | 13 | 20 | 1 | 72 | 0 | 110 | 0 | 397 | 1200 |
| | C2 | nd | nd | nd | nd | 892 | 145 | 12 | 18 | 0 | 54 | 0 | 79 | 0 | 308 | 1200 |
| | C3 | nd | nd | nd | nd | 981 | 59 | 43 | 38 | 2 | 62 | 0 | 14 | 0 | 219 | 1200 |
| | C4 | nd | nd | nd | nd | 936 | 142 | 45 | 7 | 0 | 60 | 0 | 7 | 2 | 264 | 1200 |
| | C5 | nd | nd | nd | nd | 1016 | 141 | 2 | 4 | 0 | 9 | 0 | 27 | 0 | 184 | 1200 |
| | C6 | nd | nd | nd | nd | 1047 | 97 | 15 | 8 | 0 | 21 | 0 | 10 | 2 | 153 | 1200 |
| | C7 | nd | nd | nd | nd | 703 | 315 | 4 | 28 | 0 | 59 | 0 | 89 | 2 | 497 | 1200 |
| | C8 | nd | nd | nd | nd | 977 | 110 | 12 | 12 | 1 | 29 | 0 | 55 | 4 | 223 | 1200 |
| | C9 | nd | nd | nd | nd | 853 | 279 | 22 | 11 | 2 | 24 | 0 | 5 | 4 | 347 | 1200 |
| | C10 | nd | nd | nd | nd | 995 | 180 | 6 | 2 | 0 | 16 | 0 | 0 | 1 | 205 | 1200 |
| | C11 | nd | nd | nd | nd | 920 | 189 | 12 | 11 | 1 | 29 | 0 | 37 | 1 | 280 | 1200 |
| | Average | nd | nd | nd | nd | 920 | 167 | 17 | 14 | 1 | 40 | 0 | 39 | 1 | 280 | 1200 |
| | Minimum | nd | nd | nd | nd | 703 | 59 | 2 | 2 | 0 | 9 | 0 | 0 | 0 | 72 | 1200 |
| | Maximum | nd | nd | nd | nd | 1047 | 315 | 45 | 38 | 2 | 72 | 0 | 110 | 4 | 586 | 1200 |
| Chinampas N-37 well | CH12002.0 | nd | nd | nd | nd | 1066 | 0 | 34 | 52 | 6 | 29 | 0 | 13 | 0 | 134 | 1200 |
| | CH12003.1 | nd | nd | nd | nd | 1090 | 0 | 41 | 54 | 1 | 10 | 0 | 4 | 0 | 110 | 1200 |
| | CH12004.1 | nd | nd | nd | nd | 1059 | 0 | 17 | 64 | 6 | 34 | 0 | 19 | 1 | 141 | 1200 |
| | CH12006.0 | nd | nd | nd | nd | 1141 | 0 | 19 | 30 | 1 | 6 | 0 | 3 | 0 | 59 | 1200 |
| | CH12006.6 | nd | nd | nd | nd | 1160 | 0 | 13 | 13 | 1 | 6 | 0 | 7 | 0 | 40 | 1200 |
| | CH12007.9 | nd | nd | nd | nd | 1152 | 0 | 36 | 6 | 1 | 3 | 0 | 2 | 0 | 48 | 1200 |
| | CH12008.8 | 716 | 52 | 155 | 223 | 1147 | 0 | 23 | 21 | 2 | 3 | 0 | 3 | 1 | 53 | 1200 |
| | CH12010.4 | 734 | 83 | 111 | 172 | 1100 | 0 | 54 | 32 | 3 | 8 | 0 | 3 | 0 | 100 | 1200 |
| | CH12011.2 | nd | nd | nd | nd | 1152 | 0 | 12 | 19 | 1 | 8 | 0 | 4 | 4 | 48 | 1200 |
| | Average | 726 | 67 | 133 | 198 | 1119 | 0 | 28 | 32 | 2 | 12 | 0 | 6 | 1 | 81 | 1200 |
| | Minimum | 716 | 52 | 111 | 173 | 1059 | 0 | 12 | 6 | 1 | 3 | 0 | 2 | 0 | 40 | 1200 |
| | Maximum | 735 | 83 | 155 | 223 | 1160 | 0 | 54 | 64 | 6 | 34 | 0 | 19 | 4 | 141 | 1200 |

*Others include hornblende, pyroxene, staurolite, epidote, zoisite. Micas (biotite and muscovite) are excluded from counting.
R, C and CH are sandstones and RC are conglomerates; nd, not determined.

**Appendix III.** *Chemical composition (wt%) and end members (%, for garnets of the Wolfville Formation*

| No. | SiO$_2$ | TiO$_2$ | Al$_2$O$_3$ | Cr$_2$O$_3$ | FeO | MnO | MgO | CaO | Na$_2$O | K$_2$O | F | Total % | Spesartine | Almandine | Grossular | Pyrope |
|---|---|---|---|---|---|---|---|---|---|---|---|---|---|---|---|---|
| C2-G1 | 35.68 | 0.15 | 22.01 | 0.11 | 9.22 | 30.50 | 0.30 | 2.85 | 0.02 | 0.03 | 0.05 | 100.91 | 69.74 | 20.81 | 8.25 | 1.20 |
| C2-G2 | 37.67 | 0.10 | 22.27 | 0.00 | 9.27 | 29.40 | 0.28 | 3.79 | 0.00 | 0.02 | 0.00 | 102.79 | 67.07 | 20.88 | 10.94 | 1.11 |
| C2-G3 | 36.67 | 0.10 | 22.16 | 0.01 | 9.72 | 28.97 | 0.20 | 3.64 | 0.01 | 0.01 | 0.00 | 101.49 | 66.54 | 22.05 | 10.59 | 0.83 |
| C2-G4 | 37.12 | 0.11 | 22.07 | 0.01 | 13.28 | 25.68 | 0.30 | 3.94 | 0.00 | 0.02 | 0.00 | 102.52 | 57.99 | 29.59 | 11.25 | 1.18 |
| C2-G5 | 36.60 | 0.21 | 22.17 | 0.00 | 19.85 | 18.57 | 0.52 | 4.01 | 0.00 | 0.02 | 0.00 | 101.96 | 42.05 | 44.39 | 11.48 | 2.09 |
| C2-G6 | 37.39 | 0.21 | 22.31 | 0.00 | 9.21 | 29.77 | 0.28 | 3.69 | 0.00 | 0.02 | 0.00 | 102.88 | 67.60 | 20.65 | 10.61 | 1.14 |
| C2-G7 | 37.33 | 0.05 | 22.22 | 0.00 | 16.84 | 22.57 | 0.44 | 2.90 | 0.02 | 0.03 | 0.00 | 102.39 | 51.74 | 38.11 | 8.40 | 1.76 |
| C2-G8 | 37.42 | 0.16 | 22.27 | 0.00 | 7.81 | 31.12 | 0.42 | 3.48 | 0.00 | 0.02 | 0.00 | 102.69 | 70.77 | 17.53 | 10.01 | 1.69 |
| C2-G9 | 37.39 | 0.13 | 22.16 | 0.00 | 13.86 | 26.75 | 0.26 | 2.43 | 0.00 | 0.01 | 0.00 | 102.99 | 60.84 | 31.12 | 6.98 | 1.06 |
| C2-G10 | 37.61 | 0.07 | 22.49 | 0.00 | 11.08 | 26.92 | 0.20 | 4.47 | 0.01 | 0.01 | 0.00 | 102.86 | 61.38 | 24.94 | 12.89 | 0.79 |
| C2-G11 | 36.18 | 0.09 | 22.36 | 0.02 | 11.37 | 27.61 | 0.37 | 3.01 | 0.04 | 0.02 | 0.00 | 101.07 | 63.79 | 25.94 | 8.78 | 1.49 |
| C2-G12 | 36.35 | 0.11 | 22.33 | 0.01 | 10.09 | 28.73 | 0.36 | 3.29 | 0.02 | 0.01 | 0.00 | 101.32 | 66.06 | 22.92 | 9.58 | 1.45 |
| C5-G1 | 37.25 | 0.04 | 22.40 | 0.03 | 24.18 | 15.56 | 0.77 | 2.74 | 0.01 | 0.02 | 0.00 | 103.00 | 35.17 | 53.94 | 7.84 | 3.05 |
| C5-G2 | 37.01 | 0.08 | 22.23 | 0.03 | 19.74 | 20.21 | 0.54 | 2.69 | 0.00 | 0.01 | 0.00 | 102.54 | 45.89 | 44.25 | 7.72 | 2.14 |
| C5-P3 | 37.12 | 0.03 | 22.41 | 0.00 | 31.21 | 9.87 | 1.21 | 1.16 | 0.03 | 0.01 | 0.00 | 103.06 | 22.29 | 69.59 | 3.32 | 4.81 |
| C5-G4 | 36.75 | 0.05 | 22.07 | 0.00 | 16.29 | 22.94 | 0.37 | 3.06 | 0.02 | 0.02 | 0.00 | 101.59 | 52.69 | 36.93 | 8.89 | 1.50 |
| C5-G5 | 37.34 | 0.23 | 22.28 | 0.01 | 8.16 | 30.24 | 0.42 | 3.84 | 0.00 | 0.02 | 0.00 | 102.53 | 68.88 | 18.36 | 11.07 | 1.69 |
| C5-G6 | 36.84 | 0.18 | 21.62 | 0.00 | 12.80 | 23.11 | 0.26 | 5.03 | 0.01 | 0.02 | 0.00 | 99.87 | 54.28 | 29.68 | 14.95 | 1.09 |
| C5-G7 | 37.71 | 0.07 | 22.27 | 0.00 | 13.38 | 24.62 | 0.31 | 4.55 | 0.00 | 0.03 | 0.00 | 102.95 | 55.79 | 29.94 | 13.05 | 1.23 |
| C5-P8 | 37.58 | 0.04 | 22.24 | 0.00 | 31.78 | 9.20 | 1.49 | 0.89 | 0.00 | 0.04 | 0.03 | 103.27 | 20.75 | 70.79 | 2.54 | 5.93 |
| C5-P9 | 37.53 | 0.00 | 22.34 | 0.00 | 30.29 | 10.15 | 1.40 | 1.12 | 0.02 | 0.03 | 0.00 | 102.87 | 23.10 | 68.08 | 3.22 | 5.61 |
| C5-G10 | 37.54 | 0.22 | 22.33 | 0.00 | 10.49 | 28.09 | 0.38 | 4.16 | 0.00 | 0.03 | 0.03 | 103.26 | 63.30 | 23.35 | 11.86 | 1.50 |
| C5-G11 | 37.22 | 0.12 | 22.12 | 0.00 | 11.02 | 28.02 | 0.23 | 3.80 | 0.00 | 0.03 | 0.01 | 102.57 | 63.54 | 24.66 | 10.89 | 0.91 |
| C5-G12 | 37.26 | 0.32 | 22.26 | 0.00 | 18.67 | 20.32 | 0.50 | 2.98 | 0.02 | 0.03 | 0.00 | 102.37 | 46.83 | 42.46 | 8.67 | 2.04 |
| C5-G13 | 37.51 | 0.21 | 22.27 | 0.00 | 14.50 | 23.58 | 0.33 | 4.39 | 0.01 | 0.03 | 0.00 | 102.84 | 53.54 | 32.52 | 12.62 | 1.33 |
| C5-G14 | 37.59 | 0.08 | 22.22 | 0.00 | 20.40 | 18.72 | 0.52 | 3.62 | 0.00 | 0.04 | 0.01 | 103.19 | 42.20 | 45.40 | 10.32 | 2.08 |
| C5-P1 | 37.90 | 0.01 | 22.60 | 0.01 | 34.18 | 3.63 | 3.53 | 1.19 | 0.00 | 0.00 | 0.00 | 103.05 | 8.05 | 74.84 | 3.34 | 13.76 |
| C5-P2 | 37.63 | 0.00 | 22.70 | 0.00 | 34.08 | 3.63 | 3.55 | 1.23 | 0.00 | 0.02 | 0.00 | 102.84 | 8.05 | 74.63 | 3.46 | 13.87 |
| C8-G1 | 37.48 | 0.10 | 22.42 | 0.14 | 20.80 | 18.54 | 0.64 | 2.26 | 0.01 | 0.04 | 0.04 | 102.44 | 43.05 | 47.70 | 6.64 | 2.61 |
| C8-G2 | 37.38 | 0.04 | 22.46 | 0.09 | 20.94 | 19.46 | 0.64 | 2.10 | 0.00 | 0.03 | 0.06 | 103.18 | 44.31 | 47.08 | 6.06 | 2.55 |
| C8-G3 | 35.48 | 0.18 | 21.79 | 0.10 | 14.90 | 24.69 | 0.24 | 2.20 | 0.00 | 0.02 | 0.08 | 100.64 | 57.95 | 34.53 | 6.53 | 0.99 |
| C8-G4 | 37.51 | 0.13 | 22.34 | 0.11 | 15.02 | 22.73 | 0.31 | 5.05 | 0.00 | 0.02 | 0.04 | 103.23 | 51.09 | 33.33 | 14.35 | 1.22 |
| C8-G5 | 37.41 | 0.15 | 22.29 | 0.15 | 11.55 | 28.39 | 0.22 | 3.00 | 0.00 | 0.03 | 0.06 | 103.23 | 64.55 | 25.93 | 8.62 | 0.90 |
| C8-G6 | 37.00 | 0.18 | 22.10 | 0.13 | 13.70 | 26.16 | 0.34 | 2.22 | 0.00 | 0.05 | 0.02 | 101.88 | 60.69 | 31.38 | 6.52 | 1.41 |
| C8-G7 | 43.17 | 0.19 | 18.63 | 0.09 | 13.02 | 22.00 | 0.28 | 2.00 | 0.00 | 0.02 | 0.02 | 99.42 | 58.09 | 33.95 | 6.67 | 1.29 |
| C8-G8 | 37.35 | 0.17 | 22.23 | 0.09 | 10.69 | 27.16 | 0.20 | 4.61 | 0.01 | 0.04 | 0.04 | 102.56 | 61.87 | 24.04 | 13.30 | 0.80 |
| C8-G9 | 37.31 | 0.11 | 22.11 | 0.00 | 13.60 | 26.41 | 0.31 | 2.47 | 0.00 | 0.03 | 0.00 | 102.34 | 60.72 | 30.86 | 7.19 | 1.23 |
| C8-G10 | 37.25 | 0.05 | 22.36 | 0.00 | 9.23 | 28.88 | 0.52 | 3.40 | 0.01 | 0.02 | 0.00 | 101.72 | 66.84 | 21.09 | 9.96 | 2.11 |
| C8-G11 | 37.25 | 0.10 | 22.32 | 0.00 | 16.42 | 20.54 | 0.43 | 5.13 | 0.00 | 0.01 | 0.00 | 102.19 | 46.68 | 36.83 | 14.75 | 1.74 |
| R5-G1 | 38.93 | 0.13 | 19.54 | 0.00 | 11.53 | 25.20 | 0.28 | 3.30 | 0.00 | 0.01 | 0.01 | 98.92 | 61.10 | 27.61 | 10.12 | 1.17 |
| R5-G2 | 36.35 | 0.10 | 19.79 | 0.00 | 13.64 | 25.44 | 0.31 | 2.50 | 0.00 | 0.02 | 0.04 | 98.17 | 59.71 | 31.61 | 7.42 | 1.26 |

*(Continued)*

**Appendix III.** *Continued*

| No. | $SiO_2$ | $TiO_2$ | $Al_2O_3$ | $Cr_2O_3$ | FeO | MnO | MgO | CaO | $Na_2O$ | $K_2O$ | F | Total % | Spesartine | Almandine | Grossular | Pyrope |
|---|---|---|---|---|---|---|---|---|---|---|---|---|---|---|---|---|
| R5-G3 | 35.41 | 0.04 | 19.49 | 0.00 | 16.92 | 20.86 | 0.48 | 3.09 | 0.00 | 0.01 | 0.01 | 96.30 | 49.29 | 39.47 | 9.22 | 2.01 |
| R5-G4 | 37.11 | 0.03 | 20.46 | 0.00 | 12.32 | 25.62 | 0.44 | 3.33 | 0.00 | 0.01 | 0.00 | 99.32 | 59.91 | 28.43 | 9.86 | 1.79 |
| R5-G5 | 37.36 | 0.14 | 20.74 | 0.00 | 9.41 | 28.55 | 0.32 | 3.87 | 0.00 | 0.01 | 0.01 | 100.42 | 65.94 | 21.47 | 11.31 | 1.29 |
| R5-G6 | 37.32 | 0.04 | 20.64 | 0.00 | 12.78 | 24.87 | 0.36 | 3.82 | 0.01 | 0.01 | 0.01 | 99.85 | 57.89 | 29.39 | 11.25 | 1.47 |
| R5-G7 | 36.67 | 0.03 | 20.38 | 0.00 | 13.28 | 24.28 | 0.49 | 2.73 | 0.02 | 0.01 | 0.01 | 97.89 | 58.23 | 31.44 | 8.28 | 2.05 |
| R5-G8 | 36.54 | 0.05 | 20.25 | 0.00 | 11.11 | 26.39 | 0.23 | 4.00 | 0.02 | 0.00 | 0.00 | 98.59 | 61.64 | 25.61 | 11.81 | 0.94 |
| R5-G9 | 37.22 | 0.02 | 20.89 | 0.00 | 10.76 | 26.40 | 0.43 | 3.36 | 0.04 | 0.01 | 0.00 | 99.13 | 62.82 | 25.28 | 10.11 | 1.79 |
| R5-G10 | 37.53 | 0.02 | 20.81 | 0.00 | 16.19 | 20.34 | 0.34 | 4.83 | 0.02 | 0.00 | 0.00 | 100.07 | 47.27 | 37.15 | 14.20 | 1.39 |
| R5-P | 37.75 | 0.00 | 21.08 | 0.00 | 29.16 | 8.29 | 2.25 | 1.64 | 0.00 | 0.01 | 0.03 | 100.21 | 19.23 | 66.77 | 4.82 | 9.17 |
| R8-G1 | 37.22 | 0.00 | 20.83 | 0.00 | 18.37 | 19.45 | 0.56 | 2.40 | 0.06 | 0.00 | 0.00 | 98.88 | 46.75 | 43.59 | 7.30 | 2.36 |
| R8-G2 | 37.05 | 0.05 | 20.83 | 0.00 | 11.87 | 24.76 | 0.23 | 4.44 | 0.02 | 0.01 | 0.00 | 99.25 | 58.25 | 27.57 | 13.22 | 0.96 |
| R8-G3 | 37.03 | 0.14 | 20.43 | 0.00 | 9.16 | 28.38 | 0.26 | 3.62 | 0.03 | 0.00 | 0.00 | 99.04 | 66.85 | 21.30 | 10.79 | 1.06 |
| R8-G4 | 37.43 | 0.01 | 20.91 | 0.00 | 12.82 | 23.59 | 0.29 | 4.64 | 0.01 | 0.00 | 0.00 | 99.69 | 55.34 | 29.68 | 13.78 | 1.20 |
| R8-G5 | 37.51 | 0.00 | 20.99 | 0.00 | 15.43 | 21.30 | 0.40 | 4.27 | 0.01 | 0.00 | 0.00 | 99.90 | 49.96 | 35.75 | 12.66 | 1.63 |
| R8-G6 | 37.11 | 0.69 | 20.53 | 0.00 | 15.72 | 20.82 | 0.35 | 3.93 | 0.01 | 0.01 | 0.00 | 99.17 | 49.65 | 37.01 | 11.87 | 1.47 |
| Average G | 37.26 | 0.12 | 21.59 | 0.02 | 13.65 | 24.66 | 0.37 | 3.50 | 0.01 | 0.02 | 0.01 | 101.20 | 57.08 | 31.19 | 10.23 | 1.50 |
| Average R | 37.59 | 0.01 | 22.23 | 0.00 | 31.78 | 7.46 | 2.24 | 1.21 | 0.01 | 0.02 | 0.01 | 102.55 | 16.91 | 70.78 | 3.45 | 8.86 |
| Average all | 37.29 | 0.11 | 21.66 | 0.02 | 15.59 | 22.82 | 0.57 | 3.25 | 0.01 | 0.02 | 0.01 | 101.34 | 52.78 | 35.43 | 9.50 | 2.29 |

C, Cambridge Cove; R, Rainy Cove; G, grey garnets (almandine-rich spessartine); P, pink garnets (spessartine-rich almandine).

**Appendix IV.** *Chemical composition (wt%) for tourmalines of the Wolfville Formation*

| Sample no. | $B_2O_3$ | $SiO_2$ | $TiO_2$ | $Al_2O_3$ | $FeO_t$ | MnO | MgO | CaO | $Na_2O$ | $K_2O$ | Total % |
|---|---|---|---|---|---|---|---|---|---|---|---|
| C2-1 | 10.96 | 37.13 | 0.38 | 35.89 | 6.86 | 0.78 | 5.40 | 0.15 | 1.63 | 0.00 | 99.18 |
| C2-2 | 10.71 | 35.80 | 0.39 | 35.13 | 6.40 | 0.89 | 5.80 | 0.32 | 1.66 | 0.00 | 97.10 |
| C2-3 | 10.73 | 35.41 | 0.66 | 35.17 | 5.86 | 0.03 | 6.45 | 0.74 | 1.98 | 0.02 | 97.05 |
| C2-4 | 10.65 | 35.89 | 0.43 | 34.26 | 8.16 | 0.00 | 5.40 | 0.16 | 2.06 | 0.02 | 97.03 |
| C2-5 | 10.81 | 36.09 | 0.45 | 34.51 | 7.61 | 0.00 | 6.33 | 0.99 | 1.67 | 0.00 | 98.46 |
| C2-6 | 10.63 | 36.53 | 1.02 | 30.65 | 10.39 | 0.00 | 6.49 | 0.25 | 2.48 | 0.00 | 98.44 |
| C5-1 | 10.82 | 36.25 | 0.52 | 35.76 | 7.51 | 0.00 | 5.13 | 0.41 | 1.71 | 0.04 | 98.15 |
| C5-2 | 10.81 | 36.23 | 0.54 | 35.71 | 7.43 | 0.00 | 5.18 | 0.40 | 1.69 | 0.03 | 98.02 |
| C5-3 | 10.89 | 36.28 | 0.56 | 35.98 | 7.63 | 0.00 | 5.33 | 0.42 | 1.76 | 0.04 | 98.89 |
| C8-1 | 10.75 | 35.99 | 0.23 | 33.28 | 6.14 | 0.01 | 8.13 | 1.05 | 1.99 | 0.02 | 97.59 |
| C8-2 | 10.75 | 35.78 | 0.20 | 33.51 | 6.26 | 0.03 | 8.12 | 1.06 | 1.99 | 0.02 | 97.72 |
| R5 | 9.83 | 33.83 | 0.00 | 32.04 | 11.76 | 0.13 | 1.61 | 0.04 | 1.93 | 0.01 | 91.18 |
| R8 | 9.98 | 34.42 | 0.23 | 32.50 | 11.03 | 0.03 | 2.07 | 0.01 | 1.68 | 0.01 | 91.96 |
| CH12010.4-1 | 10.38 | 35.72 | 0.92 | 32.39 | 10.83 | 0.06 | 3.25 | 0.36 | 1.98 | 0.03 | 95.91 |
| CH12010.4-2 | 10.34 | 36.31 | 0.22 | 28.81 | 10.70 | 0.00 | 6.48 | 1.23 | 2.02 | 0.03 | 96.13 |
| CH12010.4-3 | 10.51 | 36.71 | 1.01 | 30.13 | 6.44 | 0.00 | 7.76 | 0.45 | 2.59 | 0.03 | 95.63 |
| CH12010.4-4 | 10.28 | 35.90 | 0.59 | 28.38 | 9.67 | 0.00 | 6.85 | 1.93 | 1.64 | 0.03 | 95.27 |
| CH12010.4-5 | 10.48 | 36.25 | 0.33 | 32.09 | 8.02 | 0.10 | 5.94 | 0.31 | 2.13 | 0.03 | 95.68 |
| CH12010.4-6 | 10.43 | 36.49 | 0.72 | 28.83 | 7.46 | 0.00 | 8.38 | 1.00 | 2.23 | 0.03 | 95.56 |
| CH12010.4-7 | 10.59 | 36.03 | 0.77 | 33.99 | 7.33 | 0.01 | 4.99 | 0.73 | 1.64 | 0.04 | 96.12 |
| CH12010.4-8 | 10.59 | 35.98 | 0.74 | 34.10 | 7.25 | 0.01 | 4.94 | 0.78 | 1.69 | 0.05 | 96.13 |
| CH12010.4-9 | 10.48 | 36.33 | 0.34 | 31.28 | 7.71 | 0.00 | 6.70 | 0.79 | 2.06 | 0.03 | 95.72 |
| CH12010.4-10 | 10.32 | 35.78 | 0.40 | 28.81 | 9.24 | 0.04 | 7.25 | 1.92 | 1.75 | 0.03 | 95.54 |
| CH12010.4-11 | 10.53 | 36.92 | 3.38 | 26.97 | 0.74 | 0.00 | 11.74 | 1.53 | 2.18 | 0.02 | 94.02 |
| CH12010.4-12 | 10.50 | 36.18 | 0.60 | 32.21 | 8.31 | 0.02 | 5.54 | 0.49 | 2.05 | 0.04 | 95.93 |
| CH12010.4-13 | 10.37 | 35.53 | 0.48 | 33.71 | 10.39 | 0.05 | 2.64 | 0.18 | 1.93 | 0.04 | 95.31 |
| CH12010.4-14 | 10.53 | 36.76 | 0.37 | 30.31 | 6.49 | 0.00 | 8.20 | 0.91 | 2.26 | 0.01 | 95.85 |
| CH12003.08-1 | 10.54 | 35.75 | 0.93 | 33.58 | 8.38 | 0.07 | 4.49 | 0.62 | 1.85 | 0.04 | 96.25 |
| CH12003.08-2 | 10.62 | 36.42 | 0.74 | 33.80 | 6.20 | 0.00 | 5.68 | 0.68 | 1.63 | 0.04 | 95.80 |
| CH12003.08 3 | 10.76 | 36.65 | 0.93 | 34.28 | 5.47 | 0.02 | 6.25 | 0.58 | 1.80 | 0.05 | 96.79 |
| CH12003.08-4 | 10.64 | 36.47 | 0.74 | 31.37 | 6.85 | 0.00 | 7.64 | 1.18 | 2.22 | 0.04 | 97.14 |
| CH12003.08-5 | 10.04 | 35.09 | 0.57 | 27.58 | 15.66 | 0.49 | 3.34 | 0.66 | 2.51 | 0.09 | 96.02 |
| CH12003.08-6 | 10.71 | 36.55 | 0.42 | 33.32 | 5.92 | 0.04 | 7.06 | 0.94 | 1.83 | 0.05 | 96.85 |
| CH12003.08-7 | 10.60 | 35.93 | 1.00 | 33.80 | 7.88 | 0.06 | 4.85 | 0.46 | 1.94 | 0.06 | 96.58 |
| CH12003.08-8 | 10.33 | 35.60 | 0.15 | 32.67 | 13.17 | 0.34 | 1.90 | 0.15 | 2.18 | 0.07 | 96.56 |
| CH12003.08-9 | 10.52 | 35.80 | 1.03 | 33.69 | 9.92 | 0.13 | 3.47 | 0.39 | 1.62 | 0.05 | 96.62 |
| CH12003.08-10 | 10.38 | 35.68 | 0.21 | 34.57 | 13.21 | 0.11 | 0.50 | 0.08 | 1.52 | 0.05 | 96.31 |
| CH12003.08-11 | 10.51 | 35.27 | 0.55 | 32.71 | 8.20 | 0.05 | 6.05 | 1.26 | 1.67 | 0.08 | 96.35 |
| CH12003.08-12 | 10.36 | 35.57 | 1.28 | 27.10 | 9.48 | 0.05 | 8.39 | 3.14 | 1.16 | 0.08 | 96.61 |
| CH12003.08-13 | 10.65 | 37.02 | 0.41 | 31.86 | 5.52 | 0.00 | 7.85 | 0.34 | 2.48 | 0.03 | 96.16 |
| CH12003.08-14 | 10.82 | 37.22 | 0.21 | 35.96 | 6.31 | 0.04 | 4.72 | 0.16 | 1.66 | 0.03 | 97.13 |
| Average of all | 10.54 | 36.04 | 0.63 | 32.50 | 8.19 | 0.09 | 5.71 | 0.71 | 1.91 | 0.03 | 96.36 |

C, Cambridge Cove; R, Rainy Cove; CH, Chinampas Well.

# References

ALLEN, B. D. & CLARKE, D. B. 1981. Occurrence and origin of garnets in the South Mountain Batholith, Nova Scotia. *Canadian Mineralogist*, **19**, 19–24.

ANONYMOUS 1983. *Assessment Report on Geochemical, Geological, Diamond Drilling and Trenching NTS 21A-4B and 20P-13C, Pearl Lake and Peter Lake Claim Groups: Nova Scotia.* Esso Minerals Canada.

ANONYMOUS 2008. Triangle Petroleum announces analytical results for its two exploratory wells in the Windsor Basin of Nova Scotia, Canada. *The Oil & Gas Magazine Online*, June 24, http://www.wallstreet-online. de/diskussion/1137920-neustebeitraege/news-news-triangle-pct

ANTHONY, J. W., BIDEAUX, R. A., BLADH, K. W. & NICHOLS, M. C. (eds) 2005. *Handbook of Mineralogy.* Mineralogical Society of America, Chantilly, VA.

BARON, M. & PARNELL, J. 2007. Relationships between stylolites and cementation in sandstone reservoirs: Examples from the North Sea, U.K. and East Greenland. *Sedimentary Geology*, **194**, 17–35.

BARR, S. M., RAESIDE, R. P. & WHITE, C. E. 1998. Geological correlations between Cape Breton Island and Newfoundland, northern Appalachian orogeny. *Canadian Journal of Earth Sciences*, **35**, 1252–1270.

BATEMAN, R. M. & CATT, J. A. 1985. Modification of heavy mineral assemblages in English coversand by acid pedochemical weathering. *Catena*, **12**, 1–21.

BELL, W. A. 1958. *Possibility for Occurrence of Petroleum in Nova Scotia.* Department of Mines, Province of Nova Scotia.

BJØRKUM, P. A. 1996. How important is pressure in causing dissolution of quartz in sandstones? *Journal of Sedimentary Research*, **66**, 147–154.

CALDER, J. H., BOEHNER, R. C., BROWN, D. E., GIBLING, M. R., MUKHOPPADHYAY, P. K., RYAN, R. J. & SKILLITER, D. M. 1998. *Classic Carboniferous Sections of the Minas and Cumberland Basins in Nova Scotia with Special Reference to Organic Deposits*. Society for Organic Petrology Annual Meeting Field Trip, 29–30 July, 1998. Nova Scotia Natural Resources Open File Report ME **1998-5**.

CAMERON, B. I. & ZENTILLI, M. 1997. Geochemical characterization of the mineralized transition between the Goldenville and Halifax formations and the interaction with adjacent granitoid intrusions of the Liscomb Complex, Nova Scotia. *Atlantic Geology*, **33**, 143–155.

CHAREST, M. H., FARLEY, E. J. & CLARKE, D. B. 1985. *The Northeastern Part of the New Ross–Vaughan Complex*. Nova Scotia Department of Mines & Energy Open File Report **85-3**.

CHATTERJEE, A. K. 1977. *Massive Sulfide Mineralization near Blue Mountain, Cape Breton County, Southeast Cape Breton*. Nova Scotia Department Mines & Energy Report **77–1**.

CLARKE, D. B. & BOGUTYN, P. A. 2003. Oscillatory epitactic-growth zoning in biotite and muscovite from the Lake Lewis leucogranite, South Mountain Batholith, Nova Scotia, Canada. *Canadian Mineralogist*, **41**, 1027–1047.

CLARKE, D. B. & CARRUZZO, S. 2007. Assimilation of country-rock ilmenite and rutile in the South Mountain Batholith, Nova Scotia, Canada. *Canadian Mineralogist*, **45**, 31–42.

CLARKE, D. B., REARDON, N. C., CHATTERJEE, A. K. & GREGOIRE, D. C. 1989. Tourmaline composition as a guide to mineral exploration: a reconnaissance study from Nova Scotia using discriminant function analysis. *Economic Geology*, **84**, 1921–1935.

DEWER, T. & HAJASH, A. 1995. Rate laws for water-assisted compaction and stress-induced water–rock interaction in sandstones. *Journal of Geophysical Research*, **100**, 13093–13112.

DICKINSON, W. R. 1985. Interpreting provenance relations from detrital modes of sandstones. *In*: ZUFFA, G. G. (ed.) *Provenance of Arenites*. Reidel Publishing Company, Dordrecht, 331–361.

DICKINSON, W. R. & SUCZEK, C. A. 1979. Plate tectonics and sandstone compositions. *American Association of Petroleum Geologists Bulletin*, **63**, 2164–2182.

DICKINSON, W. R., INGERSOLL, R. V., COWAN, D. S., HELMOLD, K. P. & SUCZEK, C. A. 1982. Provenance of Franciscan graywackes in coastal California. *Geological Society of America Bulletin*, **93**, 95–107.

DICKINSON, W. R., BEARD, L. S. *ET AL.* 1983. Provenance of North American Phanerozoic sandstones in relation to tectonic setting. *Geological Society of America Bulletin*, **94**, 222–235.

DOSTAL, J., KONTAK, D. J. & CHATTERJEE, A. K. 2009. Trace element geochemistry of scheelite and rutile from metaturbidite-hosted quartz vein gold deposits, Meguma Terrane, Nova Scotia, Canada: genetic implications. *Mineralogy and Petrology*, **97**, 95–109.

DOUMA, S. L. 1988. *The mineralogy, petrology and geochemistry of the Port Mouton Pluton, Nova Scotia, Canada*. MSc thesis, Dalhousie University.

ERDMANN, S. 2006. *Country-rock contamination and assimilation in the South Mountain Batholith*. PhD thesis, Dalhousie University.

ERDMANN, S., JAMIESON, R. A. & MACDONALD, M. A. 2009. Evaluating the origin of garnet, cordierite, and biotite in granitic rocks: a case study from the South Mountain Batholith, Nova Scotia. *Journal of Petrology*, **50**, 1477–1503, http://dx.doi.org/10.1093/petrology/egp038

FENSOME, R. A. & WILLIAMS, G. L. (eds) 2001. *The Last Billion Years: A Geological History of the Maritime Provinces of Canada*. Atlantic Geoscience Society, Nimbas Publishing, Halifax.

FOLK, R. L. 1968. *Petrology of Sedimentary Rocks*. Hemphill Publishing Company, Austin, TX.

GALBRAITH, C. G., CLARKE, D. B., TRUMBULL, R. B. & WIEDENBECK, M. 2009. Assessment of tourmaline compositions as an indicator of emerald mineralization at the Tsa da Glisza prospect, Yukon Territory, Canada. *Economic Geology*, **104**, 713–731.

GILES, P. S. & CHATTERJEE, A. K. 1986. Peraluminous granites of the Liscomb Complex. *Nova Scotia Department of Mines & Energy, Information Series*, **12**, 83–89.

GILES, M. R., INDRELID, S. L., BEYNON, G. V. & AMTHOR, J. 2000. The origin of large-scale quartz cementation: evidence from large data sets and coupled heat-fluid mass transport modelling. *In*: WORDEN, R. H. & MORAD, S. (eds) *Quartz Cementation in Sandstones*. International Association of Sedimentologists, Special Publications, **29**, 21–38.

GREENSMITH, J. T. 1978. *Petrology of the Sedimentary Rocks*, 6th edn. George Allen & Unwin, London.

HALE, S. C. 2007. *Garnet of the South Mountain Batholith, Nova Scotia*. BA thesis, Department of Geology, Smith College, Northampton, MA.

HALLSWORTH, C. R. & CHISHOLM, J. I. 2008. Provenance of late Carboniferous sandstones in the Pennine Basin (UK) from combined heavy mineral, garnet geochemistry and palaeocurrent studies. *Sedimentary Geology*, **203**, 196–212.

HAMBLIN, A. P. 2004. *Regional Geology and Hydrogeological Potential of the Triassic/Jurassic Fundy Group, Annapolis–Cornwallis Valley, Nova Scotia*. Geological Survey of Canada Open File **4678**.

HAYNES, S. J. 1983. *Typomorphism of Turbidite-Hosted Auriferous Quartz Veins, Southern Guysborough County*. Nova Scotia Department of Mines & Energy Report **83-1**.

HENRY, D. J. & DUTROW, B. 1992. Tourmaline in a low grade clastic metasedimentary rock; an example of the petrogenetic potential of tourmaline. *Contributions to Mineralogy and Petrology*, **112**, 203–218.

HENRY, D. J. & GUIDOTTI, C. V. 1985. Tourmaline as petrogenic indicator mineral: an example from the staurolite-grade metapelites of Northern Maine. *American Mineralogist*, **70**, 1–15.

HENRY, D. J., NOVÁK, M. (CHAIR), *ET AL.* 2011. Nomenclature of the tourmaline-supergroup minerals. *American Mineralogist*, **96**, 895–913.

HICKS, R. J. 1996. *Low-grade metamorphism in the Meguma Group, Southern Nova Scotia*. MSc thesis, Dalhousie University.

HICKS, R. J., JAMIESON, R. A. & REYNOLDS, P. H. 1999. Detrital and metamorphic $^{40}Ar/^{39}Ar$ ages from muscovite and whole-rock samples, Meguma Super-group, southern Nova Scotia. *Canadian Journal of Earth Sciences*, **36**, 23–32.

HOFMANN, A., TOURANI, A. & GAUPP, R. 2000. Cyclicity of Triassic to Lower Jurassic continental red beds of the Argana Valley, Morocco: implications for palaeoclimate and basin evolution. *Palaeogeography, Palaeoclimatology, Palaeoecology*, **161**, 229–266.

HUBERT, J. F. & FORLENZA, M. F. 1988. Sedimentology of braided-river deposits in Upper Triassic Wolfville redbeds, southern shore of Cobequid Bay, Nova Scotia, and Canada. *In*: MANSPEIZER, W. (ed.) *Triassic–Jurassic Rifting: Continental Breakup and the Origin of the Atlantic Ocean and Passive Margins*. Elsevier, Amsterdam, 231–247.

HUTCHEON, I. 1990. Aspects of the diagenesis of coarse-grained siliciclastic rocks. *In*: MCLLREATH, I. A. & MORROW, D. W. (eds) *Diagenesis*. Geoscience Canada, Reprint Series, **4**, 165–175.

HUTCHINSON, H. E. 1982. *Geology, geochemistry and genesis of the Brazil Lake Pegmatites, Yarmouth County, Nova Scotia*. BSc thesis, Dalhousie University.

INGERSOLL, R. V. & SUCZEK, C. A. 1979. Petrology and provenance of Neogene sand from Nicobar and Bengal fans, DSDP sites 211 and 218. *Journal of Sedimentary Petrology*, **49**, 1217–1228.

KEPPIE, J. D. 1979. *Geological Map of the Province of Nova Scotia*. Nova Scotia Department of Mines, Scale 1:500 000.

KEPPIE, J. D. 2000. *Geological Map of the Province of Nova Scotia*. Map ME 2000-1, 1:500 000. NS Department of Natural Resources, Minerals and Energy Branch.

KETTANAH, Y., KETTANAH, M. & WACH, G. 2008. Reservoir quality, diagenetic history and provenance of the Late Triassic sandstones of the Wolfville Formation, Bay of Fundy, Nova Scotia, Canada. *In*: *Atlantic Conjugate Margin Conference*, Halifax, 2008, Program & Abstracts, 48.

KLEIN, G. DE V. 1962. Triassic sedimentation, Maritime Provinces, Canada. *Geological Society of America Bulletin*, **73**, 1127–1146.

KONTAK, D. J. & SMITH, P. K. 1993a. A metaturbidite-hosted lode gold deposit: Beaver Dam deposit. Nova Scotia. I. Vein paragenesis and mineral chemistry. *Canadian Mineralogist*, **31**, 471–522.

KONTAK, D. J. & SMITH, P. K. 1993b. An overview of Meguma gold deposits in the Meguma Terrane of southern Nova Scotia. *Exploration and Mining Geology*, **2**, 418–421.

KONTAK, D. J. & KERRICH, R. 1995. Geological and geochemical studies of a metaturbidite-hosted lode gold deposit; the Beaver Dam Deposit, Nova Scotia; II. Isotopic studies. *Economic Geology*, **90**, 885–901.

KONTAK, D. J. & KYSER, T. K. 2009. Nature and origin of an LCT-suite pegmatite with late-stage sodium enrichment, Brazil Lake, Yarmouth County, Nova Scotia, II. Implications of stable isotopes ($\delta^{18}$O, $\delta$D) for magma source, internal crystallization and nature of sodium metasomatism. *The Canadian Mineralogist*, **47**, 745–764, http://dx.doi.org/10.3749/canmin.47.4.745

KONTAK, D. J., DE WOLFE DE YOUNG, Y. & DOSTAL, J. 2002a. Late-stage crystallization history of the Jurassic North Mountain Basalt, Nova Scotia, I. Textural and chemical evidence for pervasive development of silicate-liquid immiscibilty. *The Canadian Mineralogist*, **40**, 1287–1311.

KONTAK, D. J., DOSTAL, J., KYSER, T. K. & ARCHIBALD, D. A. 2002b. A petrological, geochemical, isotopic and fluid-inclusion study of 370 Ma pegmatite-aplite sheets, Peggys Cove, Nova Scotia, Canada. *The Canadian Mineralogist*, **40**, 1249–1286.

KROGH, T. E. & KEPPIE, J. D. 1990. Age of detrital zircon and titanite in the Meguma Group, southern Nova Scotia, Canada: clues to the origin of the Meguma Terrane. *Tectonophysics*, **177**, 307–323.

KRYNINE, P. D. 1948. The megascopic study and field classification of sedimentary rocks. *Journal of Geology*, **56**, 130–165.

LACKEY, H. S., LACKEY, J. S. & MENDELSON, C. 2007. Nova Scotia: evolution of a continental margin. *In*: *20th Annual Keck Symposium*, 225–231, http://keckgeology.org/files/pdf/symvol/20th/novascotia/lackey.pdf

LACKEY, J. S., ERDMANN, S., HARK, J. S., NOWAK, R. M., MURRAY, K. E. & CLARKE, D. B. 2011. Tracing garnet origins in granitoid rocks by oxygen isotope analysis: examples from the South Mountain Batholith, Nova Scotia. *The Canadian Mineralogist*, **49**, 417–439, http://dx.doi.org/10.3749/canmin.49.2.417

LELEU, S. & HARTLEY, A. J. 2010. Controls on the stratigraphic development of the Triassic Fundy Basin, Nova Scotia: implications for the tectonostratigraphic evolution of the Triassic Atlantic rift basins. *Journal of the Geological Society, London*, **167**, 437–454.

LELEU, S., HARTLEY, A. J. & WILLIAMS, B. P. J. 2009. Large scale alluvial architecture and correlation within a Triassic pebbly braided river system, lower Wolfville Formation (Fundy Basin, Nova Scotia, Canada). *Journal of Sedimentary Research*, **79**, 292–313.

LYNCH, G. & ORTEGA, J. 1997. Hydrothermal alteration and tourmaline–albite equilibria at the Coxheath porphyry Cu–Mo–Au deposit, Nova Scotia. *The Canadian Mineralogist*, **35**, 79–94.

MACDONALD, M. A. 2001. *Geology of the South Mountain Batholith, Southwestern Nova Scotia*. Department of Natural Resources – Mineral Resources Branch Open File Report **ME 2001-002**.

MACDONALD, A. S. & SCOTT, F. 1980. *Blue Mountain Area*. Nova Scotia Department of Mines & Energy Report **80 1**.

MACDONALD, M. A., COREY, M. C., HAM, L. J. & HORNE, R. J. 1987. *South Mountain Batholith Project*. Nova Scotia Department of Mines & Energy Report **87-5**.

MANGE, M. A. & MAURER, F. W. 1992. *Heavy Minerals in Colour*. Chapman and Hall, London.

MANGE, M. & MORTON, A. C. 2007. Geochemistry of heavy minerals. *In*: MANGE, M. & WRIGHT, D. (eds) *Heavy Minerals in Use*. Developments in Sedimentology. Elsevier, Amsterdam, **58**, 345–391.

MARTEL, A. T. & GIBLING, M. R. 1994. Combined-flow generation of sole structures, including recurved groove casts, associated with Lower Carboniferous lacustrine storm deposits in Nova Scotia, Canada. *Journal of Sedimentary Research*, **A64**, 508–517.

MARTEL, A. T. & GIBLING, M. R. 1996. Stratigraphy and tectonic history of the Upper Devonian to Lower Carboniferous Horton Bluff Formation, Nova Scotia. *Atlantic Geology*, **32**, 13–38.

MCKENZIE, C. B. & CLARKE, D. B. 1975. Petrology of the South Mountain Batholith, Nova Scotia. *Canadian Journal of Earth Sciences*, **12**, 1209–1218.

MERTZ, K. A. & HUBERT, J. F. 1989. Cycles of sand-flat sandstone and playa-mudstone in the Triassic–Jurassic Blomidon redbeds, Fundy rift basin, Nova Scotia: implications for tectonic and climatic controls. *Canadian Journal of Earth Sciences*, **27**, 442–451.

MILLER, B. V., DUNNING, G. R., BARR, S. M., RAESIDE, R. P., JAMIESON, R. A. & REYNOLDS, P. H. 1996. Magmatism and metamorphism in a Grenvillian fragment: U–Pb and $^{40}$Ar/$^{39}$Ar ages from the Blair River Complex, northern Cape Breton Island, Nova Scotia, Canada. *Geological Society of America Bulletin*, **108**, 127–140.

MORTON, A. C. 1984. Stability of detrital heavy minerals in Tertiary sandstones of the North Sea Basin. *Clay Minerals*, **19**, 287–308.

MORTON, A. C. & HALLSWORTH, C. R. 1999. Processes controlling the composition of heavy mineral assemblages in sandstones. *Sedimentary Geology*, **124**, 3–29.

MORTON, A. C. & HALLSWORTH, C. R. 2007. Stability of detrital heavy minerals during burial diagenesis. *In*: MANGE, M. A. & WRIGHT, D. T. (eds) *Heavy Minerals in Use*. Developments in Sedimentology. Elsevier, Amsterdam, **58**, 215–245.

MORTON, A. C., BOYD, J. D. & EWEN, D. F. 2002. Evolution of Palaeocene sediment dispersal systems in the Foinaven Sub-Basin, West of Shetland. *In*: JOLLEY, D. W. & BELL, B. R. (eds) *The North Atlantic Igneous Province: Stratigraphy, Tectonics, Volcanic and Magmatic Processes*. Geological Society, London, Special Publications, **197**, 69–93.

MORTON, A. C., WHITHAM, A. G., FANNING, C. M. & CLAOUÉ-LONG, J. C. 2005*a*. The role of East Greenland as a source of sediment to the Vøring Basin during the Late Cretaceous. *In*: WANDÅS, B. T. G., EIDE, E. A., GRADSTEIN, F. & NYSTUEN, J. P. (eds) *Onshore–Offshore Relationships on the North Atlantic Margin*. Norwegian Petroleum Society, Special Publications, **12**, 83–110.

MORTON, A. C., WHITHAM, A. G. & FANNING, C. M. 2005*b*. Provenance of Late Cretaceous–Paleocene submarine fan sandstones in the Norwegian Sea: integration of heavy mineral, mineral chemical and zircon age data. *Sedimentary Geology*, **182**, 3–28.

MORTON, A. C., HERRIES, R. & FANNING, M. 2007. Correlation of Triassic sandstones in the Strathmore Field, West of Shetland, using heavy mineral provenance signatures. *In*: MANGE, M. A. & WRIGHT, D. T. (eds) *Heavy Minerals in Use*. Developments in Sedimentology, **58**, Amsterdam, Elsevier, 1037–1072.

MUKHOPADHYAY, P. K. 2008. Mississippian lacustrine Horton Formation Source Rocks from Nova Scotia and New Brunswick, Eastern Canada: A Major Shale Gas and Oil Shale Resource Plays. Abstract, Search and Discovery Article #90078 (2008). *American Association of Petroleum Geologists Annual Convention*, San Antonio, Texas, 20–23 April 2008.

MUKHOPADHYAY, P. K., HATCHER, P. & CALDER, J. H. 1991. Hydrocarbon generation of coal and coaly shale from fluvio-deltaic and deltaic environments of Nova Scotia and Texas. *Organic Geochemistry*, **17**, 765–783.

MUKHOPADHYAY, P. K., MACDONALD, D. J., HARVEY, P. J., BOEHNER, R. C., CALDER, J. H. & RYAN, R. 2000. Petroleum system of the Carboniferous sediments of onshore Nova Scotia. Extended Abstract. *International Journal of Coal Geology*, **43**, 137–139, http://www.sciencedirect.com/science/article/pii/S0166516299000713

MURPHY, J. B. & KEPPIE, J. D. 2005. The Acadian orogeny in the Northern Appalachians. *International Geology Review*, **47**, 663–687.

NICKEL, E. 1973. Experimental dissolution of light and heavy minerals in comparison with weathering and intrastratal dissolution. *Contributions to Sedimentology*, **1**, 1–68.

OELKERS, E. H., BJØRKUM, P. A. & MURPHY, W. M. 1996. A petrographic and computational investigation of quartz cementation and porosity reduction in North Sea sandstones. *American Journal of Science*, **296**, 420–452.

OELKERS, E. H., BJØRKUM, P. A., WALDERHAUG, O., NADEAU, P. H. & MURPHY, W. M. 2000. Making diagenesis obey thermodynamics and kinetics: the case of quartz cementation in sandstones from offshore mid-Norway. *Applied Geochemistry*, **15**, 295–309.

OLDALE, H. R. 1966. *Geology and Mineralization of a Dioritemonzonite Pluton in the Coxheath Hills, Cape Breton, Nova Scotia*. Nova Scotia Department of Mines and Energy Open-File Report **067**.

OLSEN, P. E. 1997. Stratigraphic record of the Early Mesozoic breakup of Pangea in the Laurasia–Gondwana rift system. *Annual Review of Earth and Planetary Sciences*, **25**, 337–401.

OLSEN, P. E. & SCHLISCHE, R. W. 1990. Transtensional arm of the early Mesozoic Fundy rift basin: penecontemporaneous faulting and sedimentation. *Geology*, **18**, 695–698.

OLSEN, P. E., KENT, D. V., FOWELL, S. J., SCHLISCHE, R. W., WITHJACK, M. O. & LETOURNEAU, P. M. 2000. Implications of a comparison of the stratigraphy and depositional environments of the Argana (Morocco) and Fundy (Nova Scotia, Canada) Permian–Jurassic basins. *In*: OUJIDI, M. & ET-TOUHAMI, M. (eds) *Le Permien et le Trias du Maroc, Actes de la Premièr Réunion su Groupe Marocain du Permien et du Trias*. Hilal Impression, Oujda, 165–183.

O'REILLY, G. A. 1986. Studies of mineral deposits associated with the South Mountain batholith. *Nova Scotia Department of Mines Energy Reprint*, **86-1**, 183–185.

O'REILLY, G. A., FARLEY, E. J. & CHAREST, M. H. 1982. *Metasomatic-Hydrothermal Mineral Deposits of the New Ross-Mahone Bay Area*. Nova Scotia Department of Mines & Energy Paper **82-2**.

OWEN, J. V. R., CORNEY, R., DOSTAL, J. & VAUGHAN, A. 2010. Significance of gneissic rocks in the Liscomb Complex, Nova Scotia. *Canadian Journal of Earth Sciences*, **47**, 927–940, http://dx.doi.org/10.1139/E10–019

PATTISON, D. R. M., SPEAR, F. S. & CHENEY, J. T. 1999. Polymetamorphic origin of muscovite plus cordierite plus staurolite plus biotite assemblages: implications for the metapelitic petrogenetic grid and for *P–T* paths. *Journal of Metamorphic Geology*, **17**, 685–703, http://dx.doi.org/10.1046/j.1525-1314.1999.00225.x

PE-PIPER, G. 1988. Calcic amphiboles of mafic rocks of the Jeffers Brook plutonic complex, Nova Scotia, Canada. *American Mineralogist*, **73**, 993–1006.

PE-PIPER, G. 2007. Relationship of amphibole composition to host-rock geochemistry: the A-type gabbro-granite Wentworth pluton, Cobequid shear zone, eastern Canada. *European Journal of Mineralogy*, **19**, 29–38.

PETTIJOHN, F. J., POTTER, P. E. & SIEVER, R. 1973. *Sand and Sandstone*. Springer-Verlag, Berlin.

PITTMAN, E. D., LARSE, R. E. & HEALD, M. T. 1992. Clay coats: occurrence and relevance to preservation of porosity in sandstones. *In*: HOUSEKNECHT, D. W. & PITTMAN, E. D. (eds) *Origin, Diagenesis and Petrophysics of Clay Minerals in Sandstone*. SEPM Special Publications, Tulsa, **47**, 241–255.

POOLE, J. 1984. *An example of structural deformation of the Halifax Formation in the Whitehead area, Meguma Terrane, Eastern Nova Scotia*. BSc thesis, St Francis Xavier University, Antigonish, NS.

PRESTON, J., HARTLEY, A., MANGE-RAJETZKY, M., HOLE, M., MAY, G., BUCK, ST. & VAUGHAN, L. 2002. The provenance of Triassic continental sandstones from the Beryl Field, Northern North Sea: mineralogical, geochemical and sedimentological constraints. *Journal of Sedimentary Research*, **72**, 18–29.

RAESIDE, R. P., HILL, J. D. & EDDY, B. G. 1988. Metamorphism of Meguma Group metasedimentary rocks, Whitehead Harbour area, Guysborough Caounty, Nova Scotia. *Maritime Sediments and Atlantic Geology*, **24**, 1–9.

RAST, N., KENNEDY, M. J. & BLACKWOOD, R. F. 1976. Comparison of some tectono-stratigraphic zones in the Appalachians of Newfoundland and New Brunswick. *Canadian Journal of Earth Sciences*, **13**, 868–875.

ROMER, R. L. & KRONER, U. 2012. Reply to the discussion by J.W.F. Waldron and C.E. White on 'Geochemical signature of Ordovician Mn-rich sedimentary rocks on the Avalonian shelf'. *Canadian Journal of Earth Sciences*, **49**, 775–780, http://dx.doi.org/10.1139/E2012-006

ROMER, R. L., KIRSCH, M. & KRONER, U. 2011. Geochemical signature of Ordovician Mn-rich sedimentary rocks on the Avalonian shelf. *Canadian Journal of Earth Sciences*, **48**, 703–718, http://dx.doi.org/10.1139/E10-092

RUB, M. G., PALOV, V. A., GLADKOV, N. G. & GRISHINA, N. V. 1977. Accessory garnet as an indicator of the genesis of ore-bearing granitoid rocks. *Doklady Akademii Nauk SSSR*, **235**, 1397–1400.

SABEEN, H. M., RAMANUJAM, N. & MORTON, A. C. 2002. The provenance of garnet: constraints provided by studies of coastal sediments from southern India. *Sedimentary Geology*, **152**, 279–287.

SALVINO, J. F. & VELBEL, M. A. 1989. Faceted garnets from sandstones of the Munising Formation, (Cambrian), northern Michigan: petrographic evidence for origin by intrastratal dissolution. *Sedimentology*, **36**, 371–379.

SCALLION, K.-L., REBECCA, A., JAMIESON, R. A., BARR, S. M., WHITE, C. E. & ERDMANN, S. 2011. Texture and composition of garnet as a guide to contamination of granitoid plutons: an example from the Governor Lake area, Meguma Terrane, Nova Scotia. *The Canadian Mineralogist*, **49**, 441–458, http://dx.doi.org/10.3749/canmin.49.2.441

SCHENK, P. E. 1997. Sequence stratigraphy and provenance on Gondwana's margin: the Meguma Zone (Cambrian to Devonian) of Nova Scotia, Canada. *Geological Society of America Bulletin*, **109**, 395–409.

SCHILLER, E. A. & TAYLOR, F. C. 1965. Spessartine–quartz rocks (coticules) from Nova Scotia. *The American Mineralogist*, **50**, 1477–1481.

SCHLISCHE, R. W. & OLSEN, P. E. 1990. Quantitative filling model for continental extensional basins with applications to early Mesozoic rifts of eastern North America. *Journal of Geology*, **98**, 135–155.

SCHOLLE, P. A. 1979. *A Color Illustrated Guide to Constituents, Textures, Cements and Porosities of Sandstones and Associated Rocks*. American Association of Petroleum Geologists Memoirs, **28**.

SEAL, R. R., CLARK, A. H. & MORRISSY, C. J. 1987. Stockwork tungsten (scheelite)-molybdenum mineralization, Lake George, Southwestern New Brunswick. *Economic Geology*, **82**, 1259–1282.

SHELDON, H. A., WHEELER, J., WORDEN, R. H. & CHEADLE, M. J. 2003. An analysis of the roles of stress, temperature, and pH in chemical compaction of sandstones. *Journal of Sedimentary Research*, **73**, 64–71.

SMALE, D. & MORTON, A. C. 1987. Heavy mineral suites of core samples from the McKee Formation (Eocene–Lower Oligocene), Taranaki: implications for provenance and diagenesis. *New Zealand Journal of Geology and Geophysics*, **30**, 299–306.

SMITH, P. K. 1983. *Geology of the Cochrane Hill Gold Deposit, Guysborough County, Nova Scotia*. Nova Scotia Department of Mines & Energy Report **83-1**.

SMITH, P. K. & KONTAK, D. J. 1986. Meguma gold studies: advances in geological insight as an aid to gold exploration. *In*: BATES, J. L. & MACDONALD, D. R. (eds) *Tenth Annual Open House and Review of Activities, Program and Summaries*. Nova Scotia Department of Mines and Energy Information Series, **12**, 105–113.

SMITH, P. K. & KONTAK, D. J. 1987. *Meguma Gold Studies: Some Basic Observations*. Nova Scotia Department of Mines & Energy Report **87-5**.

SMITH, P. K. & KONTAK, D. J. 1988. *Meguma gold studies II: Vein morphology classification, and information: A new interpretation of 'crack-seal' quartz veins*. Nova Scotia Department of Mines and Energy, Report **88-1**, Pt B, 61–76.

SMITHSON, F. 1941. The alteration of detrital minerals in the Mesozoic rocks of Yorkshire. *Geological Magazine*, **78**, 97–112.

SMOOT, J. P. & CASTENS-SEIDELL, B. 1994. Sedimentary features produced by efflorescent salt crusts, Saline Valley and Death Valley, California: sedimentology and geochemistry of modern and ancient Saline Lakes. *In*: RENAUT, R. W. & LAST, W. M. (eds) *Sedimentology and Geochemistry of Modern and Ancient Saline Lakes*. SEPM Special Publications, Tulsa, **50**, 73–90.

SMOOT, J. P. & OLSEN, P. E. 1988. Massive mudstones in basin analysis and palaeoclimatic interpretation of the

Newark Supergroup. *In*: MANSPEIZER, W. (ed.) *Triassic–Jurassic Rifting*. Developments in Geotectonics, Elsevier, Amsterdam, **22**, 249–274.

SOBOLEV, N. V. 1964. Classification of rock-forming garnets. *Doklady Akademii Nauk SSSR*, **157**, 353–356.

ST PETER, C. 1993. Maritimes Basin evolution: key geologic and seismic evidence from the Moncton Subbasin of New Brunswick. *Atlantic Geology*, **29**, 233–270.

TANNER, L. H. 2000. Triassic–Jurassic lacustrine deposition in the Fundy Rift Basin, Eastern Canada. *In*: GIERLOWSKI-KORDESCH, E. H. & KELTS, K. R. (eds) *Lake Basins through Space and Time*. American Association of Petroleum Geologists, Studies in Geology, **46**, 159–166.

VAN LOON, A. J. (TOM) & MANGE, M. A. 2007. 'In situ' dissolution of heavy minerals through extreme weathering, and the application of the surviving assemblages and their dissolution characteristics to correlation of Dutch and German silversands. *In*: MANGE, M. A. & WRIGHT, D. T. (eds) *Heavy Minerals in Use*. Developments in Sedimentology. Elsevier, Amsterdam, **58**, 189–213.

VELBEL, M. A. 1984. Natural weathering mechanisms of almandine garnet. *Geology*, **12**, 631–634.

VELBEL, M. A. 1993a. Formation of protective surface layers during silicate-mineral weathering under well-leached, oxidizing conditions. *American Mineralogist*, **78**, 408–417.

VELBEL, M. A. 1999. Bond strength and the relative weathering rates of simple orthosilicates. *American Journal of Science*, **299**, 679–696.

VELBEL, M. A., MCGUIRE, J. T. & MADDEN, A. S. 2007. Scanning electron microscopy of garnet from southern Michigan soils: etching rates and inheritance of pre-glacial and pre-pedogenic grain-surface textures. *In*: MANGE, M. A. & WRIGHT, D. T. (eds) *Heavy Minerals in Use*. Developments in Sedimentology. Elsevier, Amsterdam, **58**, 413–432.

WADE, J. A., BROWN, D. E., TRAVERSE, A. & FENSOME, R. A. 1996. The Triassic–Jurassic Fundy Basin, eastern Canada: regional setting, stratigraphy, and hydrocarbon potential. *Atlantic Geology*, **32**, 189–231.

WALDRON, J. W. F. & WHITE, C. E. 2012. Discussion of 'Geochemical signature of Ordovician Mn-rich sedimentary rocks on the Avalonian shelf'. *Canadian Journal of Earth Sciences*, **49**, http://dx.doi.org/10.1139/E2012-004

WHITE, C. E. 2003. *Preliminary Bedrock Geology of the Area between Chebogue Point, Yarmouth County, and Cape Sable Island, Shelburne County, Southwestern Nova Scotia*. Mineral Resources Branch, Report of Activities 2002, Nova Scotia Department of Natural Resources Report **2003-1**.

WHITE, C. E. 2008. Defining the stratigraphy of the Meguma Supergroup in southern Nova Scotia; where do we go from here? *Atlantic Geology*, **44**, 48–49.

WHITE, C. E. & BAR, S. M. 2012. Meguma Terrane revisited: stratigraphy, metamorphism, paleontology, and provenance. *Geoscience Canada*, **39**, 8–12.

WHITE, C. E., BARR, S. M., JAMIESON, R. A. & REYNOLDS, P. H. 2001. Neoproterozoic high pressure/low-temperature metamorphic rocks in the Avalon Terrane, southern New Brunswick, Canada. *Journal of Metamorphic Geology*, **19**, 519–530.

WHITE, C. E., BARR, S. M., REYNOLDS, P. H., GRACE, E. & MCMULLIN, W. A. 2006. The Pocologan metamorphic suite: high-pressure metamorphism in a Silurian fore-arc complex, Kingston Terrane, southern New Brunswick. *The Canadian Mineralogist*, **44**, 905–927.

WILLIAMS, G. L., FYFFE, L. R., WARDLE, R. J., COLMAN-SADD, S. P. & BOEHNER, R. C. 1985. *Lexicon of Canadian Stratigraphy, volume VI, Atlantic Region*. Canadian Society of Petroleum Geologists, Calgary.

WOLFSON, I. 1983. *A study of the tin mineralization and lithogeochemistry in the area of the Wedgeport Pluton, Southwestern Nova Scotia*. MSc thesis, Dalhousie University.

WRIGHT, W. I. 1938. The composition and occurrence of garnets. *American Mineralogist*, **23**, 436–449.

YU, J.-M. & JIANG, S.-Y. 2003. Chemical composition of tourmaline from the Yunlong tin deposit, Yunnan, China: implications for ore genesis and mineral exploration. *Mineralogy and Petrology*, **77**, 67–84, http://dx.doi.org/10.1007/s00710-002-0195-2

# Along-strike variations in the composition of sandstones derived from the uplifting western Greater Caucasus: causes and implications for reservoir quality prediction in the Eastern Black Sea

STEPHEN J. VINCENT[1]*, FIONA HYDEN[2] & WILLIAM BRAHAM[3]

[1]*CASP, University of Cambridge, 181a Huntingdon Road, Cambridge, CB3 0DH, UK*

[2]*Oil Quest, 12 Abbey Court, Cerne Abbas, Dorchester, Dorset DT2 7JH, UK*

[3]*11 Corner Hall, Hemel Hempstead, Hertfordshire HP3 9HN, UK*

**Corresponding author (e-mail: stephen.vincent@casp.cam.ac.uk)*

**Abstract:** Oligo-Miocene outcrops along the southern margin of the western Greater Caucasus preserve a record of sediments shed from the range into the northern and central parts of the Eastern Black Sea. Sandstones in the Russian western Caucasus are significantly more quartz-rich than those located farther SE in western Georgia. The latter contain appreciably more mudstone and volcanic rock fragments. Oligo-Miocene turbidite systems derived from the Russian western Caucasus in the Tuapse Trough and central Eastern Black Sea may therefore form better-quality reservoirs at shallow to moderate depths than sediments derived from west Georgian volcaniclastic sources in the easternmost part of the basin. Palynomorph analysis indicates sediment derivation predominantly from Jurassic and Cretaceous strata in the Russian western Caucasus and from Eocene strata, and an increasing proportion of Cretaceous strata upsection, in western Georgia. An Eocene volcaniclastic source is proposed for the increased rock fragment component in west Georgian sandstones. Eocene volcaniclastic rocks are no longer exposed in the Greater Caucasus, but similar rocks form the inverted fill of the Adjara–Trialet Basin farther south in the Lesser Caucasus. The former presence of a northern strand of this basin in the west Georgian Caucasus is supported by earlier thermochronological work.

**Supplementary material:** A sample data table, petrographic data table, petrographic key, QFL sandstone compositional plot and palynomorph reworking Stratabugs[TM] charts are available at www.geolsoc.org.uk/SUP18662

The Eastern Black Sea is one of the few remaining underexplored hydrocarbon basins in Europe. Two deep-water wells, Hopa-1 and Sürmene-1, have been drilled in Turkish waters (Fig. 1). Water depths typically in excess of 2 km hamper exploration efforts. The basin contains up to 8 km of Cenozoic sedimentary rocks (Minshull *et al.* 2005). These include a world-class source rock interval, the Oligocene to Early Miocene Maykop Series (Robinson *et al.* 1996), and turbiditic sandstones that, in the Oligocene and younger part of the succession, form important potential reservoir targets (e.g. Meisner *et al.* 2009). This paper documents along-strike variations in the composition, and therefore likely reservoir potential, of Oligo-Miocene sandstones shed from the evolving Greater Caucasus mountain belt into the Eastern Black Sea. It proposes that variations in the age and composition of earlier sedimentary rocks reworked during the inversion of the Greater Caucasus Basin are the principal control on the observed variations in composition and potential reservoir quality.

## Geological background

The Eastern Black Sea, Greater Caucasus and Adjara–Trialet basins formed as a result of Mesozoic to earliest Cenozoic extension to transtension at the southern margin of Eurasia during the northerly subduction and closure of the Tethys Ocean to the south (Dercourt *et al.* 1986, 2000; Şengör & Natal'in 1996; Nikishin *et al.* 1998, 2001, Golonka 2004; Kaz'min & Tikhonova 2006a, b; Saintot *et al.* 2006, Barrier & Vrielynck 2008). Collision of the Arabian promontory with the assembled Tethyside orogenic collage resulted in the inversion of the Greater Caucasus and Adjara–Trialet basins from the Late Eocene–earliest Oligocene onwards to form the southern slope of the Greater Caucasus and the Adjara–Trialet Belt (Saintot *et al.* 2006; Vincent *et al.* 2007, 2011; Allen & Armstrong 2008, and references therein) (Fig. 1). These regions are major contributors to the fill of the Eastern Black Sea, while the Greater Caucasus also formed a barrier to more distantly

*From*: Scott, R. A., Smyth, H. R., Morton, A. C. & Richardson, N. (eds) 2014. *Sediment Provenance Studies in Hydrocarbon Exploration and Production.* Geological Society, London, Special Publications, **386**, 111–127.
First published online September 17, 2013, http://dx.doi.org/10.1144/SP386.15

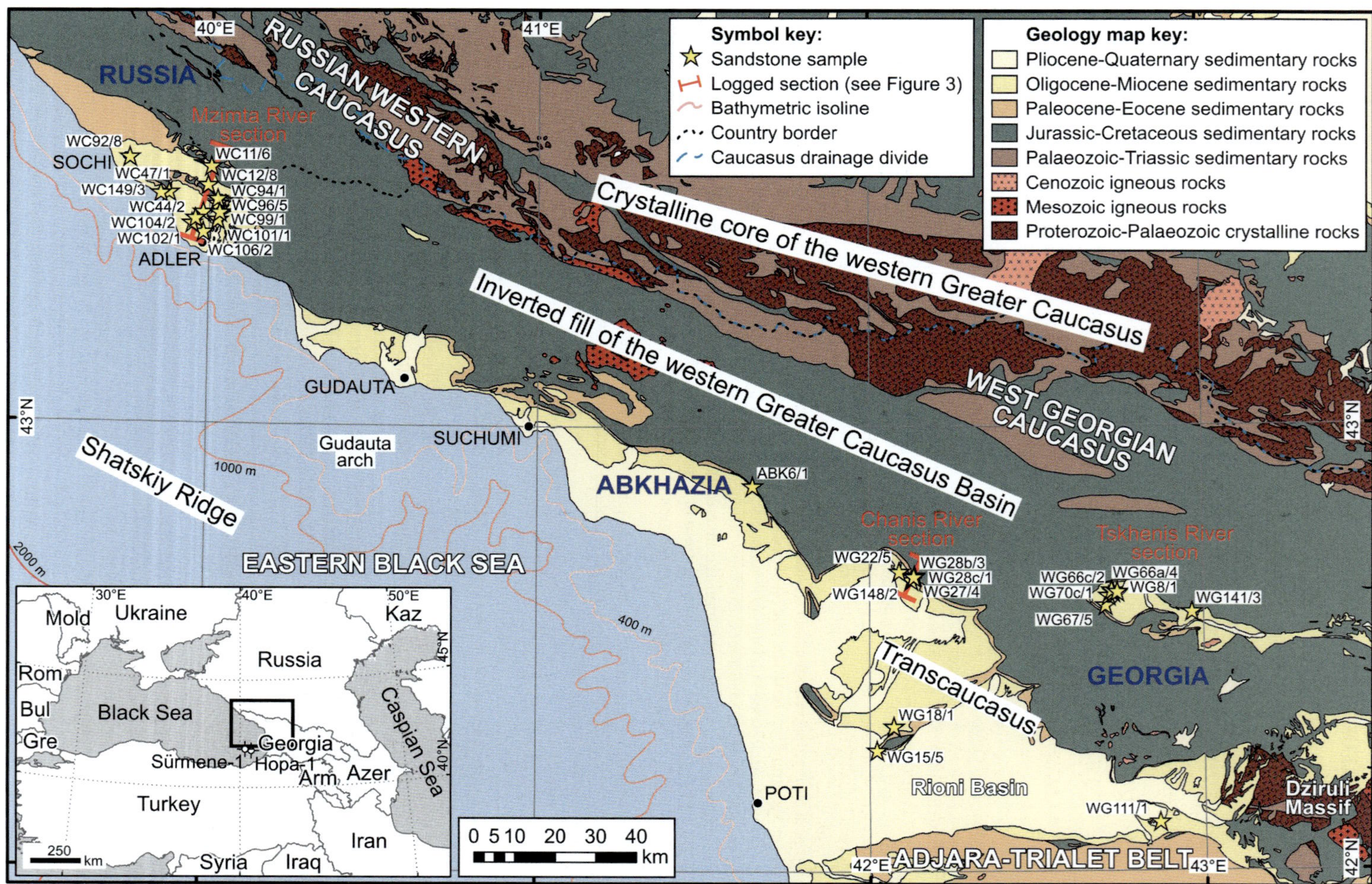



**Fig. 1.** Simplified geological map of part of the western Greater Caucasus with sample locations marked. The inset shows the position of the study area in its wider geographical context.

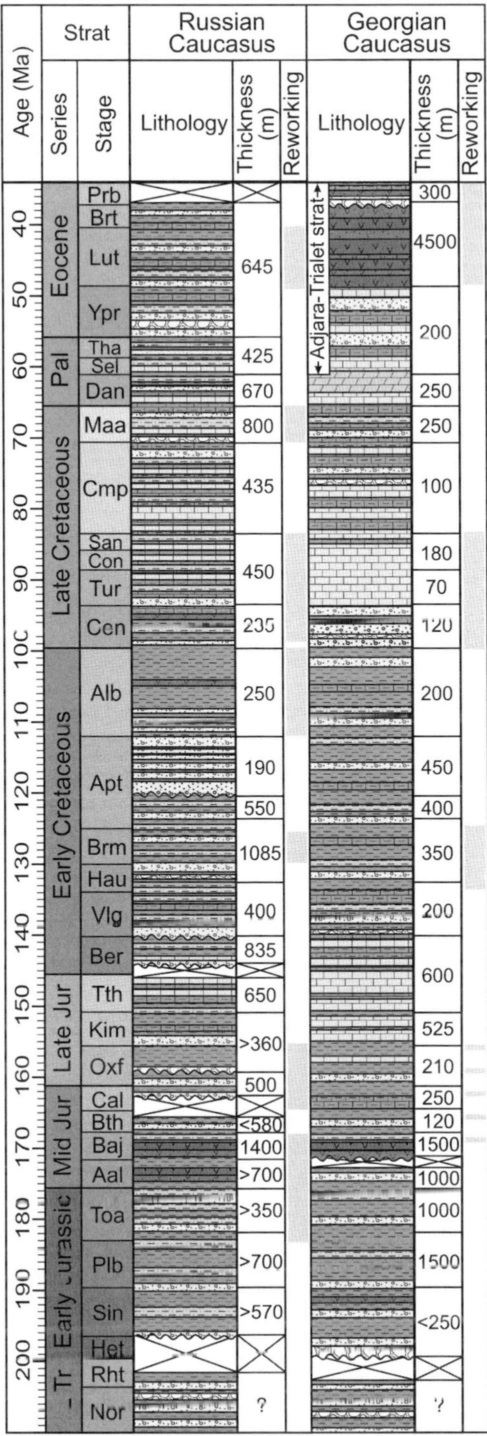

**Fig. 2.** Schematic stratigraphies through the Russian and west Georgian sectors of the western Greater Caucasus Basin. Data are compiled from local 1:200 000 geological maps, with additional information from Adamia *et al.* (1992) and McCann *et al.* (2010). Selandian to Priabonian strata are not preserved in the west Georgian sector of the Greater Caucasus Basin, and its stratigraphy is inferred from the Adjara–Trialet Belt further south (see text for discussion). See Figure 3 for the lithological key. The main intervals reworked into Oligo-Miocene sedimentary rocks are shown by shaded grey boxes; these data are derived from Figure 3. The geological timescale is from Gradstein *et al.* (2004).

sourced, Russian Platform-derived, sediment dispersal systems (Vincent *et al.* 2013). The oceanic nature of the Eastern Black Sea Basin (Starostenko *et al.* 2004; Minshull *et al.* 2005) means that it has largely resisted deformation and inversion.

Outcrops along the southern margin of the western Greater Caucasus preserve a record of the sediments that fed into the northern and central parts of the Eastern Black Sea during the Oligo-Miocene. They were derived from two main rock units within the western Greater Caucasus: the crystalline core of the range, and along its southern slope, the Jurassic to Eocene fill of the Greater Caucasus Basin (Fig. 1). The crystalline core of the range comprises Gondwana-derived, predominantly Early to middle Palaeozoic crystalline protolith and middle Palaeozoic island-arc and ophiolitic rocks that were metamorphosed and intruded during their Variscan accretion to the southern margin of Laurasia (Zonenshain *et al.* 1990; Hanel *et al.* 1992, 1993; Somin *et al.* 2006, 2007; Zakariadze *et al.* 2007; Somin *et al.* 2009; Treloar *et al.* 2009). The crystalline core is unconformably overlain by either Late Carboniferous to Permian, Triassic or Jurassic sedimentary successions (Somin 2000; Lavrishchev *et al.* 2002), which would suggest that these crystalline rocks have been at or near the surface for much of the Late Palaeozoic to Cenozoic time. Fission track thermal modelling indicates a maximum of *c.* 2 km of post-Middle Jurassic burial (Vincent *et al.* 2011).

Jurassic to Eocene sedimentary rocks to the south of the crystalline core of the range were deposited in the western segment of an extensional, or more likely transtensional, basin termed the Greater Caucasus Basin (Nikishin *et al.* 1998, 2001; Saintot *et al.* 2004, 2006; Kaz'min & Tikhonova 2006*a*; McCann *et al.* 2010). The crystalline core of the range formed the northern shoulder of this basin, whereas its southern margin is represented by the Shatskiy Ridge and its onshore equivalent, the Transcaucasus (including the Dziruli Massif, Rioni Basin, and outcrops in southern Abkhazia and the southern part of the Mzimta River catchment in Russia) (Fig. 1). The fill of the basin is characterized by turbidite, calciturbidite

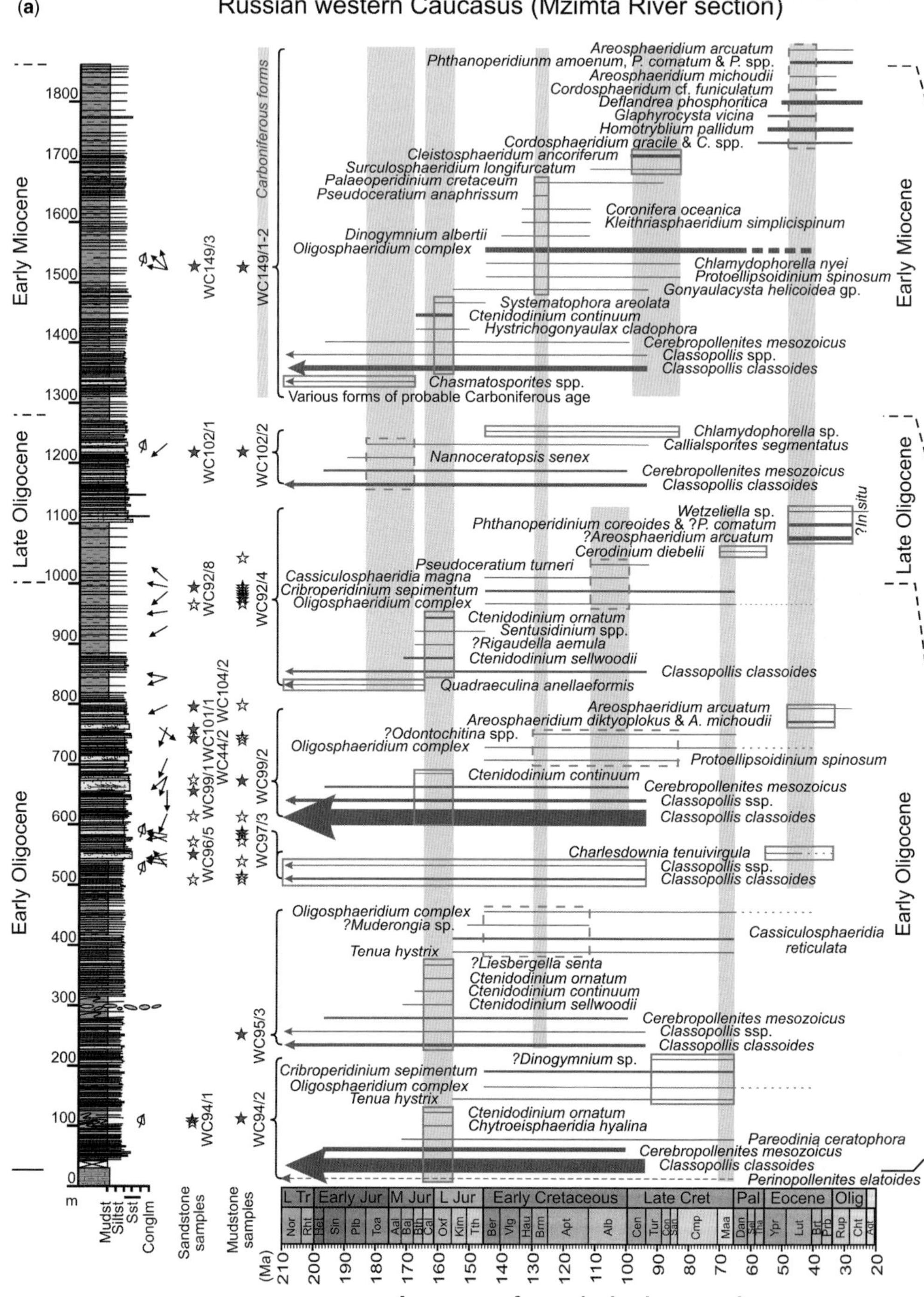



**Fig. 3.** Summary logs and palynological reworking data for latest Eocene to Middle Miocene strata in representative (**a**) Russian and (**b**) west Georgian sections on the southern side of the western Caucasus. The sections are located on Figure 1. The age range of reworked palynomorphs is shown by the horizontal lines, where the line thickness is proportional to their abundance. For each sample, the most likely ages of the rocks from which these forms

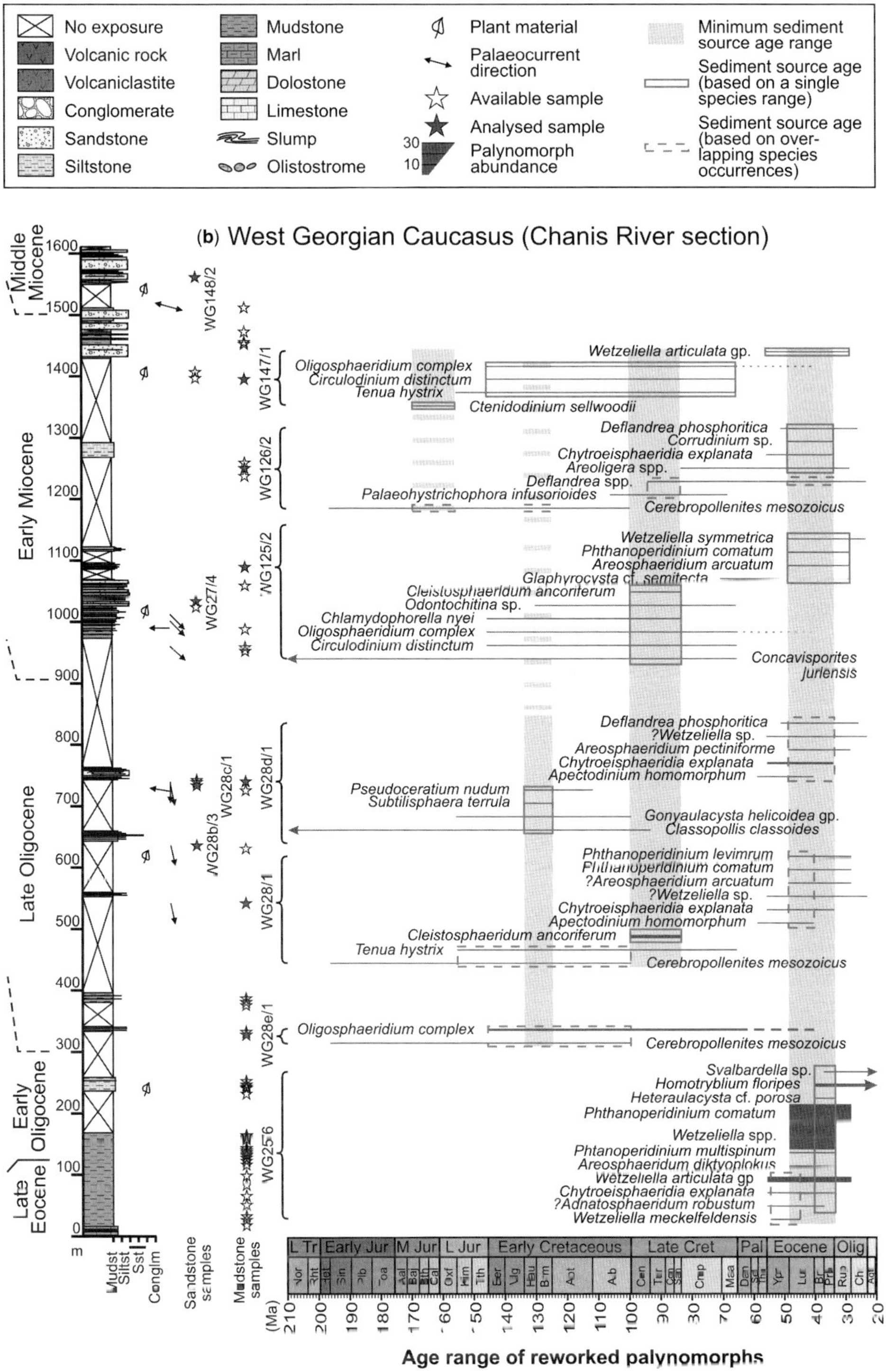

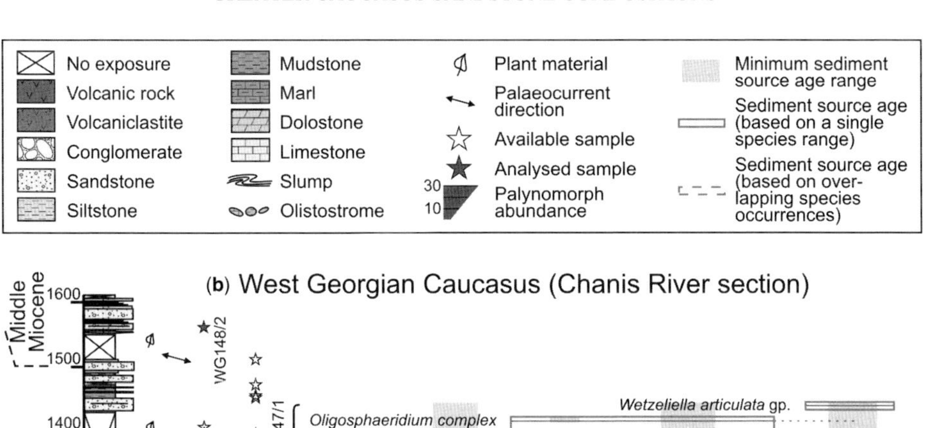



## (b) West Georgian Caucasus (Chanis River section)



**Age range of reworked palynomorphs**

**Fig. 3.** (*Continued*) are derived are shown by the boxes. The grey background shading indicates the age range of the minimum number of rock units needed to provide the range of reworked palynomorphs identified, based on their occurrence throughout each section; these are plotted on Figure 2. The geological timescale is from Gradstein *et al.* (2004).

and hemipelagic mudstone and marl deposits, although Early–Middle Jurassic and Cretaceous volcanic rocks are also present (Fig. 2).

## Sample dataset

In this dataset we have only considered sandstone and associated mudstone samples derived from the western Greater Caucasus. For a wider discussion of the provenance of sandstones entering the Eastern Black Sea, see Vincent *et al.* (2013). Determining which samples have a western Greater Caucasus provenance was an iterative process combining multiple factors, including (1) the presence of plant remains and associated conglomeratic facies (indicating a local sediment source), (2) palaeocurrent indicators and proximal to distal facies trends (indicating the direction of the sediment source, broadly from the north and NE), and (3) evidence for contemporaneous depositional systems with divergent palaeocurrent patterns and similar heavy mineral compositions on either side of the range (indicating a subaerial proto-Caucasus sediment source area from Oligocene time onward) (Vincent *et al.* 2007). Samples derived from the Adjara–Trialet Belt to the south of the west Georgian Caucasus were easily identified and excluded from the study by their rock fragment-rich and unstable heavy mineral-rich compositions and northerly palaeocurrent indicators (Vincent *et al.* 2013).

Sandstone-bearing Oligocene and younger sedimentary rocks on the southern side of the Russian western Caucasus are restricted to the region between Sochi and Adler, close to the Abkhazian border. Russian samples analysed in this study (prefix WC) come principally from exposures close to the Mzimta River (Fig. 1) that are dominated by mudstones and thin-bedded sandstones deposited by low-density turbidites and hemipelagic settling (Fig. 3a). Thick, massive, cross-bedded and laminated sandstones, interpreted as high-density turbidites, are present in the Early Oligocene part of the section, as are a number of olistostrome (debris flow) intervals.

Sandstone-bearing Oligocene and younger sedimentary rocks crop out more widely in western Georgia; sandstone samples analysed in this study (prefix WG) come from the Chanis and Tskhenis river valleys on the southern slope of the Greater Caucasus and from the margins of the Rioni Basin (Fig. 1). Mudstone samples come exclusively from the Chanis River section. West Georgian sedimentary rocks represent a greater variety of facies and depositional environments than those from the Sochi–Adler region in Russia. A broadly coarsening- and shallowing-upward trend is typical.

For instance, in the Chanis River section, the Early Oligocene interval consists of hemipelagic mudstones (Fig. 3b). Sandstones first occur in the Late Oligocene interval, are thin-bedded and were deposited by low-density turbidites. The section is capped by Early to Middle Miocene mudstones and thick deltaic sandstones (Fig. 3b). Shoreline sandstones and conglomerates are developed in Middle Miocene sedimentary rocks elsewhere (e.g. the Tskhenis River section). A single sample was analysed from the autonomous region of Abkhazia between Russia and Georgia (sample prefix ABK; Fig. 1); this was taken from an existing collection in Tbilisi due to a lack of field access.

Petrographic analyses were carried out on 27 sandstone samples in order to characterize the composition of sandstones derived from the Caucasus. Palynological analyses were carried out on 15 mudstone samples (one of which was barren) interbedded with these sandstones in order to determine the age range of earlier sedimentary rocks that were reworked into the Oligo-Miocene section. Sample preparation, analytical methodologies and results are reported below. Conventional heavy mineral and single mineral geochemical analysis, and SHRIMP U–Pb dating of zircons were carried out, but do not affect the conclusions of this study and so for brevity's sake are not reported here (see Vincent *et al.* 2013 instead).

## Sample preparation and analysis

### Sandstones

Sandstone samples were thin-sectioned, vacuum-impregnated with blue-dyed epoxy resin and then etched and stained with hydrofluoric acid and sodium cobaltinitrite, respectively, to identify alkali feldspars. Further etching with dilute hydrochloric acid and the application of a combined Alizarin-Red S and potassium ferricyanide stain facilitated identification of carbonate minerals.

Thin sections were point counted (300 points per thin section) using a Nikon Optiphot 2 microscope and Prior point-count stage. Counts were recorded via a bespoke (Oil Quest) point-count programme (P2P) that used an iteratively developed series of 77 predetermined categories to take account of both the traditional QFR method, where R stands for 'rock fragments', and the 'Gazzi-Dickinson' QFL method, where L stands for 'aphanitic lithic fragments'. The main difference between these methods is the procedure used for the classification of sand- and gravel-grade grains within rock fragments. The former method records these grains as rock fragments of a particular type (e.g. plutonic), whilst the latter records them as specific sand- or gravel-grade mineral species (e.g. alkali feldspar). The QFR data are presented here, because of the additional rock fragment provenance information they contain. The point-count data and a QFL plot are available in the supplementary material. Mudstone rock fragments are aphanitic siliciclastic sediment; carbonate mudstones are included in the

carbonate rock fragment category. All point-count percentages were expressed with reference to total rock volume, in other words, the sum of total detrital components, authigenic phases and visible porosity. The grain solid method was used, whereby intragranular voids, matrix or cement within grains were counted separately from the enclosing grains (see

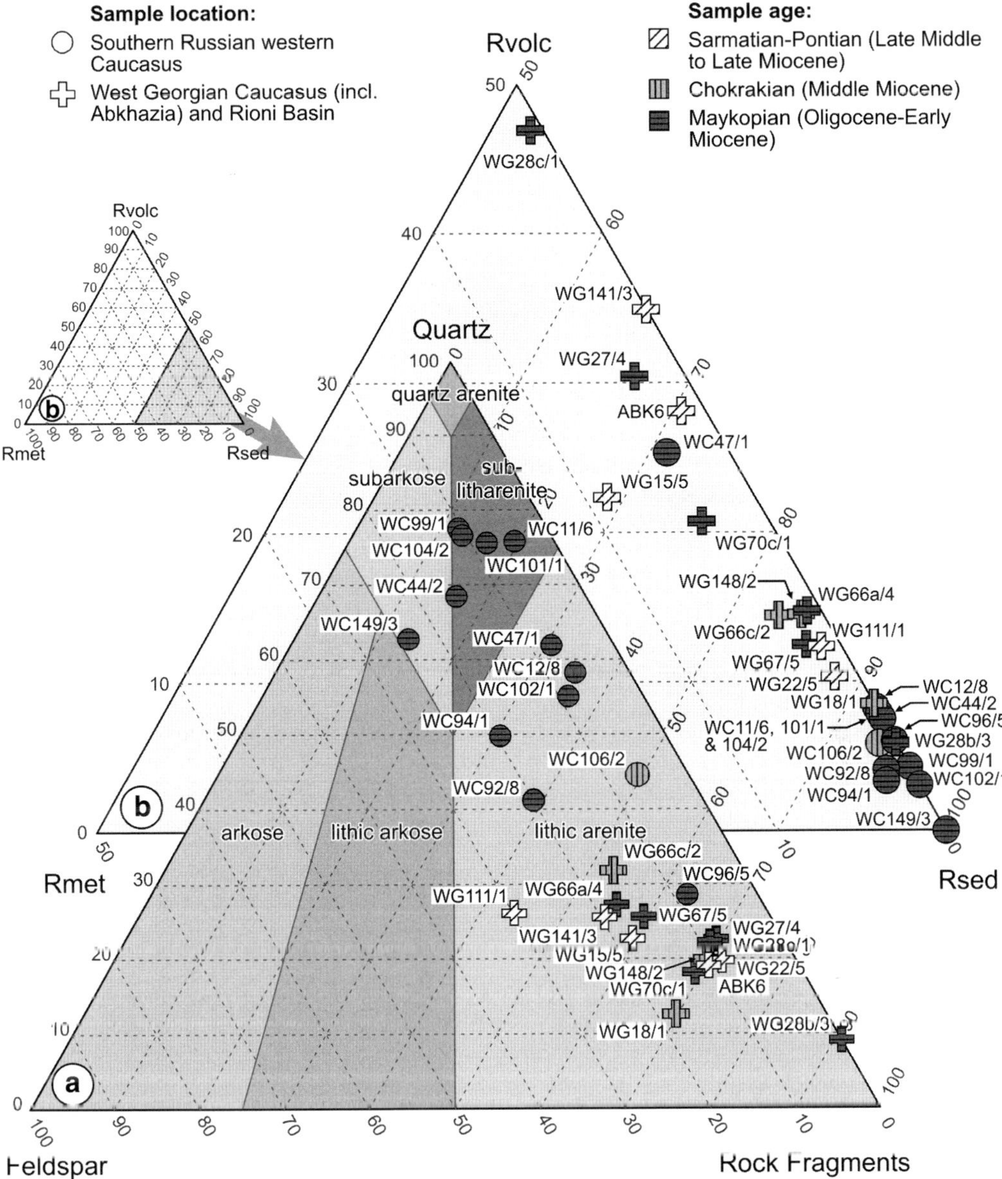


**Fig. 4.** (**a**) Quartz–feldspar–rock fragment (QFR) and (**b**) volcanic–metamorphic–sedimentary rock fragment (Rvolc–Rmet–Rsed) plots of Oligo-Miocene sandstone samples derived from the western Caucasus on the southern side of the range. Sample locations are shown on Figure 1. Note that all samples are rich in sedimentary rock fragments and only this part of the Rvolc–Rmet–Rsed plot is shown. Compositional fields are from Pettijohn *et al.* (1987).

Jaanusson 1972). The point-count distance selected for each sample was, where possible, always greater than the largest grain diameter. Reproducibility of data is dependent on the abundance of elements to be counted and heterogeneity of the sample. Errors of +4–5% are likely for 300 counts (van der Plas & Tobi 1965). Modal grain size and sorting data were based on an assessment of the whole of each thin section using industry-standard visual comparator charts (e.g. Harrell 1964).

### Mudstones

Mudstone samples were processed by standard palynological techniques involving gentle crushing to *c.* 1 mm then successive HCl (35%) and HF (60%) treatments followed by sieving at 15 μm and a further treatment in hot HCl (36%) to remove neoformed fluorides. No oxidation was performed. If amorphous organic matter (AOM) was present, a tunable ultrasonic probe was used to break up the AOM, which was then removed by sieving. The palynomorphs were mounted in Elvacite 2044 on a glass slide protected by a coverslip.

Microscopic examination was carried out on a single slide using a Leitz Dialux 20 transmitted light microscope. A series of traverses were made across the centre of the slide in order to count, where possible, a minimum of 200 palynomorphs. Once this count was reached the remainder of the slide was scanned for forms not encountered during the counting process and a note made of their presence. In instances where the preparation did not yield 200 palynomorphs, the whole slide was counted. The resulting palynomorph abundance data were then entered into Stratabugs™. Stratabugs charts are available as supplementary material.

The assemblages recovered comprise a mixture of *in situ* and reworked palynomorphs. In many

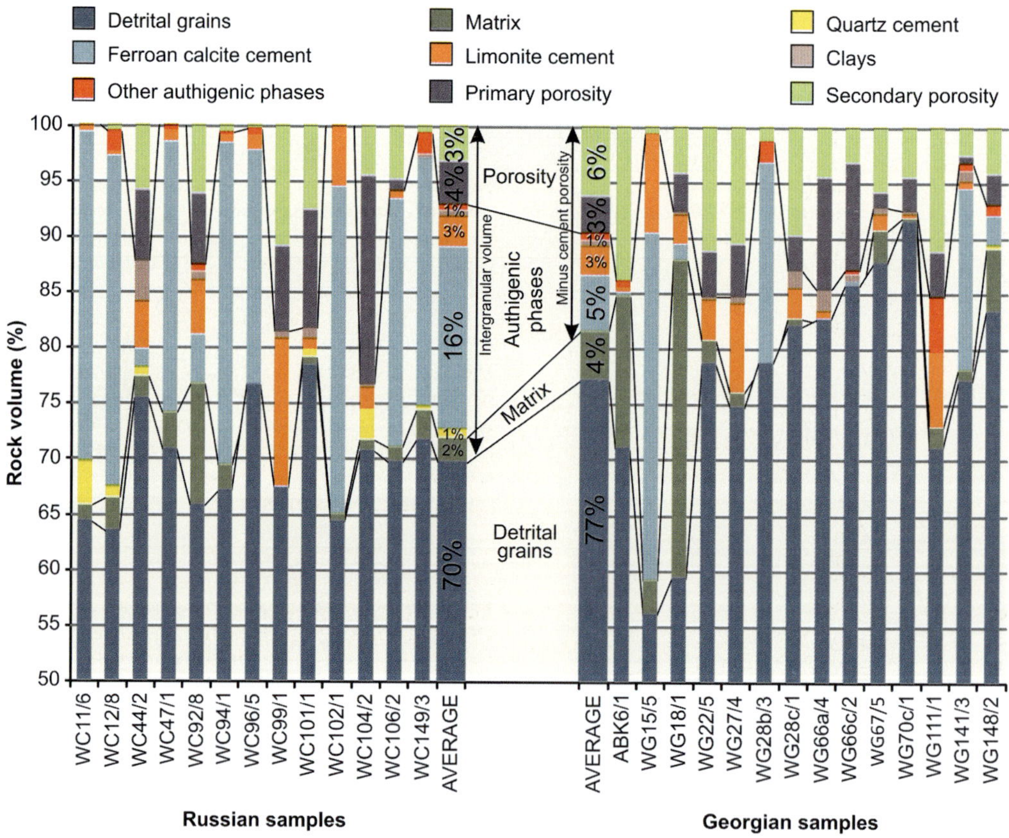


**Fig. 5.** Intergranular composition variability plot of Caucasus-derived Oligo-Miocene sandstone samples on the southern side of the range. Note the larger amounts of intergranular volume and minus cement porosity in the Russian samples and the larger amounts of secondary porosity in the Georgian samples.

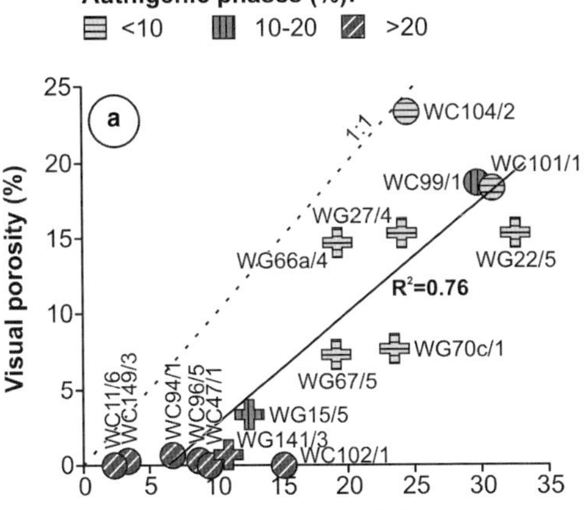



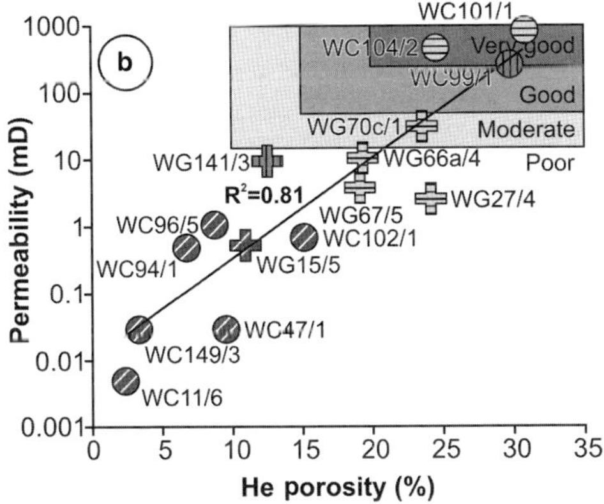



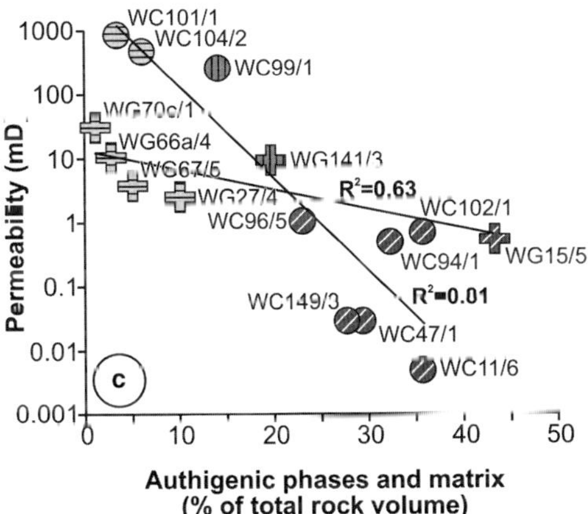



instances, determination of whether a particular palynomorph was reworked was made on the basis of exotic presence (i.e. taxa typical of Eocene sedimentary rocks being recorded within the Miocene) or, in the case of long-ranging forms, such as bisaccate pollen or pteridophyte spores, by whether they display unusually high levels of physical abrasion or thermal maturity. The age range of reworked palynomorphs in this study is used as a direct proxy for the age of the eroded sedimentary rocks contributing sediment to the studied Oligo-Miocene samples. It was not possible to distinguish between palynomorphs that have undergone single or multiple cycles of reworking, such that the palynomorph ages provide a maximum sediment source age. However, the only major episode of inversion in the Greater Caucasus Basin prior to Late Eocene–earliest Oligocene Arabia–Eurasia collision was in the latest Aalenian to Bathonian (the Mid Cimmerian orogeny; Zonenshain & Le Pichon 1986; Nikishin *et al.* 1998; Saintot *et al.* 2006; Nikishin *et al.* 2012), so any forms that have been reworked more than once are likely to be Early to Middle Jurassic in age or older.

## Analytical results

### Sandstone petrography

Sandstone compositions range from subarkose and sublitharenite to lithic arenite (Fig. 4a). Rock fragments are dominated by sediment, with lesser volcanic and negligible metamorphic types (Fig. 4b). Authigenic phases are dominated by ferroan calcite (up to 31%) and minor limonite (up to 8%) (Fig. 5). Other authigenic phases make up, on average, less than 2% of the total rock volume. The majority of samples have undergone moderate to high levels of compaction, leading to concavo-convex grain contacts and a marked reduction in primary porosity. $T_{max}$ values of adjacent mudstones indicate that they are thermally immature to

**Fig. 6.** Reservoir performance crossplots for a subset of Oligo-Miocene samples derived from the western Caucasus for which He porosity and permeability data are available. Reservoir quality categories are shown in (b). Note that visual porosity underestimates measured porosity due to the presence of microporosity (**a**), the strong correlation between measured porosity and permeability (**b**), and the moderate to strong inverse correlation between permeability and authigenic phases and matrix volume (**c**). This would imply that although rigid grain content controls compactional porosity loss and therefore minus cement porosity (Fig. 10), it is the amount of authigenic phases and matrix that controls reservoir performance at outcrop.

transitional early mature (415–435 °C). Limited spore colour and vitrinite reflectance analyses are consistent with these findings (SCI = 2–3–5–6; $V_R$ = 0.24–0.40). A lack of detrital apatite fission track annealing (Vincent *et al.* 2011) and limited or no (early) quartz cementation would suggest that the sandstones have not been heated beyond 80 °C during burial (Walderhaug 1994; Primmer *et al.* 1997; Worden & Morad 2000). Maximum recorded overburden thicknesses are *c.* 3 km. Feldspars have undergone varying degrees of dissolution and are typically replaced by calcite. In some instances dissolution at outcrop cannot be discounted and may result in a limited artificial reduction in the arkosic nature of the sandstones and an increase in secondary porosity. Visual porosity is up to 23% (average 8%). He porosity and permeability data were obtained for a subset of the samples. Visual porosity determinations underestimate measured porosity due to the presence, on average, of *c.* 10% microporosity (Fig. 6a). There

is a good correlation between He porosity and log permeability (Fig. 6b).

Large-scale spatial differences in the composition of the sandstones occur along the length of the range. Those from the Russian western Caucasus are significantly more quartz-rich and rock fragment-poor than those farther SE in western Georgia (incl. Abkhazia) (Fig. 4a). The largest differences in provenance-diagnostic grain type between the regions occur in the mudstone and volcanic rock fragment categories (Fig. 7). Caucasus-derived Oligo-Miocene sandstones in western Georgia contain, on average, *c.* 10% more volcanic rock fragments and *c.* 19% more mudstone rock fragments (relative to total extrabasinal detrital grains) than those from the southern side of the Russian western Caucasus (Fig. 7). West Georgian samples also typically contain a higher proportion of volcanic rock fragments relative to other rock fragment types (Fig. 4b). In addition to differences in detrital grain composition, samples from the Russian

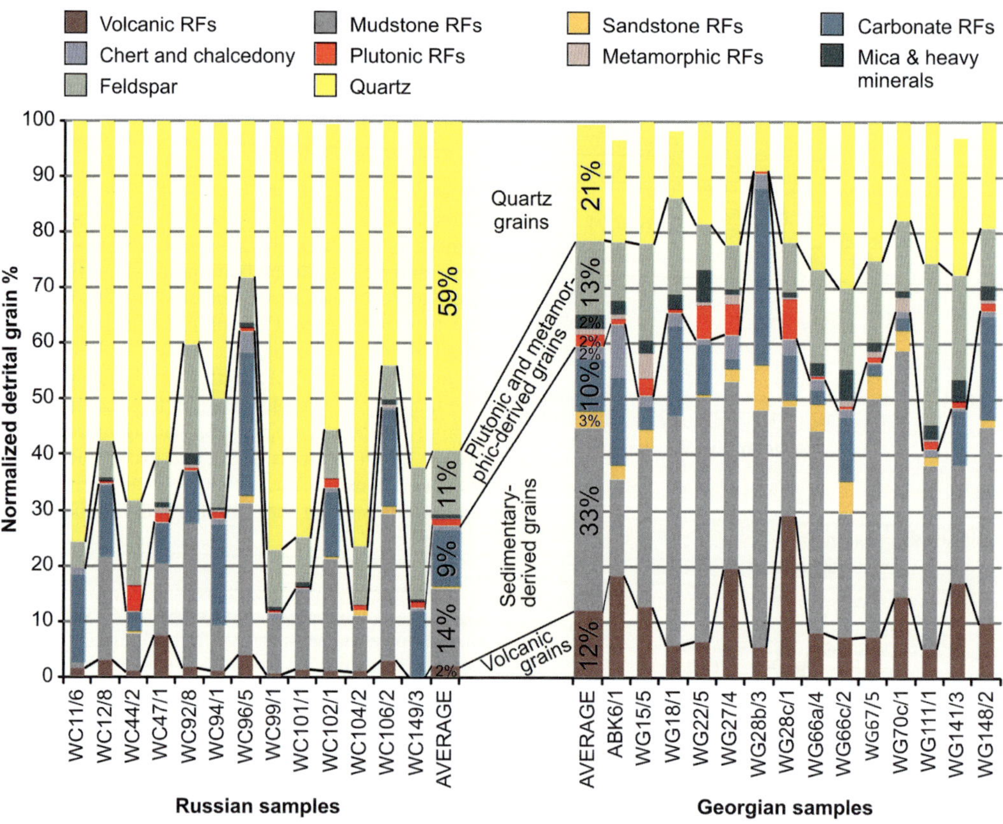


**Fig. 7.** Detrital grain variability plot of Caucasus-derived Oligo-Miocene sandstone samples on the southern side of the range. Totals do not reach 100% where unidentifiable grains were encountered. Note the higher proportion of quartz in the Russian samples and the larger amounts of mudstone and volcanic rock fragments in the Georgian samples.

western Caucasus have larger intergranular volumes and minus cement porosities. This difference is mainly due to an increase in ferroan calcite cement (Fig. 5). Volumes of other intergranular components are typically similar between the regions, although some Russian western Caucasus samples contain minor amounts of quartz cement. West Georgian samples contain larger amounts of secondary porosity (Fig. 5). This is likely to reflect the larger quantities of unstable grains within these sandstones (Fig. 7).

## Palynology data

The age range of palynomorphs contained in the mudstone samples from the Mzimta River section in the Russian western Caucasus and the Chanis River section in western Georgia are graphically presented in Figure 3. There are marked spatial differences in the age of the sedimentary rocks reworked during the uplift and erosion of the Greater Caucasus Basin. In the Russian Mzimta River section, Jurassic and Cretaceous forms are present throughout, with a lesser, variable contribution from Paleocene and Eocene sedimentary rocks. The age of the oldest sedimentary rocks being reworked increases through time and includes Late Palaeozoic forms in the youngest sample (Fig. 3a). In the west Georgian Chanis River section, reworked palynomorphs are dominated by Eocene forms, with Cretaceous forms becoming more abundant upsection (Fig. 3b).

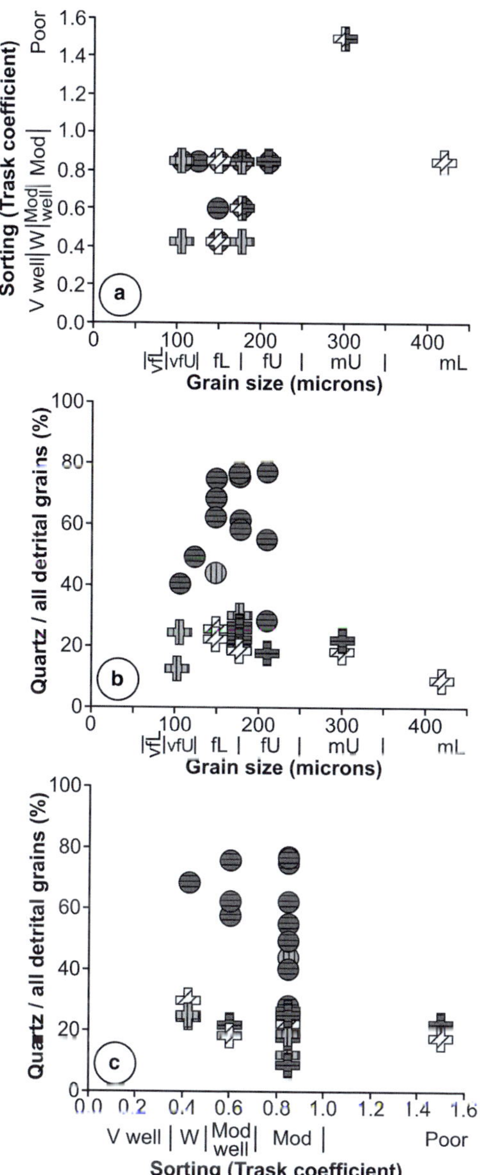


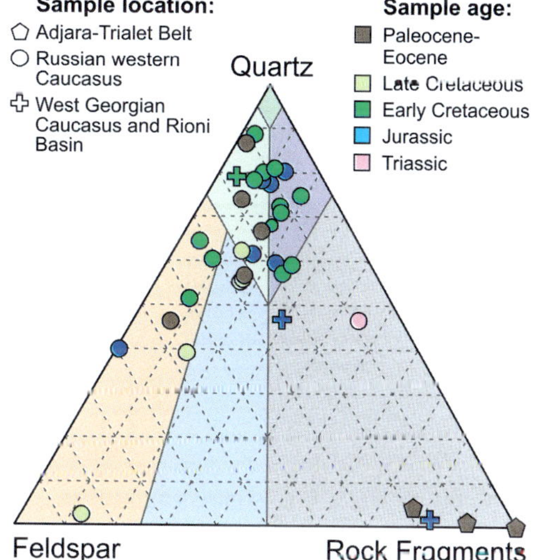


**Fig. 8.** Crossplots of (**a**) sorting v. grain size, and (**b, c**) quartz amount, as a proportion of total detrital grains, v. (b) grain size and (c) sorting. See Figure 4 for the key. Note that for any given grain size or sorting value, samples derived from the Russian western Caucasus (circles) typically have a higher proportion of quartz than those from west Georgia (crosses).

**Fig. 9.** Quartz–feldspar–rock fragment (QFR) plot of pre-Oligocene sandstones from the Caucasus region. Note the marked compositional difference between Paleocene–Eocene samples from the Russian western Caucasus and those from the Adjara–Trialet Belt (the compositional analogue for inferred sandstones from the west Georgian Caucasus).

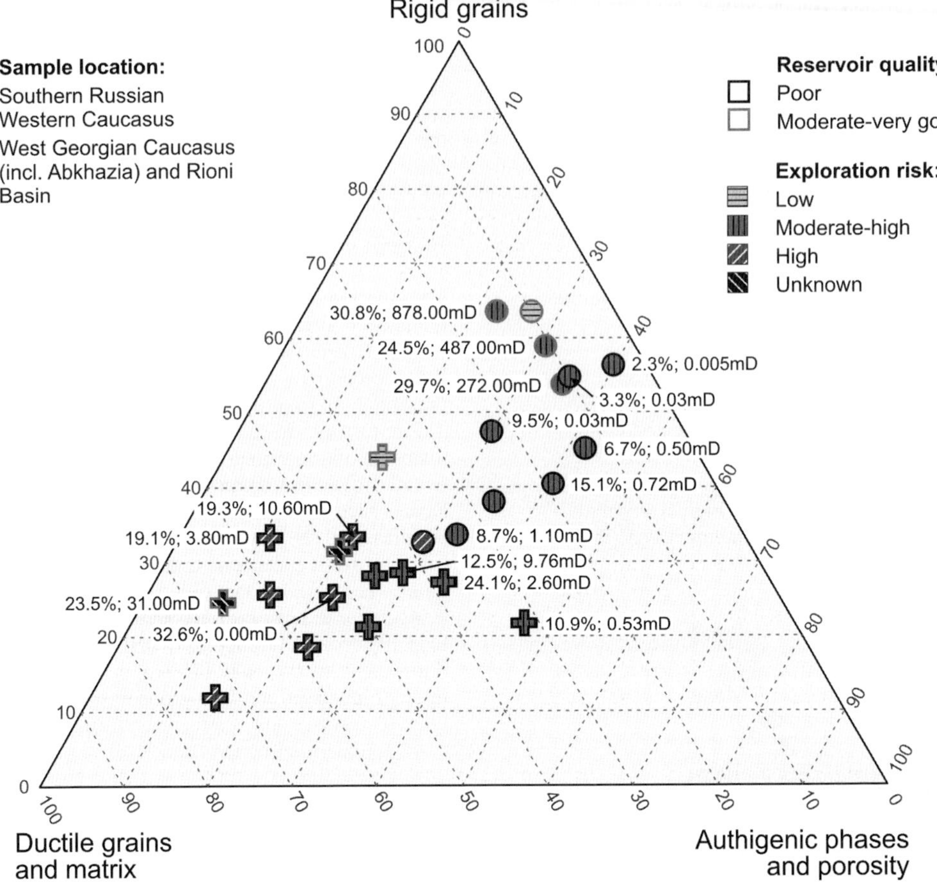



**Fig. 10.** Rigid grain–ductile grain and matrix–authigenic phases and porosity (minus cement porosity) plot of Oligo-Miocene samples derived from the western Caucasus on the southern side of the range. Rigid grains are plutonic, medium-grade metamorphic, sandstone and chert and chalcedony rock fragments, quartz and heavy minerals. Ductile grains are low-grade metamorphic, volcanic and mudstone rock fragments, micas, peloids, intraclasts, carbonaceous flakes, glauconite/berthierine/limonite pellets and phosphate. Feldspar grains, carbonate rock fragments, bioclasts and ooids vary in their mechanical behaviour and so are not included in either category. Samples have been annotated with He porosity and permeability data where available. Note (1) the general trend of decreasing volumes of authigenic phases and porosity, indicative of increasing amounts of compactional porosity loss, with increasing amounts of ductile grains and matrix, and (2) the segregation of the Russian and west Georgian samples due to their differing mechanical properties. Reservoir-grade samples are defined as those with greater than 15% He porosity and 15 mD permeability or, where these data are not available, greater than 10% visual porosity and less than 10% of authigenic phases and matrix (a proxy for permeability; see Fig. 6c). The exploration risk levels were derived using the outcrop sample appraisal methodology of Tobin (1997).

## Controls on the along-strike variability of the composition of sandstones derived from the Caucasus

### *Surface process controls*

Samples from the Russian western Caucasus were deposited by turbidity currents. As a consequence, they may have been transported farther by marine processes than samples from western Georgia that were deposited in a variety of shallow-marine shoreface and deltaic environments. However, the distance from the present-day Caucasus drainage divide to the sample sites would suggest that overall transport distances were broadly similar (up to 110 km; Fig. 1). The majority of sandstones are fine-grained and moderately to well sorted (Fig. 8a). Typically, where surface processes are an important factor in controlling sandstone composition, this is reflected in a correlation between physical and compositional maturity. This is not apparent in the Caucasus dataset (Fig. 8b, c). However, when sandstones with similar grain sizes or sorting parameters are compared, those from the Russian western Caucasus are almost always more quartz-rich than those from western Georgia (Fig. 8b, c). This instead demonstrates the overriding control of source area provenance on Oligo-Miocene sandstone composition, typical in situations like the western Greater Caucasus where sediment transport lengths are short, and relief and sedimentation rates are high (e.g. Suttner *et al.* 1981; Garzanti 1986).

*Sediment reworking controls*

Minor spatial differences in the amount of sediment derived from igneous and metamorphic rocks in the crystalline core of the Caucasus are apparent in the petrographic data (Fig. 7). Zircon U–Pb age dating is ongoing in an attempt to quantify the proportion of first-cycle sediment within the Oligo-Miocene sedimentary pile. The main along-strike variations in composition, however, occur in the amount of sedimentary and volcanic grains eroded during the inversion of the Greater Caucasus Basin (Fig. 7). Although mudstone rock fragments show

the greatest variation, it is impossible to match these to specific mudstone intervals within the Greater Caucasus Basin stratigraphy. Constraining controls on volcanic rock fragment variability is more informative and is discussed in the following text.

Volcanic intervals are present in the Middle Jurassic, mid Cretaceous and Middle Eocene stratigraphy of the Greater Caucasus and Adjara–Trialet Belt (Fig. 2). A large contribution from a Middle Jurassic volcanic source in western Georgia is not supported by the palynological data. Although mid Cretaceous volcanic rocks are present in the Transcaucasus (and Russian western Caucasus), they are not present in the west Georgian sector of the Greater Caucasus Basin (Fig. 2; Adamia *et al.* 1992). Consequently, the reworking of Cretaceous strata cannot be the cause for the higher proportion of volcanic rock fragments in Oligo-Miocene sandstones derived from the west Georgian Caucasus. The Eocene fill of the Greater Caucasus Basin is not exposed in western Georgia. (Thin accumulations of Eocene sediments in the west Georgian Caucasus represent shallow-water facies deposited to the south of the Greater Caucasus Basin on the northern margin of the Transcaucasus.) However, large volumes of volcanic and volcaniclastic rocks

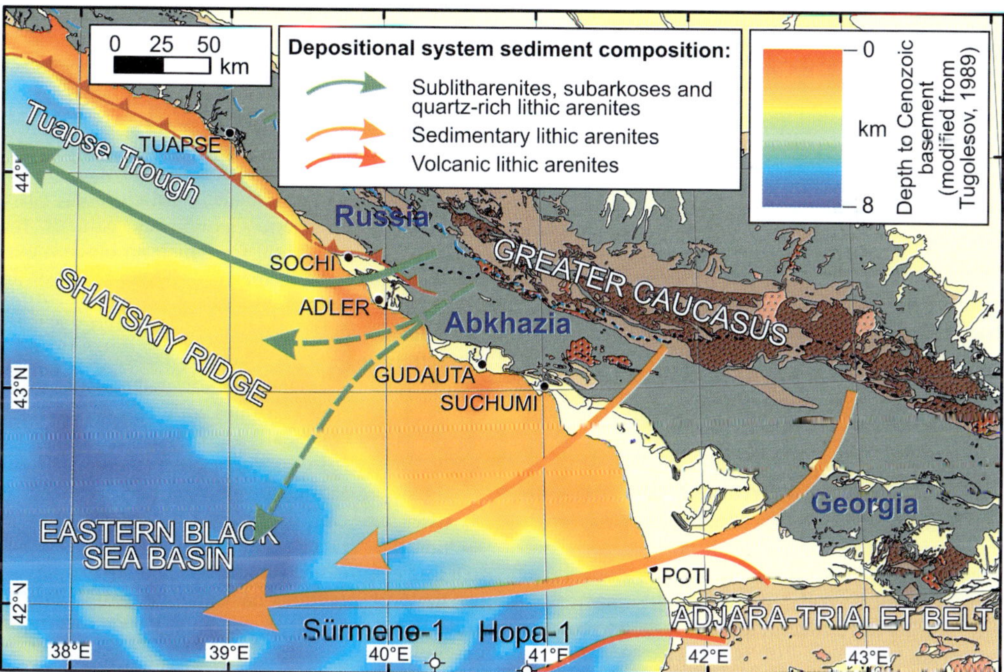


**Fig. 11.** Map showing variations in the composition of sandstones deposited by Oligocene and younger sedimentary systems derived from the western Greater Caucasus in the Eastern Black Sea. Sediment dispersal routes are interpreted from unpublished seismic data. See Figure 1 for the key to the background geological map.

do occur in the Adjara–Trialet Belt to the south of the west Georgian Caucasus (Adamia *et al.* 1974; Lordkipanidze & Zakariadze 1974; Kazmin *et al.* 1986; Yılmaz *et al.* 1997) (Fig. 1). Similar rocks are also present farther east in the Talysh region of Azerbaijan (Kazmin *et al.* 1986; Vincent *et al.* 2005). Detailed mapping indicates that these narrow east–west elongated outcrops represent the fill of rapidly subsiding Middle Eocene transtensional basins (Vincent *et al.* 2005). Eocene volcanism is also common across large parts of Iran, Turkey and Armenia (Kazmin *et al.* 1986; Vincent *et al.* 2005; Keskin *et al.* 2008; Verdel *et al.* 2011).

Palaeocurrent and facies mapping indicate that Eocene volcanic rocks in the Adjara–Trialet Belt were not the source of the sandstones in the west Georgian Caucasus; they were derived from farther north within the Caucasus. We conclude therefore that a narrow volcanic and volcaniclastic-rich sub-basin, similar to the Adjara–Trialet Basin, may have previously existed in the west Georgian sector of the Greater Caucasus Basin and that this has subsequently been completely removed during basin inversion. This conclusion is supported by young detrital apatite fission track populations in western Georgia that have age profiles best explained by temporary storage within a transient Eocene Caucasus basin (Vincent *et al.* 2011, their fig. 7). The lithologies within this sub-basin would have been markedly different from those Eocene strata that still crop out farther to the NW in the Russian sector of the Greater Caucasus Basin that are non-volcanic in nature (Fig. 2) and include sub-arkosic, lithic arkosic and arkosic turbidite sandstones (Fig. 9). The presence of this former volcanic and volcaniclastic-rich sub-basin can be tested by dating mudstone and volcanic clasts in conglomerates within the Oligo-Miocene stratigraphy of the west Georgian Caucasus.

## Implications for reservoir properties in the Eastern Black Sea

The limited He porosity and permeability data available in this study would suggest that reservoir performance at outcrop is linked to authigenic phase and matrix volume (Fig. 6c). This is because pervasive ferroan calcite cementation has filled much of the post-compactional minus cement porosity (Fig. 5); only four samples from the Russian western Caucasus and three samples from western Georgia have moderate to very good reservoir properties (Figs 6b & 10). It is unclear whether this will also be the case for equivalent sandstones in the subsurface, as the prediction of authigenic phase composition and volume is beyond the scope of this study. This requires a detailed understanding of the time-temperature history of the potential target interval. This uncertainty is reflected in the reservoir assessment methodology of Tobin (1997), which suggests that most of the samples in this study, if situated in the subsurface, would carry a moderate to high exploration risk (Fig. 10).

That being said, numerous studies indicate that sandstone composition is a major control on reservoir quality (e.g. Nagtegaal 1978; Bloch 1991; Pittman & Larese 1991; Lundegard 1992; Dutton *et al.* 2012; Tobin & Schwarzer 2013). Data presented here provide clear evidence that lateral variations in sandstone composition will be present within the Oligo-Miocene fill of the Eastern Black Sea. The larger quantities of ductile grains derived from western Georgia will result in greater sandstone compactional volume loss, and a decrease in intergranular volume and minus cement porosity in sandstones in the subsurface relative to those derived from the Russian western Caucasus (Figs 5 & 10). Despite uncertainties over expected levels of carbonate cementation, it is reasonable to assume, therefore, that the quartz-rich depositional systems derived from the Russian western Caucasus will have the potential to form better-quality reservoirs at shallow and moderate burial depths than their lithic-rich counterparts derived from western Georgia (e.g. Tobin & Schwarzer 2013). At greater depths, the preferential development of quartz cements in quartz-rich sandstones may mitigate against this trend.

## Conclusions

The most striking conclusion of this work is the possibility that an Eocene transtensional basin developed in western Georgia as part of the Greater Caucasus Basin for which no outcrop evidence remains. As in the Adjara–Trialet Basin to the south, it would appear that this basin was filled by volcanic and volcaniclastic rocks. Sediments derived from the inversion of these basins may significantly reduce the reservoir potential of Oligocene and younger sandstones within the easternmost part of the Black Sea (Fig. 11). Thus, although a crude estimate of potential sandstone compositions within young sedimentary basins may be made by an assessment of the current lithologies present within the river catchments of these basins, extreme caution needs to be taken when extrapolating these observations back in time, particularly in areas that have undergone recent episodes of rapid exhumation.

Oligo-Miocene sedimentary rocks derived from the inversion of the Greater Caucasus Basin in the Russian western Caucasus include sediment reworked from Jurassic strata, as well as the

Eocene and Cretaceous intervals also reworked into west Georgian samples. Jurassic, Early Cretaceous and Eocene sedimentary rocks in the Russian western Caucasus are relatively sandstone-rich (Fig. 2), with these sandstones being dominated by subarkoses and sublitharenites (Fig. 9). Sediment derivation from these relatively quartz-rich sandstones is likely to be a major reason why Oligo-Miocene sedimentary rocks on the southern margin of the Russian western Caucasus are also quartz-rich. These sedimentary rocks form the up-dip components of turbidite systems, which form potential hydrocarbon targets in the Tuapse Trough (e.g. Mityukov *et al.* 2011), and possibly in the central Eastern Black Sea (Fig. 11).

It is not possible to accurately constrain the boundary between the quartz- and rock fragment-rich depositional systems entering the Eastern Black Sea because of the paucity of data from the disputed Abkhazian region. It is possible that the Middle Jurassic granitoids that intrude the fill of the Greater Caucasus Basin north of Suchumi delimit the westward extent of the west Georgian volcanic- and volcaniclastic-rich Eocene Caucasus basin. Equally, the submarine Gudauta arch may have formed the bathymetric divide between lithic-rich, potentially poor-reservoir-quality depositional systems shed into the easternmost part of the Eastern Black Sea, and quartz-rich, potentially better-reservoir-quality depositional systems feeding the Tuapse Trough and, during lowstands, possibly the central part of the Eastern Black Sea (Fig. 11).

This work builds on over 12 years of research carried out by CASP and its collaborators in the Caucasus region. The research would not have been possible without the field assistance of M. Simmons, S. Inger and L. Voronova (all formerly CASP), T. Barabadze (Georgian Oil, Georgia), T. Pinchuk (Krasnodarneftgas, Russia) and V. Lavrishchev (Kavkazgeols'emka, Russia) or the funding of CASP's industrial sponsors. The analytical assistance of A. Morton (HM Associates, UK), A. Carter (UCL, UK) and S. Gibbs (NOCS, UK) during earlier phases of this research is gratefully acknowledged. WB would like to thank S. Akbari for his palynological preparation and J. Marshall (both NOCS, UK) for his useful discussions. This paper has been improved by comments from J. Omma, reviewers R. Tobin and S. Andó, and editor H. Smyth. It is dedicated to the memories of R. Scott and T. Elliott. This is Cambridge Earth Science contribution esc.2676.

# References

ADAMIA, S. A., AKHVLEDIANI, K. T., KILASONIA, V. M., NAIRN, A. E. M., PAPAVA, D. & PATTON, D. K. 1992. Geology of the Republic of Georgia: a review. *International Geology Review*, **34**, 447–476.

ADAMIA, S. A., GAMKRELIDZE, I. P., ZAKARIADZE, G. S. & LORDKIPANIDZE, M. B. 1974. Adjaria–Trialeti trough and problem of the Black Sea origin. *Geotektonica*, **1**, 78–94 [in Russian].

ALLEN, M. B. & ARMSTRONG, H. A. 2008. Arabia–Eurasia collision and the forcing of mid-Cenozoic global cooling. *Palaeogeography, Palaeoclimatology, Palaeoecology*, **265**, 52–58.

BARRIER, E. & VRIELYNCK, B. 2008. *Palaeotectonic Maps of the Middle East*. MEBE, Paris, 1:18 500 000.

BLOCH, S. 1991. Empirical prediction of porosity and permeability in sandstones. *AAPG Bulletin*, **75**, 1145–1160.

DERCOURT, J., GAETANI, M. ET AL. 2000. *Atlas Peri-Tethys, Palaeogeographical maps*. CCGM, CGMW, Paris.

DERCOURT, J., ZONENSHAIN, L. P. ET AL. 1986. Geological evolution of the Tethys belt from the Atlantic to the Pamirs since the Lias. *Tectonophysics*, **123**, 241–315.

DUTTON, S. P., LOUCKS, R. G. & DAY-STIRRAT, R. J. 2012. Impact of regional variation in detrital mineral composition on reservoir quality in deep to ultradeep lower Miocene sandstones, western Gulf of Mexico. *Marine and Petroleum Geology*, **35**, 139–153.

GARZANTI, E. 1986. Source rock versus sedimentary control on the mineralogy of deltaic volcanic arenites (Upper Triassic, northern Italy). *Journal of Sedimentary Research*, **56**, 267–275.

GOLONKA, J. 2004. Plate tectonic evolution of the southern margin of Eurasia in the Mesozoic and Cenozoic. *Tectonophysics*, **381**, 235–273.

GRADSTEIN, F. M., OGG, J. G. & SMITH, A. G. 2004. *A Geologic Time Scale 2004*. Cambridge University Press, Cambridge.

HANEL, M., GURBANOV, A. G. & LIPPOLT, H. J. 1992. Age and genesis of granitoids from the Main-Range and Bechasyn zones of the western Great Caucasus. *Neues Jahrbuch Fur Mineralogie-Monatshefte*, **1992**, 529–544.

HANEL, M., LIPPOLT, H. J., KOBER, B., GURBANOV, A. G. & BORSUK, A. M. 1993. On the Early Paleozoic age of metagranodiorites in the Main Range Zone of the Greater Caucasus. *Petrology*, **1**, 487–498.

HARRELL, J. 1964. A visual comparator for degree of sorting in thin and plane section. *Journal of Sedimentary Petrology*, **54**, 646–650.

JAANUSSON, V. 1972. Constituent analysis of an Ordovician limestone from Sweden. *Lethaia*, **5**, 217–237.

KAZ'MIN, V. G. & TIKHONOVA, N. F. 2006a. Evolution of Early Mesozoic back-arc basins in the Black Sea–Caucasus segment of a Tethyan active margin. In: ROBERTSON, A. H. F. & MOUNTRAKIS, D. (eds) *Tectonic Development of the Eastern Mediterranean Region*. Geological Society, London, Special Publications, **260**, 179–200.

KAZ'MIN, V. G. & TIKHONOVA, N. F. 2006b. Late Cretaceous–Eocene marginal seas in the Black Sea–Caspian region: paleotectonic reconstructions. *Geotectonics*, **40**, 169–182.

KAZMIN, V. G., SBORTSHIKOV, I. M., RICOU, L.-E., ZONENSHAIN, L. P., BOULIN, J. & KNIPPER, A. L. 1986. Volcanic belts as markers of the Mesozoic–Cenozoic active margin of Eurasia. *Tectonophysics*, **123**, 123–152.

KESKIN, M., GENÇ, S. C. & TÜYSÜZ, O. 2008. Petrology and geochemistry of post-collisional Middle Eocene volcanic units in North-Central Turkey: evidence for magma generation by slab breakoff following the closure of the Northern Neotethys Ocean. *Lithos*, **104**, 267–305.

LAVRISHCHEV, V. A., PRUTSKIY, N. I. & SEMENOV, V. M. 2002. *National Geological Map of the Russian Federation, Caucasus Series Sheet K-37-V (Krasnar Polyana)*. St. Petersberg Cartographic Enterprise of VSEGEI, Moscow, 1:200 000 [in Russian].

LORDKIPANIDZE, M. B. & ZAKARIADZE, G. S. 1974. Paleogene volcanism of Archara. Problems of geology of Achara–Trialeti. *Proceedings of Institute of Geology of GSSR*, **44**, 74–86 [in Russian].

LUNDEGARD, P. D. 1992. Sandstone porosity loss; a 'big picture' view of the importance of compaction. *Journal of Sedimentary Research*, **62**, 250–260.

McCANN, T., CHALOT-PRAT, F. & SAINTOT, A. 2010. The Early Mesozoic evolution of the Western Greater Caucasus (Russia): Triassic–Jurassic sedimentary and magmatic history. *In*: SOSSON, M., KAYMAKCI, N., STEPHENSON, R. A., BERGERAT, F. & STAROSTENKO, V. (eds) *Sedimentary Basin Tectonics from the Black Sea and Caucasus to the Arabian Platform*. Geological Society, London, Special Publications, **340**, 181–238.

MEISNER, A., KRYLOV, O. & NEMCOK, M. 2009. Development and structural architecture of the Eastern Black Sea. *The Leading Edge*, **28**, 1046–1055.

MINSHULL, T. A., WHITE, N. J. *ET AL.* 2005. Seismic data reveal Eastern Black Sea Basin structure. *EOS*, **86**, 413, 419.

MITYUKOV, A., AL'MENDINGER, O., MYASOEDOV, N., NIKISHIN, A. & GAIDUK, V. 2011. The sedimentation model of the Tuapse Trough (Black Sea). *Doklady Earth Sciences*, **440**, 1245–1248.

NAGTEGAAL, P. J. C. 1978. Sandstone-framework instability as a function of burial diagenesis. *Journal of the Geological Society, London*, **135**, 101–105.

NIKISHIN, A. M., CLOETINGH, S., BRUNET, M.-F., STEPHENSON, R. A., BOLOTOV, S. N. & ERSHOV, A. V. 1998. Scythian Platform, Caucasus and Black Sea region: Mesozoic–Cenozoic tectonic history and dynamics. *In*: CRASQUIN-SOLEAU, S. & BARRIER, E. (eds) *Peri-Tethys Memoir 3: Stratigraphy and Evolution of Peri-Tethyan Platforms*, Mémoire. Muséum national d'Histoire naturelle, Paris, **177**, 163–176.

NIKISHIN, A. M., ZIEGLER, P. A. *ET AL.* 2001. Mesozoic and Cainozoic evolution of the Scythian Platform–Black Sea–Caucasus domain. *In*: ZIEGLER, P. A., CAVAZZA, W., ROBERTSON, A. H. F. & CRASQUIN-SOLEAU, S. (eds) *Peri-Tethys Memoir 6: Peri-Tethyan Rift/Wrench Basins and Passive Margins*, Mémoire. Muséum national d'Histoire naturelle, Paris, **186**, 295–346.

NIKISHIN, A. M., ZIEGLER, P. A., BOLOTOV, S. N. & FOKIN, P. A. 2012. Late Palaeozoic to Cenozoic evolution of the Black Sea–southern Eastern European region: a view from the Russian Platform. *Turkish Journal of Earth Sciences*, **21**, 571–634.

PETTIJOHN, F. J., POTTER, P. E. & SIEVER, R. 1987. *Sand and Sandstone*. New York, Springer Verlag.

PITTMAN, E. D. & LARESE, R. E. 1991. Compaction of lithic sands; experimental results and applications. *AAPG Bulletin*, **75**, 1279–1299.

PRIMMER, T. J., CADE, C. A. *ET AL.* 1997. Global patterns in sandstone diagenesis: their application to reservoir quality prediction for petroleum exploration. *In*: KUPECZ, J. A., GLUYAS, J. & BLOCH, S. (eds) *Reservoir Quality Prediction in Sandstones and Limestones*. American Association of Petroleum Geologists, Memoirs, **69**, 61–77.

ROBINSON, A. G., RUDAT, J. H., BANKS, C. J. & WILES, R. L. F. 1996. Petroleum geology of the Black Sea. *Marine and Petroleum Geology*, **13**, 195–223.

SAINTOT, A., CHALOT-PRAT, F., McCANN, T., FOKIN, P., KORSAKOV, S. & STEPHENSON, R. 2004. *The Early Middle Jurassic basins of the Western Greater Caucasus. In*: AAPG European Region Conference with GSA, Program & Abstracts, Prague, 101–102.

SAINTOT, A., BRUNET, M.-F. *ET AL.* 2006. The Mesozoic–Cenozoic tectonic evolution of the Greater Caucasus. *In*: GEE, D. & STEPHENSON, R. (eds) *European Lithosphere Dynamics*. Geological Society, London, Memoirs, **32**, 277–289.

ŞENGÖR, A. M. C. & NATAL'IN, B. A. 1996. Paleotectonics of Asia: fragments of a synthesis. *In*: YIN, A. & HARRISON, T. M. (eds) *The Tectonic Evolution of Asia*. Cambridge University Press, Cambridge, 486–640.

SOMIN, M. L. 2000. Structure of axial zones in the Central Caucasus. *Doklady Earth Sciences*, **375**, 1371–1374.

SOMIN, M. L., KOTOV, A. B., SAL'NIKOVA, E. B., LEVCHENKOV, O. A., PIS'MENNYI, A. N. & YAKOVLEVA, S. Z. 2006. Paleozoic rocks in infrastructure of the metamorphic core, the Greater Caucasus Main Range Zone. *Stratigraphy and Geological Correlation*, **14**, 475–485.

SOMIN, M. L., LEPEKHINA, E. N. & KONILOV, A. N. 2007. Age of the high-temperature gneiss core of the Central Caucasus. *Doklady Akademii Nauk*, **414**, 1–5 [in Russian].

SOMIN, M. L., POTAPENKO, Y. Y. & SMUL'SKAYA, A. I. 2009. Chuchkhur xenoliths and tectonic position of Middle Paleozoic volcano-sedimentary sequences in the Peredovoi Range, Northern Caucasus. *Doklady Earth Sciences*, **428**, 1097–1099.

STAROSTENKO, V., BURYANOV, V. *ET AL.* 2004. Topography of the crust–mantle boundary beneath the Black Sea Basin. *Tectonophysics*, **381**, 211–233.

SUTTNER, L. J., BASU, A. & MACK, G. H. 1981. Climate and the origin of quartz arenites. *Journal of Sedimentary Petrology*, **51**, 1235–1246.

TOBIN, R. C. 1997. Porosity prediction in frontier basins: a systematic approach to estimating subsurface reservoir quality from outcrop samples. *In*: KUPECZ, J. A., GLUYAS, J. & BLOCH, S. (eds) *Reservoir Quality Prediction in Sandstones and Limestones*. American Association of Petroleum Geologists, Memoirs, **69**, 1–18.

TOBIN, R. C. & SCHWARZER, D. 2013. Effects of sandstone provenance on reservoir quality preservation in the deep subsurface: experimental modeling of deep-water sand in the Gulf of Mexico. *In*: SCOTT, R. A., SMYTH, H. R., MORTON, A. C. & RICHARDSON, N. (eds) *Sediment Provenance Studies in Hydrocarbon Exploration and Production*. Geological Society, London, Special

Publications, **386**. First published online September 12, 2013, http://dx.doi.org/10.1144/SP386.17

TRELOAR, P. J., MAYRINGER, F., FINGER, F., GERDES, A. & SHENGALIA, D. 2009. New age data from the Dzirula Massif, Georgia: implications for Variscan evolution of the Caucasus. *In*: DALKILIÇ, M. (ed.) *2nd International Symposium on the Geology of the Black Sea Region*. General Directorate of Mineral Research and Exploration (MTA), Ankara, Turkey, Abstracts, 204–205.

TUGOLESOV, D. A. 1989. *Album of Structural and Thickness Maps of Black Sea Basin Cenozoic Sediments; 1:1 500 000*. Main Administration of Geodesy and Cartography under the Council of Ministers of the USSR, Moscow, **71** [in Russian].

VAN DER PLAS, L. & TOBI, A. C. 1965. A chart for judging the reliability of point counting results. *American Journal of Science*, **263**, 87–90.

VERDEL, C., WERNICKE, B. P., HASSANZADEH, J. & GUEST, B. 2011. A Paleogene extensional arc flare-up in Iran. *Tectonics*, **30**, 20.

VINCENT, S. J., ALLEN, M. B., ISMAIL-ZADEH, A. D., FLECKER, R., FOLAND, K. A. & SIMMONS, M. D. 2005. Insights from the Talysh of Azerbaijan into the Paleogene evolution of the South Caspian region. *Geological Society of America Bulletin*, **117**, 1513–1533.

VINCENT, S. J., MORTON, A. C., CARTER, A., GIBBS, S. & BARABADZE, T. G. 2007. Oligocene uplift of the Western Greater Caucasus; an effect of initial Arabia–Eurasia collision. *Terra Nova*, **19**, 160–166.

VINCENT, S. J., CARTER, A., LAVRISHCHEV, V. A., RICE, S. P., BARABADZE, T. G. & HOVIUS, N. 2011. The exhumation of the western Greater Caucasus: a thermochronometric study. *Geological Magazine*, **148**, 1–21.

VINCENT, S. J., MORTON, A. C. & HYDEN, F. 2013. The composition, provenance and reservoir potential of Cenozoic siliciclastic depositional systems supplying the northern margin of the eastern Black Sea. *Marine and Petroleum Geology*, **45**, 331–348.

WALDERHAUG, O. 1994. Temperatures of quartz cementation in Jurassic sandstones from the Norwegian continental shelf – evidence from fluid inclusions. *Journal of Sedimentary Research*, **A64**, 311–323.

WORDEN, R. H. & MORAD, S. 2000. Quartz cementation in oil field sandstones: a review of the key controversies. *In*: WORDEN, R. H. & MORAD, S. (eds) *Quartz Cementation in Sandstones*. Wiley-Blackwell, Oxford, IAS Special Publication, **29**, 1–20.

YILMAZ, A., ADAMIA, S., ENGIN, T. & LAZARASHVILI, T. 1997. *Geoscientific Studies of the Area Along Turkish–Georgian Border*. MTA-SDG-GIN, Ankara.

ZAKARIADZE, G. S., DILEK, Y., ADAMIA, S. A., OBERHÄNSLI, R. E., KARPENKO, S. F., BAZYLEV, B. A. & SOLOV'EVA, N. 2007. Geochemistry and geochronology of the Neoproterozoic Pan-African Transcaucasian Massif (Republic of Georgia) and implications for island arc evolution of the late Precambrian Arabian–Nubian Shield. *Gondwana Research*, **11**, 92–108.

ZONENSHAIN, L. P. & LE PICHON, X. 1986. Deep basins of the Black Sea and Caspian Sea as remnants of Mesozoic back-arc basins. *Tectonophysics*, **123**, 181–211.

ZONENSHAIN, L. P., KUZMIN, M. I. & NATAPOV, L. M. 1990. *Geology of the USSR: A Plate Tectonic Synthesis*. American Geophysical Union, Geodynamics Series, Washington, **21**.

# Understanding the provenance and reservoir quality of the Sele Formation sandstones of the UK Central Graben utilizing detrital garnet suites

BEN KILHAMS[1]*, ANDREW MORTON[2,3], RICCARDO BORELLA[1], ANNE WILKINS[1] & ANDREW HURST[1]

[1]*Geology and Petroleum Geology, Meston Building, University of Aberdeen, Aberdeen AB24 3UE, UK*

[2]*HM Research Associates, 2 Clive Road, Balsall Common, West Midlands CV7 7DW, UK*

[3]*CASP, University of Cambridge, 181a Huntingdon Road, Cambridge CB3 0DH, UK*

**Corresponding author (e-mail: b.kilhams@shell.com)*

**Abstract:** Detrital garnet suites have been demonstrated to be reliable indicators of the mineralogical and lithological characteristics of sediment source areas. This study applies garnet analysis to the Paleocene to Eocene Sele Formation deep-water sandstone units of the central North Sea. These stratigraphic units are economically important as they represent one of the main hydrocarbon reservoir intervals in this mature basin. The routing of turbidity currents into the Central Graben has been demonstrated to be related to axial fans (ultimately sourced from Lewisian and Moine basement rocks and Triassic sandstones to the NW) and lateral fans (ultimately sourced from the Dalradian basement rocks to the west). Garnet analysis suggests the majority of samples can be attributed to the axial fan system and that the lateral system contributed little to sandstone deposition east of the Gannet Fields. This contradicts previous seismic mapping work, which suggested that the lateral fan system dominated sedimentation as far east as the Merganser Field. This reinterpretation is potentially important for our understanding of sediment routing and its impact on the distribution of reservoir quality, particularly as this is believed to relate directly to proximity to the shelf.

The development of the Palaeogene deep-water sandstone reservoirs of the UK Central Graben over the last half-century has been important to the sustained growth of the hydrocarbon industry in Europe (e.g. Ahmadi *et al.* 2003). However, the maturity of this basin means that a greater understanding of these depositional systems, their stratigraphic architecture and potential reservoir quality distribution is essential for locating the remaining hydrocarbons (e.g. Erratt *et al.* 2010). Characterization of these systems relies on understanding not only the basinal stratigraphy but also how these sandstones were routed from the hinterland and across the shelf. Provenance studies provide an opportunity to unlock some of these problems. This paper makes use of detrital garnet suites to assess the routing of Sele Formation (*c.* 56.8–54 Ma) turbidites into the Central Graben of the North Sea and the potential impact different provenance concepts may have on reservoir quality modelling.

## Palaeogene stratigraphy of the North Sea

The North Sea Palaeogene is divided into lithostratigraphic units (sand-rich turbiditic deposits bounded by mud-rich hemipelagic intervals) that have been further refined using biostratigraphic techniques (e.g. Deegan & Scull 1977; Carman & Young 1981; Knox *et al.* 1981; Stewart 1987; Knox & Holloway 1992; Mudge & Copestake 1992*a*, 1992*b*; Schröder 1992; Vining *et al.* 1993; Neal *et al.* 1994; Mudge & Bujak 1996*a*, 1996*b*; Neal 1996). The consensus suggests that each of these sand-rich units is related to successive relative sea-level lowstands that enabled routing of sediments from the shelf to the basin floor (e.g. White & Lovell 1997; Jennette *et al.* 2000; Kilhams *et al.* 2012).

Three major cycles of deep-water sedimentation occur during the Paleocene and Early Eocene. The two stratigraphically lowest units are the Maureen Sandstone Member within the Maureen Formation (*c.* 63 to *c.* 59.8 Ma, see Fig. 1) and the Mey Sandstone Member within the Lista Formation (*c.* 56.8–59.8 Ma), which together make up the Montrose Group (Mudge & Copestake 1992*a*). Stratigraphically above this is the Moray Group, which includes the Sele Formation (*c.* 56.8–54 Ma) and its associated sandstones as the third major depositional cycle (Forties, Bittern, Cromarty and Gannet members). This is capped by a tuffaceous unit

*From*: SCOTT, R. A., SMYTH, H. R., MORTON, A. C. & RICHARDSON, N. (eds) 2014. *Sediment Provenance Studies in Hydrocarbon Exploration and Production*. Geological Society, London, Special Publications, **386**, 129–142.
First published online October 3, 2013, http://dx.doi.org/10.1144/SP386.16

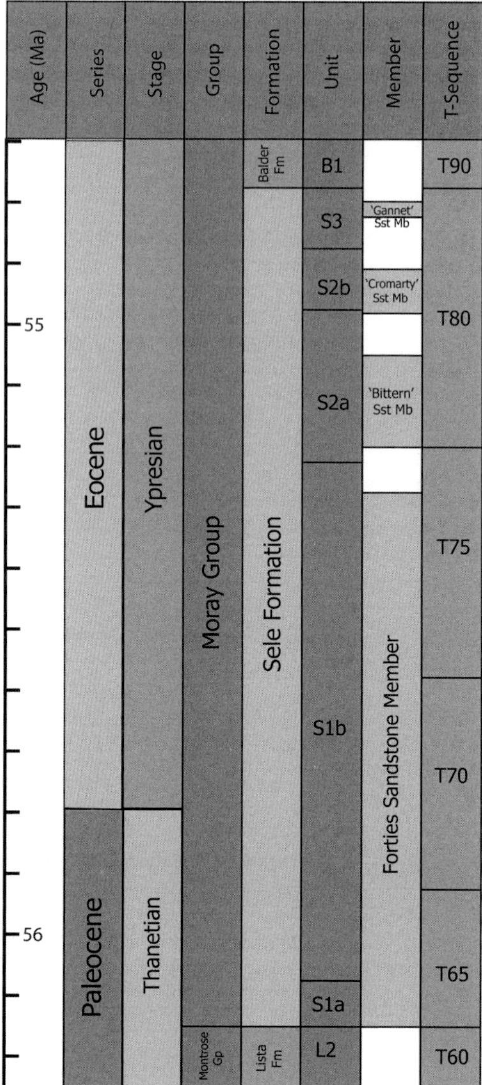

**Fig. 1.** Stratigraphic column illustrating the context of the Sele Formation sandstones within the latest Paleocene and early Eocene. Ages (Ma), series and stages from Ogg *et al.* (2008), groups from Mudge & Copestake (1992*a*), formations, units and members from Knox & Holloway (1992), T (Tertiary)-sequences after Milton *et al.* (1990) and Shell UI Europe.

known as the Balder Formation (Knox & Holloway 1992; Mudge & Copestake 1992*a*). Further smaller deep-water sandstone deposits have been recognized within the mid to late Eocene, such as the Grid and Tay sandstone members of the Horda Formation (e.g. Jennette *et al.* 2000; Kilhams *et al.* 2011).

The Sele Formation has been divided into a number of different subunits (Fig. 1). Knox &

Holloway (1992) defined three components based on mudstone-dominated sections: S1 (*c.* 56.8–56 Ma), S2 (*c.* 56–55.8 Ma) and S3 (*c.* 55.8–55.7 Ma). The S1 division contains the Forties Sandstone Member and the S2 the Cromarty Sandstone Member (Knox & Holloway 1992), with the former being the main reservoir unit in many Central Graben hydrocarbon fields (e.g. Hempton *et al.* 2005; Gill & Shepherd 2010; Scott *et al.* 2010). Less formal subdivisions have also been recognized by Shell Upstream International (UI) Europe within the S2 and S3 units describing localized sandstone members around the Gannet and Bittern fields (Figs 1 & 2). These sandstone units can be further subdivided based on detailed lithostratigraphy, with biostratigraphic support from Schröder (1992), into a series of T (Tertiary) cycles (e.g. Davis *et al.* 2009). Figure 1 illustrates that the Forties Sandstone Member can be divided into three distinct units (T65, T70 and T75). The Cromarty Sandstone Member (and associated Gannet and Bittern divisions) is associated with the T80 cycle. These T-cycles are equivalent to the units described by Jones & Milton (1994) and Dixon & Pearce (1995), which were developed synchronously but have a different naming convention (T30–40, see Ahmadi *et al.* 2003). As these T-cycles are defined on sand-rich intervals, it is considered that each one represents a period of sediment input from the shelf into the basin (Galloway 1989).

Studies of the Sele Formation (mainly based on regional three-dimensional seismic data) have concluded that there are two main sediment routes into the Central Graben: an along-graben axial system trending roughly NW to SE, and a cross-cutting lateral system trending approximately west to east (Figs 3 & 4; Morton 1987; Kulpecz & Van Geuns 1990; Den Hartog Jager *et al.* 1993; Kantorowicz *et al.* 1999; Jennette *et al.* 2000; Hempton *et al.* 2005; Davis *et al.* 2009). The lateral units may also be sourced from at least three discrete sources along the western margin of the Central Graben, allowing for the development of localized Bittern and Gannet Field sandstones within the S2 and S3 units (Fig. 3; Hempton *et al.* 2005). The largest lateral system enters the basin within the Guillemot/Gannet Field area (Kantorowicz *et al.* 1999), but there is some ambiguity as to how far east the related flows travelled. Hempton *et al.* (2005), for example, used regional seismic isochron mapping (Fig. 4; see also Jennette *et al.* 2000; Scott *et al.* 2010; Kilhams *et al.* 2012) to suggest that deposits related to this lateral unit can be found as far east as the Merganser and Scoter fields. However, these isochron maps are based on the entire Montrose and Moray Group interval and simply show thickness variations rather than precise

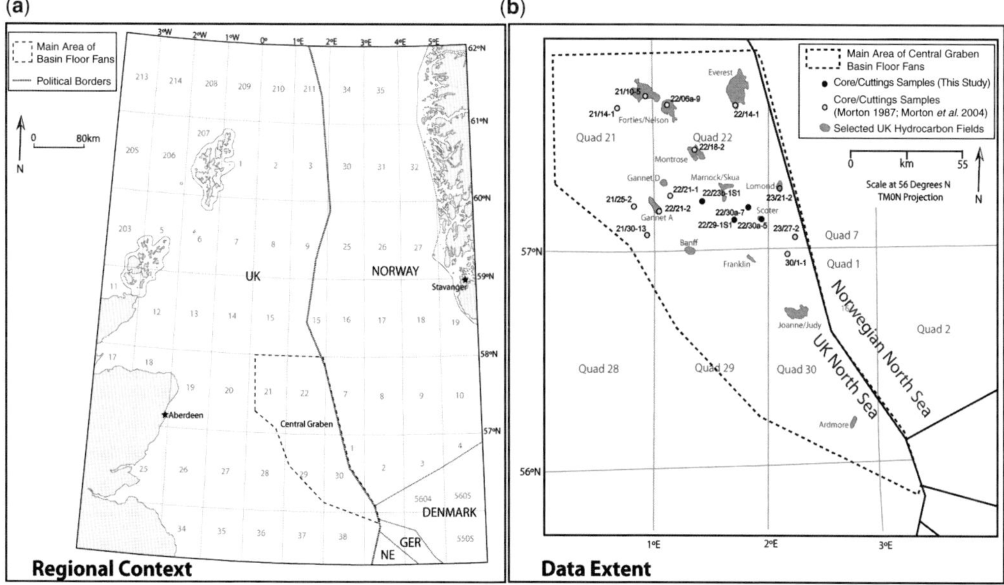


**Fig. 2.** Wider context of study locality. (**a**) Main area of basin-floor sedimentation into the Central Graben (highlighted in (b) and in Figure 3, based on Kantorowicz *et al.* (1999) and Hempton *et al.* (2005). (**b**) Study area, including selected hydrocarbon fields and the approximate locations of all wells utilized in this study and in Morton (1987) and Morton *et al.* (2004). After Kilhams *et al.* (2012).

sedimentological routing. As such, they do not illustrate the relative impact of laterally or axially derived turbidity currents within each T-cycle. Although some well correlation work has made interpretations based on the relative importance of axial and lateral units, it has not been possible to define these systems based on petrophysical response or core facies character (Kilhams *et al.* 2012). Defining these systems is potentially important because of the link made between proximity to the shelf, grain size and sandstone quality (porosity/ permeability) (Hempton *et al.* 2005; Kilhams *et al.* 2012). It has been demonstrated that more proximal axially derived deposits have a higher average grain size (fine to medium) and, therefore, greater porosity (typically over 23%) and permeability (typically over 100 mD) values, while more distal deposits have lower average grain sizes (very fine to fine) and are subsequently lower in quality (porosities *c.* 15–23% and permeabilities of 1– 100 mD) (Hempton *et al.* 2005). The routing of lateral fans into the basin has the potential to modify these relationships. For example, Figure 3 illustrates that the main 'Gannet lateral deposits' occur in a medial to distal part of the axial trend and, therefore, could result in potentially higher grain sizes and porosity/permeability values (Jennette *et al.* 2000; Ahmadi *et al.* 2003; Hempton *et al.* 2005). Therefore, understanding these source areas

has the potential to modify our understanding of regional reservoir quality. As these relationships are difficult to understand using seismic, petrophysical or core logging methods, alternative techniques are required.

## Detrital garnet suites and sediment provenance

Heavy mineral studies have been used widely to indicate provenance (e.g. Mange & Maurer 1992), with extensive applications in the North Sea from the Late Palaeozoic to the Late Palaeogene (e.g. Allen & Mange-Rajetzky 1992; Morton & Berge 1995; Hallsworth *et al.* 1996; Preston *et al.* 1998). More detailed work on the Palaeogene sandstones by Morton (1979, 1982) recognized that different source areas could be determined within these deposits related to subtle variations in heavy mineral characteristics of the sediment source areas on the East Shetland Platform and northern Scotland. Although these heavy mineral studies showed that there was some variation between source areas, they revealed little about axial and lateral depositional trends within the basin and did not enable a high-resolution discrimination of potential provenance areas. This is at least partly because heavy mineral suites show an erroneous signal (compared

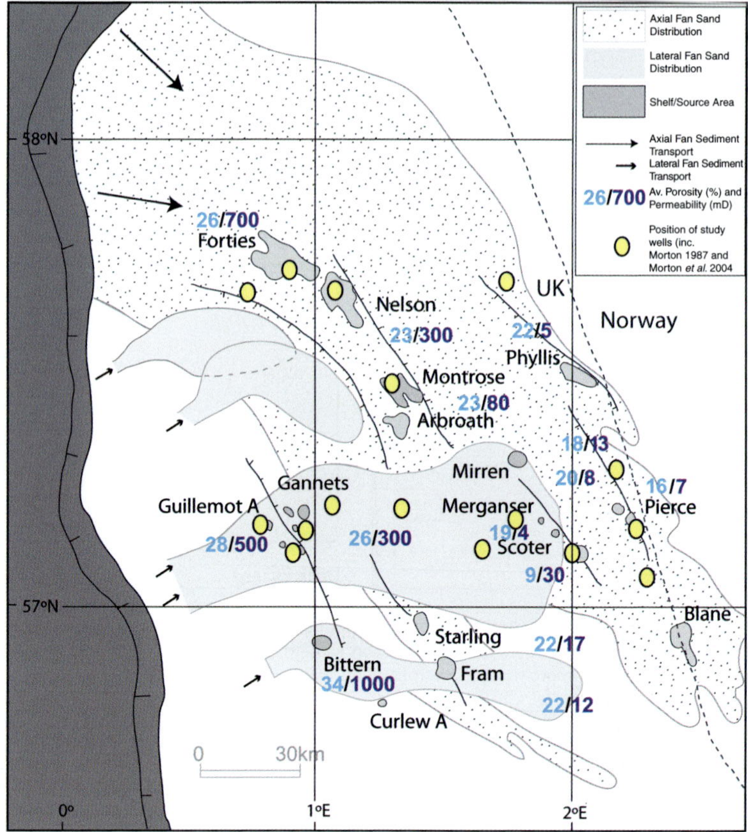



**Fig. 3.** Map showing the interpreted extent of the axial and lateral Sele Formation basin-floor fan units. Yellow dots show the positions of the data utilized in this study, including those from Morton (1987) and Morton *et al.* (2004). Approximate porosity (%; blue) and permeability (mD; purple) data are also shown. Modified from Hempton *et al.* (2005).

to the original provenance composition) when overprinted by processes such as weathering, hydrodynamics and diagenesis (Morton 1984; Morton & Hallsworth 1999; Morton *et al.* 2004). As such, it is preferable to conduct studies on a single stable mineral group using differences in composition to discriminate between provenance areas (Morton & Hallsworth 1994; Morton *et al.* 2004).

Morton (1985) described one such approach using electron microprobe analysis of detrital garnets from the Middle Jurassic of the northern North Sea. In this method the proportions of Mg-rich, Fe + Mn-rich and Ca-rich garnet compositions are plotted to show variations in composition and, therefore, provenance lithologies and mineralogies. Detrital garnet studies have since been used widely as a tool for provenance and correlation (Haughton & Farrow 1988; Tebbens *et al.* 1995; Hutchison & Oliver 1998; Hallsworth & Chisholm 2000). Within the Sele Formation, Morton (1987) suggested that

two different garnet signatures could be defined associated with the 'Forties' and 'Gannet' sandstone members.

The application of garnet geochemical analysis to evaluate the provenance of North Sea sandstones was limited by the lack of a detailed hinterland source compositional dataset until Morton *et al.* (2004) undertook a detailed study of garnet compositions from modern river sediment across northern Scotland and the Shetland Islands. This showed that four main compositional zones can be defined across this geographical area resulting from the dominant local lithologies: area S1 (Dalradian of the Grampian Highlands), area S2 (Moine of the Northern Highlands), area S3 (including the Lewisian of NW Scotland and the Hebrides and local groups around South Harris and Loch Maree) and area S4 (Moine/Dalradian of the Shetland Islands) (Fig. 5). Differences in garnet provenance are illustrated in terms of the relative abundance of garnet

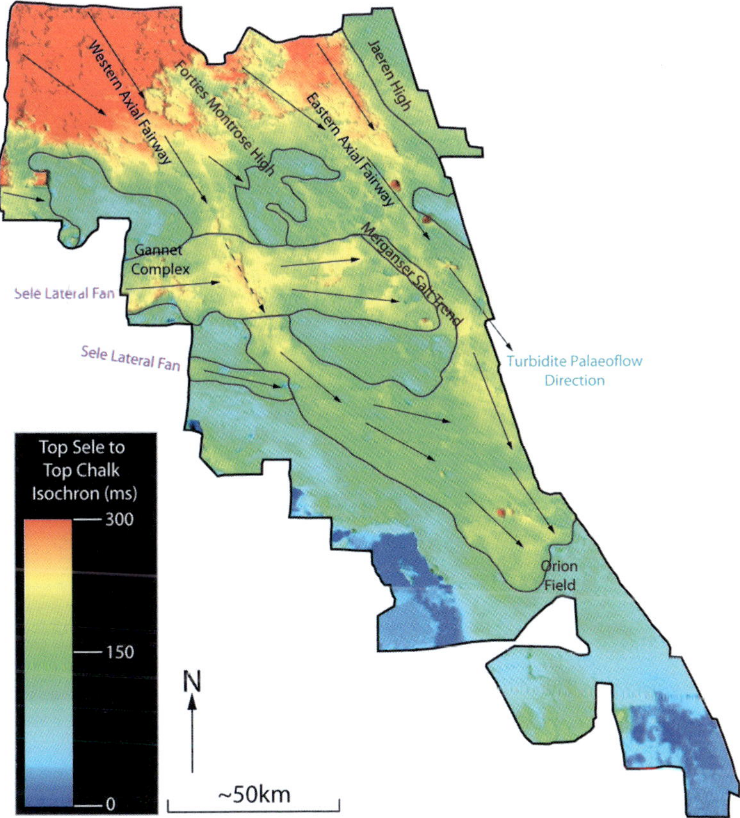


**Fig. 1.** Example seismic map and interpretation of the Sele Formation, which makes up the basis of the interpretations presented by Hempton *et al.* (2005) in Figure 3. Image shows a Top Chalk to Top Sele isochron believed to be dominated by the Sele Formation signal. Note the distinct western and eastern axial fairways and at least two lateral systems. Modified from Kilhams *et al.* (2012) and Scott *et al.* (2010).

types A (high-Mg, low-Ca), B (low-Mg, variable Ca) and C (high-Mg, high-Ca) (Figs 5 & 9), each of which represents derivation from different source-rock lithologies. Morton *et al.* (2004) utilize data from Sele and Lista Formation sandstones to propose that the North Sea Palaeogene axial systems consist of detritus derived from areas S2, S3 and S4 together with recycled Triassic detritus, with sediment routed into the basin from the northern Moray Firth and East Shetland Platform. These sandstones have a mixture of garnet compositions including types A, B and C, and garnet assemblages are therefore relatively Mg-rich. The lateral systems have a different signature because they are sourced directly from the Grampian area (area S1) with little or no mixing, consistent with the Tertiary sediment transport patterns proposed by Hall & Bishop (2002). As such, these sandstones have garnet assemblages dominated by Type B (low-Mg) (Figs 5 & 9). This technique

may therefore be used to assign individual Sele Formation sandstones (T-units) to either axial or lateral input into the basin, and could help to lessen uncertainty about reservoir quality distribution across the Central Graben.

## Locality, aims and limitations of study

In this study we utilize the methods and provenance areas set out by Morton (1985) and Morton *et al.* (2004) to study four Central Graben wells that penetrate the Sele Formation (Fig. 2). This was designed as a pilot study in order to demonstrate, alongside existing datasets (e.g. Morton 1987; Morton *et al.* 2004, Figs 2 & 9), the probable extent of the Sele Formation lateral input systems and to determine if these vary temporally between T-sequences. As such it should be possible to make predictions about how important these lateral units are in

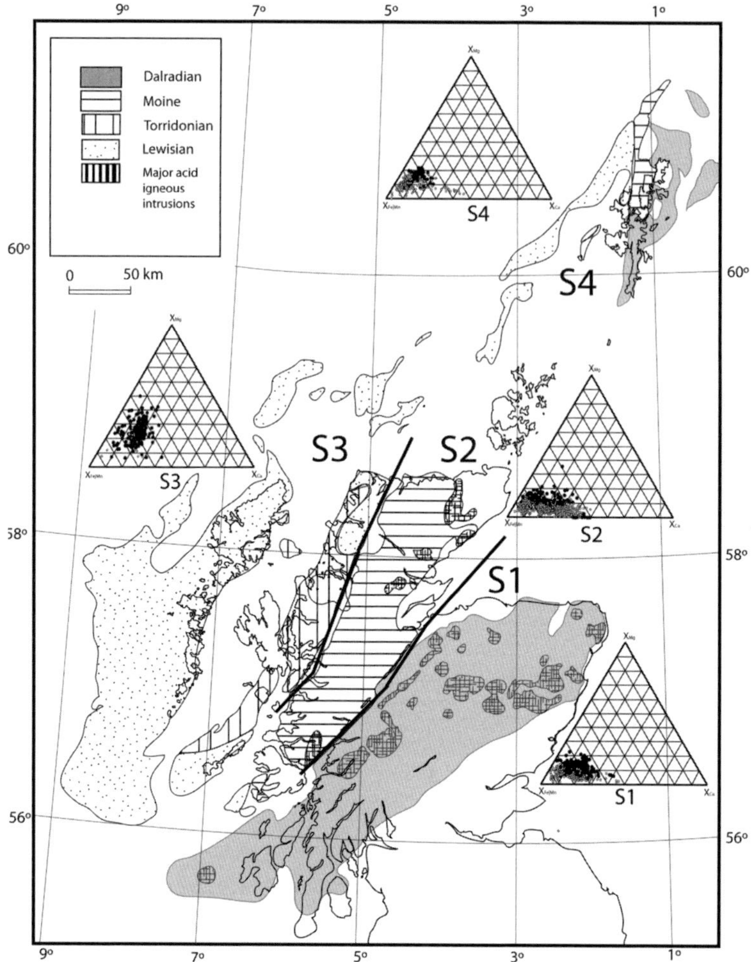



**Fig. 5.** Summary geological map of northern Scotland showing the different source areas (areas S1–S4) that contribute to the detrital garnet signatures seen in the Central Graben. Triangular plots illustrate the relative abundance of Fe + Mn (left), Mg (top) and Ca (right) in the garnet formula for lithologies typical of each area (see also Figs 7 & 9). After Morton *et al.* (2004).

constraining sandstone porosity/permeability relationships across the Central Graben.

However, there are some data limitations and unknowns that should be recognized. First, detrital garnets, particularly those rich in Ca, are impacted by burial diagenesis, especially below *c.* 2500 m (*c.* 8200 ft) in the central North Sea (Morton 1987). The precise impact this process has had on the samples used is described below. However, because the key indicator to discriminate between axial and lateral sediment input systems is the Mg content, burial diagenesis should not drastically alter any interpretations about provenance area and sediment routing. Second, the precise nature of the stratigraphic interaction between the lateral and axial

sediment systems is unknown. For example, it is unclear if these systems are of slightly different ages or if they operate at exactly the same time and, therefore, interfinger. The authors recognize that if the latter is true then some of the interpretations given in the following may be oversimplified, and further high-resolution analysis within single wells would be required to fully understand their relationships.

Therefore, although this study is currently limited in scope, the main aim is to demonstrate a powerful concept that can be developed with further work. This has the potential to substantially modify our understanding of prospectivity in the Central Graben Palaeogene.

**Table 1.** *Samples studied in this paper including T-sequences (defined by Shell UI Europe), depths (AHRT) and interpretation of axial or lateral sediment routing based on detrital garnet analysis*

| Well | T-sequence of sample | AHRT depth (m) | AHRT depth (ft) | Burial depth (m) | Interpretation |
|---|---|---|---|---|---|
| 22/23b-1S1 | T75 | 2746.2 | 9010.0 | 2625.0 | Axial |
| 22/23b-1S1 | T75 | 2773.7 | 9100.0 | 2653.0 | Axial |
| 22/23b-1S1 | T70 | 2837.7 | 9310.0 | 2717.0 | Axial |
| 22/29-1S1 | T75 | 3011.4 | 9880.0 | 2894.0 | Axial |
| 22/29-1S1 | T70 | 3060.2 | 10 040.0 | 2942.0 | Axial |
| 22/30a-5 | T75 | 2952.5 | 9686.6 | 2841.0 | Axial |
| 22/30a-5 | T75 | 2954.2 | 9692.3 | 2843.0 | Axial |
| 22/30a-5 | T75 | 2957.7 | 9703.6 | 2846.0 | Axial |
| 22/30a-5 | T70 | 2966.4 | 9732.4 | 2855.0 | Axial |
| 22/30a-5 | T70 | 3001.5 | 9847.6 | 2890.0 | Axial |
| 22/30a-5 | T70 | 3047.1 | 9997.0 | 2936.0 | Axial |
| 22/30a-5 | T65 | 3067.9 | 10 065.3 | 2956.0 | Lateral? |
| 22/30a-7 | T75 | 3038.3 | 9968.3 | 2923.0 | Axial |
| 22/30a-7 | T75 | 3066.2 | 10 059.8 | 2950.0 | Axial |
| 22/30a-7 | T70 | 3084.6 | 10 120.0 | 2969.0 | Axial |
| 22/30a-7 | T65 | 3133 | 10 279.0 | 3015.0 | Axial |

Only one sample is associated with the lateral system (22/30a-5) and, as discussed in the text, this may be reworked. Results are plotted in Figure 8 and mapped in Figure 10. AHRT, along hole rotary table.

**Table 2.** *Samples studied by Morton (1987) and Morton et al. (2004) updated with T-sequences (defined by Shell UI Europe) and depths (AHRT)*

| Well | Study | T-sequence of sample | AHRT depth (m) | AHRT depth (ft) | Interpretation |
|---|---|---|---|---|---|
| 21/10-5 | Morton (1987) | T75 | 2198 | 7211.3 | Axial |
| 21/14-1 | Morton (1987) | T80 | 2170 | 7119.5 | Axial |
| 21/25-2 | Morton (1987) | T75 | 1859 | 6099.2 | Lateral |
| 21/25-2 | Morton (1987) | T70 | 1911 | 6269.8 | Lateral |
| 22/14-1 | Morton (1987) | Undiff. S1 | 2669.3 | 8757.5 | Axial |
| 22/18-2 | Morton (1987) | T75 | 2503.6 | 8213.9 | Axial |
| 22/18-2 | Morton (1987) | T75 | 2538.6 | 8328.7 | Axial |
| 22/18-2 | Morton (1987) | T75 | 2522.6 | 8374.6 | Axial |
| 22/21-1 | Morton (1987) | Undiff. S1 | 2659.9 | 8726.6 | Axial |
| 22/21-1 | Morton (1987) | Undiff. S1 | 2672.9 | 8769.2 | Axial |
| 22/21-2 | Morton (1987) | T75 | 2447.7 | 8030.6 | Lateral |
| 23/21-2 | Morton (1987) | Undiff. S1 | 2835.3 | 9302 | Axial |
| 23/27-2 | Morton (1987) | T75 | 3121.2 | 10 240.3 | Axial |
| 23/27-2 | Morton (1987) | T75 | 3156.2 | 10 355.1 | Axial |
| 30/1-1 | Morton (1987) | Undiff. S1 | 3101.3 | 10 175 | Axial |
| 30/1-1 | Morton (1987) | Undiff. S1 | 3126.3 | 10 257 | Axial |
| 21/30-13 | Morton et al. (2004) | Undiff. S1 | 2261.3 | 7419 | Lateral |
| 21/30-13 | Morton et al. (2004) | Undiff. S1 | 2279.6 | 7479 | Lateral |
| 21/30-13 | Morton et al. (2004) | Undiff. S1 | 2298.3 | 7540.5 | Lateral |
| 21/30-13 | Morton et al. (2004) | Undiff. S1 | 2334.5 | 7659 | Lateral |
| 22/6a-9 | Morton et al. (2004) | T75 | 2247.6 | 7374 | Axial |
| 22/6a-9 | Morton et al. (2004) | T75 | 2259.2 | 7412 | Axial |
| 22/6a-9 | Morton et al. (2004) | T75 | 2272.6 | 7456 | Axial |
| 22/6a-9 | Morton et al. (2004) | T75 | 2278.7 | 7476 | Axial |
| 22/6a-9 | Morton et al. (2004) | T70 | 2285.1 | 7497 | Axial |
| 22/6a-9 | Morton et al. (2004) | T70 | 2294.2 | 7527 | Axial |
| 22/6a-9 | Morton et al. (2004) | T70 | 2305.5 | 7564 | Axial |
| 22/6a-9 | Morton et al. (2004) | T65 | 2317.7 | 7604 | Axial |

Interpretation of axial or lateral sediment routing based on detrital garnet analysis. Results are plotted in Figure 9 and mapped in Figure 10. AHRT, along hole rotary table.

## Datasets and methodology

Access to core and cuttings material was provided by Shell UI Europe. Sixteen cuttings samples were taken from four different wells across the Central Graben (22/23b-1S1, 22/29-1S1, 22/30a-5, 22/30a-7) with the aim of sampling each of the different T-sequences of the Forties Sandstone Member (Table 1). These particular wells were chosen from the material available because, combined with those analysed by Morton (1987) and Morton *et al.* (2004), they gave a good spatial and temporal distribution of data across the potential lateral fan system. The samples were taken from the middle of each T-sequence, although no particular grain size or sedimentary features were favoured. For each sample, garnet grains were picked at random with a needle from the dry residues during optical examination under a polarizing microscope, then mounted on double-sided adhesive tape, coated with carbon, and analysed at Aberdeen University using a Link Systems AN10000 energy-dispersive X-ray analyser attached to a Cambridge Instruments Microscan V electron microprobe. The quality of each result was monitored to ensure that the stoichiometrically determined formulae corresponded to ideal garnet compositions. Garnet compositions are expressed in terms of the relative abundance of the Mg, $Fe^{2+}$, Ca and Mn end members. Fifty garnet analyses were acquired from each sample, with garnet assemblage compositions shown using ternary diagrams with proportions of $Fe^{2+} + Mn$, Mg and Ca in the garnet formula as poles, calculated assuming all Fe is present as $Fe^{2+}$. Note that the methodology used here is similar to that of Morton (1987) and Morton *et al.* (2004), and the resulting dataset has been further populated with the results of this previous work (correlated to specific T-sequences using Shell UI Europe in-house studies) (Fig. 1; Table 2).

## Data analysis

It is evident from Figure 6 that garnet grains have undergone extensive corrosion in the four studied

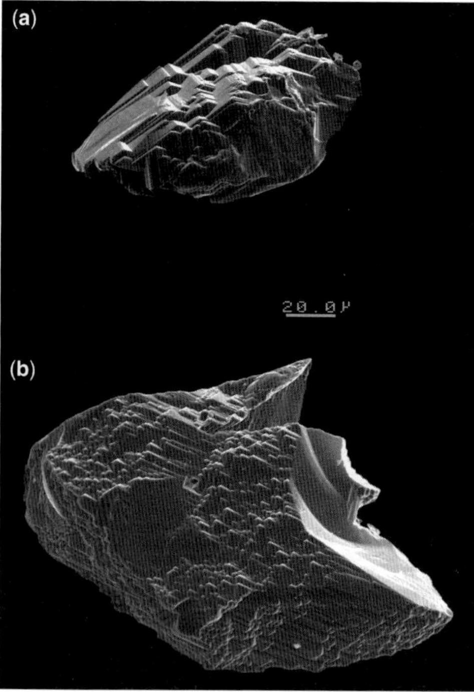

**Fig. 6.** Scanning electron micrographs of garnet grains from 22/23b-1S1, 9100 ft (2773.7 m), showing the presence of large-scale etch facets indicative of advanced grain dissolution similar to those illustrated by Turner & Morton (2007).

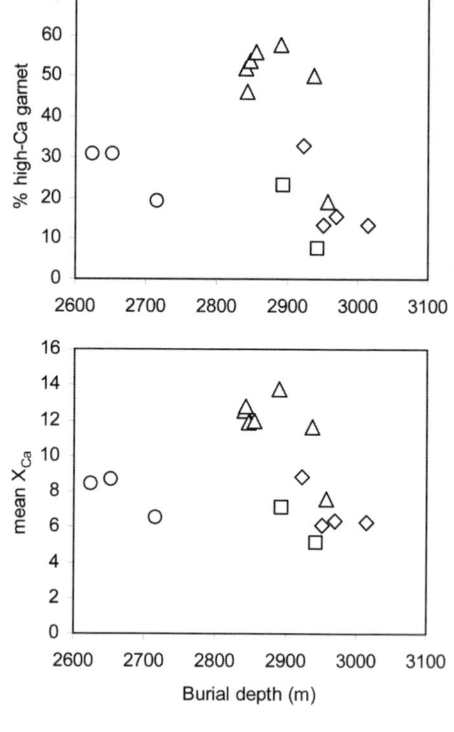



**Fig. 7.** Relationships between garnet Ca content and burial depth in T65, T70 and T75 sandstones. In wells 22/29-1S1, 22/30a-5 and 22/30a-7, there is a clear decrease in Ca content with increasing depth (total vertical depth subsea, TVDSS), both in terms of mean $X_{Ca}$ and in terms of the proportions of high-Ca garnet ($X_{Ca} > 10\%$) in the garnet assemblages.

wells. Garnet grain surfaces show moderate to intense etch features, similar to those illustrated by Turner & Morton (2007). Given the evidence for advanced garnet dissolution, it is possible that burial diagenesis may have had an effect on garnet assemblage compositions. Previous studies on garnet dissolution during burial diagenesis have shown that garnet stability is controlled by Ca content, with Ca-rich garnets being less stable than Ca-poor varieties (Morton 1987; Morton & Hallsworth 2007). As a result, garnet assemblages show a general reduction in diversity with increasing burial depth, manifested by a decrease in Ca contents.

The possibility of a diagenetic control on garnet compositions in the samples described in this paper was investigated by plotting two parameters, mean $X_{Ca}$ content and the proportion of garnets with $X_{Ca} > 10\%$, against burial depth (Fig. 7). The diagrams show that for three wells (22/29-1S1, 22/30a-5 and 22/30a-7), there is a broad linear relationship between Ca content and burial depth, with

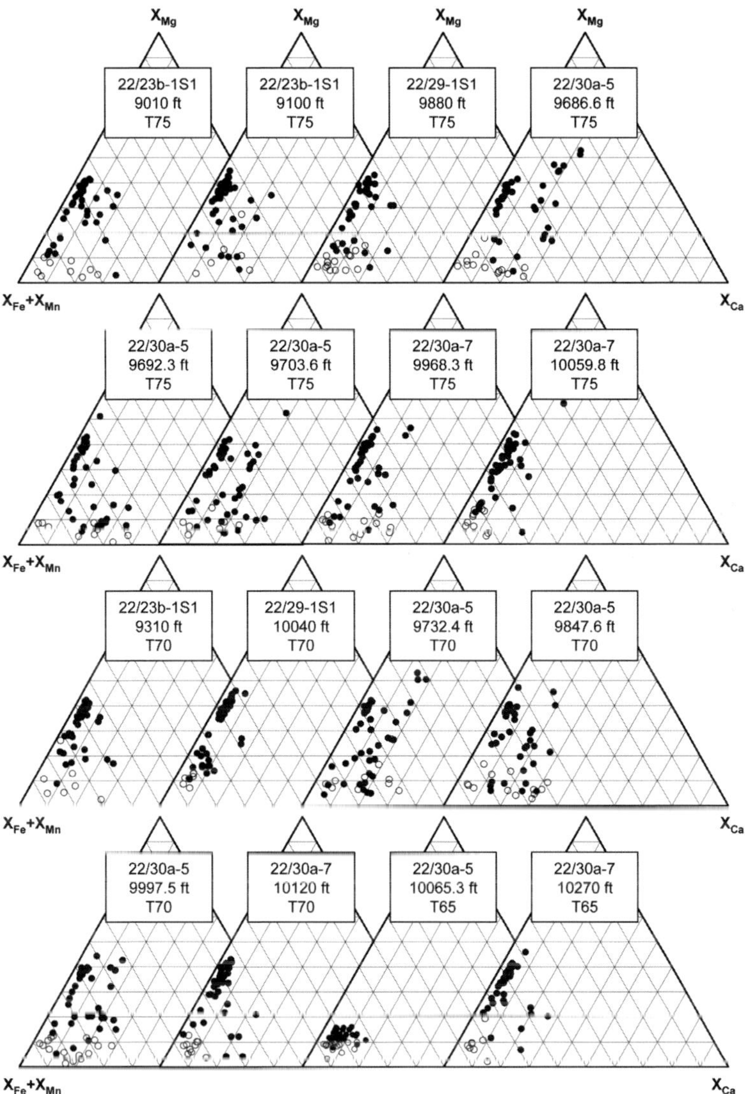

**Fig. 8.** Garnet assemblages in T65, T70 and T75 sandstones in wells between the Gannet Complex and Merganser Field, expressed in terms of Fe + Mn, Mg and Ca. $X_{Fe}$, $X_{Mg}$, $X_{Ca}$, $X_{Mn}$ = proportions of Fe, Mg, Ca and Mn in the garnet formula. All Fe calculated as $Fe^{2+}$. Filled circles, $X_{Mn} < 5\%$; open circles, $X_{Mn} > 5\%$.

garnet populations in the deeper sandstones having lower Ca. However, 22/23b-1S1 is anomalous, because Ca contents are lower than in the other samples, even though the sandstones are from shallower depths. This is not a provenance effect because the garnet populations in this well have similar relative proportions of types A, B and C to all the other samples (with the notable exclusion of the sample from 22/30a-5, 10 065.3 ft/3067.9 m), indicating they were deposited by the axial fan system. The data therefore imply that garnet dissolution is more advanced in the 22/23b-1S1 sandstones compared with the other wells, despite the fact that they are less deeply buried.

The most likely reason for this anomaly lies in the nature of the fluids present in the sandstones in the four wells. The sandstones in 22/23b-1S1 are water-bearing, whereas those in 22/29-1S1, 22/30a-5 and 22/30a-7 contain hydrocarbons (oil/condensate). Previous case studies (Yurkova 1970; Morton 1984; Morton & Hallsworth 2007) have shown that heavy mineral dissolution is inhibited in the presence of hydrocarbons, and this is the preferred explanation for the apparent anomaly in this study. The sandstones in 22/23b-1S1 remain water-bearing and therefore open to dissolution, whereas hydrocarbon migration into the 22/29-1S1, 22/30a-5 and 22/30a-7 sandstones caused dissolution to either cease or be seriously inhibited. Given that the

sandstones in 22/29-5S1, 22/30a-5 and 22/30a-7 are hydrocarbon-bearing, the downhole decrease in Ca contents in these wells also requires explanation. The implication of the observed profiles is that, within any one well, the deeper sandstones remained water-wet for a longer period than those at shallower depth. Hence, the garnet dissolution patterns in these wells are most likely to be a result of the hydrocarbon charge history.

Variations in garnet Ca content are important in terms of constraining our understanding of the diagenetic history of these samples. However, it is the Mg content of the garnets that reveals the most information about provenance areas and enables an interpretation about sediment routing (axial and lateral fans) to be made. All but one of the new samples contain diverse garnet assemblages with Type A (high-Mg, low-Ca), Type B (low-Mg, variable-Ca) and Type C (high-Mg, high-Ca) components all represented (Table 1; Fig. 8). Assemblages such as these suggest a mixed provenance including input from the Lewisian (area S3 in Fig. 5) and reworked Triassic detritus from the Orkney–Shetland Platform (Morton *et al.* 2004), consistent with axial fan deposition. This includes samples from the T75, T70 and T65 intervals. The only exception to this rule is found within the garnet assemblage in the T65 interval of well 22/30a-5 (Scoter field area), which, as well as

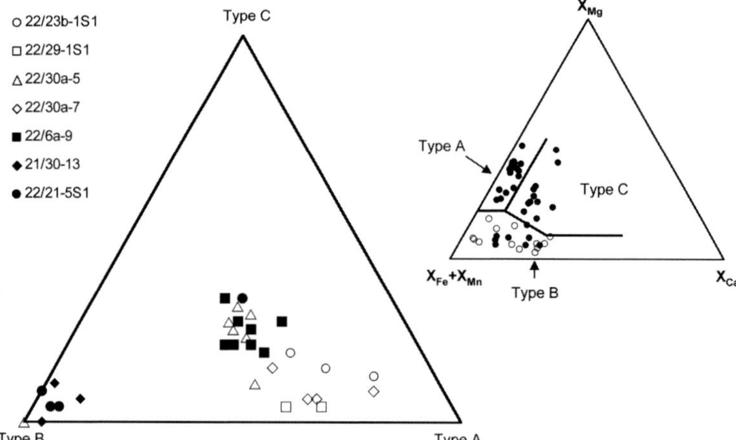


**Fig. 9.** Comparison of garnet assemblage compositions in axial and lateral fan systems of the central North Sea, expressed in terms of the relative abundance of garnet types A, B and C. Differentiation of garnet types follows Morton *et al.* (2004), as illustrated using the assemblage from 22/30a-5, 9847.6 ft. With one exception, garnet assemblages in 22/30a-5 are directly comparable to those in Nelson Field well 22/6a-9 (from Morton *et al.* 2004), located in the more proximal area of the axial fan system. Assemblages in 22/23b-1S1, 22/29-1S1 and 22/30a-7 fall between the area defined by 22/30a-5 (and 22/6a-9) and the Type A pole, owing to burial depth-related depletion of high-Ca garnet, which causes a reduction in the proportions of both Type B and Type C garnet. One sample from 22/30a-5 falls at the Type B pole, closely comparable to Sele Formation and Lista Formation sandstones in the Gannet area (wells 21/30-13 and 22/21b-5S1), as described by Morton *et al.* (2004).

exhibiting enhanced Ca dissolution, also has low Mg content. Such garnet compositions are usually attributed to a provenance area in the Grampian Highlands (Figs 5 & 9) and lateral fan deposition. It should be noted, however, that the T65 sandstones in this interval appear to show evidence of reworking with centimetre-scale angular chalk fragments and low textural maturity (medium to coarse grains and poor sorting). This is consistent with reworking from the salt high associated with the Scoter field (e.g. Hempton *et al.* 2005). It is possible, therefore, that this garnet signature is not truly representative of turbidite routing within the Sele interval, because the sandstones may have originally been deposited as Maureen or Lista-age deposits. This possibility is supported by evidence for lateral sediment input in older Paleocene sandstones (Lista Formation) in well 22/21-5S1 (Gannet Field area), most of which have garnet assemblages

dominated by the Type B component (Morton *et al.* 2004) directly comparable to the sample in 22/30a-5 (Fig. 9).

## Discussion and implications

### Sediment routing

According to seismic attribute studies (Fig. 4), and the work of Hempton *et al.* (2005) (Fig. 3), the Sele Formation sandstones of the Central Graben (especially between the Gannet and Merganser fields) should contain a mixed provenance signal with axial fans sourced from the north meeting sandstones sourced from the Grampian Highlands to the west. In particular, Hempton *et al.* (2005) suggested that the lateral fans in this area are extensive. However, the results of this study illustrate that the dominant provenance signal in the Central

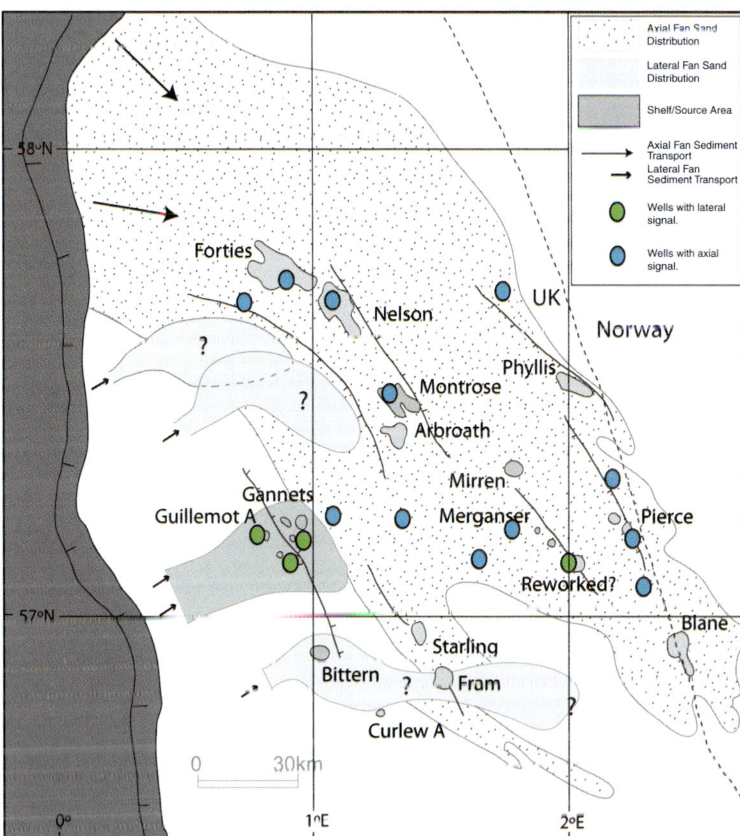


**Fig. 10.** Reinterpreted map showing the wells utilized in this study alongside those from Morton (1987) and Morton *et al.* (2004). Those with axial garnet signatures are shown in blue and those with lateral signatures are shown in green. Based on the distribution of these results a new limit for the main lateral fan system (in the area of the Gannet Fields) has been proposed. More work is required to understand the precise extent of the lateral fans to the north and south. Modified from Hempton *et al.* (2005).

Graben is associated with axial fan systems. It is also possible to update the data presented by Morton (1987) and Morton *et al.* (2004) with T-sequence picks and present a larger combined dataset. This work suggests that in each unit (T65, T70, T75 and T80) the majority of samples can be attributed to a provenance area to the north and, therefore, to the lateral fan system. There are only three confirmed exceptions to this rule (wells 21/25-2, 21/30-13 and 22/21-2), which can be attributed to the axial fan system with a high degree of confidence. This is of particular interest because these wells are all confined towards the west of the study area (Fig. 2), suggesting that the lateral fan systems did not deposit sandstones far to the east of the Gannet fields.

Therefore, seismic attribute maps (e.g. Kilhams *et al.* 2012), as well as the published interpretations based on such work (e.g. Hempton *et al.* 2005), may not be truly representative of turbidite routing within the Sele Formation. Figure 10 illustrates the likely limit of lateral fan routing based on this provenance analysis work.

### *Implications for reservoir quality understanding and the hydrocarbon industry*

As further discoveries within the mature North Sea basin become harder to make it is important to develop a detailed understanding of the sedimentological relationships within reservoir units. For example, it is imperative to comprehend the provenance areas and distributions of the Forties Sandstone Member because proximity to the shelf has been demonstrated to be related to reservoir quality (e.g. Jennette *et al.* 2000; Hempton *et al.* 2005; Kilhams *et al.* 2012). Therefore, the considerable reinterpretation of sediment distribution presented in this paper may result in the reassessment of both exploration prospects and existing fields across the Central Graben.

### *Limitations and further work*

This work has presented a potentially high-impact reinterpretation of existing assumptions about sediment provenance areas in the Sele Formation of the Central Graben based on garnet analysis. However, as suggested in the study aims, the dataset used is relatively small when compared to the number of wells drilled over the last 40 years of hydrocarbon exploration and production. As a result, it is not yet possible to understand the precise temporal relationship between the axial and lateral fan systems. For example, do the sandstones interfinger or are they entirely separate? It is also only possible to define an approximate limit of the lateral fan

system and it is not clear if this limit varies significantly within the different T-sequences. It may also be possible that the Lista and Maureen Formation sandstones show similar trends. These limitations, and related questions, can be solved only with further, more extensive and higher-resolution sampling of core and cuttings.

## Conclusions

The Sele Formation of the central North Sea contains a series of deep-water sandstones that form one of the most important hydrocarbon reservoirs in the world. These sandstones have been demonstrated, by garnet analysis, to be sourced from two main areas. The main axial fan system was derived from northern and northwestern Scotland, with lateral fan systems supplied from the Grampian Highlands of eastern Scotland to the west. Previous seismic mapping suggested that the lateral fan system was extensive within the Central Graben from the Gannet Complex in the west to the Merganser Field area in the east. Detailed garnet analysis on four wells in this area, combined with previously published data from a further eleven wells, suggests that the lateral fan system is not as extensive as suggested. Instead, the majority of samples, especially east of the Gannet Complex, can be associated with the axial fan system. This is of particular importance because the proximity of sandstone deposits to their provenance areas, and related shelf, is likely to determine reservoir quality (grain size, sorting, porosity, permeability).

This study formed a component of R. Borella's MSc degree, with further work conducted by B. Kilhams during his PhD. Both projects were generously supported by Shell Upstream International (UI) Europe. Special thanks should go to C. Davis, S. Drake, G. Lynn, J. Marshall, T. McKie and P. Watt for organizing core viewings, data support and permissions. A. Hartley and M. Huuse should be thanked for their guidance, without which these projects would not have happened. We are indebted to J. Still for undertaking the microprobe analysis. Finally, we would like to thank two anonymous reviewers, whose comments improved the manuscript greatly.

## References

Ahmadi, Z. M., Sawyers, M., Kenyon-Roberts, S., Stanworth, C. W., Kugler, K. A., Kristensen, J. & Fugelli, E. M. 2003. Paleocene. *In*: Evans, D., Graham, C., Armour, A. & Bathurst, P. (eds) *The Millennium Atlas: Petroleum Geology of the Central and Northern North Sea*. Geological Society, London, 548–597.

Allen, P. A. & Mange-Rajetzky, M. A. 1992. Devonian–Carboniferous sedimentary evolution of the Clair area, offshore northwestern UK: impact of changing provenance. *Marine and Petroleum Geology*, **9**, 29–52.

CARMAN, G. J. & YOUNG, R. 1981. Reservoir geology of the Forties oil field. *In*: ILLING, L. V. & HOBSON, G. D. (eds) *Petroleum Geology of the Continental Shelf of Northwest Europe*. Geological Society, London, 371–379.

DAVIS, C., HAUGHTON, P., MCCAFFREY, W., SCOTT, E., HOGG, N. & KITCHING, D. 2009. Character and distribution of hybrid sedimentary gravity flow deposits from the outer Forties Fan, Palaeocene central North Sea, UKCS. *Marine and Petroleum Geology*, **26**, 1919–1939.

DEEGAN, C. & SCULL, B. J. 1977. *A Standard Lithostratigraphic Nomenclature for the Central and Northern North Sea*. Institute of Geological Sciences Report 77/25.

DEN HARTOG JAGER, D., GILES, M. R. & GRIFFITHS, G. R. 1993. Evolution of Paleogene submarine fans of the North Sea in space and time. *In*: PARKER, J. R. (ed.) *Petroleum Geology of Northwest Europe: Proceedings of the 4th Conference*. Geological Society, London, 59–71.

DIXON, R. J. & PEARCE, J. 1995. Tertiary sequence stratigraphy and play fairway definition, Bruce–Beryl Embayment, Quadrant 9, UKCS. *In*: STEEL, R. J., FELT, V. L., JOHANNESSEN, E. P. & MATHIEU, C. (eds) *Sequence Stratigraphy on the Northwest European margin*. Norwegian Petroleum Society, Special Publications, **5**, 443–469.

ERRATT, D., THOMAS, G. M., HARTLEY, N. R., MUSUM, R., NICHOLSON, P. H. & SPISTO, Y. 2010. North Sea hydrocarbon systems: some aspects of our evolving insights into a classic hydrocarbon province. *In*: VINING, B. A. & PICKERING, S. C. (eds) *Petroleum Geology: From Mature Basins to New Frontiers – Proceedings of the 7th Petroleum Geology Conference*. Geological Society, London, 37–56.

GALLOWAY, W. E. 1989. Genetic sequence stratigraphic sequences in basin analysis I: architecture and genesis of flooding-surface bounded depositional units. *American Association of Petroleum Geologists Bulletin*, **73**, 125–142.

GILL, C. & SHEPHERD, M. 2010. Locating the remaining oil in the Nelson Field. *In*: VINING, B. A. & PICKERING, S. C. (eds) *Petroleum Geology: From Mature Basins to New Frontiers – Proceedings of the 7th Petroleum Geology Conference*. Geological Society, London, 349–368.

HALL, A. & BISHOP, P. 2002. Scotland's denudational history: an integrated view of erosion and sedimentation at an uplifted passive margin. *In*: DORÉ, A. G., CARTWRIGHT, J. A., STOKER, M. S., TURNER, J. P. & WHITE, N. (eds) *Exhumation of the North Atlantic Margin: Timing, Mechanisms and Implications for Petroleum Exploration*. Geological Society, London, Special Publications, **196**, 271–290.

HALLSWORTH, C. R. & CHISHOLM, J. I. 2000. Stratigraphic evolution of provenance characteristics in Westphalian sandstones of the Yorkshire coalfield. *Proceedings of the Yorkshire Geological Society*, **53**, 43–72.

HALLSWORTH, C. R., MORTON, A. C. & DORE, G. 1996. Contrasting mineralogy of Upper Jurassic sandstones in the Outer Moray Firth North Sea: implications for

the evolution of sediment dispersal patterns. *In*: HURST, A., JOHNSON, H. D., BURLEY, S. D., CANHAM, A. C. & MACKERTICH, D. S. (eds) *Geology of the Humber Group: Central Graben and Moray Firth, UKCS*. Geological Society, London, Special Publications, **114**, 131–144.

HAUGHTON, P. D. W. & FARROW, C. M. 1988. Compositional variation in lower old red sandstone detrital garnets from the midland valley of Scotland and the Anglo-Welsh basin. *Geological Magazine*, **126**, 373–396.

HEMPTON, M., MARSHALL, J., SADLER, S., HOGG, N., CHARLES, R. & HARVEY, C. 2005. Turbidite reservoirs of the Sele Formation, central North Sea: geological challenges for improving production. *In*: PARKER, J. R. (ed.) *Petroleum Geology of Northwest Europe and Global Perspectives – Proceedings of the 6th Petroleum Geology Conference*. Geological Society, London, 449–459.

HUTCHISON, A. R. & OLIVER, G. J. H. 1998. Garnet provenance studies, juxtaposition of Laurentian marginal terranes and timing of the Grampian Orogeny in Scotland. *Journal of the Geological Society, London*, **155**, 541–550.

JENNETTE, D. C., GARFIELD, T. R., MOHRIG, D. C. & CAYLEY, G. T. 2000. The interaction of shelf accommodation, sediment supply and sea level in controlling the facies, architecture and sequence stacking patterns of the Tay and Forties/Sele basin-floor fans, central North Sea. *In*: WEIMER, P., SLATT, R. M., COLEMAN, J., ROSEN, N. C., NELSON, H., BOUMA, A. H., STYZEN, M. J. & LAWRENCE, D. T. (eds) *Deep-Water Reservoirs of the World. Proceedings of the GCSSEPM Foundation 20th Annual Research Conference*. The Write Enterprise, Houston, 402–421.

JONES, R. W. & MILTON, N. J. 1994. Sequence development during uplift: Palaeogene stratigraphy and relative sea-level history of the Outer Moray Firth, UK North Sea. *Marine and Petroleum Geology*, **11**, 157–165.

KANTOROWICZ, J. D., ANDREWS, I. J. ET AL. 1999. Innovation and risk management in a small subsea-tieback: Arkwright Field, central North Sea. *In*: FLEET, A. J. & BOLDY, S. A. R. (eds) *Petroleum Geology of Northwest Europe: Proceedings of the 5th Conference*. Geological Society, London, 1125–1134.

KILHAMS, B. A., GODFREY, S., HARTLEY, A. & HUUSE, M. 2011. An integrated 3D seismic, petrophysical and analogue core study of the Mid Eocene Grid channel complex in the greater Nelson Field area, UK central North Sea. *Petroleum Geoscience*, **17**, 127–142.

KILHAMS, B. A., HARTLEY, A., HUUSE, M. & DAVIS, C. 2012. Characterising the Paleocene Turbidites of the North Sea: the Mey Sandstone Member, Lista Formation, UK Central Graben. *Petroleum Geoscience*, **18**, 337–354.

KNOX, R. W. O'B. & HOLLOWAY, S. 1992. Paleogene of the central and northern North Sea. *In*: KNOX, R. W. O'B. & CORDEY, W. G. (eds) *Lithostratigraphic Nomenclature of the UK North Sea*. British Geological Survey, Nottingham, 3–34.

KNOX, R. W. O'B., MORTON, A. C. & HARLAND, R. 1981. Stratigraphical relationships of Palaeocene Sands in the UK Sector of the central North Sea. *In*: ILLING,

L. V. & HOBSON, G. D. (eds) *Petroleum Geology of the Continental Shelf of Northwest Europe.* Geological Society, London, 267–281.

KULPECZ, A. A. & VAN GEUNS, L. C. 1990. Geological modeling of a turbidite reservoir, Forties Field, North Sea. *In*: BARWIS, J. H., MCPHERSON, J. G. & STUDLICK, J. R. J. (eds) *Sandstone Petroleum Reservoirs.* Springer-Verlag, New York, 489–508.

MANGE, M. A. & MAURER, H. F. W. 1992. *Heavy Minerals in Colour.* Chapman & Hall, London.

MILTON, N. J., BERTRAM, G. T. & VANN, I. R. 1990. Early Palaeogene tectonics and sedimentation in the central North Sea. *In*: HARDMAN, R. F. P. & BROOKS, J. (eds) *Tectonic Events Responsible for Britain's oil and Gas Reserves.* Geological Society, London, Special Publications, **55**, 339–351.

MORTON, A. C. 1979. The provenance and distribution of the Paleocene sands of the central North Sea. *Journal of Petroleum Geology*, **2**, 11–21.

MORTON, A. C. 1982. Lower Tertiary sand development in Viking Graben, North Sea. *American Association of Petroleum Geologists Bulletin*, **66**, 1542–1559.

MORTON, A. C. 1984. Stability of detrital heavy minerals in Tertiary sandstones of the North Sea Basin. *Clay Minerals*, **19**, 287–308.

MORTON, A. C. 1985. A new approach to provenance studies: electron microprobe analysis of detrital garnets from Middle Jurassic sandstones of the northern North Sea. *Sedimentology*, **32**, 553–566.

MORTON, A. C. 1987. Influences of provenance and diagenesis on detrital garnet suites in the Forties sandstone, Paleocene, central North Sea. *Journal of Sedimentary Petrology*, **57**, 1027–1032.

MORTON, A. C. & BERGE, C. 1995. Heavy mineral suites in the Statfjord and Nansen formations of the Brent Field, North Sea: a new tool for reservoir subdivision and correlation. *Petroleum Geoscience*, **1**, 355–364.

MORTON, A. C. & HALLSWORTH, C. R. 1994. Identifiying provenance-specific features of detrital heavy mineral assemblages in sandstones. *Sedimentary Geology*, **90**, 241–256.

MORTON, A. C. & HALLSWORTH, C. R. 1999. Processes controlling the composition of heavy mineral assemblages in sandstones. *Sedimentary Geology*, **124**, 3–29.

MORTON, A. C., HALLSWORTH, C. & CHALTON, B. 2004. Garnet compositions in Scottish and Norwegian basement terrains: a framework for interpretation of North Sea sandstone provenance. *Marine and Petroleum Geology*, **21**, 393–410.

MORTON, A. C. & HALLSWORTH, C. R. 2007. Stability of detrital heavy minerals during burial diagenesis. *In*: MANGE, M. & WRIGHT, D. T. (eds) *Heavy Minerals In Use.* Developments in Sedimentology. Elsevier, Amsterdam, **58**, 215–245.

MUDGE, D. C. & BUJAK, J. P. 1996*a*. An integrated stratigraphy for the Paleocene and Eocene of the North Sea. *In*: KNOX, R. W. O'B., CORFIELD, R. M. & DUNAY, R. E. (eds) *Correlation of the Early Paleogene in Northwest Europe.* Geological Society, London, Special Publications, **101**, 91–113.

MUDGE, D. C. & BUJAK, J. P. 1996*b*. Palaeocene biostratigraphy and sequence stratigraphy of the UK central North Sea. *Marine and Petroleum Geology*, **13**, 295–312.

MUDGE, D. C. & COPESTAKE, P. 1992*a*. A revised Lower Palaeogene lithostratigraphy for the Outer Moray Firth. *Marine and Petroleum Geology*, **9**, 53–69.

MUDGE, D. C. & COPESTAKE, P. 1992*b*. Lower Palaeogene stratigraphy of the northern North Sea. *Marine and Petroleum Geology*, **13**, 287–301.

NEAL, J. E. 1996. A summary of Paleogene sequence stratigraphy in Northwest Europe and the North Sea. *In*: KNOX, R. W. O'B., CORFIELD, R. M. & DUNAY, R. E. (ed.) *Correlation of the Early Paleogene in Northwest Europe.* Geological Society, London, Special Publications, **101**, 15–42.

NEAL, J. E., STEIN, J. A. & GAMBER, J. H. 1994. Graphic correlation and sequence stratigraphy in the Paleogene of NW Europe. *Journal of Micropalaeontology*, **13**, 55–80.

OGG, J. G., OGG, G. & GRADSTEIN, F. M. (eds) 2008. *The Concise Geologic Time Scale.* Cambridge University Press, Cambridge.

PRESTON, J., HARTLEY, A., HOLE, M., BUCK, S., BOND, J., MANGE, M. & STILL, J. 1998. Integrated whole-rock trace element geochemistry and heavy mineral chemistry studies: aids to the correlation of continental red-bed reservoirs in the Beryl Field, UK North Sea. *Petroleum Geoscience*, **4**, 7–16.

SCHRÖDER, T. 1992. A palynological zonation for the Paleocene of the North Sea Basin. *Journal of Micropalaeontology*, **11**, 113–126.

SCOTT, E. D., GELIN, F., JOLLEY, S. J., LEENARTS, E., SADLER, S. P. & ELSINGER, R. J. 2010. Sedimentological control of fluid flow in deep marine turbidite reservoirs: Pierce Field, UK central North Sea. *In*: JOLLEY, S. J., FISHER, Q. J., AINSWORTH, R. B., VROLIJK, P. J. & DELISLE, S. (eds) *Reservoir Compartmentalization.* Geological Society, London, Special Publications, **347**, 113–132.

STEWART, I. J. 1987. A revised stratigraphic interpretation of the Early Palaeogene of the central North Sea. *In*: BROOKS, J. & GLENNIE, K. W. (eds) *Petroleum Geology of Northwest Europe – Proceedings of the 3rd Petroleum Geology Conference.* Geological Society, London, 557–576.

TEBBENS, L. A., KROONENBERG, S. B. & VANDENBURGH, M. W. 1995. Compositional variation of detrital garnets in Quaternary Rhine, Meuse and Baltic river sediments in the Netherlands. *Geologie en Mijnbouw*, **74**, 213–224.

TURNER, G. & MORTON, A. C. 2007. The effects of burial diagenesis on detrital heavy mineral grain surface textures. *In*: MANGE, M. & WRIGHT, D. T. (eds) *Heavy Minerals in Use.* Developments in Sedimentology. Elsevier, Amsterdam, **58**, 393–412.

VINING, B. A., IOANNIDES, N. S. & PICKERING, K. T. 1993. Stratigraphic relationships of some Tertiary lowstand depositional systems in the central North Sea. *In*: PARKER, J. R. (ed.) *Petroleum Geology of Northwest Europe: Proceedings of the 4th Conference.* Geological Society, London, 17–29.

WHITE, N. & LOVELL, B. 1997. Measuring the pulse of a plume with the sedimentary record. *Nature*, **387**, 888–891.

YURKOVA, R. M. 1970. Comparison of post-sedimentary alteration of oil- and gas-, and water-bearing rocks. *Sedimentology*, **15**, 53–68.

# Heavy mineral provenance of the Mesozoic succession in Andøya, northern Norway: implications for sand transport in the Vøring Basin


ANDREW MORTON[1,2]*, MARK FANNING[3] & JOHN R. BERRY[4]

[1]*HM Research Associates, 2 Clive Road, Balsall Common, West Midlands CV7 7DW, UK*

[2]*CASP, Department of Earth Sciences, University of Cambridge, 181a Huntingdon Road, Cambridge CB3 0DH, UK*

[3]*Research School of Earth Sciences, The Australian National University, Canberra, ACT 0200, Australia*

[4]*BP Norge AS, Godesetdalen 8, PO Box 197, 4065 Stavanger, Norway*

**Corresponding author (e-mail: heavyminerals@hotmail.co.uk)*



**Abstract:** A heavy mineral, mineral chemical and detrital zircon study of Jurassic–Cretaceous (Bathonian–Valanginian) sandstones of the Andøya B borehole, Lofoten–Vesterålen, northern Norway, has revealed the existence of significant differences within the succession. These are related partly to changes in source and partly to variations in the extent of weathering during alluvial storage. Three mineralogical units have been identified. The main change takes place within the Bathonian, and is interpreted as marking a switch from eastern (West Troms) to western (Andøya–Lofoten High) sourcing, consistent with previously published sedimentological models. U–Pb age data indicate that most of the zircons were derived from Palaeoproterozoic rocks (*c.* 1750–1860 Ma), with a subordinate Archaean group (*c.* 2600–2800 Ma) and a small early Palaeozoic group (mostly in the 435–446 Ma range). These groups can all be tied back to lithological components of the Lofoten–Vesterålen and West Troms regions, including Palaeozoic rocks hosted in Caledonian allochthons.

The provenance characteristics of the Andøya succession have no counterpart in Cretaceous and Paleocene sandstones of the Vøring Basin. This suggests that sediment fed into the basin from Lofoten–Vesterålen was of minor importance, and that prospective Cretaceous–Paleocene hydrocarbon reservoir sandstones in the Vøring Basin were mainly derived from either northern Nordland or northern East Greenland.

**Supplementary material:** Zircon isotopic compositions and ages are available at http://www.geolsoc.org.uk/SUP18616


The Mesozoic succession of Andøya is of crucial importance for palaeogeographic reconstructions of the Norwegian Sea, being the only known outcrop of Jurassic–Cretaceous sediment in northern Norway (Dalland 1975, 1981). The succession is located in a small fault-bounded basin in the eastern part of the island of Andøya, which is the northernmost island in the Lofoten–Vesterålen archipelago (Fig. 1). It is believed to be a remnant of a more extensive Mesozoic sedimentary basin located along the western margin of northern Norway (Vajda & Wigforss-Lange 2009), preserved from later Cretaceous and Tertiary erosion by virtue of its downthrown structural location. The sediments preserved on Andøya therefore provide a unique opportunity to acquire provenance data from this region. Such data are important not only because they place constraints on the sediment source region for the Andøya succession, but also because they help to establish the nature of sediment putatively fed into the Vøring Basin from the NE during the Cretaceous and Paleocene. Transport from the NE has been inferred to account for several submarine fan systems at several stratigraphic levels within the Cretaceous–Paleocene interval in the northern part of the Vøring Basin (Vergara *et al.* 2001).

This article presents the results of an integrated heavy mineral provenance study of Jurassic and Early Cretaceous sandstones from the Andøya succession, involving conventional heavy mineral analysis and provenance-sensitive ratio determinations, major element analysis of garnet and tourmaline populations, trace element mineral chemical analysis of rutile populations, and U–Pb detrital zircon age dating. This approach proved successful in characterizing the provenance of the Jurassic, Cretaceous and Paleocene sandstones of


*From*: SCOTT, R. A., SMYTH, H. R., MORTON, A. C. & RICHARDSON, N. (eds) 2014. *Sediment Provenance Studies in Hydrocarbon Exploration and Production.* Geological Society, London, Special Publications, **386**, 143–161.
First published online June 11, 2013, http://dx.doi.org/10.1144/SP386.3

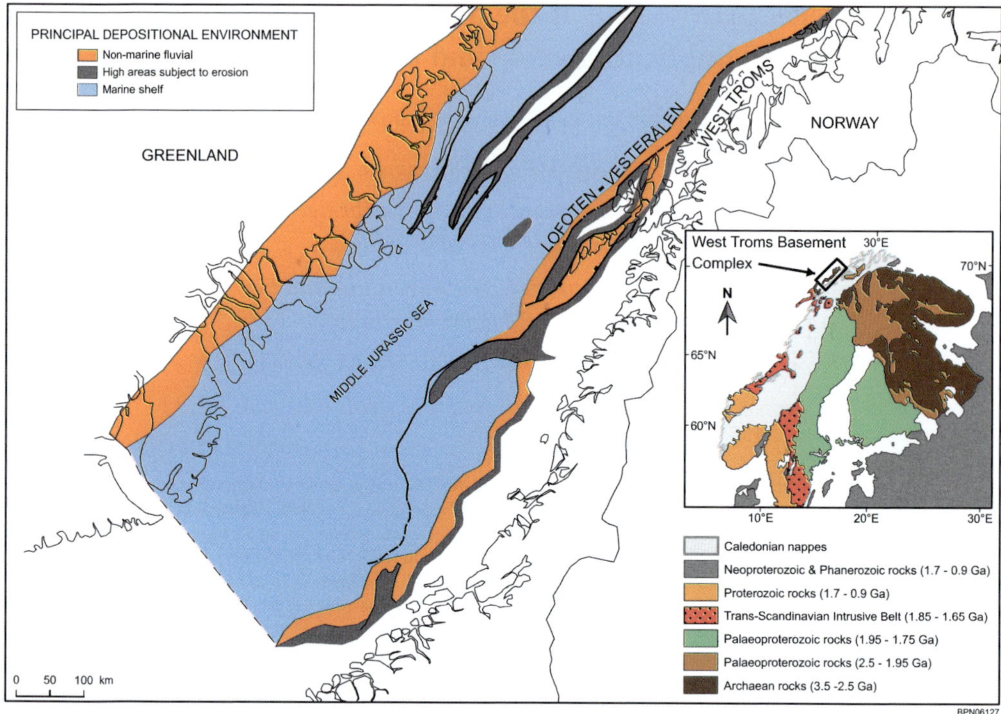



**Fig. 1.** Location of Andøya, northern Norway, shown on a generalized Middle Jurassic (Bajocian–Bathonian) palaeogeographic reconstruction. Inset location map from Bergh *et al.* (2010). Sand types MN1 and MN2 have been identified in wells from the Vøring Basin; for detailed locations see Morton *et al.* (2005*a*).

the Vøring and Møre basins to the south (Morton *et al.* 2005*a*, *b*, 2009). The samples analysed in the study are from cores taken in the Andøya B borehole, drilled by Norminol in 1972–1973. This borehole was analysed in preference to samples from outcrop, because the borehole provides a complete geological record of the succession, whereas the exposures are patchy and generally small (Dalland 1975). Analysis of borehole samples also avoids the possibility of surficial weathering modifying the composition of the heavy mineral assemblages.

## Geological setting

The Mesozoic succession on Andøya lies in a small fault-bounded basin landward of the Lofoten and Vesterålen margin of the NE Atlantic. Multistage rifting and basin development along the Lofoten and Vesterålen margin have been described in detail by Hansen *et al.* (2012). They document rifting in the Early Triassic, succeeded by a period of tectonic quiescence during the Mid–Late Triassic to Early Jurassic, during which time deltaic to shallow marine sediments were deposited. Renewed rifting commenced in the Middle Jurassic

(Bajocian), leading to uplift and erosion followed by transgression and deposition in continental and shallow marine settings. The climax of the rifting in Callovian to Tithonian times led to rapid transgression and deposition in pro-delta/shelf environments. This was followed by tectonic quiescence, uplift and erosion over most of the margin, and then by renewed rifting in the earliest Cretaceous (Valanginian), leading to deposition of marine shales and turbidites. This was again followed by quiescence, until a final rift event began in Campanian/Maastrichtian times, which ultimately led to separation and passive margin development in the Eocene.

The Mesozoic succession of Andøya (Fig. 2), which has been described in detail by Dalland (1975, 1981), is an important constituent of the record of rifting along the Lofoten and Vesterålen margin compiled by Hansen *et al.* (2012). The succession overlies a Precambrian basement complex comprising granite and gabbro (Dalland 1975). The basement rocks of the Lofoten–Vesterålen area comprise Archaean gneisses (Griffin *et al.* 1978; Jacobsen & Wasserburg 1978) intruded by plutons of an anorthosite–mangerite–charnockite–granite (AMCG) suite, ranging in age from

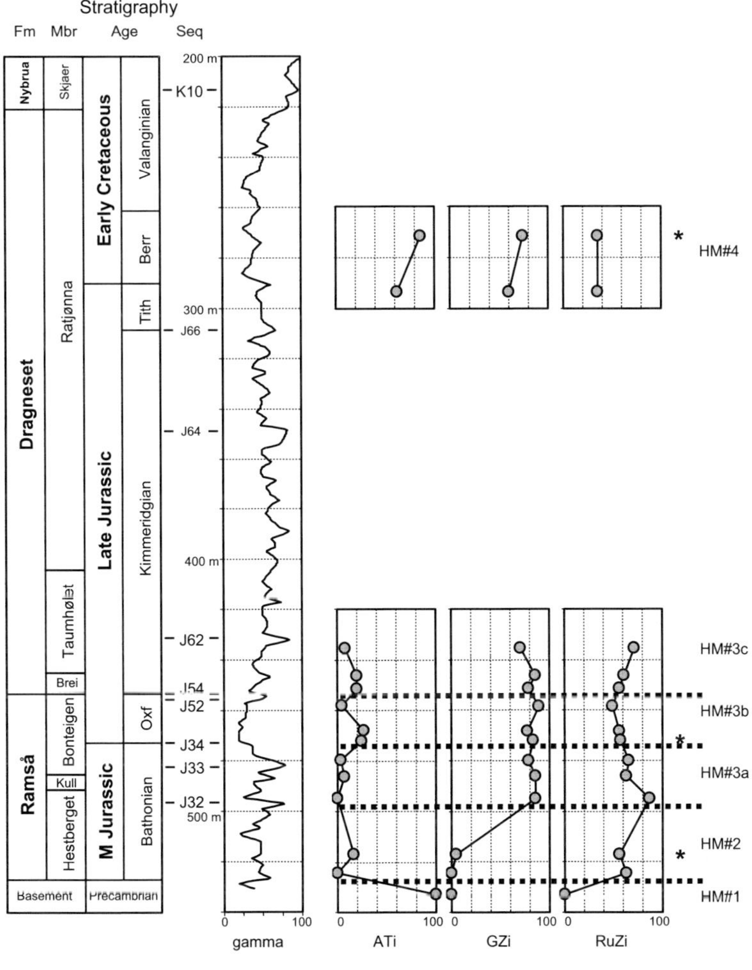



**Fig. 2.** Lithostratigraphy, age, sequence stratigraphy and heavy mineral stratigraphy of the Andøya B borehole. Only the interval from 200 m to total depth (TD) (540 m) is shown. ATi, GZi and RuZi are apatite:tourmaline, garnet:zircon and rutile:zircon indices, as defined by Morton & Hallsworth (1994). Skjaer, Skjaermyrbekken Member; Brei, Breisanden Member; Kull, Kullgrøfta Member.

1870 Ma to 1770 Ma (Corfu 2004*a*). In addition, there are allochthonous metasediments of Neoproterozoic to Ordovician age belonging to the Caledonian Nappe Domain (Leknes Group), together with Early Palaeozoic granitoids (Corfu 2004*b*).

In the southern part of the outcrop, an *in situ* kaolinitic weathering profile is preserved above the granitic basement. Sturt *et al.* (1979) interpreted K–Ar age determinations from this profile to indicate it formed during the Late Devonian to Early Carboniferous, with partial resetting of ages during subsequent diagenesis. The kaolinitic weathering profile is locally overlain by a non-marine limestone (Holen Formation), believed to be of

Carboniferous age (Dalland 1975, 1981). The weathering profile and the Holen Formation are not preserved in the Andøya B borehole.

The lowermost Jurassic sediments in the Andøya succession belong to the Hestberget Member of the Ramså Formation, and were deposited in a fluviolacustrine setting (Dalland 1975, 1981). In the Andøya B borehole, the Hestberget Member comprises siltstones with interbeds of fine to coarse sandstones and conglomerates, and is Early to Middle Bathonian in age. The Hestberget Member is overlain by the Kullgrøfta Member, which comprises carbonaceous claystones, siltstones and coals deposited in a lacustrine or lagoonal setting

(Dalland 1975, 1981). The Kullgrøfta Member in Andøya B is Middle Bathonian in age. The lower part of the Bonteigen Member (the topmost unit of the Ramså Formation) consists of stacked fining-upward coarse to fine-grained sandstones, siltstones and mudstones, with minor interbeds of coal, representing deposition in a fluvial setting (Dalland 1975, 1981). The upper part of the Bonteigen Member is dominated by medium- to coarse-grained sandstones that contain marine palynomorphs, and was deposited in a high-energy, very shallow marine shoreface environment (Dalland 1975, 1981). The Bonteigen Member is of Middle Bathonian to Oxfordian age in Andøya B.

The lowermost part of the overlying Dragneset Formation (Breisanden Member: Dalland 1975, 1981) mainly comprises fine- to medium-grained sandstones deposited in a shallow marine setting. In Andøya B, the Breisanden Member is Kimmeridgian in age. The Breisanden Member is overlain by the Taumhølet Member (Kimmeridgian), which comprises very fine-grained bioturbated sandstones, deposited in a low-energy distal offshore environment (Dalland 1975, 1981). Overlying the Taumhølet Member is the Ratjønna Member (Kimmeridgian–Valanginian), which mainly comprises siltstones and mudstones deposited in relatively deep water, but contains some sandy beds in the upper part (Dalland 1975, 1981).

The Dragneset Formation is overlain by the Nybrua Formation, which comprises two members (Leira and Skjaermyrbekken). The Leira Member consists of calcareous sandstones that were deposited in shallow marine conditions, and is of Valanginian age. The Leira Member does not appear to be present in the Andøya B succession. The overlying Skjaermyrbekken Member consists of slightly calcareous red siltstones and mudstones of Valanginian–Hauterivian age, deposited below wave base (Dalland 1975, 1981).

The trend of increasing water depth continues through the later part of the Early Cretaceous succession (Skarstein Formation). The lower part of the Skarstein Formation comprises bioturbated silty and sandy mudstones (Nordelva Member), deposited in intermediate depth marine conditions, possibly of pro-deltaic origin (Dalland 1975, 1981), during the Hauterivian–Barremian. The topmost unit in the Andøya succession (Hellneset Member, Aptian) comprises mudstones with calcareous sandstone interbeds, representing relatively deep water conditions. The sandstone interbeds are interpreted as turbidites (Dalland 1975, 1981).

On the basis of grain size trends and onlap patterns, Dalland (1975, 1981) considered that the majority of the sediment was shed from an exposed area of basement (Andøya–Lofoten High) to the west, although some of the sediment

in the lower part of the Hestberget Member (Early Bathonian) was interpreted as having a source to the east.

In regional palaeogeographic reconstructions (Fig. 1), the Andøya area is located on the eastern margin of a narrow seaway between Norway and East Greenland that connected the Boreal and Tethyan oceans (Doré 1991, 1992). Hence, the Andøya succession appears to have been deposited in an embayment on the eastern margin of this narrow seaway, between the Norwegian landmass and the Andøya–Lofoten High. It is uncertain whether the embayment opened to the north, into the Harstad Basin, or to the south, into the Vestfjorden Basin, although Dalland (1981) considered the latter to be more likely.

# Integrated heavy mineral analysis: procedures

An integrated approach to provenance studies builds a comprehensive picture of source area characteristics and overcomes limitations associated with individual methods. For example, bulk heavy mineral assemblages not only reflect provenance characteristics, but also the effects of other processes that operate during the sedimentary cycle, such as hydrodynamics, weathering and diagenesis (Komar 2007; Garzanti *et al.* 2008; Morton 2012). Studies of individual mineral phases such as garnet (Morton 1985), tourmaline (Henry & Guidotti 1985), rutile (Zack *et al.* 2004) and zircon (Rainbird *et al.* 1992) all help to overcome the effects of the processes that operate during the sedimentary cycle, but offer insights only into the provenance of the individual mineral component. An approach that combines conventional heavy mineral analysis with mineral chemical and isotopic studies of several phases optimizes the amount of provenance information that can be gathered. This approach proved successful in previous studies of Mesozoic and Tertiary sandstones in the Norwegian Sea region (Morton *et al.* 2005*a*, 2009), and was therefore adopted in the current study.

## Conventional analysis

The core samples from the Andøya B borehole were gently disaggregated by use of a pestle and mortar, avoiding grinding action. Chemical pretreatment was not used, thereby avoiding the possibility of modifying assemblages in the laboratory. Following disaggregation, the samples were immersed in water and cleaned by ultrasonic probe to remove and disperse any clay that was adhering to grain surfaces. The samples were then washed through a 63 μm sieve and resubjected to ultrasonic treatment until no more clay passed into suspension. Following this, the samples were wet-sieved through the 125 μm and 63 μm sieves, and the resulting >125 μm and 63–125 μm fractions were dried in an oven at 80 °C. The 63–125 μm fraction was placed in bromoform with a measured specific gravity of 2.8. Heavy minerals were allowed to separate under gravity, with frequent stirring to ensure complete separation. The heavy mineral residues were mounted under Canada Balsam for optical study using a polarizing microscope. Where possible, a split was retained for mineral chemical analysis.

Heavy mineral proportions (Table 1) were estimated by counting 200 non-opaque detrital grains using the ribbon method described by Galehouse (1971). Identification was made on the basis of optical properties, as described for grain mounts by Mange & Maurer (1992). A qualitative assessment was also made of other components, such as diagenetic minerals, opaques and mica. Provenance-sensitive mineral ratios (Table 2), as defined by Morton & Hallsworth (1994), were determined on the basis of a 200 grain count per mineral pair (where heavy mineral recovery allowed).

## Garnet and tourmaline major element analysis

Mineral grains were picked with a needle from dry residues during optical examination under the polarizing microscope, placed on double-sided adhesive tape, coated with carbon, and analysed using a Link Systems AN 10/55S energy-dispersive X-ray analyser attached to a Cambridge Instruments Microscan V electron microprobe at the University of Aberdeen. The count time was 30 s for each grain, and populations were characterized by analysis of approximately 50 grains per sample. Data reduction used the ZAF4 programme. Results from poorly oriented or rough grain surfaces were identified by low analytical totals and/or deviations from ideal stoichiometry, and were discarded. Garnet compositions are expressed in terms of the relativeproportions of Mg, $Fe^{2+}$, Ca and Mn in the garnet molecule, with all Fe calculated as $Fe^{2+}$, as recommended by Droop & Harte (1995). Tourmaline provenances are discriminated on the basis of their Al, Fe and Mg contents, as proposed by Henry & Guidotti (1985).

## Rutile trace element analysis

Rutile grains were picked with a needle from dry residues during optical examination under the polarizing microscope, placed on double-sided adhesive tape, and analysed using a Thermo Elemental X(7) series inductively coupled plasma-mass spectrometer (ICP-MS) coupled to a New Wave Research UP213 Nd:YAG 213 nm UV laser in the School of Earth, Ocean and Planetary Sciences at Cardiff University. The laser beam diameter was 30 μm and the laser repetition rate was set at 4 Hz. Helium gas was used for ablation initial transport from the laser cell, and this was combined with argon outside the cell as the sample

**Table 1.** *Relative abundance of non-opaque detrital heavy minerals (HM) in samples from Andøya B, expressed as frequency (%) of the non-opaque detrital component in the 63–125 μm size fraction on a grain count of 200*

| Depth (m) | HM unit | At | Ap | Ca | Cr | Ct | Ep | Gh | Gt | Mo | Ru | To | Zr |
|---|---|---|---|---|---|---|---|---|---|---|---|---|---|
| 271.7 | #4 | 0.5 | 49.5 | | | | | | 30.0 | R | 4.5 | 7.5 | 8.0 |
| 293.7 | #4 | 0.5 | 21.5 | | | | | | 34.0 | | 13.5 | 9.0 | 21.5 |
| 435.7 | #3c | 1.0 | 1.5 | | | | | | 33.0 | R | 27.0 | 28.5 | 9.0 |
| 446.4 | #3c | R | 4.0 | | | | | | 60.0 | R | 12.0 | 17.0 | 7.0 |
| 451.6 | #3c | 0.5 | 3.0 | | | 0.5 | | R | 50.5 | 0.5 | 19.5 | 14.0 | 11.5 |
| 458.3 | #3b | 1.5 | R | | | 1.0 | | R | 69.5 | 1.5 | 12.0 | 7.5 | 7.0 |
| 468.1 | #3b | 0.5 | 3.0 | R | 0.5 | 0.5 | | | 46.0 | 3.5 | 18.0 | 17.0 | 11.0 |
| 472.2 | #3b | R | 7.0 | | | 2.5 | | | 51.0 | R | 11.5 | 19.0 | 9.0 |
| 479.9 | #3a | R | R | | | | | R | 52.0 | 1.0 | 27.0 | 5.5 | 14.5 |
| 486.2 | #3a | R | R | | | | | R | 60.0 | 2.0 | 21.0 | 7.5 | 9.5 |
| 494.9 | #3a | R | | | | 0.5 | | | 31.0 | R | 32.5 | 31.5 | 4.5 |
| 517.1 | #2 | 0.5 | 0.5 | | | | R | | 1.5 | 0.5 | 50.0 | 3.5 | 43.5 |
| 524.3 | #2 | | | | | | | | | | 58.0 | 10.0 | 32.0 |
| 542.8 | #1 | | 74.0 | 0.5 | | 0.5 | | | 0.5 | | R | | 24.5 |

In addition to the detrital non-opaque phases, diagenetic carbonate is abundant at 271.7 m and 293.7 m, and diagenetic $TiO_2$ (anatase and brookite) is common at 524.3 m. Micas (muscovite and biotite) and chlorite are also common.
*Abbreviations*: At, anatase; Ap, apatite; Ca, calcic amphibole; Cr, chrome spinel; Ct, chloritoid; Ep, epidote; Gh, gahnite; Gt, garnet; Mo, monazite; Ru, rutile; To, tourmaline; Zr, zircon; R, rare (<0.5%).

**Table 2.** *Provenance-sensitive heavy mineral (HM) ratio data in samples from Andøya B*

| Depth (m) | HM unit | ATi | Total | GZi | Total | RuZi | Total | MZi | Total | CZi | Total |
|---|---|---|---|---|---|---|---|---|---|---|---|
| 271.7 | #4 | 85.5 | 200 | 73.8 | 271 | 35.7 | 171 | 0.9 | 111 | 0.0 | 110 |
| 293.7 | #4 | 61.9 | 118 | 60.2 | 211 | 35.4 | 130 | 0.0 | 84 | 0.0 | 84 |
| 435.7 | #3c | 8.0 | 200 | 70.1 | 214 | 71.2 | 281 | 7.4 | 216 | 0.0 | 200 |
| 446.4 | #3c | 19.5 | 200 | 85.5 | 234 | 60.8 | 255 | 0.5 | 201 | 0.0 | 200 |
| 451.6 | #3c | 19.5 | 200 | 78.4 | 255 | 56.7 | 231 | 4.3 | 209 | 0.0 | 200 |
| 458.3 | #3b | 4.5 | 200 | 88.9 | 225 | 49.5 | 200 | 9.5 | 221 | 0.0 | 200 |
| 468.1 | #3b | 27.5 | 200 | 77.8 | 257 | 56.9 | 232 | 11.1 | 225 | 2.9 | 206 |
| 472.2 | #3b | 25.0 | 200 | 83.3 | 240 | 58.2 | 239 | 3.6 | 196 | 0.0 | 189 |
| 479.9 | #3a | 4.0 | 200 | 79.4 | 252 | 66.0 | 294 | 11.5 | 226 | 0.0 | 200 |
| 486.2 | #3a | 6.5 | 107 | 85.5 | 234 | 63.5 | 274 | 15.3 | 236 | 0.0 | 200 |
| 494.9 | #3a | 0.0 | 200 | 85.8 | 233 | 87.1 | 255 | 1.1 | 178 | 0.0 | 176 |
| 517.1 | #2 | 17.0 | 200 | 4.8 | 210 | 56.5 | 230 | 1.5 | 203 | 0.0 | 200 |
| 524.3 | #2 | 0.0 | 69 | 0.0 | 200 | 64.1 | 234 | 0.0 | 200 | 0.0 | 200 |
| 542.8 | #1 | 100.0 | 200 | 0.5 | 201 | 0.5 | 201 | 0.0 | 200 | 0.0 | 200 |

For definition of index measurements, see Morton & Hallsworth (1994).
*Abbreviations*: ATi, apatite–tourmaline index; GZi, garnet–zircon index; RuZi, rutile–zircon index; MZi, monazite–zircon index; CZi, chrome spinel–zircon index.

was transported to the ICP-MS. Thermo Elemental Plasmalab time-resolved analysis (TRA) data acquisition software was used with a total acquisition time of 60 s per analysis, allowing about 30 s for background followed by 25 s for laser ablation. Plasmalab was used for initial data reduction with post-processing in Excel. The calibration made use of BIR-1G, BHVO-2G and BCR-2G (USGS basalt glass standards) to produce a four-point (including the origin) calibration curve. Data were normalized to Ti (98% $TiO_2$) and adjusted accordingly. Instrumental drift was monitored by repeat analysis of BHVO-2G after every 25–30 grains. Approximately 50 rutiles were analysed per sample.

The great majority of rutiles are derived from either metamafic or metapelitic sources (Force 1980; Zack *et al.* 2004). Discrimination of rutiles derived from these two sources is achieved using their Cr and Nb contents (Zack *et al.* 2004; Triebold *et al.* 2007; Meinhold *et al.* 2008). In this article, we have classified rutiles with >800 ppm Nb and Cr < Nb as metapelitic, and all others as metamafic, following the criteria proposed by Meinhold *et al.* (2008). Rutile formation temperatures have been estimated using the Zr content following the formula proposed by Watson *et al.* (2006).

## Zircon dating

Zircon grains were separated from heavy residues using heavy liquid (density 3.3 g cm$^{-3}$) and paramagnetic procedures. The zircon-rich heavy mineral concentrates were poured onto double-sided adhesive tape, mounted in epoxy together with chips of the reference zircons (Temora, FC1 and SL13), sectioned approximately in half, and polished. Reflected light photomicrographs, transmitted light photomicrographs and cathodoluminescence (CL) scanning electron microscope (SEM) images were prepared for all zircon grains. The CL images were used to decipher the internal structures of the sectioned grains and to ensure that the *c.* 20 µm SHRIMP spot was wholly within a single age component (usually the youngest) within the sectioned grains.

The U–Th–Pb analyses were made using SHRIMP II and SHRIMP RG at the Research School of Earth Sciences, The Australian National University, Canberra, Australia, following the method described by Williams (1998 and references therein). For each sample, zircons on the mount were analysed sequentially and randomly until a total of at least 60 grains was reached for each sample. Each analysis consisted of four or five scans through the mass range, with a reference zircon analysed for every five unknown zircon analyses. The data were reduced using the SQUID Excel Macro of Ludwig (2001). The U–Pb ratios were normalized relative to a value of 0.0668 for the Temora reference zircon, equivalent to an age of 417 Ma (Black *et al.* 2003), or 0.1859 for the FC1 reference zircon, equivalent to an age of 1099 Ma (Paces & Miller 1993). Uncertainties given for individual analyses (ratios and ages) are at the 1σ level. Correction for common Pb was made using the measured $^{204}Pb/^{206}Pb$ ratio, except for grains younger than *c.* 800 Ma (or those low in U and therefore low in radiogenic Pb), where the $^{207}Pb$ correction method was used (Williams 1998). When the $^{207}Pb$ correction is applied, it is not possible to determine radiogenic $^{207}Pb/^{206}Pb$ ratios or ages. In general, for areas younger than *c.* 800 Ma (and for areas that are low in U and therefore low in radiogenic Pb), the radiogenic $^{206}$Pb–$^{238}$U age has been used for the probability density plots. The $^{207}$Pb/$^{206}$Pb age is used for grains older than 800 Ma, or for younger grains enriched in U. Tera & Wasserburg (1972) concordia plots and probability density plots with stacked histograms were generated using ISOPLOT/EX (Ludwig 2003). Zircon isotopic compositions and ages are available in the Supplementary Publication material.

## Mineralogical and isotopic characteristics

The heavy mineral assemblages in the sandstones of the Andøya B borehole (Table 1) have relatively low diversity. Twelve species have been identified in total, most of which (anatase, apatite, chrome spinel, gahnite spinel, monazite, tourmaline and zircon) are known to be essentially stable during burial diagenesis (Morton & Hallsworth 2007). These minerals account for *c.* 30–100% of the assemblages. The remaining *c.* 0–70% of the assemblages consists of garnet and chloritoid, which are the two most stable heavy minerals known to undergo intrastratal dissolution (Morton & Hallsworth 2007). Garnet grains show well-developed etch facets similar to those figured by Turner & Morton (2007), indicating relatively advanced dissolution. The low diversity of the heavy mineral assemblages is therefore attributed to depletion of unstable minerals during burial diagenesis. The almost complete absence of minerals with lower stability than garnet (such as pyroxene, amphibole, epidote, titanite, kyanite and staurolite) is therefore considered to be due to dissolution rather than reflecting the nature of the source. The degree of modification to the Andøya B assemblages is considerably greater than that predicted for the present burial depth, and a phase of burial and subsequent uplift is therefore inferred. The heavy mineral assemblage recovered from the basement also has low diversity, comprising mainly apatite and zircon, but in this case the low diversity is a function of the basement lithology (granitic gneiss). It is noteworthy that the basement sample also contains unstable phases (amphibole and epidote), although they are present only in very low abundances.

Variations in the provenance-sensitive heavy mineral indices apatite:tourmaline (ATi), garnet: zircon (GZi) and rutile:zircon (RuZi) (Table 2) enable a subdivision of the succession in Andøya B into four heavy mineral units, #1 to #4 (Fig. 2). Unit #3 is further subdivided into #3a, #3b and #3c on the basis of smaller-scale changes in ATi.

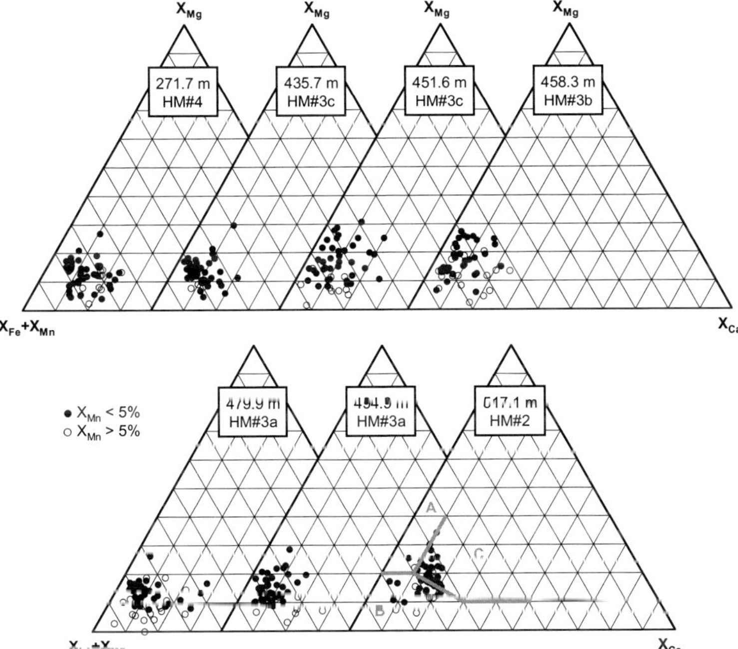


**Fig. 3.** Garnet compositions in sandstones from Andøya B, expressed in terms of the proportions of Mg, Fe$^{2+}$, Ca and Mn in the garnet molecule, normalized to total Fe + Mg + Ca + Mn, with all Fe calculated as Fe$^{2+}$, as recommended by Droop & Harte (1995). Subdivision of the ternary plot into Types A, B and C is taken from Morton *et al.* (2004).

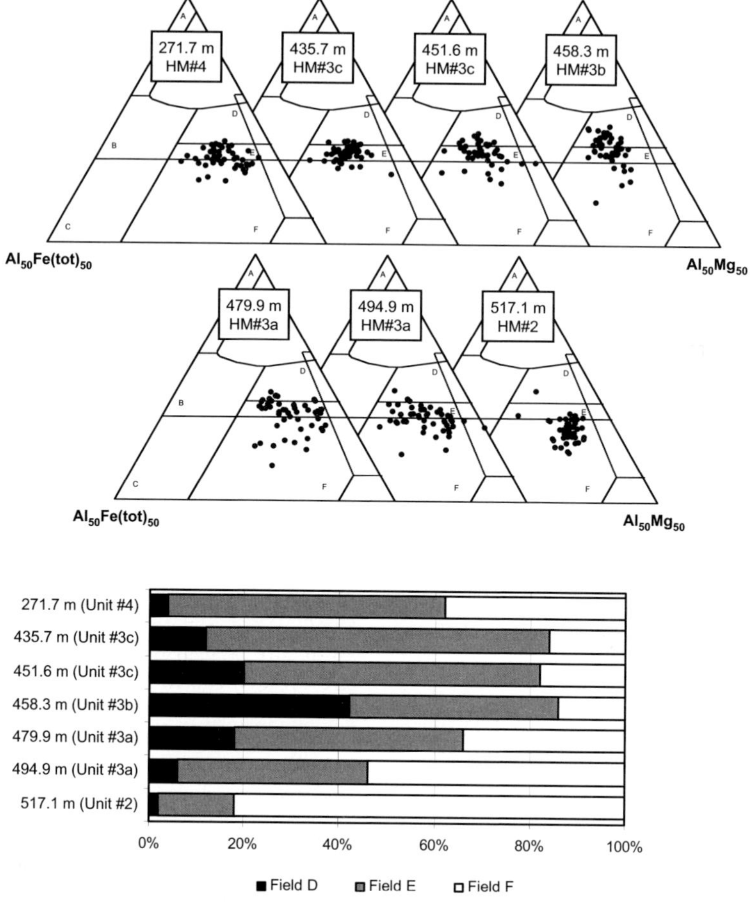



**Fig. 4.** Tourmaline compositions in Andøya B sandstones plotted on the provenance-discriminant Al–Fe–Mg ternary diagram devised by Henry & Guidotti (1985), and stratigraphic variations in the relative abundance of tourmaline types D, E and F. Field A: Li-rich granitoids, pegmatites and aplites. Field B: Li-poor granitoids, pegmatites and aplites. Field C: hydrothermally altered granitic rocks. Field D: Aluminous metapelites and metapsammites. Field E: Al-poor metapelites and metapsammites. Field F: Fe$^{3+}$-rich quartz-tourmaline rocks, calc-silicates and metapelites.

## Unit #1

Unit #1 represents the Proterozoic gneiss basement. It has a low-diversity heavy mineral assemblage dominated by apatite and zircon, and therefore has extremely high ATi (100) and extremely low GZi and RuZi (both <1). The assemblage in the basement sample is typical of a granite or acidic gneiss.

## Unit #2

Unit #2 represents the lower part of the Hestberget Member (Early Bathonian) of the Ramså Formation. The top of the unit is closely coincident with the J32 flooding surface (Fig. 2). Unit #2 is characterized by high RuZi (57–64) in conjunction with low ATi (0–17) and GZi (0–5). The

garnet population falls mainly in Field C (high-Mg, high-Ca) as defined by Morton *et al.* (2004), although the population straddles the boundary with Field B (Fig. 3). The tourmaline population (Fig. 4) is distinctive in that the great majority of grains fall in Field F, considered to represent derivation from Al-poor metapelites, Fe$^{3+}$-rich quartz-tourmaline rocks or calc-silicates (Henry & Guidotti 1985). The rutile assemblage (Fig. 5) is also distinctive, with 87% of the rutiles having metamafic compositions. Temperature estimates suggest the rutiles formed under lower amphibolite or eclogite facies conditions (main group falling in the 550–650 °C range). Although Zack *et al.* (2004) caution against using the Zr content to determine the formation temperatures of metamafic rutiles, subsequent studies (Meinhold *et al.* 2008;

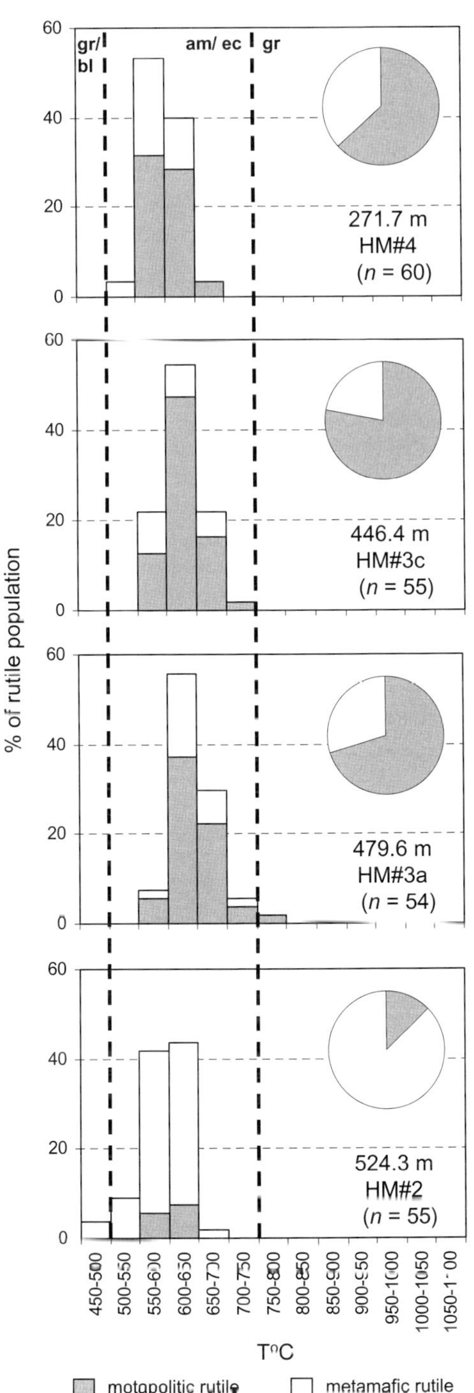



**Fig. 5.** Rutile geochemistry in sandstones from Andøya B, showing variations in source lithotype (metamafic and metapelitic) and in metamorphic grade. Rutiles with >800 ppm Nb and Cr < Nb are classed as metapelitic, and all others as metamafic, following Meinhold *et al.*

Morton & Chenery 2009) have shown that metamafic rutiles yield similar temperatures to those in metapelitic rutiles from the same samples, and therefore appear to be reliable in many cases.

Most of the zircons have U–Pb compositions that are >90% concordant, although the Archaean group shows a greater degree of discordance (Fig. 6). The zircon age spectrum (Fig. 7) shows the presence of one main peak at *c.* 1780–1820 Ma, together with a more diffuse group of Archaean grains between *c.* 2600–2800 Ma and a small number of older Palaeoproterozoic zircons at *c.* 1840–1860 Ma. There is also a small number of scattered isolated zircons with ages of uncertain significance.

## Unit #3

Unit #3 is Bathonian–Kimmeridgian in age. It consists of the upper part of the Hestberget Member, the Kullgrøfta Member and Bonteigen Member of the Ramså Formation, together with the Breisanden Member and lower part of the Taumhølet Member of the Dragneset Formation. The top is essentially defined by the disappearance of sand-size detritus. The unit has consistent heavy mineral characteristics, with relatively low ATi (0–28), high GZi (70–89) and high RuZi (50–87) throughout. The subunits (#3a, #3b and #3c) are based on variations in ATi. Subunit #3a has low ATi thoughout, with the base of subunit #3b being marked with a rise coincident with the J34 flooding surface. ATi values then fall through the J34 sequence, before rising immediately above the J54 flooding surface, which marks the base of subunit #3c. The ATi trend in subunit #3c mimics that of subunit #3b, showing an upward decrease.

Garnet compositions in Unit #3 differ from those in Unit #2, being dominated by the Type B component, with subordinate Types A and C (Fig. 3). Tourmaline compositions also differ markedly from Unit #2 (Fig. 4), being richer in Field E (Al-poor metasedimentary source) and Field D (Al-rich metasedimentary source). There appear to be differences between the subunits, with subunit #3a having higher proportions falling in Field F, subunit #3b having higher proportions falling in Field D, and subunit #3c having predominantly Field E tourmaline (Fig. 4). Rutile geochemistry (Fig. 5) also confirms a marked change in provenance characteristics between Unit #2 and Unit #3,

(2008). Rutile formation temperatures have been estimated using the Zr content as proposed by Watson *et al.* (2006). gr/bl, greenschist/blueschist; am/ec, amphibolite/eclogite; gr, granulite.

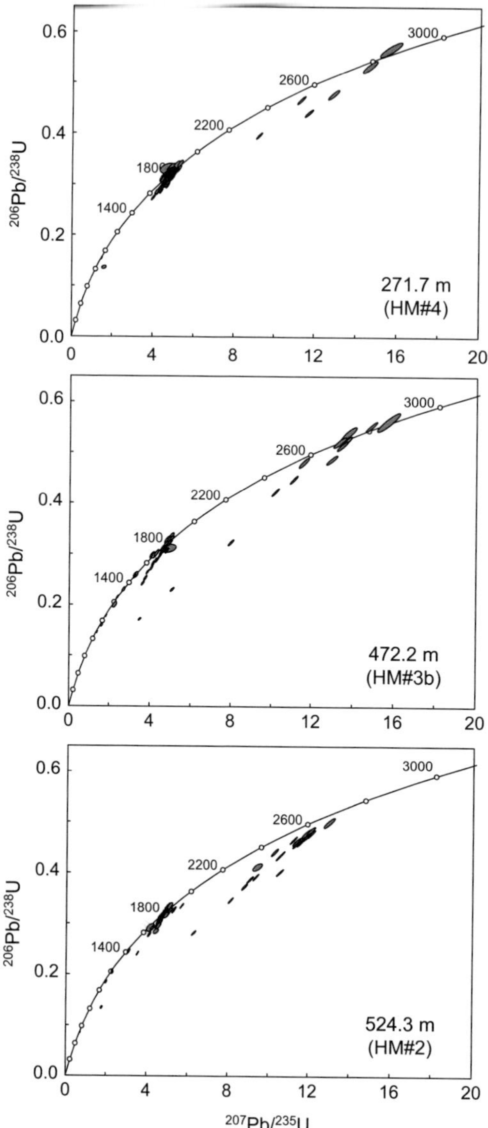



**Fig. 6.** U–Pb isotopic compositions of detrital zircon in sandstones from Andøya B, plotted on Wetherill concordia digrams.

with rutile populations in Unit #3 being dominated by the metapelitic component (70–78% of the populations, compared with 13% in Unit #2). Formation temperatures are similar to those of Unit #2, corresponding to amphibolite/eclogite facies conditions, although there does appear to be a slight increase in the proportion of upper amphibolite facies rutiles (>650 °C).

Zircons were analysed from the basal sample in subunit #3b. They mostly have >90% concordant compositions but there is evidence for Pb loss from some Proterozoic and Archaean grains (Fig. 6). The zircon age spectrum (Fig. 7) is more diverse than in Unit #2, although the main two groups are similar. The majority of zircons fall in the c. 1730–1820 Ma age range, with an apparent bimodal structure peaking at c. 1760–1770 Ma and c. 1780–1790 Ma. Archaean zircons in the c. 2600–2900 Ma bracket form a large group, and there is a small number of older Palaeoproterozoic zircons at c. 1850–1870 Ma. The remainder of the spectrum comprises scattered Proterozoic grains, the youngest being 880 Ma, together with a small but conspicuous group of Early Palaeozoic zircons, most of which have model ages between 435 Ma and 446 Ma (Late Ordovician–Early Silurian).

## Unit #4

Unit #4 consists of Tithonian–Berriasian sandstones that occur in the upper part of the Ratjønna Member of the Dragneset Formation (Fig. 2). The sandstones in Unit #4 have similar GZi values to those of Unit #3, but ATi values are much higher (62–85) and RuZi is lower (36). Garnet compositions are similar to those of Unit #3, dominated by grains falling in Field B (Fig. 3). Tourmalines mainly fall in Field E, similar to those found in subunits #3a and #3c (Fig. 4). Rutiles are also similar to those of Unit #3, being rich in the metapelitic component (63%) and most having lower amphibolite/eclogite facies formation temperatures (Fig. 5).

The zircon population in Unit #4 (Fig. 7) is dominated by zircons dated as c. 1760–1820 Ma, with a slightly older group (c. 1840–1890 Ma) also evident. Archaean zircons are also present, but are less abundant than in either the Unit #2 or Unit #3 samples.

## Provenance

The variations in provenance-sensitive ratios, mineral chemistry and zircon age data in the Andøya B succession are interpreted as resulting from a combination of changes in sediment source and the extent of weathering during the sedimentary cycle. The changes in provenance through the succession are summarized in Table 3.

The most profound change in provenance takes place in the Bathonian, between Unit #2 and Unit #3, and is recorded by a major increase in GZi, changes in garnet, tourmaline and rutile geochemistry and, to a lesser extent, a difference in the zircon age spectrum. Unit #2 has rutile and garnet geochemical characteristics indicating that

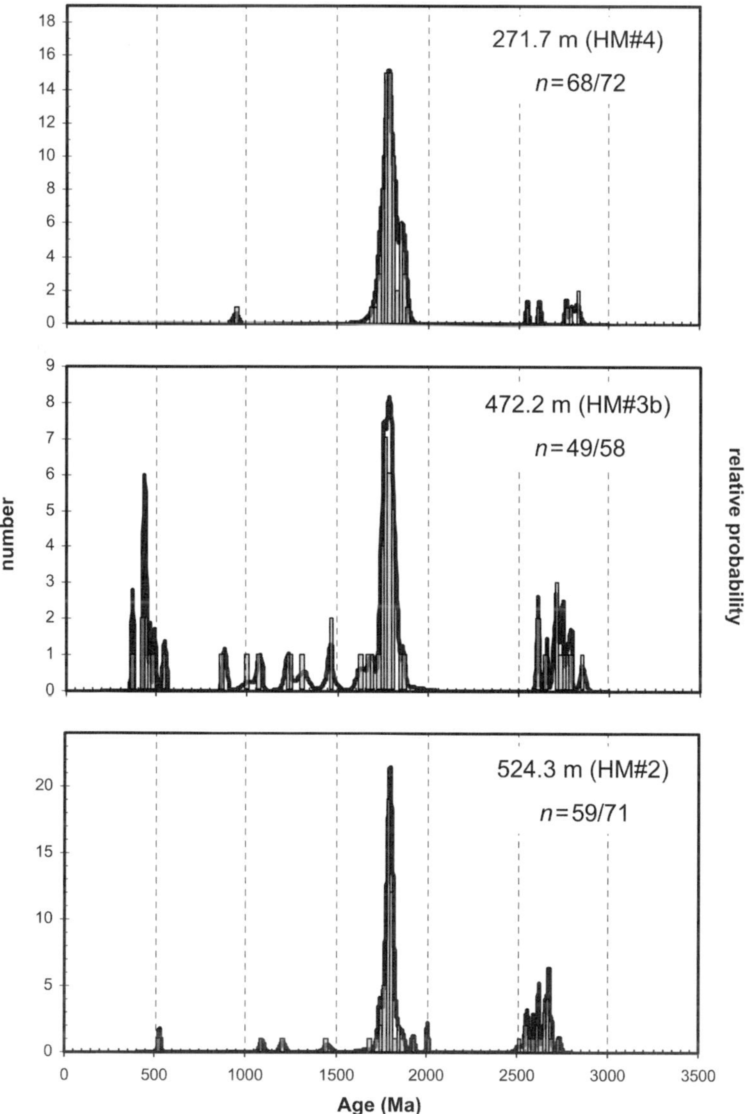



**Fig. 7.** Combined relative probability–histogram plots of detrital zircon ages in sandstones from Andøya B. *n* refers to the number of zircon analyses with > 90% concordance in the total number of analyses. Note that for zircons younger than *c.* 800 Ma, discordance cannot be measured because correction for common Pb is made using the $^{207}$Pb correction method. When the $^{207}$Pb correction is applied, it is not possible to determine radiogenic $^{207}$Pb/$^{206}$Pb ratios or ages, thereby precluding the use of the discordance formula.

amphibolite/eclogite facies metamafic rocks were an important source of detritus. The tourmaline data suggest involvement of Al-poor metapelites, $Fe^{3+}$ rich quartz tourmaline rocks or calc-silicates. The tourmaline population in Unit #2 is closely comparable to that found in sandstones derived from the Western Gneiss Region further south in Norway (Morton *et al.* 2005*a*), which also have

garnet and rutile characteristics indicating sourcing from moderate to high-grade metamafic rocks. The zircon population in Unit #2 indicates derivation from both Archaean and Palaeoproterozoic terranes. The Archaean ages correspond closely with those found in the Lofoten–Vesterålen area (Griffin *et al.* 1978; Jacobsen & Wasserburg 1978; Corfu 2007). The main Palaeoproterozoic peak is

**Table 3.** *Summary of the provenances of heavy mineral (HM) units #2 to #4 in the Andøya B borehole, Lofoten–Vesterålen area*

| HM unit | Lithostratigraphy | Age | Provenance interpretation |
| --- | --- | --- | --- |
| Unit #4 | Ratjønna Member (Dragneset Formation) | Tithonian–Berriasian | AMCG suite of the Andøya–Lofoten High; minor input from metasediments and Archaean basement |
| Unit #3 | Upper Hestberget, Kullgrøfta, Bonteigen and lower Taumhølet members (Ramså and Dragneset formations) | Middle Bathonian–Kimmeridgian | Archaean basement, AMCG suite, metasediments (Leknes Group), and Caledonian granites of the Andøya–Lofoten High |
| Unit #2 | Lower Hestberget Member (Ramså Formation) | Early Bathonian | West Troms Basement Complex east of Andøya |

AMCG suite, anorthosite–mangerite–charnockite–granite in the Lofoten and Vesterålen area.

at *c.* 1780–1820 Ma (Fig. 8), which corresponds to the age range of intrusions in the Lofoten–Vesterålen AMCG suite (Corfu 2004*a*).

The heavy mineral, mineral chemical and zircon age data for Unit #2 can therefore be reconciled with a source on the Andøya–Lofoten High, as proposed by Dalland (1975, 1981). However, the geological history of the West Troms Basement Complex, located to the east of Andøya, has close similarities with that of Lofoten–Vesterålen, particularly in terms of Archaean and Palaeoproterozoic events (Corfu *et al.* 2003; Bergh *et al.* 2010), and derivation from the east is therefore equally possible. The model proposed by Dalland (1975, 1981) for the Bathonian Hestberget Member included input from both east and west, but at higher stratigraphic levels sediment input is believed to be confined to the western (Andøya–Lofoten High) source. Given the strong contrast in provenance characteristics between Unit #2 and Unit #3, it is considered most likely that Unit #2 had a source to the east, in the West Troms Basement Complex.

The low ATi in Unit #2 is probably attributable to loss of apatite as a result of weathering during the sedimentary cycle. Apatite is one of the least stable minerals in humid weathering conditions (Morton 2012; Morton *et al.* 2012), and the fluviolacustrine depositional environment and warm humid climatic conditions (Vajda & Wigforss-Lange 2009) of Unit #2 are likely to have promoted apatite dissolution. Furthermore, it is unlikely that either of the potential source regions (West Troms Basement Complex and the Andøya–Lofoten High) inherently lack apatite. Garnet is also relatively unstable under weathering conditions, although less so than apatite (Morton 2012), and weathering may therefore also account for the low garnet abundances in Unit #2.

The provenance characteristics of Unit #3, particularly garnet and rutile compositions, indicate there was a significant difference in source lithology compared with Unit #2. Unit #3 garnets mainly have Type B compositions, which are typical of amphibolite-facies metasediments, although they can also be derived from a number of other metamorphic lithologies (Morton *et al.* 2004; Mange & Morton 2007). The rutiles in Unit #3 were predominantly derived from amphibolite or eclogite facies metapelites, consistent with the evidence from the garnets, although metamafic rutiles are also present in relatively high amounts. Tourmaline assemblages are dominated by Types D and E, providing further evidence for the presence of metasediments with both Al-rich and Al-poor compositions in the source area.

The zircon population in Unit #3 is more diverse than in Unit #2. Although there are Archaean and Palaeoproterozoic components that are closely comparable to those in Unit #2, there is a significant number of younger Proterozoic and Early Palaeozoic zircons, unlike the situation in Unit #2. The younger Proterozoic grains are likely to represent recycling from metasediments of the Caledonian Nappes, as these are known to contain a wide range of Precambrian zircons representing a diversity of ultimate basement sources (Bingen & Solli 2009). The Leknes Group of Lofoten–Vesterålen is a potential candidate to supply such grains (Corfu 2004*b*). Metasediments of the Caledonian Nappe Domain are the most likely source of the metapelitic rutile, Type B garnet, and the metasedimentary tourmalines found in Unit #2. The Leknes Group is also a potential candidate for the Early Palaeozoic zircons found in Unit #3, although they could also have been derived from Caledonian granites in the Upper and Uppermost Allochthons (Corfu 2004*b*; Bingen & Solli 2009).

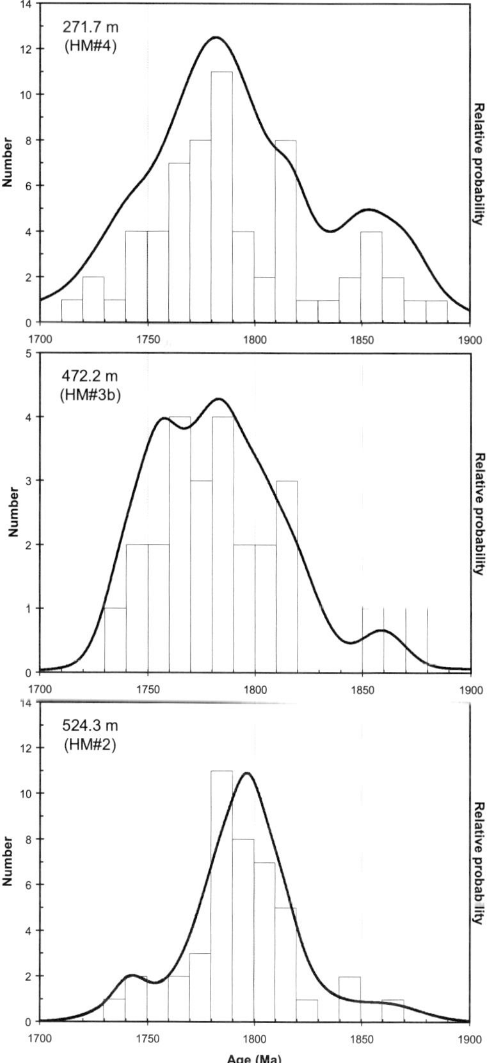

**Fig. 8.** Detail of the Palaeoproterozoic of the Andøya B zircon age spectra, showing that the Unit #2 sample (524.3 m) contains fewer zircons in the c. 1740–1780 Ma age range compared with Unit #3 and Unit #4, and that zircons dated in the c. 1840–1870 Ma interval are most common in Unit #4.

The Palaeoproterozoic group is also different in detail compared with Unit #2. The main group evident in Unit #2 (c. 1780–1820 Ma) is also present in Unit #3. However, Unit #3 also has a slightly younger group at c. 1760–1770 Ma (Fig. 8), which corresponds to ages obtained from monazite in the Torset granite, zircon and titanite in the Borge granite pegmatite, and titanite in the Eidsfjord monzonite (Corfu 2004a), representing the late- to

post-magmatic stage of activity in Lofoten–Vesterålen.

The heavy mineral, mineral chemical and zircon age data from Unit #3 are therefore consistent with derivation from the Andøya–Lofoten High, as proposed by Dalland (1975, 1981), with evidence for input from Archaean rocks, the full range of plutons in the Lofoten–Vesterålen AMCG suite, metasediments of the Leknes Group, and Palaeozoic granites hosted in the Caledonian Nappes.

The low ATi found in Unit #3 is most likely attributable to weathering during the sedimentary cycle, as discussed for Unit #2. It is noteworthy that changes in ATi take place across maximum flooding surfaces, with increases occurring immediately above the J34 and J54 MFS (Fig. 2). The same pattern has been seen in the Upper Jurassic of the North Sea, where it was attributed to changes in the extent of floodplain storage during periods of sea-level change (Morton *et al.* 2012). During sea-level highstands, the amount of exposed floodplain is decreased, reducing the potential for floodplain storage and consequent apatite dissolution; the reverse is true for lowstand events. The same scenario is considered to be the most likely explanation for the observed ATi pattern in Andøya B.

The provenance characteristics of Unit #4 differ somewhat from those of Unit #3. RuZi values are significantly lower, implying greater input from zircon-bearing lithologies (granitoids being most likely). GZi values are also slightly lower, probably for the same reason. Garnet, rutile and tourmaline compositions do not differ greatly from Unit #3, implying that metasedimentary sources remained important. The main change in mineralogy and mineral chemistry therefore appears to be increased supply from igneous sources. The zircon age data support this hypothesis, because the abundance of post-1700 Ma zircons is much lower than in Unit #3, and there is a corresponding increase in the abundance of the Palaeoproterozoic component. The Palaeoproterozoic zircons ages are very similar to those present in Unit #3, with ages corresponding to both the main phase and the late- to post-magmatic stage of activity in Lofoten–Vesterålen. However, there is an increase in abundance of an older group of zircons (c. 1850–1870 Ma), which is present in Unit #2 and Unit #3 in low abundances. These correspond to the age of early magmatism in the AMGC complex, which took place at 1860–1870 Ma and is represented by the Lodingen granite and the Hopen pluton (Corfu 2004a). Archaean zircons also appear to be less abundant than at lower stratigraphic levels.

The provenance of the Unit #4 sandstones is therefore considered to have lain on the Andøya–Lofoten High, as with Unit #3, the differences relating essentially to changes in relative input from the lithological units exposed in the source, possibly through a change in sediment routing or by unroofing.

The other main difference between Unit #4 and the underlying sandstones relates to the ATi value, which is much higher in Unit #4. This is again interpreted as related to differences in the extent of floodplain weathering. Unit #4 is found in the Tithonian–Berriasian interval within the Ratjønna Member of the Dragneset Formation, which represents the deepest marine conditions in the Jurassic of the area (Dalland 1975, 1981). It was therefore a time of particularly high sea-level, which would have significantly reduced the area of floodplain available for alluvial storage and weathering.

## Comparison with Cretaceous and Tertiary sandstones of the Vøring Basin

Cretaceous and Tertiary deep-water submarine fan sandstones are important targets for hydrocarbon exploration in the Vøring and Møre basins, with discoveries in Ormen Lange (Gjelberg *et al.* 2001), Nyk-Luva (Kittilsen *et al.* 1999), and more recently in Ellida and Midnattsol. On the basis of an integrated provenance study, Morton *et al.* (2005*a*) considered that the sandstones in the Vøring Basin were sourced from both the east (Scandinavia) and west (East Greenland), a view shared by Fonneland *et al.* (2004). The sands sourced from the east (sand type MN1 as defined by Morton *et al.* 2005*a*) are characterized by high RuZi values (>35), whereas those sourced from the west (MN2a and MN2b) have lower RuZi, mostly <35

(Fig. 9). Garnet geochemistry does not effectively discriminate between MN1 and MN2, both being dominated by Type B garnet (Fig. 10), but tourmaline compositions are more useful (Fig. 11). MN1 sandstones have a relatively simple zircon age spectrum dominated by a single main peak at *c.* 1780–1800 Ma, whereas MN2 sandstones have complex zircon age spectra including Archaean, Palaeoproterozoic, Mesoproterozoic, Early Palaeozoic, Permian–Triassic and mid-Cretaceous elements (Fig. 12).

On the basis of facies mapping, Vergara *et al.* (2001) concluded that in addition to sediment fed from East Greenland and the Norwegian landmass, the Vøring Basin also received sediment from the NE, in the direction of Lofoten–Vesterålen and West Troms, and that submarine fans derived from this area were an important contributor to the reservoir succession in Nyk. The Lofoten–Vesterålen and West Troms region is the only likely source of Archaean zircons on the Norwegian margin (Bingen & Solli 2009), one of the key features used by Morton *et al.* (2005*a, b*) to distinguish Norwegian and Greenland sourcing in the Vøring Basin. The data acquired from the Andøya B succession have established the mineralogical attributes of sediment shed from Lofoten–Vesterålen and possibly the West Troms area during the Jurassic and Early Cretaceous. These attributes can be compared with those acquired from the Vøring Basin sandstones in order to determine if any of them have characteristics suggestive of input from the NE.

Most of the sandstones in Andøya B have much higher RuZi values than those found in sand type MN2, and many have RuZi higher even than MN1

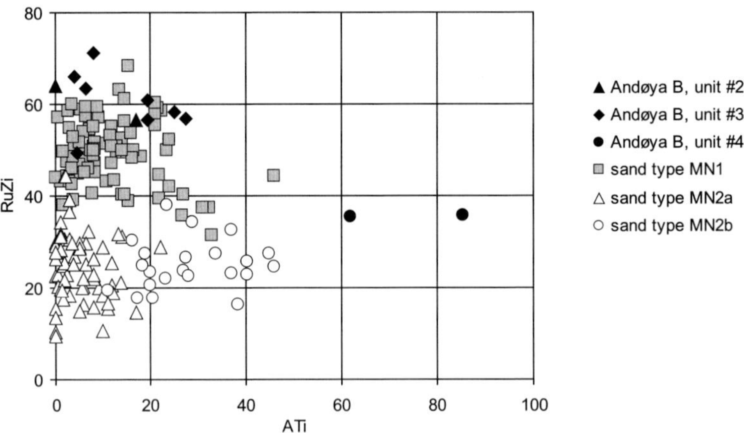


**Fig. 9.** Heavy mineral ratio data from Andøya B compared with those characterizing sand types MN1, MN2a and MN2b from the Mesozoic and Paleocene of the Vøring Basin. MN1, MN2a and MN2b data are from Morton *et al.* (2005*a*).

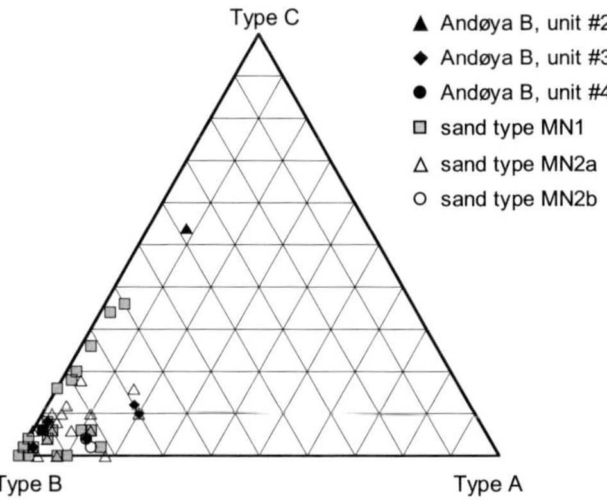



**Fig. 10.** Relative abundance of garnet types A, B and C (see Fig. 3 for definition) in the sandstones of Andøya B, compared with sand types MN1, MN2a and MN2b from the Mesozoic and Paleocene of the Vøring Basin (from Morton *et al.* 2005*a*).

are also rich in Type B garnet and are hence indistinguishable from Andøya B in terms of garnet composition. The only exception is the garnet population in Unit #2, which is rich in Type C garnet and therefore plots close to the C pole (Fig. 11). A small number of MN1 sandstones, from the Paleocene in wells 6610/3-1 and 6710/10-1 (Morton *et al.* 2005*a*), have elevated Type C garnet contents (Fig. 10). It is possible that these were supplied from the same (or similar) source that supplied the sediment in Andøya B Unit #2. This possibility requires further testing by means of rutile geochemistry, because the rutiles in Andøya B Unit #2 have distinctive metamafic compositions. Zircon dating may also provide useful information given the presence of a small Archaean component in the Andøya B provenance.

(Fig. 9). The lowest RuZi values in Andøya B are found in Unit #4: these fall at the lower end of the range shown by MN1, but differ from both MN1 and MN2 by virtue of their higher ATi (Fig. 9). In terms of conventional heavy mineral data, therefore, sandstones in Andøya B have different characteristics from those found in both MN1 and MN2, but, of the two, they have greater affinities with MN1.

Most of the garnet populations in Andøya B are rich in the Type B component, and therefore plot close to the B pole on the garnet ternary summary diagram (Fig. 10). Both MN1 and MN2 sandstones

The tourmaline populations in MN1 are distinguished from those of MN2 on the basis of the abundance of granitic types (B and C) and, to a lesser extent, on the relative abundance of Al-rich metasedimentary (Type D) and Al-poor metasedimentary (E and F) tourmaline (Fig. 11). The tourmaline populations in Andøya B are radically different to those in MN2, having lower abundances of granitic types and higher abundances of Al-poor metasedimentary types (Fig. 11). They compare better with MN1, in that granitic types are scarce, but most samples fall outside the range shown by MN1 because they contain a higher proportion of Types E and F. The tourmaline data therefore indicate that the Andøya sandstones differ from both MN1 and MN2, but have greater affinities with MN1.

The zircon spectrum in Andøya B compares reasonably well with the spectrum found in the

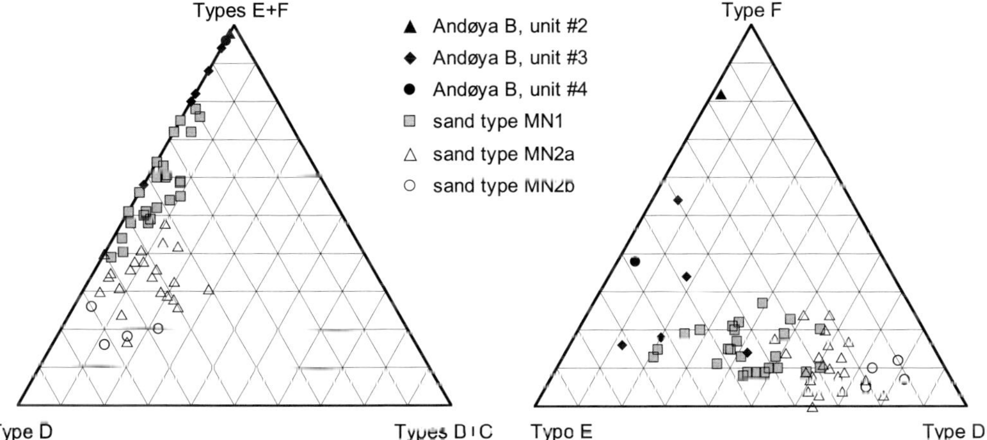



**Fig. 11.** Tourmaline populations in Andøya B compared with those in sand types MN1, MN2a and MN2b from the Mesozoic and Paleocene of the Vøring Basin (from Morton *et al.* 2005*a*). Tourmaline types B, C, D, E and F are as defined by Henry & Guidotti (1985), as shown in Figure 4.

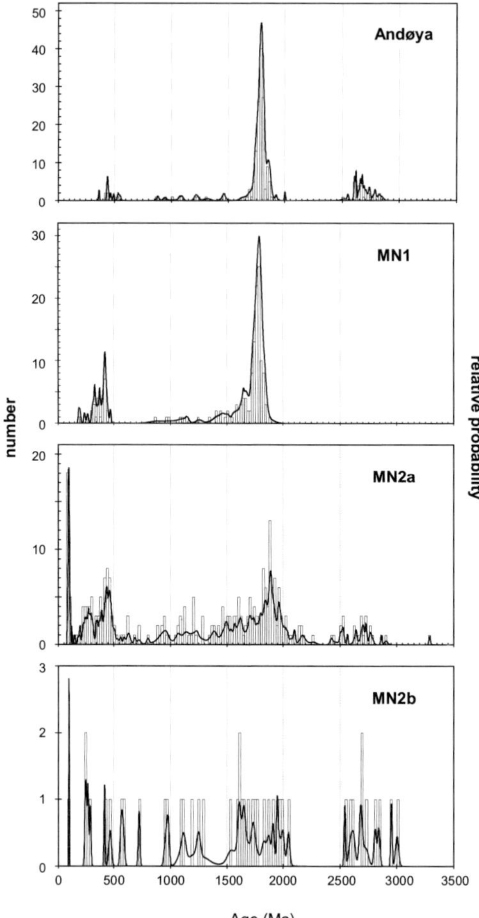



**Fig. 12.** Zircon age spectrum from Andøya B (all three samples combined), compared with those from sand types MN1, MN2a and MN2b. The MN1 spectrum is a composite using data from 6710/10-1, 1459.2 m (from Morton *et al.* 2005*a*) and 6610/3-1, 2300.0 m and 6507/2-2, 3281.0 m (from Morton *et al.* 2005*b*). The MN2a spectrum combines data from 6707/10-1, 3002.8 m, 6607/5-2, 4172.0 m, 6507/2-2, 2830.1 m and 6505/10-1, 3711.6 m (from Morton *et al.* 2005*b*). The MN2b spectrum is from 6704/12-1, 2574.6 m (Morton *et al.* 2005*a*).

MN1 sand type (Fig. 12), with both displaying prominent peaks at *c.* 1780–1800 Ma. The only significant difference is that Archaean zircons are relatively common in Andøya B, but are absent in MN1. By contrast, although both MN2a and MN2b contain an Archaean group similar to that found in Andøya B, the Proterozoic and Phanerozoic parts of the MN2a and MN2b spectra bear no resemblance to the Andøya spectrum. In particular, the

prominent *c.* 1780–1800 Ma peak in Andøya B is not evident in either MN2a or MN2b.

MN1 and MN2 sandstones therefore have different mineralogical and mineral chemical attributes to those found in Andøya B. Consequently, there is no available evidence to suggest that sands fed from the Lofoten–Vesterålen and West Troms area entered the Vøring Basin during the Cretaceous and Paleocene. The only possible input identified to date occurs in the Paleocene of 6610/3-1 and 6710/10-1, where elevated Type C garnet contents (Morton *et al.* 2005*a*) indicate supply from high-grade metamafic rocks similar to those that sourced the sediment in Andøya B unit #2. The differences between the Andøya B and MN2 sandstones are especially profound, consistent with the evidence for westerly (Greenland) sourcing for MN2 (Morton *et al.* 2005*a*, *b*). There are greater similarities between Andøya B and MN1 sandstones, consistent with the inferred juxtaposition of their respective source areas. The results of this study show that sandstones fed from the Lofoten–Vesterålen and West Troms area have a distinctive provenance signature (high RuZi, tourmaline populations dominated by Al-poor metasedimentary types, and a zircon spectrum comprising a major Palaeoproterozoic peak at *c.* 1780–1800 Ma with a subsidiary, more scattered Neoarchaean group). Identification of sand fed into the Vøring Basin from the NE should therefore be a straightforward issue, but, as yet, sands with these characteristics have not been observed.

## Conclusions

An integrated heavy mineral, mineral chemical and zircon age dating study of sandstones drilled in the Andøya B borehole, northern Norway, has identified three mineralogically distinct units (#2 to #4) overlying Precambrian granitic basement (Unit #1).

Unit #2 sandstones (Bathonian, lower Hestberget Member, Ramså Formation) have characteristics indicating that lower amphibolite or eclogite facies metamafic rocks were an important component of the source area. Zircon age data indicate a major crust-forming episode at *c.* 1780–1820 Ma as well as supply from Archaean rocks (*c.* 2600–2800 Ma).

There is a marked change in provenance characteristics between Unit #2 and Unit #3, expressed by differences in garnet, rutile and tourmaline geochemistry and by a more subtle change in zircon age data. The provenance of Unit #3 is interpreted as including amphibolite facies metasediments with subordinate metamafics. Zircon age data indicate a major crust-forming event at *c.* 1760–1820 Ma age range, with subordinate supply from Archaean (c. 2600–2900 Ma) and Early Palaeozoic (435–446 Ma) rocks and scattered representation through the mid-Proterozoic.

Dalland (1975, 1981) inferred input from both the west (Andøya–Lofoten High) and east (West Troms region) for the Hestberget Member, with the overlying sandstones being sourced exclusively from the Andøya–Lofoten High. Although the provenance characteristics of Unit #2 can be reconciled with both source regions, the marked change between Unit #2 and Unit #3 suggests derivation from the east is more likely for Unit #2.

Sandstones in Unit #3 (Bathonian–Kimmeridgian, upper part of the Ramså Formation and lower part of the Dragneset Formation) were derived from a more complex source area than Unit #2. Rutile, tourmaline and garnet data indicate input from amphibolite/eclogite facies metasedimentary rocks of variable composition, and zircon data indicate input from Archaean gneisses, Proterozoic rocks with ages that correspond to those found in the AMGC suite of Lofoten–Vesterålen, and Early Palaeozoic (Caledonian) granitoids. These data are consistent with derivation from the Andøya–Lofoten High to the west, as proposed by Dalland (1975, 1981). A similar source is diagnosed for Unit #4 sandstones (upper part of the Dragneset Formation, Tithonian–Berriasian), although there appears to have been less input from the Archaean and Early Palaeozoic at this time.

Variations in ATi appear to be independent of changes in provenance, and were probably caused by differences in the extent of weathering during alluvial storage, similar to that seen in Jurassic sandstones further south in the North Sea (Morton et al. 2012).

The sandstones in Andøya B were therefore derived either from Lofoten–Vesterålen or the West Troms region of northern Norway, and their characteristics can therefore be regarded as representative of sediment that might have entered the Vøring Basin from the NE during the Jurassic and Cretaceous. However, sandstones with these characteristics have not yet been identified in hydrocarbon exploration wells from the area, and it would therefore appear that sediment supply from the direction of Lofoten–Vesterålen was of minor importance. Most of the sandstones were derived either from northern Nordland (sand type MN1, as defined by Morton et al. 2005a, b) or northern East Greenland (sand type MN2). Sand type MN1 has closer affinities with the sandstones in Andøya B, consistent with the inferred juxtaposition of their respective source regions. They differ principally in that Andøya B sandstones have higher RuZi, tourmaline populations that are richer in Types E and F, and contain a significant Archaean group. Some MN1

sandstones of Paleocene age have elevated Type C garnet contents, similar to those seen in Unit #2 of Andøya B, and it is possible that the source of Unit #2 contributed sediment to the northeastern part of the Vøring Basin during the Paleocene. This possibility requires testing by rutile geochemical analysis and zircon age dating.

We are grateful to BP Norway, who funded much of the analytical work discussed in the paper, and to the NPD, for allowing access to the core from the Andøya B borehole. Thanks are also due to P. Milner for his support and encouragement of the study, and for his constructive comments on an earlier version of this manuscript.

# References

BERGH, S. G., KULLERUD, K., ARMITAGE, P. E. B., ZWAAN, K. B., CORFU, F., RAVNA, E. J. K. & MYHRE, P. I. 2010. Neoarchaean to Svecofennian tectono-magmatic evolution of the West Troms Basement Complex, North Norway. *Norwegian Journal of Geology*, **90**, 21–48.

BINGEN, B. & SOLLI, A. 2009. Geochronology of magmatism in the Caledonian and Sveconorwegian belts of Baltica: synopsis for detrital zircon provenance studies. *Norwegian Journal of Geology*, **89**, 267–290.

BLACK, L. P., KAMO, S. L., ALLEN, C. M., ALEINIKOFF, J. N., DAVIS, D. W., KORSCH, R. J. & FOUDOULIS, C. 2003. TEMORA 1: a new zircon standard for Phanerozoic U–Pb geochronology. *Chemical Geology*, **200**, 155–170.

CORFU, F. 2004a. U–Pb age, setting and tectonic significance of the anorthosite–mangerite–charnockite–granite suite, Lofoten–Vesterålen, Norway. *Journal of Petrology*, **45**, 1799–1819.

CORFU, F. 2004b. U–Pb geochronology of the Leknes Group: an exotic Early Caledonian metasedimentary assemblage stranded on Lofoten basement, northern Norway. *Journal of the Geological Society, London*, **161**, 619–627.

CORFU, F. 2007. Multistage metamorphic evolution and nature of the amphibolite-granulite facies transition in Lofoten–Vesterålen, Norway, revealed by U–Pb in accessory minerals. *Chemical Geology*, **241**, 108–128.

CORFU, F., ARMITAGE, P. E. B., KULLERUD, K. & BERGH, S. G. 2003. Preliminary U–Pb geochronology in the West Troms Basement Complex, North Norway: Archaean and Palaeoproterozoic events and younger overprints. *Norges Geologiske Undersøkelse Bulletin*, **441**, 61–72.

DALLAND, A. 1975. The Mesozoic rocks of Andøy, northern Norway. *Norges Geologiske Undersøkelse Bulletin*, **316**, 271–287.

DALLAND, A. 1981. Mesozoic sedimentary succession at Andøy, northern Norway, and relation to structural development of the North Atlantic area. *In*: KERRAND, J. W. & FERGUSSON, A. J. (eds) *Geology of the North Atlantic Borderlands*. Canadian Society of Petroleum Geologists Memoirs, **7**, 563–584.

DORÉ, A. G. 1991. The structural foundation and evolution of the Mesozoic seaway between Europe and the

Arctic. *Palaeogeography, Palaeoclimatology and Palaeoecology*, **87**, 441–491.

DORÉ, A. G. 1992. Synoptic palaeogeography of the North East Atlantic Seaway: Late Permian to Cretaceous. *In*: PARNELL, J. (ed.) *Basins on the Atlantic Seaboard*. Geological Society, London, Special Publications, **62**, 421–448.

DROOP, G. T. R. & HARTE, B. 1995. The effect of Mn on the phase relations of medium grade pelites: constraints from natural assemblages on petrogenetic grid topology. *Journal of Petrology*, **36**, 1549–1578.

FONNELAND, H. C., LIEN, T., MARTINSEN, O. J., PEDERSEN, R. B. & KOSLER, J. 2004. Detrital zircon ages: a key to understanding the deposition of deep marine sandstones in the Norwegian Sea. *Sedimentary Geology*, **164**, 147–159.

FORCE, E. R. 1980. The provenance of rutile. *Journal of Sedimentary Petrology*, **50**, 485–488.

GALEHOUSE, J. S. 1971. Point-counting. *In*: CARVER, R. E. (ed.) *Procedures in Sedimentary Petrology*. Wiley-Interscience, New York, 385–407.

GARZANTI, E., ANDÒ, S. & VEZZOLI, G. 2008. Settling equivalence of detrital minerals and grain-size dependence of sediment composition. *Earth and Planetary Science Letters*, **273**, 138–151.

GJELBERG, J. G., ENOKSEN, T., KJAERNES, P., MANGERUD, G., MARTINSEN, O. J., ROE, E. & VÅGNES, E. 2001. The Maastrichtian and Danian depositional setting along the eastern margin of the Møre Basin (mid-Norwegian Shelf): implications for reservoir development of the Ormen Lange Field. *In*: MARTINSEN, O. J. & DREYER, T. (eds) *Sedimentary Environments Offshore Norway – Palaeozoic to Recent*. Norwegian Petroleum Society, Special Publications, **10**, 421–449.

GRIFFIN, W. L., TAYLOR, P. N. *ET AL*. 1978. Archaean and Proterozoic crustal evolution in Lofoten–Vesterålen, N. Norway. *Journal of the Geological Society, London*, **135**, 629–647.

HANSEN, J. A., BERGH, S. G. & HENNINGSEN, T. 2012. Mesozoic rifting and basin evolution on the Lofoten and Vesterålen Margin, North-Norway; time constraints and regional implications. *Norwegian Journal of Geology*, **91**, 203–228.

HENRY, D. J. & GUIDOTTI, C. V. 1985. Tourmaline as a petrogenetic indicator mineral: an example from the staurolite-grade metapelites of NW Maine. *American Mineralogist*, **70**, 1–15.

JACOBSEN, S. B. & WASSERBURG, G. J. 1978. Interpretation of Nd, Sr and Pb isotope data from Archaean migmatites in Lofoten–Vesterålen, Norway. *Earth and Planetary Science Letters*, **41**, 245–253.

KITTILSEN, J. E., OLSEN, R. R., MARTEN, R. F., HANSEN, E. K. & HOLLINGSWORTH, R. R. 1999. The first deepwater well in Norway and its implications for the Upper Cretaceous play, Vøring Basin. *In*: FLEET, A. J. & BOLDY, S. A. R. (eds) *Petroleum Geology of Northwest Europe: Proceedings of the 5th Conference*. Geological Society, London, 275–280.

KOMAR, P. D. 2007. The entrainment, transport and sorting of heavy minerals by waves and currents. *In*: MANGE, M. A. & WRIGHT, D. T. (eds) *Heavy Minerals in Use*. Developments in Sedimentology, Elsevier, Amsterdam, **58**, 3–48.

LUDWIG, K. R. 2001. *SQUID 1.02, A User's Manual*. Berkeley Geochronology Center Special Publication, Berkeley, CA 94709, no. 2.

LUDWIG, K. R. 2003. *Isoplot/Ex Version 3.0: A Geochronological Toolkit for Microsoft Excel*. Berkeley Geochronology Center Special Publication, Berkeley, CA 94709, no. 4.

MANGE, M. A. & MAURER, H. F. W. 1992. *Heavy Minerals in Colour*. Chapman and Hall, London.

MANGE, M. A. & MORTON, A. C. 2007. Geochemistry of heavy minerals. *In*: MANGE, M. & WRIGHT, D. T. (eds) *Heavy Minerals In Use*. Developments in Sedimentology, Elsevier, Amsterdam, **58**, 345–391.

MEINHOLD, G., ANDERS, B., KOSTOPOULOS, D. & REISCHMANN, T. 2008. Rutile chemistry and thermometry as provenance indicator: an example from Chios Island, Greece. *Sedimentary Geology*, **203**, 98–111.

MORTON, A. C. 1985. A new approach to provenance studies: electron microprobe analysis of detrital garnets from Middle Jurassic sandstones of the northern North Sea. *Sedimentology*, **32**, 553–566.

MORTON, A. C. 2012. Value of heavy minerals in sediments and sedimentary rocks for provenance, transport history and stratigraphic correlation. *In*: SYLVESTER, P. (ed.) *Quantitative Mineralogy and Microanalysis of Sediments and Sedimentary Rocks*. Mineralogical Association of Canada Short Course Series, **42**, 133–165.

MORTON, A. C. & CHENERY, S. 2009. Detrital rutile geochemistry and geothermometry as guides to provenance of Jurassic–Paleocene sandstones of the Norwegian Sea. *Journal of Sedimentary Research*, **79**, 540–553.

MORTON, A. C. & HALLSWORTH, C. R. 1994. Identifying provenance-specific features of detrital heavy mineral assemblages in sandstones. *Sedimentary Geology*, **90**, 241–256.

MORTON, A. C. & HALLSWORTH, C. R. 2007. Stability of detrital heavy minerals during burial diagenesis. *In*: MANGE, M. & WRIGHT, D. T. (eds) *Heavy Minerals in Use*. Developments in Sedimentology, Elsevier, Amsterdam, **58**, 215–245.

MORTON, A. C., HALLSWORTH, C. R. & CHALTON, B. 2004. Garnet compositions in Scottish and Norwegian basement terrains: a framework for interpretation of North Sea sandstone provenance. *Marine and Petroleum Geology*, **21**, 393–410.

MORTON, A. C., WHITHAM, A. G. & FANNING, C. M. 2005a. Provenance of Late Cretaceous–Paleocene submarine fan sandstones in the Norwegian Sea: integration of heavy mineral, mineral chemical and zircon age data. *Sedimentary Geology*, **182**, 3–28.

MORTON, A. C., WHITHAM, A. G., FANNING, C. M. & CLAOUÉ-LONG, J. C. 2005b. The role of East Greenland as a source of sediment to the Vøring Basin during the Late Cretaceous. *In*: WANDÅS, B. T. G., EIDE, E. A., GRADSTEIN, F. & NYSTUEN, J. P. (eds) *Onshore–Offshore Relationships on the North Atlantic Margin*. Elsevier, Amsterdam, NPF (Norsk Petroleumsforening) Special Publications, **12**, 83–110.

MORTON, A. C., HALLSWORTH, C. R., STROGEN, D., WHITHAM, A. G. & FANNING, C. M. 2009. Evolution of provenance in the NE Atlantic rift: the Early–Middle Jurassic succession in the Heidrun Field,

Halten Terrace, offshore Mid Norway. *Marine and Petroleum Geology*, **26**, 1100–1117.

MORTON, A. C., MUNDY, D. J. C. & BINGHAM, G. 2012. High-frequency fluctuations in heavy mineral assemblages from Upper Jurassic sandstones of the Piper Formation, UK North Sea: relationships with sea level change and floodplain residence. *In*: RASBURY, E. T., HEMMING, S. R. & RIGGS, N. R. (eds) *Mineralogical and Geochemical Approaches to Provenance*. Geological Society of America Special Papers, **487**, 163–176.

PACES, J. B. & MILLER, J. D . 1993. Precise U–Pb ages of Duluth Complex and related mafic intrusions, northeastern Minnesota: geochronological insights to physical, petrogenetic, paleomagnetic, and tectono-magmatic process associated with the 1.1 Ga Midcontinent Rift System. *Journal of Geophysical Research*, **98**, 13 997–14 013.

RAINBIRD, R. H., HEAMAN, L. M. & YOUNG, G. M. 1992. Sampling Laurentia: detrital zircon geochronology offers evidence for an extensive Neoproterozoic river system originating from Grenville orogen. *Geology*, **20**, 351–354.

STURT, B., DALLAND, A. & MITCHELL, J. G. 1979. The age of the sub Mid-Jurassic tropical weathering profile of Andøya, northern Norway, and the implications for the Late Paleozoic paleogeography in the North Atlantic region. *Geologische Rundschau*, **68**, 523–542.

TERA, F & WASSERBURG, G. 1972. U–Th–Pb systematics in three Apollo 14 basalts and the problem of initial Pb in lunar rocks. *Earth and Planetary Science Letters*, **14**, 281–304.

TRIEBOLD, S., VON EYNATTEN, H., LUVIZOTTO, G. L. & ZACK, T. 2007. Deducing source rock lithology from detrital rutile geochemistry: an example from the Erzgebirge, Germany. *Chemical Geology*, **244**, 421–436.

TURNER, G. & MORTON, A. C. 2007. The effects of burial diagenesis on detrital heavy mineral grain surface textures. *In*: MANGE, M. & WRIGHT, D. T. (eds) *Heavy Minerals in Use*. Developments in Sedimentology, Elsevier, Amsterdam, **58**, 393–412.

VAJDA, V. & WIGFORSS-LANGE, J. 2009. Onshore Jurassic of Scandinavia and related areas. *Geologiska Föreningens Förhandlingar*, **131**, 5–23.

VERGARA, L., WREGLESWORTH, I., TRAYFOOT, M. & RICHARDSON, G. 2001. The distribution of Cretaceous and Paleocene deep-water reservoirs in the Norwegian Sea basins. *Petroleum Geoscience*, **7**, 395–408.

WATSON, E. B., WARK, D. A. & THOMAS, J. B. 2006. Crystallization thermometers for zircon and rutile. *Contributions to Mineralogy and Petrology*, **151**, 413–433.

WILLIAMS, I. S. 1998. U–Th–Pb geochronology by ion microprobe. *In*: MCKIBBEN, M. A., SHANKS, W. C., III & RIDLEY, W. I. (eds) *Applications of Microanalytical Techniques to Understanding Mineralising Processes*. Society of Economic Geologists, Reviews in Economic Geology, **7**, 1–35.

ZACK, T., VON EYNATTEN, H. & KRONZ, A. 2004. Rutile geochemistry and its potential use in quantitative provenance studies. *Sedimentary Geology*, **171**, 37–58.

# Tying catchment to basin in a giant sediment routing system: a source-to-sink study of the Neogene–Recent Amur River and its delta in the North Sakhalin Basin


UISDEAN NICHOLSON[1,2]*, SARAH POYNTER[3], PETER D. CLIFT[4] & DAVID I. M. MACDONALD[5]

[1]*Shell International Exploration and Production, PO Box 162, 2501 AN, The Hague, The Netherlands*

[2]*Formerly School of Geosciences, University of Aberdeen, Aberdeen AB24 3UE, UK*

[3]*Eni Australia, Eni House, 226 Adelaide Terrace, Perth, Western Australia, 6000, Australia*

[4]*Department of Geology and Geophysics, Louisiana State University, Baton Rouge, LA 70803, USA*

[5]*School of Geosciences, University of Aberdeen, Aberdeen AB24 3UE, UK*

**Corresponding author (e-mail: uisdean.nicholson@shell.com)*



**Abstract:** This paper uses an extensive dataset from more than 200 samples to provide a comprehensive source-to-sink analysis of the Amur River and its delta in the Russian Far East. The majority of sand-sized sediment in the Amur River and its former delta comes from upstream of the Lesser Khingan Ridge, shown by uniformity of sediment composition in the lower 1700 km of the river. Stable mineral ratios, U–Pb age spectra and garnet geochemistry show little stratigraphic provenance-specific variation in the Neogene delta. This renders Miocene–Pliocene drainage capture models unlikely.

The onset of uplift in the delta is marked by a decrease in the apatite–tourmaline index (ATi) in Upper Pliocene offshore well samples, caused by dissolution of apatite as sediments were uplifted and eroded onshore Sakhalin. These wells also show variable ATi and garnet–zircon index (GZi) values in Lower Miocene samples, which could potentially be used for stratigraphic correlation.

A positive correlation between GZi values and distance from the river mouth is attributed to hydrodynamic sorting across the delta system. This has negative implications for the use of this stable mineral index and others of a similar hydraulic equivalence as regional correlation tools on a basin scale ($>100$ km).

**Supplementary material:** Heavy mineral data, petrographic data, geochronometric data, sample locations available at www.geolsoc.org.uk/SUP18643


The North Sakhalin Basin in the Russian Far East is one of the world's most prolific hydrocarbon provinces, with more than 6000 MMboe discovered to date (Lindquist 2000). The basin covers an area of 84 000 km², extending from the Tatar Strait in the west to the Deryugin Basin in the central Sea of Okhotsk in the east (Fig. 1; Lindquist 2000). Throughout the Neogene, the basin was supplied with clastic sediment by the Amur River, one of the ten largest rivers on Earth, with a catchment area of more than 1.8 M km² and a trunk stream more than 4500 km long (Fig. 2; Hovius 1998) Much of the basin sediments are now exposed onshore in the northern part of Sakhalin. Regional transpression along the Sakhalin–Hokkaido Shear Zone (Worrall *et al.* 1996), which separates the

Eurasian and Amurian plates in the west from the Okhotsk microplate in the east (Apel *et al.* 2006; Hindle *et al.* 2006), has resulted in active exhumation and recycling of the Neogene sediments into offshore areas of the basin.

This paper aims to provide a comprehensive source-to-sink study for the Amur sediment routing system, with particular emphasis on recycling of Neogene deltaic deposits from Sakhalin into the offshore basin. We use heavy mineral analysis and associated petrographic and single-grain analytical techniques to look in detail at the provenance of the palaeo-Amur deposits, which could have a significant impact on their reservoir quality, as well as constraining alternative sources of sediments from onshore Sakhalin. In order to accurately


*From*: Scott, R. A., Smyth, H. R., Morton, A. C. & Richardson, N. (eds) 2014. *Sediment Provenance Studies in Hydrocarbon Exploration and Production.* Geological Society, London, Special Publications, **386**, 163–193. First published online July 15, 2013, http://dx.doi.org/10.1144/SP386.7

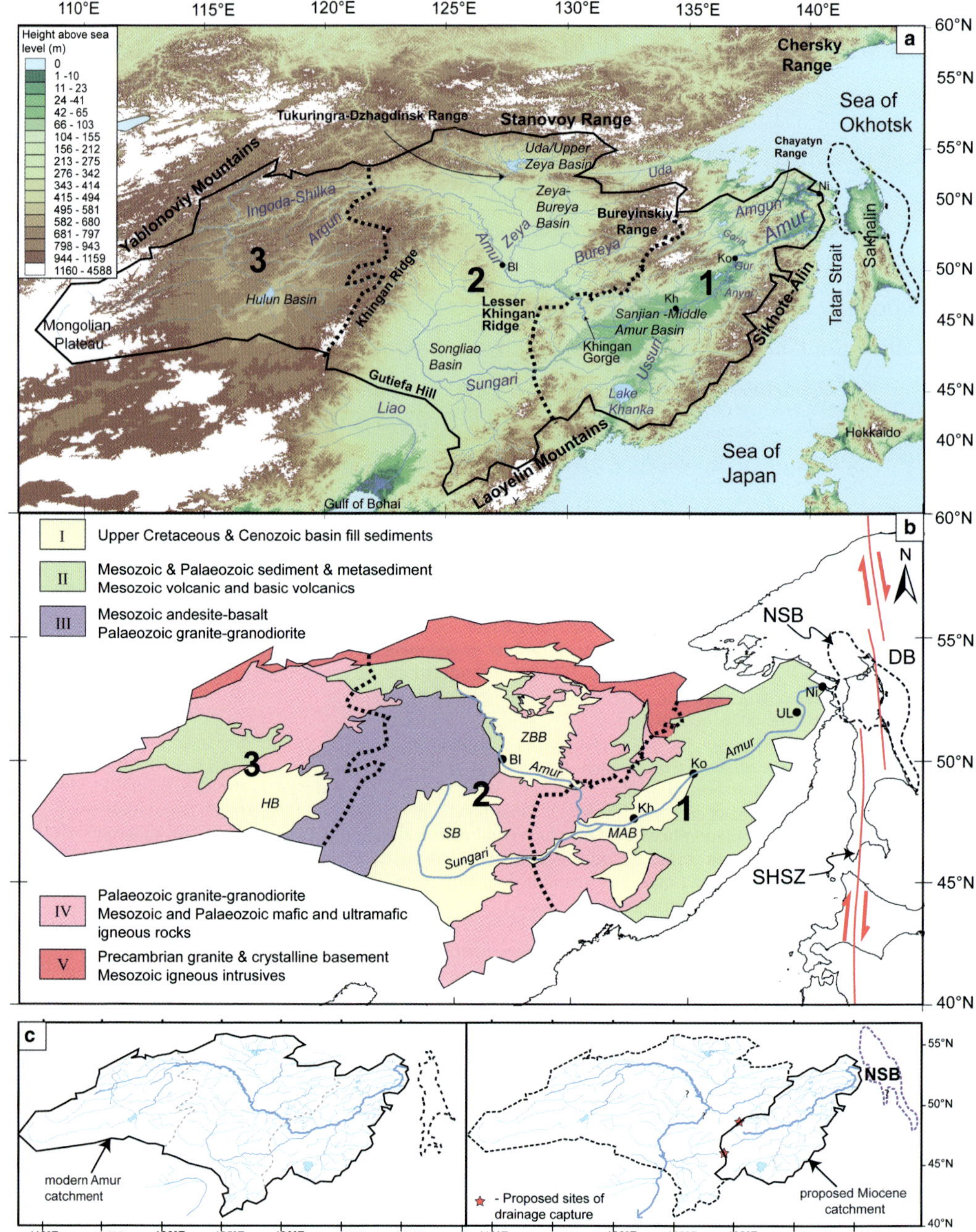


**Fig. 1.** The Amur drainage basin in the Russian Far East. (**a**) Topographic features and drainage pattern of the Amur catchment. Dashed lines show position of main physiographic divides in the catchment, which separate the drainage basin into sites of distinct relief and elevation (1 to 3). Topographic data from USGS (1996). (**b**) Simplified bedrock geology for the Amur catchment, divided into five lithological groups. The most important rock type in each unit is noted in the key, followed by important secondary types. (**c**) Model for Late Neogene drainage capture in the Amur catchment. Modern drainage configuration on the left and proposed drainage configuration in the Amur Basin prior to the Pliocene, as proposed by Makhinov (2004) and Sorokin & Artyomenko (2003), on the right. In this scenario, the Upper Amur River flows southward through the proto-Lesser Khingan Ridge and Songliao Basin into the Gulf of Bohai. Drainage capture occurs by headward erosion by the Lower Amur at the NE and SE of the Lesser Khingan Ridge, to

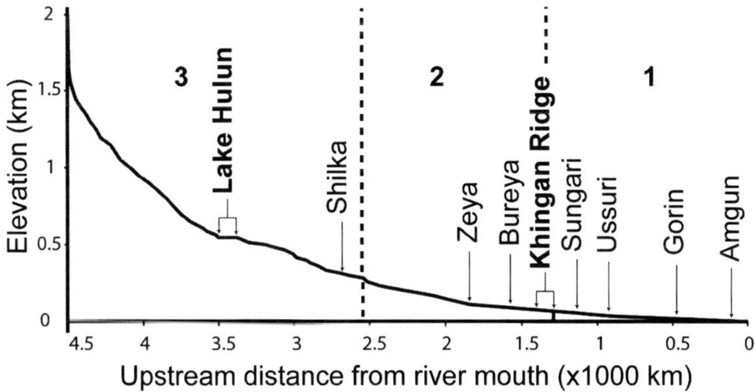


**Fig. 2.** Longitudinal river profile of the Argun and Amur River. Note the change in gradient upstream of the Zeya River confluence, and in the upper catchment of the Amur, upstream of the confluence of the Argun and Shilka rivers. There is no resolvable knickpoint at the Khingan Ridge, which separates the lower and middle physiographic regions of the catchment. The mouth of the Zeya River is 81 m asl and the mouth of the Ussuri River is 20 m asl.

constrain the source-to-sink history of these sediments, the modifying processes that affect mineral composition in this area must be deconvolved. These datasets will be used to test hypotheses of drainage evolution of the Amur River and the tectonic history of the North Sakhalin Basin.

*Models of catchment evolution and sediment production*

Despite several focused studies, the evolution of the Amur River since the Miocene remains poorly constrained. In particular, it is not certain whether the limits of the river catchment have been continuous throughout the lifetime of the river (Velichko & Spasskaya 2002), or whether a series of major drainage capture events have taken place during the Neogene (Sorokin & Artyomenko 2003; Makhinov 2004; Fig. 3). We test these competing hypotheses for the age and catchment evolution of the river by looking at the petrography and mineralogy of Amur River sediments and of the continuous Lower Miocene–Recent deltaic sequence penetrated in offshore wells. Significant drainage capture events should result in a pronounced difference in sediment supply to the basin because of the change in the size of the catchment and the different lithologies being eroded. This should be discernible in the stratigraphic record of the North Sakhalin Basin.

*Models of evolution of the North Sakhalin Basin: discriminating intrabasinal sources*

The geodynamic evolution of the North Sakhalin Basin during the time of deltaic deposition remains poorly constrained. Deltaic deposition on Sakhalin initiated in the Early Miocene, during a time of regional extension (Worrall *et al.* 1996). The uplift of the main north–south-trending mountain belts, the East Sakhalin Mountains and West Sakhalin Mountains (Fig. 3), as well as inversion of the North Sakhalin Basin, was a late Neogene event associated with basin-wide transpression (Rozhdestvenskiy 1982). This event is believed to have begun at some point between the Late Miocene and Pleistocene (Rozhdestvenskiy 1982; Worrall *et al.* 1996), causing exhumation and re-working of older Neogene sediments. Inversion would be expected to be marked by a flux of recycled deltaic material from Sakhalin into contemporaneous deposits offshore. We attempt to constrain the timing of this tectonic event by looking for evidence of such material.

---

**Fig. 1.** (*Continued*) connect the upper and lower Amur and the Sungari River to the lower Amur (positions highlighted by red stars). Such events would increase the size of the Amur catchment by a factor of three, resulting in a significant change in the composition and volume of sediment reaching the North Sakhalin Basin. There would also be a corresponding decrease in sediment supply to the Liao River delta in the Gulf of Bohai. Bl, Blagovoshchenk; Kh, Khabarovsk; Ko, Khomsomolsk-na-Amure; Ni, Nikolayevsk-na-Amure; UL, Udyl Lake; DB, Deryugin Basin; NSB, North Sakhalin Basin; SHSZ, Sakhalin–Hokkaido Shear Zone; ZBB, Zeya-Bureya Basin; SB, Songliao Basin; HB, Hulun Basin; A, Field A; B, Field B. Modified from Poynter (2003).

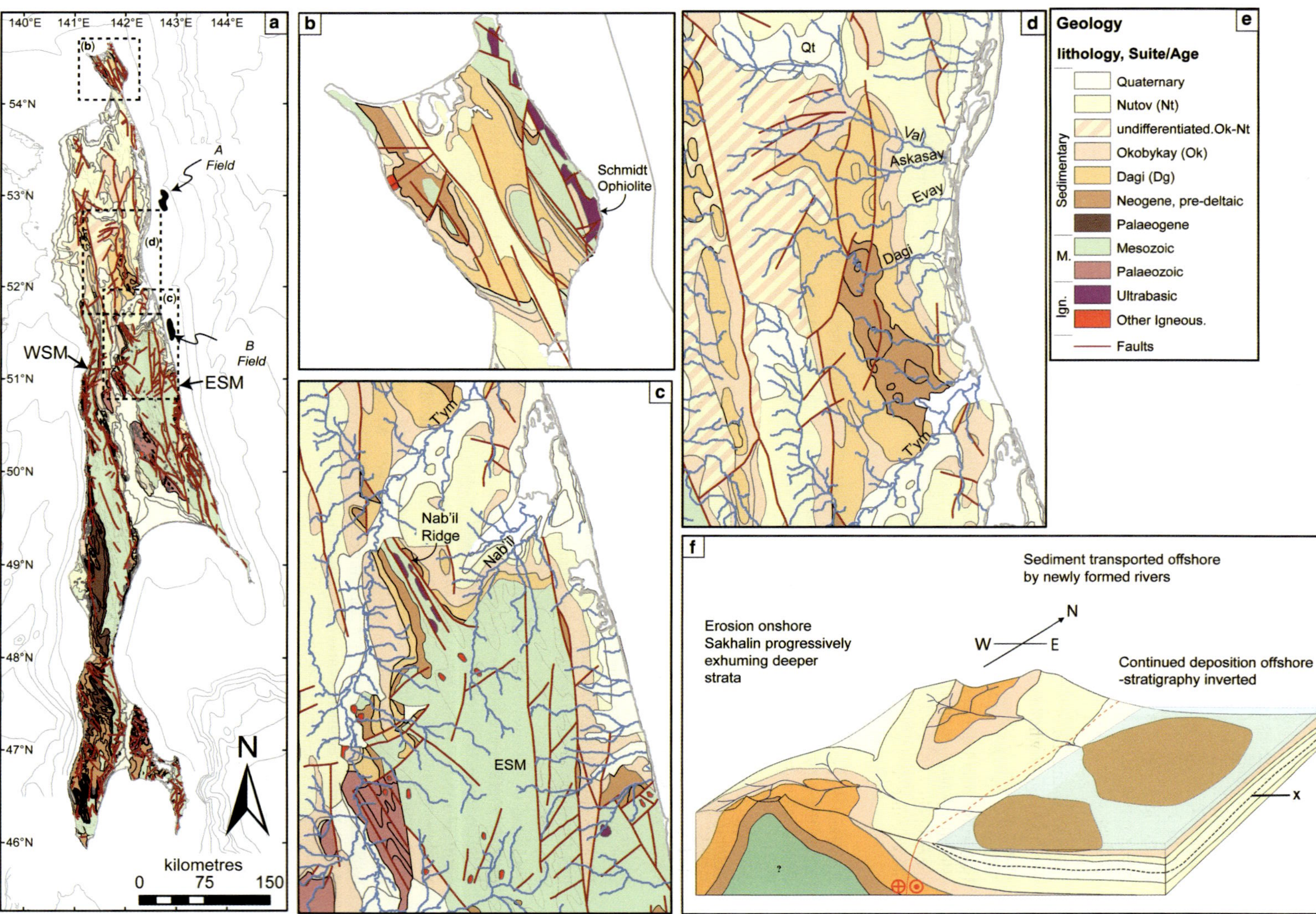
a
140°E 141°E 142°E 143°E 144°E
54°N
A Field
53°N
(d)
52°N
(c)
WSM
B Field
51°N
ESM
50°N
49°N
48°N
47°N
46°N
N
kilometres
0 75 150
b
Schmidt Ophiolite
c
Tym
Nab'il Ridge
Nab'il
ESM
d
Qt
Val
Askasay
Evay
Dagi
Tym
e
Geology
lithology, Suite/Age
Sedimentary
Quaternary
Nutov (Nt)
undifferentiated.Ok-Nt
Okobykay (Ok)
Dagi (Dg)
Neogene, pre-deltaic
Palaeogene
M.
Mesozoic
Palaeozoic
Ign.
Ultrabasic
Other Igneous.
Faults
f
Sediment transported offshore
by newly formed rivers
Erosion onshore
Sakhalin progressively
exhuming deeper
strata
N
W E
Continued deposition offshore
-stratigraphy inverted
x
?

*Heavy mineral stratigraphy*

The use of heavy mineral data as a tool for stratigraphic correlation in otherwise barren sandstones is an important economic application of heavy mineral analysis (e.g. Morton *et al.* 2007; Morton & Milne 2012). This method has also been carried out in key stratigraphic intervals in the onshore North Sakhalin Basin (Yurkova 1970), with key index minerals (particularly chromian spinel) being used as marker beds for correlation between wells on a local scale in the onshore basin (N.A. Derevskova, Sakhalinmorneftegaz, pers. comm. 2006). We investigate whether such a stratigraphic correlation scheme is possible in offshore hydrocarbon fields in this basin. If so, this could represent an important economic application of these analyses. We also investigate whether there is any basis for more basin-wide stratigraphic correlation based on heavy mineral data in the palaeo-Amur sequence.

## Sediment source areas

*Amur drainage basin in Far-East Asia*

The bedrock geology of the Amur drainage basin in the Russian Far East is complex. The area consists of a number of discrete tectonic terranes accreted to Eurasia between the Palaeozoic and the Cenozoic (Sengor & Natal'in 1996). These comprise a diverse range of intrusive and extrusive igneous rocks, and metasedimentary and sedimentary rocks ranging in age from Precambrian to Recent. For the purposes of this paper, the geology can be simplified into five groups, based on the age and lithology of the most abundant rocks in a wider area (Poynter 2003; Fig. 1b). These can be assessed in terms of their relative contribution to each of the main physiographic regions that form the Amur catchment (Fig. 1a).

In the lower (eastern) catchment (Area 1 in Fig. 1), from the Sikhote-Alin Mountains to the Lesser Khingan Ridge, the drainage basin is characterized by a large contrast in relief between the Mid-Amur Basin and the surrounding mountain belts. The bedrock is dominated from east to west by Mesozoic volcanic and volcaniclastic sedimentary rocks of the Sikhote-Alin volcanic arc and fold belt; Neogene–Quaternary sedimentary rocks of the Mid-Amur Basin; and Palaeozoic (and subsidiary Mesozoic and Precambrian) intermediate-acidic intrusive rocks of the eastern Khingan Ridge and Bureyinskiy Mountains. Sedimentary and metasedimentary rocks are dominant. Cretaceous sedimentary rocks in the Udyl Lake area (Fig. 1b) consist of a series of volcaniclastic greywackes, arkoses and cherts, mostly derived from the Cretaceous Sikhote-Alin arc (Markevich *et al.* 2007). Because of their distinct provenance, arc-derived volcanic minerals dominate all these sequences, mainly clino- and orthopyroxenes, calcic amphibole, magnetite and chromite (Nechaev *et al.* 1997); chromite represents up to 54% of the greywackes, suggesting an ultramafic source. If the catchment area of the Amur River was restricted to this area during the Miocene then the Miocene stratigraphy in the North Sakhalin Basin should be dominated by arc-derived sediment and metasedimentary detritus.

The central catchment area (Area 2 in Fig. 1) is dominated by the western Bureyinsky Mountains and the Khingan Ridge in the east; Cretaceous–Neogene sedimentary rocks in the Zeya-Bureya Basin; Cretaceous–Quaternary sediments and sedimentary rocks in the Songliao Basin (Ren *et al.* 2002) (Fig. 2); Precambrian basement and Mesozoic intrusive rocks of the Stanovoy Range (part of the southern margin of the Siberian Craton); and Mesozoic–Palaeogene extrusive (and subsidiary intrusive) intermediate-basic igneous rocks in the Khingan Mountains. Cenozoic deposits in the Zeya-Bureya Basin are considered to be a particularly important potential sediment source rock because of the petrographic and mineralogical similarities to the sediments of the modern Amur River (Poynter 2003). These sediments were predominantly derived from granitic and granodioritic sources, with subsidiary sources including metasedimentary rocks. Sediments in the Zeya-Bureya Basin have experienced mild–moderate burial dissolution,

---

**Fig. 3.** Potential local sources of sediment from onshore Sakhalin. (**a**) Geological map modified from Vereshchagin *et al.* (1969) with locations of (b), (c) and (e). WSM, West Sakhalin Mountains; ESM, East Sakhalin Mountains. (**b**) Schmidt Peninsula in northern Sakhalin showing position of Cretaceous ophiolite, which is exposed along much of the east coast. (**c**) Northern East Sakhalin Mountains in central Sakhalin, showing position of Nab'il Ridge. The Nab'il and T'ym rivers drain this area and flow northward into the southern North Sakhalin Basin. (**d**) Neogene deltaic sedimentary rocks in the central–southern part of the North Sakhalin Basin. Uplift of this area (and other Neogene deposits on Sakhalin) provides recycled deltaic sandstone into adjacent areas of the North Sakhalin Basin. (**e**) Key to geological maps. (**f**) Hypothetical schematic representation of Neogene stratigraphy of the North Sakhalin Basin. Assuming uplift is initiated at point X (shown by dashed line), the offshore basin will begin to receive the erosional products of the uplifted island at that time, resulting in the formation of an inverted stratigraphy. If diagnostic stratigraphic marker minerals are present onshore then the first appearance of these in the offshore detrital record indicates the time at which that unit was incised onshore.

as well as acid weathering of garnet and apatite grains (Poynter 2003). Heavy mineral assemblages derived from this basin should be relatively enriched in stable minerals such as zircon, tourmaline and rutile (Table 1).

The upper (western) catchment (Area 3 in Fig. 1) is significantly more elevated (mean elevation >500 m) than the lower catchments, with a corresponding increase in the concavity of the river profile (Fig. 2). The area is dominated by the Khingan Mountains in the east; Neogene–Quaternary sediments of the Hulun Basin; and Palaeozoic intrusive igneous and metasedimentary rocks of the western Stanovoy Range, Yablonoviy Mountains and Mongolian Plateau on the west and northwestern margin of the catchment. Because the catchment is dominated by crystalline igneous rocks, the heavy mineral assemblage should be dominated by minerals derived from crystalline continental crust. However, as metasedimentary rocks in the catchment are likely to produce more heavy minerals per unit area than granitic rocks; the heavy mineral assemblage may be relatively enriched in minerals from these subordinate lithologies (Poynter 2003).

## The North Sakhalin Basin

The Neogene stratigraphy of the North Sakhalin Basin can be divided into six main units and their lateral equivalents: the Daekhurin, Uynin, Dagi, Okobykay, Nutov and Pomyr suites (Figs 3 & 4). Sediments of the Oligocene–lowermost Miocene Daekhurin and Uynin suites were derived from local sources, whereas those from the Lower Miocene Dagi Suite and younger units are inferred to have been derived from the palaeo-Amur River.

Potential secondary sources for detritus in the North Sakhalin Basin are located on the margins of the basin, or within the basin itself (Fig. 3). However, it is uncertain how significant a contribution these have had in terms of sediment supply to the basin. This is partly because of lack of constraints on the timing of subaerial exposure and erosion of these areas in the geological past. The potential source areas within or proximal to the North Sakhalin Basin are (i) Permian–Cretaceous greenschist–blueschist facies metamorphic rocks and subsidiary ultramafic rocks in the East Sakhalin Mountains; (ii) Cretaceous ultramafic and metamorphic rocks on the Schmidt Peninsula and the now-submerged Schmidt Ridge; (iii) Cretaceous–Palaeogene sedimentary rocks in the West Sakhalin Mountains; and (iv) reworked deltaic sedimentary rocks in the North Sakhalin Basin (Fig. 3).

Ophiolites exposed on the Nab'il Ridge and Schmidt Peninsula (Fig. 3b) are of particular importance because they are likely to provide a characteristic heavy mineral suite distinct from that of the Amur River (Table 2). Ultramafic rocks are rare in the Amur catchment, and ultramafic minerals such as chrome spinel, enstatite and olivine are typically rare or absent in the trunk stream of the upper and middle Amur (Poynter 2003). The ophiolite exposed on Schmidt Peninsula is only part of the much more laterally extensive Schmidt Ridge, which is currently a submarine feature. This feature would have been subaerially exposed during eustatic lowstands (glacial maxima) during the Late Pliocene and Pleistocene (Nicholson 2009), potentially acting as a sediment source for the northeastern part of the North Sakhalin Basin.

The East Sakhalin Mountains (Fig. 3c) predominantly comprise greenschist (and locally blueschist)

**Table 1.** *Characteristic heavy mineral assemblage of the main lithological provinces in the Amur catchment*

| Lithology | Characteristic heavy minerals | Occurrence in Amur catchment |
| --- | --- | --- |
| Basic/ultrabasic igneous rocks | Calcic amphibole, clino- and ortho-pyroxene, chrome-spinel, olivine | Province II, III and IV – predominantly lower and middle catchment – Sikhote–Alin, Khingan Mountains, Cenozoic Mid-Amur rift volcanic |
| Continental crust (high-grade metamorphics + acid plutons) | Monazite, rutile, tourmaline, zircon– granite; andalusite, garnet, kyanite, sillimanite, staurolite–metamorphic | Province IV and V – upper catchment – Stanovoy Range, Bureyinskiy Mountains, Yablonoviy Mountains |
| Low–medium grade sedimentary and metamorphosed basic igneous rocks | Axinite, calcic amphibole, chloritoid, epidote, prehnite, pumpellyite | Province II – lower catchment – Badzhal Range and Sikhote Alin Range |
| Recycled mineralogically mature sandstone | Rutile, tourmaline, zircon | Province II – Zeya–Bureya Basin, Songliao Basin |

Provinces are shown in Figure 1 (Mange-Rajetzky 1983; Thornburg & Kulm 1987; Morton & Smale 1990; Morton & Johnsson 1993; Nechaev & Isphording 1993; Garzanti *et al.* 2002). Adapted from Poynter (2003).

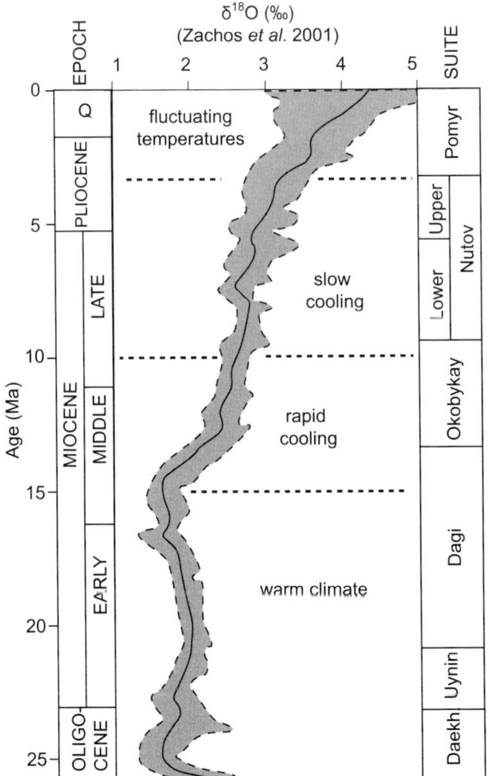



**Fig. 4.** Late Palaeogene–Neogene stratigraphy of the North Sakhalin Basin. Geological timescale and local stratigraphic names are from Gladenkov *et al.* (2002) and Vereshchagin *et al.* (1969). Note that the Russian term 'svita' is translated as 'suite' rather than the incorrect 'Formation'. The feminine adjectival ending – skaya – has been omitted for simplicity. The oxygen isotope curve from Zachos *et al.* (2001) in the central column is shown as a proxy for global climatic conditions during the time of Amur Delta deposition. Climatic conditions have cooled significantly since the Mid-Miocene climatic optimum (15–17 Ma).

facies metamorphic rocks, with minor ultramafic units. Greenschist–blueschist facies metamorphic rocks typically yield a range of provenance-specific index minerals (Table 2), some of which are rare or absent in the Amur River. As with the Schmidt Ophiolite, ultramafic rocks in the Nab'il Ridge are likely to have experienced metamorphism up to greenschist facies along with the surrounding metasedimentary rocks.

As with sediments derived from the Zeya-Bureya Basin, recycled Neogene sedimentary rocks on Sakhalin (Fig. 3d) are likely to yield heavy mineral assemblages that are relatively enriched in stable minerals, depending on the extent of intrastratal dissolution that has occurred during burial.

## Analytical techniques

Sandstone samples were collected in the field, after removing any weathered crust. River samples were collected from exposed sand bars where present, and were collected from river beds otherwise. Samples of fine to very fine sand were sampled where possible in order to reduce the effects of hydraulic segregation on heavy mineral assemblages. River samples were collected upstream of roads and bridges to avoid contamination from construction aggregates. Samples were disaggregated and wet-sieved to retrieve the 63–125 μm grain size fraction and were then immersed in water and cleaned by ultrasonic probe to remove clay from grain surfaces. These samples were then dried and the heavy minerals were extracted by gravity settling in a separating funnel using bromoform (tribromethane). After separation a fraction of grains was mounted and evenly dispersed on a glass slide using Canada Balsam (refractive index of 1.5) as a mounting medium. Grain counting was carried out using the Ribbon counting technique (Mange & Maurer 1992). When the total number of grains was very low all grains on the slide were counted to maximize the quantity of data collected. At least 400 non-opaque grains were counted per slide for the basic count, excluding micas and authigenic phases, and an additional 200 grains were counted per mineral-pair for stable mineral ratios (Morton *et al.* 2007). Stable mineral ratios were used where significant modification of grain assemblages had occurred, most prominently acid weathering during erosion and sediment storage and diagenesis during burial (Morton & Hallsworth 1994). Where grains were difficult to identify by optical methods alone, grains were identified using a scanning electron microscope (SEM).

Electron microprobe analyses (EMPA) of garnets were carried out at the University of Aberdeen and at Moscow State University. A minimum of 50 individual grains were hand-picked from the samples, mounted on double-sided tape attached to a plain glass slide, and coated with 20–25 nm of carbon. Fifty grains is considered a minimum number of grains for a statistically robust result (Ruhl & Hodges 2005). Samples were analysed by the electron microprobe on a MICROSCAN MK5 made by Cambridge Scientific Instruments. Each grain was analysed for 30 s, allowing quantitative analysis of major element compositions. For garnet these are Si, Ti, Al, Cr, Fe, Mg, Mn, Ca and O. Data were plotted on ternary diagrams to show the relative abundances of the major elements present and to allow determination of the mineral species within the group. The electron microprobe used does not distinguish between $Fe^{2+}$ and $Fe^{3+}$, so the relative proportions of these two types of Fe

**Table 2.** *Characteristic heavy mineral assemblage of main lithological provinces in Sakhalin*

| Lithology | Characteristic heavy minerals | Occurrence in North Sakhalin |
|---|---|---|
| Ultrabasic igneous rocks | Clino- and ortho-pyroxene, chrome-spinel, olivine | Schmidt Peninsula, Nab'il Ridge |
| Low–medium grade (greenschist facies) metasedimentary rocks | Axinite, calcic amphibole, chloritoid, epidote, prehnite, pumpellyite | East Sakhalin Mountains, Schmidt Peninsula |
| Medium–high grade (blueschist facies) metasedimentary rocks | Lawsonite, glaucophane, pumpellyite, sodic-pyroxene | East Sakhalin Mountains |
| Recycled mineralogically mature sandstone | Rutile, tourmaline, zircon | North Sakhalin Basin |

were calculated using the technique outlined by Droop (1987). This is only necessary for garnets belonging to the ugrandite series where substitution of $Fe^{3+}$ occurs in the trivalent site; in garnets of the pyralspite series Fe typically only occurs in the bivalent site, and is therefore assumed to be all $Fe^{2+}$ (Mange & Morton 2007). Ugrandite garnets were subdivided on the basis of the $Fe^{3+}/Al$ ratio, with values of <0.1, 0.1–1, 1–10 and >10 correspond to <10, 10–50, 50–90 and >90% andradite respectively. Almandine and spessartine were similarly distinguished using cutoff values of 5 and 50% spessartine to subdivide the group.

U–Pb dating of zircon grains was carried out at the thermochronometry laboratory in the Department of Earth Sciences at University College, London. Samples were sieved to obtain the 63–250 μm grain size and zircons were separated by submersion in methylene iodide (di-iodomethane); the remaining grains were subsequently removed using a magnetic separator. Samples were analysed by laser ablation inductively coupled plasma mass spectrometry (LA-ICPMS) using a New Wave 213 aperture imaged frequency quintupled laser ablation system (213 nm) coupled to an Agilent 750a quadrupole-based ICP–MS. Real-time data were processed using GLITTER. Repeated measurements of external zircon standard PLESOVIC (TIMS reference age $337.13 \pm 0.37$ Ma; (Slama *et al.* 2008) and NIST 612 silicate glass (Pearce *et al.* 1997) were used to correct for instrumental mass bias and depth-dependent interelement fractionation of Pb, Th and U. Data were subsequently filtered using standard discordance tests with a $\pm 10\%$ cutoff. The $^{206}Pb-^{238}U$ ratio was used to determine ages younger than 1000 Ma and the $^{207}Pb/^{206}Pb$ ratio for older grains. Common Pb was determined by the $^{208}Pb$ method assuming a common Pb composition from the age-dependent Pb model of Cumming & Richards (1975). Data were processed using Isoplot (Ludwig 2003), and plotted graphically as probability density plots,

which take into account age uncertainties as well as peak ages (Vermeesch 2004). In order to resolve all the major source terranes within the tectonically complex and geologically diverse catchment of the Amur River, at least 100 grains were analysed per sample, if recovery was sufficient. This allowed resolution of all grain populations representing over 6% of the sample at a 95% ($2\sigma$) confidence level (Vermeesch 2004).

Petrographic analysis was carried out on the medium (250–500 μm) sand fraction of 38 samples from two fields in the offshore North Sakhalin Basin (Fig. 1; named A and B for confidentiality reasons), using the Gazzi-Dickinson technique (quartz-feldspar-lithics, QFL: Gazzi 1966; Dickinson 1970). This differs slightly from techniques previously used on Sakhalin where the whole grains were counted (quartz-feldspar-rock fragments, QFR: Poynter 2003). However, the two methods converge for medium-sand-grade rocks, so this is not viewed as a serious methodological difference. Samples were selected from both core and ditch cuttings, necessitating the use of different preparation procedures. Care should therefore be taken when comparing samples from ditch cuttings and core, as mechanical abrasion during drilling and subsequent sample preparation can cause the destruction of mechanically weak grains (Morton *et al.* 2007). Cuttings samples are also susceptible to contamination by sediment from overlying intervals falling into the drill hole, whereas core samples are free from such contamination.

## Results of heavy mineral analysis: modern rivers and beaches

*Modern Amur River composition*

The heavy mineral composition and petrography of the modern Amur River and its tributaries have been investigated by Poynter (2003). That study focused on the area between Blagoveshchensk, at the confluence of the Amur and Zeya Rivers, and Khomsomolsk-na-Amure, at the northern end of the Mid-Amur Basin (Fig. 1). However, it did not include samples of the modern Amur River downstream of Khomsomolsk-na-Amure, so the lower 550 km stretch of the river remained unconstrained. Because of this, the detrital contribution of the numerous smaller tributaries draining NW Sikhote-Alin and the last major tributary, the Amgun River, which drains the eastern Bureyinsky and Chayatyn ranges (Fig. 1), remained unaccounted for. In our study, samples obtained from the trunk stream of the Amur between Khomsomolsk-na-Amure and the confluence of the Amur with the Amgun were also analysed (Fig. 5b), allowing the composition of the whole modern Amur River to be determined, including the detritus contributed by all its major tributaries. Stable mineral ratios were also calculated in order to allow comparison with potentially modified palaeo-Amur assemblages in the North Sakhalin Basin.

The heavy mineral composition of the trunk stream of the Amur River changes very little between Blagoveshchenk and its confluence with the Amgun River, a distance of more than 1700 km (Fig. 5). The heavy mineral fraction of the river is consistently dominated by calcic amphibole and epidote (60–85%), with subsidiary pyroxene, sphene, apatite, tourmaline, garnet and zircon. The heavy mineral suite and petrography of river samples show a close relationship to rivers draining the Zeya–Bureya Basin (Zeya and Bureya Rivers) and the Khingan Ridge (Arkhara, Uril and Bira Rivers). However, they show less similarity to the rivers draining the Sikhote–Alin (Anyui and Gur) and Badzhal ranges (Gorin). The latter tributaries are rich in epidote and calcic amphibole, but they also have an elevated component of pyroxene (up to 25% in the Gorin River), reflecting the common occurrence of basic volcanic rocks in their catchments (Markevich *et al.* 2007). They also have a higher chrome-spinel component (0.2–1.0%), a mineral that is rare or absent in the Amur River (<0.2%). There is a slight increase in pyroxene content in the trunk stream *c.* 600 km upstream of the river mouth, and from 500 km there is a relative decrease in epidote and increase in calcic amphibole, which may reflect some contribution from tributaries draining the Sikhote–Alin and Chayatyn ranges (Fig. 5). However, the Amur does not show any substantial change in provenance along its course and remains stable after the confluence of the large Sungari and Ussuri rivers. As these rivers were not sampled, it is impossible to tell if this is because the two rivers have similar bulk mineralogy to the Amur, or whether they contribute a relatively small volume of

bedload sediment to the fluvial system. As with the confluence of the other major tributaries, sediment input from the Amgun River, 100 km upstream of the river mouth, appears to make little or no difference to the bulk heavy mineral population of the sediments in the Amur River. The pyroxene abundance, which was observed to increase slightly immediately upstream of Khomsomolsk-na-Amure, drops in this downstream section to values similar to those present in the upper Amur.

Stable mineral ratios between Khomsomolsk-na-Amure and the Amgun confluence are also fairly consistent, although there are significant variations in the GZi and, to a lesser degree, in the RZi (rutile–zircon index). It is evident that the slight density differential of these mineral pairs results in the preferential settling of denser zircon grains in coarser sediment, resulting in a decrease of both the GZi and the RZi. The ATi shows very little variation downstream over this 500 km stretch. This suggests that rates of chemical weathering rates are low, because apatite is highly susceptible to chemical weathering during sediment storage in floodplain environments or bar systems (Morton & Hallsworth 1994). Apatite and tourmaline have a much closer degree of hydraulic equivalence (average of 3% difference) than rutile and zircon (9% difference) or garnet and zircon (12% difference). It is therefore the least susceptible of these mineral pairs to hydrodynamic segregation, and therefore the most accurate indicator of provenance, if chemical dissolution of apatite is minimal (Morton & Hallsworth 1994).

## Streams on Sakhalin

The drainage system in NE Sakhalin is characterized by a series of subparallel, east-flowing rivers that incise the Neogene stratigraphy as the island is being uplifted under the current tectonic regime (Fig. 6a). The depth of the incision generally increases southward, with the greatest exhumation in the Dagi River catchment, where over 4 km of stratigraphic section is exposed (Davies *et al.* 2003; Fig. 3).

Samples were collected from a total of 21 rivers, including all of the large north- and east-flowing rivers in northern Sakhalin. North of Kydalani the rivers are short, with small catchment areas restricted to Nutov Suite bedrock. Heavy mineral assemblages are dominated by epidote and calcic amphibole, which account for over 60% of the total grain population, a similar composition to the modern Amur River (Fig. 6). By contrast, rivers south of Kydalani (Fig. 6), which incise older stratigraphy, have significantly lower calcic amphibole epidote content (Ca + Ep values typically 30–40%). The ZTR index (the percentage

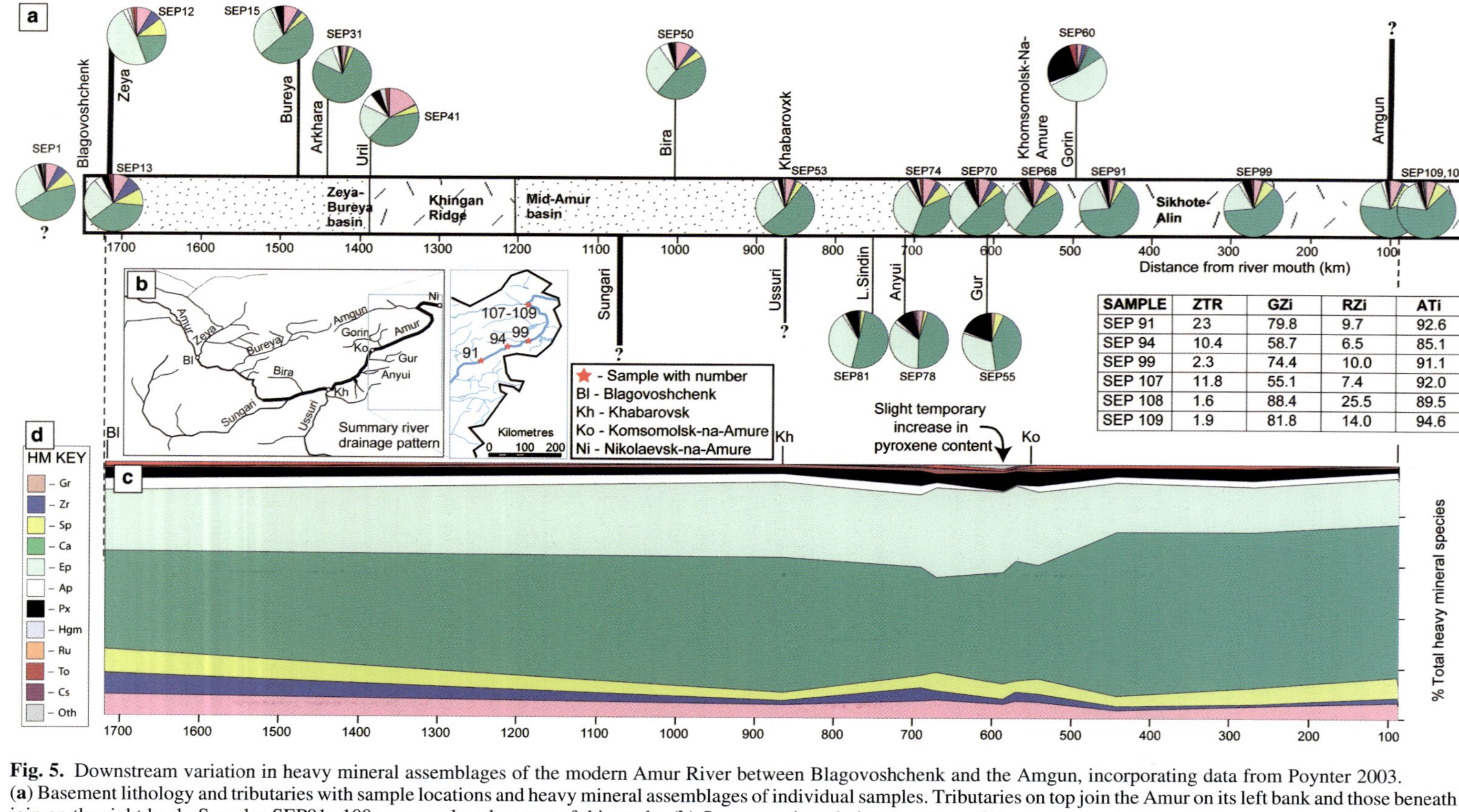


| SAMPLE | ZTR | GZi | RZi | ATi |
|---|---|---|---|---|
| SEP 91 | 23 | 79.8 | 9.7 | 92.6 |
| SEP 94 | 10.4 | 58.7 | 6.5 | 85.1 |
| SEP 99 | 2.3 | 74.4 | 10.0 | 91.1 |
| SEP 107 | 11.8 | 55.1 | 7.4 | 92.0 |
| SEP 108 | 1.6 | 88.4 | 25.5 | 89.5 |
| SEP 109 | 1.9 | 81.8 | 14.0 | 94.6 |

**Fig. 5.** Downstream variation in heavy mineral assemblages of the modern Amur River between Blagovoshchenk and the Amgun, incorporating data from Poynter 2003. (**a**) Basement lithology and tributaries with sample locations and heavy mineral assemblages of individual samples. Tributaries on top join the Amur on its left bank and those beneath join on the right bank. Samples SEP91–109 were analysed as part of this study. (**b**) Summary river drainage pattern of the study area. (**c**) Cumulative percentage plot of heavy minerals in the trunk stream of the Amur. (**d**) Key to heavy mineral plots: Gr, garnet; Zr, zircon; Sp, sphene; Ca, calcic amphibole; Ep, epidote; Ap, apatite; Px, pyroxene; Hgm, high grade metamorphics (alumino-silicates + staurolite); Ru, rutile; To, tourmaline; Cs, chrome-spinel; Oth, other. Modified from Poynter (2003). Table shows stable mineral indices calculated for samples SEP91–SEP109. ZTR, zircon–tourmaline–rutile index; GZi, garnet–zircon index; RZi, rutile–zircon index; ATi, apatite–tourmaline index. Chrome spinel-zircon (CZi) indices all have a value of 0 and are not shown on this table.

of zircon, tourmaline and rutile in the total transparent heavy mineral assemblage; Hubert 1962) of these samples is also relatively high, with values in the range of 12–32%, implying more extensive modification of the heavy mineral suite. The Dagi River has the most mature heavy mineral assemblage (Ca + Ep value of only *c.* 25% and a ZTR of 31.4). (Maturity is defined here as the percentage of unstable minerals in the sample; it is often quantified with the ZTR index, but can also be qualitatively assessed by the absence of unstable phases in a sample.) This river incises most deeply into the Neogene stratigraphy, with sediments belonging to the Daekhurin and Machigar Suites exposed in the core of the Dagi anticline. The main exception to this regional trend is the Val River, which is dominated by calcic amphibole and epidote, similar to the rivers draining the area north of Kydalani. However, the Val River drains a large area of Quaternary deposits, which has a much lower relief than surrounding regions and consequently the depth of incision is less.

Stable mineral ratios show moderate variability in these rivers. ATi values in particular are variable (4–44%) but very low in all river samples in northern Sakhalin compared to the modern Amur River (>80%) (Fig. 7). Because apatite is stable with burial but susceptible to dissolution in acid-weathering conditions, it is likely that significant apatite dissolution has taken place as sedimentary rocks are uplifted and recycled in northern Sakhalin. This low ATi value is an important proxy when looking for evidence of sediment recycling in the detrital record.

The T'ym and Nab'il rivers (Fig. 6) drain areas of different bedrock lithology to the streams discussed above. The T'ym is the largest river draining Sakhalin, with a drainage area exceeding 6000 km$^2$, including extensive Cretaceous to Palaeogene outcrops in the East and West Sakhalin Mountains, as well as the Cenozoic sedimentary rocks of the Central Depression. The heavy mineral assemblage is broadly similar to many of the rivers draining only the Neogene deltaic sequence. The total heavy mineral suite, however, is much more diverse than these rivers. It contains a relatively large component of lawsonite, which is restricted to high-pressure, low-temperature metamorphic rocks of the blueschist facies; pumpellyite, a low-medium grade metamorphic index mineral; and gedrite, an Fe–Mg amphibole often associated with reaction zones around ultramafic rocks, as well as a broad range of regionally metamorphosed rocks (Mange & Maurer 1992).

The Nab'il River drains much of the area to the east of the Nab'il Ridge, which marks the northern limit of the East Sakhalin Mountains. The heavy mineral signature of this river is significantly different to the rest of those sampled in northern Sakhalin. The heavy mineral content is dominated by epidote, calcic amphibole and pyroxene in approximately equal abundance, as well as a significant component (4%) of chrome-spinel. Most of the grains in the pyroxene group belong to the orthopyroxene enstatite, unlike those found in the Amur River and the rivers draining the palaeo-Amur deposits samples, which are predominantly augite, diopside and hypersthene. The chrome spinel-enstatite association in the Nab'il River is characteristic of ultramafic rocks (Mange & Maurer 1992), so it is likely that these are derived from the ophiolites, which crop out in the Nab'il Ridge (Fig. 3c). It is thus important to consider these two minerals in particular when looking for evidence of detritus derived from this area in Neogene rocks in this region. In particular, the chrome spinel–zircon index (CZi) of 76.9 is clearly distinguishable from that of the Amur River and from rivers draining the North Sakhalin Basin.

## Beaches on Sakhalin

A number of samples were collected from beaches along a coastal transect from the SE Schmidt Peninsula to Piltun Bay (Fig. 6c). This was done primarily to quantify the sediment contribution from Cretaceous metamorphic and ophiolitic rocks exposed along the east coast of the Schmidt Peninsula (Vereshchagin *et al.* 1969), and to see whether any mineralogical signal would be discernible in the Neogene detrital record. In addition, samples were collected from a number of the back barrier bays and lagoons that separate the drainage network in Sakhalin from the coast in order to assess the role of these systems in controlling the sediment flux to the offshore basin.

There is a clear southward decrease in the total heavy mineral content of the sands along the east coast of Sakhalin, south of Schmidt Peninsula. This corresponds with a change in the colour and texture of the beach sediment itself. Beach sediment at the northernmost section sampled, within 5 km of the exposed ophiolite, consists predominantly of cobble-size clasts of olivine-rich ultramafic rocks (Fig. 6d), but farther south the sediment becomes progressively finer-grained and lighter in colour. Heavy mineral assemblages in these sands are dominated by olivine, enstatite, chrome spinel and garnet (Fig. 6c). Serpentine is also abundant in these beach samples, and serpentinized enstatite (bastite) and olivine grains are common. Enstatite is the principal rock-forming mineral in ultramafic rocks, and this mineral association is typical of an ophiolitic source. Serpentine is an index mineral of serpentinites, an altered ultramafic rock present in the eastern Schmidt Peninsula. Many of the

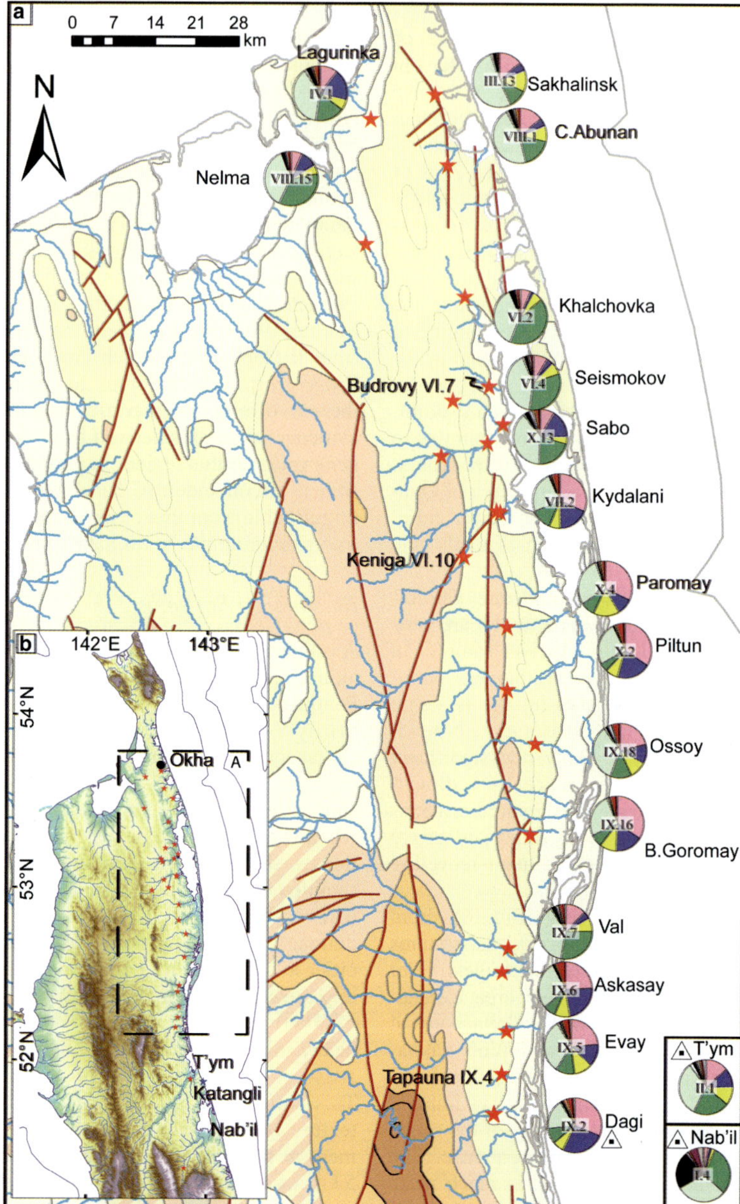


**Fig. 6.** Heavy mineral assemblages of Recent sediments in NE Sakhalin. Key to heavy mineral plots shown in Figure 5. (**a**) Streams. There is a clear distinction between the northernmost rivers draining only the Nutov Suite and the more southerly rivers, which drain large areas of Okobykay-equivalent and older sediment. (**b**) location of Area A and of the Nab'il and T'ym rivers. Sample sites are shown by red stars. (**c**) Beaches and bays. Beach samples are shown outside the coastline whereas bay samples are shown inside. There is a progressive southward decrease in ultramafic orthopyroxene, chrome-spinel and olivine on the east coast with increasing distance from the Schmidt Peninsula. Bay samples have occasional concentration of ultramafic grains, which suggests that these bays and enclosing bar–spit systems are transient features with sediment mixing occurring between sand transported by rivers and marine processes. (**d–f**) Photographs showing change in composition of beach sand southward from Schmidt Peninsula.

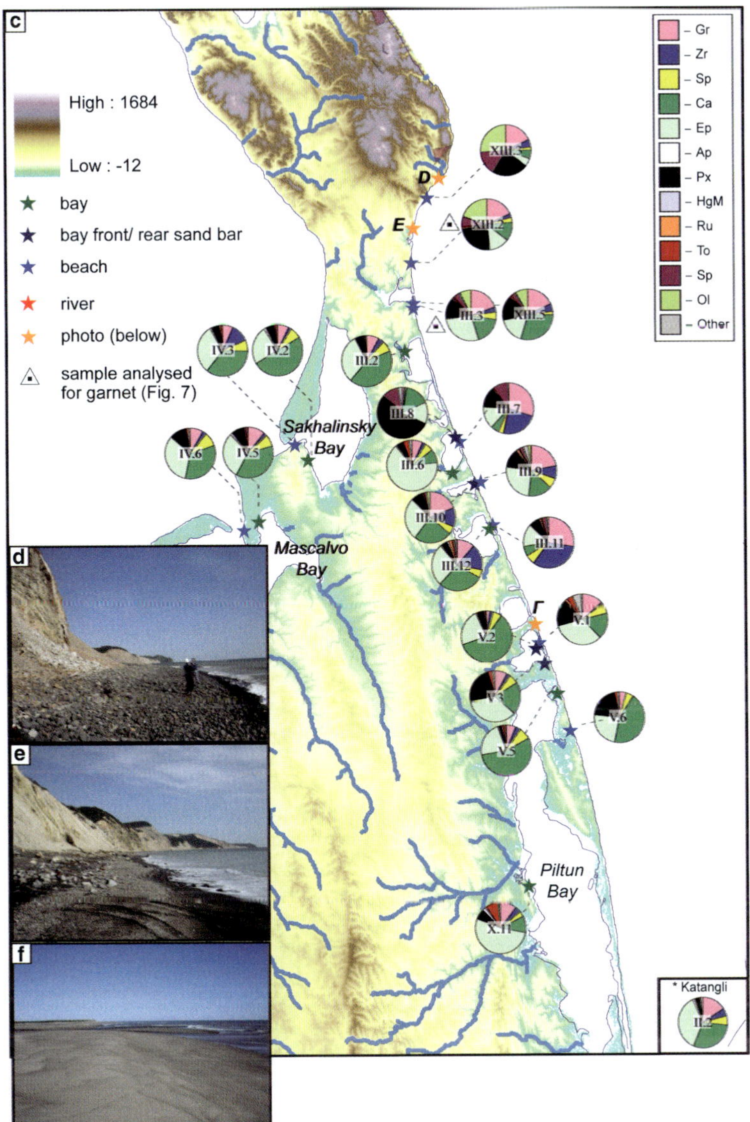


**Fig. 6.** (*Continued*)

garnet grains ($>10\%$) in these samples are brown or green, suggesting that these may be andradite (Mange & Maurer 1992). Such grains are very rare or absent in the modern Amur River.

Farther south, in beach samples east of Piltun Bay, heavy mineral assemblages are dominated by calcic amphibole and epidote, although these still contain a substantial component of enstatite. This suggests that the ultramafic component in the sand is reduced as it is diluted by sand derived from reworked palaeo-Amur sediments from the onshore North Sakhalin Basin. Although there is a consistent decrease in the proportion of ultramafic minerals towards the south, the rate at which these different minerals decline varies. Samples XIII.3 and XIII.2 at the southern margin of Schmidt Peninsula (5 km from the source) contain 20–25% olivine, but this is reduced to $<10\%$ in samples XIII.5 and III.3 (20 km from the source) and only trace amounts are present in samples farther south. Because olivine is one of the least stable heavy minerals (Turner & Morton 2007), it is likely that a combination of aggressive chemical weathering and mechanical abrasion due to wave and tidal energy are the main reasons for its absence in more distal samples. Pyroxene also decreases to the south, although it is still common in samples 70 km from the inferred source. Although pyroxene is also highly chemically unstable, its relatively low density (3.2 g cm$^{-3}$) and bladed or prismatic morphology means that it is easily entrained and transported and concentrates in finer-grained sediment relative to denser grains such as zircon (Komar 2007). Chrome-spinel likewise decreases with distance south from the Schmidt Peninsula and is present only in trace amounts at 70 km from the source. However, chrome-spinel is highly stable, and its disappearance is unlikely to be caused by chemical dissolution or mechanical abrasion (Morton & Hallsworth 2007). It is more likely that the relative abundance of chrome-spinel decreases to the south because of dilution by reworked palaeo-Amur derived sediment, in which chrome-spinel is absent or very rare. In beach samples east of Piltun Bay, heavy mineral assemblages are dominated by calcic amphibole and epidote, although these still contain a substantial component of enstatite. Because of its mechanical and chemical stability, chrome-spinel is the most useful mineral for distinguishing an ultramafic source, such as the Schmidt ophiolite, in the detrital record, particularly in moderate to deeply buried samples in which intrastratal dissolution may have removed unstable minerals.

Sediments collected from bays in northern Sakhalin have a mixed heavy mineral signature, in some places reflecting the composition of the rivers feeding them from the west (Fig. 6a) and in other places reflecting the composition of the beach sands to the east (Fig. 6c). Many samples at the bay fronts (east side) in particular have significant proportions of enstatite, a mineral that is not present in river samples draining the North Sakhalin Basin or in the modern Amur River. Conversely, most samples from the west (landward) sides of the bays do not contain enstatite and other minerals indicative of input from an ultramafic source. The variability of the heavy mineral signatures in bay samples in eastern Sakhalin implies sediment mixing of longshore-transported beach sands and river sands from the North Sakhalin Basin.

There is a pronounced difference in the pyroxene assemblages of bays and beaches on the west coast of Sakhalin from those on the east. As mentioned above, samples collected on the east coast of Sakhalin contain a significant component of orthopyroxenes, mostly enstatite. The ratio of orthopyroxene to clinopyroxene is typically at least 2:1, and can be significantly higher. In contrast, samples collected along the NW coast of Sakhalin, in Sakhalinsky and Mascalvo bays, all have abundant augite (clinopyroxene) grains, which generally outnumber orthopyroxenes.

The most likely cause of the difference in pyroxene compositions between NE and NW Sakhalin is the presence of enstatite-bearing ultramafic rocks on the northeastern Schmidt Peninsula, in contrast to the west where these rocks are absent. Pyroxenes species in the modern Amur River are predominantly augite, hypersthene and diopside, whereas enstatite is completely absent. However, as the pyroxene group minerals on West Schmidt are dominated by titanian augite, there may also be a more proximal source for these. Possible sources include Cretaceous intrusive rocks exposed on the west Schmidt Peninsula or another unknown source in the lowermost Amur River (northern Sikhote–Alin Ridge), downstream of the confluence with the Amgun River.

## Results of heavy mineral analysis: detrital record in the North Sakhalin Basin

### Onshore North Sakhalin Basin

There is a great deal of spatial and stratigraphic variation in heavy mineral assemblages in Neogene sandstones exposed in the North Sakhalin Basin (Poynter 2003; Nicholson 2009); however, some general trends can be observed. The youngest stratigraphic samples, from the Upper Nutov Suite, are compositionally very similar to the modern Amur River, being typically dominated by calcic amphibole and epidote. By contrast, older stratigraphic samples have a less diverse heavy mineral

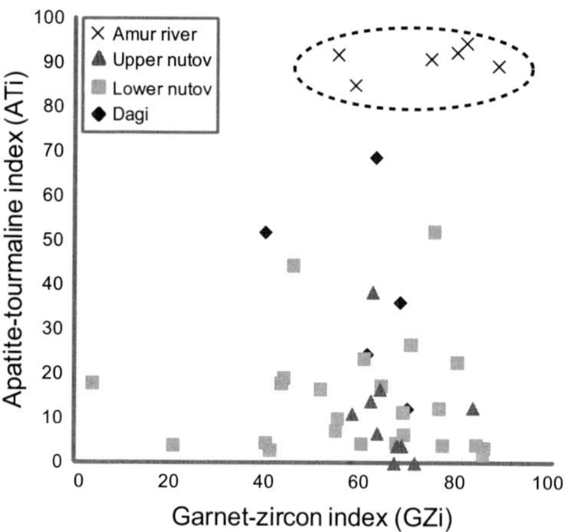


**Fig. 7.** Stable mineral ratios for onshore Neogene samples from the North Sakhalin Basin. There is a clear difference in values from modern Amur samples, particularly the ATi. Amur compositional field highlighted within dashed circle.

suite, with a decrease in both calcic amphibole and epidote. Stable mineral indices also show considerable variation, both spatially and stratigraphically, and are all significantly different to the modern Amur (Fig. 7). In particular, ATi values are typically less than 40, similar to the Sakhalin stream samples described above.

## Offshore North Sakhalin Basin

In order to look at a continuous stratigraphic sequence it is necessary to use samples recovered from offshore wells in two hydrocarbon fields in the North Sakhalin Basin. The main reservoir in Field A is Upper Miocene Lower Nutov Suite sandstone (Figs 3 & 4), with a shale unit at the base of the Upper Nutov Suite providing the top seal. Sands are generally moderately to well-sorted and very fine- to medium-grained. The main reservoir in Field B is the Lower–Middle Miocene Dagi Suite, with overlying Middle Miocene Okobykay Suite shales providing a regional seal. The reservoir consists of fine- to medium-grained, well-sorted to very well-sorted sands, and subsidiary silts and muds.

Cores from a number of wells were made available from the two fields (Fig 3). All measurements described below are measured depth below rotary table (MBRT). The sedimentary rocks that form the reservoirs in these two fields have been buried to a similar depth (1800–2500 m), so the effects of post-burial dissolution should be similar. This allows a more direct comparison between the mineral assemblages in sedimentary rocks of the two main deltaic pulses of the Lower–Middle Miocene Dagi Suite and the Upper Miocene Lower Nutov Suite than is possible from onshore samples.

Cuttings samples were also made available for one well in each field (A-5 and B-2) in order to extend the database from the cored intervals. Because core samples are considered to be more reliable and less susceptible to contamination than ditch cuttings, core samples were used for a detailed, high-resolution study of the reservoir intervals and for the main geochemical analyses where possible. However, as core samples were only available over limited stratigraphic intervals, ditch cuttings were used to constrain compositional changes over the whole stratigraphic sequence.

Samples were analysed for heavy mineral analysis and petrographic analysis from regular (c. 15 m) intervals in key reservoir intervals in the available cores, as well as from every 200 m from ditch cuttings. This was done to test the hypothesis that onshore uplift can be tracked by the presence of key index minerals in the offshore stratigraphic sequences, either from recycled deltaic sedimentary rocks from the onshore North Sakhalin Basin, or from other local sources.

## Petrography

Results of petrographic analyses are presented graphically in Figure 8 and numerically in the Supplementary Material. Core samples analysed from the Lower Nutov Suite (Upper Miocene) of Field A and the Dagi Suite of Field B all plot in the lithic arkose field of Folk (1974) and in the recycled orogen field on a QFL provenance plot (Dickinson & Suczek 1979; Fig. 8) with average values of c. 60% quartz, 25% feldspar and 15% lithics. Samples from cuttings also mainly plot in these fields, although they generally plot closer to the QmQp and Q end-members. This is likely to be an artefact of drilling or the sample preparation process, because lithic grains and feldspars are susceptible to mechanical abrasion (particularly if the grains are already weakened by chemical dissolution/replacement). Samples from the Pomyr Suite (Pliocene) have a relatively elevated component of lithic sedimentary grains in comparison to others.

The overall proportion of lithic grains is similar for samples from both the Lower Nutov and Dagi Suites, although the types of lithic grains are different (Fig. 8e, f). Sedimentary/metasedimentary grains are more abundant in the Dagi Suite, whereas (felsic) volcanic lithic grains are dominant in the Lower Nutov Suite.

## Heavy mineralogy

Heavy mineral analysis was carried out on 73 core samples and 32 cuttings samples from the two fields. The results of these analyses are presented numerically in the Supplementary Material. As with other detrital grains, samples taken from ditch cuttings are susceptible to some contamination from overlying units. The effects of mechanical abrasion during sample preparation are likely to be uniform however, as the same method of sample preparation is used for core and cuttings samples.

Cuttings samples were analysed from two wells (A-5 and B-6) in order to look at stratigraphic changes that occur throughout the palaeo-Amur sequence. Shallow samples are dominated by calcic amphibole and epidote (c. 85%), with relatively minor components of garnet, zircon, tourmaline, apatite and sphene (Fig. 9). Minerals associated with local source terranes in Sakhalin, such as chrome-spinel, enstatite, olivine, glaucophane, lawsonite or pumpellyite, are not present in samples from wells A-5 or B-6 at any stratigraphic level. There is a progressive change in mineral assemblages downhole, with samples at the base of the well dominated by garnet, zircon, tourmaline and apatite. Bulk non-opaque heavy mineral assemblages from core samples covering the

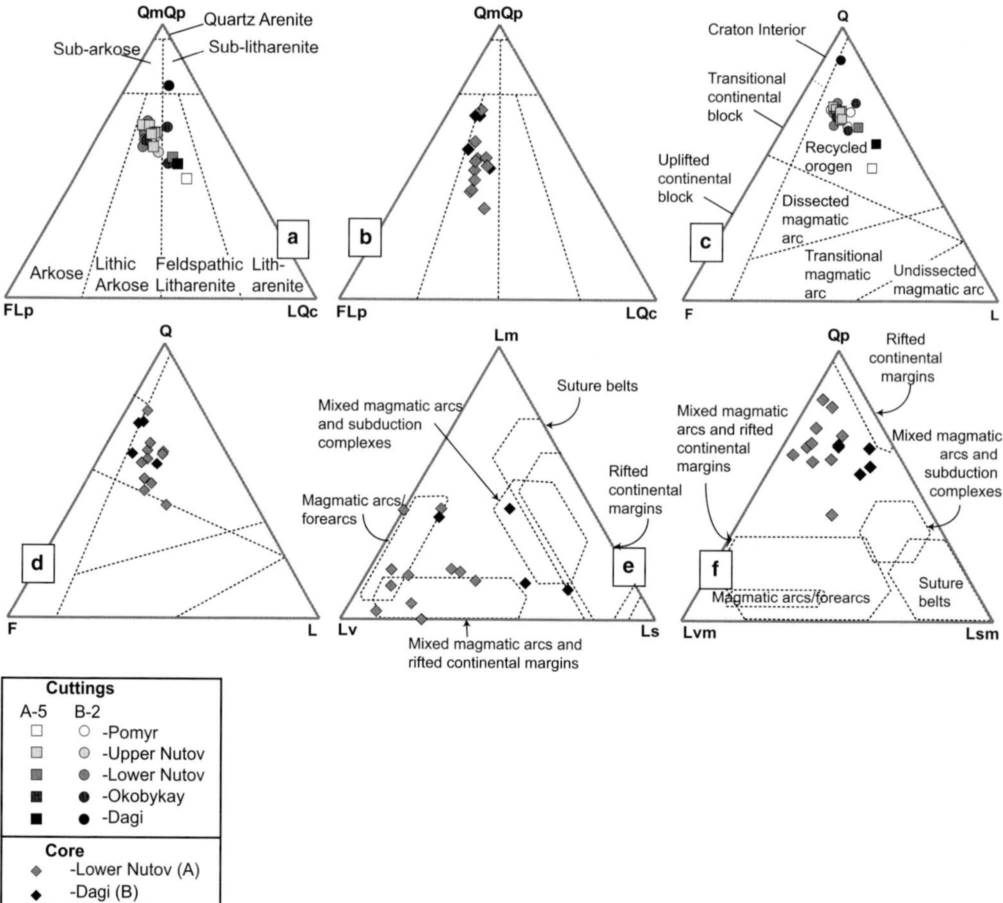



**Fig. 8.** Detrital petrographic analyses of palaeo-Amur samples from offshore wells A-5 and B-2. Core samples and cuttings samples are plotted separately due to the different methods used for sample preparation. (**a, b**) Sandstone classification plots of Folk (1974). Samples mostly plot in the lithic arkose fields. (**c**) QFL plots of cuttings samples. Samples all plot in the recycled orogen plot of Dickinson & Suczek (1979). The Pomyr sample from A-5 plots closer to the lithic axis due to an elevated Ls component. (**d**) QFL plot of core samples of Dagi and Lower Nutov age, all plotting in the recycled orogen field. (**e, f**) LmLvLs and QpLvmLsm plots for core samples of Dagi and Lower Nutov suites. The percentage of volcanic lithic grains is consistently higher in Nutov samples than in Dagi samples. Qm, monocrystalline quartz; Qp, polycrystalline quartz; F, feldspar; Lp, lithic plutonic; L, lithics (all); Qc, microcrystalline quartz; Q, quartz (all); Lm, lithic metamorphic; Lv, lithic volcanic; Ls, lithic sedimentary; Lvm, Lithic volcanic + metavolcanic; Lsm, lithic sedimentary + metasedimentary.

reservoir intervals in Field A (Lower Nutov Suite) and Field B (Dagi Suite) fields are similar, dominated by garnet, zircon, apatite, tourmaline and rutile, with variable proportions of sphene and rare epidote in some samples. High-grade metamorphic minerals (mainly staurolite and kyanite) are also present in both, but are more common in the Dagi Suite (average of 1.3%) than in the Lower Nutov Suite (0.4%).

Interpretation of provenance using heavy mineral data from diagenetically altered sandstones must take into account the progressive elimination of unstable mineral species that occurs during sediment burial (Morton & Hallsworth 1994, 2007; Milliken 2007). To account for this process, it is necessary to look for evidence of corrosive textures, such as etch pits, facets and denticles, on individual grain surfaces (Velbel 2007). If the disappearance of individual species from heavy mineral populations is associated with such grain dissolution features, it can reasonably be assumed that intrastratal dissolution, rather than a change in provenance, is responsible for the change in mineral population (Morton & Hallsworth 2007). Surface textures of the main mineral species were observed by SEM, using backscattered electron imaging and secondary electron imaging, as well as observations made with transmitted light microscopy.

Figure 9 shows the extent of surface corrosion on individual mineral species at different depths in well A-5. There is a clear inverse relationship in this well between mineral diversity and burial depth. Moreover, there is a clear association between extensive mineral dissolution observed on surface textures and the disappearance of these mineral species downhole. Therefore, it is highly likely that the decrease in diversity downhole is caused primarily by progressive dissolution of unstable heavy mineral species. An order of the relative stabilities for the main minerals present can be ascertained from this well. From least to most stable these are pyroxene, amphibole, epidote, sphene and garnet. This corresponds closely with the order of stability described by Morton & Hallsworth (2007). Zircon, tourmaline and rutile are all ultrastable and show no evidence of dissolution in these wells at any depth. Apatite grains show evidence of grain-surface dissolution, but, unlike the unstable minerals described above, these features do not show any systematic increase with depth of burial. Apatite is highly susceptible to acid weathering near the surface, so it is likely that dissolution features in these grains are more closely controlled by surface processes and the environment of deposition than by intrastratal dissolution.

*Stable minerals*

Stable mineral indices for well A-5 are shown in Figure 10. Chromite is absent from all cuttings samples except for two individual samples with one grain, so CZi values are not shown. The ATi value is significantly higher than stratigraphic equivalents onshore Sakhalin (Nicholson 2009), and is comparable to values for the modern Amur River, indicating that significant surface weathering is the reason why ATi values are so low in outcrop samples. The ZTR index is an indicator of overall maturity and there is a strong positive correlation with depth. There is a pronounced increase in the ZTR below 2500 m, corresponding to a similar decrease in GZi. This is likely to be caused by intensive garnet dissolution below this depth, consistent with that recorded by Morton (1987). Corresponding decreases in the RZi and ATi values are, however, unlikely to be related to dissolution during deep burial, because these minerals are all ultrastable during such conditions (Morton 2007). All four curves fluctuate significantly below 3000 m. This area represents a condensed sequence of Oligocene–Middle Miocene Okobykay, Dagi, Uynin and Daekhurin Suites. The GZi, RZi and ATi values all show a moderate increase between 1890 and 1260 m, corresponding to the lower part of the Upper Nutov Suite. GZi and ATi values decrease most significantly in the Pomyr Suite, at 210 and 180 m. This is especially pronounced in the ATi index, in which values decrease from >80 at 420 m to <60 at 180 m. Extensive grain-surface etching on many of the apatite grains in these samples indicates that the change in ATi is caused by dissolution of apatite grains, most likely in response to pedochemical weathering. The similarity of the ATi value in the Pomyr Suite to weathered samples and to stream sediments onshore Sakhalin suggests that these may be derived from recycled palaeo-Amur sandstone rather than directly from the Amur River. A comparison of all samples from the North Sakhalin Basin offshore and onshore shows a significant difference in ATi in particular (Fig. 11). GZi and ATi of the offshore Dagi and Nutov suite samples display a linear relationship, which is not apparent in the onshore samples.

The GZi and ATi values also show marked variations in core samples on the scale of individual reservoir units in both the Dagi (Field B) and Lower Nutov (Field A) suites (Fig. 12). GZi values in the Dagi Suite of Field B vary between 40 and 88 in the two wells studied. Based on these values, the two wells can be divided into packages of high and low GZi values. Garnet and apatite are both susceptible to acid weathering during alluvial storage, which may explain their linear relationship (e.g. Velbel 1984; Morton & Hallsworth 1994), so these packages may represent units with variable degrees of acid dissolution. Both the GZi and the ATi values decrease significantly below *c.* 2000 m in both wells, and increase again below 2063 m in well B-1. This allows the sequence encountered in these cored intervals to be divided into three provisional units, D1–D3.

Wells A-1 and A-2 from Field A are also tentatively divided into three units, N1–N3, on the basis of their heavy mineral data. N1 consists of very high ATi and GZi values (>90), which overlap on this scale. N2 consists of a highly variable sequence with values in both wells ranging from 70 to 95 for both the ATi and GZi. N3a is subdivided into two units in A-1, but this is speculative given the limited sample set at this depth.

# Garnet geochemistry and U–Pb zircon data

## Garnet geochemistry

EMPA of garnet grains was used to determine the geochemistry of individual grains and to distinguish between different source terranes in the Amur River basin and Sakhalin, and to compare to Neogene

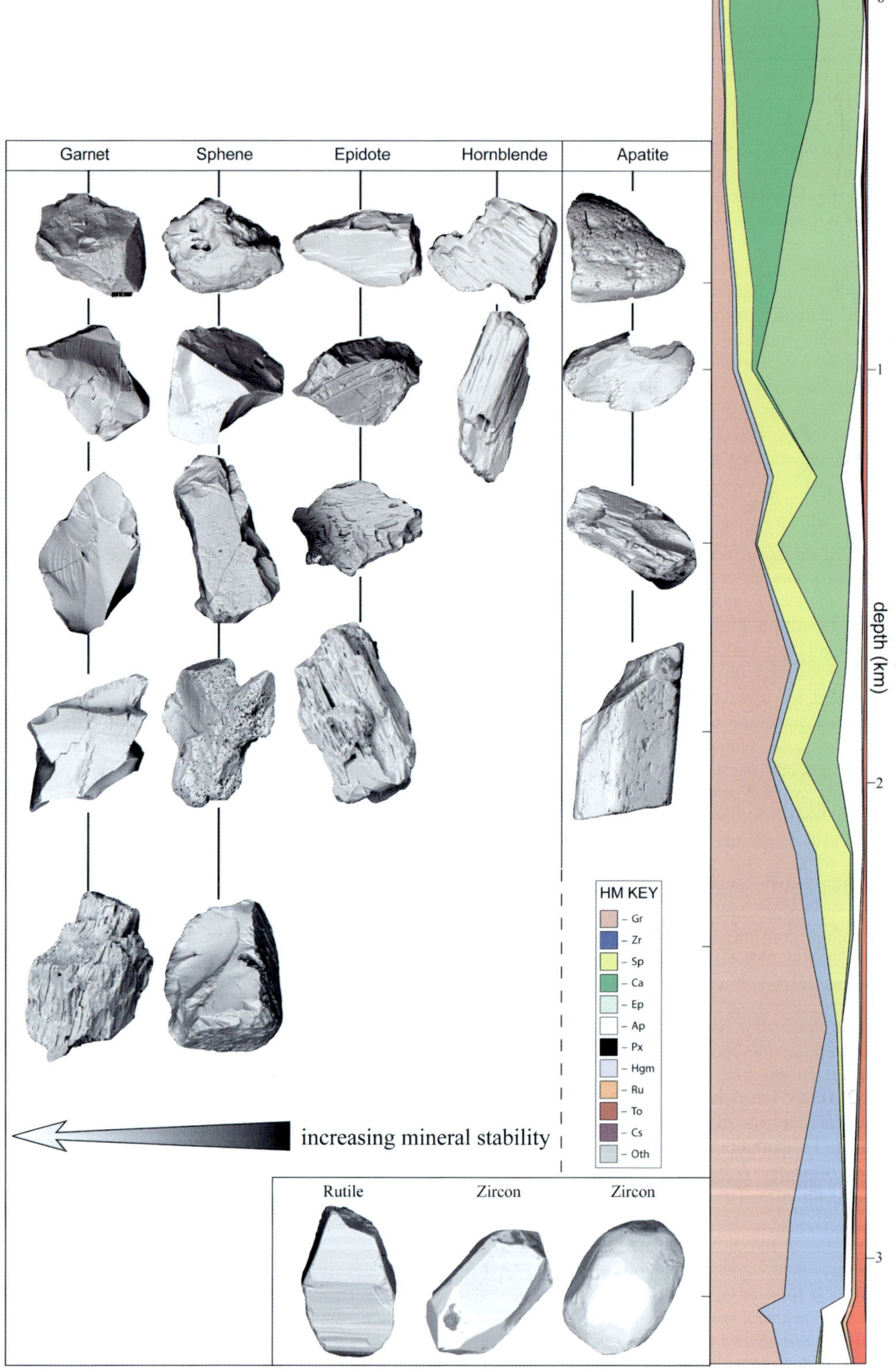


Garnet
Sphene
Epidote
Hornblende
Apatite
increasing mineral stability
HM KEY
Gr
Zr
Sp
Ca
Ep
Ap
Px
Hgm
Ru
To
Cs
Oth
Rutile
Zircon
Zircon
depth (km)
0
1
2
3

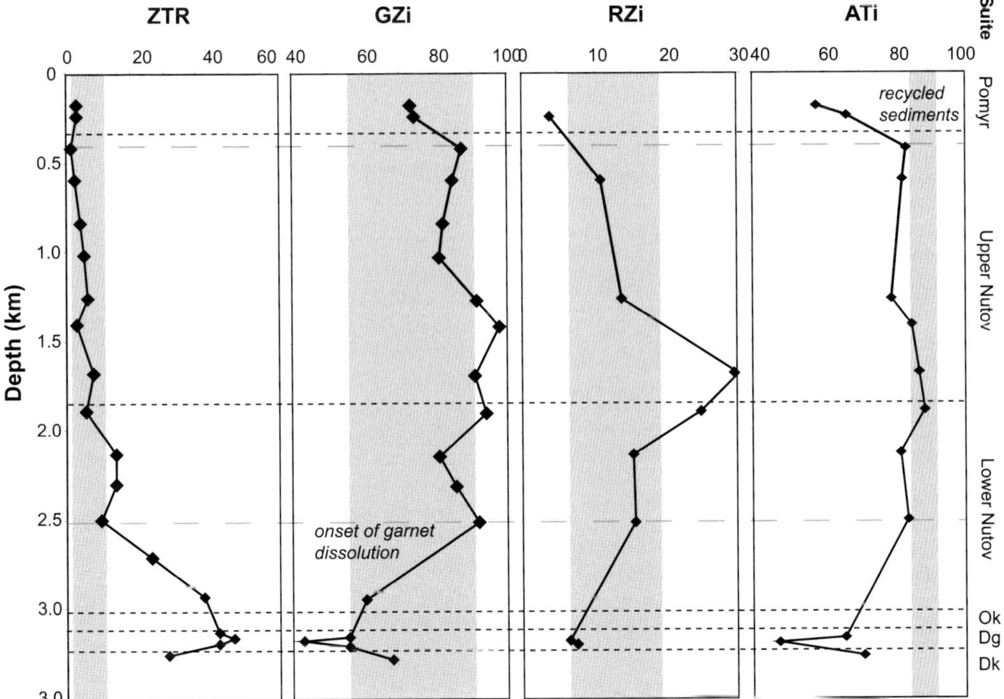



**Fig. 10.** Stable mineral ratios of cuttings samples from A-5 plotted against depth. Typical Amur River ranges shown in grey. Only samples with 50 grains or more per mineral pair are included. Shallow samples of the Pomyr Suite show a pronounced decrease in ATi, together with a decrease in GZi and RZi, and a slight increase in ZTR. There is a significant increase in RZi between 1260 and 1890 m, corresponding to high GZi values. The GZi decreases significantly below 2500 m, probably due to garnet dissolution (see Fig. 9). The area between 3130 and 3260 m represents a condensed sequence of Okobykay–Dagi–Daekhurin age and the heavy mineral assemblages vary accordingly. Low GZi and ATi vales in the Dagi Suite may also represent acid weathering profiles. Ok, Okobykay Suite; Dg, Dagi Suite; Dk, Daekhurin Suite.

sedimentary rocks deposited in the North Sakhalin Basin.

## Constraining source area geochemistry: river and beach sediments

Six samples were identified as potential end-member sources for detrital garnet grains, including the Amur River (SEP108), the Schmidt Peninsula (XIII.2, III.3), the exhumed North Sakhalin Basin (IX.2, II.1) and the East Sakhalin Mountains (I.4, II.1) (Fig. 13; sample locations shown in Figs 5 & 6).

The modern Amur River is dominated by Type A, Bi and Bii (pyralspite) garnets, with a significant proportion (18%) in the Type D field, typical of metasomatic or low-grade metabasic rocks (Deer *et al.* 1992). A large proportion of the pyralspite garnets are spessartine-rich, whereas the ugrandite (Type D) garnets are predominantly grossular. Beach samples from SE Schmidt Peninsula and Rybach'ye are dominated by Type A almandine garnets, typical of very high grade (granulite facies) metamorphic rocks. However, they also contain a significant component (>15%) of andradite garnets, many of which are relatively magnesium-rich. Samples from the Dagi and T'ym rivers are dominated by Type A and Type Bi garnets, with small components of Type Bii and Type D garnets

---

**Fig. 9.** Downhole heavy mineral composition and dissolution features of main mineral groups from A-5. The order of increasing stability of these grains can be seen to be hornblende–epidote–sphene–garnet–rutile + zircon. Apatite shows no systematic downhole increase in dissolution features, although surface etching is variably present at all depths. Zircon and rutile are free of surface corrosion textures at >3000 m. Key for the cumulative % plot is shown in Figure 5. The data are also presented numerically in the Supplementary Material.

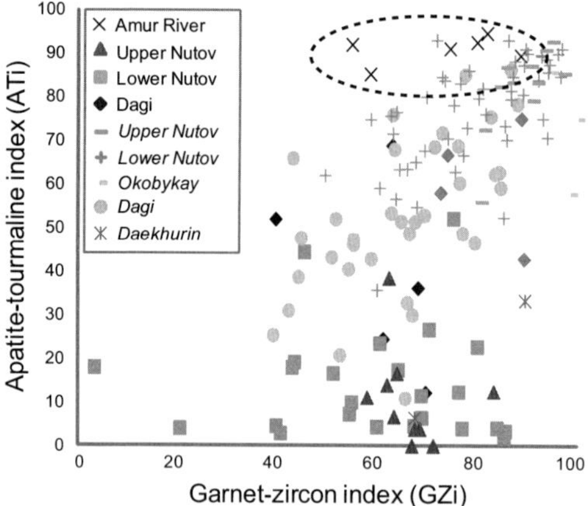



**Fig. 11.** Stable minerals of Neogene offshore well samples (italics), together with those from the onshore Sakhalin Basin and modern Amur River shown in Figure 7. Offshore samples plot much nearer to the modern Amur compositional field (highlighted within dashed ellipse) than those from the onshore basin.

(mostly grossular). Sample 1.4 from the Nab'il River is dominated almost exclusively by Type A and Type Ci garnets.

The garnet compositions of samples near Schmidt Peninsula are very distinctive as a result of the high magnesium content of grains in the Type D field. It is unusual for andradite grains to contain magnesium, with ugrandite garnets typically lying on the Fe + Mn–Ca axis on ternary plots. These Fe–Mg-rich ugrandite grains are likely equivalent to the green-brown grains identified optically, and their abundance is strongly correlated with ultramafic minerals in these samples. Although andradite is typically associated with metamorphosed calcic rocks and associated metasomatic rocks, such as skarns (Deer *et al.* 1992, Mange & Maurer 1992), it has also been observed in association with metabasic rocks (Coombs *et al.* 1977; Morton *et al.* 2003). However, ultrabasic rocks are usually associated with Type Cii (Mg-rich) garnets (Morton *et al.* 2004). Based on its juxtaposition with medium-grade metasedimentary rocks, the ophiolite exposed in Schmidt Peninsula is likely to have experienced moderate levels of regional metamorphism, which may explain the paragenesis of these Mg-rich andradite grains.

Garnet compositions in the T'ym and Dagi rivers are virtually indistinguishable, despite the variation in catchment lithologies. The Dagi River drains an area almost exclusively comprising Neogene palaeo-Amur-derived siliciclastic sedimentary rocks. By contrast, the T'ym River drains a large area of central Sakhalin, including the East Sakhalin Mountains, the West Sakhalin Mountains and the Central Depression. It is possible, however, that most of the garnets in the T'ym River are being derived from palaeo-Amur deposits in the Central Depression.

Type A garnets are abundant in Amur and palaeo-Amur sandstones, but Ci garnets are not (Poynter 2003). Type Ci garnets are typically associated with high-grade metabasic rocks. The Nab'il River drains a large area of the East Sakhalin Mountains and Nab'il Ridge, which is comprised primarily of greenschist–blueschist facies metasedimentary rocks, with ophiolites also exposed along the Nab'il Ridge (Fig. 3). As with the ophiolite on the Schmidt Peninsula, those on the Nab'il Ridge are likely to have experienced low-grade metamorphism. As there are no other known metabasites in the area, the Type Ci garnets were probably derived from these ultramafic rocks.

Garnet geochemistry also varies considerably between different sources and can be used to distinguish between different sources in the palaeo-Amur sequence. In particular, Ci garnets are diagnostic of a source in the East Sakhalin Mountains, and Mg-rich Type D garnets are diagnostic of a source in Schmidt Peninsula.

## Offshore well samples

Garnet grains from 21 core and cuttings samples were analysed to identify changes in provenance between the main stratigraphic units and to see whether any garnets derived from local source areas on Sakhalin are present. Eight cuttings samples from well A-5 and one from well B-2 were analysed along with six core samples from the Lower Nutov Suite (wells A-5 and A-6) and five from the Dagi Suite (well B-2).

Cuttings from well A-5 sample over 3000 m of stratigraphy, including the Pomyr, Lower and Upper Nutov, Dagi and Daekhurin suites (Fig. 14). All samples are dominated by Mn-poor Type A garnets, with a diffuse range of Type B and Type Ci garnets. The number of Ca-rich Type D (ugrandite) garnets is highly variable (0–38%), with the relative abundance of this group having an inverse relationship with depth. This suggests that these may be susceptible to intrastratal dissolution. Ca-rich garnets are known to be more susceptible to intrastratal dissolution than Ca-poor varieties (Morton & Hallsworth 2007).

Core samples from wells in Field A and Field B are from the Lower Nutov and Dagi suites, respectively. Garnets in both suites plot predominantly in the Type A and Type Bi compositional fields, with a broad range of garnets also plotting in the Type Bii and Ci fields (Fig. 14). The proportion of Mn-rich garnets is similar in both samples (17.3%

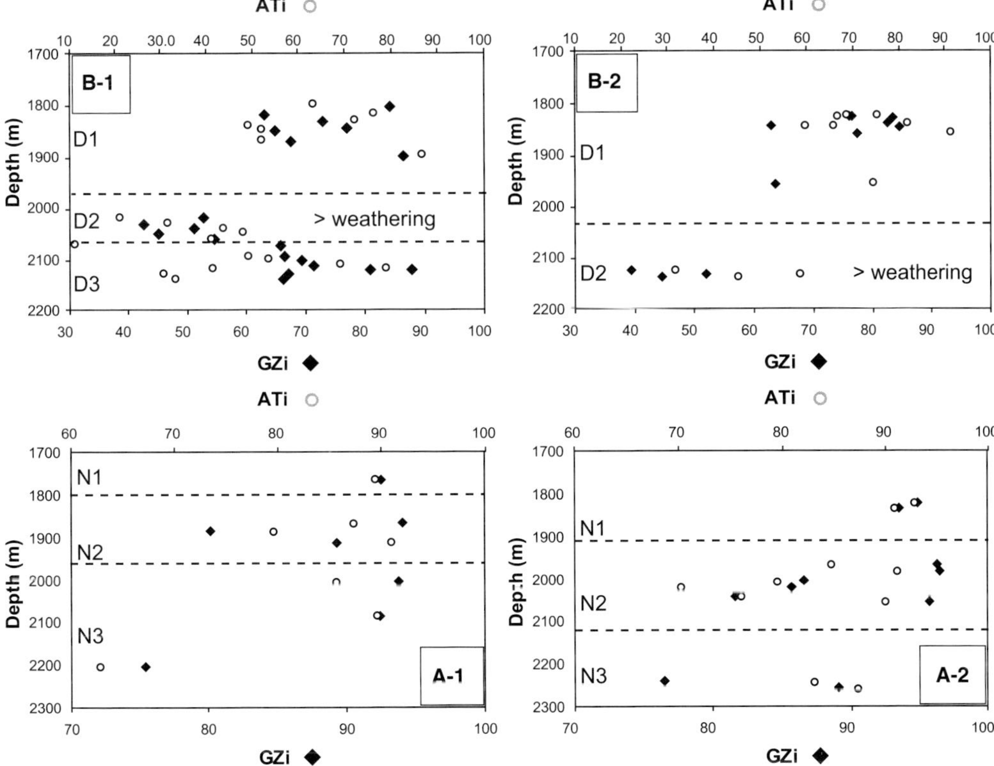

**Fig. 12.** GZi and ATi values for core samples from the Dagi reservoir in two wells in Field B, and from the Nutov reservoir in two wells in Field A. Packages D1, D2 and D3 in Field B are based on stable mineral ratios and may form correlatable units for this field. N1, N2 and N3 in Field A may also form correlatable units. The close relationship between the ATi and GZi trends suggests that the stratigraphic changes in values are caused by acid weathering during sediment transport and storage rather than reflecting changes in provenance or intrastratal dissolution.

in the Dagi, 24.6% in the Nutov). Type D garnets are very rare in both suites.

Samples from the Lower Nutov and Dagi suites are compositionally similar, but are significantly different to those from the modern river. Sands from the modern river contain a much higher proportion of Type D garnets (18%) and Mn-rich garnets (49%). Type D garnets are susceptible to intrastratal dissolution, as described above. Although this decrease may be partially a result of intrastratal dissolution, resulting in the relative enrichment of Fe-rich garnets with increasing depth and age, Mn-rich garnets are relatively stable, and are not expected to have experienced such pervasive dissolution at such a depth (Morton & Hallsworth 2007).

*Detrital geochronology: U–Pb zircon data*

U–Pb dating of zircon grains was carried out on four samples from the palaeo-Amur delta and two samples from the modern Amur River near the confluence with the Amgun River. Palaeo-Amur samples were selected from core rather than ditch cuttings to avoid the possibility of contamination. The zircon age spectra of all four palaeo-Amur samples are similar, being dominated by a broad range of Phanerozoic grains (62–78% of total) with variable quantities of Palaeoproterozoic grains also present in all the samples (Fig. 16). The Palaeoproterozoic component is, however, lower in the modern river than in the palaeo-Amur deposits. Although the zircon data in the Phanerozoic are also somewhat variable, three dominant peak ages can be extracted at c. 100, 200 and 250 Ma, with numerous minor peaks. The absence of Mesoproterozoic and Neoproterozoic grains in all samples is notable, and Archaean grains are also rare or absent from all samples. The Palaeoproterozoic grains in most samples have dominant age ranges between 1.8 and 2.0 Ga, which is a typical crustal generation peak in East Asia and elsewhere (Hawkesworth & Kemp 2006; Safanova *et al.* 2010).

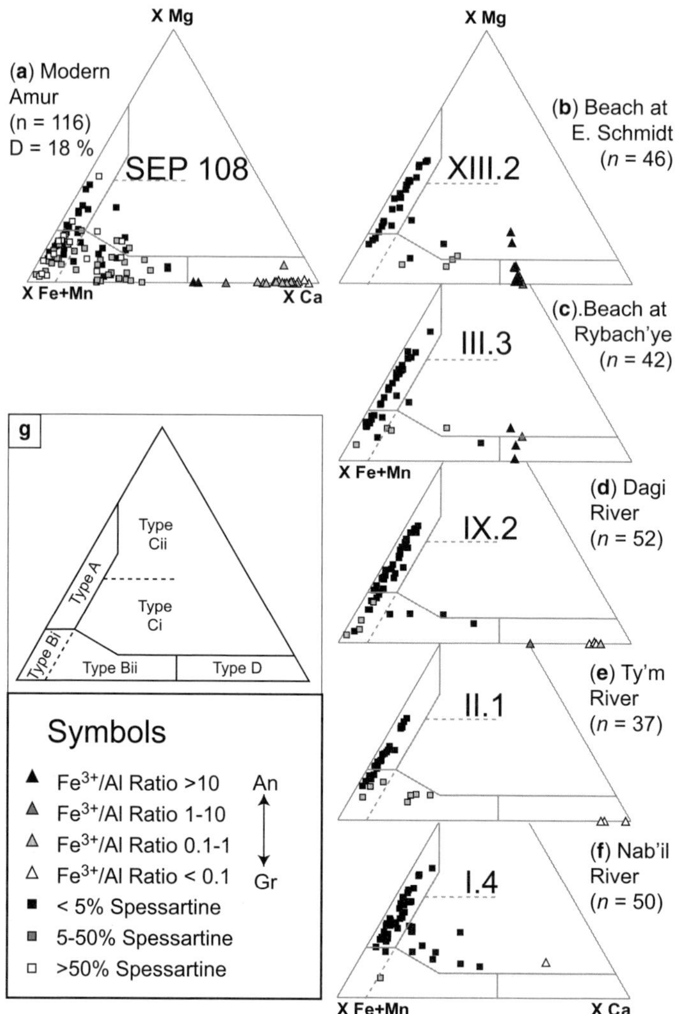



**Fig. 13.** Garnet geochemistry data of different source endmembers (sample locations shown in Fig. 6). (**a**) Modern Amur River sample, downstream of the confluence with the Amgun. (**b, c**) Beach samples on the SE of Schmidt Peninsula. (**d**) Dagi River, incising the full palaeo-Amur sequence. (**e**) T'ym River, draining the Central Depression, with tributaries in both the East and West Sakhalin Mountains. (**f**) Nab'il River, draining almost exclusively from the Nab'il Ridge and East Sakhalin Mountains, consisting of both greenschist-facies metamorphic rocks and ultramafic rocks. (**g**) Compositional fields of Mange & Morton (2007): Type A, granulite-grade metasediments and charnokites, also deeply sourced intermediate-acidic igneous rocks; Type B, amphibolite facies metasediments; Bi, intermediate-acidic igneous rocks; Type C, high-grade metabasic rocks – Cii garnets are derived from ultramafic rocks; Type D, metasomatic rocks, very low-grade metabasic rocks, ultrahigh-temperature calc-silicate granulites. *n*, total number of grains analysed.

## Discussion

### *Downstream stability of Amur sediment composition*

The lack of significant downstream variation in the trunk stream of the Amur has two implications for this river system. First, many of the large tributaries

that join this river downstream of the Khingan Ridge (including the Sungari, Ussuri, Anyui, Gorin and Amgun rivers) make little difference to the overall sediment composition in the trunk stream. This is surprising, as these tributaries drain a substantial area of the Sikhote–Alin Mountains. This area has relatively high relief and a relatively wet climate due to its proximity to the coast, so rates of sediment production might be expected to be high (e.g. Bookhagen & Burbank 2006). In addition, the source area in the Sikhote–Alin is dominated by metasedimentary, basic and ultrabasic rocks, and volcaniclastic sediments (Fig. 1, Table 1), all of which contribute a relatively high proportion of heavy minerals when compared with other areas (largely acidic-intermediate continental crust and sedimentary rocks) in the Amur catchment. It appears that much of this sediment is not being incorporated into the Amur sediment load in the Mid Amur Basin, at least into the sand-size fraction. Poynter (2003) suggested that much of this locally derived sediment may be trapped by tributary mouth lakes, which form behind sediment build-up at stream confluences. These lakes may, in places, form traps for sediment, but their presence also indicates that subsidence rates are very high throughout much of the basin, outpacing sedimentation rates. Because the Mid Amur Basin (Fig. 1) is actively aggradational, it is possible that much of the coarse sediment being derived from the Sikhote-Alin Mountains is being deposited in the distributary fan systems on the basin margins, with Amur-derived material from the upper catchment being deposited in the basin centre. However, the relatively thin sedimentary succession in the Mid-Amur Basin (<1 km; Varnavskiy 1971) suggests that this process is unlikely to have been active throughout the lifetime of the Amur River. Alternatively, the much larger catchment area and high elevation of the upper Amur River may swamp the sediment signature from the Sikhote–Alin Mountains, despite being dominated by crystalline magmatic and sedimentary rocks. Below Khomsomolsk-na-Amure, the trunk stream leaves the Mid Amur Basin and flows through a confined river valley. In this lower 500 km of the river system, the proportion of calcic amphibole increases, suggesting a contribution from tributaries draining the northern Sikhote–Alin and Chayatyn ranges, which is not seen in the Mid Amur Basin.

The second implication is that mechanical abrasion in river channels and chemical weathering in floodplain environments appears to be insignificant in the river, with unstable minerals such as apatite remaining unaffected by such processes. This is consistent with geochemical observations suggesting only superficial chemical weathering of sediment in other East Asian rivers, including the Lena River (Huh et al. 1998). It is also noteworthy that the sediment of the Amur trunk stream is compositionally very similar to the Zeya River. Poynter (2003) suggested that much of the lack of down stream variation in the river can be attributed to the fact that a large component of the sediment budget of the river is coming from the Zeya–Bureya Basin, where the Amur, Zeya and Bureya

rivers are actively incising Cenozoic sedimentary basins that are currently experiencing tectonically driven rock uplift (Ren et al. 2002).

However, sediments from a series of Quaternary Amur terraces upstream of the confluence of the Zeya are also compositionally almost identical to those recorded from downstream of the catchment (Poynter 2003), whereas those from the Zeya River, by comparison, are depleted in calcic amphibole. Although the Quaternary samples are taken from within the confines of the Zeya–Bureya Basin and a component of the sediment may be from there, the lack of major tributaries draining the basin upstream of the Zeya confluence suggests that it is more likely that this sediment has come from the much larger Upper Amur catchment (Fig. 1). Furthermore, as significant apatite dissolution is common in the Zeya–Bureya and Mid-Amur Basins (Poynter 2003), the consistently high ATi index in Amur sediments downstream makes it unlikely that the majority of sediment in the modern river is coming from these sources. Conversely, the consistency in ATi values may be the product of the limited residence time of sediment in floodplains and intracatchment sedimentary basins.

## Provenance of palaeo-Amur samples

Neogene sediments exposed in the offshore North Sakhalin Basin are mainly derived from the palaeo-Amur River (Davies et al. 2005; Macdonald & Flecker 2007), but little work has been done previously to determine the stability of the palaeo-Amur drainage system through time.

Bulk heavy mineral populations of diagenetically unaltered Neogene samples of the North Sakhalin Basin are almost identical to those from the modern Amur River (Poynter 2003; Nicholson 2009; Figs 5 & 9). Pliocene and younger samples from both onshore and offshore areas are dominated by calcic amphibole and epidote, as are those from the modern river. Stable mineral ratios are also broadly consistent with the modern river throughout the Amur sequence (Fig. 10), particularly in samples from the offshore wells where samples are unaffected by the second-stage acid weathering seen in surface samples of the onshore North Sakhalin Basin.

It is clear that most of the sedimentary rocks in the North Sakhalin Basin are derived directly from the Amur River, rather than local sources. Offshore samples in particular have very consistent stable mineral ratios and appear to be too distant from local source areas to have received any significant sediment input. It is therefore possible to use a combination of petrographic, heavy mineral and geochronological data to provide

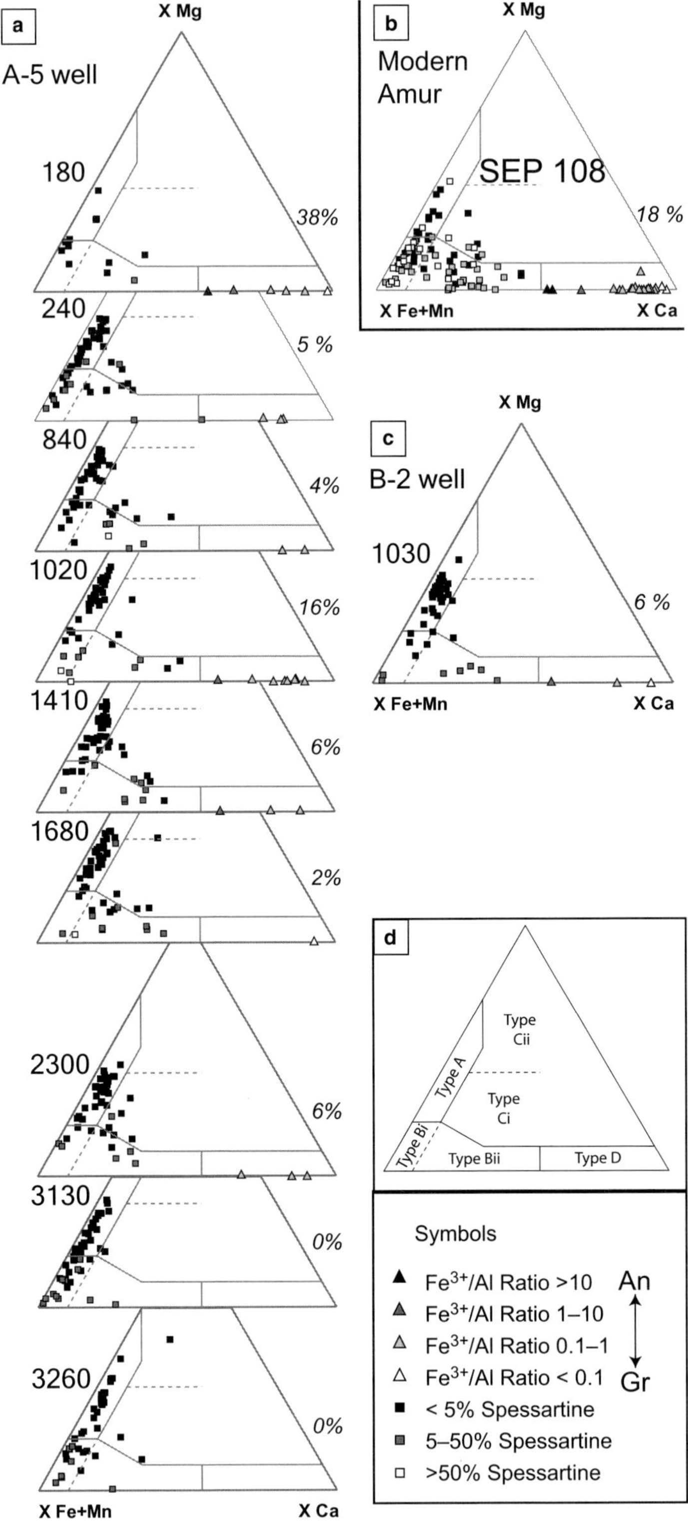
a
A-5 well
X Mg
180
38%
240
5 %
840
4%
1020
16%
1410
6%
1680
2%
2300
6%
3130
0%
3260
0%
X Fe+Mn
X Ca

b
X Mg
Modern
Amur
SEP 108
18 %
X Fe+Mn
X Ca

c
X Mg
B-2 well
1030
6 %
X Fe+Mn
X Ca

d
Type Cii
Type A
Type Ci
Type Bi
Type Bii
Type D

Symbols
Fe$^{3+}$/Al Ratio >10
Fe$^{3+}$/Al Ratio 1–10
Fe$^{3+}$/Al Ratio 0.1–1
Fe$^{3+}$/Al Ratio < 0.1
An
Gr
< 5% Spessartine
5–50% Spessartine
>50% Spessartine

some constraints on catchment evolution through time.

When the effects of post-burial dissolution are taken into account, the heavy mineral assemblages of sandstone samples from the Dagi and Nutov suites are broadly similar. However, there are some key differences between these two units. Heavy mineral assemblages from the Dagi Suite typically have lower and more variable GZi and ATi values than those from the Lower Nutov Suite. In addition, they contain a significantly higher component of grains derived from a high-grade metamorphic source (alumino-silicates and staurolite) than those in the Lower Nutov Suite. By contrast, sandstones from the Lower Nutov Suite contain a significantly higher proportion of volcanic lithic grains (Fig. 8), which are not present in the Dagi Suite, as well as a significant component of swelling clays (most likely smectite) in core samples.

Other data suggest that the sediment being supplied by the palaeo-Amur River remained fairly constant between the two main phases of deltaic sedimentation. Garnet geochemistry shows some variability between individual samples, but an overall comparison of the Dagi and Nutov suites at similar burial depths (Fig. 15) suggests that the source of these minerals remained broadly consistent through time.

Mn-rich (spessartine) garnets are much more common in the modern river than they are anywhere in the palaeo-Amur sequence (Figs 13 & 14). However, garnet samples in the modern river upstream of Khomsomolsk-na-Amure are more similar to those in the palaeo-Amur (Poynter 2003). It is possible that the large component of spessartine in Sample SEP108 is of local significance, particularly in consideration of its proximity to the confluence with the Amgun River. In particular, these may be derived from plutonic rocks exposed in the eastern Bureyinsky Mountains (Fig. 1), at the headwaters of the Amgun River.

U–Pb dates of zircon grains also show relatively little difference between the Dagi Suite, the Lower Nutov Suite and the modern river (Fig. 16). All samples are dominated by Phanerozoic grains that date at less than 500 Ma, with peaks discernible in most samples at 100, 200 and 250 Ma. These peaks may relate to active margin magmatism at these times, during the closure of the Mongol–Okhotsk Ocean, and subsequent rifting events near the Mongol–Okhotsk Suture Zone (e.g. Faure et al. 1995; Ren et al. 2002; Tomurtogoo et al. 2005). There are numerous Mesozoic–Palaeozoic

granitoid complexes in the upper catchment of the Amur in particular, such as the vast 290–330 Ma Angara–Vitim batholith (Rytsk et al. 2011) and the 195–225 Ma Khentei batholith (Yarmolyuk et al. 2002; Kovalenko et al. 2003), both now exposed in the Yablonoviy Mountains (Fig. 1), which may correlate with these ages (Safanova et al. 2010), in addition to possible sources in the Bureyinsky Ranges and Khingan Ridge (Fig. 1b).

There does, however, appear to be a decrease in Proterozoic zircons in the modern river sample, which may suggest less input from Precambrian sources to the modern river than in the ancient river. This may be caused by reactivation of the ranges from which younger sediments are eroded (possibly including the Sikhote–Alin, Bureyinsky, Khingan and Yablonoviy ranges) or a decrease in erosion rates in the Stanovoy Range and Mongolian Plateau, where most Proterozoic basement rocks are exposed (Fig. 1). Zircons analysed by Safanova et al. (2010) from the Amur River near Khomsomolsk-na-Amure did, however, yield abundant U–Pb ages in the 1.8–2.0 Ga age range, as well as a large peak at c. 2.5 Ga, consistent with the palaeo-Amur samples analysed in this study. Safanova et al. (2010) interpret these Proterozoic grains as having been derived from the North China Craton terranes, now exposed in the Mongolian Plateau in the upper Amur catchment (Fig. 1), where peak ages of 1.6–2.0 Ma are dominant, with an additional strong peak at 2.5 Ga (Rojas-Agramonte et al. 2011). The presence of these in Lower to Upper Miocene samples indicates that this area was integrated in the Amur catchment at that time.

When taking all different datasets into consideration, it seems unlikely that any major basin-scale drainage capture events have affected the Amur River from the Early Miocene to the present. Sediment supply has been relatively stable during this time period, with consistency of composition more significant than the minor differences. However, a smaller-scale drainage capture event in the Songliao Basin cannot be ruled out. The presence of barbed tributaries in the Liao and Sungari rivers, and the overall centripetal drainage pattern in the basin (Fig. 1), strongly suggest that capture occurred at some point in the geological past. However, as the Sungari River does not significantly alter the composition of the Amur trunk stream downstream of their confluence, and until analysis of sediment composition and sediment volumes in the Songliao and Liao basins of northern China is carried out, this hypothesis remains difficult to test.

---

**Fig. 14.** Garnet geochemistry for well samples from offshore fields. (**a**) Cuttings samples from A-5 showing a decrease in Type D (ugrandite garnets) with depth. Percent of Type D garnets is shown in italics. (**b**) Modern Amur River composition. (**c**) Ditch cuttings sample from B-2, 1030 m. (**d**) Compositional fields from Mange & Morton (2007).

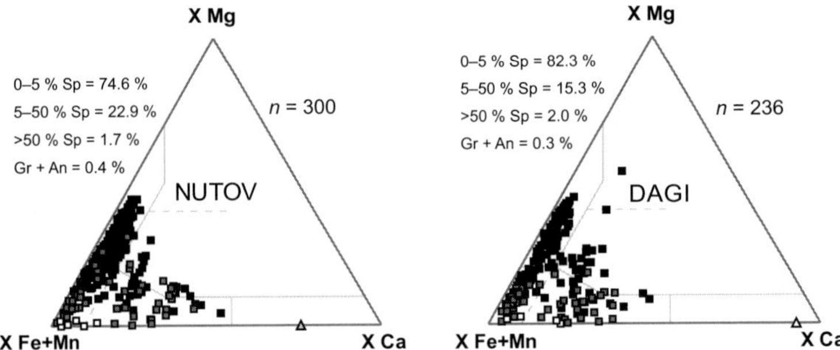



**Fig. 15.** Combined garnet geochemistry plots for the Lower Nutov (wells A-5 and A-6) and Dagi (B-2) from core samples only. Both sets of samples are dominated by Type A and Type Bi garnets, with a broad range of subsidiary Type B and Type Ci. Ugrandite garnets (Type D) are very rare (<0.5%), despite being relatively common in modern Amur sands and in some palaeo-Amur samples onshore (Poynter 2003).

## Sediment recycling and local source input

Several strands of evidence suggest that there was some input from local sources to the North Sakhalin Basin during deposition of the palaeo-Amur delta sequence. Pebble clast assemblages, where present, remain consistent throughout most of the Dagi and Nutov suites, but become markedly more diverse at the top of the Upper Nutov Suite in the coastal sections of NE Sakhalin (Nicholson 2009). Clasts here include those from a wide variety of igneous and metamorphic rocks. Most significantly, there are also rounded clasts of sandstone that are compositionally identical to that of the matrix and underlying sand units. This suggests that uplift of older deltaic deposits was occurring simultaneously with sedimentation in this area, the first conclusive stratigraphic evidence for uplift of deltaic sedimentary rocks either onshore or offshore in northern Sakhalin. Clasts of opoka (partially recrystallized diatomite) are present in the Pomyr Suite of Schmidt Peninsula (Poynter 2003), providing further evidence that uplift was active at this time.

One of the main purposes of analysing samples from offshore wells in Sakhalin was to look for evidence of sediment being recycled from the uplifting North Sakhalin Basin. The rivers draining the North Sakhalin Basin have heavy mineral assemblages that appear to be controlled by the depth of incision and the stratigraphic levels exposed in their catchments. GZi and ATi values are significantly lower in weathered surface samples onshore Sakhalin and in modern rivers draining the island than those of the modern Amur River or unweathered palaeo-Amur samples encountered in offshore wells (Fig. 11). Apatite is highly susceptible to weathering in acidic conditions (Morton 1986), so the low ATi values in the rivers draining Sakhalin and in Pliocene–Recent samples offshore likely represent dissolution and depletion of these grains at the surface and during sediment storage.

Samples from the reservoir sections (1800–2500 m) in both the Dagi Suite (Field B) and Lower Nutov Suite (Field A) age show no evidence of sediment recycling or of input from local sources. Shallow ditch cuttings (<250 m) however, do contain some petrographic evidence that sediment was being reworked from the uplifting island to the west (Figs 8 & 9). Pomyr Suite samples from both fields contain an elevated proportion of siltstone and mudstone clasts that appear to be detrital in origin. We therefore interpret the decrease in ATi and GZi values in Upper Pliocene samples offshore as representing the first flux of reworked deltaic sediment from the uplifting island of Sakhalin into the offshore basin. This marks the onset of transpression along the Sakhalin–Hokkaido Shear Zone in this area of the basin, consistent with the observations of Rozhdestvenskiy (1982).

Sediment derived from other local sources surrounding the North Sakhalin Basin was not evident in the well samples, but this is probably due to the relatively large distances of the offshore locations from any of these sources (Fig. 3). There is some evidence from more proximal sources for local sediment input from the East Sakhalin Mountains during the Early–Middle Miocene (Nicholson 2009), implying that this area was subaerially exposed as early as that point. However, this may have been a relict high from an earlier phase of deformation and is not necessarily indicative of transpressive uplift during this time.

## Heavy minerals: a tool for stratigraphic correlation

Correlation schemes based on heavy mineral data have previously been published by Morton *et al.*

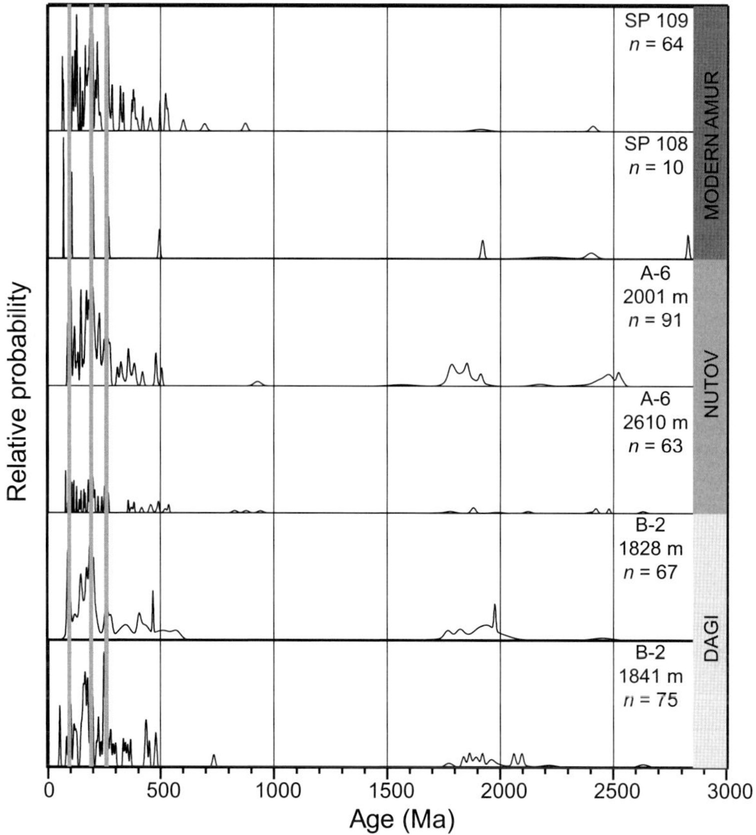



**Fig. 16.** Probability density plots of best U–Pb zircon ages for samples from the Dagi Suite, Lower Nutov Suite and modern Amur River. All samples are dominated by a broad range of Phanerozoic zircons with shared peaks at c.100, 200 and 250 Ma (highlighted in grey). SEP 109 contains a longer 'tail' of older zircons in the Late Neoproterozoic/Early Palaeozoic, but less Palaeoproteroic zircons in comparison to palaeo-Amur derived samples. n, number of zircons plotted (<10% discordant).

(2007) for biostratigraphically barren fluvial sedimentary rocks in the North Sea, which can be used at the well site to guide drilling decisions in 'real time' (Morton & Milne 2012). The construction of such a field-scale heavy mineral stratigraphy for the North Sakhalin Basin could be particularly useful as there is no high-resolution biostratigraphic framework for the North Sakhalin Basin at present. The ATi and GZi of samples from offshore wells do appear to fluctuate systematically, particularly in the Dagi Suite sands of Field B (Fig. 12). The simultaneous changes in GZi and ATi values indicate that the packages of low GZi and ATi represent relatively intensively chemically weathered horizons, most likely prior to deposition (Morton & Berge 1995; Morton et al. 1996). If the decrease in the GZi values was because of intrastratal dissolution of garnet during burial, then the ATi would remain unaffected, as

apatite and tourmaline are both ultrastable during burial (Morton 1986; Morton & Hallsworth 2007). These zones of increased chemical weathering may correlate with lowstand systems tracts, when sediment on the shelf would have been subaerially exposed and subject to weathering (Mitchum & Vail 1977; Posamentier & Allen 1993). In this case, Unit D1 and D2 may be stratigraphic equivalents in the two wells. Alternatively, these units may have been transported during periods with a relatively warm and humid climate, resulting in the intensification of chemical weathering on floodplains (West et al. 2005). This may be more pronounced in the Middle Miocene Dagi Suite than in the Lower Miocene Nutov Suite because of the warmer, and potentially wetter, climatic conditions in eastern Eurasia during the Middle Miocene (Fig. 4; Zachos et al. 2001). Comparison of the stable mineral data with core sedimentary facies

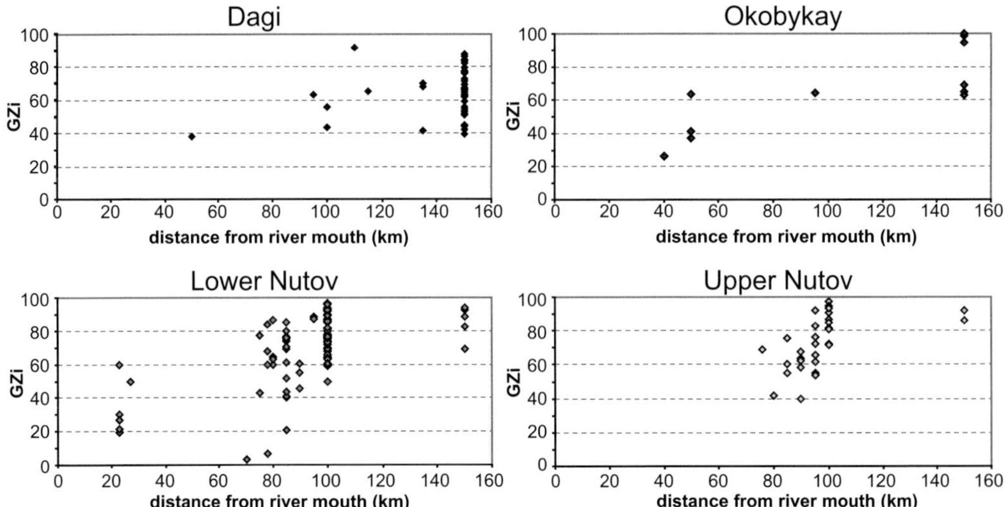

**Fig. 17.** Summary of spatial and stratigraphic variations in GZi, incorporating data from Nicholson (2009). Data from each main stratigraphic suite (or lateral equivalents) are plotted as GZi v. distance from modern river mouth in order to look at the effects of grain sorting. GZi values show a positive correlation with distance from the river mouth, indicating pronounced hydrodynamic segregation of garnet and zircon grains on the delta plain; the 12% density differential resulted in denser zircon grains being deposited closer to the sediment point source at the Amur River mouth.

analysis may assist in understanding the relationship between heavy mineral ratios and climatic and eustasic conditions.

Despite the stable mineral stratigraphic correlation that may be made at a field scale, there is a strong geographic variation in stable mineral ratios on a more regional scale, particularly for GZi and RZi, between palaeo-Amur sediments encountered in offshore wells and chronostratigraphically equivalent formations onshore Sakhalin (Fig. 17). In the onshore North Sakhalin Basin, the GZi is typically significantly lower than that in the modern river (Nicholson 2009), unlike the offshore North Sakhalin Basin, where the values are broadly similar to the river until depths greater than *c*. 2.5 km. There is a clear east–west trend in stable mineral ratios in the North Sakhalin Basin, with RZi and GZi values increasing to the east. Both of these stable mineral indices have moderate density differentials (9% and 12%, respectively), resulting in the preferential settling of denser zircon grains in coarser grained deposits. This implies that hydrodynamic segregation and mineral sorting was occurring along the delta plain during deposition of these samples. However, the degree to which these ratios vary over *c*. 150 km is unexpected. This casts doubt on the practice of employing some of these commonly used heavy mineral indices for stratigraphic correlation over wide (>100 km) geographic areas.

## Conclusions

- The majority of the sand-sized sediment in the Amur River came from upstream of the Lesser Khingan Ridge, as can be seen from the uniformity of sediment composition in the lower 1700 km of the river. The Mid-Amur basin is actively subsiding and it is possible that much of the sediment derived from the Sikhote–Alin Range is trapped within this basin and did not enter the trunk stream.
- The sediment of the lower Amur River is mineralogically immature, with little evidence for significant dissolution of chemically susceptible grains such as apatite, pyroxene and calcic amphibole. This implies that residence time in floodplain deposits or bar systems is limited in this fluvial system, or that the cold climatic conditions in the basin inhibit chemical dissolution.
- There are only minor variations in the palaeo-Amur sediment composition from the Early Miocene to the present day. The consistency of the petrographic signature in the palaeo-Amur delta argues against Miocene–Pliocene drainage capture models involving the Zeya–Bureya Basin and the Upper Amur.
- A pronounced decrease in GZi and ATi values is evident in samples from the Pliocene Pomyr Suite (≥3.6Ma) of Sakhalin and was caused by

dissolution of garnet and apatite grains as sedimentary rocks were uplifted and exposed to pedochemical process. This trend also corresponds with an increase in Type D garnets and sedimentary rock fragments in cuttings samples and marks the onset of uplift on Sakhalin associated with transpression. These complementary datasets imply active recycling of deltaic sediments from Sakhalin during Pliocene deposition of the Pomyr Suite.

- A positive correlation between GZi and horizontal distance from the river mouth is attributed to hydrodynamic sorting laterally across the delta system, with the lighter garnet grains being transported farther than denser zircons. This has negative implications for the use of certain stable mineral ratios as basin-wide correlation tools, in particular GZi and RZi, which are commonly used for these purposes.
- Intensively weathered heavy mineral zones within the basin stratigraphy characterized by low ATi and GZi values may be used as a basis for stratigraphic correlation if they can be linked with sequence stratigraphic systems tracts.

This work was carried out as part of a PhD programme for U. Nicholson, which was funded by the Natural Environment Research Council (NERC). We extend our thanks to A. Morton, R. Willan and J. Still, who gave assistance with heavy mineral analysis techniques, petrographic interpretation and garnet geochemical analysis, respectively, at the University of Aberdeen, as well as to Andy Carter of Birkbeck College for carrying out ICP-MS analysis. Many thanks also to Nina Alekseevena Derevskova and other staff members of SakhalinMorNefteGaz (SMNG) in Okha for assistance with fieldwork and sample preparation, and to M. Ivanov, D. Korost and colleagues at Moscow State University for providing the facilities to carry out heavy mineral analysis over an extended period of time.

# References

APEL, E. V., BURGMANN, R., STEBLOV, G., VASILENKO, N., KING, R. & PRYTKOV, A. 2006. Independent active microplate tectonics of northeast Asia from GPS velocities and block modelling. *Geophysical Research Letters*, **33**, L11303, http://dx.doi.org/10.1029/2006GL026077

BOOKHAGEN, B. & BURBANK, D. W. 2006. Topography, relief, and TRMM-derived rainfall variations along the Himalaya. *Geophysical Research Letters*, **33**.

COOMBS, D. S., KAWACHI, Y., HOUGHTON, B. F., HYDEN, G., PRINGLE, I. J. & WILLIAMS, J. G. 1977. Andradite and andradite-grossular solid solutions in very low-grade regionally metamorphosed rocks in Southern New Zealand. *Contributions to Mineralogy and Petrology*, **63**, 229–246.

CUMMING, G. L. & RICHARDS, J. R. 1975. Ore lead isotope ratios in a continuously changing earth. *Earth and Planetary Science Letters*, **28**, 155–171.

DAVIES, C. E., POYNTER, S. P. *ET AL.* 2005. Facies analysis of the Neogene delta of the Amur River, Sakhalin, Russian Far East: controls on sand distribution. *In*: GIOSAN, L. & BHATTACHARYA, J. P. (eds) *River Deltas – Concepts, Models and Examples*. SEPM (Society for Sedimentary Geology) Special Publications, Tulsa, **83**, 207–229.

DEER, W. A., HOWIE, R. A. & ZUSSMAN, J. 1992. *An Introduction to the Rock-Forming Minerals*, 2nd edn. Prentice-Hall, Englewood Cliffs, NJ.

DICKINSON, W. R. 1970. Interpreting detrital modes of greywackes and arkose. *Journal of Sedimentary Petrology*, **40**, 695–707.

DICKINSON, W. R. & SUCZEK, C. A. 1979. Plate tectonics and sandstone compositions. *American Association of Petroleum Geologists Bulletin*, **63**, 2164–2182.

DROOP, G. T. R. 1987. A general equation for estimating $Fe^{3+}$ concentrations in ferromagnesian silicates and oxides from microprobe analyses, using stoichiometric criteria. *Mineralogical Magazine*, **51**, 431–435.

FAURE, M., NATAL'IN, B. A., MONIE, P., VRUBLEVSKY, A. A., BORUKAIEV, C. & PRIKHODKO, V. 1995. Tectonic evolution of the Anuy metamorphic rocks (Sikhote Alin, Russia) and their place in the Mesozoic geodynamic framework of East Asia. *Tectonophysics*, **241**, 279–301.

FOLK, R. L. 1974. *Petrology of Sedimentary Rocks*. Hemphill Publishing Company, Austin, TX.

GARZANTI, E., CANCLINI, S., FOGGIA, F. M. & PETRELLA, N. 2002. Unraveling magmatic and orogenic provenance in modern sand: the back-arc side of the Apennine thrust belt, Italy. *Journal of Sedimentary Research*, **72**, 2–17.

GAZZI, P. 1966. *I minerali pesanti nei flysch arenacei fra Monte Ramaceto e Monte Molinatico (Appennino settentrionale)*. Istituto di Mineralogia e Petrografia dell'Universita di Bologna, Bologna.

GLANDENKOV, Y. B., BAZHENOVA, O. K., GRECHIN, V. I., MARGULIS, L. S. & SALNIKOV, B. A. 2002. *Cenozoic geology and oil and gas of Sakhalin (Cainozoye Sakhalin e yego neftegazonosnost)*. Geological Institute of the Russian Academy of Sciences, Moscow.

HAWKESWORTH, C. J. & KEMP, A. I. S. 2006. Evolution of the continental crust. *Nature*, **443**, 811–817.

HINDLE, D., FUJITA, K. & MACKEY, K. 2006. Current deformation rates and extrusion of the northwestern Okhotsk plate, northeast Russia. *Geophysical Research Letters*, **33**.

HOVIUS, N. 1998. Controls on sediment supply by large rivers. *In*: SHANLEY, K. W. & McCABE, P. J. (eds) *Relative Role of Eustasy, Climate and Tectonism in Continental Rocks*. SEPM (Society for Sedimentary Geology) Special Publications, Tulsa, **59**, 3–16.

HUBERT, J. F. 1962. A zircon-tourmaline-rutile maturity index and the interdependence of the composition of heavy mineral assemblages with the gross composition and texture of sandstones. *Journal of Sedimentary Petrology*, **32**, 450–440.

HUH, Y., PANTELEYEV, G., BABICH, O., ZAITSEV, A. & EDMOND, J. M. 1998. The fluvial geochemistry of the rivers of Eastern Siberia: II. Tributaries of the Lena, Omoloy, Yana, Indigirka, Kolyma, and Anadyr draining the collisional/accretionary zone of the Verkhoyansk and Cherskiy ranges. *Geochimica et Cosmochimica Acta*, **62**, 5063–5075.

KOMAR, P. D. 2007. The entrainment, transport and sorting of heavy minerals by waves and currents. *In*: MANGE, M. A. & WRIGHT, D. T. (eds) *Heavy Minerals in Use.* Elsevier, Amsterdam, 3–48.

KOVALENKO, V. I., YARMOLYUK, V. K. *ET AL.* 2003. Sources of igneous rocks and genesis of the Early Mesozoic tectonomagnetic area of the Mongolia–Transbaikalia magmatic region:1. *Geology and Isotope Geochronology, Petrology*, **11**, 147–160.

LINDQUIST, S. J. 2000. *The North Sakhalin Neogene Total Petroleum System of Eastern Russia.* US Geological Survey Open File Report **99-50-O**.

LUDWIG, K. 2003. *Isoplot 3.0: A Geochronological toolkit for Microsoft Excel.* Berkeley Geochronological Center Special Publication No. 4.

MACDONALD, D. I. M. & FLECKER, R. 2007. Injected sand sills in a strike–slip fault zone: a case study from the Pil'sk Suite (Miocene), SE Schmidt Peninsula, Sakhalin. *In*: HURST, A. & CARTWRIGHT, J. (eds) *Sand Injectites: Implications for Hydrocarbon Exploration and Production.* American Association of Petroleum Geologists, Memoirs, **87**, 253–263.

MAKHINOV, A. N. 2004. Natural and anthropogenic factors of the Amur River runoff formation. *Far Eastern Branch of Russian Academy of Sciences Bulletin*, **2**, 41–46.

MANGE, M. A. & MAURER, H. F. W. 1992. *Heavy Minerals in Colour*, 1st edn. Chapman & Hall, London.

MANGE, M. A. & MORTON, A. C. 2007. Geochemistry of heavy minerals. *In*: MANGE, M. A. & WRIGHT, D. T. (eds) *Heavy Minerals in Use.* Elsevier, Amsterdam, 345–391.

MANGE-RAJETSKY, M. A. 1983. Sediment dispersal from source to shelf on an active continental margin, S. Turkey. *Marine Geology*, **52**, 1–26.

MARKEVICH, P. V., MALINOVSKY, A. I., TUCHKOVA, M. I., SOKOLOV, S. D. & GRIGORYEV, V. N. 2007. The use of heavy minerals in determining the provenance and tectonic evolution of Mesozoic and Caenozoic sedimentary basins in the continent–Pacific Ocean transition zone: examples from Sikhote-Alin and Koryak-Kamchatka regions (Russian Far East) and Western Pacific. *In*: MANGE, M. A. & WRIGHT, D. T. (eds) *Heavy Minerals in Use.* Elsevier, Amsterdam, 789–822.

MILLIKEN, K. L. 2007. Provenance and diagenesis of heavy minerals, Cenozoic Units of the Northwestern Gulf of Mexico sedimentary basin. *In*: MANGE, M. A. & WRIGHT, D. T. (eds) *Heavy Minerals in Use.* Elsevier, Amsterdam, 247–261.

MITCHUM, R. M. & VAIL, P. R. 1977. Seismic stratigraphy and global changes of sea level; Part 6, Stratigraphic interpretation of seismic reflection patterns in depositional sequences. *In*: PAYTON, C. E. (ed.) *Seismic Stratigraphy – Applications to Hydrocarbon Exploration.* American Association of Petroleum Geologists, Memoirs, **26**, 117–133.

MORTON, A. C. 1986. Dissolution of apatite in North Sea Jurassic sandstones: implications for the generation of secondary porosity. *Clay Minerals*, **21**, 711–733.

MORTON, A. C. 1987. Detrital garnets as provenance and correlation indicators in North Sea reservoir sandstones. *In*: BROOKS, J. & GLENNIE, K. W. (eds) *Petroleum Geology of North West Europe: Proceedings of the 3rd Conference on Petroleum Geology of North West Europe*, Graham & Trotman, London, 991–995.

MORTON, A. C. 2007. The role of heavy mineral analysis as a geosteering tool during drilling of high-angle wells. *In*: MANGE, M. A. & WRIGHT, D. T. (eds) *Heavy Minerals in Use.* Elsevier, Amsterdam, 1123–1142.

MORTON, A. C. & BERGE, C. 1995. Heavy mineral suites in the Statfjord and Nansen Formations of the Brent Field, North Sea: a new tool for reservoir subdivision and correlation. *Petroleum Geoscience*, **1**, 355–364.

MORTON, A. C. & HALLSWORTH, C. 1994. Identifying provenance-specific featues of detrital heavy mineral assemblages in sandstones. *Sedimentary Geology*, **90**, 241–256.

MORTON, A. C. & HALLSWORTH, C. 2007. Stability of detrital heavy minerals during burial diagenesis. *In*: MANGE, M. A. & WRIGHT, D. T. (eds) *Heavy Minerals in Use.* Elsevier, Amsterdam, 215–245.

MORTON, A. C. & JOHNSSON, M. J. 1993. Factors influencing the composition of detrital heavy mineral suites in Holocene sands of the Apure River drainage basin, Venezuela. *In*: JOHNSSON, M. J. & BASU, A. (eds) *Processes Controlling the Composition of Clastic Sediments.* Geological Society of America Special Papers, **284**, 171–185.

MORTON, A. C. & MILNE, A. 2012. Heavy mineral stratigraphic analysis on the Clair Field, UK, west of Shetlands: a unique real-time solution for red-bed correlation while drilling. *Petroleum Geoscience*, **18**, 115–128.

MORTON, A. C. & SMALE, D. 1990. The effects of transport and weathering on heavy minerals from the Cascade River, New Zealand. *Sedimentary Geology*, **68**, 117–123.

MORTON, A. C., CLAOUÉ-LONG, J. C. & BERGE, C. 1996. SHRIMP constraints on sediment provenance and transport history in the Mesozoic Statfjord Formation, North Sea. *Journal of the Geological Society, London*, **153**, 915–929.

MORTON, A., ALLEN, M. *ET AL.* 2003. Provenance patterns in a neotectonic basin: Pliocene and Quaternary sediment supply to the South Caspian. *Basin Research*, **15**, 321–337.

MORTON, A., HALLSWORTH, C. & CHALTON, B. 2004. Garnet compositions in Scottish and Norwegian basement terrains: a framework for interpretation of North Sea sandstone provenance. *Marine and Petroleum Geology*, **21**, 393–410.

MORTON, A. C., HERRIES, R. & FANNING, M. 2007. Correlation of Triassic Sandstones in the Strathmore Field, West of Shetland, using heavy mineral provenance signatures. *In*: MANGE, M. A. & WRIGHT, D. T. (eds) *Heavy Minerals in Use.* Elsevier, Amsterdam, 1037–1072.

NECHAEV, V. P. & ISPHORDING, W. C. 1993. Heavy mineral assemblages of continental margins as indicators of plate tectonic environments. *Journal of Sedimentary Petrology*, **63**, 1110–1117.

NECHAEV, V. P., MARKEVICH, P. V., MALINOVSKIY, A. I., FILLIPOV, A. N. & VYSOTESKIY, S. V. 1997. Heavy-mineral assemblages in the Cretaceous sediments of the lower Amur Region, Russia's Far East: implications for the geodynamic environments of deposition. *Geology of the Pacific Ocean*, **13**, 471–486.

NICHOLSON, U. 2009. *Landscape evolution and sediment routing across a strike–slip plate boundary*. PhD thesis. University of Aberdeen.

PEARCE, N. J. G., PERKINS, W. T., WESTGATE, J. A., GORTON, M. P., JACKSON, S. E., NEAL, C. R. & CHERERY, S. P. 1997. A compilation of new and published major and trace element data for NIST SRM 610 and NIST SRM 612 glass reference materials. *Geostandards Newsletter*, **21**, 115–144.

POSAMENTIER, H. W. & ALLEN, G. P. 1993. Variability of the sequence stratigraphic model: effects of local basin factors. *Sedimentary Geology*, **86**, 91–109.

POYNTER, S. 2003. *The Neogene evolution of the Amur River and its delta in the Russian Far-East*. PhD thesis. Darwin College, University of Cambridge.

REN, J., TAMAKI, K., LI, S. & JUNXIA, Z. 2002. Late Mesozoic and Cenozoic rifting and its dynamic setting in Eastern China and adjacent areas. *Tectonophysics*, **344**, 175–205.

ROJAS-AGRAMONTE, Y., KRONER, A., DEMOUX, A., XIA, X., DONSKAYA, T., LIU, D. & SUN, M. 2011. Detrital and xenocrystic zircon ages from Neoproterozoic to Palaeozoic arc terranes of Mongolia: significance for the origin of crustal fragments in the Central Asian Orogenic Belt. *Gondwana Research*, **19**, 751–763.

ROZHDESTVENSKIY, V. S. 1982. The role of wrench faults in the structure of Sakhalin. *Geotectonics*, **16**, 323–332.

RUHL, K. W. & HODGES, K. V. 2005. The use of detrital mineral cooling ages to evaluate steady state assumptions in active orogens: an example from the central Nepalese Himalaya. *Tectonics*, **24**, TC4015, http://dx.doi.org/10.1029/2004TC001712

RYTSK, E. Y., KOVACH, V. P., YARMOLYUK, V. V., KOVALENKO, V. I., BOGOMOLOV, E. S. & KOTOV, A. B. 2011. Isotopic structure and evolution of the continental crust in the East Transbaikalian segment of the Central Asian Foldbelt. *Geotectonics*, **45**, 349–377.

SAFANOVA, I., MARUYAMA, S., HIRATA, T., KON, Y. & RINO, S. 2010. LA ICP MS U–Pb ages of detrital zircons from Russia largest rivers: implications for major granitoid events in Eurasia and global episodes of supercontinent formation. *Journal of Geodynamics*, **50**, 134–153.

SENGOR, A. M. C. & NATAL'IN, B. A. 1996. Palaeotectonics of Asia: fragments of a synthesis. *In*: YIN, M. (ed.) *The Tectonic Evolution of Asia*. Cambridge University Press, Cambridge, 486–640.

SLAMA, J., KOSLER, J. *ET AL.* 2008. Plezovice zircon; a new natural reference material for U–Pb and Hf isotopic microanalysis. *Chemical Geology*, **249**, 1–35.

SOROKIN, A. P. & ARTYOMENKO, T. V. 2003. Structural evolution of the Eastern Margin of Eurasia in Late Mesozoic and Cenozoic. *Journal of Geoscientific Research in Northeast Asia*, **6**, 150–160.

THORNBURG, T. M. & KULM, L. D. 1987. Sedimentation in the Chile Trench: depositional morphologies, lithofacies, and stratigraphy. *Geological Society of America Bulletin*, **98**, 33–52.

TOMURTOGOO, O., WINDLEY, B. F., KRONER, A., BADARCH, G. & LIU, D. Y. 2005. Zircon age and occurrence of the Adaatsag ophiolite and Muron shear zone, central Mongolia: constraints on the evolution of the Mongol–Okhotsk Ocean, suture and orogen. *Journal of the Geological Society, London*, **162**, 125–134.

TURNER, G. & MORTON, A. C. 2007. The effects of burial diagenesis on detrital heavy mineral grain surface textures. *In*: MANGE, M. A. & WRIGHT, D. T. (eds) *Heavy Minerals in Use*. Elsevier, Amsterdam, 393–412.

VARNAVSKIY, V. G. 1971. *Paleogene and Neogene sediments of the Sredne–Amursk Basin*. Nauka, Moscow.

VELBEL, M. A. 1984. Natural weathering mechanisms of almandine garnet. *Geology*, **12**, 631–634.

VELBEL, M. A. 2007. Surface textures and dissolution processes of heavy minerals in the sedimentary cycle: examples from pyroxene and amphiboles. *In*: MANGE, M. A. & WRIGHT, D. T. (eds) *Heavy Minerals in Use*. Elsevier, Amsterdam, 113–150.

VELICHKO, A. & SPASSKAYA, I. 2002. Climatic change and the development of landscapes. *In*: SHAHGEDANOVA, M. (ed.) *The Physical Geography of Northern Eurasia*. Oxford University Press, Oxford, 36–69.

VERESHCHAGIN, V. H., KOVTUNOVICH, U. M. & MIKHAILOV, K. M. 1969. *Geolgecheskaya karta Sakhalina* [Geological map of Sakhalin]. Ministry of Geology of the USSR, Moscow.

VERMEESCH, P. 2004. How many grains are needed for a provenance study? *Earth and Planetary Science Letters*, **224**, 441–451.

WEST, A. J., GALY, A. & BICKLE, M. 2005. Tectonic and climatic controls on silicate weathering. *Earth and Planetary Science Letters*, **235**, 211–228.

WORRALL, D. M., KUNST, F., KRUGLYAK, V. & KUZNETSOV, V. 1996. Tertiary tectonics of the Sea of Okhotsk, Russia: far-field effects of the India–Eurasia collision. *Tectonics*, **15**, 813–826.

YARMOLYUK, V. V., KOVALENKO, V. I., SAL'NIKOVA, E. B., BUDNIKOV, C. V. & KOVACH, V. P. 2002. Tectono-magmatic zoning, magma sources and geodynamics of the Early Mesozoic Mongolia–Transbaikal Province. *Geotectonics*, **4**, 293–311.

YURKOVA, R. M. 1970. Comparison of post-sedimentary alteration of oil, gas and waterbearing rocks. *Sedimentology*, **15**, 53–68.

ZACHOS, J., PAGANI, H., SLOAN, L., THOMAS, E. & BILLUPS, K. 2001. Trends, rhythms, and aberrations in global climate 65 Ma to present. *Science*, **292**, 686–693.

# Provenance study on Eocene–Miocene sandstones of the Rakhine Coastal Belt, Indo-Burman Ranges of Myanmar: geodynamic implications


T. T. NAING[1,4], D. A. BUSSIEN[1], W. H. WINKLER[1]*, M. NOLD[2] & A. VON QUADT[3]

[1]*Department of Earth Sciences, Geological Institute, ETH-Zentrum, Sonneggstrasse 5, ETH Zurich, 8092 Zurich, Switzerland*

[2]*Dr Mario Nold Think Tank, Project Management & Consulting, World Trade Center, Leutschenbachstrasse 98, 8050 Zurich, Switzerland*

[3]*Department of Earth Sciences, Institute of Geochemistry and Petrology, ETH-Zentrum, Clausiusstrasse 25, ETH Zurich, 8092 Zurich, Switzerland*

[4]*Present address: No.1050, Ye'baw 3rd Street, Phawkan Quarter, Insein Township, Yangon, Myanmar*

**Corresponding author (e-mail: wilfried.winkler@erdw.ethz.ch)*



**Abstract:** The Indo-Burman Ranges (IBR) represent an accretionary wedge, which is the result of subduction of the Indian plate beneath the Asian plate. In the Rakhine Coastal Belt it comprises a thick stack of Cretaceous to Neogene turbiditic sediments and localized thrust sheets of oceanic plate mafics and pelagic sediments. We investigate Eocene–Miocene sandstones, aiming to reveal the provenance of the detrital material using modal framework grain, heavy mineral and detrital zircon analysis (U–Pb laser ablation ICP-MS dating, Hf isotope geochemistry and typology). The results show a predominant derivation of the clastic material from: (i) Late Cretaceous to Oligocene igneous rocks, which are often bimodal with a low number of zircons spanning the Cretaceous–Palaeogene boundary, and (ii) recycled orogenic terrane sources comprising ophiolitic rocks. Age corrected Hf isotope ratios confirm subduction-related mixed mantle-crust sources. We also observe minor reworking of older magmatic zircons. By comparing our obtained petrographic parameters and zircon characteristics with potential Himalayan, Indian continent and Burman margin sources we conclude a Burman margin and arc origin provenance. With regard to hydrocarbon exploration in the IBR, a forearc and trench basin system model linked with the Burman arc appears more appropriate for evaluating the petroleum system.

**Supplementary material:** sandstone modal composition, heavy mineral contents, detrital zircon U–Pb LA-ICPMS dating and hafnium results are available at http://www.geolsoc.org.uk/SUP18651


Myanmar (Burma) is located on the eastern edge of the zone of Himalayan convergence, south of the Eastern Himalayan Syntaxis where the west–east-striking Himalayan structures rapidly change to a north–south orientation (e.g. Brunnschweiler 1974; Bender 1983; Barley *et al.* 2003; Searle *et al.* 2007; Allen *et al.* 2008) (Fig. 1a). Myanmar is the region of transition between the main Himalayan collision belt and the Andaman arc, where, since the Cretaceous, the Indian plate has been subducting under Asia (e.g. Curray *et al.* 1979; Mitchell 1993; Acharyya 1998). The main territory of Myanmar is situated on the Burma micro-plate between the Asian plate to the east and the Indian plate to the west (Fig. 1b) (Fitch 1972; Curray *et al.* 1979; Pivnik *et al.* 1998; Bertrand & Rangin 2003). Due

to oblique, NE-directed subduction of the Indian plate under the Asian continent, a northward displacement of about 300–500 km of the Burma micro-plate has occurred since the Late Oligocene (e.g. Curray *et al.* 1979; Mitchell 1993; Pivnik *et al.* 1998; Curray 2005). This subduction is also driving the spreading in the Andaman Sea to the south (Stephenson & Marshall 1984). Since the Late Cretaceous the Burma micro-plate is assumed to have moved northward, relative to SE Asia, about 1000 km (Mitchell 1993). Chakraborty & Khan (2009) call it a sliver plate forming through right-lateral displacement between the subduction zone and the Sagaing fault, and its southward continuation in the Andaman Sea and offshore Sumatra. The Burma micro-plate constitutes part of the larger


*From*: SCOTT, R. A., SMYTH, H. R., MORTON, A. C. & RICHARDSON, N. (eds) 2014. *Sediment Provenance Studies in Hydrocarbon Exploration and Production.* Geological Society, London, Special Publications, **386**, 195–216. First published online July 22, 2013, http://dx.doi.org/10.1144/SP386.10

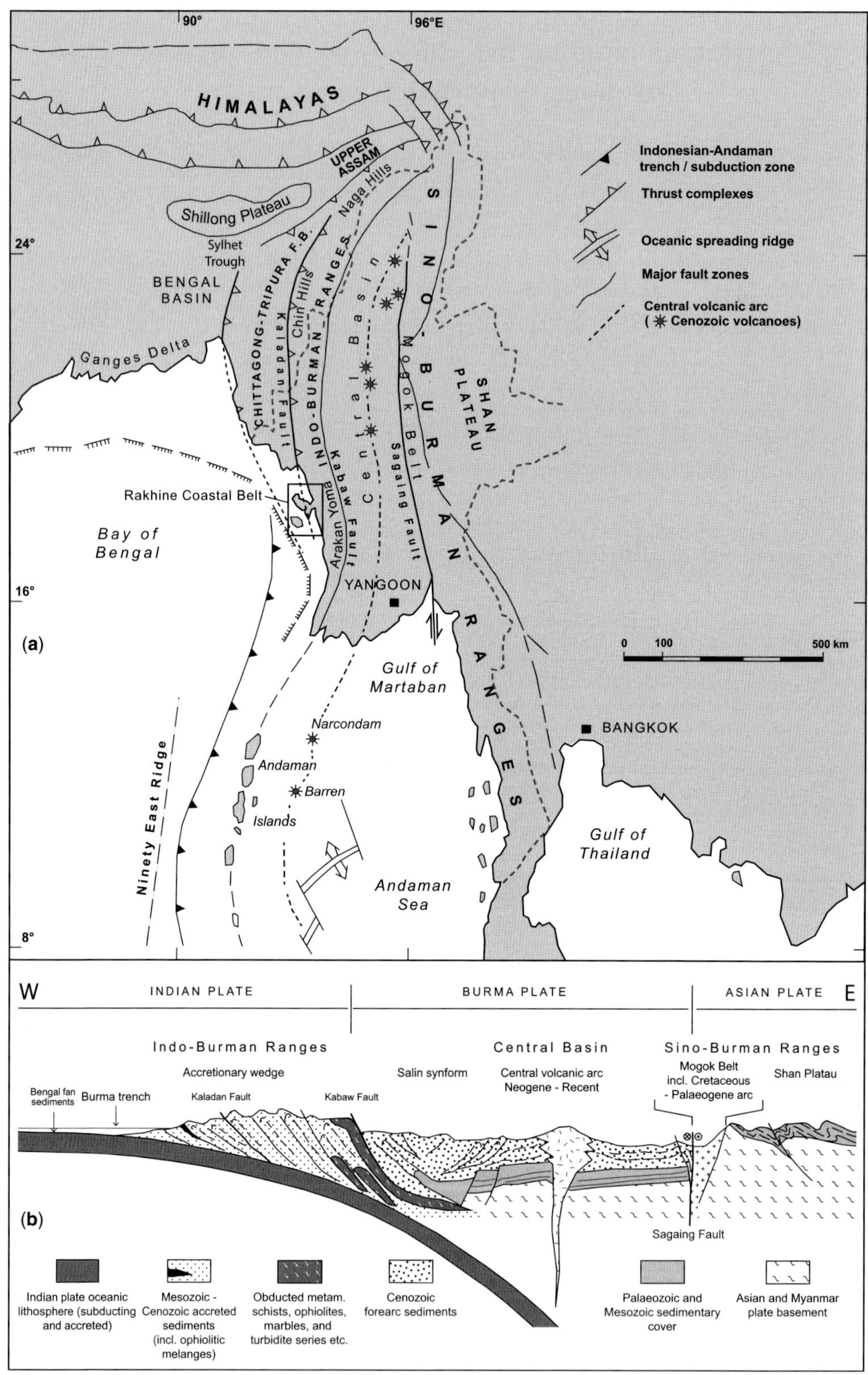
90°
96°E
HIMALAYAS
UPPER ASSAM
Shillong Plateau
Naga Hills
24°
Sylhet Trough
BENGAL BASIN
SINO-BURMAN RANGES
Ganges Delta
CHITTAGONG-TRIPURA F.B.
Kaladan Fault
Chin Hills
INDO-BURMAN RANGES
Central Basin
Mogok Belt
SHAN PLATEAU
Rakhine Coastal Belt
Arakan Yoma
Kabaw Fault
Sagaing Fault
Bay of Bengal
16°
YANGOON
(a)
Gulf of Martaban
Narcondam
Andaman
Barren
Islands
Ninety East Ridge
Andaman Sea
BANGKOK
Gulf of Thailand
8°
Indonesian-Andaman trench / subduction zone
Thrust complexes
Oceanic spreading ridge
Major fault zones
Central volcanic arc ( ✳ Cenozoic volcanoes)
0   100          500 km

W   INDIAN PLATE          BURMA PLATE          ASIAN PLATE   E
Indo-Burman Ranges          Central Basin          Sino-Burman Ranges
Accretionary wedge
Salin synform
Central volcanic arc Neogene - Recent
Mogok Belt incl. Cretaceous - Palaeogene arc
Shan Platau
Bengal fan sediments
Burma trench
Kaladan Fault
Kabaw Fault
(b)
Sagaing Fault
Indian plate oceanic lithosphere (subducting and accreted)
Mesozoic - Cenozoic accreted sediments (incl. ophiolitic melanges)
Obducted metam. schists, ophiolites, marbles, and turbidite series etc.
Cenozoic forearc sediments
Palaeozoic and Mesozoic sedimentary cover
Asian and Myanmar plate basement

Indo-Burma–Andaman micro-continent, which rifted and drifted apart from Australia in the Late Jurassic and Early Cretaceous (Acharyya & Lahiri 1991; Acharyya 1998). It was brought into collision with the Asian continent in the Late Oligocene, and the northern tip of the micro-plate collided with India in the Mio-Pliocene (Acharyya 1998).

The Indo-Burman Ranges (IBR) occupy western Myanmar, and geographically are subdivided into the Naga Hills, the Chin Hills and the Arakan Yoma Range (Fig. 1a). They formed along the western margin of the Burma micro-plate buttress (Fig. 1). The IBR represent an active, west-verging accretionary complex encroaching on the Bengal basin from the east due to subduction of the Indian plate beneath the Burman micro-plate (e.g. Bender 1983; Sengupta et al. 1990; Mitchell 1993). Lithologically, the IBR consist of a complex tectonic collage of obducted metamorphic schists, marbles, ophiolites, Mesozoic and Cenozoic oceanic sediment, basin pelagic as well as turbiditic series (e.g. Brunnschweiler 1974; Bender 1983; Mitchell 1985; Pivnik et al. 1998).

The origin of the Cenozoic syn-accretionary turbiditic material of the IBR is a matter of debate. A current interpretation is that the west-verging range consists of off-scraped material of a proto-Bengal fan, which has been continuously incorporated into the IBR (Curray et al. 1979; Bender 1983; Hutchinson 1989; Curray 2005) since the Eocene. Hence, the clastic material could have been supplied from early collision events of the Indian plate with the Eurasian plate, dating back nearly 60 Ma sourced from the Himalaya (e.g. Curray et al. 1979; Alam et al. 2003). In Burma, evidence for that Himalayan-sourced sediment is poor and based only on preliminary field observations given by Brunnschweiler (1974) and Bender (1983). Allen et al. (2008) also found a source for the Palaeogene turbiditic sandstones in the Myanmar orogenic belt and arcs.

After recent large gas discoveries (e.g. Shwe discovery), western Myanmar and the Rakhine coastal basin have attracted renewed interest in hydrocarbon exploration. It is now one of the last unexplored deep-water basin systems in the world. In earlier days, when exploration focused close to the onshore, the lack of reservoirs was a main cause of failure, and the long distance to the Bengal Fan sandstones was used to explain this unfavourable situation. The results of the present study open new opportunities for exploration with reservoirs which did not form from Himalayan clastic material, but a close link of the petroleum system with the Burman arc and forearc, including the Myanmar Central Basin, appears promising.

The present work aims to increase our knowledge of the Cenozoic geological history of Myanmar by investigating the provenance of Eocene–Miocene sandstones tectonically included in the accretionary wedge in the Rakhine coastal area. The recognition of the detrital source rocks of the sandstones is of great importance for tracing the geodynamic evolution of the eastern Himalayan area in general. We investigate the modal framework grain composition of the sandstones and their heavy mineral associations to characterize the lithologies present in the source areas. Detrital zircon ICP-MS laser ablation U–Pb dating is used to evaluate the crystallization age ranges in the source regions and/or recycling of older crustal elements. Hf isotope ratios of the dated detrital zircons are applied to infer the origin of the syn-sedimentary magmas. Additional zircon typology arguments are used to constrain magma types present in the source regions. Finally, we are able to discuss the sourcing of the detrital material from Burman arc and continental plate rocks. Himalayan-sourced rocks, which one would expect to find in pre-Bengal fan deposits, were not identified.

## Geological framework of Myanmar

Tectonically as well as geomorphologically, Myanmar can be subdivided into four major, north–south-orientated provinces (Fig. 1). These are, from east to west: (1) the Eastern Highland or Shan Plateau and Scarp, (2) the Central Lowlands or Myanmar Central Basin, (3) the Indo-Burman Ranges, and (4) the Rakhine Coastal Belt.

The Eastern Highland is a part of the Shan–Thai Block, a large continental domain of the Asian plate that connects to the Pacific plate in the east. It represents the southern continuation of the Quiangtang and Lhasa blocks of Central Tibet (Searle et al. 2007). In central and southern Myanmar the Shan Scarps and Mogok Metamorphic Belt (MMB) occupy the western margin of the highlands. It is a high grade metamorphic belt of Precambrian to Palaeozoic rocks (marbles, schists and gneisses), which show evidence of Cretaceous to Miocene magmatism and metamorphism (e.g. Mitchell 1993; Barley et al. 2003; Searle et al. 2007). Age and petrological features suggest that the MMB represents the remainder of an Andean-type subduction



**Fig. 1.** (**a**) Structural map of Myanmar and adjacent regions adapted and modified from Bender (1983), Pivnik et al. (1998) and Alam et al. (2003). (**b**) Schematic tectonic section (not to scale) across Myanmar at about latitude 20°N adapted and modified from Pivnik et al. (1998), including results of Barley et al. (2003) and Bertrand & Rangin (2003).

margin, which developed prior and during collision of India with the Eurasian continent (Barley *et al.* 2003). In the Shan Scarps and Plateau the unconformable contact between Proterozoic sandstone and shale formations, with overlying Cambrian to Early Cretaceous series, is observed (Mitchell 1993).

The Myanmar Central Basin (MCB) is located between the major dextral Sagaing Fault in the east and the Kabaw Fault and IBR in the west (Fig. 1). It hosts thick series of mixed marine and fluvial sediments of Late Cretaceous through Plio-Pleistocene age. They unconformably overlie underplated oceanic rocks in the west (belonging to the IBR) and metamorphic/crystalline continental basement and cover of the Burma plate (Fig. 1b). This lowland hosts a series of *en echelon* depocentres with Cenozoic basins presently affected by active tectonic inversion (Pivnik *et al.* 1998). According to Bertrand & Rangin (2003), basin inversion started in the Middle to Late Miocene. The central volcanic line is interpreted as the modern volcanic arc separating the thick forearc basin in the west from the relatively thin back-arc basin in the east. The back-arc basin is underlain by Mid- to Late Cenozoic non-marine sediments, and the forearc basin is filled by thick latest Cretaceous to Quaternary marine and non-marine sediments (Mitchell 1993; Pivnik *et al.* 1998).

The IBR is a prominent geotectonic element extending from the Naga Hills in the north to the Arakan Yoma Range in the south. From the Cape of Negrais southwards the IBR is submerged in the Bay of Bengal. The IBR of western Myanmar is considered an active accretionary wedge linked to the eastward subduction of the Bengal basin oceanic lithosphere beneath the Burmese platelet (e.g. Sengupta *et al.* 1990; Mitchell 1993; Searle *et al.* 2007). It terminates northward at the Eastern Himalayan Suture and extends southward to Sumatra. The IBR is composed mainly of Late Cretaceous–Neogene turbiditic and hemipelagic series. Along its eastern margin, the IBR includes a prominent range composed of Upper Triassic turbiditic ('flysch-like') sandstones and marbles in tectonic contact with a belt of mica schists (Mt Victoria). The tectonically associated ophiolitic rocks are thought to be the southern continuation of the Indus Tsangpo suture zone (Mitchell 1993; Searle *et al.* 2007). In the IBR major tectonic and subduction events have been identified from late Early Cretaceous to Middle Miocene and Quaternary periods (Mitchell 1993; Allen *et al.* 2008). The western edge of the IBR is represented by the Kaladan thrust fault along the Naga Hills in the north and the Arakan Yoma Hills in the south (Fig. 1a). All or most of the IBR sediments are widely interpreted either as accreted sediments of

the early Bengal Fan, derived from the India–Asia collision belt (Mitchell 1974; Curray *et al.* 1979; Bender 1983; Hutchinson 1989; Curray 2005), or as sediments shed from the Indian margin (Sengupta *et al.* 1990), or from the Burman active volcanic arc (Allen *et al.* 2008).

The Rakhine Coastal Belt (RCB), including several offshore islands, is geologically part of the IBR because of a similar origin, but with an increased presence of Miocene sediments west of the Kaladan Fault (Fig. 1). The RCB developed between the subduction zone in the west and the main Arakan Yoma ranges in the east (IBR). In the southern part of the area, structural deformation is greater than further north. The Miocene sequences in this area are frequently tilted and faulted, locally folded and overthrust. The pre-Miocene turbiditic formations occurring below the transgressive Miocene beds show the intense folded structural style of the IBR (Bender 1983). Further north, the Miocene series exhibit relatively minor tectonic deformation. In Bangladesh, the RCB has its continuation in the Chittagong–Tripura Fold Belt.

## Samples and methods

We undertook provenance analysis on Eocene to Miocene sand and sandstone samples from the onshore section of the Rakhine Coastal area. GPS and geographical locations, chronostratigraphic age correlations, lithologies of the samples and methods applied are given in Table 1.

Modal framework grain analysis of the sandstones was performed on thin sections stained for feldspars and carbonates (Dickson 1966; Norman 1974). Monomineralic grains (quartz and feldspars) and lithic rock fragments were distinguished according to the standard method of Dickinson & Suczek (1979) and Dickinson (1985). At least 300 grains were point-counted in the feldspar-stained thin sections. The percentages of various combinations of grains are plotted in standard triangular diagrams used to classify the sandstones (Folk 1974) and to interpret the geodynamic setting of the source terranes (Dickinson 1985).

To obtain the transparent heavy mineral fractions, the sandstones were crushed by SelFrag laboratory batch equipment using high voltage pulse power technology to liberate morphologically intact minerals. In the <2 mm sieved fraction the carbonate was dissolved in warm (60–70 °C) 10% acetic acid. The heavy minerals were extracted in separation funnels (Mange & Maurer 1992) from the 0.063–0.4 mm sieve fraction using bromoform (density 2.88). The bulk heavy mineral fractions were mounted in piperine (Martens 1932) between glass slab and cover. Mineral identification and

**Table 1.** *Geographical location and characteristics of the analysed samples in the Rakhine Coastal Belt*

| Sample no. | Latitude (deg: min: sec) | Longitude (deg: min: sec) | Location | Formation | Stratigraphical age | Lithology | Analysis |
|---|---|---|---|---|---|---|---|
| 10TTN01 | 18°27'11" | 94°17'44" | Near Thandwe airport | Ngapali | M. Eocene | Turbiditic shale, dark-grey, moderately hard | HM |
| 10TTN02 | 18°27'58" | 94°17'_7" | Sabarkyi village | Ngapali | M. Eocene | Turbiditic sandstone, brownish-grey, associated with carbonaceous shales | HM, MFA |
| 10TTN03 | 18°29'_8" | 94°16'_8" | Gaw village | Ngapali | M. Eocene | Turbiditic sandstone, dark-grey, medium- grained | DZA, HM |
| 10TTN04 | 18°22'11" | 94°23'01" | Andrew Bay (Abe hill) | Ngapali | M. Eocene | Turbiditic sandstone, dark-grey, medium-grained | DZA, HM, ZT |
| 10TTN05 | 18°22'07" | 94°23'02" | Andrew Bay | Ngapali | M. Eocene | Turbiditic sandstone, dark-grey, medium-grained | HM, MFA |
| 10TTN06 | 18°18'31" | 94°19'38" | Aungkyawmaw Chaing | Ngapali | M. Eocene | Very coarse turbiditic sandstone, brownish-grey to grey | DZA, HM, ZT |
| 10TTN09 | 17°59'37" | 94°29'44" | Thandwe-Gwa Rd. (MP 51/4) | Yenandaung | M. Miocene | Turbiditic sandstone, light-grey | DZA, ZT, HM, MFA |
| 10TTN10 | 13°08'20" | 94°26'37" | Htudu Island | Yechangyi | Oligocene | Turbiditic sandstone, reddish-brown, medium-grained | DZA, ZT, HM |
| 10TTN12 | 13°00'_3" | 94°26'16" | Kyeintali North | Yechangyi | Oligocene | Turbiditic sandstone, light-grey, medium-grained | HM, MFA |
| 10TTN13 | 13°00'11" | 94°26'14" | Kyeintali North | Yechangyi | Oligocene | Turbiditic sandstone, yellowish-brown, medium-grained | DZA, ZT, HM |
| 10TTN14 | 13°00'11" | 94°26'14" | Kyeintali North | Yechangyi | Oligocene | Turbiditic sandstone, yellowish-brown, medium-grained | HM, MFA |
| 10TTN16 | 19°52'55" | 93°12'05" | East Bayonga Island | Yechangyi | Oligocene | Turbiditic sandstone, medium-grained | DZA, ZT, HM, MFA |
| 10TTN18 | 19°20'27" | 93°29'34" | Kalaba to Gortu | Leikkammaw | L. Miocene | Sandstone, soft | DZA, ZT, HM, MFA |
| 10TTN20 | 18°54'01" | 93°57'16" | Kyauknimaw | Yechangyi | Oligocene | Turbiditic sandstone, fine- to medium-grained | DZA, ZT, HM, MFA |

Analysis key: DZA, detrital zircon age, ZT, zircon typology, HM, heavy minerals, MFA, modal framework grain analysis.

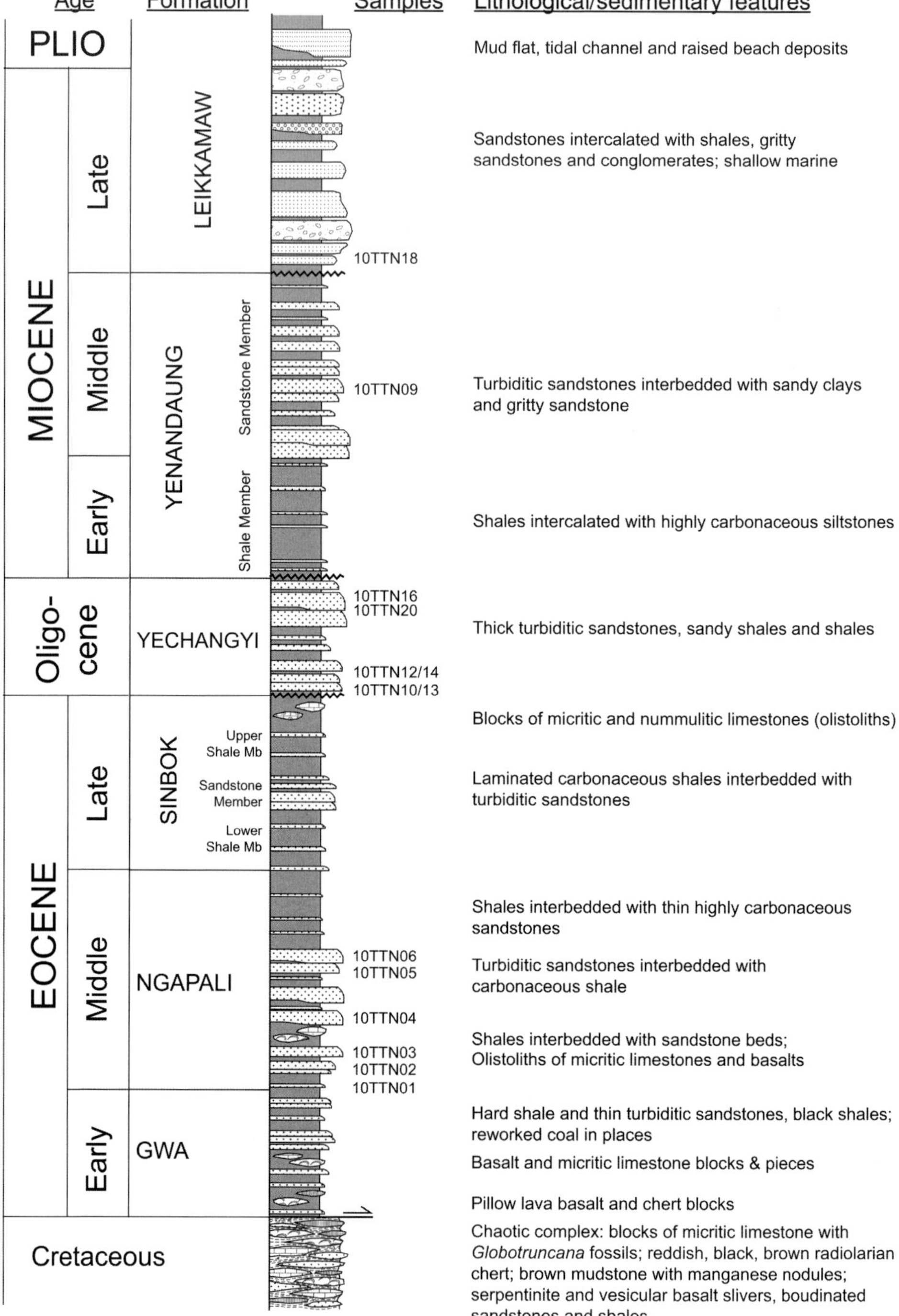



**Fig. 2.** Idealized stratigraphic section (not to scale) of the Rakhine coastal area with comments on lithological and sedimentary features (see also Htut *et al.* 2010). The sedimentary series are overthrusting an ophiolitic mélange, including oceanic plate pelagic sediments. In the turbiditic formations a general decrease in tectonic deformation is observed upsection, including several unconformities. The Late Miocene shallow-marine and coastal Leikkammaw Formation unconformably overlies the stack of turbiditic and hemipelagic sediments.

quantification was carried out under the petrographic microscope (e.g. Mange & Maurer 1992) using the mid-point ribbon and fleet counting methods, and a minimum of 200 grains was counted per sample.

Detrital zircons were extracted from *c.* 2 kg of sandstone using standard mineral separation techniques, including SelFrag high voltage fragmentation, dissolution of carbonate cement in cold hydrochloric acid and heavy liquid (methylene iodide, density 3.32) separation. Handpicked zircons were mounted in epoxy blocks and polished down to expose their core. Cathodoluminescence (CL) imaging was carried out on all analysed zircons in order to determine the internal structures of the grains prior to isotopic analysis. For typological analysis (Pupin 1980), only euhedral zircon crystals were investigated. Zircon classification was based on the zircon typology method of Pupin (1980). Each crystal was described in terms of relative development of the crystal surfaces. The distributions of morphological types are shown on the typology diagram for each sample.

In laser ablation ICP-MS U–Pb dating of detrital zircons, only magmatic domains were investigated, and we did not date inherited cores. Laser ablation ICP-MS analyses were performed on an Elan 6100 DRC instrument coupled to an in-house-built 193 nm Excimer laser at the Institute of Geochemistry and Petrology, ETH Zurich. Helium gas (1.1 l/min) was used as the carrier gas in the ablation cell. The laser was run at pulse rate of 10 Hz with energy of 0.5 mJ/pulse and a spot size of 40 μm. The accuracy and reproducibility within each run of analysis were monitored by periodic measurements of the GJ-1 external standard, with $^{207}$Pb/$^{206}$Pb age of 608.5 $\pm$ 0.4 Ma (Jackson *et al.* 2004). Data reduction was performed using the GLITTER software (Van Achterbergh *et al.* 2001) to calculate the relevant isotopic ratios, ages and errors. Concordia and frequency probability diagrams were performed using ISOPLOT v.3.0 (Ludwig 2003).

The *in-situ* Hf isotopic analysis on dated zircons was carried out using the Nu plasma MC-ICP-MS (Nu instrument Ltd) attached to a 193 nm UV ArF excimer laser, at the Institute of Geochemistry and Petrology, ETH Zurich. Laser repetition rate of 5 Hz, spot size of 60 μm and He carrier gas (0.8–1.11 l/min) were applied. The energy density used in this study was 10–20 J cm$^{-2}$. Standard Monastery zircon was used for external correction. Each ablation was preceded by a 30 s-on mass background measurement, and ablated zircon was measured within 60 s. For the accurate measurement of Hf isotope ratios in zircon, the isobaric interference of $^{176}$Yb and $^{176}$Lu on $^{176}$Hf were corrected by measuring $^{171}$Yb ($^{176}$Yb/$^{171}$Yb = 0.897145) and $^{175}$Lu ($^{176}$Lu/$^{175}$Lu = 0.026549),

respectively. The calculation of epsilonHf$_{(T)}$ (time-corrected) values was based on zircon U–Pb ages and the chondritic values ($^{176}$Hf/$^{177}$Hf = 0.282772, $^{176}$Lu/$^{177}$Hf = 0.0332; Blichert-Toft & Albarède 1997).

## Lithostratigraphy and sedimentology

The present lithostratigraphic subdivision and chronostratigraphic correlations apply the scheme developed in Htut *et al.* (2010). The Eocene–Miocene sediment sequence regionally is thrust over a chaotic complex (tectonic mélange) comprising single blocks of variable size, which are embedded in a thoroughly deformed and slightly metamorphosed matrix of shales and boudinated sandstone beds (Fig. 2). The blocks comprise sediments of Cretaceous age (Brunnschweiler 1974; Bender 1983) and include white and red *Globotruncana*-bearing micritic limestones, reddish-brown to black radiolarian cherts and shales, and brown manganesoferous shales. Radiolarian cherts are also found to be intruded by vesicular basalts. Volcanic elements of the basal complex are blocks of dark-brown basalts with pillow and columnar structures.

The Gwa Formation (Fig. 2) is the lowest exposed formation in the surveyed area. It may reach a thickness of 3400 m and even more in the Rakhine Yoma area. The lower part of this formation is faulted in many places; boudinated sandstone beds and pencil structures are characteristic. The unit is composed mainly of indurated, carbonaceous bluish-grey shales, occasionally intercalated by thin grey to dark grey turbiditic sandstone beds. Flute, groove and load casts are common along bedding planes. Blocks of micritic limestones, radiolarian chert and (pillow) basalts are assembled at various basal levels of the formation.

The Ngapali Formation (Middle Eocene) has a compound thickness of *c.* 5500 m. In its lower part, it comprises fine- to medium-grained, hard, brown to grey bedded turbiditic sandstones, which are intercalated by laminated, grey to greenish-grey and highly carbonaceous shales. Occasionally, limestones, conglomerates and thin sandstones are intercalated into the generally massive sandstone-rich lower part. The middle and upper parts of the formation are dominated by grey to pinkish-grey, very hard and compact siltstone and thin highly carbonaceous turbiditic sandstone beds.

The Sinbok Formation (Late Eocene), with an estimated sediment thickness of about 2200 m, consists of three members. The Lower Shale Member is made up of highly fractured carbonaceous blue shales with some intercalated *Nummulite*-bearing turbiditic sandstone beds. The middle Sandstone

Member comprises fine- to medium-grained yellowish-brown to dark brown thick sandstone beds (up to 13 m) with minor shale intercalations. The Upper Shale Member is composed of thinly to thickly laminated dark grey shales, which contain brown sandstone and limestone blocks with *Nummulite* and *Globotruncana* fossils, respectively. The shale members represent mainly pelagic sedimentation with intercalations of low-density turbidite deposits.

The Yechangyi Formation (Oligocene) measures about 2000 m and unconformably overlies the Sinbok. It consists of fine- to medium-grained, thick-bedded light grey turbiditic sandstones interbedded with sandy shale and shale beds. The shales are grey to greenish-grey, fairly soft, and micaceous. This formation clearly shows less tectonic deformation than the underlying ones.

The Yenandaung Formation (Early–Middle Miocene) amounts to about 850–2000 m in thickness. In turn it unconformably overlies the Yechangyi Formation in most locations, and is faulted in only a few places. It consists of a lower Shale and Siltstone Member and an upper Sandstone and Shale Member. The lower member comprises blue and light grey thick shale beds and highly carbonaceous siltstone beds of turbiditic facies. The upper member is made up of thick sandstone beds associated with sandy shales and gritty sandstone beds.

The Leikkamaw Formation (Late Miocene), with a sediment thickness of about 700 m, again unconformably overlies the Yenandaung Formation. It comprises an irregular succession of fine- to medium-grained, hard, well-bedded yellowish-brown sandstones intercalated with shale and gritty sandstone beds. Individual sand-rich sequences are up to 120 m thick. The formation was deposited in a near-shore environment. Pliocene and Pleistocene mud flat, tidal channel and beach sediments conformably overlie the Leikkamaw Formation.

Thrusts and folds in the sediment series show generally west-vergence with an obvious decreasing trend of tectonic deformation going from older to younger formations (Htut *et al.* 2010). Eocene and Oligocene sediments are highly folded and thrusted, whereas the Miocene series are gently folded or only tilted. Turbiditic and hemipelagic sedimentation is characteristic for the Eocene, Oligocene and Early–Middle Miocene formations ('flysch-type' sediments; Bender 1983; Allen *et al.* 2008). The Late Miocene and Pliocene show a shallowing trend from shelfal to coastal facies, generally classed as 'Molasse-type' facies by Bender (1983) and Htut *et al.* (2010). The presence of the ophiolite-bearing mélange at the base of the folded and thrusted, west-verging turbiditic sediment pile suggests an accretionary wedge tectonic environment (e.g. Curray *et al.* 1979; Mitchell 1981;

Alam *et al.* 2003; Allen *et al.* 2008), which comprises oceanic basement and pelagic cover, deformed trench and slope basin deposits, which finally were topped by shallow shelf and coastal deposits (Htut *et al.* 2010).

## Results

The modal framework grain analysis classifies the sandstones mostly as feldspathic litharenite and litharenite within the QFL diagram of Folk (1974) (Fig. 3a). However, the amount of quartz grains is variable, showing that Oligocene and Miocene sandstones tend to have higher quartz content. Plagioclase is the dominating feldspar (Fig. 3b) but all Oligocene sandstones in addition reveal *c.* 5–15% K-feldspar. Comprising at least *c.* 50% of the sandstones, volcanic hypabyssal rock fragments prevail over metamorphic and sedimentary lithic ones (Fig. 3c). Also with regard to the inferred plate tectonic setting of the source terranes (e.g. Dickinson & Suczek 1979; Dickinson 1985) some trend is visible. A recycled orogenic nature of the source terranes is inferred for the majority of Oligocene and Miocene sandstones (Fig. 3d). An exception is represented by the Middle Eocene sample 10TTN03, which also shows particularly heavy mineral and detrital zircon age distributions compatible with recycling of older crustal elements (see below). Otherwise, most Eocene and parts of Oligocene and Miocene sandstones plot in the dissected and transitional magmatic arc fields.

Heavy mineral distributions show varying amounts of the ultrastable minerals zircon, tourmaline and rutile (ZTR, 23–60%) (Fig. 4). A variable contribution from continental crust rocks (including granitoids and volcanics) and recycled sandstones is implied (Mange & Maurer 1992). Garnet, epidote, zoisite and chloritoid grains show a wide range, from 2% up to 75%, which suggests variable detrital sources in medium-grade metamorphic rocks, or alternatively in garnet-bearing leucocratic intrusives. Oligocene sandstones generally show increased amounts of metamorphic minerals (up to 60–70%). The abundance of chromian spinel (up to 40%) suggests partly important sourcing of the detrital material from exhumed ophiolitic rocks (e.g. serpentinites and ultrabasic rocks). The Middle Eocene Ngapali Formation. turbidites, and the Middle and Late Miocene sandstones (10TTN09, 10TTN18) show highest chromian spinel occurrences (Fig. 4). The opposite trends of ultrastable mineral ZTR and the chromian-spinel/zircon index (Morton & Hallsworth 1999; Fig. 5), and the low to moderate ZTR index (20–60) implies that the observed heavy mineral distributions are due less to weathering but rather are source-rock

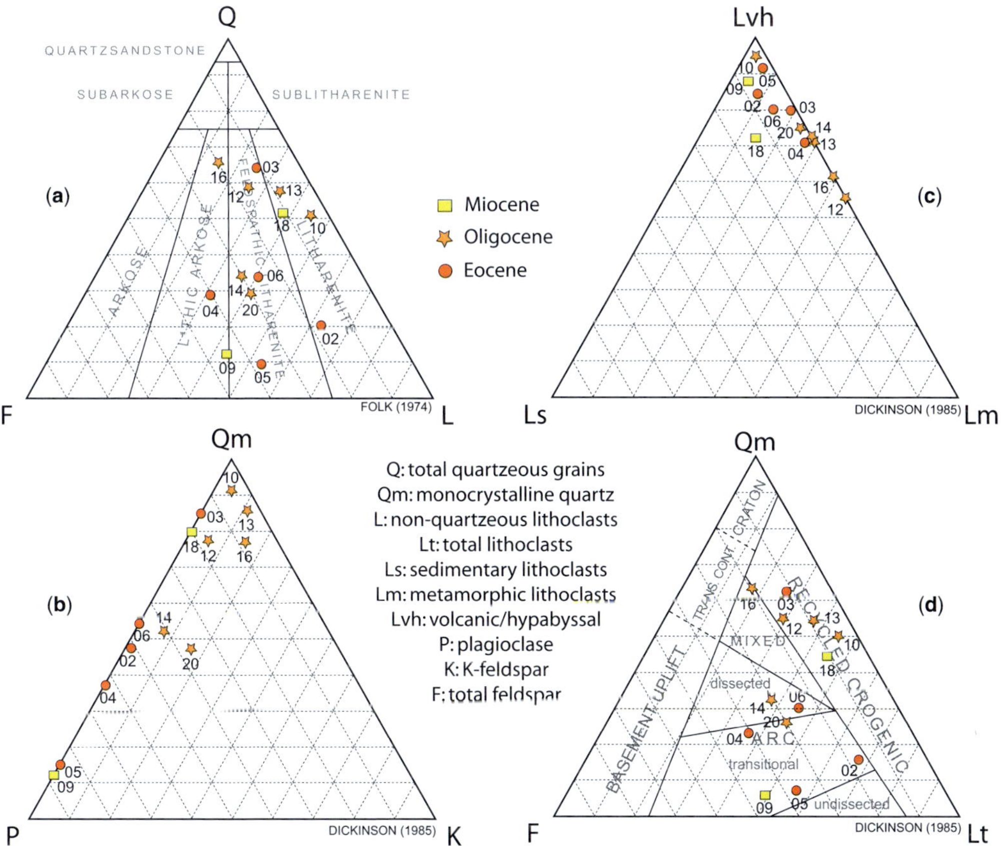



**Fig. 3.** Detrital composition and classification of the sandstones examined in the Rakhine coastal sections.

controlled. However, with regard to the marked surface etching of the garnets, the strong variations of the garnet–zircon index (Fig. 5) may partly be derived from temperature/burial controlled intrastratal dissolution (Morton & Hallsworth 1999).

Detrital zircon laser ablation ICP-MS U–Pb dating work in the Rakhine area revealed, in the majority of sandstone samples (8 out of 9), a dominant age distribution peaking in the mid and Late Cretaceous, and in the Cenozoic (Fig. 6). Only concordant and quasi concordant ages were used, as indicated by the presented concordia diagrams with each age probability curve (Fig. 6). The Cretaceous peaks are located between 123 and 66 Ma. With Palaeogene peaks ranging from 54 to 30 Ma, a nearly continuous supply of detrital zircons into the Cenozoic is documented. However, the Cretaceous–Cenozoic transition is often accentuated by a diminished number of grains. With two exceptions (10TTN16 and 10TTN03), only sparse reworking of Early Mesozoic, Palaeozoic and Precambrian zircons is observed. The Oligocene sandstone

10TTN16 reveals a prominent Cambrian–Early Ordovician population. The Middle Eocene sandstone 10TTN03 is characterized by a very broad and flat pre-Cambrian zircon distribution, together with rare Cambrian, Ordovician and Triassic grains. Even with the low number of zircons (54) available for dating in this sample, the absence of Cretaceous and Palaeogene age zircons in the source area is apparent, with similar cases of lower statistical rates still showing clear peaks for these periods. Together with the very high >90% ZTR amount, prominent recycling of older sediments in the source area of this sandstone is corroborated.

Where possible, depending on the grain size of the zircons and their internal growth structure (supported by CL imaging), Hf isotope ratios were measured only on dated zircons. Hf isotopes in zircon of known U–Pb ages may provide information on the source of the magma from which the zircon crystallized (e.g. Griffin *et al.* 2004). The $\varepsilon Hf_{(T)}$ data from zircons older than Permian

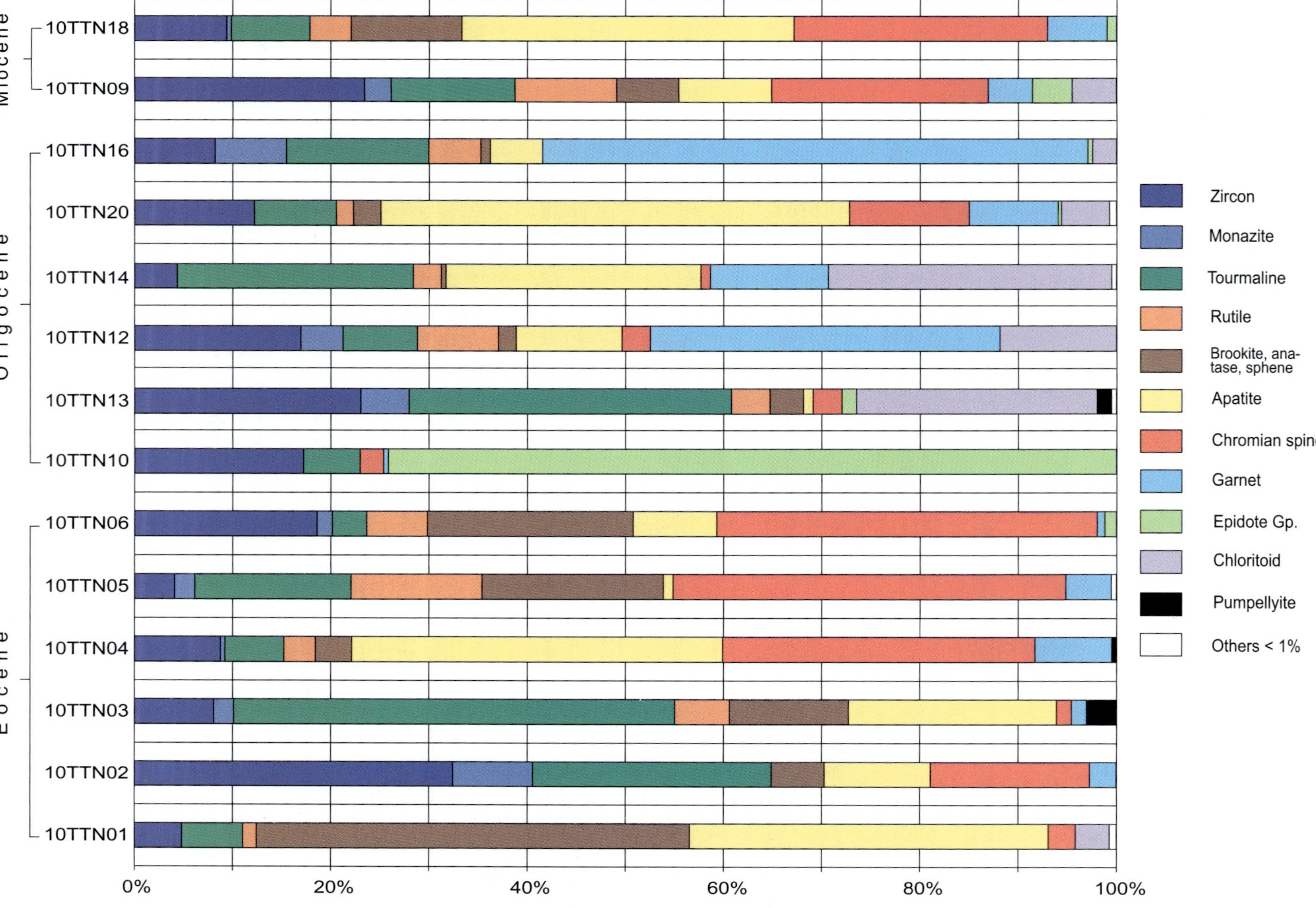


**Fig. 4.** Heavy mineral composition of the sandstones of the Rakhine coastal area arranged in reconstructed stratigraphical order.

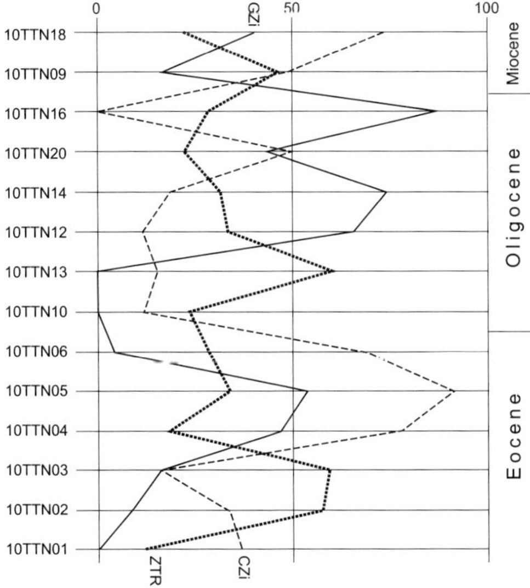



**Fig. 5.** Provenance-sensitive heavy mineral ratios (Morton & Hallsworth 1999) calculated for the sandstones in the Rakhine Coastal Belt. Garnet–zircon index, GZi; zircon–tourmaline–rutile index, ZTR; chromian spinel–zircon index, CZi.

are not shown in Figure 7 because of their low relevance for the present purpose. In this plot, a majority of the $\varepsilon Hf_{(T)}$ values are located between the Depleted Mantle and Chondritic Uniform Reservoir (CHUR) lines, whereas few grains fall below the CHUR line. They show a large spread in $\varepsilon Hf_{(T)}$ from +15.7 to −2.6. Close to depleted mantle values are noticeable in Cretaceous (mean +12.6 $\varepsilon Hf_{(T)}$) and Palaeogene zircons (mean +13 $\varepsilon Hf_{(T)}$). The Hf isotopic composition for detrital zircons, clustering at 145–66 Ma shows $\varepsilon Hf_{(T)}$ values of +15.7 to +3.8 and that of 65–32 Ma age have positive $\varepsilon Hf_{(T)}$ values of +14.5 to +0.97 with a few negative ones down to −2.6. In general, Cretaceous and Palaeogene detrital zircons scattered $\varepsilon Hf_{(T)}$ values suggest a mixed mantle-crust origin as for subduction-related magmatism (Patchett & Tatsumoto 1980; Patchett 1983). There are no visible strong trends of change in $\varepsilon Hf_{(T)}$ correlated with age. With some caution a negative trend of $\varepsilon Hf_{(T)}$ values from the Cretaceous to the Palaeogene may be inferred for samples 10TTN04 and 10TTN09, which had detrital sources in magmatic arc environments (see above). A slight trend to increased continental crust involvement in magma generation may be inferred.

Aimed at determining the type of magmatic melts from which the detrital zircons were derived, a typology study (Pupin 1980) of euhedral zircons was carried out (Fig. 8), although a small number of grains was analysed. With this approach we assume that this analysis grasps in particular idiomorphic Late Cretaceous and Palaeogene zircons and, therefore, gives information on the types of arc magmas in which the reworked zircons have crystallized. In the basic typology diagram (Fig. 8) comparing the alkalinity (I.A) and temperature (I.T) indices, the zircon crystals show broader distributions with some clustering. The zircon suites contain mainly S-type crystals, with {100}, {110} prisms and {101}, {211} pyramids. They cluster particularly in the S7, S12, S13, S17, S18 and S19 fields, which characterize zircons derived from diorites, quartz gabbros, tonalites and granodiorites. According to the petrogenetic classification scheme for non-granitic rocks, zircons of the dominant populations (see Fig. 9) fall into the field of tonalites (T) and calc-alkaline series rhyolites (C.A.R), which figure at the commencement of hybrid granite development. The typology of the detrital zircons indicates that the parental magmas had a hybrid origin, which occurs by mixing of mantle and crustal material in the melts (Pupin 1980). This observation corroborates the $\varepsilon Hf_{(T)}$ results, which were carried out on dated Cretaceous–Palaeogene zircons.

## Discussion

The provenance of the detrital material in the Indo-Burman accretionary range in the larger framework of Himalayan orogeny has been a subject of research for decades. Often, the sediments of the IBR were reported to represent accreted deposits of the palaeo-Bengal fan (e.g. Brunnschweiler 1974; Bender 1983; Curray *et al.* 2003). DSDP and ODP drillings have penetrated the *in-situ* Bengal fan sediments, finding the last 15–17 Ma recorded (e.g. Thompson 1974; Yokoyama *et al.* 1990; Amano & Taira 1992; Curray *et al.* 2003) above a major hiatus with the pre-fan pelagic sediments of the Indian oceanic plate. The missing sediment record of the palaeo-Bengal fan was suggested to have been scraped off by eastward subduction and incorporated into the Indo-Burman accretionary wedge (IBR) and Andaman–Nicobar ridge (Curray *et al.* 2003). This is also a current model for inferring earlier, Eocene and younger Himalayan collisional events between the north-drifting Indian plate and Eurasia.

However, different potential sediment sources in this orogenic suture at the eastern termination of the Himalayan range can be considered: (1) the eastern Himalaya thrust belt situated on the Indian plate, (2) the Cretaceous–Cenozoic Trans-Himalayan volcanic arc situated on the Asian plate, or (3) the Burmese active (subduction) continental margin

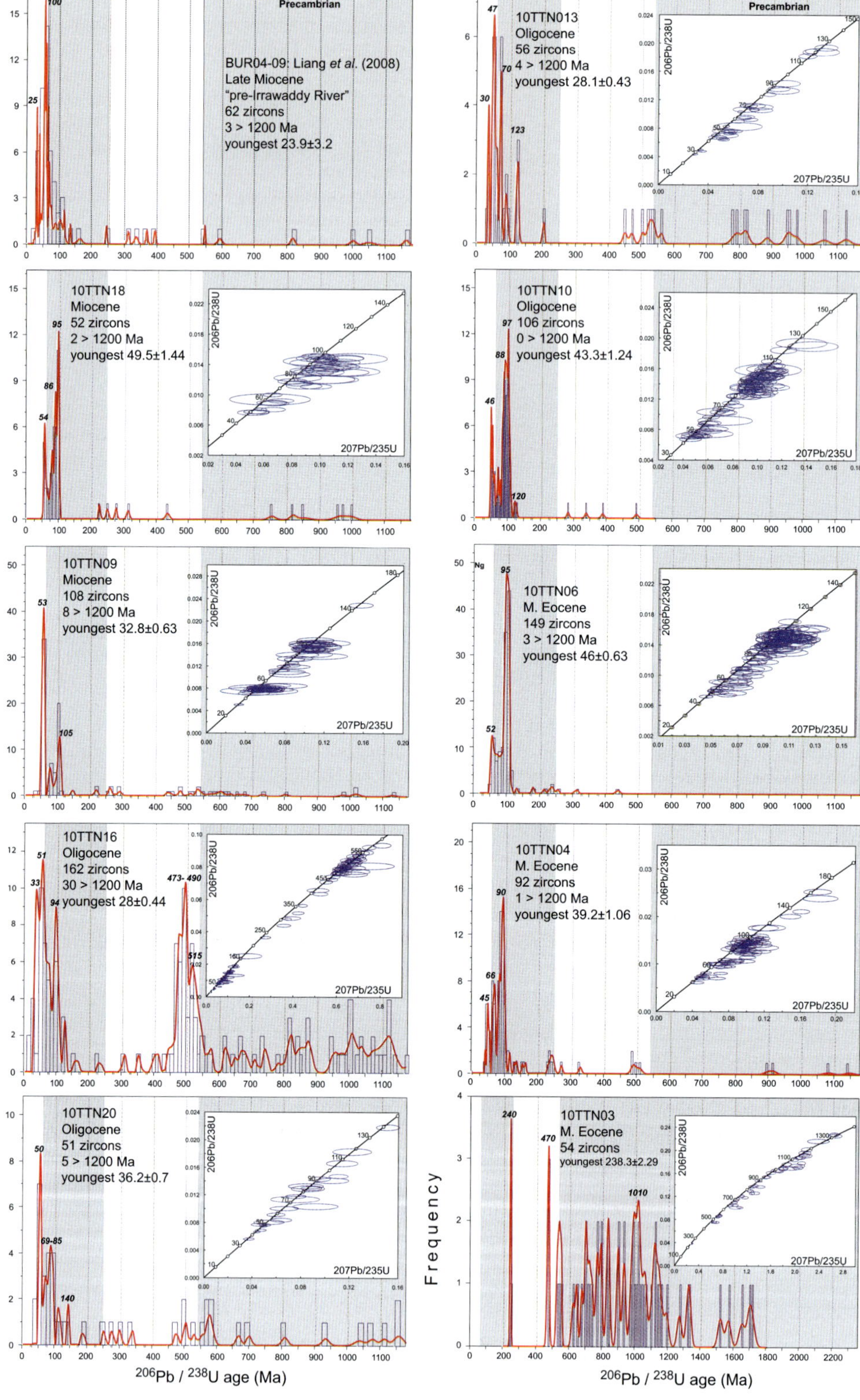
Precambrian
BUR04-09: Liang et al. (2008)
Late Miocene
"pre-Irrawaddy River"
62 zircons
3 > 1200 Ma
youngest 23.9±3.2
100
25
10TTN013
Oligocene
56 zircons
4 > 1200 Ma
youngest 28.1±0.43
47
70
30
123
Precambrian
10TTN18
Miocene
52 zircons
2 > 1200 Ma
youngest 49.5±1.44
95
86
54
10TTN10
Oligocene
106 zircons
0 > 1200 Ma
youngest 43.3±1.24
97
88
46
120
10TTN09
Miocene
108 zircons
8 > 1200 Ma
youngest 32.8±0.63
53
105
10TTN06
M. Eocene
149 zircons
3 > 1200 Ma
youngest 46±0.63
95
52
Ng
10TTN16
Oligocene
162 zircons
30 > 1200 Ma
youngest 28±0.44
51
33
94
473-490
515
10TTN04
M. Eocene
92 zircons
1 > 1200 Ma
youngest 39.2±1.06
90
66
45
10TTN20
Oligocene
51 zircons
5 > 1200 Ma
youngest 36.2±0.7
50
59-85
140
10TTN03
M. Eocene
54 zircons
youngest 238.3±2.29
240
470
1010
Frequency
206Pb / 238U age (Ma)
206Pb / 238U age (Ma)

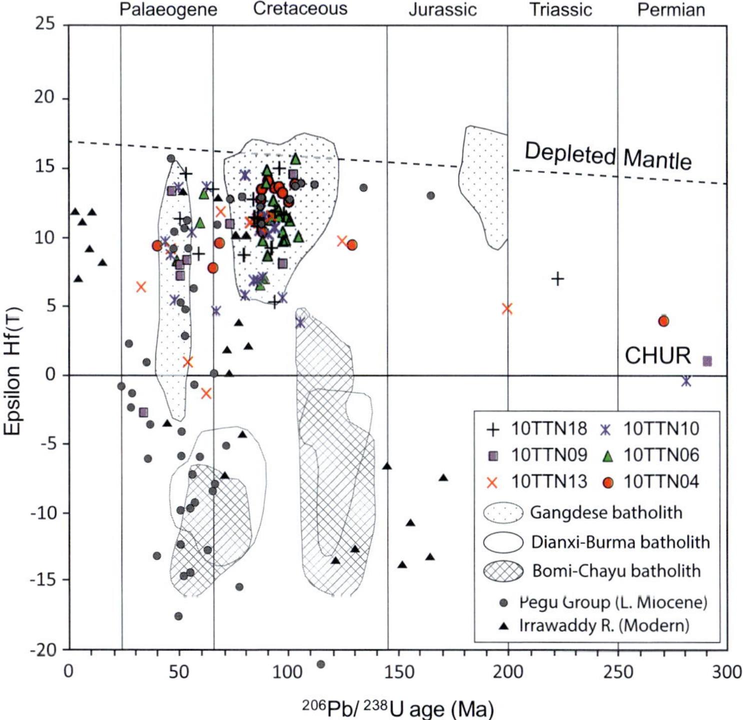



**Fig. 7.** Diagram of epsilon Hf$_{(T)}$ (time-corrected) v. $^{206}$Pb/$^{238}$U zircon age for sandstones of the Rakhine coastal area. Corrections in function of age are based on chondritic values (CHUR) from Blichert-Toft & Alabarède (1997), which become the reference value. The depleted mantle evolution trend is from Griffin *et al.* (2002) values. The results are compared with relevant data derived from Transhimalayan batholiths, pre-Irrawaddy River (Late Miocene) and Modern Irrawaddy River sediments, as reported by Chu *et al.* (2006) and Liang *et al.* (2008). Timescale after Ogg *et al.* (2008).

comprising Cretaceous–Neogene magmatic arc rocks in the Mogok metamorphic belt, Shan Scarps and in the Central Basin (see also discussions in Najman *et al.* 2008; Allen *et al.* 2008). The present results need to be evaluated and compared with provenance data from coeval sediments in these various settings (Table 2).

Contemporaneous Eocene to Miocene Himalayan foreland basin series (Table 2) show common Himalaya-derived heavy mineral associations (zircon, tourmaline, epidote, zoisite, garnet) in addition to high-grade metamorphic minerals staurolite, kyanite and sillimanite at the detriment of chromian spinel (Szulc *et al.* 2006; Ravikant *et al.* 2011). Detrital zircon U–Pb age results from the Eocene–Pleistocene Subathu sub basin in NW India reveal a dominant population ranging from

1000–900 Ma, with subordinate populations of 650–500, 1900–1800 and 2500 Ma (Ravikant *et al.* 2011). In the Nepalese thrust belt, the coeval Bhainskati and Dumri foreland basin formations are characterized by prominent detrital zircon age populations peaking at 500–600, 1000–1300, *c.* 1800 and *c.* 2500 Ma, which are correlated with sources in the Tethyan, Greater and Lesser Himalayan units, respectively (DeCelles *et al.* 2004). The three Neogene–Holocene megasequences of the Hatia Trough and Chittagong Hill tracts in Bangladesh with a presumed Himalayan provenance (Najman *et al.* 2012) are characterized by quartz-rich sandstones of recycled orogenic sources and, similar to the Siwlaik foreland basin series, showing high grade metamorphic detrital grains of staurolite, kyanite and sillimanite.

**Fig. 6.** Detrital zircon laser ablation ICP-MS U–Pb age distributions (histograms and relative-age-probability curves, red), and concordia diagrams measured in sandstones of the Rakhine Coastal Belt. Concordant to slightly discordant age data are plotted. Also given is the number of zircons with ages >1200 Ma and the youngest dated grain in samples. For comparison (upper left) the detrital zircon ages of a Late Miocene–Pliocene transition sandstone from the Pegu Group in the Myanmar Central Basin (Liang *et al.* 2008). Timescale after Ogg *et al.* (2008).

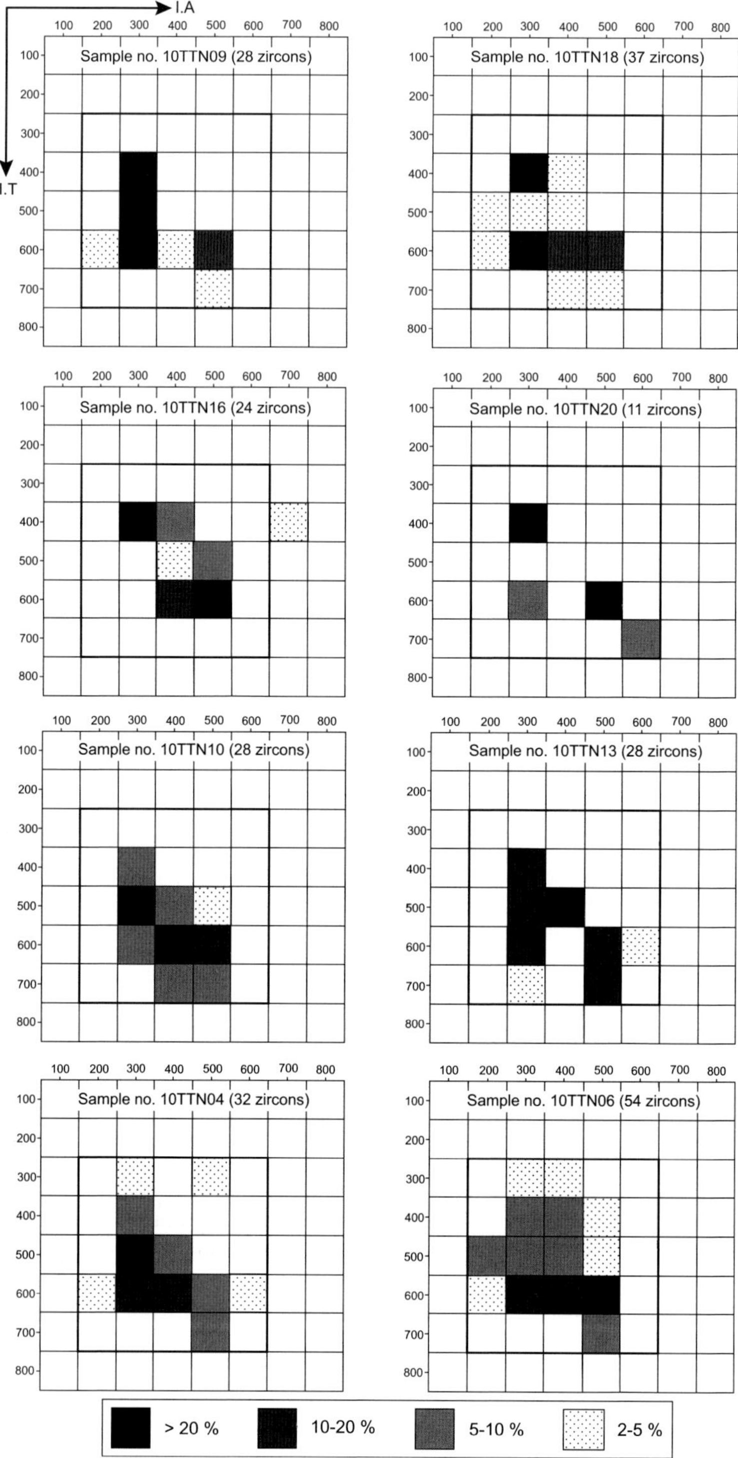



**Fig. 8.** Typological distribution of zircons from sandstones of the Rakhine Coastal Belt, according to the alkalinity (IA) and temperature index (IT) of Pupin (1980). See also Figure 9.

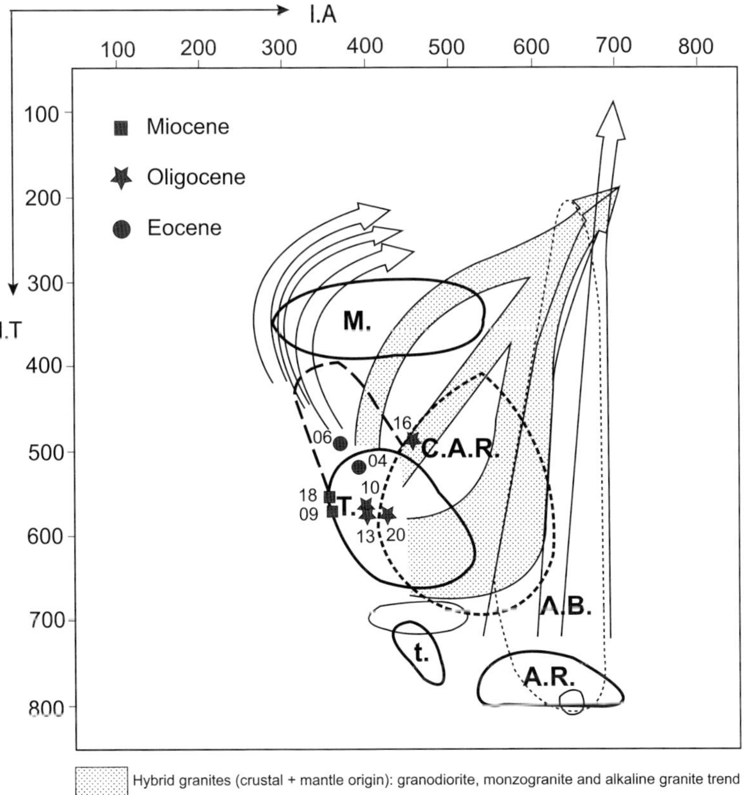



**Fig. 9.** Summary typological distribution of zircons of the Rakhine Coastal Belt in the Typological Evolutionary Trend (TET) model of Pupin (1980). The mean points of the samples plot in the hybrid field of tonalites and calc-alkaline series rhyolites of mixed crust and mantle origin.

In the Bengal Basin, including the Syleth, Surma, Assam and Haita sub-basins, Palaeogene and Neogene sandstones exhibit variable sources (Table 2). The Eocene Kopili and Oligocene Barail formations in the Syleth Trough contain low amounts of the stable minerals tourmaline, garnet, rutile, zircon and chromian spinel, leading Uddin & Lundberg (1998) to infer Indian craton derivation and intense chemical weathering. However, due to the coeval presence of pronounced *c.* 500 Ma and *c.* 1000 Ma detrital zircon age populations, and the prevalence of ZTR heavy mineral associations, Najman *et al.* (2008) inferred a major Trans-Himalayan arc source for the Kopili and Barail for mations of the Surma Basin since about 38 Ma. In the Miocene Surma Group and younger formations, detrital staurolite, kyanite, sillimanite and orthopyr-oxene grains clearly suggest a prevailing Himalayan source (Uddin *et al.* 2007; Najman *et al.* 2012).

Autochthonous Bengal fan turbidites were ana-lysed for heavy mineral contents in various sites of DSDP Leg 22 (218, 217, 219) and ODP Leg 116 (717–719). The results document Himalayan material reworking since 17–15 Ma (Langhian, Middle Miocene) (Table 2). In Leg 22 the main mineral phase is calcic amphibole (20–60%) with a general increasing trend towards younger ages, associated with epidote, garnet, ZTR, scarce tremo-lite, and clinopyroxene in the youngest part (Thompson 1974; Yokoyama *et al.* 1990). Stauro-lite, kyanite and sillimanite were observed in minor proportions and chromian spinel occurs very spora-dically, thus correlating with other results from the Bengal Basin (Table 2). In the Bengal fan ODP sites 117–119 from the Middle Miocene upwards (*c.* 15 Ma to present day), the variations in calcic amphibole amounts (*c.* 25% and higher) and other minerals allow discrimination of Higher Himalaya from Indian subcontinent detrital sources (Amano & Taira 1992).

The eastern Trans-Himalayan batholiths in Lhasa and South China blocks like the Gangdese,

**Table 2.** *Compilation of provenance indicators from the Indo-Burman Ranges and contemporaneous basins in the larger Himalayan system*

| Units | Indo-Burman Ranges | Indo-Burman Ranges | Central Burma Basin | Bengal Basin | Haita Trough and Chittagong Hills | Bengal Fan (DSDP/ODP drill sites) | Himalayan Foreland Basins |
|---|---|---|---|---|---|---|---|
| Authors | Present study | Allen *et al.* (2008) | Liang *et al.* (2008) | 1. Syleth Trough Uddin & Lundberg (1998), 2. Surma Basin Najman *et al.* (2008), 3. Assam Basin Uddin *et al.* (2007) | Najman *et al.* (2012) | 1. Thompson (1974), 2. Yokoyama *et al.* (1990), 3. Amano & Taira (1992) | 1. India/Siwalik: Szulc *et al.* (2006); 2. India/Subathu: Ravikant *et al.* (2011); 3. Nepal/ Bhainskati-Dumri fms: DeCelles *et al.* (2004) |
| Sample ages | Middle Eocene–Miocene | 1. Palaeogene bed rock and river sands; 2. Neogene bed rock and river sands | Late Miocene (*c.* 10–5 Ma) | 1. Late Eocene–Plio/ Pleistocene, 2. Late Eocene–Early Miocene, 3. Oligocene–Plio/ Pleistocene | Pliocene–Recent (4–0 Ma) | 1. last 12 Ma (Late Miocene–Recent), 2 and 3. last 17 Ma (Early Miocene–Recent) | 1. ≈16–1 Ma (Middle Miocene–Pliocene); 2. ≈55–12 Ma (Eocene–M. Miocene); 3. Eocene–Early Miocene (≈55–15 Ma) |
| Sandstone provenance (Dickinson 1985) and heavy minerals | Dissected/ transitional magmatic arc and recycled orogenic provenance; HM: ZTR, chromian spinel, garnet, chloritoid, epidote group | 1. Magmatic arc provenance (minority); 2. Recycled orogenic provenance (majority); 1. and 2. HM: ZTR, epidote, garnet, chromian spinel (particularly in Palaeogene), traces of staurolite and amphibole (in Neogene) | | 1. HM: ZTR, garnet, AKS, epidote, amphibole; 2. Recycled orogen source; 3. HM: pre-Miocene ZTR, Miocene and younger ZTR, AKS, garnet, epidote, hornblende, chloritoid | Recycled orogen provenance (quartz-rich); HM: ZTR, garnet staurolite, kyanite, amphibole, epidote, chloritoid | 1. HM: hornblende dominates and increases upsection, pyroxene, minor kyanite and sillimanite; 2. HM: calcic amphibole, epidote, garnet, clinopyroxene, ZTR; 3. HM: calcic amphibole, epidote, garnet, AKS, ZTR, pumpellyite | 1. Recycled orogenic provenance; HM: garnet, ZTR, epidote group, chloritoid, staurolite, kyanite and sillimanite, minor chloritoid; 2. no data available; 3. Cratonal and recycle orogen provenance |
| Detrital zircons – concordant ages (U–Pb) | Major populations: 123–66 Ma (Cretaceous) and 54–30 Ma | 1. *c.* 110–40 Ma populations dominate, minor 500–650 Ma, | Nearly continuous Cretaceous–Cenozoic record: 110–70 Ma and | 2. Cretaceous–Cenozoic (younging upward) combined with minor peaks at ≈500–600 | Early Palaeozoic–Precambrian zircons dominate with prominent | | 1. no data available; 2. population 1000–900 Ma most prominent, others 650– |

| | | | | | | | |
|---|---|---|---|---|---|---|---|
| | (Eocene–Oligocene) older grains >125 Ma randomly distributed; occasionally a 550–450 Ma population | >650 Ma random; 2. c. 100–10 Ma population present, majority of grains >450 Ma with occasional populations at c. 450–600 and 700–1200 Ma | 65–23 Ma; minor (22%) >200 Ma randomly distributed | and 1000–1200 Ma, and older grains | peak at ± 500 Ma; in Plio-Pleistocene a Late Jurassic–Miocene (≈150–10 Ma) population is present | | 500, 1900–1800 and 2500 Ma; 3. populations ≈600–500, ≈1300–1000, ±1800, ± 2500–Ma |
| Zircon Hf isotope ratios | General range: $\varepsilon$Hf(T) c. 16–1; 145–66 Ma $\varepsilon$Hf(T) c. +15.7 to +3.8, 65–32 Ma: $\varepsilon$Hf(T) c. +14.5 to +0.97 | | Large spread $\varepsilon$Hf(T) c. 15 to −18, correlating with Gangdese, Boni–Chayn and Dianxi–Burma batholiths in the Lhasa Block and eastern Trans-Himalayan zone respectively (see Fig. 7) | | | | |
| Authors' interpretation of provenance | Burman Late Cretaceous–Palaeogene magmatic arc, accreted ophiolitic series and continental margin basement and cover (i.e Mogok Metamorphic Belt, Shan Plateau) | 1. Palaeogene: significant Burmese arc-derived input; 2. Neogene: Himalayan source predominates | Palaeo-Irrawaddy River draining the Trans-Himalayan region | 1. pre-Miocene from Indian continent, Miocene and younger from Himalayan orogenic front; 2. pre-Late Eocene from Indian continent, Late Eocene and younger: Himalayan source with a Burman arc component; 3. eastern Himalayas and Indo-Burman Ranges | Predominating Himalayan source | 1. low–high-grade metamorphic rocks and acidic igneous rocks in Himalayas; 2. Himalayan source with strong affinity with Siwalik Group; 3. Tethys and Lower Himalaya (i.e. pumpellyite), Higher Himalaya (i.e. calcic amphibole) and subordinate Deccan Traps sources | 1. Greater Himalaya corresponding with rapid exhumation from 20–15 Ma; 2. Tethys Himalaya, Higher Himalaya, and Lesser Himalaya (i.e. 1900–1800 Ma detrital zircons); 3. Tethys, Greater and Lesser Himalaya, pre-Eocene Indian Craton provenance |

*Abbreviations*: Heavy minerals, HM; ultrastable heavy mineral assemblage zircon–tourmaline–rutile (may also include brookite, anatase and sphene), ZTR; high-grade metamorphic association andalusite–kyanite–sillimanite, AKS.

Bomi–Chayu and Dianxi–Burma batholiths are characterized by prominent zircon age populations of Jurassic, Cretaceous and early Palaeogene (up to 45 Ma) age (Liang *et al.* 2008); Bomi-Chayu age ranges: 140–105 Ma, 65–50 Ma; Dianxi–Burma batholith: 140–110 Ma, 90–50 Ma; Gangdese batholith: 210–180 Ma, 115–75 Ma and 55–45 Ma (Chu *et al.* 2006). These batholith age clusters show some incomplete overlap with the present detrital zircon ages of the Rakhine Coastal Belt, and closest similarity exists with the Gangdese Cretaceous and Palaeogene zircons with respect to the reported $\varepsilon Hf_{(T)}$ values (Chu *et al.* 2006) (Fig. 7). However, the Jurassic component is much more developed in that batholith than in the detrital zircons described here. The Bomi–Chayu and Dianxi-Burma granitoids certainly can be excluded as important sources, because these batholiths systematically reveal low and negative $\varepsilon Hf_{(T)}$ values between +5 and −15 in Cretaceous and Palaeogene zircons (Fig. 7; Liang *et al.* 2008). With regard to the detrital zircon populations documented in the Rakhine Coastal Belt, there is an affinity observed with the age distributions reported by Liang *et al.* (2008) from a Late Miocene Pegu Group sandstone in the Myanmar Central Basin (Fig. 6; Table 2). However, Palaeogene (*c.* 65–23 Ma) zircons from this sandstone reveal a strong negative $\varepsilon Hf_{(T)}$ component ($\approx$ −4 to −18; Liang *et al.* 2008), and a similar igneous rock sourcing of coeval sandstones in the Rakhine coastal area (Fig. 7) can be excluded. Presumably, this sandstone of the Pegu Group is younger than our youngest sample (base of Late Miocene, 10TTN18) and the detrital contribution from the Dianxi–Burma and Bomi–Chayu batholiths, as inferred by Liang *et al.* (2008), occurred later, or the river drainage pattern was more complex.

In Myanmar, Jurassic to Neogene subduction-related magmatic activity is documented in the MMB, in the Shan Scarps and in the Myanmar Central Basin (Mitchell 1993; Barley *et al.* 2003; Mitchell *et al.* 2007; Searle *et al.* 2007). Unfortunately, few occurrences have been dated with modern geochronological methods yet (Mitchell *et al.* 2007). The MMB represents a middle to lower crustal section giving evidence of magmatization and magmatic intrusions at different times, and final high-grade metamorphism lasting at least from *c.* 43–29 Ma (Searle *et al.* 2007). The MMB was exhumed prior to displacement along the right-lateral, brittle Sagaing fault, which initiated between 22 and 16 Ma (Barley *et al.* 2003; Bertrand & Rangin 2003; Searle *et al.* 2007). Mid-Cretaceous to Eocene I-type granitoids (*c.* 120–50 Ma) in the MMB suggest that in eastern Myanmar and southern Eurasia in general (from Pakistan to Sumatra), an Andean-type subduction margin existed during that time (Barley *et al.* 2003). Oligocene zircons were identified in the Mandalay Hill area (*c.* 30 Ma) and Yesin dam area (*c.* 25 Ma; Barley *et al.* 2003). During the youngest metamorphic event (*c.* 45–24 Ma), hybrid mantle-derived hornblende syenites (35–23 Ma) and crust-derived leucogranite melts (24.5 Ma) were produced (Searle *et al.* 2007).

South of Mandalay, a granitic belt cuts the Shan Scarps. The plutons intrude Late Jurassic–Early Cretaceous sediment series and provided age ranges from Early Cretaceous to Early Miocene (Darbyshire & Swainbank 1988). Cobbing *et al.* (1992) concluded hybrid magma sources from mantle and crust for these intrusions. In the Myanmar Central Basin (Banmauk area) the allochthonous Mawgyi andesites and ophiolitic suites (Salingyi) are intruded by *c.* 106–91 Ma old (K–Ar data) granodioritic/tonalitic plutons. According to Mitchell (1993), intrusive rocks covering the Cretaceous–Cenozoic age transition (60–70 Ma) are rather rare. This author suggests that subduction and magmatism ceased during ophiolite obduction in the Indo-Burman Ranges during this period. Subduction and magmatic activity resumed in the Palaeogene. Eocene (*c.* 50–53 Ma, K–Ar) magmatic activity (andesites and quartzdiorites) is documented in the Shangalon porphyry copper near the Banmauk granodioritic plutons (Mitchell 1993). In the Myanmar Central Basin an Oligocene age (*c.* 38 Ma) of a granodiorite intrusion is also reported from the area south of Banmauk. Neogene–Quaternary volcanic arc activity is represented by the porphyry copper at Monywa north of Salingyi and the Mt Popa volcano (Mitchell 1993).

According to the various provenance aspects, the investigated sandstones of the Rakhine coastal area show nearly coherent results. With one exception (Middle Eocene 10TTN03), a dominant sourcing of sediment from Late Cretaceous and Palaeogene (tonalitic/calc-alkaline rhyolitic) magmatic rocks is indicated. The bimodal occurrence of Late Cretaceous and Palaeogene detrital zircons (Fig. 6) is in agreement with Burman arc age distributions as suggested by Mitchell (1993). A mixed mantle–crust origin of the magmas is inferred by the measured Hf isotope ratios and the typology of the zircons. Consequently, a Late Cretaceous and Palaeogene subduction-related volcanic arc environment for the sediment source areas is evident. Ultrastable heavy mineral associations (ZTR), chromian spinel and variable medium-grade metamorphic mineral spectra show that the magmatic rocks were associated with sources that comprised continental and oceanic basement and older sedimentary rocks. This is in agreement with the development of the arcs in the Mogok Metamorphic Belt, the Shan Scarps and Central Basin (Mitchell

1993; Barley *et al.* 2003; Mitchell *et al.* 2007; Searle *et al.* 2007). The nearly ubiquitous chromian spinel in the heavy mineral fractions implies additional sources from accreted ophiolitic rocks in the Indo-Burman Ranges. Unless some heavy mineral grains, such as garnet or epidote, stem from leucocratic intrusions (frequent in the MMB; Barley *et al.* 2003) or from the alteration of basic volcanic rocks, the heavy minerals in general indicate the erosion of metamorphic basement as exposed today at the eastern edge of the Shan Plateau. Discriminating key heavy minerals indicative for Himalayan sources, such as calcic amphiboles, clinopyroxene, staurolite, andalusite, kyanite and sillimanite, were not detected by us in sandstones of the Rakhine coastal area. Almost all Middle Eocene to Late Miocene sandstones show only subordinate – in the range of single grains – reworking of older Mesozoic, Palaeozoic and Precambrian zircons (Fig. 6). One Oligocene sandstone (10TTN16) sample indicates a strong sourcing in Cambrian–Ordovician rocks or their secondary reworking. Further noticeable Proterozoic detrital zircon populations, as typically reported from Himalayan foreland basins (Table 2 and references therein), are not obvious.

Two sandstones reveal a different composition and need further discussion. The Middle Eocene sample 10TTN03 shows a high recycling index in the heavy minerals (ZTR >60) and a broad Early Palaeozoic–Proterozoic detrital zircon age distribution, including one Triassic zircon, with only one to two grains per age bin (bin width is 15 Ma; Fig. 6). A dominant source area in exhumed Triassic flysch-like turbidite series (e.g. Mitchell 1993) in the IBR or in Palaeozoic sandstone formations in the Shan Plateau and Scarps could explain these very restricted grain and age spectra. The other example, the Oligocene sandstone 10TTN16, is characterized by the co-existence of prominent bimodal Burman arc zircon populations with a strong Cambrian peak and a broad Proterozoic age spectrum. A *c.* 500–650 Ma age population, together with a strong Proterozoic component, could be indicative of a Himalayan source, as revealed in the foreland basins (DeCelles *et al.* 2004; Ravikant *et al.* 2011). Therefore, the origin of the detritus of this sandstone may be the confluence of Burman and Himalayan material in a turbiditic source. However, with respect to the overall picture, a secondary supply from reworking Cambrian sediments and Proterozoic basement in the Shan Plateau and Scarps (Mitchell 1993) could be an explanation.

The presence of detrital zircon populations of 450–600 and 700–1200 Ma in river sands and bedrocks in the Arkhan coastal area was used by Allen *et al.* (2008) to infer a predominant Himalayan

sourcing from the Neogene onwards. However, these heavy mineral assemblages (available from Allen *et al.* 2008, supplements) do not correlate with such ones reported from coeval sediments in the larger Bengal Basin and Himalayan foreland basin occurrences (Table 2), instead showing qualitatively similar assemblages, as in the present study. Finally, we arrive at the similar conclusion that during the Palaeogene the dominating source of detrital material was the Burman arc and crystalline basement with sedimentary cover. The studied Neogene deposits do not yet show strong influx from the Himalayan range, but some river confluence and consequent mixing of provenances may have occurred.

The results of the present study have interesting implications for hydrocarbon exploration in the larger area of Myanmar, because they suggest reservoirs were not only derived from the Himalayas but predominantly from the Burman active margin, and a connection with the development of the Myanmar Central Basin may be established. There exist good correlations of thick sediment packages of pre-Eocene, Eocene and post-Eocene age in the Rakhine Coastal Belt with those of the MCB and deep-water offshore areas. Further investigation of the stratigraphy and the history of deposition in the MCB will be fundamental. This in the context of an already proven petroleum system along the Rakhine coastal area, with significant hydrocarbon discoveries in stratigraphic traps or combined structural–stratigraphic traps, infallible indications of gas seepages from mud volcanoes, and local oil production from hand-dug wells.

## Conclusions

The present work evaluated the provenance of detrital material in Middle Eocene to Late Miocene sandstones comprised in the Indo-Burman Ranges distal accretionary wedge and slope series in the Rakhine Coastal Belt. This area has large-scale significance for the reconstruction of the collisional history of the Indian with the Eurasian plate and Himalayan orogenesis in general. Our results do not confirm opinions that the Rakhine detrital sediments were supplied from Himalayan sources, and therefore, could be proxies of the palaeo-Bengal fan and early (≤55 Ma, Eocene and younger) collisional events in the Himalaya.

Our results, from framework grain, heavy mineral and detrital zircon analyses suggest a principal sourcing of the detrital material from the Late Cretaceous and Palaeogene magmatic arc and basement formations related to the Indian plate subduction beneath the Burman margin, as earlier suggested by Mitchell (1993). Ophiolitic series in the Indo-Burman Ranges, and metamorphic complexes in the larger Mogok Metamorphic Belt and Shan Plateau area represent secondary sources.

With regard to the sediment routing from the Burman active margin, we infer that, from Middle Eocene to Middle Miocene, the erosional scree presumably was transported by rivers and subsequent turbiditic flows into the Burman trench and slope basins to be accreted and uplifted in the external Indo-Burman Ranges. Unconformable Late Miocene shallow shelf sediment shows a similar volcanic arc signature. However, in the light of the limited geographical and stratigraphic distribution of our sample material, we cannot exclude the merging of Burman arc and Himalayan sources in places closer to the Himalayan Range to the north and in younger times.

A better understanding of the provenance of the sandstones in the Rakhine Coastal Belt may also contribute to the reassessment of the petroleum system in the Indo-Burman Ranges and Myanmar Central Basin.

The Swiss Federal Grant Commission for Foreign Students funded this work to T. T. Naing, grant no. ESKAS-2010. 0424. The authors gratefully acknowledge this support. The authors would like to thank the geoscience team (MPRL E & P Pte Ltd) for efficient support during fieldwork and express their sincere thanks and appreciation to all those who have been directly and indirectly involved in this research work. Y. Najman, A. Morton and an unknown reviewer are thanked for constructive comments on an earlier manuscript version.

# References

ACHARYYA, S. K. 1998. Break-up of the greater Indo-Australian continent and accretion of blocks framing south and east Asia. *Journal of Geodynamics*, **26**, 149–170.

ACHARYYA, S. K. & LAHIRI, T. C. 1991. Cretaceous palaeogeography of the Indian subcontinent: a review. *Cretaceous Research*, **12**, 3–26.

ALAM, M., ALAM, M. M., CURRAY, J. R., CHOWDHURY, M. L. R. & GANI, M. R. 2003. An overview of the sedimentary geology of the Bengal Basin in relation to the regional tectonic framework and basin-fill history. *Sedimentary Geology*, **155**, 179–208.

ALLEN, R., NAJMAN, Y. *ET AL.* 2008. Provenance of the Tertiary sedimentary rocks of the Indo-Burman Ranges, Burma (Myanmar): Burman arc or Himalayan-derived? *Journal of the Geological Society, London*, **165**, 1045–1057.

AMANO, K. & TAIRA, A. 1992. Two-phase uplift of Higher Himalayas since 17 Ma. *Geology*, **20**, 391–394.

BARLEY, M. E., PICKARD, A. L., ZAW, K., RAK, P. & DOYLE, M. G. 2003. Jurassic to Miocene magmatism and metamorphism in the Mogok metamorphic belt and the India–Eurasia collision in Myanmar. *Tectonics*, **22**, 1019, http://dx.doi.org/10.1029/2002TC001398

BENDER, F. 1983. *Geology of Burma*. Gebr. Borntraeger, Berlin–Stuttgart.

BERTRAND, G. & RANGIN, C. 2003. Tectonics of the western margin of the Shan plateau (central Myanmar): implication for the India-Indochina oblique convergence since the Oligocene. *Journal of Asian Earth Sciences*, **21**, 1139–1157.

BLICHERT-TOFT, J. & ALBARÈDE, F. 1997. The Lu–Hf geochemistry of the chondrites and the evolution of the mantle crust system. *Earth and Planetary Science Letters*, **148**, 243–258.

BRUNNSCHWEILER, R. O. 1974. Mesozoic–Cenozoic orogenic belts. *In*: SPENCER, A. M. (ed.) *Mesozoic Orogenic—Cenozoic Belts: Data for Orogenic Studies*. Geological Society, London, Special Publications, **4**, 279–299.

CHAKRABORTY, P. P. & KHAN, P. K. 2009. Cenozoic geodynamic evolution of the Andaman–Sumatra subduction margin: current understanding. *Island Arc*, **18**, 184–200.

CHU, M. F., CHUNG, S. L., SONG, B., LIU, D., O' REILLY, S. & PEARSON, N. J. 2006. Zircon U–Pb and Hf isotope constrains on the Mesozoic tectonics and crustal evolution in southern Tibet. *Geology*, **34**, 745–748.

COBBING, E. J., PITFIELD, P. E. J., DARBYSHIRE, D. P. F. & MALLIK, D. I. J. 1992. The granites of the South-East Asian tin belt. *British Geological Survey Overseas Memoir*, **10**, 369.

CURRAY, J. R. 2005. Tectonics and history of the Andaman Sea region. *Journal of Asian Earth Sciences*, **25**, 187–232.

CURRAY, J. R., MOORE, D. G., LAWVER, L. A., EMMEL, F. J., RAITT, R. W., HENRY, M. & KIECKHEFER, R. 1979. Tectonics of the Andaman Sea and Burma. *In*: WATKINS, J., MONTADERT, L. & DICKERSON, P. W. (eds) *Geological and Geophysical Investigations of Continental Margins*. American Association of Petroleum Geologists, Memoirs, **29**, 189–198.

CURRAY, J. R., EMMEL, F. J. & MOORE, D. G. 2003. The Bengal Fan: morphology, geometry, stratigraphy, history and processes. *Marine and Petroleum Geology*, **19**, 1191–1223.

DARBYSHIRE, D. P. F. & SWAINBANK, I. G. 1988. *Geochronology of Thai granites. Geochronology of a selection of granites from Burma*. NERC Isotope Geology Centre Report, **88**.

DeCELLES, P. G., GEHRELS, G. E., NAJMAN, Y., MARTIN, A. J., CARTER, A. & GARZANTI, E. 2004. Detrital geochronology and geochemistry of Cretaceous – Early Miocene strata of Nepal: implications for timing and diachroneity of initial Himalayan orogenesis. *Earth and Planetary Science Letters*, **227**, 313–330.

DICKINSON, W. R. 1985. Interpreting provenance relations from detrital modes of sandstones. *In*: ZUFFA, G. G. (ed.) *Provenance in Arenites*. NATO Advanced Study Institutes Series C, Mathematical and Physical Science. Springer, Berlin, **148**, 333–361.

DICKINSON, W. R. & SUCZEK, C. 1979. Plate tectonics and sandstone compositions. *American Association of Petroleum Geologists Bulletin*, **63**, 2164–2182.

DICKSON, J. A. D. 1966. Carbonate identification and genesis as revealed by staining. *Journal of Sedimentary Petrology*, **36**, 491–505.

FITCH, T. J. 1972. Plate convergence, transcurrent faults, and internal deformation adjacent to southeast Asia and western Pacific. *Journal of Geophysical Research*, **77**, 4432–4460.

FOLK, R. I. 1974. *Petrology of Sedimentary Rocks*. Hemphill, Austin, Texas.

GRIFFIN, W. L., WANG, X., JACKSON, S. E., PEARSON, N. J., O'REILLY, S. Y., XU, X. & ZHOU, X. 2002. Zircon chemistry and magma mixing, SE China: in-situ analysis of Hf isotopes, Tonglu and Pingtan igneous complexes. *Lithos*, **61**, 237–269.

GRIFFIN, W. L., BELOUSOVA, E. A., SHEE, S. R., PEARSON, N. J. & O'REILLY, S. Y. 2004. Archean crustal evolution in the northern Yilgarn Craton: U–Pb and Hf-isotope evidence from detrital zircons. *Precambrian Research*, **131**, 231–282.

HTUT, T., MYINT, K., KO, K., MYINT, A. Z., WIN, K. S. & NAING, T. T. 2010. *Evidence of petroleum systems on the coast of Rakhine and implications for offshore petroleum exploration*. Myanmar Geological Society (MGS).

HUTCHINSON, C. S. 1989. *Geological Evolution of South-East Asia*. Oxford University Press, New York.

JACKSON, S. E., PEARSON, N. J., GRIFFIN, W. L. & BELOUSOVA, E. A. 2004. The application of laser ablation-inductively coupled plasma-mass spectrometry to in situ U–Pb zircon geochronology. *Chemical Geology*, **21**, 47–69.

LIANG, Y.-H., CHUNG, S.-L. ET AL. 2008. Detrital zircon evidence from Burma for reorganization of the eastern Himalayan river system. *American Journal of Science*, **308**, 618–638.

LUDWIG, K. R. 2003. Isoplot 3.00 – a geochronological toolkit for Microsoft Excel. Berkeley Geochronological Center, Berkeley, CA, Special Publications, **4**, 71.

MANGE, M. A. & MAURER, H. F. W. 1992. *Heavy Minerals in Colour*. Chapman & Hall, London.

MARTENS, J. H. C. 1932. Piperine as an immersion medium in sedimentary petrography. *American Mineralogist*, **17**, 198–199.

MITCHELL, A. H. G. 1974. Flysch–ophiolite successions: polarity indicators in arc and collision type orogens. *Nature*, **248**, 747–749.

MITCHELL, A. H. G. 1981. Phanerozoic plate boundaries in mainland SE Asia, the Himalayas and Tibet. *Journal of the Geological Society, London*, **130**, 109–122.

MITCHELL, A. H. G. 1985. Collision-related fore-arc and back-arc evolution of the northern Sunda Arc. *Tectonophysics*, **116**, 323–334.

MITCHELL, A. H. G. 1993. Cretaceous-Cenozoic Tectonic events in the western Myanmar (Burma) – Assam region. *Journal of the Geological Society, London*, **150**, 1089–1102.

MITCHELL, A. H. G., HTAY, M. T., HTUN, K. M., WIN, M. N., OO, T. & HLAING, T. 2007. Rock relationships in the Mogok metamorphic belt, Tatkon to Mandalay, central Myanmar. *Journal of Asian Earth Sciences*, **29**, 891–910.

MORTON, A. C. & HALLSWORTH, C. R. 1999. Processes controlling the composition of heavy mineral assemblages in sandstones. *Sedimentary Geology*, **124**, 3–30.

NAJMAN, Y., BICKLE, M. ET AL. 2008. The Paleogene record of Himalayan erosion: Bengal basin, Bangladesh. *Earth and Planetary Science Letters*, **273**, 1–14.

NAJMAN, Y., ALLEN, R. ET AL. 2012. The record of Himalayan erosion preserved in the sedimentary rocks of the Hatia Trough of the Bengal Basin and the Chittagong Hill Tracts, Bangladesh. *Basin Research*, **24**, http://dx.doi.org/10.1111/j.1365–2117.2011.00540.x

NORMAN, M. B. 1974. Improved technique for selective staining feldspar and other minerals using amaranth. *Journal of Research US Geological Survey*, **2**, 73–79.

OGG, J., OGG, G. & GRADSTEIN, F. 2008. *The Concise Geological Time Scale*. Cambridge University Press, Cambridge.

PATCHETT, P. J. 1983. Importance of the Lu-Hf isotopic system in studies of planetary chronology and chemical evolution. *Geochimica et Cosmochimica Acta*, **47**, 81–91.

PATCHETT, P. J. & TATSUMOTO, M. 1980. Hafnium isotope variations in oceanic basalts. *Geophysical Research Letters*, **7**, 1077–1080.

PIVNIK, D. A., NAHM, J., TUCKER, R. S., SMITH, G. O., NYEIN, K., NYUNT, M. & MAUNG, P. H. 1998. Polyphase deformation in a fore-arc/back-arc basin, Salin Subbasin, Myanmar (Burma). *American Association of Petroleum Geologists Bulletin*, **82**, 1837–1856.

PUPIN, J. P. 1980. Zircon and granite petrology. *Contributions to Mineralogical Petrology*, **73**, 207–220.

RAVIKANT, V., WU, F. Y. & JI, W. Q. 2011. U–Pb age and Hf isotopic constraints of detrital zircons from the Himalayan foreland Subathu sub-basin on the Tertiary palaeogeography of the Himalaya. *Earth and Planetary Science Letters*, **304**, 356–368.

SEARLE, M. P., NOBLE, S. R., COTTLE, J. M., WATERS, D. J., MITCHELL, A. H. G., HLAING, T. & HORSTWOOD, M. S. A. 2007. Tectonic evolution of the Mogok metamorphic belt, Burma (Myanmar) constrained by U–Th–Pb dating of metamorphic and magmatic rocks. *Tectonics*, **26**, http://dx.doi.org/10.1029/2006TC002083

SENGUPTA, S., RAY, K. K., ACHARYYA, S. K. & DESMITH, J. B. 1990. Nature of ophiolite occurrence along the eastern margin of the Indian plate and their tectonic significance. *Geology*, **18**, 439–442.

STEPHENSON, D. & MARSHALL, T. R. 1984. The petrology and mineralogy of Mt. Popa volcano and the nature of the late Cainozoic Burma volcanic arc. *Journal of the Geological Society, London*, **141**, 747–762.

SZULC, A. G., NAJMAN, Y. ET AL. 2006. Tectonic evolution of the Himalaya constrained by detrital $^{40}$Ar $^{39}$Ar, Sm–Nd and petrographic data from the Siwalik foreland basin succession, SW Nepal. *Basin Research*, **18**, 375–391.

THOMPSON, R. W. 1974. Mineralogy of sands from the Bengal and Nicobar Fans, Site 218, 211, Eastern Indian Ocean. *In*: VON DER BORCH, C. C. & SCLATER, J. C. (eds) *Initial reports of the Deep Sea Drilling Project, Leg 22*. US Government Printing Office, Washington, DC, 711–713.

UDDIN, A. & LUNDBERG, N. 1998. Unroofing history of the Eastern Himalaya and the Indo-Burman Ranges: heavy mineral study of Cenozoic sediments from the Bengal Basin, Bangladesh. *Journal of Sedimentary Research*, **68**, 465–472.

UDDIN, A., KUMAR, P., SARMA, J. N. & AKHTER, S. H. 2007. Heavy mineral constraints on the provenance of Cenozoic sediments from the foreland and basins of Assam and Bangladesh: Erosional history of the eastern Himalayas and the Indo-Burman Ranges. *Developments in Sedimentology*, **58**, 823–847.

VANACHTERBERGH, E., RYAN, C. G., JACKSON, S. E. & GRIFFIN, W. L. 2001. Data reduction software for LA-ICP-MS. *In*: SYLVESTER, P. J. (ed.) *Laser Ablation ICP-Mass Spectrometry in the Earth Sciences: Principles and Applications*. Mineralogical Association of Canada, Quebec, Short Course Series, **29**, 239–243.

YOKOYAMA, K., AMANO, K., TAIRO, A. & SAITO, Y. 1990. Mineralogy of silts from the Bengal Fan. *In*: COCHRAN, J. R., STOW, D. A. V. *ET AL.* (eds) *Proceedings of the Ocean Drilling Project, Scientific Results*, College Station, TX, **116**, 59–73.

# Heavy mineral record of Andean uplift and changing sediment sources across the NE margin of South America: a case study from Trinidad and Barbados

HASLEY VINCENT[1,*], GRANT WACH[1] & YAWOOZ KETANNAH[2]

[1]*Department of Earth Sciences, Dalhousie University, Life Sciences Centre, Halifax, Nova Scotia, Canada B3H 4J1*

[2]*Department of Geology, Salahaddin University, Erbil, Iraq*

**Corresponding author (e-mail: hvincent@dal.ca)*

**Abstract:** The heavy mineral compositions of sandstones in Trinidad and Barbados record the onset of Andean-related erosion and a reduction of craton-derived sediments into NE South America. The changing provenance was deduced by comparing heavy mineral assemblages interpreted from ancient sandstones with associations recognized in modern sands that can be reasonably correlated to existing tectonic domains. The impact of the Andean orogeny across the margin was to introduce a suite of minerals characteristic of low-temperature metamorphism that today is prevalent adjacent to the Caribbean Mountain belt and differs from the zircon-rich assemblage produced within cratonic plains. Twenty-one Paleocene–Late Pliocene sandstone samples from Trinidad revealed systematic changes in mineral diversity and maturity that recorded this provenance transition, and suggests Andean erosion during deposition of the Late Oligocene Nariva Formation. Similar to Palaeogene sandstones of Trinidad, four Eocene Scotland Formation samples from Barbados support craton derivation, but with additional evidence of minor Andean input probably due to the proximity of the Scotland Formation delivery systems to an earlier uplift episode. By the Late Miocene, most of the sediments delivered into Trinidad basins were supplied from the Andean orogeny as suggested by the relative abundance of minerals of this affinity. The heavy mineral records of Trinidad and Barbados are similar to that described across northern South America from both modern and ancient environments that collectively mark the uplift of the Andean mountain belt, with its strong influence on drainage patterns and reservoir provenance along this sector of the continental margin.

**Supplementary material:** Sample location coordinates, sample and outcrop photographs, and summary outcrop sections are available at http://www.geolsoc.org.uk/SUP18728.

The ability for spot sampling coupled with the geological discrimination derived from heavy mineral analysis makes this scientific tool ideal to elucidate sandstone reservoir provenance in structurally complex areas. The islands of Trinidad and Barbados are ideal case studies given their locations on active plate margins, the relatively limited rock exposures and their status as hydrocarbon producers. Both islands are part of the prolific Upper Cretaceous petroleum system that exists across northern South America (Rodrigues 1988; Requejo *et al.* 1994; Hill & Schenk 2005). Trinidad lies at a tectonic divide along the NE South American continental margin where, to the south of the island, the continental coastline continues as a passive margin to Brazil and to the west; the margin is extensively deformed by the northern extension of the Andean mountain belt (Fig. 1). Here, the Andean mountain belt consists of a number of accreted terrains that includes the Merida Andes of western Venezuela and Colombia, and the Caribbean Mountain belt that extends from western Venezuela through the Northern Range of Trinidad to the North Coast Schists of Tobago (Fig. 1) (Dengo 1953; Kugler 1953; Gonzales de Juana *et al.* 1968; Maresch 1974; Frost & Snoke 1989; Donovan 1994; Pindell *et al.* 2005). Barbados lies approximately 290 km NE of Trinidad and is considered a 'type' exposure of an accretionary prism complex (Speed 1985) because of its location on the Barbados Ridge, a tectonic melange formed from the subduction of Atlantic oceanic crust beneath the Caribbean Plate (Speed 1985; Brown & Westbrook 1987). The Ridge continues south into the deep-water area just east of Trinidad. To the west of the Barbados Ridge is the Tobago Trough forearc basin, the Lesser Antilles island arc and the Grenada back-arc basin, all associated with the subducting Atlantic Plate (Fig. 1).

Plate tectonic models for the Caribbean Plate infer either an in inter-American origin (James

*From:* SCOTT, R. A., SMYTH, H. R., MORTON, A. C. & RICHARDSON, N. (eds) 2014. *Sediment Provenance Studies in Hydrocarbon Exploration and Production.* Geological Society, London, Special Publications, **386**, 217–241. First published online February 6, 2014, http://dx.doi.org/10.1144/SP386.19

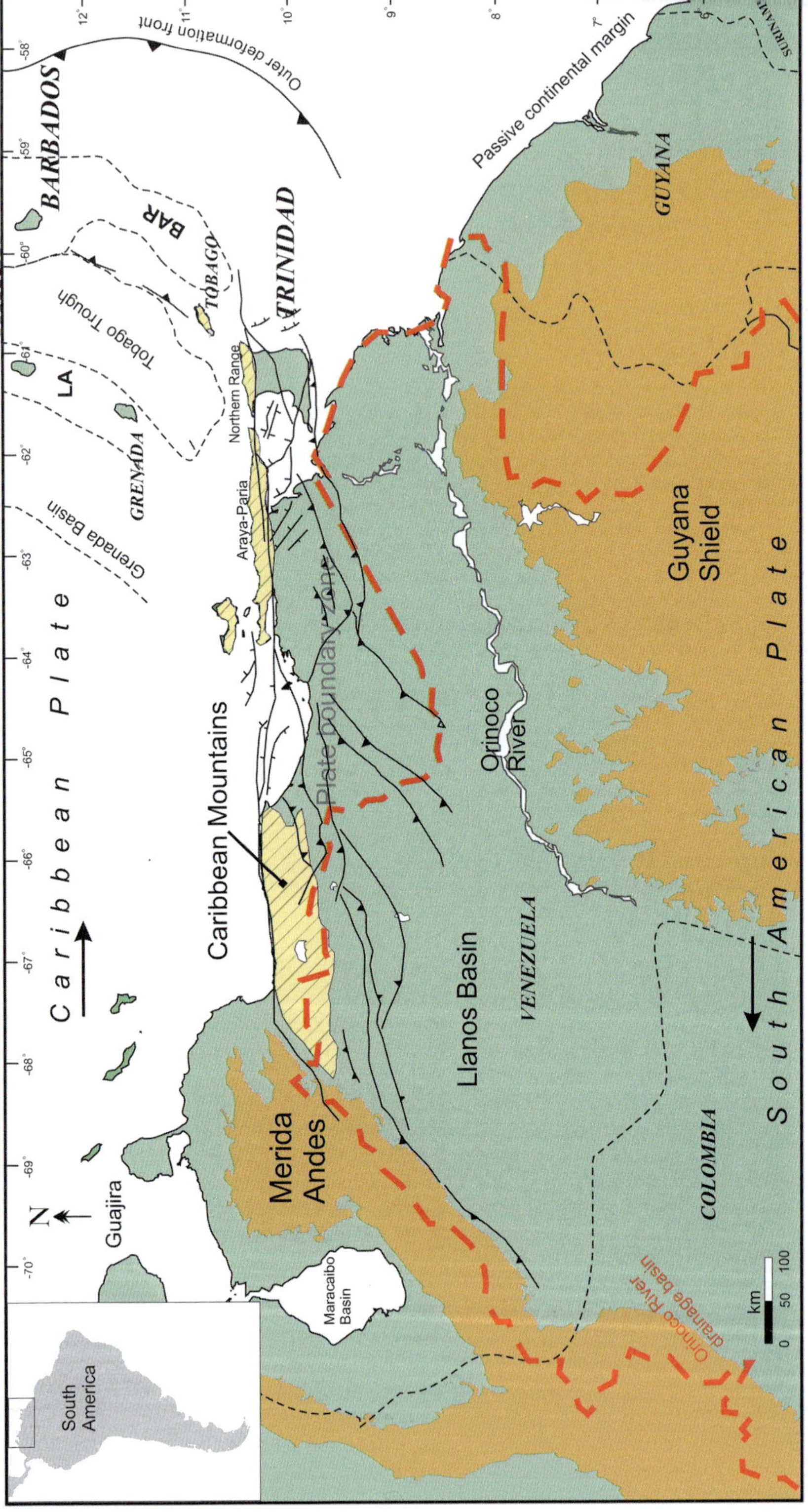



**Fig. 1.** Map of northern South America showing the locations of Trinidad and Barbados relative to major structural and geomorphic features that include the Caribbean Mountain belt, the Llanos Basin and the Guyana Shield, which are discussed in the text. The northern margin is extensively deformed by a wide plate boundary zone that separates the Caribbean and South American plates. The arrows depict the relative direction of plate motion. The extent of the Orinoco River drainage basin is after Warne *et al.* (2002). BAR, Barbados Ridge; LA, Lesser Antilles volcanic arc.

2006) or far-travelled eastwards migration from the Pacific, with related convergence and right lateral strike-slip components (Burke *et al.* 1984; Duncan & Hargraves 1984; Pindell 1994; Weber *et al.* 2001*a*; Pindell & Kennan 2007; Pindell *et al.* 2009). The latter appears to be the more commonly referenced model (Meschede & Frisch 1998) despite the numerous arguments for an inter-American origin. The various plate tectonic models- tend to converge on a relative eastwards plate migration since the Oligocene (Meschede & Frisch 1998; James 2006) at an approximate velocity of 20 mm a$^{-1}$ (Weber *et al.* 2001*a*).

The differing plate migration models variably explain the nature and timing of deformation across the northern margin of South America. Earlier pre-plate tectonic models for the evolution of the NE margin hypothesized Paleocene highs along the northern South American margin that were the sediment source for Eocene deep-water turbidites. This was largely based on lithological relationships (including the assumption of a coarse-grained 'flysch' and 'wildflysch' facies proximity to this uplift) and heavy mineral differences in contemporaneous sands (Senn 1940; Hedberg 1950; Kugler 1953; Kugler & Saunders 1967). Similarly, early plate tectonic models envisaged Late Cretaceous–Early Cenozoic arc–continent collision and uplift of oceanic basement within the western Caribbean Mountains (Maresch 1974). Since then, the recurrent interpretation of Paleocene–Eocene highs was based on a variety of data and techniques including plate motion reconstructions, and well, seismic and field-based studies (Morris *et al.* 1990; Tyson & Ali 1991; Tyson *et al.* 1991; Wielchowsky *et al.* 1991; Pindell *et al.* 2005; Higgs 2009). Alternatively, some models suggest the perpetuation of passive margin basin settings in the Trinidad area to the Oligocene, with Caribbean Plate deformation thereafter (Pindell & Barrett 1990, 1998; Erikson & Pindell 1993; Pindell 1994; Algar 1998; Pindell *et al.* 2005). More complex hybrid models invoke multiple phases of deformation across the NE margin including Paleocene–Eocene 'proto-Caribbean' and Oligo-Miocene 'Caribbean'-involved deformation (Kennan & Pindell 2007; Pindell & Kennan 2007; Pindell *et al.* 2009). These models invoke a Paleocene–Eocene passive to active margin collision, followed by later emplacement of the 'Caribbean' allochthon (that equates to the modern Andean belt).

The various models differ in their timing for the passive to active margin transition, and this has implications for palaeogeographical reconstructions and the prediction of reservoir distribution across the NE margin. It is well known that the Andean uplift led to a major reorganization of depositional systems (Hoorn *et al.* 1995; Diaz de Gamero

1996; Aslan *et al.* 2003). Geo- and thermochronology studies across the Andean mountain belt provided some resolution to the timing of uplift. Apatite fission track (AFT) cooling ages vary from 1.4 to 24 Ma (Early Miocene–Pleistocene) in the vicinity of the Merida Andes (Kohn *et al.* 1984*a*, *b*; Erikson *et al.* 2012), 'Late Oligocene' in north-central Venezuela (Pérez de Armas 2005) and 6.1–12 Ma (Miocene) in the eastern Caribbean Mountains (Kohn *et al.* 1984*b*; Weber *et al.* 2001*b*). These cooling ages generally agree with white mica Ar/Ar crystallization ages of 25 Ma (Early Miocene) for the Northern Range (Speed & Foland 1990; Weber *et al.* 2001*b*) and collectively suggest Neogene deformation across the margin. An even earlier initiation of uplift is, however, suggested by Middle Eocene AFT cooling identified in western Venezuela (Pérez de Armas 2005) and Tobago (Cerveny & Snoke 1993), and Late Eocene (35 Ma) AFT cooling in the eastern Serrania del Interior (Locke & Garver 2005), and implies that there still remains some ambiguity regarding the uplift history across the margin.

Locally, Trinidad comprises a series of tectonic uplifts and basins juxtaposed across strike-slip faults. The highest elevation and oldest rocks (Jurassic) occur in the Northern Range metamorphic belt, which is bounded to the north and south by a series of deformed basins with Cretaceous and younger sedimentary fill. These basins have cumulatively produced commercial volumes of hydrocarbons for more than a century and include the Northern Basin, Southern Basin, Columbus Basin, Gulf of Paria, Central Range and its offshore extension (Fig. 2). Most of the hydrocarbon production to date was produced from Late Miocene and younger deltaic reservoirs, often attributed to a 'proto-Orinoco' depositional system (e.g. Michelson 1976; Strainforth 1978; Leonard 1983; Wood 2000). Deltaic lobes of this system have been variably interpreted to originate from Guyana Shield tributaries to the SW and SE (e.g. Barr *et al.* 1958; Michelson 1976; Stainforth 1978; Leonard 1983) and Andean tributaries to the north and NW (Michelson 1976; Babb & Mann 1999; Kugler *et al.* 2001; Pindell *et al.* 2009). Babb & Mann (1999) additionally suggested the Central Range as a source for some of the Late Pliocene sands present in the Gulf of Paria. Today, the Orinoco drainage basin comprises approximately 50% Llanos plains, 35% Guyana Shield and 15% Andean mountains sources (Warne *et al.* 2002).

Hydrocarbon production in Barbados occurs only from the Woodbourne Trough where the Eocene Scotland Formation sandstones comprise the single reservoir unit. The proposed source of these clastics is the South American continent with contributions from metamorphic and

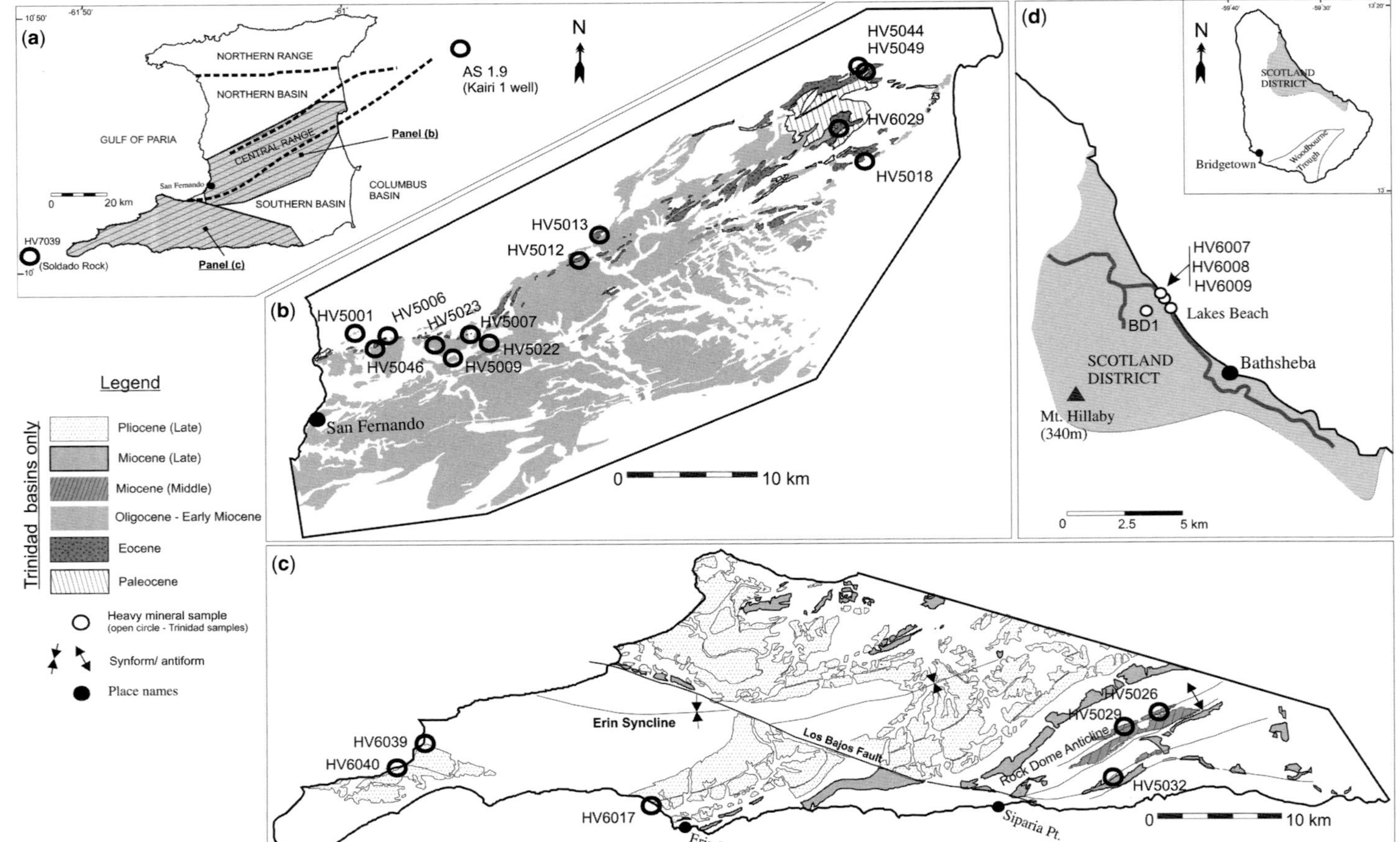


**Fig. 2.** Sample locations across major basins in (**a**)–(**c**) Trinidad and (d) Barbados. (a) Primary basins across Trinidad and the location of two offshore samples from Soldado Rock (HV7039) and Kairi 1 well (AS 1.9). The locations of panels (b) and (c) are also shown. (b) Samples and their related outcrop ages across the Central Range and (c) the Southern Basin of Trinidad. (**d**) The oil-producing Woodbourne Trough of Barbados and the Scotland District to the north, where the Eocene reservoirs are exposed; the locations of samples are shown. Trinidad geology after Saunders *et al.* (1998).

quartz-rich sources (Larue & Speed 1983; Pudsey 1985). These reservoirs occur among a mass of tectonically deformed Eocene–Early Miocene sedimentary rocks that lie beneath Pleistocene coral terraces (Fig. 3) and are uplifted up to 325 m above sea-level (Saunders 1979). Stratigraphic age relationships among these units are constrained through biostratigraphy of claystone units such as the Oceanic Formation (Fig. 3).

This paper describes heavy mineral suites in coarse clastics across Trinidad and Barbados basins, and outlines some of the implications for provenance and general basin configuration across the NE continental margin (Figs 2 & 3). The paper utilizes well-known provenance relationships established from modern sands sampled in rivers and basins around Trinidad, the Guyana shelf and alluvial plains across northern South America, and attempts to demonstrate the replication of these relationships in the ancient rock record. Confidence in these provenance relationships and an appreciation for their tectonic significance were obtained from a more profound understanding of heavy mineral sorting and dissolution (Morton 1979; Garzanti & Andò 2007b; Komar 2007) and, overall, it will be demonstrated that the heavy mineral record suggests a relatively simple margin evolution compared to current models.

In addition, mainstream literature rarely alludes to the history of heavy mineral research across the region that began in the early twentieth century (e.g. Illing 1928; Matley 1932; Senn 1940; J. C. Griffiths in Suter 1960; Harry 1992; Kugler 2001) and, with the exception of Pindell et al. (2009), geological constraints using heavy mineral compositions are generally absent from palaeogeographical reconstructions across the NE margin. The need for an updated study has become increasingly important given the general inaccessibility of older studies, and the fact that ages and formation names used in these studies have become obsolete or modified (Kugler 1956, 1959; Carr-Brown & Frampton 1979; Carr-Brown 1998; Saunders et al. 1998). The heavy mineral attributes of at least one unit, the Angostura sandstone, has not before been published. This is an oil-producing

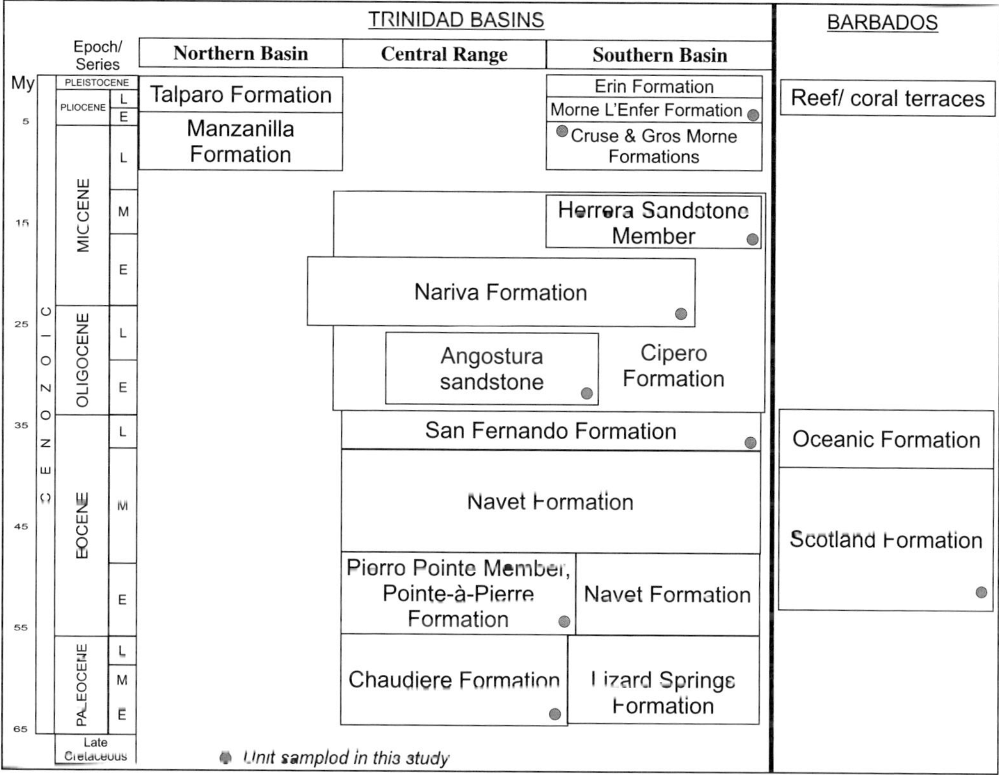


**Fig. 3.** Simplified stratigraphic table for Trinidad and Barbados showing the relative superposition of lithological units that are discussed in the text. Those sampled for this study are also highlighted.

sandstone in the offshore Central Range (Fig. 2a), which has been informally dated as Oligocene (Carr-Brown *et al.* 2002).

## Heavy mineral analysis and results

### Sample collection and preparation

Twenty-one heavy mineral separates were made from Cenozoic sandstone outcrops and core plugs across Trinidad, and four separates from outcrops of the Eocene Scotland Formation in Barbados (Figs 2 & 3). All samples are from litho- and biostratigraphical rock units as designated by Kugler (1959), Saunders *et al.* (1998) and Kugler (2001), although there is some uncertainty concerning the age of sample HV5018 (Plum Mitan) collected near the contact between Eocene and Late Oligocene formations. For each separate, 120 g of the sample were disaggregated using a ceramic mortar and pestle, and wet-sieved in distilled water into four grain-size categories (>1 mm; 0.25–1 mm; 0.0625–0.25 mm; and <0.0625 mm). Samples were then oven dried with low heat (60 °C) for 12–16 h. The 0.0625–0.25 mm sand fraction (fine to very fine) was chosen for gravity separation following Morton (1985), and the mineral separation procedure was guided by Mange & Maurer (1992). Following gravity separation in sodium polytungstate at 2.89 g cm$^{-3}$ (Callahan 1987; Mange & Maurer 1992), each sample was weighed and optical thin sections then mounted in Canada balsam. A few samples were selected for polished thin sections to conduct microprobe analysis. Sample BD-2 of the Scotland Formation was originally intended for fission track counts, and was disaggregated using a jawcrusher and disc-mill followed by Wilfley table density separation, sieving

**Table 1.** *Total heavy mineral counts of samples from the Cenozoic of Trinidad and Eocene of Barbados*

| | Sample No. | Formation | Age | Location (see Fig. 2) | Zircon | Tourmaline | Rutile | Garnet | Chloritoid | Chlorite | Kyanite |
|---|---|---|---|---|---|---|---|---|---|---|---|
| | HV6017 | Morne L'Enfer | Pliocene | Puerto Grande Bay | 6 | 48 | 16 | 3 | 12 | – | 9 |
| | HV6039 | Morne L'Enfer | Pliocene | Esperanza Bay | 8 | 34 | 12 | 10 | 19 | – | – |
| | HV6040 | Morne L'Enfer | Pliocene | Cedros Bay | 30 | 31 | 7 | 13 | 21 | – | 3 |
| | HV5032 | Cruse | Late Miocene | Moreau Road (2) | 33 | 43 | 5 | 11 | 23 | 111 | – |
| | HV5026 | Herrera Sandstone | Middle Miocene | Rock River road, Moruga | 84 | 53 | 10 | 17 | 4 | 1 | – |
| | HV5029 | Herrera Sandstone | Middle Miocene | Moreau road | 109 | 29 | 18 | 23 | 6 | 7 | – |
| | HV5018 | Cipero/Nariva | Early Miocene | Plum Mitan | 84 | 181 | 32 | 5 | – | – | – |
| Trinidad | HV5007 | Nariva | Late Oligocene | Esmeralda Junction | 97 | 128 | 33 | 7 | 28 | 50 | – |
| | HV5009 | Nariva | Late Oligocene | Kelly Hill | 85 | 86 | 52 | 8 | 74 | – | – |
| | HV5022 | Nariva | Late Oligocene | Corbeaux Hill | 154 | 93 | 61 | 3 | 65 | – | – |
| | HV5023 | Nariva | Late Oligocene | Sandstone Trace | 161 | 36 | 54 | 33 | 67 | – | – |
| | AS1.9 | Angostura | Oligocene | Kairi 1 well - 5264 ft | 194 | 19 | 8 | 3 | – | – | – |
| | HV5001 | San Fernando | Middle Eocene | Plaisance conglomerate | 184 | 92 | 28 | 4 | – | – | – |
| | HV7039 | San Fernando | Late Eocene | Soldado Rock | 86 | 93 | 8 | 43 | – | – | – |
| | HV5006 | Pointe-à-Pierre | Middle Eocene | Caratal Road | 233 | 57 | 29 | 4 | – | – | – |
| | HV5012 | Pointe-à-Pierre | Middle Eocene | Tabaquite sawmill | 253 | 37 | 22 | – | – | – | – |
| | HV5013 | Pointe-à-Pierre | Middle Eocene | Allen Road | 237 | 57 | 32 | – | – | – | – |
| | HV5046 rec | Pointe-à-Pierre | Middle Eocene | Fabien Road quarry | 242 | 53 | 16 | 3 | – | – | – |
| | HV6029 | Pointe-à-Pierre | Middle Eocene | Chaudière River, Mount Harris | 160 | 91 | 63 | 4 | – | – | – |
| | HV5044 | Chaudiere | Paleocene | Forest Park ('Growing rock') | 264 | 45 | 5 | – | – | – | – |
| | HV5049 | Chaudiere | Paleocene | Forest Park ('Growing rock') | 259 | 33 | 7 | – | – | – | – |
| Barbados | BD1 | Chalky Mount | Middle Eocene | Chalky Mount | 264 | 65 | 25 | 4 | – | 20 | 3 |
| | HV6007 | Chalky Mount | Middle Eocene | Lakes Beach coast section | 130 | 51 | 24 | 36 | 33 | 51 | – |
| | HV6008 | Chalky Mount | Middle Eocene | Lakes Beach coast section | 127 | 106 | 44 | 6 | 30 | – | 4 |
| | HV6009 | Chalky Mount | Middle Eocene | Lakes Beach coast section | 200 | 69 | 28 | 12 | 6 | 1 | 6 |

The Trinidad samples are arranged in stratigraphic order.

to 0.25 mm and liquid fractionation as described above.

Grain counts sampled at least 250 translucent grains per slide, with most samples attaining over 300 counts. Total counts exceeded 500 for most samples and up to a maximum of 1100 counts in one case, with the majority of grains being opaque. The line counting method (Mange & Maurer 1992) was used on a Leitz polarizing microscope with 10× oculars, up to 50× objectives and a mechanical point-counter. Muscovite and biotite were excluded in the heavy mineral counts, although chlorite was included as its sudden appearance, abundance and mineral association was considered significant for provenance determinations; it is likely that the counts underestimated their abundance. The entire slide was examined for minerals that may have been missed during the counts and these were classified as 'accessories'. The identification of minerals was guided by standard texts (Kerr & Rogers 1977; MacKenzie & Guilford 1980; Mange & Maurer 1992; Nesse 2004; Perkins & Henke 2004).

## Summary of heavy mineral assemblages

The heavy minerals of Cenozoic sandstones are listed in Table 1, graphically portrayed in Figure 4 and some of the more common minerals are illustrated in Figure 5. Zircon is a dominant mineral in most sandstones and, together with tourmaline and rutile, is present in all samples. For simplification, the characteristic heavy minerals were grouped into three assemblages based on relative mineral abundance (Fig. 4). The first assemblage ('1') is dominated by the ultrastable minerals, with zircon, tourmaline and rutile comprising >90% of translucent minerals. Secondary minerals include epidote,

| Andalusite | Sillimanite | Epidote Group | Staurolite | Actinolite | Apatite | ? Brookite | ? Limonite | Unrecognized | Opaque | Total count | Total transluscent | Accessories |
|---|---|---|---|---|---|---|---|---|---|---|---|---|
| 16 | 10 | 170 | – | 14 | – | – | – | 16 | 287 | 607 | 320 | Amphibole?, biotite, apatite, limonite, staurolite |
| 6 | 2 | 173 | – | – | 2 | – | – | 4 | 205 | 475 | 270 | |
| 23 | 2 | 201 | – | – | 6 | – | – | 6 | 453 | 796 | 343 | Hornblende |
| – | – | 6 | – | – | 39 | – | – | 16 | 253 | 540 | 287 | Staurolite |
| – | – | 11 | – | – | 9 | – | 111 | 5 | 802 | 1107 | 305 | |
| 1 | – | – | – | – | 5 | 1 | 100 | 5 | 854 | 1158 | 304 | Kyanite, epidote, ?sphene |
| – | – | 1 | – | – | – | 5 | 30 | 4 | 509 | 851 | 342 | ?sphene (2), sillimanite, ?pumpeyllite |
| – | – | 7 | – | – | – | 1 | 13 | 7 | 594 | 965 | 371 | |
| – | – | 2 | – | – | – | 4 | 9 | 6 | 432 | 758 | 326 | |
| – | – | 11 | – | – | – | – | 8 | 1 | 212 | 608 | 396 | |
| – | – | 9 | – | – | – | 1 | – | 10 | 223 | 594 | 371 | ?Sillimanite |
| – | – | 13 | – | – | – | – | 3 | 17 | 722 | 979 | 257 | |
| – | – | – | – | – | – | – | – | 1 | 279 | 588 | 309 | Sillimanite, epidote–clinozoisite |
| 1 | – | 4 | 46 | – | 29 | – | – | 8 | 536 | 854 | 318 | |
| – | – | 12 | – | – | – | – | 10 | 4 | 456 | 805 | 349 | Staurolite, andalusite, chlorite |
| – | – | – | – | – | – | 1 | 1 | 6 | 246 | 566 | 320 | |
| – | – | – | – | – | – | – | – | – | 123 | 449 | 326 | |
| – | – | 2 | 5 | – | – | – | – | – | 379 | 700 | 321 | |
| – | – | – | – | – | – | – | 3 | 1 | 379 | 701 | 322 | Clinozoisite–zoisite |
| – | – | – | – | – | – | – | – | 5 | 272 | 591 | 319 | Sillimanite, ?garnet |
| – | – | – | – | – | – | 13 | – | 5 | 208 | 525 | 317 | Saussurite, ?sphene, uralite, tremolite–actinolite, chloritoid, epidote, ?kyanite |
| 10 | 3 | – | 9 | – | – | – | – | 3 | 398 | 806 | 408 | |
| 11 | – | 4 | – | – | – | – | – | 6 | 368 | 714 | 346 | Kyanite, limonite, ?apatite |
| 2 | 1 | 11 | – | – | – | – | – | 8 | 447 | 786 | 339 | ?Anatase |
| 8 | 1 | 2 | 6 | – | – | – | – | 2 | 483 | 824 | 341 | |

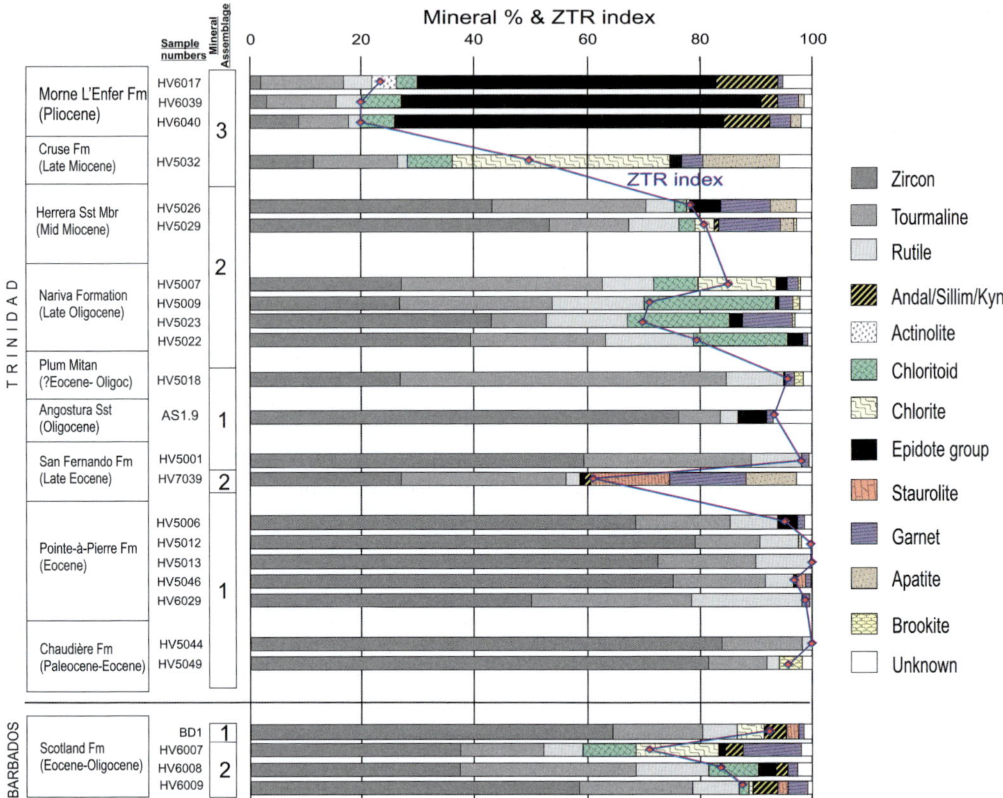



**Fig. 4.** Composite bar chart showing actual counts of translucent heavy minerals and their relative contribution to the total fraction. The ZTR index for each sample is also shown and highlights the increasing maturity with age of Trinidad rock units, which are displayed in stratigraphic order (except for HV5018, for which there is some age uncertainty). All samples have been grouped into three mineral assemblages based on the relative amount of ultrastable minerals to allow comparisons between ages and locations.

garnet, staurolite and brookite, with sillimanite, andalusite, chloritoid, chlorite and possibly kyanite as accessories. This assemblage is characteristic of Palaeogene samples from Trinidad basins (including sample HV5018 from Plum Mitan) and one sample from the Scotland Formation of Barbados. A second assemblage ('2') is characteristic of Late Oligocene–Middle Miocene samples from Trinidad basins and also includes one Late Eocene sample. Three of four samples from the Scotland Formation of Barbados also fall into this category. This assemblage is similarly dominated by ultrastable minerals, but also contains 'common' minerals such as chloritoid, staurolite, apatite, chlorite and garnet occurring collectively in significant proportions (>20% of translucent grains). Garnet and staurolite account for up to 29% of translucent minerals in a sample from the Late Eocene San Fernando Formation (Fig. 4, sample HV 7039), while chloritoid alone averaged 16% ($n = 4$) in samples

from the Late Oligocene Nariva Formation and 7% from the Scotland Formation ($n = 3$). Detrital chlorite is also locally significant (13% of translucent minerals in one sample from the Nariva Formation and 15% in one sample from the Scotland Formation: Fig. 4). Two samples from the Herrera Sandstone Member also fall into this assemblage, with ultrastables comprising less than 70% of translucent minerals. Accessory minerals include kyanite and sphene. The mineralogy of the Scotland Formation samples are noteworthy since they are largely classified within a different assemblage from contemporaneous Trinidad samples (Pointe-à-Pierre Formation) and demonstrate a greater mineralogical similarity to the Late Oligocene Nariva Formation (Trinidad). This is discussed in more detail below.

The third assemblage ('3') is the most diverse, and characteristic of Late Miocene and Pliocene samples in Trinidad basins, and is not recognized from the Barbados samples. It is characterized by ultrastable minerals accounting for less than 50% of translucent grains, with abundant epidote (40% of translucent minerals, $n = 4$), detrital chlorite (39% of one sample), and a variety of other minerals, including actinolite, sillimanite, kyanite, andalusite, hornblende, apatite, staurolite, chloritoid and biotite. Biotite is recorded only from this assemblage, and apatite was found in three of four samples and accounted for up to 14% of translucent minerals in the Late Miocene sample HV5032 (the most recorded in the dataset) (Figs 4 & 5).

With the exception of one Late Eocene sample in the Trinidad dataset, the three assemblages occur in sequence with decreasing age and suggest a systematic variation in heavy mineral composition throughout the Trinidad Cenozoic. This is further evident in the decreasing zircon–tourmaline–rutile (ZTR) index with decreasing age (Fig. 4). Before provenance implications are explored, the impact of grain dissolution is considered in the following subsection as it is well established that heavy minerals are susceptible to chemical dissolution, the effects of which vary with factors such as the age of the rock, mineral species, stage of sedimentary cycle, relief, burial depth and sediment permeability (Pettijohn 1941; Hubert 1962; Nickel 1973; Morton 1979, 1985; Morton & Johnsson 1993; Garzanti & Andò 2007a). The sequential changes in mineralogy may also be attributed to varying sediment sources and it will be demonstrated that there are some similarities between the assemblages described above and mineral groups recognized in modern sands. This will also be elaborated before provenance interpretations are explored.

*Impact of dissolution on mineral assemblage*

The effect of grain dissolution was evaluated using the combined ZTR and heavy mineral concentration (HMC) indices (Hubert 1962; Garzanti & Ando 2007a). The ZTR index is an indicator of the relative concentration of ultrastable heavy minerals and overall compositional maturity, with higher ZTR values suggesting greater maturity. The HMC is a measure of the relative yields of transparent heavy minerals after corrections are applied for carbonates or phyllosilicates (e.g. chlorite is not included in the assessment) (Garzanti & Andò 2007b). It is the abundance of all heavy minerals within the fine- to very-fine-grained sand portion and the weighted average densities of total framework and total dense grains, expressed as a ratio. Its underlying principle is that different sources can be associated with low and high heavy mineral concentrations (e.g. chert or limestone v. ultramafic sources,

respectively), and a declining HMC with increasing age of samples is suggestive of grain dissolution. Together, both indices provide a measure for the effect of dissolution on a grain population (Garzanti & Andò 2007a), assuming a common source throughout time.

Within the Trinidad samples, the ZTR index suggests decreasing compositional maturity in younger sandstones (Fig. 4), and this may be attributed to either dissolution of mineral species in stratigraphically older sandstones, or the addition of newer mineral species in the younger, related to provenance. The HMC index suggests that the addition of newer mineral species is more likely. The HMC index for Cenozoic sandstone ranges from 0.2 to 4.0 and is indicative of generally low heavy mineral yields (Fig. 6). There is no difference in the range of values with increasing age, which would be expected if dissolution were the main determinant on heavy mineral abundance (e.g. Garzanti & Andò 2007a). Similarly, when considered against the ZTR index, there is no difference in yields with increasing compositional maturity; that is, a similar quantity of heavy minerals per unit rock volume was derived from both stable and diverse assemblages. Some of the best yields were even derived from older samples (e.g. Fig. 6, Herrera Sandstone Member and Pointe-à-Pierre Formation), whereas the opposite would be expected if grain dissolution were significant. In general, very similar yields were derived from all samples, regardless of age, which suggests that there was a constant supply of heavy minerals from low-yielding source rocks throughout the Cenozoic, without appreciable loss of minerals in the receiving basin either by dissolution or sediment reworking. For Scotland Formation samples, the heavy mineral yield, as indicated by the HMC index, is relatively low and similar to those of the Trinidad sandstones.

Collectively, it was therefore assumed that mineral dissolution was not detrimental in changing the grain population within the receiving basin and the observed heavy mineral assemblages are representative of their 'depositional' character. This assumption is partially supported by the similarity between modern and ancient mineral associations (discussed in the next subsection) that suggests modern associations are being successfully preserved in the rock record. The likely preservation of a 'source' signature in older samples is further supported by the common occurrence of apatite (a mineral susceptible to acidic weathering) within a sample from the Eocene San Fernando Formation. Indirect supporting evidence is gained from the appreciable amounts of feldspar that are preserved in Cretaceous sandstones of Trinidad (Barr 1962; Vincent 2008), demonstrating that some of the

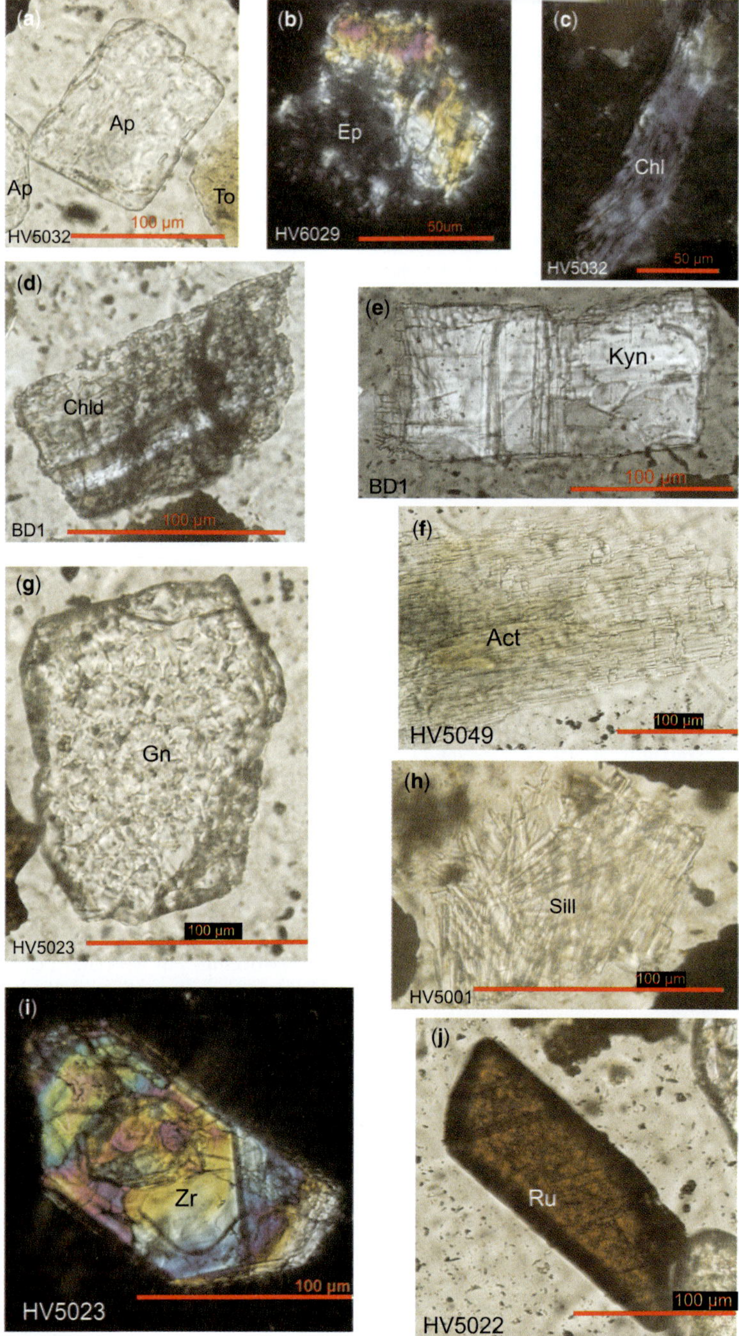

**Fig. 5.** Typical heavy minerals found in Trinidad and Barbados Cenozoic sandstones. (**a**) Colourless, prismatic apatite (plane polarized light, PPL) common in Neogene samples; (**b**) anhedral epidote showing multiple interference colours (cross-polarized light, XPL); (**c**) anomalous interference colour of chlorite, common in Late Miocene samples, XPL; (**d**) *c*-axis-parallel chloritoid, showing twinning, is abundant in Late Oligocene and younger samples PPL; (**e**) orthogonal cleavage and pale interference colours of kyanite, XPL; (**f**) actinolite showing bladed habit and pale-green colour, PPL; (**g**) colourless garnet crystal displaying typical high relief, occurs throughout the Cenozoic, PPL; (**h**) colourless and fibrous sillimanite fragment, PPL; (**i**) euhedral zircon crystal with high relief, zoned and

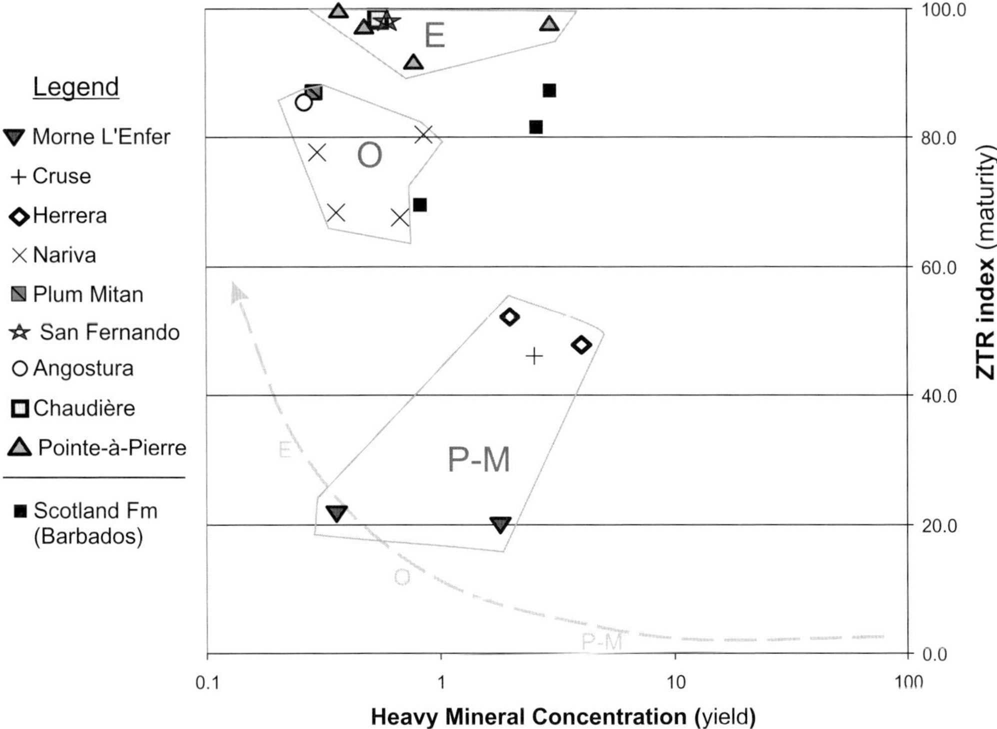


**Fig. 6.** Heavy mineral concentration (HMC) plotted against ZTR index for Cenozoic sandstones showing the relative heavy mineral yields against sample maturity. There is no apparent change in heavy mineral yield with changing mineralogical maturity, as would be expected if grains were lost to dissolution (trend shown by the dashed arrow). Dashed polygons encircle Pliocene–Late Miocene (PM), Oligocene (O) and Eocene–Paleocene (E) Trinidad samples.

more susceptible minerals remain available for provenance study.

## Sources for Cenozoic sediments

Late Cretaceous and Cenozoic sediment sources into NE South American basins are considered relative to two general tectonic regimes; those associated with a relatively stable craton; and those associated with active deformation, including any potential volcanic arc sources.

The Guyana Shield (Fig. 1) is a long-lived cratonic source throughout the Phanerozoic development of the margin (e.g. van Andel & Postma 1954; Barr et al. 1958; Nota 1958; van Andel 1958; Jones 1968; Baldwin et al. 1986; Goldstein et al. 1997; Algar 1998; Pindell et al. 2005). The

Shield comprises approximately 35 000 km$^2$ of exposed Archaean, Precambrian and Palaeozoic crystalline continental basement rock at an estimated thickness greater than 2400 m across portions of eastern Venezuela, Guyana, Suriname, French Guyana and northern Brazil, including the basement complex below the Barinas-Apure–Llanos basins (Lopez et al. 1942; Edmond et al. 1995). Rock types consist of numerous intrusive and extrusive igneous rocks of both continental and oceanic origin, variably metamorphosed in successive deformation events, intercalated with continental sandstones and quartzites (Almeida 1978; Santos et al. 2000). The Shield has been an emergent feature since the Middle Precambrian and has historically been drained by several large rivers (Hoorn et al. 1995; Diaz de Gamero 1996).

**Fig. 5.** (*Continued*) multiple high-order interference colours is ubiquitous, XPL; (**j**) rutile crystal with very high relief and dark brown colour is also ubiquitous, PPL. Minerals (a)–(g) are typical of mineral assemblages 2 and 3, while assemblage 1 is dominated by zircon and rutile (together with tourmaline, not illustrated).

The geological longevity, drainage patterns and various rock types present across the Shield combine for an established sediment source to the NE South American basins, including the Trinidad area.

Studies involving modern sands across northern South America have highlighted the role of extensive, 'transport-limited' (Stallard *et al.* 1990) cratonic lowland plains, such as the Barinas-Apure–Llanos basins, as important sediment storage centres and modifiers of sediment composition by enhanced chemical weathering. Although these depocentres cannot be considered parent sediment sources, the warm–wet climate and the development of thick soil horizons in these areas are arguably the most important determinants in the composition of sands delivered to the coasts from craton and foreland regions (Franzinelli & Potter 1983; Johnsson *et al.* 1988; Johnsson 1990; Stallard *et al.* 1990). These lowland plains act as temporary transport 'sinks' and sediment 'filters' producing increasingly mature, first-cycle sediments that are ultimately delivered to river mouths and continental shelves. Compositional links to specific source rocks, both within the Shield and from the Cenozoic uplifts, become ambiguous in this domain (e.g. van Andel & Postma 1954).

Sediment sources associated with a deforming continental margin include the Caribbean Mountain and Merida Andes uplifts. Today these uplifts are a significant source of detritus into South American basins (Stallard *et al.* 1990; Morton & Johnsson 1993; Potter 1994; Warne *et al.* 2002), with initiation suggested by thermochronological and radiometric data varying between the Late Eocene and Miocene (Maresch 1974; Beets *et al.* 1984; Kohn *et al.* 1984*a, b*; Speed & Foland 1990; Cerveny & Snoke 1993; Smith *et al.* 1999; Weber *et al.* 2001*b*). The Caribbean Mountain belt is the more proximal of these uplifts to Trinidad and Barbados, and generally consists of several east–west-trending, fault-bounded tectonic units of metamorphic rocks, with metamorphic grade generally increasing in a northerly direction in Venezuela and decreasing towards the east of Trinidad (Maresch 1974; Frey *et al.* 1988; Bellizzia & Dengo 1990; Smith *et al.* 1999; Weber *et al.* 2001*b*). The belt mostly comprises greenschist-facies metamorphic rocks, although with widely varying lithology. Within the Cordillera de la Costa alone (including the Northern Range of Trinidad), rock types comprise slates and chloritic phyllites, sericite–epidote and quartz–mica schists, blueschists, amphibolites, eclogites, serpentinites, limestone, marble, feldspathic gneiss, conglomerates, zeolite–prehnite–pumpellyite metamorphic facies, and gypsum beds (Dengo 1953; Gonzales de Juana *et al.* 1968; Potter 1968; Maresch 1974; Algar 1993; Smith *et al.* 1999).

The deformed basement complex is also varied, and includes granitic gneiss of the Sebastopol Group (Dengo 1953) to the west of the mountain belt and hypothetical oceanic crust below the Northern Range of Trinidad to the east (Pindell & Kennan 2007).

North of the Caribbean Mountains, Mesozoic plutonic and volcanic basement rocks are exposed across Tobago (Tobago Volcanic Group) and include mafic volcaniclastic breccias, ultramafic rocks, gabbro-diorite and biotite–hornblende–tonalite (Maxwell 1948; Snoke *et al.* 2001). The basement thermal history suggests that these rocks have occupied upper-crustal levels since the mid-Cretaceous (Cerveny & Snoke 1993), although they were probably too far west to influence mineralogical signatures at that time (Pindell & Kennan 2007). Today, they are exposed to erosion and were probably a source of clastics into the Trinidad basins during the later Cenozoic. Abundant pyroxenes within Eocene turbidites on the island of Grenada (Saunders *et al.* 1985) (see Fig. 1 for the location) suggest active volcanic arc sources in basins north of South America at least since that time. The Lesser Antilles island-arc complex represents the modern equivalent that has been active at least since 21 Ma (Duncan & Hargraves 1984), and is a highly probable source for volcanic detritus into Trinidad and Barbados basins.

The Central Range of Trinidad is a modern topographical feature with initiation of uplift variably interpreted as Middle Miocene (Erlich *et al.* 1993) and Plio-Pleistocene, and has been cited as a source for Pliocene sediments (Babb & Mann 1999). Uplift should, theoretically, be reflected in the erosional record as the recycling of older sedimentary particles. Various other hypothetical Late Cretaceous–Eocene uplifts have been suggested for the Trinidad area which would have acted as potential active margin sediment sources (e.g. Senn 1940; Hedberg 1950; Kugler & Saunders 1967; Tyson & Ali 1991; Pindell *et al.* 2005; Higgs 2009; Pindell *et al.* 2009). Pindell *et al.* (2009), for example, suggested the existence of a Late Cretaceous–Eocene submarine ridge north of the present-day location of the Caribbean Mountains, which was attributed to inversion tectonics along the margin at that time. It is assumed that the erosional history of these ridges would be recorded in the sedimentary mineral suites.

The results of studies involving modern sands across the northern South American continent provided a basis for differentiating between the sources mentioned above. These studies have shown that the fill of many basins (e.g. Barinas-Apure, Llanos and Eastern Venezuelan) is strongly reflective of the proximity to major tectonic or provenance domains, and, for the most part, a differentiation can be made between the craton and uplift sources (Franzinelli & Potter 1983; Johnsson *et al.* 1988; DeCelles & Hertel 1989; Stallard *et al.* 1990; Potter 1994). At least three groups are apparent in the modern distribution of particular minerals relative to the major tectonic domains described above (Fig. 7). Chloritoid is characteristic of the first group (here termed the CM petrological association) as it is commonly found in basins adjacent to the Caribbean Mountain uplift and, together with lawsonite and glaucophane, is particularly distinctive due to a relative absence in cratonic-sourced sands. Epidote, where occurring as the most abundant mineral, may also be diagnostic of this group when compared to other domains (Fig. 7, charts 1 & 5). A second group is characterized by ultrastables, epidote, hornblende and other minerals found along the Orinoco drainage basin and river mouth (Fig. 7, OR petrological association). Although ubiquitous, zircon is clearly the most abundant mineral in this category. A third group is found along the Guyana shelf just SE of Trinidad (Fig. 7, GS petrological association), where NW flowing littoral currents distribute sands from adjacent Guyana Shield sources (Nota 1958; Wong *et al.* 1998; Aslan *et al.* 2003). Within this group, sillimanite and andalusite appear distinctive as they occur only as accessory minerals within the other groups. These three petrological association, apparent from modern sands, provide a first approximation for provenance determination in the ancient rock record.

Distinctive mineral signatures can also be gauged from the rock compositions within the potential source domains. The Caribbean Mountains contain low-pressure, low-temperature (LP–LT) metamorphosed rocks with minerals that are characteristic of low-grade metamorphism such as chlorite, epidote, chloritoid and actinolite (Winkler 1974). Frey *et al.* (1988) lists chlorite as a primary mineral in samples from the Northern Range of Trinidad, and noted the occurrence of actinolite, chloritoid and epidote among the metamorphic mineral suite. Caribbean Mountain zones of high-pressure–low-temperature (HP–LT) metamorphism are the blueschists of the Villa de Cura and Cordillera de la Costa, characterized by the additional presence of glaucophane and lawsonite, with kyanite being the stable aluminosilicate (Maresch 1974). These minerals were all identified in modern sands of the Apure drainage basin of Venezuela and used to differentiate Caribbean Mountain from Merida Andes-derived sands (Morton & Johnson 1993) (Fig. 7, chart 1). Local sources of epidote exist within the Caribbean Mountain greenschist-facies rocks such as the Tacagua Formation of the Caracas Group (Dengo 1953) and San Souci metabasalts of eastern Northern

Range (van Andel & Postma 1954, their table D-1). These sources may account for the epidote flood characteristic of the 'CM' petrological association in the Gulf of Paria (Fig. 7, chart 5).

The abundant ultrastable minerals of the Orinoco (OR) petrological association found in modern sands at the mouth of the Orinoco River are reasonably well explained within a transport-limited 'source' framework and its distinctive mineralogical signature, with dominant zircons, tourmaline, rutile and monocrystalline quartz. The minerals of this group have long been associated with Llanos basin plains and intense chemical weathering (van Andel & Postma 1954). Within this 'source' domain, minerals readily susceptible to weathering, such as apatite, are unlikely to be present in significant numbers (Morton & Johnsson 1993).

The common andalusite and sillimanite in sediments along the Guyana and Suriname shelf is not surprising given its stability in LP–HT metamorphic rocks that are present in the Guyana Shield. Within ancient rocks, sillimanite has been recorded in abundance in regionally metamorphosed granites of the Kanuku Group in southern Guyana and granite gneiss of the Imataca Province of the Guyana Shield (Kalliokoski 1965; Singh 1968). It is notably absent from the Caribbean Mountains (Maresch 1974). In the Apure Basin of northern South America, Morton & Johnson (1993) used sillimanite to distinguish modern river sands sourced from the Merida Andes from those derived from Caribbean Mountain sources. It is apparent that the presence of sillimanite in Cenozoic sediments indicates derivation from Shield and other high-grade metamorphic sources; the likelihood is increased when found in association with other high-grade metamorphic minerals such as andalusite, garnet and staurolite (GS petrological association of Fig. 7).

Hornblende is ubiquitous in modern sands around the NE margin of South America and was probably derived from multiple sources. This mineral is widespread in Orinoco-derived sands, which may have originated from amphibolites or similar rocks of the Imataca Province (northern Guyana Shield). Hornblende was also identified in river sands that drain the Northern Range of Trinidad (van Andel & Postma 1954), where volcanic beds in the Maracas Formation are a potential source (Jackson *et al.* 1991). Amphiboles and pyroxenes are scarce in both modern and ancient sands (Fig. 7) (Matley 1932; Senn 1940; van Andel & Postma 1954; Suter 1960). Within modern sands, some contribution comes from the Guyana Shield (GS association) and metabasalts of the San Souci volcanics that crops out to the NE of the Northern Range (van Andel & Postma 1954).

H. VINCENT *ET AL.*

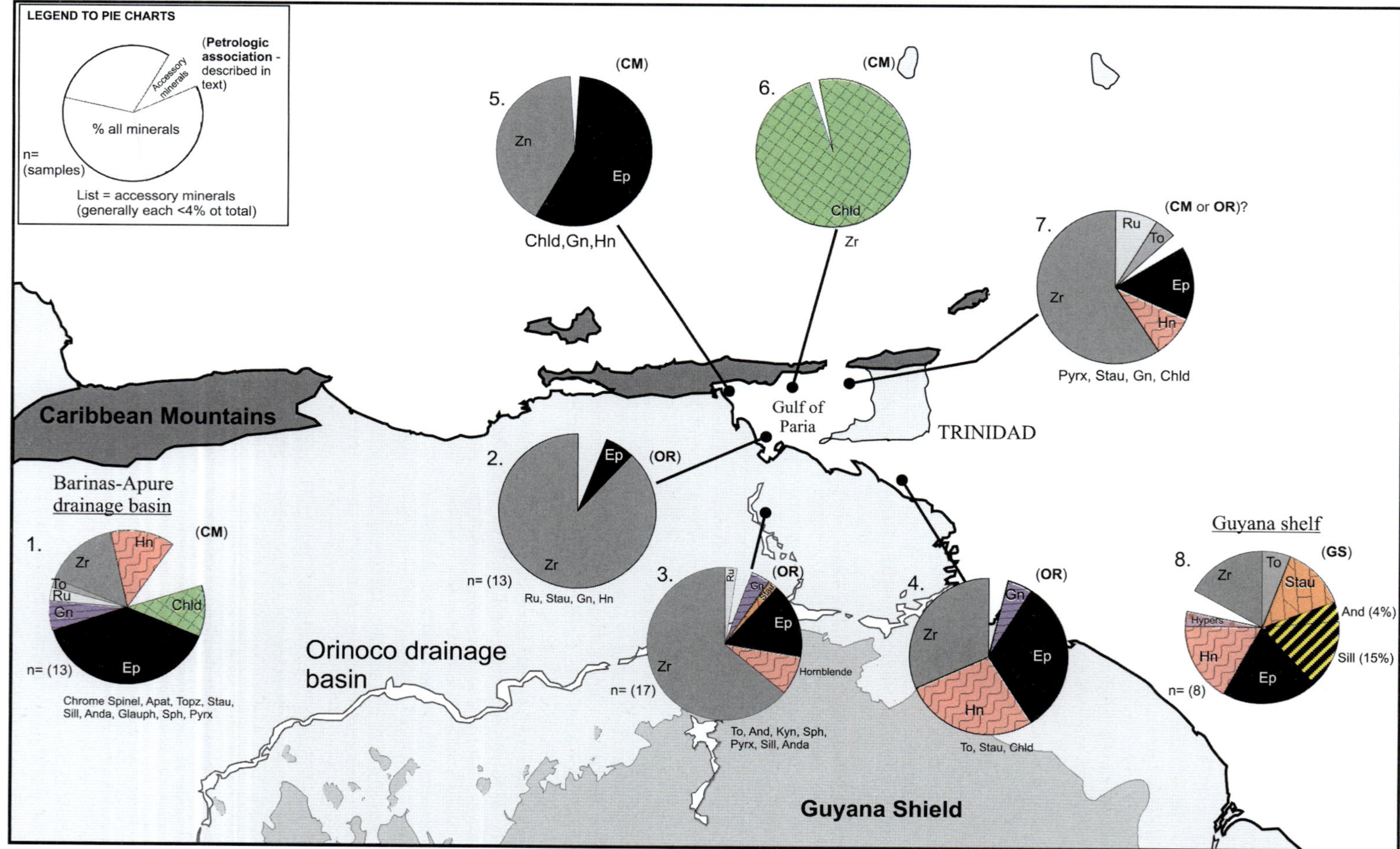



**Fig. 7.** Compilation of heavy mineral percentages from modern sands across the NE margin of South America. The charts highlight the relative abundance of chloritoid and epidote adjacent to the Caribbean Mountains (CM petrological association), and dominance of zircon at the mouth of the Orinoco river (OR petrological association). The Guyana shelf with sediments supplied from the Guyana Shield contains common staurolite, andalusite and sillimanite (GS association). The data were compiled from van Andel & Postma (1954, tables 3 & D1); Nota (1958, table 3); Wong *et al.* (1998); Morton & Johnsson (1993, table 1, 'Caribbean Mountains').

## Sediment provenance interpretations

The preceding discussion established that, first, mineral assemblages identified from the rock record are representative of their depositional diversity and, secondly, diagnostic minerals can be identified from the modern distribution of heavy minerals with respect to sediment sources. The mineralogical maturity of assemblage '1', typical of the Palaeogene Chaudiere, Pointe-à-Pierre, Angostura, Scotland and Plum Mitan sandstones (Fig. 4), is correlated to the zircon-rich sands that are currently being deposited at the mouth of the Orinoco River (OR petrological association) and this suggests inheritance during particle transit across broad cratonic alluvial plains. The occurrence of andalusite, sillimanite and staurolite as accessories in this assemblage (Table 1) suggests parent rock affiliation to an area of medium- to high-grade metamorphism of Al-rich pelites (Winkler 1974; Nockolds *et al.* 1978), the signature for which is now largely removed. Parent rocks may have originated from within both the Guyana Shield and Merida Andes. There are also minor epidote and other accessory minerals that are typical of the CM petrological association (actinolite, chloritoid, chlorite and probable kyanite, Fig. 5f) present in the Trinidad samples and, similar to the Eocene Scotland Formation, hint at an orogenic source at that time. This correlation has not previously been made as these accessories were not recorded in previous heavy mineral studies (Pindell *et al.* 2009).

Chloritoid is typical of assemblage '2' and, together with the presence of chlorite and minerals of the epidote group (epidote, zoisite and clinozoisite), suggests a low-grade metamorphic input for Nariva and Herrera sandstones. Supportive evidence is gained from the correlation of this assemblage with the CM petrological association seen in modern sands adjacent to the Caribbean Mountain belt and this is proposed as the affiliated source. The common chloritoid and chlorite present in the Scotland Formation samples also suggest this unit is affiliated with this source, although samples are dominated by ultrastables with accessory medium- to high-grade metamorphic minerals such as andalusite, sillimanite, staurolite and garnet. Although relatively insignificant (average 2%, $n = 4$), these minerals hint at a contribution from high-grade metamorphic sources. A similar high-grade metamorphic affiliation is suggested by the common garnet and staurolite within the Late Eocene sample of the San Fernando Formation (Fig. 4, sample HV7039), although garnet is not restricted to that rock type (Winkler 1974; Nesse 2004).

The diversity of minerals of assemblage '3' that includes the Late Miocene and younger rocks

in Trinidad basins suggest numerous sources, including high-grade metamorphic (sillimanite–kyanite–andalusite–staurolite–hornblende), low-grade metamorphic (chloritoid–chlorite–actinolite) and igneous (hornblende) rocks. Elements of the three petrological associations are present and, together with the common occurrence of apatite, suggests that sediment sources other than the cratonic plains, with its intense chemical weathering, were active.

In order to assess the relative contributions from the various sediment sources, sample mineralogy was viewed on a ternary plot whereby each pole represented one of the petrological domains interpreted from modern sands and suggested from sample counts (Fig. 8a). Each pole is the sum of index minerals for transport-limited, high- and low-grade metamorphic petrological domains (TL–HG–LG), normalized to 100%. Samples with higher ultrastable suites will plot closer to the 'transport-limited/reworked sedimentary' pole, which emphasizes the contribution of source domains with an abundance of mature heavy minerals, almost to the exclusion of all other minerals (i.e. high ZTR index). This pole is interpreted to represent intracratonic basins, craton lowlands or 'stable' basins within the Guyana Shield or Llanos Basin. In addition, it may represent areas of uplift and erosion of mature sedimentary rocks, and other criteria will be required to differentiate between these two basin settings (e.g. textural immaturity). A second index emphasizes the importance of LT–HP/LP metamorphic sources, such as the Caribbean Mountains, and is sensitive to low-grade, greenschist-facies metamorphic mineral assemblages. Index minerals for this pole include those characteristic of Caribbean Mountain detritus plus chlorite, and compare to the 'CM' association of heavy minerals (Fig. 7). The third index emphasizes the role of HT metamorphic minerals for which sources will include the Guyana Shield and Merida Andes of Venezuela; the minerals in this index compare best with the 'GS' association for modern sands (Fig. 7).

Samples from the Paleocene–Eocene of Trinidad, characterized by assemblage '1' minerals, distinctively cluster around the transport-limited pole on the TL–LG–HG ternary plot (Fig. 8a), suggesting derivation almost exclusively from the cratonic plains where dissolution during transport have largely eliminated ties to any parent rock within the Shield or elsewhere. Accessory minerals only hint at high- and low-grade metamorphic rocks as found today on the Guyana shelf and Caribbean Mountains, respectively.

Similar to the Paleocene–Eocene sandstones, ultrastable heavy minerals dominate the heavy mineral suite in both the San Fernando Formation

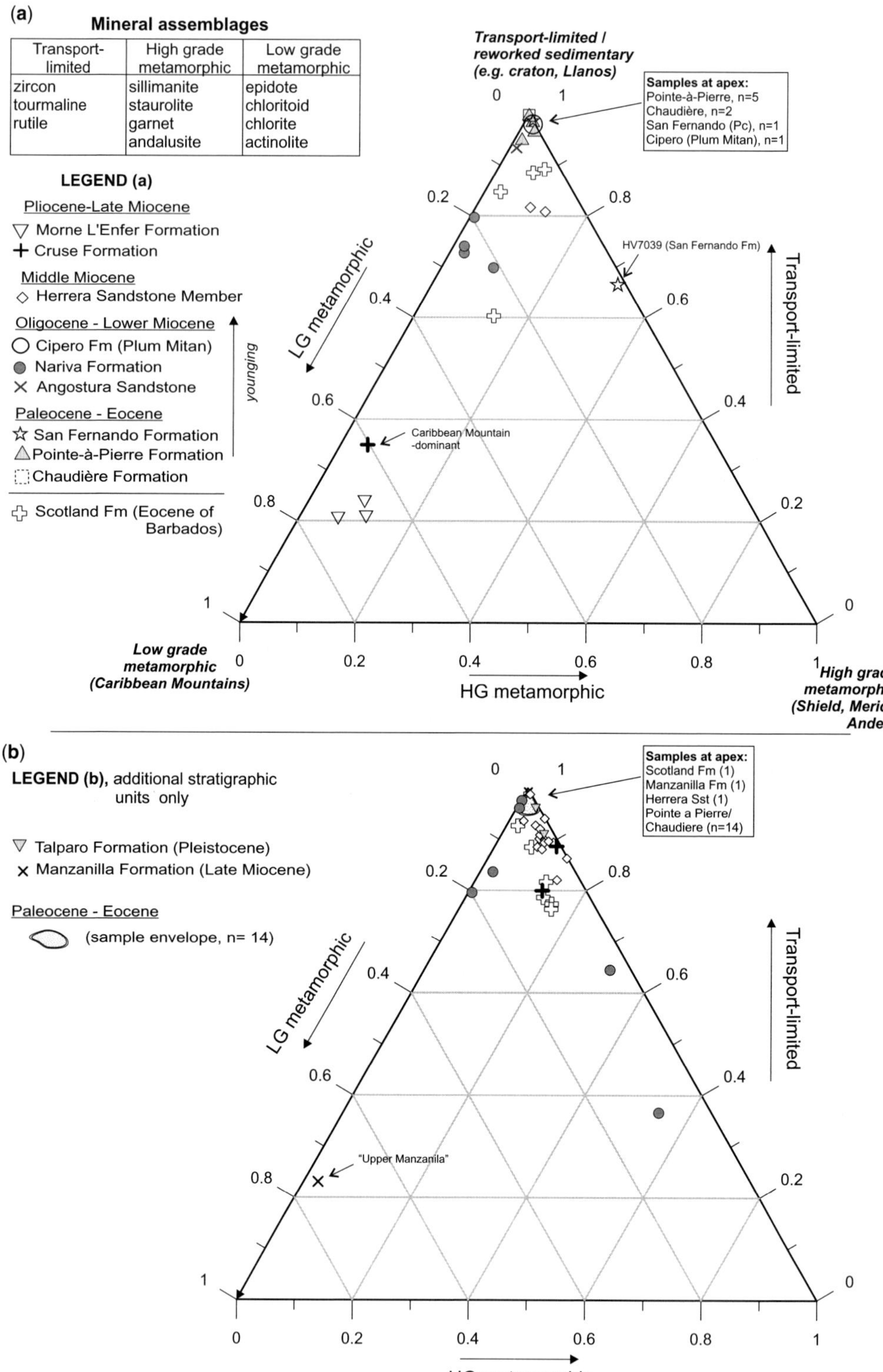
(a)
Mineral assemblages
Transport-limited | High grade metamorphic | Low grade metamorphic
zircon | sillimanite | epidote
tourmaline | staurolite | chloritoid
rutile | garnet | chlorite
| andalusite | actinolite

LEGEND (a)
Pliocene-Late Miocene
Morne L'Enfer Formation
Cruse Formation
Middle Miocene
Herrera Sandstone Member
Oligocene - Lower Miocene
Cipero Fm (Plum Mitan)
Nariva Formation
Angostura Sandstone
Paleocene - Eocene
San Fernando Formation
Pointe-à-Pierre Formation
Chaudière Formation
Scotland Fm (Eocene of Barbados)

Transport-limited /
reworked sedimentary
(e.g. craton, Llanos)
0   1
Samples at apex:
Pointe-à-Pierre, n=5
Chaudière, n=2
San Fernando (Pc), n=1
Cipero (Plum Mitan), n=1
0.2   0.8
HV7039 (San Fernando Fm)
0.6
younging
LG metamorphic
0.4
Caribbean Mountain-dominant
0.6   0.4
Transport-limited
0.8   0.2
1   0
Low grade
metamorphic
(Caribbean Mountains)
0   0.2   0.4   0.6   0.8   1
HG metamorphic
High grade
metamorphic
(Shield, Merida Andes)

(b)
LEGEND (b), additional stratigraphic units only
Talparo Formation (Pleistocene)
Manzanilla Formation (Late Miocene)
Paleocene - Eocene
(sample envelope, n= 14)

0   1
Samples at apex:
Scotland Fm (1)
Manzanilla Fm (1)
Herrera Sst (1)
Pointe a Pierre/
Chaudiere (n=14)
0.2   0.8
LG metamorphic
0.4   0.6
0.6   0.4
Transport-limited
"Upper Manzanila"
0.8   0.2
1   0
0   0.2   0.4   0.6   0.8   1
HG metamorphic

(Late Eocene) and the Angostura sandstone sample (Early Oligocene), and these samples plot adjacent to the transport-limited pole of the TL–HG–LG plot, indicative of a similar sediment source as the Chaudière/Pointe-à-Pierre sandstones. Common garnet and staurolite within the San Fernando Formation sample cause a departure from the transport-limited pole towards the high-grade metamorphic pole and suggest an additional contribution from this source. These samples suggest that compositionally mature sediments were still being derived from the South American craton during the Late Eocene–Middle Oligocene.

The prominence of ultrastable minerals within Late Oligocene samples of the Nariva Formation also suggests that most of the coarse clastics were being derived from transport-limited sources. The assemblage '2' heavy mineral proportions of chloritoid and chlorite, however, divert samples away from the transport-limited pole and towards the 'Caribbean Mountain' pole in the TL–HG–LG ternary plot (Fig. 7), suggesting that an eroding greenschist-facies metamorphic orogen was then supplying influential amounts of detritus to these sandstones; this is proposed to be a segment of the Caribbean Mountains.

Ultrastable minerals continue to dominate the heavy mineral assemblage within the Middle Miocene Herrera Sandstone Member, causing samples to plot near the transport-limited pole in the TL–HG–LG ternary plot. Similar to the Nariva Formation sandstones, this suggests a supply primarily of mature cratonic sands that occurred together with a continued supply of low-grade metamorphic detritus of Caribbean Mountain affinity, as indicated by the presence of chloritoid, chlorite and the accessory kyanite.

Assemblage '3' samples from the Cruse and Morne L'Enfer formations plot closest to the 'Caribbean Mountains' pole on the discriminatory ternary diagram attributed to the relative abundance of 'CM' index minerals (epidote + chlorite + chloritoid + actinolite + kyanite). This suggests the uplift was the dominant source of sediments into the Southern Basin from the Late Miocene onwards. The continental sources that supplied mature sands throughout most of the Cenozoic are now suppressed, as reflected in the low ZTR index

values, although their presence is still reflected in the ultrastables (zircons) and high-grade metamorphic minerals (andalusite, sillimanite, garnet) that are present in these samples.

A systematic variation in the relative contribution between cratonic and Caribbean Mountain areas is apparent in the collective Trinidad dataset, with successively younger formations showing a greater degree of divergence from the transport limited pole. This trend is interrupted only by the samples from the Herrera Sandstone Member, with a greater proportion of ultrastables than the older Nariva Formation (Figs 4 & 8). Unlike the Palaeogene sandstones, the framework portion of these samples is texturally immature, consisting of a greater quantity of sedimentary lithic fragments (Vincent 2008), suggesting increased sedimentary erosion and reworking at this time. This reworking is also invoked to explain the 'reversal' in the heavy mineral trend and probably resulted in a dilution of Caribbean Mountain detritus within the Herrera Sandstone Member, being replaced with abundant zircons and opaque minerals (the two samples examined also contained over 70% opaque heavy minerals, the highest recorded in all samples). It is also noteworthy that some of the highest heavy mineral yields were obtained from the two samples of this sandstone (Fig. 6), probably attributed to the sedimentary reworking of the older mineral species and the redeposition of the more resistant opaque minerals and zircons. The Herrera sandstones originated from multiple sources: (1) proximal, reworked sedimentary rocks; (2) the Caribbean Mountain; and (3) the lowlands of the South American craton.

Samples from the Scotland Formation of Barbados plot adjacent to the transport-limited pole on the TL–HG–LG ternary diagram (Fig. 8) and a cratonic source is implied. The divergence and spread from the transport-limited pole is a departure from their chronostratigraphic equivalent within Trinidad basins (Pointe-à-Pierre) attributed to the presence of chlorite and chloritoid, with common sillimanite, andalusite and kyanite. This divergence is similar to that of the Late Oligocene Nariva Formation samples, suggesting a similar input of the Caribbean Mountain detritus for both sandstones. Overall, a mixed sediment source signature is apparent for

---

**Fig. 8.** (a) Ternary plot of the relative abundance of minerals ascribed to the 'transport-limited', 'high-grade metamorphic' and 'low-grade metamorphic' provenance groups for Trinidad and Barbados sandstones. The minerals used to discern these groups are shown in the table. The plot highlights a deviation from the 'transport-limited' pole from the Nariva Formation onwards, and suggests a dominant clastic contribution from 'low-grade' sources (Caribbean Mountains) for the Cruse and Morne L'Enfer formations. (b) Equivalent-aged samples from Pindell *et al.* (2009, supplementary tables) are shown on a similar plot for comparison, and suggests a greater amount of reworking in younger samples, which plot adjacent to the transport-limited pole. Only one sample, 'upper Manzanilla', clearly differentiates the Caribbean Mountain clastic influence. Charuma silt and 'probable' samples are excluded from plot.

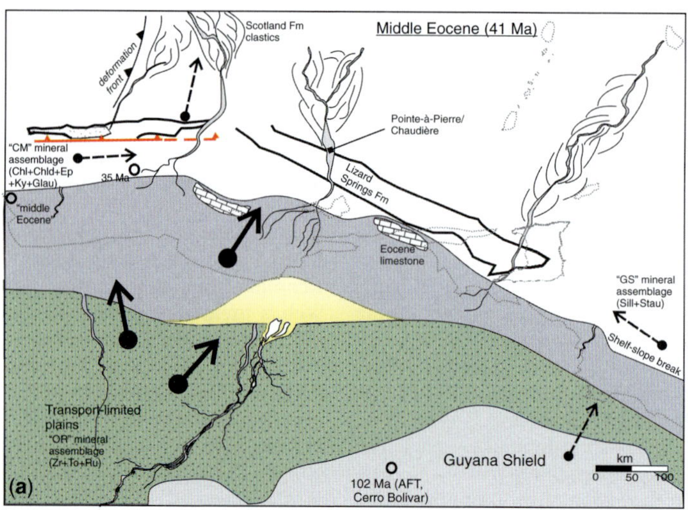


Middle Eocene (41 Ma)
Scotland Fm clastics
deformation front
Pointe-à-Pierre/ Chaudière
Lizard Springs Fm
"CM" mineral assemblage (Chl+Chld+Ep +Ky+Glau)
35 Ma
"middle Eocene"
Eocene limestone
"GS" mineral assemblage (Sill+Stau)
Shelf-slope break
Transport-limited plains
"OR" mineral assemblage (Zr+To+Ru)
Guyana Shield
102 Ma (AFT, Cerro Bolivar)
km
0 50 100
(a)


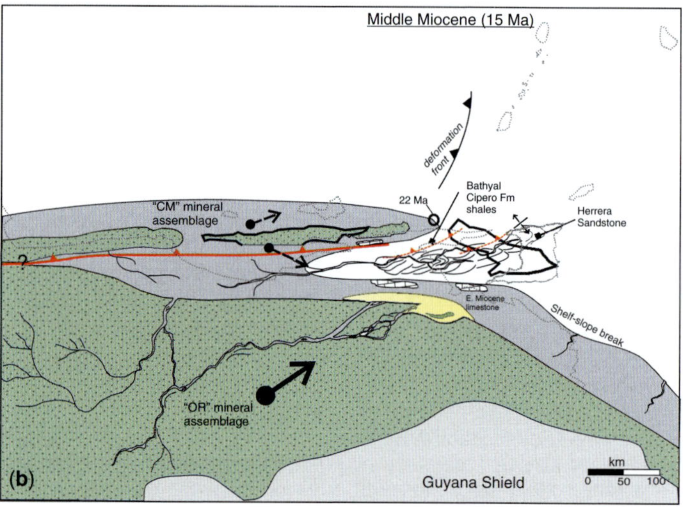


Middle Miocene (15 Ma)
deformation front
22 Ma
Bathyal Cipero Fm shales
Herrera Sandstone
"CM" mineral assemblage
?
E. Miocene limestone
Shelf-slope break
"OR" mineral assemblage
Guyana Shield
km
0 50 100
(b)


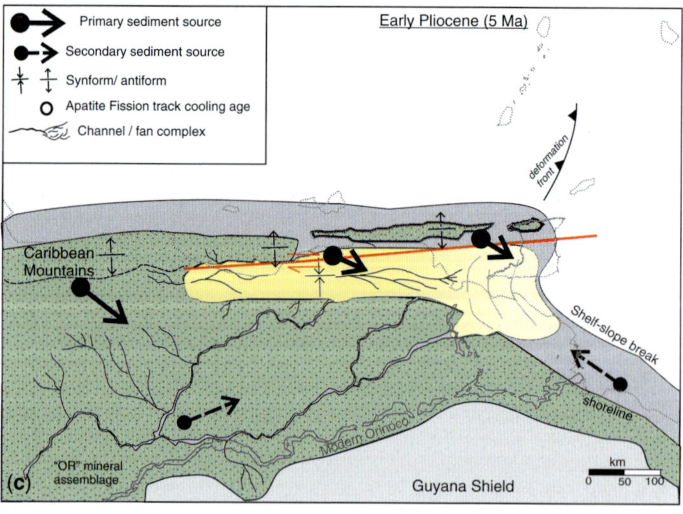


Primary sediment source
Secondary sediment source
Synform/ antiform
Apatite Fission track cooling age
Channel / fan complex
Early Pliocene (5 Ma)
deformation front
Caribbean Mountains
Shelf-slope break
shoreline
Modern Orinoco
"OR" mineral assemblage
Guyana Shield
km
0 50 100
(c)

the Scotland Formation, overwhelmingly dominated by cratonic detritus.

## Discussion

The temporal trends that are evident within samples from this study suggest an overall mixing or dilution of the craton-derived sands that was incipient from the Eocene, clearly evident by the Late Oligocene, until most of the detritus was derived from the Caribbean Mountains during the Pliocene. This trend is interpreted to record the imprint of Andean tectonics on a former passive continental margin, and coincides with apatite reset dates and metamorphic mica ages across the Caribbean Mountain system. Supportive evidence was seen in the framework sandstone constituent of the Nariva Formation, where the changing diversity was coincident with a clear increase in the presence of lithic schists relative to older sandstones (Vincent 2008).

The heavy mineral characteristics and trends are not unique, and conform to observations across the northern South American region that collectively record the influence of Andean erosion around the Late Oligocene–Early Miocene. Erikson *et al.* (2012, p. 149) cited a similar change in heavy mineral diversity along the foothills of the Venezuelan Andes and attributed this to a Miocene unroofing sequence. Hoorn (1993) recognized the Andean sediment input from the Miocene in NW Brazil by a similar change from a mature heavy mineral assemblage to a diverse chloritoid–epidote–garnet assemblage. Along the eastern flank of the Colombian and Venezuelan Andes, changes in sedimentary facies, zircon and fission track thermochronology, and U–Pb geochronology also support a Late Oligocene–Miocene basin event (Horton *et al.* 2010; Erikson *et al.* 2012).

The heavy mineral characteristics and trends also compare well with the genetic clastic domains of Pindell *et al.* (2009) that were summarized into: (1) mature South American-derived assemblages; and (2) orogen-derived assemblages associated with an allochthonous Caribbean Plate. The transition between these two assemblages occurred during the Oligocene. When normalized for the TL–HG–LG ternary plot, one of their Late Miocene samples ('Upper Manzanilla') supports a predominance of Caribbean Mountain detritus at that time owing to the abundance of epidote + clinozoisite in that sample (Fig. 8b). Their collective Cenozoic dataset, however, is less discriminatory of the Caribbean Mountain input, and suggests a greater contribution of cratonic to high-grade metamorphic detritus from the Late Oligocene and younger samples primarily because of the high garnet counts. Recycling of older rocks was proposed to account for the maturity observed in the younger sandstones (Pindell *et al.* 2009).

The heavy mineral variation between the Eocene sandstones of Trinidad and Barbados deserves attention. Similar to the Pointe-à-Pierre Formation, the dominant source of the Scotland Formation was the South American craton, with detritus delivered through meandering rivers that tapped into high-grade metamorphic rocks and meandered across the cratonic plains before eventual deposition as Eocene deep-water clastics (Pudsey & Reading 1982; Larue & Speed 1983; Vincent & Wach 2007). Scotland Formation rivers show greater proportions of Caribbean Mountain input and this has been used to support the existence of an intervening high that separates the two Eocene clastic domains (Senn 1940; Pindell *et al.* 2009). Such an interpretation is reasonable, although it lacks supportive evidence from the sedimentary record that indicates erosion or proximity to such a high. There is no recognized change in the Late Cretaceous–Palaeogene (Pindell *et al.* 2009) heavy mineral assemblage that supports its existence and, similarly, it is not supported by the mature textural attributes of Trinidad Palaeogene sandstones (Vincent 2008). If it existed, the high was either not sufficiently significant to modify the mineralogy (as occurred with the Middle Miocene Herrera sandstone 'reversal', with a supportive abundance of reworked foraminifera, brachiopod shells, chert, mudstone and siltstone fragments within the sandstone framework: Vincent 2008) or the sedimentary record is preserved elsewhere. The presence of Caribbean Mountain detritus as accessory minerals within

---

**Fig. 9.** Summary palaeogeographical maps of the NE margin of South America showing the changing sediment distribution patterns and diachronous margin deformation during (**a**) the Middle Eocene, (**b**) the Middle Miocene and (**c**) the Early Pliocene. For simplification, all structural elements are not shown. The bold polygons are the restored positions of Trinidad and northern Venezuela after Rohr (1991) and Pindell & Kennan (2007), and the position of the 'deformation front' is approximated after Pindell & Kennan (2007). Cerro Bolívar AFT after Dr A. Grist (pers. comm.) and demonstrates the relative 'stability' of the craton. (a) Middle Eocene passive margin deep-water clastics are delivered north of the margin primarily from mature cratonic sources, with relative orogenic contribution dependant on proximity to the uplift. Scotland Formation tributaries are closest to the deformation front and will be subsequently translated over 600 km eastwards with its migration. (b) Middle Miocene deformation front advanced just east of Trinidad, with an increasing supply of orogenic and reworked clastics, although intermediate to the craton-derived sands. (c) The Caribbean Mountains is now the dominant supplier of sediment to Neogene deltas.

Trinidad samples hints at a greater commonality of source than previously appreciated. The mineralogical differences between the Eocene sandstones are here attributed to the relative proximity of 'Scotland' and 'Pointe-à-Pierre' fluvial or shelf delivery systems to an incipient Andean uplift, similar to the decrease seen in modern 'CM' minerals that occurs with increasing distance from the mountain front (e.g. Johnsson 1990). The incipient uplift probably existed to the west of the margin, as suggested by Eocene AFT cooling ages in the vicinity of western Venezuela (Locke & Garver 2005; Pérez de Armas 2005). Plate reconstructions suggest approximately 600 km of eastwards translation for the Barbados Ridge since the Eocene, which means that the source of the Scotland Formation detritus lay at least that distance to the west (Fig. 9) (Pindell & Kennan 2007; Pindell *et al.* 2009).

The Andean uplift had a major impact on sediment detritus along the NE margin, and is arguably the primary reason for the occurrence of the rapid shallowing and thick deltaic sequences observed throughout the Plio-Pleistocene rock record. The mineral diversity and diagnostic assemblages demonstrated in this study are known from all of the major deltaic successions across the Southern Basin of Trinidad, including the Cruse, Forest, Morne L'Enfer and Gros Morne formations (Kugler 2001). It is reasonable to assume that a major reorganization of the palaeo-Orinoco was in effect during the Pliocene, with minor tributaries (in terms of length relative to cratonic streams) contributing significant amounts of sediment into the NE margin. The sediment yield from the Andean front to the NE margin may have greatly exceeded that of the craton, judging by the relative proportion of unstable heavy minerals found in these sediments. This is not surprising given the high estimated sediment discharge derived from smaller and lesser known tributaries that drain relatively small, but high relief terrains (Milliman & Syvitski 1992).

An absence of volcanic assemblages was not anticipated in this study given the proximity to modern and ancient arc-related rocks and plutonic suites along the Caribbean Mountains, the island of Tobago and the Guyana Shield to the south. This finding is, however, consistent with many of the heavy mineral studies on both modern and ancient sands across the margin, and signifies either limited erosion of volcanic terrains or the loss of parent volcanic signatures across the Orinoco drainage basin.

The heavy mineral trends described within Trinidad sandstones suggest a Late Oligocene transition from passive to active margin. Palaeogene clastics were deposited in deep-water palaeoenvironments directly from passive margin cratonic plains (Fig. 9a). This also includes the Angostura

sandstone and 'Plum Mitan' sample (HV5018) as their mature assemblage is similar to other Palaeogene samples, suggesting a close relationship in depositional space and time. The tributaries of the Scotland Formation were most proximal to the deformation front, and this accounts for the varying mineralogy, confinement and amalgamation of these sandstones (Pudsey & Reading 1982; the authors' own field observations). The orogenic detritus accounted for a minor contribution into the contemporaneous Trinidad sandstones, with tributaries located further from this source (Fig. 9a). The evidence for uplift and margin deformation is obvious from the Nariva Formation clastics, although the depositional site is still removed from the Andean front that lay to the west. The encroaching deformation front is evident in the composition and reworking of the Herrera clastics (Fig. 9b). A major reorganization of the basin is evident from the Late Miocene and throughout the Pliocene, with most of the clastic detritus now supplied by the uplifted Caribbean Mountains (Fig. 9c). Beginning with the Cruse Formation, thick deltaic sedimentation and rapid deposition were characteristic of the Trinidad Pliocene (Barr *et al.* 1958; Ablewhite & Higgins 1968; Wood 2000), and this was a direct response to Andean erosion that was well underway at this time and in close proximity to Trinidad basins. Overall, the gradual increase in the Andean influence throughout the Cenozoic is consistent with interpretations for a diachronous migration of the deformation front as it gradually encroached on the eastern continental margin.

Heavy minerals and other mineralogical techniques can contribute significant value to reconstructions along deformed plate margins, where it is very difficult to confidently recreate palaeocurrents and establish sandbody orientations for reservoir intervals. The heavy mineral characteristics described herein provide a good approximation for reservoir fairways and sandbody orientations among and between these two islands. There is still significant benefit to be gained from varietal studies of zircons and garnets given the relative abundance of both minerals as detailed heavy mineral characterization has proved a very useful technique (Mange & Wright 2007). The common apatite in Neogene sands may also be a used to better constrain the uplift histories of highs that bound the Southern and Columbus basins where most of Trinidad's hydrocarbons originate. This study did not include samples of Pleistocene sands, which are currently mined for their mature quartzose composition, suggesting a difference from earlier formations and a probable return to predominantly cratonic or reworked provenance sources. It would be interesting to evaluate this interval further in order to test this hypothesis.

## Conclusions

- Heavy mineral assemblages interpreted from ancient sandstones are comparable to mineral associations recognized in modern sands that were correlated with three active petrological domains: (a) mature transport-limited alluvial plains; (b) low-grade metamorphic belt of the Caribbean Mountains (and the Northern Range); and (c) high-grade metamorphic domains such as the Merida Andes and Guyana Shield.
- The impact of the Andean orogeny across the margin was to introduce a suite of minerals characteristic of low-temperature metamorphism that today is prevalent adjacent to the Caribbean Mountain belt and differs from the zircon-rich assemblage produced within cratonic plains. Andean-related erosion is recorded in clastics of Trinidad and Barbados from Eocene time, although this source remains subsidiary until the Late Miocene coincident with Cruse Formation deltaic deposition.
- The heavy mineral records of Trinidad and Barbados are similar to that described across northern South America with uplift of the Andean mountain belt occurring around the Late Oligocene.
- The changing heavy mineral attributes suggest an initial passive margin setting adjacent to extensive cratonic lowlands with a Late Miocene transition to deltaic deposition flanked by active deformation in the Trinidad area.
- Barbados clastics are primarily craton-supplied, with secondary input from low-grade metamorphic detritus, and are interpreted to be the more proximal of Eocene deep-water clastics relative to an advancing deformation front. There may be greater heavy mineral similarity between these and Trinidad Eocene clastics than previously recognized.

This research was funded primarily by the Canadian Commonwealth Fund and the Ministry of Energy and Energy Affairs, Trinidad. The authors also acknowledge grants from the American Association of Petroleum Geologists and Geological Society of America, and data and samples provided by the Petroleum Company of Trinidad and Tobago and BHP Billiton. Finally, we acknowledge L. Kennan and L. Pinto whose critique significantly improved the content and layout of this paper.

## References

ABLEWHITE, K. & HIGGINS, G. E. 1968. A review of Trinidad, West Indies, oil development and the accumulations at Soldado, Brighton Marine, Grande Ravine, Barrackpore-Penal and Guayaguayare. *In*: SAUNDERS, J. B. (ed.) *Transactions of the Fourth Caribbean Geological Conference, Port of Spain, Trinidad and Tobago*. Caribbean Printers, Arima, Trinidad & Tobago, 41–73.

ALGAR, S. T. 1993. *Structural, stratigraphic, and thermochronological evolution of Trinidad*. PhD thesis, Dartmouth College, Hanover, NH.

ALGAR, S. 1998. Tectonostratigraphic Development of the Trinidad Region. *In*: PINDELL, J. L. & DRAKE, C. L. (eds) *Paleogeographic. Evolution and Non-Glacial Eustacy Northern South America*. SEPM Special Publications, Tulsa, **58**, 87–109.

ALMEIDA, F. F. M. 1978. *Tectonic Map of South America*. 1:5,000,000, and Explanatory Notes. UNESCO, Commission for the Geological Map of the World and Ministry of Mines and Energy, Brazil.

ASLAN, A., WHITE, W. A., WARNE, A. G. & GUEVARA, E. H. 2003. Holocene evolution of the western Orinoco Delta, Venezuela. *Geological Society of America Bulletin*, **115**, 479–498.

BABB, S. & MANN, P. 1999. Structural and sedimentary development of a Neogene Transpressional Plate Boundary between the Caribbean and South America Plates in Trinidad and the Gulf of Paria. *In*: MANN, P. (ed.) *Caribbean Basins, Sedimentary Basins of the World*, Volume **4**. Elsevier Science, Amsterdam, 495–557.

BALDWIN, S. L., HARRISON, T. M. & BURKE, K. 1986. Fission track evidence for the source of accreted sandstones, Barbados. *Tectonics*, **5**, 457–468.

BARR, K. W. 1962. The geology of the Toco district, Trinidad, West Indies, Part 1. *Overseas Geology and Mineral Resources*, **8**, 379–415.

BARR, K. W., WAITE, S. T. & WILSON, C. C. 1958. The mode of oil occurrence in the Miocene of southern Trinidad, B.W.I. *In*: WEEKS, L. G. (ed.) *Habitat of Oil*. American Association of Petroleum Geologists, Tulsa, OK, 533–549.

BEETS, D. J., MARESCH, W. V., KLAVER, G. T., MOTTANA, A., BOCCHIO, R., BEUNK, F. F. & MONEN, H. P. 1984. Magmatic rock series and high-pressure metamorphism as constraints on the tectonics history of the southern Caribbean. *In*: BONINI, W. E., HARGRAVES, R. B. & SHAGAM, R. (eds) *The Caribbean–South American Plate Boundary and Regional Tectonics*. Geological Society of America, Memoirs, Boulder, **162**, 95–130.

BELLIZZIA, A. & DENGO, G. 1990. The Caribbean mountain system, northern South America; a summary. *In*: DENGO, G. & CASE, J. E. (eds) *The Caribbean Region. The Geology of North America*, Volume **H**. Geological Society of America, Boulder, CO, 167–175.

BROWN, K. M. & WESTBROOK, G. K. 1987. The tectonic fabric of the Barbados Ridge accretionary complex. *Marine and Petroleum Geology*, **4**, 71–81.

BURKE, K., COOPER, C., DEWEY, J. F., MANN, P. & PINDELL, J. L. 1984. Caribbean tectonics and relative plate motions *In*: BONINI, W. E., HARGRAVES, R. B. & SHAGAM, R. (eds) *The Caribbean–South American Plate Boundary and Regional Tectonics*. Geological Society of America, Memoirs, Boulder, **162**, 31–63.

CALLAHAN, J. 1987. A non-toxic heavy liquid and inexpensive filters for separation of mineral grains. *Journal of Sedimentary Petrology*, **57**, 765–766.

CARR-BROWN, B. 1998. Revision of the Late Neogene stratigraphy of southeast Trinidad (Abstract). *In*: ALI, W., PAUL, A. & YOUNG ON, V. (eds) *Transactions of the 3rd Geological Conference of the Geological Society of Trinidad and Tobago and the 14th Caribbean Geological Conference, Port of Spain*. Geological Society of Trinidad and Tobago, Trinidad and Tobago, 757.

CARR-BROWN, B. & FRAMPTON, J. 1979. An outline of the stratigraphy of Trinidad. *In*: *Field Guide, Trinidad, Tobago, Barbados, July 1979, 4th Latin American Geological Congress*. Key Caribbean Publications, Port of Spain, Trinidad, 7–20.

CARR-BROWN, B., JONES, G., LA BORDE, K., GREGORY, J. & FRAMPTON, J. 2002. *Biostratigraphic Report of Well Canteen-2. Trinidad East Coast Marine Area, Block 2c*. BHP Billiton Well Report. Ministry of Energy and Energy Affairs, Port of Spain, Trinidad.

CERVENY, P. F. & SNOKE, A. W. 1993. Thermochronologic data from Tobago, West Indies; constraints on the cooling and accretion history of Mesozoic oceanic-arc rocks in the southern Caribbean. *Tectonics*, **12**, 431–440.

DECELLES, P. G. & HERTEL, F. 1989. Petrology of fluvial sands from the Amazonian foreland basin, Peru and Bolivia. *Geological Society of America Bulletin*, **101**, 1552–1562.

DENGO, G. 1953. Geology of the Caracas region, Venezuela. *Geological Society of America Bulletin*, **64**, 7–39.

DIAZ DE GAMERO, M. L. 1996. The changing course of the Orinoco River during the Neogene: a review. *Palaeogeography, Palaeoclimatology, Palaeoecology*, **123**, 385–402.

DONOVAN, S. K. 1994. Northern South America. *In*: DONOVAN, S. K. & JACKSON, T. A. (eds) *Caribbean Geology, An Introduction*. University of the West Indies Publishers' Association, Kingston, 229–248.

DUNCAN, R. A. & HARGRAVES, R. B. 1984. Plate tectonic evolution of the Caribbean region in the mantle reference frame. *In*: BONINI, W. E., HARGRAVES, R. B. & SHAGAM, R. (eds) *The Caribbean–South American Plate Boundary and Regional Tectonics*. Geological Society of America, Memoirs, **162**, 81–93.

EDMOND, J. M., PALMER, M. R., MEASURES, C. I., GRANT, B. & STALLARD, R. F. 1995. The fluvial geochemistry and denudation rate of the Guyana Shield in Venezuela, Colombia, and Brazil. *Geochimica et Cosmochimica Acta*, **59**, 3301–3325.

ERIKSON, J. P. & PINDELL, J. L. 1993. Analysis of subsidence in northeastern Venezuela as a discriminator of tectonic models for northern South America. *Geology*, **21**, 945–948.

ERIKSON, J. P. & PINDELL, J. L. 1998. Sequence stratigraphy and relative sea-level history of the Cretaceous to Eocene passive margin of Northeastern Venezuela and the possible tectonic and eustatic causes of stratigraphic development. *In*: PINDELL, J. L. & KENNAN, L. (eds) *Paleogeographic Evolution and Non-glacial Eustasy, Northern South America*. SEPM Special Publications, Tulsa, **58**, 261–281.

ERIKSON, J. P., KELLEY, S. A., OSMOLOVSKY, P. & VEROSUB, K. L. 2012. Linked basin sedimentation and orogenic uplift: the Neogene Barinas basin sediments derived from the Venezuelan Andes. *Journal of South American Earth Science*, **39**, 138–156.

ERLICH, R. N., FARFAN, P. F. & HALLOCK, P. 1993. Biostratigraphy, depositional environments and diagenesis of the Tamana Formation, Trinidad: a tectonic marker horizon. *Sedimentology*, **40**, 743–768.

FRANZINELLI, E. & POTTER, P. E. 1983. Petrology, chemistry, and texture of modern river sands, Amazon River system. *Journal of Geology*, **91**, 23–39.

FREY, M., SAUNDERS, J. & SCHWANDER, H. 1988. The mineralogy and metamorphic geology of low-grade metasediments, Northern Range, Trinidad. *Journal of the Geological Society, London*, **145**, 563–575.

FROST, C. D. & SNOKE, A. W. 1989. Tobago, West Indies, a fragment of a Mesozoic oceanic island arc; petrochemical evidence. *Journal of the Geological Society, London*, **146**, 953–964.

GARZANTI, E. & ANDÒ, S. 2007a. Plate tectonic and heavy mineral suites of modern sands. *In*: MOORE, C. H. & WADE, W. J. (eds) *Carbonate Reservoirs. Porosity and Diagenesis in a Sequence Stratigraphic Framework*. Developments in Sedimentology, **58**. Elsevier, Amsterdam, 741–763.

GARZANTI, E. & ANDÒ, S. 2007b. Heavy mineral concentration in modern sands: implications for provenance interpretation. *In*: MOORE, C. H. & WADE, W. J. (eds) *Carbonate Reservoirs. Porosity and Diagenesis in a Sequence Stratigraphic Framework*. Developments in Sedimentology, **58**. Elsevier, Amsterdam, 517–545.

GOLDSTEIN, S. L., ARNDT, N. T. & STALLARD, R. F. 1997. The history of a continent from U–Pb ages of zircons from Orinoco River sand and Sm–Nd isotopes in Orinoco Basin river sediments; highlights of the *Goldschmidt meeting, in honor* of A. W. Hofmann. *Chemical Geology*, **139**, 271–286.

GONZALEZ DE JUANA, C., MUNOZ, J. N. G. & VIGNALI, C. M. 1968. On the geology of eastern Paria (Venezuela). *In*: SAUNDERS, J. B. (ed.) *Transactions of the Fourth Caribbean Geological Conference, Port of Spain, Trinidad and Tobago*. Caribbean Printers, Arima, Trinidad & Tobago, 25–29.

HARRY, B. E. 1992. *Depositional environments and burial history of the late early Pliocene, Moruga Group, south coast, Trinidad, West Indies*. PhD thesis, University of Cincinnati, Cincinnati, OH.

HEDBERG, H. D. 1950. Geology of the Eastern Venezuela Basin (Anzoategui-Monagas-Sucre-Eastern Guarico Portion). *Geological Society of America Bulletin*, **61**, 1173–1216.

HIGGS, R. 2009. Caribbean–South America oblique collision model revised. *In*: JAMES, K., LORENTE, M. A. & PINDELL, J. L. (eds) *The Origin and Evolution of the Caribbean Plate*. Geological Society, London, Special Publications, **328**, 613–657.

HILL, R. J. & SCHENK, C. J. 2005. Petroleum geochemistry of oil and gas from Barbados: implications for distribution of Cretaceous source rocks and regional petroleum prospectivity. *Marine and Petroleum Geology*, **22**, 917–943.

HOORN, C. 1993. Marine incursions and the influence of Andean tectonics on the Miocene depositional history of northwestern Amazonia; results of a palynostratigraphic study. *Palaeogeography, Palaeoclimatology, Palaeoecology*, **105**, 267–309.

HOORN, C., GUERRERO, J., SARMIENTO, G. A. & LORENTE, M. A. 1995. Andean tectonics as a cause for changing drainage patterns in Miocene northern South America. *Geology*, **23**, 237–240.

HORTON, B. K., PARRA, M. *ET AL.* 2010. Resolving uplift of the northern Andes using detrital zircon age signatures. *GSA Today*, **20**, 4–9.

HUBERT, J. F. 1962. A zircon-tourmaline-rutile maturity index and the interdependence of the composition of heavy mineral assemblages with the gross composition and texture of sandstones. *Journal of Sedimentary Petrology*, **32**, 440–450.

ILLING, V. C. 1928. Geology of the Naparima Region of Trinidad (British West Indies). *Quarterly Journal of the Geological Society, London*, **84**, 1–56.

JACKSON, T. A., SMITH, T. E. & DUKE, M. J. M. 1991. The geochemistry of a metavolcanic horizon in the Maracas Formation, Northern Range, Trinidad; evidence of ocean floor basalt activity. *In*: GILLEZEAU, K. A. (ed.) *Transactions of the Second Geological Conference of the Geological Society of Trinidad and Tobago*. Geological Society of Trinidad and Tobago, Trinidad and Tobago, 42–47.

JAMES, K. H. 2006. Arguments for and against the Pacific origin of the Caribbean Plate: discussion, finding for an inter-American origin. *Geologica Acta*, **4**, 279–302.

JOHNSSON, M. J. 1990. Tectonic versus chemical-weathering controls on the composition of fluvial sands in tropical environments. *Sedimentology*, **37**, 713–726.

JOHNSSON, M. J., STALLARD, R. F. & MEADE, R. H. 1988. First-cycle quartz arenites in the Orinoco River basin, Venezuela and Colombia. *Journal of Geology*, **96**, 263–277.

JONES, H. P. 1968. The geology of the Herrera sands in the Moruga West oilfield of south Trinidad. *In*: SAUNDERS, J. B. (ed.) *Transactions of the Fourth Caribbean Geological Conference, Port of Spain, Trinidad and Tobago*. Caribbean Printers, Arima, Trinidad & Tobago, 91–100.

KALLIOKOSKI, J. 1965. Geology of north-central Guayana shield, Venezuela. *Geological Society of America Bulletin*, **76**, 1027–1049.

KENNAN, L. & PINDELL, J. 2007. Middle Miocene orogeny in Trinidad: Record of Caribbean collision and template for younger deformation. *In*: *Transactions of the Fourth Geological Conference of the Geological Society of Trinidad and Tobago, Port of Spain, Trinidad and Tobago*. Geological Society of Trinidad and Tobago, Trinidad and Tobago.

KERR, P. F. & ROGERS, A. F. 1977. *Optical Mineralogy*. McGraw-Hill, New York.

KOHN, B. P., SHAGAM, R., BANKS, P. O. & BURKLEY, L. A. 1984a. Mesozoic–Pleistocene fission-track ages on rocks of the Venezuelan Andes and their tectonic implications. *In*: BONINI, W. E., HARGRAVES, R. B. & SHAGAM, R. (eds) *The Caribbean–South American Plate Boundary and Regional Tectonics*. Geological Society of America, Memoirs, Boulder, **162**, 365–384.

KOHN, B. P., SHAGAM, R. & SUBIETA, T. 1984b. Results and preliminary implications of sixteen fission-track ages from rocks of the western Caribbean Mountains, Venezuela. *In*: BONINI, W. E., HARGRAVES, R. B. & SHAGAM, R. (eds) *The Caribbean–South American Plate Boundary and Regional Tectonics*. Geological Society of America, Memoirs, Boulder, **162**, 415–421.

KOMAR, P. D. 2007. The entrainment, transport and sorting of heavy minerals by waves and currents. *In*: MOORE, C. H. & WADE, W. J. (eds) *Carbonate Reservoirs. Porosity and Diagenesis in a Sequence Stratigraphic Framework*. Developments in Sedimentology, **58**. Elsevier, Amsterdam, 3–48.

KUGLER, H. G. 1953. Jurassic to recent sedimentary environments in Trinidad. *Swiss Association of Petroleum Geologists and Engineers*, **20**, 27–60.

KUGLER, H. G. 1956. Trinidad. *In*: HOFFSTETTER, R. (ed.) *Lexique Stratigraphique International, Volume 5, Amérique Latine, Fascimile 2b, Antilles (sauf Cuba et Antilles vénézuéliennes)*. CNRS, Paris, 41–116.

KUGLER, H. G. 1959. *Geologic Map of Trinidad*. Petroleum Association of Trinidad, Port of Spain, Trinidad and Tobago.

KUGLER, H. G. 2001. *Treatise on the Geology of Trinidad, Part 4: Palaeocene to Holocene Formations* (BOLLI, H. M. & KNAPPERTSBUSCH, M. (eds)). Museum of Natural History, Basel.

KUGLER, H. G. & SAUNDERS, J. B. 1967. On Tertiary turbidity-flow sediments in Trinidad, W.I. *Boletin Informativo–Asociacion Venezolana De Geologia, Mineria y Petroleo*, **10**, 243–259.

LARUE, D. K. & SPEED, R. C. 1983. Quartzose turbidites of the accretionary complex of Barbados; I, Chalky Mount succession. *Journal of Sedimentary Petrology*, **53**, 1337–1352.

LEONARD, R. 1983. Geology and hydrocarbon accumulations, Columbus Basin, offshore Trinidad. *American Association of Petroleum Geologists Bulletin*, **67**, 1081–1093.

LOCKE, B. D. & GARVER, J. I. 2005. Thermal evolution of the eastern Serranía del Interior foreland fold and thrust belt, northeastern Venezuela, based on apatite fission-track analyses. *In*: AVÉ LALLEMANT, H. G. & SISSON, V. B. (eds) *Caribbean–South American Plate Interactions, Venezuela*. Geological Society of America Special Papers, **394**, 315–328.

LOPEZ, V. M., MENCHER, E. & BRINEMAN, J. J. H. 1942. Geology of southeastern Venezuela. *Geological Society of America Bulletin*, **53**, 849–872.

MACKENZIE, W. S. & GUILFORD, C. 1980. *Atlas of Rock-Forming Minerals in Thin Section*. Longman, London.

MANGE, M. A. & MAURER, H. F. W. 1992. *Heavy Minerals in Colour*. Chapman & Hall, Hong Kong.

MANGE, M. A. & WRIGHT, D. T. 2007. High Resolution Heavy Mineral Analysis (HRHMA): a brief summary. *In*: MOORE, C. H. & WADE, W. J. (eds) *Carbonate Reservoirs. Porosity and Diagenesis in a Sequence Stratigraphic Framework*. Developments in Sedimentology, **58**. Elsevier, Amsterdam, 433–436.

MARESCH, W. V. 1974. Plate tectonics origin of the Caribbean mountain system of Northern South America; discussion and proposal. *Geological Society of America Bulletin*, **85**, 669–682.

MATLEY, C. A. 1932. The old basement of Barbados, with some remarks on Barbadian geology. *Geological Magazine*, **69**, 366–373.

MAXWELL, J. C. 1948. Geology of Tobago, British West Indies. *Geological Society of America Bulletin*, **59**, 801–854.

MESCHEDE, M. & FRISCH, W. 1998. A plate-tectonic model for the Mesozoic and Early Cenozoic history of the Caribbean Plate. *Tectonophysics*, **296**, 269–291.

MICHELSON, J. E. 1976. Miocene Deltaic Oil Habitat, Trinidad. *American Association of Petroleum Geologists Bulletin*, **60**, 1502–1519.

MILLIMAN, J. D. & SYVITSKI, J. P. M. 1992. Geomorphic/tectonic control of sediment discharge to the ocean: the importance of small mountainous rivers. *Journal of Geology*, **100**, 525–544.

MORRIS, A. E. L., TANER, I., MEYERHOFF, H. A. & MEYERHOFF, A. A. 1990. Tectonic evolution of the Caribbean region; alternative hypothesis. *In*: DENGO, G. & CASE, J. E. (eds) *The Caribbean Region*. Geological Society of America, Boulder, CO.

MORTON, A. C. 1979. Depth control of intrastratal solution of heavy minerals from the Palaeocene of the North Sea. *Journal of Sedimentary Petrology*, **49**, 281–286.

MORTON, A. C. 1985. Heavy minerals in provenance studies; provenance of arenites: NATO ASI series. *Mathematical and Physical Sciences*, **148**, 249–277.

MORTON, A. C. & JOHNSSON, M. J. 1993. Factors influencing the composition of detrital heavy mineral suites in Holocene sands of the Apure River drainage basin, Venezuela. *In*: JOHNSSON, M. J. & BASU, A. (eds) *Processes Controlling the Composition of Clastic Sediments*. Geological Society of America Special Papers, **284**, 171–185.

NESSE, W. D. 2004. *Introduction to Optical Mineralogy*. Oxford University Press, Oxford.

NICKEL, E. 1973. Experimental dissolution of light and heavy minerals in comparison with weathering and intrastratal solution. *Contributions to Sedimentology*, **1**, 3–68.

NOCKOLDS, S. R., KNOX, R. W. O. & CHINNER, G. A. 1978. *Petrology for Students*. Cambridge University Press, Cambridge.

NOTA, D. J. G. 1958. *Sediments of the western Guiana shelf*. PhD thesis, Utrecht University.

PÉREZ DE ARMAS, J. 2005. Tectonic and thermal history of the western Serranía del Interior foreland fold and thrust belt and Guarico basin, north-central Venezuela: implications of new apatite fission-track analysis and seismic interpretation. *In*: AVÉ LALLEMANT, H. G. & SISSON, V. B. (eds) *Caribbean–South American Plate Interactions, Venezuela*. Geological Society of America Special Papers, **394**, 271–314.

PERKINS, D. & HENKE, K. R. 2004. *Minerals in Thin Section*. Pearson Education, Upper Saddle River, NJ.

PETTIJOHN, F. J. 1941. Persistence of heavy minerals and geologic age. *Journal of Geology*, **49**, 610–625.

PINDELL, J. 1994. Evolution of the Gulf of Mexico and Caribbean Region. *In*: DONOVAN, S. K. & JACKSON, T. A. (eds) *Caribbean Geology, An Introduction*. University of the West Indies Publishers' Association, Kingston, 13–40.

PINDELL, J. L. & BARRETT, S. F. 1990. Geological evolution of the Caribbean region; a plate tectonic perspective. *In*: DENGO, G. & CASE, J. E. (eds) *The Caribbean Region*. The Geology of North America, Volume H. Geological Society of America, Boulder, CO, 405–432.

PINDELL, J. L. & KENNAN, L. 2007. Cenozoic kinematics and dynamics of oblique collision between two convergent plate margins: The Caribbean–South America Collision in Eastern Venezuela, Trinidad

and Barbados. *In*: KENNAN, L., PINDELL, J. & ROSEN, N. C. (eds) *The Paleogene of the Gulf of Mexico and Caribbean Basins: Processes, Events, and Petroleum Systems*. 27th Annual GCSSEPM Foundation Annual Bob F. Perkins Research Conference Proceedings. SEPM, Houston, TX, 458–553.

PINDELL, J., KENNAN, L., MARESCH, W. V., STANEK, K., DRAPER, G. & HIGGS, R. 2005. Plate kinematics and crustal dynamics of circum-Caribbean arc–continental interactions: tectonic controls on basin development in Proto-Caribbean margins. *In*: AVE LALLEMANT, H. G. & SISSON, V. G. (eds) *Caribbean–South American Plate Interactions, Venezuela*. Geological Society of America Special Papers, **394**, 7–52.

PINDELL, J., KENNAN, L., WRIGHT, D. & ERIKSON, J. 2009. Clastic domains of sandstones in central/eastern Venezuela, Trinidad, and Barbados: heavy mineral and tectonic constraints on provenance and palaeogeography. *In*: JAMES, K., LORENTE, M. A. & PINDELL, J. (eds) *The Origin and Evolution of the Caribbean Plate*. Geological Society, London, Special Publications, **328**, 743–797.

POTTER, H. C. 1968. A preliminary account of the stratigraphy and structure of the eastern part of the Northern Range, Trinidad. *In*: SAUNDERS, J. B. (ed.) *Transactions of the Fourth Caribbean Geological Conference, Port of Spain, Trinidad and Tobago*. Caribbean Printers, Trinidad & Tobago, 15–20.

POTTER, P. E. 1994. Modern sands of South America; composition, provenance and global significance. *Geologische Rundschau*, **83**, 212–232.

PUDSEY, C. J. 1985. Stratigraphy and sedimentology of the Scotland Group, Barbados. *In*: *Transactions of the Fourth Latin American Geological Conference, 7–15 July 1979, Port-of-Spain, Trinidad and Tobago, Volume 2*. Ministry of Energy and Energy Industries, Arima, Trinidad & Tobago, 903–908.

PUDSEY, C. J. & READING, H. G. 1982. Sedimentology and structure of the Scotland Group, Barbados. *In*: LEGGETT, J. K. (ed.) *Trench–Forearc Geology; Sedimentation and Tectonics on Modern and Ancient Active Plate Margins*. Geological Society, London, Special Publications, **10**, 291–308.

REQUEJO, A. G., WIELCHOWSKY, C. C., KLOSTERMAN, M. J. & SASSEN, R. 1994. Geochemical characterization of lithofacies and organic facies in Cretaceous organic-rich rocks from Trinidad, East Venezuela Basin. *Organic Geochemistry*, **22**, 441–459.

RODRIGUES, K. 1988. Oil source bed recognition and crude oil correlation, Trinidad, West Indies. *Organic Geochemistry*, **13**, 365–371.

ROHR, G. M. 1991. Paleogeographic maps, Maturing Basin of E. Venezuela and Trinidad. *In*: GILLEZEAU, K. A. (ed.) *Transactions of the Second Geological Conference of the Geological Society of Trinidad, Tobago, held at Port-of-Spain, Trinidad, April 3–8 1990*. Geological Society of Trinidad and Tobago, Trinidad and Tobago, 88–105.

SANTOS, J. O. S., HARTMANN, L. A., GAUDETTE, H. E., GROVES, D. I., MCNAUGHTON, N. J. & FLETCHER, I. R. 2000. A new understanding of the provinces of the Amazon Craton based on integration of field mapping and U–Pb and Sm–Nd geochronology. *Gondwana Research*, **3**, 453–488.

SAUNDERS, J. B. 1979. Field trip Barbados. *In: Field Guide, Trinidad, Tobago, Barbados, July 1979, 4th Latin American Geological Congress.* Key Caribbean Publications, Port of Spain, Trinidad, 71–88.

SAUNDERS, J. B., BERNOULLI, D. & MARTIN-KAYE, P. H. A. 1985. Late Eocene deep water clastics in Grenada, W. I. *In: Transactions of the Fourth Latin American Geological Conference, 7–15 July 1979, Port-of-Spain, Trinidad and Tobago, Volume 2.* Ministry of Energy and Energy Industries, Trinidad & Tobago, 909–918.

SAUNDERS, J. B., ROBERTS, C., ALI, W. & EGGERTSON, E. B. 1998. *Trinidad and Tobago Geologic Map: Scale 1:100, 000.* Ministry of Energy and Energy Industries, Trinidad & Tobago.

SENN, A. 1940. Palaeogene of Barbados and its bearing on history and structure of Antillean–Caribbean region. *American Association of Petroleum Geologists Bulletin*, **24**, 1548–1610.

SINGH, S. 1968. The occurrence of sillimanite and its status as an index of metamorphic grade in Guyana Shield rocks of southern British Guiana. *In: SAUNDERS, J. B. (ed.) Transactions of the Fourth Caribbean Geological Conference, Port of Spain, Trinidad and Tobago.* Caribbean Printers, Arima, Trinidad & Tobago, 395–399.

SMITH, C. A., SISSON, V. B., AVE LALLEMANT, H. G. & COPELAND, P. 1999. Two contrasting pressure-temperature-time paths in the Villa de Cura blueschist belt, Venezuela; possible evidence for Late Cretaceous initiation of subduction in the Caribbean. *Geological Society of America Bulletin*, **111**, 831–848.

SNOKE, A. W., ROWE, D. W., YULE, J. D. & WADGE, G. 2001. *Petrologic and Structural History of Tobago, West Indies; a Fragment of the Accreted Mesozoic oceanic arc of the Southern Caribbean.* Geological Society of America Special Papers, **354**.

SPEED, R. C. 1985. New views on the geology of Barbados. *In: Transactions of the Fourth Latin American Geological Conference, 7–15 July 1979, Port-of-Spain, Trinidad and Tobago, Volume 2.* Ministry of Energy and Energy Industries, Trinidad & Tobago, 929–941.

SPEED, R. C. & FOLAND, K. A. 1990. Mid-Tertiary $^{40}Ar/^{39}Ar$ ages of metamorphism of Northern Ranges schists, Trinidad. *In:* GILLEZEAU, K. A. (ed.) *Transactions of the Second Geological Conference of the Geological Society of Trinidad, Tobago, held at Port-of-Spain, Trinidad, April 3–8 1990.* Geological Society of Trinidad and Tobago, Trinidad and Tobago,, 19–20.

STAINFORTH, R. M. 1978. Was It the Orinoco? *American Association of Petroleum Geologists Bulletin*, **62**, 303–306.

STALLARD, R. F., KOEHNKEN, L. & JOHNSSON, M. J. 1990. Weathering processes and the composition of inorganic material transported through the Orinoco River system, Venezuela and Colombia. *In:* WEIBEZAHN, F. H., ALVAREZ, H. & LEWIS, W. M., JR. (eds) *The Orinoco River as an Ecosystem.* Universidad Simon Bolivar, Caracas, 81–119.

SUTER, H. H. 1960. *The General and Economic Geology of Trinidad, B.W.I.* Overseas Geological Surveys, Mineral Resources Division. HMSO, London.

TYSON, L. & ALI, W. 1991. Cretaceous to Middle Miocene sediments in Trinidad. *In:* GILLEZEAU, K. A. (ed.) *Transactions of the Second Geological Conference of the Geological Society of Trinidad, Tobago, held at Port-of-Spain, Trinidad, April 3–8 1990.* Geological Society of Trinidad and Tobago, Trinidad and Tobago, 266–277.

TYSON, L., BABB, S. & DYER, B. 1991. Middle Miocene tectonics and its effects on Late Miocene sedimentation in Trinidad. *In:* GILLEZEAU, K. A. (ed.) *Transactions of the Second Geological Conference of the Geological Society of Trinidad, Tobago, held at Port-of-Spain, Trinidad, April 3–8 1990.* Geological Society of Trinidad and Tobago, Trinidad and Tobago, 26–40.

VAN ANDEL, Tj. H. 1958. Origin and classification of Cretaceous, Palaeocene, and Eocene sandstones of western Venezuela. *American Association of Petroleum Geologists Bulletin*, **42**, 734–763.

VAN ANDEL, Tj. H. & POSTMA, H. 1954. *Recent Sediments of the Gulf of Paria: Reports of the Orinoco Shelf Expedition: Volume 1.* North-Holland, Amsterdam.

VINCENT, H. 2008. *Cenozoic Sediment Dispersal Patterns Across Trinidad, West Indies.* PhD Dissertation, Dalhousie University, Halifax, Nova Scotia.

VINCENT, H. & WACH, G. D. 2007. Paleogene slope deposits; examples from the Chaudière, Pointe-a-Pierre and San Fernando formations, Central Range, Trinidad. *In:* KENNAN, L., PINDELL, J. & ROSEN, N. C. (eds) *The Paleogene of the Gulf of Mexico and Caribbean Basins: Processes, Events, and Petroleum Systems.* 27th Annual GCSSEPM Foundation Annual Bob F. Perkins Research Conference Proceedings. SEPM, Houston, TX, 554–585.

WARNE, A. G., MEADE, R. H. *ET AL.* 2002. Regional controls on geomorphology, hydrology, and ecosystem integrity in the Orinoco Delta, Venezuela; Geomorphology on large rivers. *Geomorphology*, **44**, 273–307.

WEBER, J. C., DIXON, T. H. *ET AL.* 2001a. GPS estimate of relative motion between the Caribbean and South American plates, and geologic implications for Trinidad and Venezuela. *Geology*, **29**, 75–78.

WEBER, J. C., FERRILL, D. A. & RODEN-TICE, M. K. 2001b. Calcite and quartz microstructural geothermometry of low-grade metasedimentary rocks, Northern Range, Trinidad. *Journal of Structural Geology*, **23**, 93–112.

WIELCHOWSKY, C. C., RAHMANIAN, V. D. & HARDENBOLD, J. 1991. A preliminary tectonostratigrahic framework for onshore Trinidad. *In:* GILLEZEAU, K. A. (ed.) *Transactions of the Second Geological Conference of the Geological Society of Trinidad, Tobago, held at Port-of-Spain, Trinidad, April 3–8 1990.* Geological Society of Trinidad and Tobago, Trinidad and Tobago, 41.

WINKLER, H. G. F. 1974. *Petrogenesis of Metamorphic Rocks.* Springer, New York.

WONG, T. E., KROOK, L. & ZONNEVELD, J. I. S. 1998. Investigations in the coastal plain and offshore area of Suriname. *In:* WONG, T. E., DE VLETTER, D. R., KROOK, L., ZONNEVELD, J. I. S. & VAN LOON, A. J. (eds) *The History of Earth Sciences in Suriname.* Netherlands Institute of Applied Geoscience TNO, Amsterdam.

WOOD, L. J. 2000. Chronostratigraphy and tectonostratigraphy of the Columbus Basin, eastern offshore Trinidad. *American Association of Petroleum Geologists Bulletin*, **84**, 1905–1928.

# Provenance of late Palaeozoic terrestrial sediments on the northern flank of the Mid North Sea High: detrital zircon geochronology and rutile geochemical constraints

ANDERS MATTIAS LUNDMARK[1]*, EDINA PÓZER BUE[1], ROY H. GABRIELSEN[1], KJERSTI FLAAT[2], TOR STRAND[2] & SVERRE E. OHM[1,2]

[1]*Department of Geosciences, University of Oslo, Sem Sælands vei 1, Oslo N-0371, Norway*

[2]*ConocoPhillips Norway, PO Box 220, Tananger N-4098, Norway*

**Corresponding author (e-mail: mattias@geologi.uio.no)*



**Abstract:** Zircon U–Pb and rutile trace element data are used to investigate the provenance of late Devonian to early Permian terrestrial sandstones in the Embla and Flora oil fields on the north flank of the Mid North Sea High, central North Sea. Two Old Red Sandstone samples (ORS 1) are dominated by 1.2–0.9 Ga Grenvillian zircons and low- to medium-grade rutile, with sparse Cambro-Ordovician Caledonian zircons (2–4%) and high-grade rutiles (0–5%). The samples are interpreted as recycled metasediments from the Scottish Caledonides. Two other Old Red Sandstone samples (ORS 2) contain a high proportion of Caledonian, mainly Silurian zircons (15–19%) and high-grade rutiles (15–18%); we propose that these components are traceable to the Krummedal sequence on East Greenland (and related sediments). We interpret the data to reflect a temporal evolution of the regional drainage system from northwestern to northeastern sources, with high grade detritus reaching the Mid North Sea High in the Famennian–early Carboniferous. A late Carboniferous and an early Permian sandstone yielded zircon and rutile signatures compatible with recycling of Palaeozoic sediments north of southernmost Scotland, probably reflecting inversion tectonics. Recycling of Mesoproterozoic to Palaeozoic sediments is thus a prominent feature of the studied late Palaeozoic sandstones.



**Supplementary material:** Electron microprobe data from rutile trace element analyses, additional rutile temperature plots and zircon LA-ICPMS U–Pb data are available at http://www.geolsoc.org.uk/SUP18617


The provenance of sediments can be constrained by comparing their heavy mineral signatures to the present-day geology of potential source areas. The inherent limitations of this approach are partly mitigated as new provenance data increase our understanding of the geographical distribution of sediments from different sources over time, and build up a record of the geological units lost to erosion.

The post-Caledonian Palaeozoic sediments in the North Sea and Britain have been the focus of numerous studies using different combinations of mineralogical, geochemical and isotopic data along with sedimentological observations to constrain source areas and transport routes (e.g. Drewery et al. 1987; Cliff et al. 1991; Hallsworth et al. 2000; Morton et al. 2001; Tyrrell et al. 2006; Hallsworth & Chisholm 2008). On the megascale, potential sediment sources include Laurentia and Baltica to the north, and Avalonia and related terranes to the south. Laurentia and Baltica consist of Archaean nuclei, modified and added to during successive events of crustal building. Unfortunately, their age signatures are, on the whole, quite similar (cf.

provenance signatures in Cawood et al. 2007; Morton et al. 2008; Andersen et al. 2010). This problem is compounded along the plate margins by terrane transfer during plate collisions (Roberts et al. 2007; Augland et al. 2011) and possible sediment transport across plate boundaries (Lamminen & Köykkä 2010). Complementary data reflecting additional source characteristics are therefore important to further constrain potential source areas to the north. Avalonia and related terranes to the south represent mainly Neoproterozoic crust formed along the northern margin of Gondwana, subsequently rifted off and accreted to Laurussia in the Palaeozoic. A distinct age signature with important Cadomian and Variscan populations can be expected in sediments of Avalonia origins (Nance & Murphy 1996; von Raumer et al. 2003).

The present study reports the first detrital zircon U–Pb and rutile trace element data from the late Palaeozoic Embla and Flora oil field reservoir sandstones on the Grensen Nose on the northern flank of the Mid North Sea High, the offshore continuation of the Southern Uplands in the central North Sea (Fig. 1). Detrital zircon ages mainly


*From*: Scott, R. A., Smyth, H. R., Morton, A. C. & Richardson, N. (eds) 2014. *Sediment Provenance Studies in Hydrocarbon Exploration and Production*. Geological Society, London, Special Publications, **386**, 243–259.
First published online June 19, 2013, http://dx.doi.org/10.1144/SP386.4

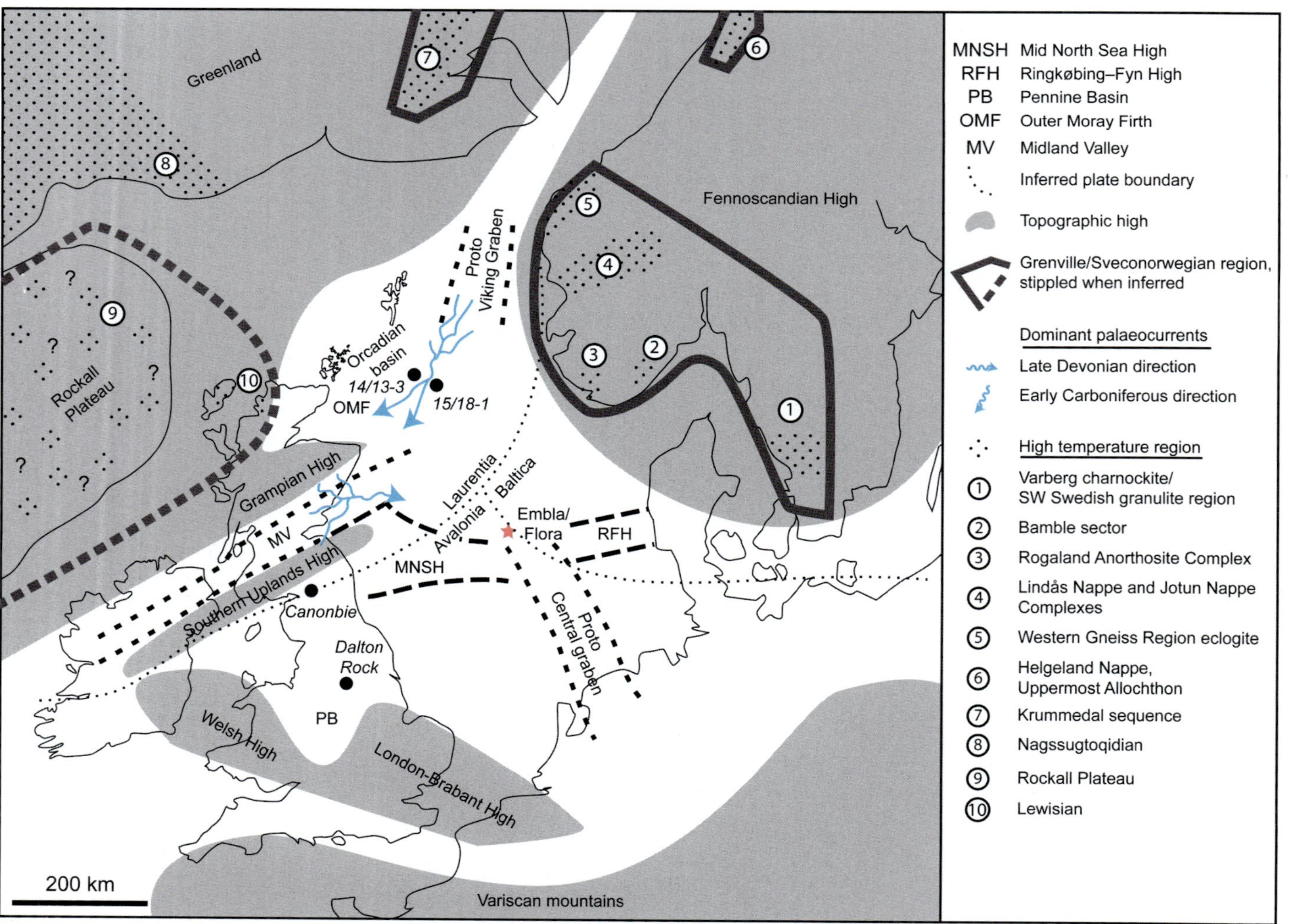

MNSH Mid North Sea High
RFH Ringkøbing–Fyn High
PB Pennine Basin
OMF Outer Moray Firth
MV Midland Valley
Inferred plate boundary
Topographic high
Grenville/Sveconorwegian region, stippled when inferred
Dominant palaeocurrents
Late Devonian direction
Early Carboniferous direction
High temperature region
1 Varberg charnockite/SW Swedish granulite region
2 Bamble sector
3 Rogaland Anorthosite Complex
4 Lindås Nappe and Jotun Nappe Complexes
5 Western Gneiss Region eclogite
6 Helgeland Nappe, Uppermost Allochthon
7 Krummedal sequence
8 Nagssugtoqidian
9 Rockall Plateau
10 Lewisian
Greenland
Fennoscandian High
Proto Viking Graben
Orcadian basin
14/13-3
OMF
15/18-1
Rockall Plateau
Grampian High
Laurentia
Avalonia · Baltica
Embla/Flora
RFH
MV
Southern Uplands High
MNSH
Proto Central graben
Canonbie
Dalton Rock
PB
Welsh High
London-Brabant High
Variscan mountains
200 km

reflect granitic (*sensu lato*) magmatism and medium- to high-grade metamorphism. Because mafic rocks tend to lack zircons (Hoskin & Schaltegger 2003), they are commonly poorly represented in zircon provenance data. Rutile typically forms during medium- to high-grade metamorphism. The trace element content can be used to differentiate between felsic and mafic source rocks, and to constrain the temperature of rutile formation (Zack *et al.* 2004; Meinhold *et al.* 2008; Meinhold 2010). Rutile is unstable under greenschist facies conditions, so it tends to record the last medium- to high-grade metamorphic event (Zack *et al.* 2004). Both zircon and rutile are highly resistant to the effects of weathering, sediment transport and diagenesis, and may therefore represent recycling of older sedimentary rocks (Morton & Hallsworth 1999).

Whereas zircon and rutile are present in all the sandstones analysed in this study, garnet, which has previously been used for provenance studies in the North Sea area and Britain (e.g. Hallsworth *et al.* 2000; Morton *et al.* 2001), is not found. Its absence may reflect dissolution during weathering and/or burial (Morton & Hallsworth 1999; Walderhaug & Porten 2007), or it could reflect source characteristics. However, garnet geochemical data are typically used to infer the source lithology and its metamorphic facies (e.g. Mange & Morton 2007), and similar information can be extracted from rutile data (e.g. Meinhold 2010), thereby allowing comparisons to be made.

This study was undertaken to test whether zircon and rutile isotopic and geochemical data from late Palaeozoic sediments on the northern flank of the Mid North Sea High vary significantly between sediments of different ages. If so, the variations could be used to constrain interwell correlations in the Embla oil field and to evaluate changes in sediment provenance.

## Geological framework

The late Silurian to early Carboniferous Old Red Sandstone sediments in northwestern Europe were largely sourced from erosion of the Caledonian orogen (Ziegler 1990) and deposited during the warm and semi-arid climate conditions typical for the newly formed continent Laurussia (Witzke 1990). In the northern and central North Sea, a late Devonian to early Carboniferous evolution from southeastward-flowing drainage systems reflecting erosion of the Scottish Caledonides to a dominantly

southwestward-directed system reflecting erosion of the Caledonides to the NE appears to have controlled sediment transport (Leeder 1988). This northeastern source area may have included Greenland, the Rockall Plateau and Baltica (Fig. 1).

From the late Tournaisian, the changing palaeogeographic setting of the North Sea region and the onset of an extensional tectonic regime combined with eustatic sea-level rise ended the deposition of Old Red Sandstone sediments (Gatliff *et al.* 1994; Besly 1998). In the Tournaisian, Visean and the Namurian (the latter spanning the Serpukhovian and lower Bashkirian, all translations from sub stages to global stages after Heckel & Clayton 2006), large volumes of siliciclastics sourced from the NE continued to spread southward into the Pennine Basin in central Britain via the Pennine River (Cliff *et al.* 1991). The long duration and large volumes of siliciclastic input have been suggested to reflect uplift in the northeastern Caledonides (Besly 1998) and channelling of sediments through a proto-Viking Graben (Fig. 1; Coward 1995).

In the early Westphalian (middle to late Bashkirian), the sediment supply from the north appears to have waned, and the Pennine Basin received detritus from a Laurentian source to the west (Glover *et al.* 1996; Hallsworth *et al.* 2000). This pattern was modified in the mid-Westphalian (Moscovian) by the advance of the Variscan mountain chain, causing influx of sediments from the south. With the aid of the detrital zircon age signature the sediments derived from the Variscan orogen have been identified as far north as southern Scotland (Canonbie in Fig. 1; Morton *et al.* 2010). The advance of the Variscan orogenic front was accompanied by inversion tectonics in the Variscan foreland (Corfield *et al.* 1996) and a gradual return to red bed facies, the 'Barren Red Group' of Besly *et al.* (1993). The continental red bed facies continued in the Cisuralian by deposition of the lower and upper Rotliegend groups. Carboniferous sediments are generally lacking on the Mid North Sea High (Maynard *et al.* 1997), but terrestrial Westphalian to Stephanian (Middle to Upper Pennsylvanian) sediments are present on its northern flank as seen in the Flora oil field (Martin *et al.* 2002).

During the late Carboniferous–early Permian, widespread and locally intense magmatism affected northwestern Europe (Neumann *et al.* 2004; Timmerman 2004). Magmatic rocks of these ages have been encountered in the central North Sea in the Argyll/Ardmore (Robson 1991), Flora (Hayward *et al.* 2002), Embla and potentially the Auk

---

**Fig. 1.** Overview of structural elements in the post-Caledonian Palaeozoic North Sea region. Greenland is restored according to Torsvik *et al.* (2002). The palaeohighs are adapted from Ziegler (1990), inferred plate boundaries after Banka *et al.* (2002) and Landes *et al.* (2005), and palaeostream flow from Leeder (1988).

(Trewin *et al.* 2003) oil fields. In the Auk, Argyll/ Ardmore and Embla oil fields, late Carboniferous–Permian sediments unconformably overlie the Old Red Sandstone (Robson 1991; Trewin *et al.* 2003). It is commonly difficult to differentiate between these sediments because they represent similar lithologies, and the Rotliegend sediments partly derive from recycling of Old Red Sandstone (Marshall & Hewett 2003). With the Permian collapse of the Variscan orogen, the northern and southern Permian basins developed, separated by the Mid North Sea High–Ringkøbing-Fyn High. The South Permian Basin continued to receive detritus from the Variscan orogen (Glennie 1972), but the northern flank of the Mid North Sea High area was likely cut off from this source (cf. Ziegler 1990).

## The late Palaeozoic of the Embla and Flora oil fields

In the Embla oil field the deepest cored beds consist of volcanic rocks interbedded with Old Red Sandstone braidplain and floodplain/lacustrine sediments (Knight *et al.* 1993). One alkali rhyolite yields a zircon U–Pb age of 374 ± 3 Ma, overlapping with the Frasnian–Famennian boundary (timescale of Kaufmann 2006). Together with sparse palynological evidence for a Givetian–Famennian age, this indicates that the sediments are timeequivalent to the Buchan Formation (Lundmark *et al.* 2012), and potentially to the Tayport Formation. The latter represents a diachronous transition to the coal measures of the Visean–early Namurian Firth Coal Formation (Cameron 1993). The volcanism has been proposed to reflect Famennian rifting taking place along a proto-Central Graben across the Mid North Sea High (Fig. 1; Lundmark *et al.* 2012). Marine plankton from cuttings taken above the volcanic rocks in well 2/ 7-21S indicate a limited marine influence (Knight *et al.* 1993) potentially associated with this rifting. The Old Red Sandstone extends into the lower Carboniferous, which may also be the case for the corresponding Embla sediments. The Old Red Sandstone is in places overlain by terrestrial sediments interbedded with highly altered mafic volcanic rocks, interpreted to be of early Permian (Rotliegend) age, and in other places by postTriassic marine sediments that cap the Palaeozoic reservoir rocks (Knight *et al.* 1993). Palynological, palaeomagnetic (Conoco-Phillips, unpublished studies) and isotopic (Lundmark *et al.* 2012) age data in the studied wells are summarized in Figure 2.

Five medium-grained sandstones were sampled in the Embla oil field. Four samples of Old Red Sandstone represent the two main reservoir sequences in well 2/7-26S (samples M09-02 and M09-03), a highly productive reservoir sequence in well 2/7-21S (M09-06), and sandstone above an apparent angular unconformity in well 2/7-D07 (M09-05). Although the latter is assigned to the Old Red Sandstone it should be noted that palaeomagnetic inclination data hint at a possible late Carboniferous–early Permian age (Fig. 2). Interwell correlations are poorly constrained due to fault compartmentalization and poor seismic continuity, but chemostratigraphical similarities (ConocoPhillips, unpublished data) suggest a correlation between the upper sandstone sampled in well 2/ 7-26S and the one sampled in well 2/7-21S. In contrast, dissimilar magnetic susceptibility patterns suggest that the sediments are of slightly different ages. The fifth sample represents early Permian sandstone in well 2/7-23S (M09-04). All the sampled intervals are sublithic to lithic arenites. Clasts in associated pebbly and conglomeratic sediments are dominantly grey to white or red (sub)rounded quartz/quartzites. The general absence of feldspar may reflect pervasive dissolution related to weathering, diagenesis or hydrothermal alteration, processes that have all been active in the Embla oil field (Lundmark *et al.* 2012). However, textural evidence for feldspar dissolution is limited, hinting that the initial feldspar content was low. Sampling depths and sedimentological interpretations are summarized in Table 1.

In the Flora oil field three units have been informally defined and described by Martin *et al.* (2002). The deepest drilled unit is the Lower Flora Sandstone, interpreted as Duckmantian–Bolsovian (Westphalian B-C) flood plain deposits from a lowsinuosity fluvial system. A Bolsovian mafic volcanic unit (Ar/Ar dated to 299 ± 9 Ma) separates it from the Bolsovian to Stephanian braided river deposits of the Upper Flora Sandstone (Martin *et al.* 2002). The arkosic to subarkosic and lithic arenitic Flora sandstone is terminated by an unconformity, proposed to correlate to the 'Asturic unconformity' (Martin *et al.* 2002), which reflects inversion tectonics in the late Westphalian–Stephanian (Maynard *et al.* 1997). Above follows the lower Rotliegend Grensen Formation, representing a fluvial system with a palaeotransport direction toward the NE (Hayward *et al.* 2002; Martin *et al.* 2002). This is in turn overlain by 299 ± 3 Ma basaltic lava (Heeremans *et al.* 2004). Only the Upper Flora Sandstone has been cored, and the mediumgrained sample M10-09 analysed in this study was collected in its upper part (Table 1), where a dominantly southeastern palaeocurrent direction has been recorded (cf. Martin *et al.* 2002, fig. 8). The sandstone is interpreted to correlate with the Ketch Member of the Schooner Formation in the southern North Sea (Martin *et al.* 2002).

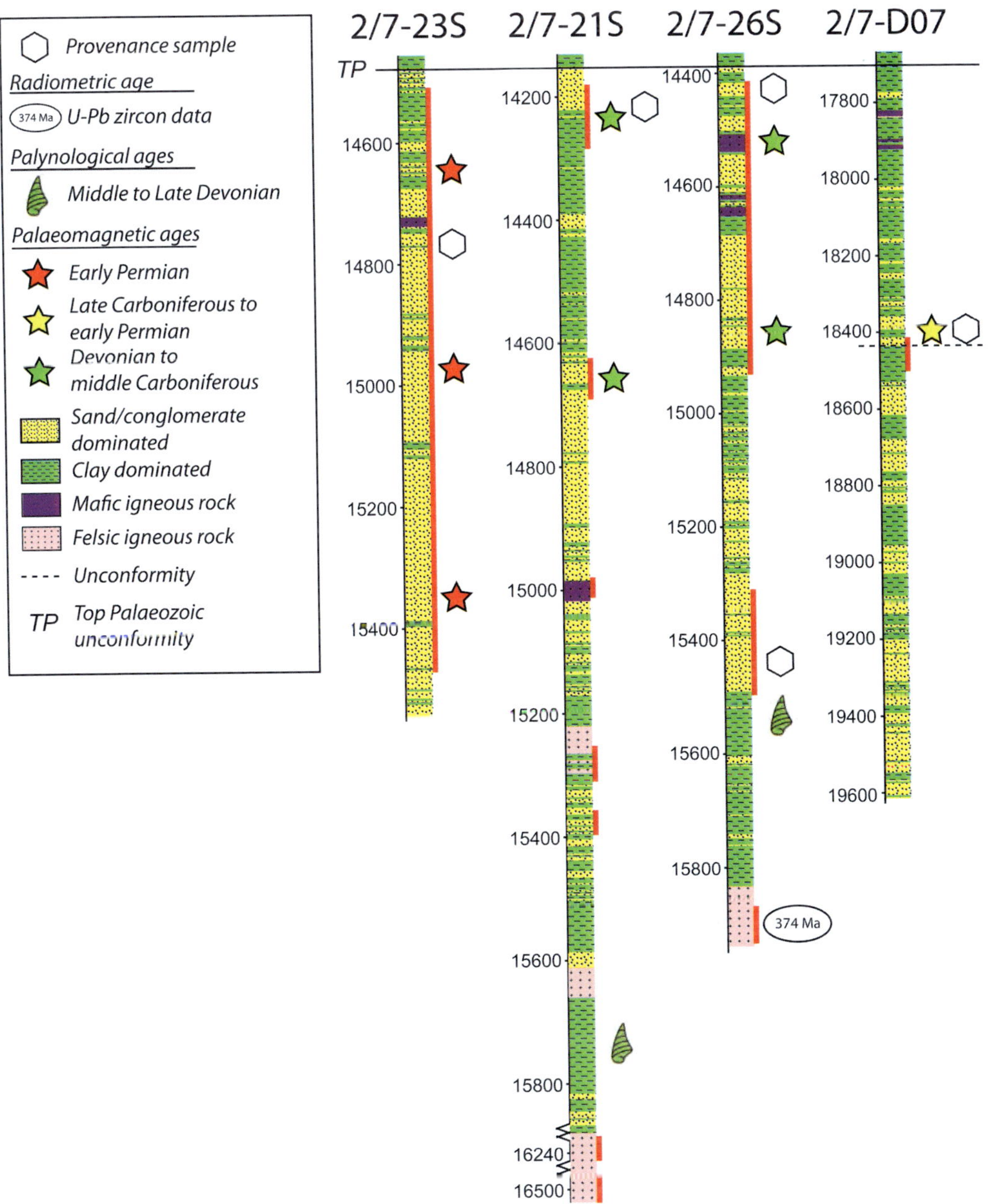



**Fig. 2.** Sampled wells in the Embla oil field flattened at top Palaeozoic. Simplified lithological logs, age constraints and provenance sample depths shown at measured depths in feet. Well 2/7-23S is interpreted as early Permian, the remaining wells as late Devonian to early Carboniferous. Cored intervals marked in red. All wells are deviated.

## Analytical methods

Zircon and rutile were separated from *c.* 100 g of sandstone for each of the six samples using standard methods, that is, crushing, gentle removal of the clay fraction by washing, followed by concentration of the non-magnetic heavy minerals using Frantz magnetic separators and sodium heteropolytungstate heavy liquid. Zircons and rutiles were randomly picked from the non-magnetic fractions, mounted in epoxy and polished using standard methods. Rutile from the magnetic fractions of

**Table 1.** *Samples in this study*

| Sample no. | Well/oil field | Measured depth (ft) | Age | Sedimentological interpretation |
|---|---|---|---|---|
| M09-02* | 2/7-26S/Embla | 15 461 | Late Devonian/early Carboniferous ORS | Braided fluvial channel (potentially proximal braid plain to distal alluvial fan) |
| M09-03* | 2/7-26S/Embla | 14 423 | Late Devonian/early Carboniferous ORS | Braided fluvial channel |
| M09-04 | 2/7-23S/Embla | 14 783 | Early Permian | Braided fluvial channel |
| M09-05[†] | 2/7-D07/Embla | 18 428 | Late Devonian/early Carboniferous ORS | Braided fluvial channel |
| M09-06[†] | 2/7-21S/Embla | 14 220 | Late Devonian/early Carboniferous ORS | Braidplain fluvial deposits (potentially braided stream outwash environment or alluvial fan) |
| M10-09 | 31/26c-13/Flora | 8991 | Late Carboniferous (Ketch Member) | Braided fluvial channel |

*ORS 1; [†]ORS 2.
*Abbreviation*: ORS, Old Red Sandstone.

two samples, M09-03 and M09-05, were mounted separately. The rutiles were analysed for the trace elements Nb, Cr, Zr and Si in wavelength dispersive mode (WDS) on a Cameca SX100 electron microprobe at the Department of Geosciences, University of Oslo. Detection limits were *c*. 75 ppm for Nb, 45 ppm for Cr, 60 ppm for Zr and 30 ppm for Si. The measured ratios from the non-magnetic fractions are compiled in the Supplementary material, and data plots are presented in Figure 3a. The results from the magnetic fractions are also summarized in the Supplementary material.

Zircon grains mounted in epoxy were imaged by cathodoluminescence on a JEOL-JSM6460LV scanning electron microscope. U–Pb isotope compositions were then analysed by laser ablation inductively coupled plasma mass spectrometry (LA-ICPMS) using a Nu Plasma HR mass spectrometer and a NewWave LUV213 laser microprobe at the Department of Geosciences, University of Oslo. The analytical protocols used are described in detail by Rosa *et al.* (2009) and Andersen *et al.* (2009).

The U–Pb data from the mass spectrometer were calibrated to reference zircons GJ-01 (Belousova *et al.* 2006), 91500 (Wiedenbeck *et al.* 1995) and in-house standard A382. The data are presented as probability–density plots (Fig. 3b, c) and Concordia diagrams (Fig. 4). Analyses requiring corrections due to large amounts of $^{204}$Pb were excluded from the figures but are listed with Stacey & Kramers' (1975) corrections in the Supplementary material. The U–Pb ages were calculated using Isoplot version 3 (Ludwig 2003). Combined zircon age probability–density and histogram plots were produced using AgeDisplay (Sircombe 2004).

# Results

## Rutile trace element results

The samples contain variably abraded (fragments of) black, red and yellow detrital rutile grains. Forty grains were analysed from each sample. The source lithologies of the samples in this study, as defined by the Nb and Cr content according to the criteria of Meinhold *et al.* (2008), are mainly metapelitic (Fig. 3a).

The Zr content in rutile is temperature-dependent in a system buffered by quartz and zircon, and can be used to determine the metamorphic grade of the source rocks at the time of rutile formation (Zack *et al.* 2004; Watson *et al.* 2006). Experimental data also indicate that the thermometer can be successfully applied to rutiles from metamafic sources (e.g. Zack & Luvizotto 2006).

The rutile crystallization temperatures in this study were calculated after Watson *et al.* (2006). In the absence of pressure data, metamorphic facies are assigned based on temperature. In Figure 3a all data below the detection limit of 60 ppm Zr, the equivalent of 528 °C, are included in the greenschist and blueschist facies part of the diagram. The temperatures indicated by data from metamafic protoliths appear to be slightly lower than those from metafelsic sources.

Rutiles from the magnetic fractions of samples M09-03 and M09-05 were analysed to test for possible bias introduced by the magnetic concentration of the rutiles. Sixty analyses from sample M09-03 and 24 analyses from M09-05 appear to reproduce the temperature signatures of the non-magnetic

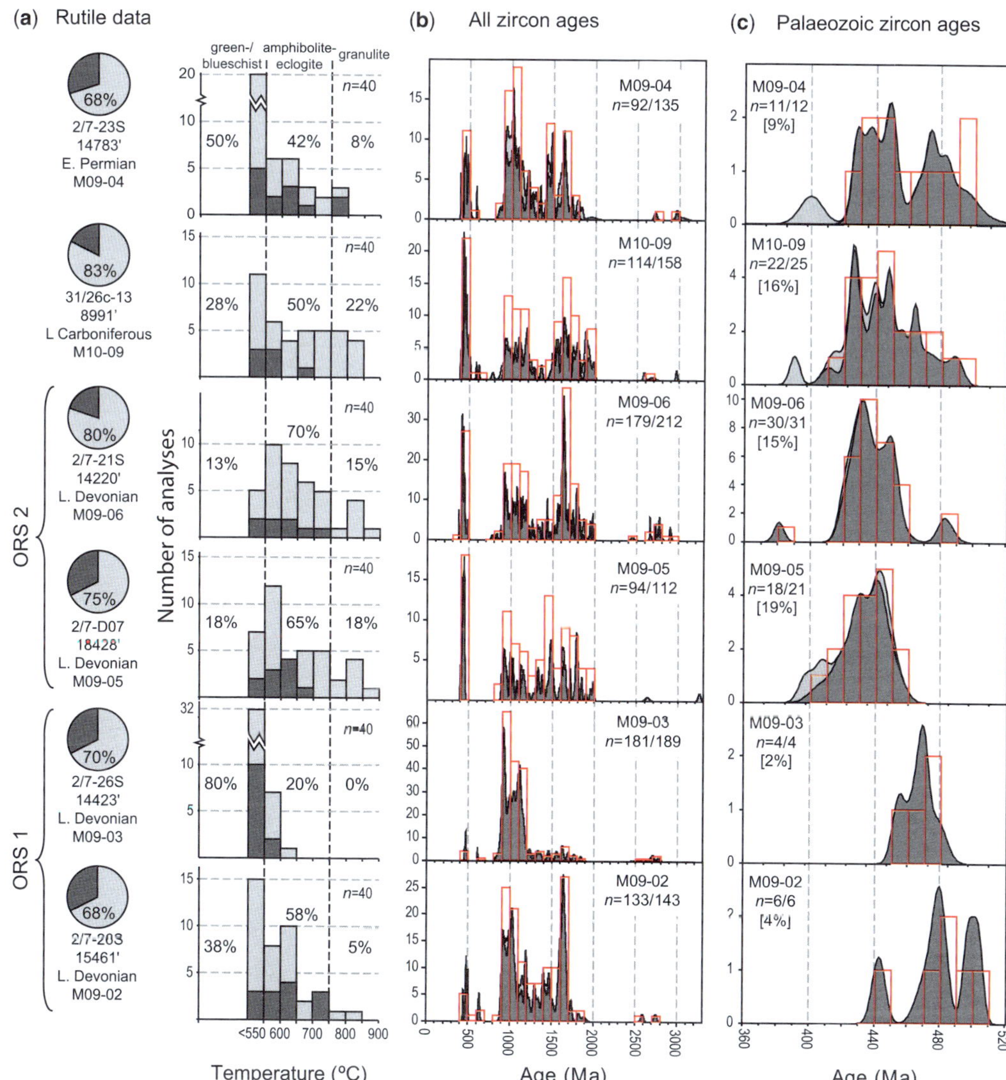



**Fig. 3.** (**a**) Relative abundance of rutiles from mafic (dark grey) and pelitic (light grey) sources (pie charts), and rutile temperature data (histograms) with % analyses of different metamorphic facies. (**b**) Combined histograms and probability–density zircon age plots for <10% discordant analyses (dark grey) and all analyses (light grey). For zircons <700 Ma, the $^{206}$Pb–$^{238}$U age is shown; the remaining data are shown at $^{207}$Pb/$^{206}$Pb ages. $n = \leq 10\%$ discordant analyses/total number of analyses. (**c**) Analysis of Palaeozoic (Caledonian) ages. [%] = 500–380 Ma analyses/total number of analyses.

fractions, whereas a slightly lower incidence of rutile from metamafic protoliths is recorded (Supplementary material).

## Zircon U–Pb results

The U–Pb results are summarized in Figures 3 and 4 and in the Supplementary material. The six samples include sparse 2.5 to 3.3 Ga Archaean grains (*c.* 1–3% of the analysed grains) and display an age gap from 2.5 to 2 Ga. In the Proterozoic part of the spectrum, grains with ages ranging between 2 and 0.9 Ga are present in all samples. The samples display a second age gap from 0.8 to 0.5 Ga, interrupted in four samples by sparse *c.* 600 Ma analyses (*c.* 1%). The Palaeozoic part of the age spectrum is dominated by grains with ages from 0.5 to 0.39 Ga (Fig. 3c).

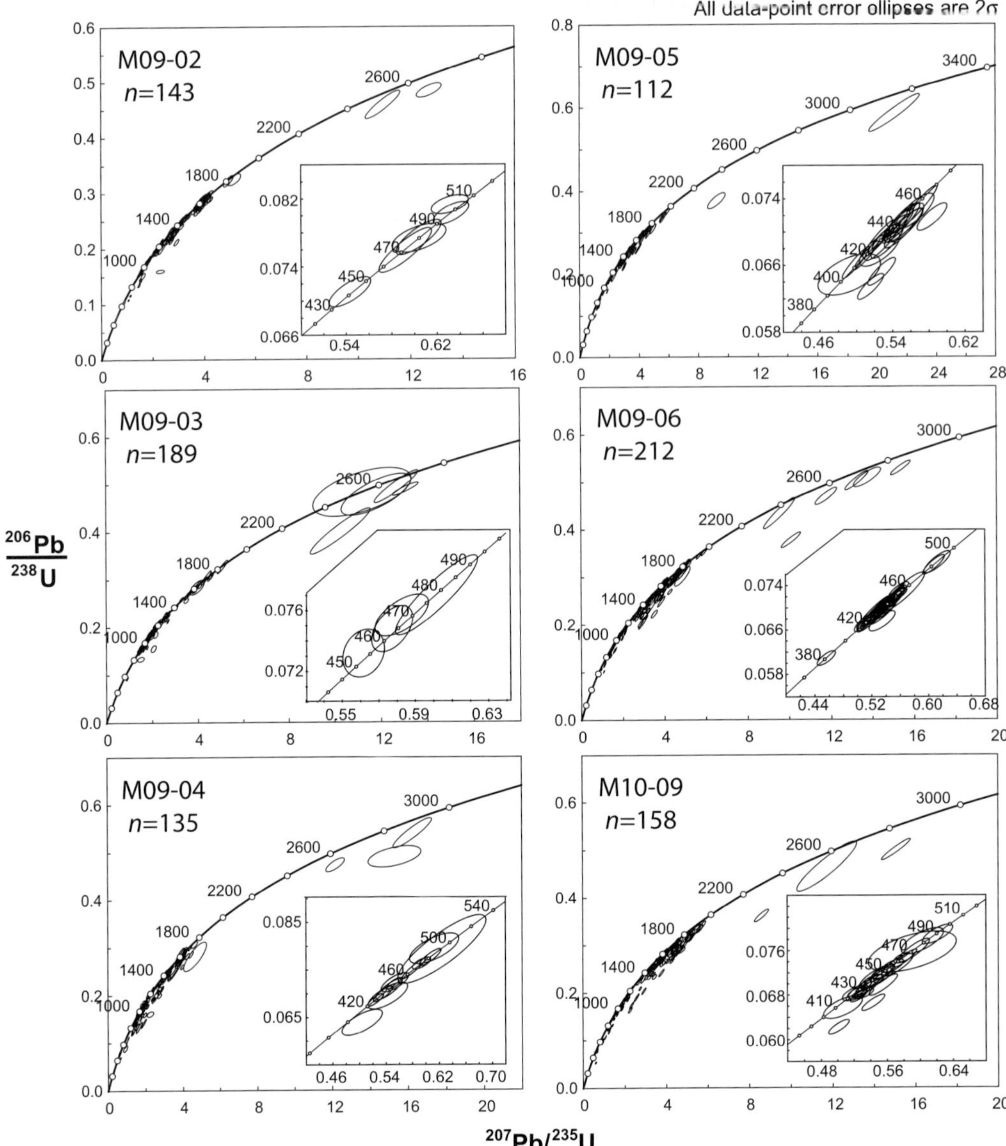



**Fig. 4.** Concordia diagrams. Insets show <500 Ma analyses.

## Discussion

### *Sample differentiation based on zircon and rutile data, and potential for testing of well correlations*

To a first approximation all the samples display similar zircon age patterns with matching peaks and gaps (Fig. 3b). The main differences between the samples are the positions and relative importance of the Palaeozoic, *c.* 500–380 Ma (Caledonian) age peaks and the relative importance of 1.25–0.9 Ga (Grenville–Sveconorwegian) ages. The rutile data show a dominantly greenschist to lower amphibolite facies signature in samples from well 2/7-26S, whereas a significant high-grade rutile population, mainly from felsic source rocks, is present in the remaining samples (Fig. 3a). The rutile signatures correlate with the zircon age patterns, indicating that variations in the relative importance of Caledonian and Grenville/Sveconorwegian age groups represent actual source characteristics (as opposed to mere random variation). Based on the zircon and rutile signatures the late Devonian samples are arranged in two groups, Old Red Sandstone (ORS) 1 and 2.

The ORS 1 group comprises samples M09-02 and M09-03 from well 2/7-26S, which yield few (≤4%) Caledonian zircons of mainly Cambro-Ordovician ages, and sparse c. 600 Ma zircons. The dominant zircon populations are of Grenville–Sveconorwegian ages (48–82%). Both samples display rutile temperature data dominated by the <650 °C component.

The ORS 2 group consists of samples M09-05 and M09-06 from wells 2/7-D07 and 2/7-21S, respectively, which yield a high proportion of Caledonian, mainly Silurian grains (15–19%). No c. 600 Ma zircons are present. The Grenville–Sveconorwegian ages (29% to 33%) represent a smaller proportion of the analyses than the 1.25–2 Ga ages (48–51%). Both samples yield rutile temperature data with significant granulite facies components (15–18%).

The zircon and rutile signatures of the late Carboniferous sample M10-09 from the Flora oil field are similar to the ORS 2 samples, yielding a high proportion of Caledonian grains, a relatively subdued Grenville–Sveconorwegian population, and a high proportion of the rutiles record granulite facies conditions. However, the sample contains both Silurian and Cambro-Ordovician zircons, and several grains yielding c. 600 Ma ages. The early Permian sample M09-04 is intermediate in character between the ORS 1 and 2 samples, both in terms of age patterns and rutile temperature data.

In summary, the provenance data suggest input from (at least) two different sources in the late Devonian–early Carboniferous, which should allow testing of correlations between different wells in the Embla oil field. For example, the correlation between the sand bodies represented by sample M09-03 (2/7-26S) and M09-06 (2/7-21S) suggested by chemostratigraphy does not find support in the new data. With the palaeomagnetic support for different sediment ages in 2/7-26S and 2/7-21S, we interpret the two samples to represent detritus from (partly) different sources, deposited at slightly different times.

*Provenance of late Palaeozoic sediments on the northern flank of the Mid North Sea High*

The samples investigated in this study contain detrital zircons with ages spanning the Archaean to the Palaeozoic, representing detritus sourced from different geological provinces either directly or by recycling of older sediments. The diverse age and temperature signatures, the inferred low primary feldspar content, and the mainly quartz/quartzite pebbles and clasts associated with the sampled sandstones hint at a polycyclic origin of the sediments.

The overall age patterns are compatible with ultimate sources in Laurentia and/or Baltica, and include detritus from both the Caledonian and the Grenville–Sveconorwegian orogens. The c. 600 Ma zircon ages in the Embla and Flora oil field sandstones overlap in age with Cadomian magmatism along the Gondwana margin, and could reflect the presence of detritus derived from the south (cf. Hallsworth *et al.* 2000). However, sparse c. 620–550Ma zircons are also present in circum-Atlantic rocks. The Dalradian Supergroup in Scotland hosts the c. 590 Ma Ben Vuirich granite (Rogers *et al.* 1989), and c. 600 Ma augen granites intrude the Moine Supergroup in Scotland (Kinny *et al.* 2003). Detrital zircon grains of this age are present in Ordovician sediments in the Southern Uplands, and are believed to have been derived from Laurentia (Waldron *et al.* 2008). Mafic dykes of this age are also known from Baltica (Bingen *et al.* 1998). The c. 600 Ma ages reflect the (mainly mafic) magmatism associated with the opening of the Iapetus Ocean (Andréasson 1994; Stølen 1994; Cawood *et al.* 2007; Oliver *et al.* 2008; Bingen & Solli 2009). The sparse c. 600 Ma zircons in samples in this study are therefore in agreement with a northern provenance.

## Old Red Sandstone: ORS 1

The zircon and rutile signatures in the ORS 1 sediments indicate an ultimate source area dominated by low- to medium-grade Grenville–Sveconorwegian rocks, and sparse Cambro-Ordovician zircons point to minor contributions from rocks related to the Iapetan realm, now parts of the Caledonides.

In Greenland, Grenville ages are restricted to detrital zircons and c. 950–920 Ma granitoids in allochthonous Neoproterozoic metasediments in the East Greenland Caledonides (Watt & Thrane 2001; Kalsbeek *et al.* 2008) and in younger sediments derived from these (Slama *et al.* 2011). Much of the metasediments yield a Caledonian granulite facies rutile signature (Gilotti *et al.* 2008; Meinhold *et al.* 2011). The high-grade source area fits garnet data from the Buchan and Tayport formations in the Outer Moray Firth situated roughly mid-way between Greenland and the Mid North Sea High (wells 14/13-3 and 15/18-1; Fig. 1), which record influx of detritus from the NE (Hallsworth *et al.* 2000; Morton *et al.* 2001) at the time of deposition of the Embla Old Red Sandstone sediments. However, the signature does not match the rutile data in ORS 1.

In Baltica, *c.* 950–1170 Ma granites are present in SW Sweden and south Norway (Andersen *et al.* 2001; Bingen *et al.* 2008). However, they represent a minor and geographically scattered part of the basement and are unlikely to dominate the regional provenance signature. In SW Norway, the allochthonous Lindås Nappe is dominated by Sveconorwegian rocks (Bingen *et al.* 2001) and is associated with the *c.* 480–490 Ma Gullfjellet ophiolite (Dunning & Pedersen 1988). However, Cretaceous and Paleocene sediments derived from west Norway yield mainly metamafic rutile (Morton & Chenery 2009), which is probably also a fair representation of the anorthosite–mangerite–charnockite–granite rocks in the Lindås Nappe.

The Scottish Lewisian basement is interpreted to be linked to the Rockall Plateau microcontinent and SE Greenland (Friend & Kinny 2001). Sparse Grenvillian metamorphic ages (Miller *et al.* 1973) and brittle deformation suggest that these rocks were situated at the margin of the Grenville belt (Sherlock *et al.* 2008 and references therein), and are unlikely sources for prolific zircons of Grenville age. Furthermore, the subordinate Archaean populations in the ORS 1 sediments are not compatible with significant contributions from Lewisian-like basement.

In contrast, Scottish Neoproterozoic metasediments were largely derived from the Grenville orogen, with minor contributions from erosion of local basement. Important Archaean zircon populations in Torridonian (Rainbird *et al.* 2001) and the Dalradian Argyll, Appin and Southern Highland groups (Cawood *et al.* 2003) do not match the signatures in the ORS 1 sediments. The Grampian Group, on the other hand, displays important late Mesoproterozoic to early Neoproterozoic populations, corresponding to the age of the Grenville orogeny, and only minor Archaean populations, as do the Moine metasediments (Friend *et al.* 2003; Cawood *et al.* 2004). Both the Northern Highlands and the Grampian Highlands host numerous *c.* 470–465 Ma 'Grampian Event' granites and *c.* 430–400 Ma 'Newer granites'. Metamorphic conditions range from greenschist to upper amphibolite facies and migmatites are locally present (Strachan *et al.* 2002 and references therein). Garnets from present-day rivers sampling the Dalradian and Moine metasediments do not yield high-grade signatures (Hallsworth *et al.* 2000; Morton *et al.* 2004).

Ordovician sandstones in the Midland Valley (Phillips *et al.* 2009) and in the Southern Uplands (Waldron *et al.* 2008) also display strong Grenville signatures. Zircons from the former yield mainly *c.* 475 Ma and *c.* 1000 Ma ages, suggested by Phillips *et al.* (2009) to reflect erosion of local Ordovician volcanic rocks and their basement during strike–slip tectonics. Detrital zircons from Ordovician sandstones in the Southern Uplands are typically dominated by Grenville populations, with 8–35% Archaean zircons, and sparse *c.* 600 Ma and Ordovician ages (Waldron *et al.* 2008). The sediments are dominantly greenschist grade or lower (Oliver 1988), and intruded by *c.* 425–390 Ma granites (Oliver *et al.* 2008).

The Grampian High, the Southern Uplands High and the Midland Valley (Fig. 1) thus include potential sources matching the ORS 1 zircon and rutile signature (Fig. 5). We regard the two palaeohighs as the most likely sources for the ORS 1 sediments.

This interpretation is consistent with Famennian palaeostream flow data, which document erosion of the Scottish Caledonides and sediment transport southeastward into the North Sea (Fig. 1; Leeder 1988; Ziegler 1990). In addition, detritus of Southern Uplands affinity were likely derived locally from the rift flanks of the late Devonian proto-Central Graben (Fig. 1). The ORS 1 sediments are thus interpreted to be polycyclic, although input from the inferred Grenville-dominated basement

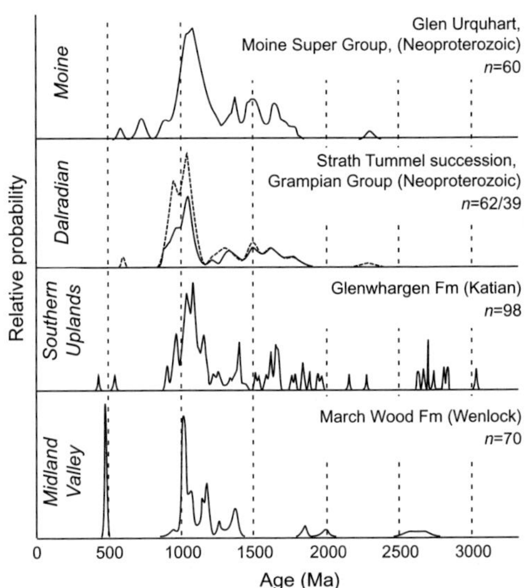


**Fig. 5.** Zircon U–Pb data from the Moine Supergroup (Cawood *et al.* 2004), the Dalradian Supergroup (Cawood *et al.* 2003), the Southern Uplands (Waldron *et al.* 2008) and the Midland Valley (Phillips *et al.* 2009), selected to highlight similarities to the ORS 1 zircon age signature in this study. Data are shown for <10% discordant analyses (black lines), and in the Dalradian Strath Tummel succession where few analyses are available, with all analyses included (dashed line). The emergence of this specific source signature in both the Neoproterozoic and the Palaeozoic hints at repeated influx from an as yet unidentified source in the Grenville orogen.

supplying sediments to the Neoproterozoic to Ordovician Scottish basins cannot be excluded.

The ORS 1 age signature puts additional constraints on the nature of the source areas; the c. 600 Ma zircons hint at erosion of Neoproterozoic granites in the Dalradian and Moine metasediments, whereas the low proportion of Caledonian zircons, and the lack of <440 Ma zircons in particular, suggest that few of the Grampian granites and none of the younger Caledonian granites were exhumed in the source areas.

## Old Red Sandstone: ORS 2

The age signatures in the ORS 2 sediments indicate a source dominated by Proterozoic rocks and an important contribution from mainly Silurian granites. The rutile data indicate high-grade felsic rocks in the source area.

High-grade terrains in the circum-Atlantic and North Sea region include variably retrogressed Archaean and Proterozoic granulites in NW Scotland, the Rockall Plateau and on SE Greenland (Fig. 1; Friend & Kinny 2001; Hitchen 2004). The sparse Archaean zircons in the samples in this study do not support a significant contribution from these rocks.

In SW Sweden, the Proterozoic basement includes minor c. 1.4 Ga variably retrograded granulite facies rocks exposed in the Varberg area (Page et al. 1996; Christoffel et al. 1999), and largely retrogressed mafic and felsic Sveconorwegian eclogites and granulites metamorphosed up to c. 800 °C (Johansson et al. 1991; Austin Hegardt et al. 2005). These rocks are situated SE of the Caledonian front, distant from known Silurian granitoids.

In south and SW Norway, pre-Caledonian mainly mafic granulite facies rocks are present in the Lindås Nappe and the related Jotun Nappe Complex, and in the autochthonous Rogaland anorthosite complex and the Bamble sector (Heier 1973). Mesozoic sediments derived from SW Norway mainly contain rutiles from mafic sources, and only sparse high-temperature rutiles (Morton & Chenery 2009). The latter may be pre-Caledonian, or could reflect Caledonian eclogite facies metamorphism in west Norway, locally reaching in excess of 800 °C (Fig. 1; Cuthbert et al. 2000). This pattern of largely mafic granulite signatures is confirmed by garnet data from present-day streams, which also indicate the Rogaland anorthosite complex as a potential source for both high-grade felsic and mafic rutile (Morton et al. 2004).

The East Greenland Krummedal sequence makes up the erosional remnants of a vast Caledonian granulite facies terrain (Gilotti et al. 2008). The metasediments yield a distinct felsic high-grade rutile and garnet signature (Morton et al. 2009;

Meinhold et al. 2011) and include Caledonian c. 466–420 Ma granites (Kalsbeek et al. 2008; Rehnström 2010). Correlated metasediments are also present in thrust sheets in the Uppermost Allochthon in west Norway (Roberts et al. 2007), and potentially off the Norwegian coast (Slagstad et al. 2011).

Given the propensity of rutile to retrograde to other titaniferous minerals during low-grade metamorphism (Zack et al. 2004), Caledonian high-grade rocks are more likely sources for the granulite facies rutile in ORS 2 than the Archaean to Proterozoic variably retrograded granulite facies terrains described above. This would also fit the association of the high-grade component with Caledonian zircons. We therefore propose that the East Greenland Krummedal sequence (and related rocks) is the best match for the mainly felsic high-grade rutile in the ORS 2.

A northeastern source matches early Carboniferous palaeostream flow-direction data demonstrating southwestward outbuilding of the Carboniferous Pennine River (Fig. 1; Leeder 1988). Further constraints on this source come from the Buchan and Tayport formations in the Outer Moray Firth (Fig. 1). Their northeastern source area is characterized by high-grade metasediments derived from mainly mid-Proterozoic rocks, Caledonian granites and minor Archaean basement. The garnet signatures indicate near exclusive (garnet) derivation from a high-grade felsic source. Of the analysed zircons 37–39% are Caledonian and yield mainly Silurian ages. No Caledonian grains are older than c. 450 Ma (Morton et al. 2001).

We propose that the high-grade felsic rutile component and the Caledonian, mainly Silurian zircons in the ORS 2 sediments signal the arrival of sediments from the NE via the Pennine River system. A comparison to the garnet and zircon signature documented in the Outer Moray Firth suggests that the sediments in the Embla field have been diluted, potentially by sediments from the NW and/or local detritus from the uplifted rift flanks of the late Devonian proto-Central Graben.

## Late Carboniferous sandstone

The zircon age signature of the mid-Bolsovian–Stephanian Upper Flora Sandstone indicates ultimate sources in Laurentia and/or Baltica, resembling those inferred for the ORS 2 sediments, including a significant contribution from Silurian granites and high-grade felsic rocks.

The sampled interval has been interpreted as part of an eastward prograding alluvial fan system (Martin et al. 2002). At the time of its deposition the Variscan foreland underwent compression with northward-progressing folding, thrusting and basin inversion, leading to considerable uplift of the Mid North Sea High (Corfield *et al.* 1996). Shifting palaeocurrents from southwestward transport directions in the lower part of the Upper Flora Sandstone to (south-)eastward in the upper part (sampled in this study), combined with an upward increase in grain size, has been interpreted as the result of hinterland rejuvenation in the west (Martin *et al.* 2002).

The Bolsovian–Asturian Ketch Member in the Schooner Formation in the southern North Sea comprises similar lithologies and is also interpreted to derive from erosion of uplifted older Palaeozoic sediments (Leeder & Hardman 1990). It has therefore been proposed to be a distal equivalent to the Upper Flora Sandstone (Martin *et al.* 2002).

Heavy mineral data from the Lower Ketch Member (well 44/18-1) show high chromite contents indicative of mafic–ultramafic rocks in the source region. This led Morton *et al.* (2001) to suggest the Ringkøbing-Fyn High, NE of Block 44, as the source. Although the zircon age signature of the Lower Ketch Member is similar to that of the Upper Flora Sandstone, the latter contains no chromite (Martin *et al.* 2002; this study). Because chromite is resistant to weathering and burial diagenesis (Mange & Morton 2007; Beinlich *et al.* 2010), its absence indicates that the two units have different sources.

The zircon and rutile data in this study are compatible with continued influx of detritus from the north in the late Carboniferous. Alternatively, and more likely given the sedimentological data and the evidence for hinterland rejuvenation, the sample represents reworking of older Palaeozoic sediments, including detritus originally from the Krummedal sequence and related rocks. The absence of a Variscan signature in the Upper Flora Sandstone despite the influx of Variscan detritus to the south of the Southern Uplands–Mid North Sea High, documented in central England in the Bolsovian (Dalton Rock; Hallsworth *et al.* 2000) and at the Canonbie Coalfields in the Asturian (Morton *et al.* 2010), indicates that the source of the reworked sediments lay north of southernmost Scotland. Given an increasing distance to the Variscan front and a presumably decreasing importance of inversion tectonics towards the north, it is appealing to attribute the sediments to differential uplift and erosion of the Southern Uplands–Mid North Sea High in the late Carboniferous, thus providing a local source for the sediments.

It is noteworthy that the influx of detritus from southern sources in the late Carboniferous provides a potential provenance criterion for distinguishing late Carboniferous and Rotliegend red beds from Old Red Sandstone south of the Southern Uplands–Mid North Sea High.

## Early Permian sandstone

The early Permian sandstone sample yields a zircon and rutile provenance signature intermediate in character between the two Old Red Sandstone groups in this study.

Age constraints for the sampled sandstone come from palaeomagnetic inclination data indicating an early Permian age, and the polarity, which indicates deposition during the Kiaman Reverse Superchron (*c.* 317–262 Ma; e.g. Opdyke & Channell 1996). Highly altered decimetre-thick mafic volcanic beds interleaved with the sediments suggest that the sediments were deposited in the context of the intense Stephanian–early Permian magmatism in northwestern Europe (Timmerman 2004). The volcanic rocks could potentially be related to 299 ± 3 Ma (Heeremans *et al.* 2004) and *c.* 310 Ma mafic volcanism in the Flora oil field, interpreted as part of the Inge Formation, Lower Rotliegend Group (Martin *et al.* 2002).

A Variscan origin for the early Permian sediments, as in the Rotliegend deserts in the southern North Sea (Glennie 1972), is ruled out by the provenance signature. The zircon and rutile data in this study are compatible with continued influx of detritus from the north in the early Permian. Alternatively, because the early Permian sediments appear to rest unconformably on Old Red Sandstone in the Embla oil field, they could represent reworking of the missing sequence. Recycling of the Old Red Sandstone (and late Carboniferous sediments) described in this study would fit the observed provenance signature. Rotliegend sediments are commonly attributed to reworking of Old Red Sandstone (Marshall & Hewett 2003), and this is tentatively proposed as the origin of the early Permian sediments in this study. As in the case for the Upper Flora sandstone, the absence of Cadomian and Variscan ages points to a source area to the north of southernmost Scotland.

## Conclusions

Zircon U–Pb ages and rutile geochemistry demonstrate that Old Red Sandstone sediments on the northern flank of the Mid North Sea High form two groups. One group (ORS 1) is dominated by Grenvillian zircons and low- to medium-grade rutiles. The main source is interpreted to be recycling of sediments from the Grampian High and the Southern Uplands High–Mid North Sea High. The second group (ORS 2) is distinguished by the appearance of high-grade rutile from a felsic source and a significant Silurian zircon population. We propose that both the high-grade and Silurian components can be traced back to the Krummedal sequence on East Greenland (and related sediments) via late Devonian–early Carboniferous sediments in the Outer Moray Firth (Morton *et al.* 2001).

The appearance of high-grade sediments at the Mid North Sea High is most easily interpreted in terms of a temporal evolution of the Caledonide drainage system from northwestern sources to increasingly important contributions from the NE through the proto-Viking Graben, reflecting the growth of the Pennine river system. This indicates that ORS 1 predates ORS 2.

Mid Bolsovian–Stephanian Upper Flora Sandstone and early Permian sandstone yield zircon and rutile signatures compatible with recycling of uplifted Palaeozoic sediments similar to the Old Red Sandstones above. Because molasse deposits from the Variscan orogen extend to the Southern Uplands High–Mid North Sea High, where they probably terminate, the lack of distinct Cadomian and Variscan signatures indicates that sediments in this study were derived from sources north of southernmost Scotland, potentially the Southern Uplands High–Mid North Sea High itself.

The provenance interpretations indicate that much of the sediments in this study represent repeated recycling of Mesoproterozoic to Palaeozoic sediments of Laurentian ancestry.

This study was financed by ConocoPhillips Norway, who also supplied core samples from the Embla oil field, and made available unpublished palynological, palaeomagnetic and geochemical data. Amerada Hess provided access to the Flora oil field core sample, and Martin Hill assisted with on-site sampling. Siri Lene Simonsen provided support for the LA-ICPMS work. Muriel Erambert assisted with microprobe work. Magnus Kristoffersen assisted with sample preparation. This is publication no. 32 from the Isotope Geology Laboratory at the Department of Geosciences, University of Oslo. The authors gratefully acknowledge the reviewers Guido Meinhold and John Marshall for their constructive comments.

# References

ANDERSEN, T., ANDRESEN, A. & SYLVESTER, A. G. 2001. Nature and distribution of deep crustal reservoirs in the southwestern part of the Baltic Shield: evidence from Nd, Sr and Pb isotope data on late Sveconorwegian granites. *Journal of the Geological Society, London*, **158**, 253–267.

ANDERSEN, T., ANDERSSON, U. B., GRAHAM, S., ÅBERG, G. & SIMONSEN, S. L. 2009. Granitic magmatism by melting of juvenile continental crust: new constraints on the source of Paleoproterozoic granitoids in Fennoscandia from Hf isotopes in zircon. *Journal of the Geological Society, London*, **166**, 233–248, http://dx.doi.org/10.1144/0016-76492007-166.

ANDERSEN, T., SAEED, A., GABRIELSEN, R. H. & OLAUSSEN, S. 2010. Provenance characteristics of the Brumunddal sandstone in the Oslo Rift derived from U–Pb, Lu–Hf and trace element analyses of detrital zircons by laser ablation ICPMS. *Norwegian Journal of Geology*, **91**, 1–19.

ANDRÉASSON, P. G. 1994. The Baltoscandian margin in Neoproterozoic–early Palaeozoic times. Some constraints on terrane derivation and accretion in the Arctic Scandinavian Caledonides. *Tectonophysics*, **231**, 1–32.

AUGLAND, L. E., ANDRESEN, A. & CORFU, F. 2011. Terrane transfer during the Caledonian orogeny: Baltican affinities of the Liverpool Land eclogite terrane in East Greenland. *Journal of the Geological Society, London*, **168**, 15–26.

AUSTIN HEGARDT, E., CORNELL, D. H., CLAESSON, L., SIMAKOV, S., STEIN, H. J. & HANNAH, J. L. 2005. Eclogites in the central part of the Sveconorwegian Eastern Segment of the Baltic Shield: support for an extensive eclogite terrane. *GFF*, **127**, 221–232.

BANKA, D., PHARAOH, T. C. & WILLIAMSON, J. P. 2002. Potential field imaging of Palaeozoic orogenic structure in northern and central Europe. *Tectonophysics*, **360**, 23–45.

BEINLICH, A., AUSTRHEIM, H., GLODNY, J., ERAMBERT, M. & ANDERSEN, T. B. 2010. $CO_2$ sequestration and extreme Mg depletion in serpentinized peridotite clasts from the Devonian Solund basin, SW-Norway. *Geochimica et Cosmochimica Acta*, **74**, 6935–6964.

BELOUSOVA, E. A., GRIFFIN, W. L. & O'REILLY, S. Y. 2006. Zircon crystal morphology, trace element signatures and Hf isotope composition as a tool for petrogenetic modelling: examples from Eastern Australian granitoids. *Journal of Petrology*, **47**, 329–353, http://dx.doi.org/10.1093/petrology/egi077

BESLY, B. M. 1998. Carboniferous. *In*: GLENNIE, K. W. (ed.) *Introduction to the Petroleum Geology of the North Sea*, 4th edn. Blackwell, Oxford, 104–136.

BESLY, B. M., BURLEY, S. D. & TURNER, P. 1993. The late Carboniferous 'Barren Red Bed' play of the Silver Pit area, Southern North Sea. *In*: PARKER, J. R. (ed.) *Petroleum Geology of Northwest Europe: Proceedings of the 4th Conference*. Geological Society, London, 727–740.

BINGEN, B. & SOLLI, A. 2009. Geochronology of magmatism in the Caledonian and Sveconorwegian belts of Baltica: synopsis for detrital zircon provenance studies. *Norwegian Journal of Geology*, **89**, 267–290.

BINGEN, B., DEMAIFFE, D. & VAN BREEMEN, O. 1998. The 616 Ma old Egersund basaltic dike swarm, SW Norway, and late Neoproterozoic opening of the Iapetus Ocean. *Journal of Geology*, **106**, 565–574.

BINGEN, B., DAVIS, W. J. & AUSTRHEIM, H. 2001. Zircon U–Pb geochronology in the Bergen Arc eclogites and their Proterozoic protoliths, and implications for the pre-Scandian evolution of the Caledonides in western Norway. *Geological Society of America Bulletin*, **113**, 640–649.

BINGEN, B., NORDGULEN, Ø. & VIOLA, G. 2008. A four-phase model for the Sveconorwegian orogeny, SW Scandinavia. *Norwegian Journal of Geology*, **88**, 43–72.

CAMERON, T. D. J. 1993. Triassic, Permian, pre-Permian of the Central, Northern North Sea. *In*: KNOX, R. W. O'B. & CORDEY, W. G. (eds) *Lithostratigraphic nomenclature of the UK North Sea.* British Geological Survey, Nottingham, 1–163.

CAWOOD, P. A., NEMCHIN, A. A., SMITH, M. & LOEWY, S. 2003. Source of the Dalradian Supergroup constrained by U/Pb dating of detrital zircon, implications for the East Laurentian margin. *Journal of the Geological Society, London,* **160**, 231–246.

CAWOOD, P. A., NEMCHIN, A. A., STRACHAN, R. A., KINNY, P. D. & LOEWY, S. 2004. Laurentian provenance and an intracratonic tectonic setting for the Moine Supergroup, Scotland, constrained by detrital zircons from the Loch Eil, Glen Urquhart successions. *Journal of the Geological Society, London,* **161**, 861–874.

CAWOOD, P. A., NEMCHIN, A. A., STRACHAN, R. A., PRAVE, A. R. & KRABBENDAM, M. 2007. Sedimentary basin and detrital zircon record along East Laurentia and Baltica during assembly and breakup of Rodinia. *Journal of the Geological Society, London,* **164**, 257–275.

CHRISTOFFEL, C. A., CONNELLY, J. N. & ÅHÄLL, K.-I. 1999. Timing and characterization of recurrent pre-Sveconorwegian metamorphism and deformation in the Varberg–Halmstad region of SW Sweden. *Precambrian Research,* **98**, 173–195.

CLIFF, R. A., DREWERY, S. E. & LEEDER, M. R. 1991. Sourcelands for the Carboniferous Pennine river system: constraints from sedimentary evidence and U–Pb geochronology using zircon and monazite. *In:* MORTON, A. C., TODD, S. P. & HAUGHTON, P. D. W. (eds) *Developments in Sedimentary Provenance Studies.* Geological Society, London, Special Publications, **57**, 137–159.

CORFIELD, S. M., GAWTHORPE, R. L., GAGE, M., FRASIER, A. J. & BESLY, B. M. 1996. Inversion tectonics of the Variscan foreland of the British Isles. *Journal of the Geological Society, London,* **153**, 17–32.

COWARD, M. P. 1995. Structural and tectonic setting of the Permo-Triassic basins of northwest Europe. *In:* BOLDY, S. A. R. (ed.) *Permian, Triassic Rifting in Northwest Europe.* Geological Society, London, Special Publications, **91**, 7–39.

CUTHBERT, S. J., CARSWELL, D. A., RAVNA, E. J. K. & WAIN, A. L. 2000. Eclogites and eclogites in the Western Gneiss Region, Norwegian Caledonides. *Lithos,* **52**, 165–195.

DREWERY, S. A., CLIFF, R. A. & LEEDER, M. R. 1987. Provenance of Carboniferous sandstones from U–Pb dating of detrital zircons. *Nature,* **325**, 50–53.

DUNNING, G. & PEDERSEN, R.-B. 1988. U/Pb ages of ophiolites and arc related plutons of the Norwegian Caledonides: implications for the development of Iapetus. *Contributions to Mineralogy and Petrology,* **98**, 13–23.

FRIEND, C. R. L. & KINNY, P. D. 2001. A reappraisal of the Lewisian Gneiss Complex: geochronological evidence for its tectonic assembly from disparate terranes in the Proterozoic. *Contributions to Mineral and Petrology,* **142**, 198–218, http://dx.doi.org/10.1007/s004100100283

FRIEND, C. R. L., STRACHAN, R. A., KINNY, P. & WATT, G. R. 2003. Provenance of the Moine Supergroup of NW Scotland: evidence from geochronology of detrital and inherited zircons from sediments, granites and

migmatites. *Journal of the Geological Society, London,* **160**, 247–258.

GATLIFF, R. W., RICHARDS, P. C., SMITHS, K., GRAHAM, C. C., MCCORMAC, M., SMITH, N. J. P. & LONG, D. 1994. *The geology of the central North Sea.* British Geological Survey UK Offshore Regional Report, **5**, Her Majesty's Stationery Office, London.

GILOTTI, J. A., JONES, K. A. & ELVEVOLD, S. 2008. Caledonian metamorphic patterns in Greenland. *In:* HIGGINS, A. K., GILOTTI, J. A. & SMITH, M. P. (eds) *The Greenland Caledonides: Evolution of the Northeast Margin of Laurentia.* Geological Society of America Memoirs, **202**, 201–205.

GLENNIE, K. W. 1972. Permian Rotliegendes of northwest Europe interpreted in light of modern desert sedimentation studies. *American Association of Petroleum Geologists Bulletin,* **56**, 1048–1071.

GLOVER, B. W., LENG, M. J. & CHISHOLM, J. I. 1996. A second major fluvial sourceland for the Silesian Pennine Basin of northern England. *Journal of the Geological Society, London,* **153**, 901–906.

HALLSWORTH, C. R., MORTON, A. C., CLAOUÉ-LONG, J. & FANNING, C. M. 2000. Carboniferous sand provenance in the Pennine basin, UK: constraints from heavy mineral and detrital zircon age data. *Sedimentary Geology,* **137**, 147–185.

HALLSWORTH, C. R. & CHISHOLM, J. I. 2008. Provenance of late Carboniferous sandstones in the Pennine Basin (UK) from combined heavy mineral, garnet geochemistry and palaeocurrent studies. *Sedimentary Geology,* **203**, 196–212, http://dx.doi.org/10.1016/j.sedgeo.2007.11.002.

HAYWARD, R., MARTIN, C. A. L., HARRISON, D., VAN DORT, G., GUTHRIE, S. & PADGET, N. 2002. The Flora Oil Field, Blocks 31/26a, 31/26c, UK North Sea. *In:* GLUYAS, J. & HICHENS, H. (eds) *Oil and Gas Fields Commemorative Millennium Volume.* Geological Society, London, Memoirs, **20**, 549–555, http://dx.doi.org/10.1144/GSL.MEM.2003.021.01.44

HECKEL, P. H. & CLAYTON, G. 2006. The Carboniferous system. Use of the new official names for the subsystems, series and stages. *Geologica Acta,* **4**, 403–407.

HEEREMANS, M., TIMMERMAN, M. J., KIRSTEIN, J. I. & FALEIDE, J. I. 2004. New constraints on the timing of late Carboniferous–early Permian volcanism in the central North Sea. *In:* WILSON, M., NEUMANN, E.-R., DAVIES, G. R., TIMMERMAN, M. J., HEEREMANS, M. & LARSEN, B. T. (eds) *Permo-Carboniferous Magmatism and Rifting in Europe.* Geological Society, London, Special Publications, **223**, 177–194.

HEIER, K. S. 1973. Geochemistry of granulite facies rocks and problems of their origin. *Philosophical Transactions of the Royal Society of London A,* **273**, 429–442, http://dx.doi.org/10.1098/rsta.1973.0009

HITCHEN, K. 2004. The geology of the UK Hatton–Rockall margin. *Marine and Petroleum Geology,* **21**, 993–1012, http://dx.doi.org/ 10.1016/j.marpetgeo.2004.05.004

HOSKIN, P. W. O. & SCHALTEGGER, U. 2003. The composition of zircon and igneous and metamorphic petrogenesis. *In:* HANCHAR, J. & HOSKIN, P. W. O. (eds) *Zircon.* Mineralogical Society of America, Washington, D.C. Reviews in Mineralogy and Geochemistry, **53**, 27–62.

JOHANSSON, L., LINDH, A. & MÖLLER, C. 1991. Late Sveconorwegian (Grenville) high-pressure granulite facies metamorphism in southwest Sweden. *Journal of Metamorphic Geology*, **9**, 283–292.

KALSBEEK, F., HIGGINS, A. K., JEPSEN, H. F., FREI, R. & NUTMAN, A. P. 2008. Granites and granites in the East Greenland caledonides. *In*: HIGGINS, A. K., GILOTTI, J. A. & SMITH, M. P. (eds) *The Greenland Caledonides: Evolution of the Northeast Margin of Laurentia*. Geological Society of America Memoirs, **202**, 227–250.

KAUFMANN, B. 2006. Calibrating the Devonian time scale: a synthesis of U–Pb ID-TIMS ages and conodont stratigraphy. *Earth Science Reviews*, **76**, 175–190, http://dx.doi.org/10.1016/j.earscircv.2006.01.001

KNIGHT, I. A., ALLEN, L. R., COIPEL, J., JACOBS, L. & SCANLAN, M. J. 1993. The Embla Oil field. *In*: PARKER, J. R. (ed.) *Petroleum Geology of Northwest Europe: Proceedings of the 4th Conference*. Geological Society, London, 1433–1444.

KINNY, P. D., STRACHAN, R. A., KOCKS, H. & FRIEND, C. R. L. 2003. U–Pb geochronology of late Neoproterozoic augen granites in the Moine Supergroup, NW Scotland: dating of rift-related, felsic magmatism during supercontinent break-up? *Journal of the Geological Society, London*, **160**, 925–934.

LAMMINEN, J. & KÖYKKÄ, J. 2010. The provenance and evolution of the Rjukan Rift Basin, Telemark, south Norway: the shift from a rift basin to an epicontinental sea along a Mesoproterozoic supercontinent. *Precambrian Research*, **181**, 129–149.

LANDES, M., RITTER, J. R. R., READMAN, P. W. & O'REILLY, B. M. 2005. A review of the Irish crustal structure and signatures from the Caledonian and Variscan Orogenies. *Terra Nova*, **17**, 111–120, http://dx.doi.org/10.1111/j.1365-3121.2004.00590.x

LEEDER, M. R. 1988. Devonian–Carboniferous river systems and sediment dispersal from the orogenic belts and cratons of NW Europe. *In*: HARRIS, A. L & FETTES, D. J. (eds) *The Caledonian–Appalachian Orogen*. Geological Society, London, Special Publications, **38**, 549–558.

LEEDER, M. R. & HARDMAN, M. 1990. Carboniferous geology of the Southern North Sea Basin and controls on hydrocarbon prospectivity. *In*: HARDMAN, R. F. P. & BROOKS, J. (eds) *Tectonic Events Responsible for Britain's Oil and Gas Reserves*. Geological Society, London, Special Publications, **55**, 87–105.

LUDWIG, K. R. 2003. *Isoplot 3.0 – a geochronological toolkit for Microsoft excel*. Berkeley Geochronology Center, Berkeley, Special Publication, **4**.

LUNDMARK, A. M., GABRIELSEN, R. H., AUSTRHEIM, H., FLAAT, K., STRAND, T. & OHM, S. E. 2012. Late Devonian rifting in the central North Sea: evidence from altered felsic volcanic rocks in the Embla oil field. *Marine and Petroleum Geology*, **29**, 204–218, http://dx.doi.org/10.1016/j.marpetgeo.2011.06.002

MAYNARD, J. R., HOFMANN, W., DUNAY, R. E., BENTHAM, P. N., DEAN, K. P. & WATSON, I. 1997. The Carboniferous of Western Europe: the development of a petroleum system. *Petroleum Geoscience*, **3**, 97–115, http://dx.doi.org/10.1144/petgeo.3.2.97

MANGE, M. A. & MORTON, A. C. 2007. Geochemistry of heavy minerals. *In*: MANGE, M. & WRIGHT, D. T. (eds) *Heavy Minerals in Use*. Elsevier, Amsterdam, Developments in Sedimentology, **58**, 345–391.n

MARSHALL, J. E. A. & HEWETT, A. J. 2003. Devonian. *In*: EVANS, D., GRAHAM, C., ARMOUR, A. & BATHURST, P. (eds) *The Millennium Atlas: Petroleum Geology of the Central and Northern North Sea*. Geological Society, London, 65–81.

MARTIN, C. A., STEWART, S. A. & DOUBLEDAY, P. A. 2002. Upper Carboniferous and Lower Permian tectonostratigraphy on the southern margin of the Central North Sea. *Journal of the Geological Society, London*, **159**, 731–749.

MEINHOLD, G. 2010. Rutile and its applications in earth sciences. *Earth-Science Reviews*, **102**, 1–28, http://dx.doi.org/10.1016/j.earscirev.2010.06.001

MEINHOLD, G., ANDERS, B., KOSTOPOULOS, D. & REISCHMANN, T. 2008. Rutile chemistry and thermometry as provenance indicator: an example from Chios Island, Greece. *Sedimentary Geology*, **203**, 98–111.

MEINHOLD, G., MORTON, A. C., FANNING, M. & WHITHAM, A. G. 2011. U–Pb Shrimp ages of detrital granulite-facies rutiles: further constraints on provenance of Jurassic sandstones on the Norwegian margin. *Geological Magazine*, **148**, 473–480, http://dx.doi.org/10.1017/S0016756810000877

MILLER, I A., MATTHEWS, D. H. & ROBERTS, D. G. 1973. Rock of Grenville age from the Rockall Bank. *Nature Physical Science*, **246**, 61.

MORTON, A. C. & CHENERY, S. 2009. Detrital rutile geochemistry, thermometry as guides to provenance of Jurassic–Paleocene sandstones of the Norwegian Sea. *Journal of Sedimentary Research*, **79**, 540–553.

MORTON, A. C. & HALLSWORTH, C. R. 1999. Processes controlling the composition of heavy mineral assemblages in sandstones. *Sedimentary Geology*, **124**, 3–29.

MORTON, A. C., CLAOUÉ-LONG, J. C. & HALLSWORTH, C. R. 2001. Zircon age and heavy mineral constraints on provenance of North Sea Carboniferous sandstones. *Marine and Petroleum Geology*, **18**, 319–337.

MORTON, A. C., HALLSWORTH, C. & CHALTON, B. 2004. Garnet compositions in Scottish and Norwegian basement terrains: a framework for interpretation of North Sea sandstone provenance. *Marine and Petroleum Geology*, **21**, 393–410.

MORTON, A. C., FANNING, C. M. & MILNER, P. 2008. Provenance characteristics of Scandinavian basement terrains: constraints from detrital zircon ages in modern river sediments. *Sedimentary Geology*, **210**, 61–85.

MORTON, A. C., HALLSWORTH, C. R., STROGEN, D., WHITHAM, A. G. & FANNING, C. M. 2009. Evolution of provenance in the NE Atlantic rift: the Early–Middle Jurassic succession in the Heidrun Field, Halten Terrace, offshore Mid Norway. *Marine and Petroleum Geology*, **26**, 1100–1117.

MORTON, A. C., FANNING, M. & JONES, N. 2010. Variscan sourcing of Westphalian (Pennsylvanian) sandstones in the Canonbie Coalfield, UK. *Geological Magazine*, **147**, 718–727, http://dx.doi.org/10.1017/S0016756810000014

NANCE, R. D. & MURPHY, J. B. 1996. Basement isotopic signatures, Neoproterozoic palaeogeography of Avalonian–Cadomian and related terranes in the circum-North Atlantic. *In*: NANCE, R. D. & THOMPSON, M. D. (eds) *Avalonian and Related Peri-Gondwanan*

*Terranes of the Circum-North Atlantic.* Geological Society of America, Special Papers, **304**, 333–346.

NEUMANN, E.-R., WILSON, M., HEEREMANS, M., DUNWORTH, E. A., OBST, K., TIMMERMAN, M. J. & KIRSTEIN, L. A. 2004. Carboniferous–Permian rifting and magmatism in southern Scandinavia, the North Sea and northern Germany: a review. *In*: WILSON, M., NEUMANN, E.-R., DAVIES, G. R., TIMMERMAN, M. J., HEEREMANS, M. & LARSEN, B. T. (eds) *Permo-Carboniferous Rifting and Magmatism in Europe.* Geological Society, London, Special Publications, **223**, 11–40, http://dx.doi.org/10.1144/GSL.SP.2004.223.01.02

OLIVER, G. J. H. 1988. Arenig to Wenlock regional metamorphism in the Paratectonic Caledonides of the British Isles: a review. *In*: HARRIS, A. L. & FETTES, D. J. (eds) *The Caledonian-Appalachian Orogen.* Geological Society, London, Special Publications, **38**, 347–363, http://dx.doi.org/10.1144/GSL.SP.1988.038.01.20

OLIVER, G. J. H., WILDE, S. A. & WAN, Y. 2008. Geochronology and geodynamics of Scottish granitoids from the late Neoproterozoic break-up of Rodinia to Palaeozoic collision. *Journal of the Geological Society, London*, **165**, 661–674.

OPDYKE, N. D. & CHANNELL, J. E. T. 1996. *Magnetic stratigraphy. International Geophysics Series 64.* Academic Press, New York.

PAGE, L. M., MÖLLER, C. & JOHANSSON, L. 1996. $^{40}Ar/^{39}Ar$ geochronology across the Mylonite Zone and the south–west Swedish Granulite Province of the Sveconorwegian orogen of S. Sweden. *Precambrian Research*, **79**, 239–259.

PHILLIPS, E. R., SMITH, R. A., STONE, P., PASHLEY, V. & HORSTWOOD, M. 2009. Zircon age constraints on the provenance of Llandovery to Wenlock sandstones from the Midland Valley terrane of the Scottish Caledonides. *Scottish Journal of Geology*, **45**, 131–146.

RAINBIRD, R. H., HAMILTON, M. A. & YOUNG, G. M. 2001. Detrital zircon geochronology and provenance of the Torridonian, NW Scotland. *Journal of the Geological Society, London*, **158**, 15–27.

REHNSTRÖM, E. 2010. Prolonged Paleozoic magmatism in the East Greenland Caledonides: some constraints from U–Pb ages, Hf isotopes. *Journal of Geology*, **118**, 447–465.

ROBERTS, D., NORDGULEN, Ø. & MELEZHIK, V. A. 2007. The Uppermost Allochthon in the Scandinavian Caledonides: from a Laurentian ancestry through Taconian orogeny to Scandian crustal growth on Baltica. *In*: HATCHER, R. D, JR, CARLSON, M. P., MCBRIDE, J. H. & MARTINEZ CATALAN, J. R. (eds) *4-D Framework of Continental Crust.* Geological Society of America, Memoirs, 357–377.

ROBSON, D. 1991. The Argyll, Duncan and Innes Fields, Blocks 30/24, 30/25a, UK North Sea. *In*: ABBOTTS, I. L. (ed) *United Kingdom Oil and Gas Fields, 25 Years Commemorative Volume.* Geological Society, London, Memoirs, **14**, 219–226.

ROGERS, G., DEMPSTER, T. J., BLUCK, B. J. & TANNER, P. W. G. 1989. A high-precision U–Pb age for the Ben Vuirich granite: implications for the evolution of the Scottish Dalradian Supergroup. *Journal of the Geological Society, London*, **146**, 789–798.

ROSA, D. R. N., FINCH, A. A., ANDERSEN, T. & INVERNO, C. M. C. 2009. U–Pb geochronology and Hf isotope ratios of magmatic zircons from the Iberian Pyrite Belt. *Mineralogy and Petrology*, **95**, 47–69.

SHERLOCK, S. C., JONES, K. A. & PARK, R. G. 2008. Grenville age pseudotachylyte in the Lewisian: laserprobe $^{40}Ar/^{39}Ar$ ages from the Gairloch region of Scotland (UK). *Journal of the Geological Society, London*, **165**, 73–84.

SIRCOMBE, K. N. 2004. AgeDisplay: an excel workbook to evaluate and display univariate geochronological data using binned frequency histograms and probability density distributions. *Computers & Geosciences*, **30**, 21–31.

SLAGSTAD, T., DAVIDSEN, B. & DALY, J. S. 2011. Age and composition of crystalline basement rocks on the Norwegian continental margin: offshore extension and continuity of the Caledonian–Appalachian orogenic belt. *Journal of the Geological Society, London*, **168**, 1167–1185, http://dx.doi.org/10.1144/0016-76492010-136

SLAMA, J., WALDERHAUG, O., FONNELAND, H., KOSLER, J. & PEDERSEN, R. B. 2011. Provenance of Neoproterozoic to upper Cretaceous sedimentary rocks, eastern Greenland: implications for recognizing the sources of sediments in the Norwegian Sea. *Sedimentary Geology*, **238**, 254–267.

STACEY, J. S. & KRAMERS, J. D. 1975. Approximation of terrestrial lead isotope evolution by a two-stage model. *Earth and Planetary Science Letters*, **26**, 207–221.

STØLEN, L. K. 1994. The rift-related mafic dyke complex of the Rohkunborri Nappe, Indre Troms. northern Norwegian Caledonides. *Norsk geologisk Tidskrift*, **74**, 35–47.

STRACHAN, R. A., SMITH, M., HARRIS, A. L. & FETTES, D. J. 2002. The Northern Highland and Grampian Terranes. *In*: TREWIN, N. (ed.) *Geology of Scotland.* Geological Society, London, 81–147.

TIMMERMAN, M. J. 2004. Timing, geodynamic setting and character of Permo-Carboniferous magmatism in the foreland of the Variscan Orogen, NW Europe. *In*: WILSON, M., NEUMANN, E.-R., DAVIES, G. R., TIMMERMAN, M. J., HEEREMANS, M. & LARSEN, B. T. (eds) *Permo-Carboniferous Magmatism and Rifting in Europe.* Geological Society, London, Special Publications, **223**, 41–74.

TORSVIK, T. H., CARLOS, D., MOSAR, J., COCKS, L. R. M. & MALME, T. 2002. Global reconstructions and North Atlantic palaeogeography 440 Ma to Recent. *In*: EIDE, E. A. (ed.) *BATLAS – Mid Norway Plate Reconstructions Atlas with Global and Atlantic Perspectives.* Geological Survey of Norway, 18–39.

TREWIN, N. H., FRYBERGER, S. G. & KREUTZ, H. 2003. The Auk Field, Block 30/16, UK North Sea. *In*: GLUYAS, J. G. & HICHENS, H. M. (eds) *United Kingdom Oil and Gas Fields', Commemorative Millennium Volume.* Geological Society, London, Memoirs, **20**, 485–496.

TYRRELL, S., HUGHTON, P. D. W., DALY, J. S., KOKFELT, T. F. & GAGNEVIN, D. 2006. The use of the common Pb isotope composition of detrital K-feldspar grains as a provenance tool and its application to Upper Carboniferous palaeodrainage, Northern England. *Journal of Sedimentary Research*, **76**, 324–345, http://dx.doi.org/10.2110/jsr.2006.023

VON RAUMER, J. F., STAMPFLI, G. M. & BUSSY, F. 2003. Gondwana-derived microcontinents – the constituents of the Variscan and Alpine collisional orogens. *Tectonophysics*, **365**, 7–22.

WALDERHAUG, O. & PORTEN, K. W. 2007. Stability of detrital heavy minerals on the Norwegian continental shelf as a function of depth and temperature. *Journal of Sedimentary Research*, **77**, 992–1002.

WALDRON, J. W. F., FLOYD, J. D., SIMONETTI, A. & HEAMAN, L. M. 2008. Ancient Laurentian detrital zircon in the closing Iapetus Ocean, Southern Uplands Terrane, Scotland. *Geology*, **36**, 527–530.

WATSON, E. B., WARK, D. A. & THOMAS, J. B. 2006. Crystallization thermometers for zircon and rutile. *Contributions to Mineralogy and Petrology*, **151**, 413–433.

WATT, G. R. & THRANE, K. 2001. Early Neoproterozoic events in East Greenland. *Precambrian Research*, **110**, 165–184.

WIEDENBECK, M., ALLE, P. *ET AL.* 1995. Three natural zircon standards for U–Th–Pb, Lu–Hf, trace element and REE analyses. *Geostandards Newsletter*, **19**, 1–23.

WITZKE, B. J. 1990. Palaeoclimatic constraints for Palaeozoic Paleolatitudes of Laurentia and Euramerica. *In*: MCKERROW, W. S. & SCOTESE, C. R. (eds) *Palaeozoic Paleogeography and Biogeography*. Geological Society, London, Memoirs, **12**, 57–73.

ZACK, T. & LUVIZOTTO, G. L. 2006. Application of rutile thermometry to eclogites. *Mineralogy and Petrology*, **88**, 69–85.

ZACK, T., VON EYNATTEN, H. & KRONZ, A. 2004. Rutile geochemistry and its potential use in quantitative provenance studies. *Sedimentary Geology*, **171**, 37–58.

ZIEGLER, P. A. 1990. *Geological Atlas of Western, Central Europe*, 2nd edn. Shell Internationale Petroleum Maatschappij B.V., The Hague.

# Constraining depositional models in the Barents Sea region using detrital zircon U–Pb data from Mesozoic sediments in Svalbard

E. PÓZER BUE* & A. ANDRESEN

*Department of Geosciences, University of Oslo, PO Box 1047, Blindern, N-0316 Oslo, Norway*

*Corresponding author (e-mail: e.p.bue@geo.uio.no)*

**Abstract:** Detrital zircon U–Pb laser ablation inductively coupled plasma mass spectrometry age data on sandstones from Mesozoic successions on Svalbard are used to investigate provenance changes over time, constrain potential source areas, and to test and refine previous interpretations of the Mesozoic filling of the Barents Sea. The zircon age data indicate a western Laurentian (North Greenland) source in the Early and Middle Triassic. The westerly derived sediments most likely include reworked older sediments with proto-sources in Canada and Greenland. Sediments reaching Svalbard in the Late Triassic display a distinct Uralide signature that demonstrates derivation from the east. Zircon age populations in Late Triassic–Early Jurassic sands suggest mixing of zircons from the Early and Middle Triassic and Late Triassic sediments; the data are interpreted to reflect reworking of older Mesozoic sands and possible renewed input of sediments from the west. The data thus demonstrate a shift from westerly to easterly sediment sources in the early Late Triassic. The Early and Middle Triassic zircon age signature in this study appears to resurface in published Early Cretaceous provenance data from Svalbard, suggesting that sediment input from the east ceased during the Jurassic, and shifted back to westerly sources.

**Supplementary material:** A summary of U–Pb isotopic results, Concordia diagrams of U–Pb age data, K–S test results and cumulative probability plots for all samples are available at www.geolsoc.org.uk/SUP18652

The Barents Sea region is of considerable commercial interest as a not yet fully explored hydrocarbon province, and has been the target of extensive geological and geophysical investigations in recent decades (e.g. Ulmishek 1982, 1985; Faleide *et al.* 1984; Johansen *et al.* 1993; Mørk 1999; Mørk & Worsley 2006; Riis *et al.* 2008; Worsley 2008; Glørstad-Clarck *et al.* 2010). The post-Caledonian depositional history of the Barents Sea region is of particular interest because it controls the distribution of hydrocarbon source and reservoir rocks.

The Svalbard Archipelago, the uplifted northwestern corner of the Barents Sea (Figs 1 & 2), is a fertile study area with excellent exposure of post-Caledonian rocks. Previous sedimentological and geophysical studies indicate that sediments reached Svalbard from western (Greenland–Canada) and eastern (Baltica and/or Uralides) source areas in the Mesozoic and Cenozoic (Mørk *et al.* 1982; Mørk 1999; Riis *et al.* 2008; Røhr & Andersen 2009; Smelror *et al.* 2009). A source area to the north has also been suggested (Lock *et al.* 1978; Mørk *et al.* 1982; Mørk 1999) but remains enigmatic because no land masses are presently found in that region other than Nordaustlandet, which makes up northeastern Svalbard (Fig. 2). Heavy mineral studies on sediments of well-constrained ages represent an avenue for testing the current depositional models of the Mesozoic sediments of the Barents Sea. These models are, to a large degree, based on geophysical data.

In this study, detrital zircons from eight Triassic and Jurassic sandstones from Svalbard, representing five different formations and one subgroup, have been analysed using *in situ* U–Pb laser ablation inductively coupled plasma mass spectrometry (LA-ICPMS). The resulting zircon age distributions are used to investigate changes in sand provenance through time, and to constrain the ultimate sources (proto-sources) of the sediments, thereby also constraining sediment transport directions.

## Geological setting

In the Mesozoic, the Barents Sea was a large epicontinental sea stretching from the Sverdrup Basin and the Norwegian–Greenland Sea in the west to Novaya Zemlya, the Pechora and the Kara Seas in the east, and from the Arctic Ocean, Svalbard and Franz Josef Land in the north to northern Norway and northwestern Russia in the south (Fig. 1). The shelf covers *c.* 1.3 million km$^2$ of the northwestern corner of the Eurasian plate (Worsley 2008). The

*From*: Scott, R. A., Smyth, H. R., Morton, A. C. & Richardson, N. (eds) 2014. *Sediment Provenance Studies in Hydrocarbon Exploration and Production*. Geological Society, London, Special Publications, **386**, 261–279.
First published online October 1, 2013, http://dx.doi.org/10.1144/SP386.14

 E. PÓZER BUE & A. ANDRESEN

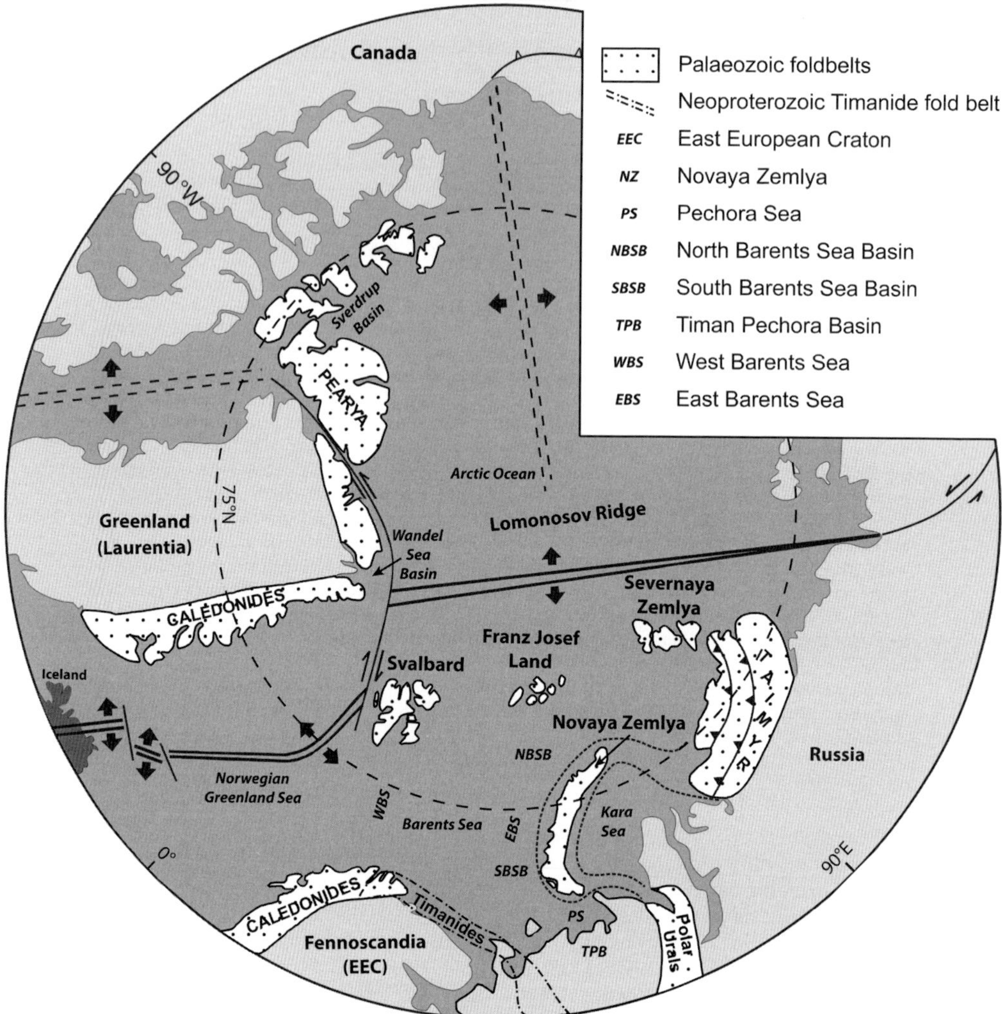


**Fig. 1.** Simplified map of the Arctic region and the Svalbard Archipelago, together with major crustal domains that are potential source areas for detrital zircons in the Triassic and Jurassic sediments on Svalbard (after Johansson *et al.* 2005, adapted from Jackson & Gunnarsson 1990).

present-day Barents Sea is one of the largest shelf areas in the world, with water depths typically less than 500 m. The depth to the pre-Devonian basement rocks exceeds 10 km throughout the Barents Sea (Faleide *et al.* 1984).

The Barents Sea can be subdivided into two major provinces based on the basin development history: the Eastern and Western Barents Sea. The boundary between the two provinces is defined by a structural high, the Central Barents Arch (also referred to as the Fersmanovskoye High), which more or less coincides with the previously disputed border area between Russia and Norway (Worsley 2008; Henriksen *et al.* 2011). The basement of the western province is interpreted to be of Caledonian origin (Faleide *et al.* 1984; Breivik *et al.* 2002; Ritzmann & Faleide 2007; Bjørlykke 2010; Henriksen *et al.* 2011) and the Caledonian orogenic front, representing the northeastward continuation of the Scandinavian Caledonides, is possibly the boundary between the eastern and western provinces of the shelf (e.g. Breivik *et al.* 2002; Roberts & Olovyanishnikov 2004; Gee *et al.* 2008). The northward-trending branch of the Caledonian Orogen is exposed on Svalbard (Henriksen *et al.* 2011). The Western Barents Sea is dominated by

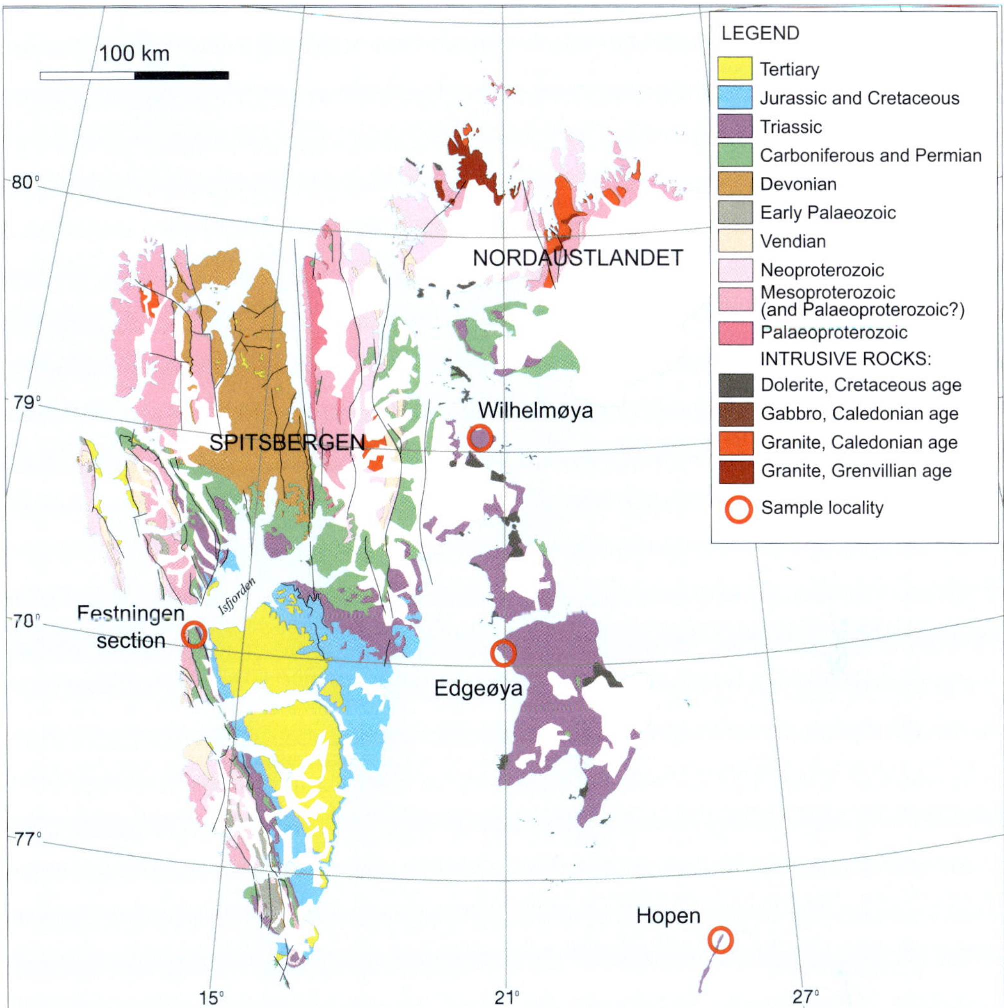


**Fig. 2.** Geological map of Svalbard with stratigraphic overview (after Dallmann *et al.* 2002) and sample localities.

basins, highs and platforms, whereas the Eastern Barents Sea part consists of fewer and larger basins, including the offshore continuation of the Timan–Pechora Basin, and the South and North Barents Sea basins.

The geological evolution of the Western Barents Sea has been discussed in detail, for example in Faleide *et al.* (1984), Doré (1991) and Worsley (2008). The basement geology is shaped by the Caledonian orogeny, and Caledonian rocks are exposed on Svalbard and East Greenland, for instance. In late- to post-Caledonian times, erosion of the mountain chain led to the deposition of molasse sediments – the Old Red Sandstone – which are preserved in rift and strike–slip basins (Worsley 2008; Smelror *et al.* 2009).

In the Devonian to Early Carboniferous, the Barents Sea region, as part of Laurussia, continued its northward drift from equatorial latitudes to its present-day Arctic position, and a carbonate platform developed in the central part of the Eastern Barents Sea (Worsley 2008). During the Early Palaeozoic, the Eastern Barents Sea and Kara Sea areas were dominated by pericratonic extension and the development of a passive margin (Ziegler 1990; Zonenshain *et al.* 1990; Doré 1991; Smelror *et al.* 2009). In the Devonian this passive margin transformed into a convergent margin when the Uralian oceanic crust was subducted westward under the East European Craton (Puchkov 1997; Smelror *et al.* 2009). The progressive subduction of the Uralian oceanic crust ended in a continent–continent collision between Baltica and Kazakhstan, resulting in the formation of the Uralian Orogen in Carboniferous and Permian times. The Uralian collisional front was oblique and moved progressively northward, ceasing in the Late Triassic. In the north the Uralian Orogen sutures the East European Craton to Siberia and includes the Polar Urals, Novaya Zemlya and the Taimyr fold belt (Ulmishek 1982; Ziegler 1989; Nikishin *et al.* 1996; Puchkov 1997; Petrov *et al.* 2008; Smelror *et al.* 2009).

In the Permian, the depositional environment in the Barents Sea changed from a warm carbonate platform to a deep-water setting with deposition of fine clastic sediments. The transition from carbonates to clastics is younging towards the west. Throughout the Triassic, large amounts of sediments were shed from the Polar Urals and deposited as thick sedimentary successions in the Barents Sea (Doré 1991; Johansen *et al.* 1993; Otto & Bailey 1995; Riis *et al.* 2008; Worsley 2008; Glørstad-Clarck *et al.* 2010).

In Carboniferous times, the Western Barents Sea region was dominated by regional extension, part of the long-lived Palaeozoic–Mesozoic pre-opening rifting events of the North Atlantic region. Carboniferous–Permian carbonate platform deposits are also extensive in this part of the Barents Sea, with evaporite deposits in graben structures becoming more important during the Late Palaeozoic, for example on Svalbard (Faleide *et al.* 1984; Worsley & Aga 1986; Worsley 2008; Smelror *et al.* 2009). A seaway opened in the Late Permian, connecting the Barents Sea via East Greenland and southwards to the North Sea, and deposition of siliciclastics became dominant (Faleide *et al.* 1984; Worsley & Aga 1986; Worsley 2008).

A major rifting episode dominated the tectonic regime in the Western Barents Sea during the Triassic. Early and Middle Triassic strata comprise clastic sediments deposited in transgressive–regressive cycles in marine, deltaic and continental settings. On Svalbard, these successions have been interpreted to reflect deltaic progradations, with a westerly source in Laurentia (Mørk *et al.* 1982, 1989; Mørk 1999). Strata of Carnian age (*c.* 228–216 Ma, Gradstein & Ogg 2009) thicken significantly towards the west, which has been interpreted to reflect clastic sediment input from Laurentia in the west and Fennoscandia in the south (Worsley 2008; Smelror *et al.* 2009).

In Permo-Triassic times, sedimentation in the eastern parts of the shelf and in the Timan–Pechora basin was controlled by the Uralian Orogeny (Johansen *et al.* 1993; Otto & Bailey 1995). From the Middle to Late Triassic, progressive uplift characterized the southern, eastern and northern Barents Sea. This uplift and erosion in the east led

to coastal progradation towards the west in the form of marine–deltaic cycles, most likely in a temperate climate (Doré 1991; Worsley 2008) and, by the Late Triassic, most of the Barents Sea was dominated by coastal-plain and continental depositional environments, with marine environments restricted to the westernmost part of the present-day shelf (Doré 1991; Johansen *et al.* 1993; Smelror *et al.* 2009).

An overall transgressive trend characterize the Early Jurassic deposits of the Timan–Pechora Basin and, in the southwestern Barents Sea, shallow-marine deposits with several transgressive pulses are recognized (Dalland *et al.* 1988; Doré 1991). On Svalbard the diminishing influx of sediments continued through the Early and Middle Jurassic (Doré 1991). During the Middle Jurassic and Early Cretaceous, Atlantic rifting affected the westernmost parts of the Barents Sea and in the northern areas extension in the Amerasia Basin became influential (Doré 1991; Worsley 2008). In the Late Jurassic a major transgressional event changed the depositional environment throughout the Barents Sea. The region once again became flooded and shallow-shelf to deep-marine sedimentation became dominant (Doré 1991; Johansen *et al.* 1993).

The tectonic regime in the Barents Sea during the Early Cretaceous was characterized by strike–slip movements and reverse-faulting (Doré 1991). A major Early Cretaceous magmatic event (Doré 1991; Faleide *et al.* 2008; Worsley 2008), forming the High Arctic Large Igneous Province (Drachev & Saunders 2006; Polteau *et al.* 2010; Tegner *et al.* 2011), led to significant uplift along the northern rim of the Barents Sea. The sedimentary successions that crop out on Svalbard and Franz Josef Land are locally affected by Cretaceous magmatism. In the southwestern part of the Barents Sea, rifting episodes led to subsidence and the development of deep basins. By Late Cretaceous times this rifting was replaced by strike–slip movements, leading to the creation of pull-apart basins (Faleide *et al.* 1993, 2008; Smelror *et al.* 2009).

The Cenozoic opening of the Atlantic Ocean led to partitioning of strain along a strike–slip zone separating Svalbard and the Barents Sea from Laurentia, and the development of a fold-and-thrust belt along the western edge of Svalbard was accompanied by regional uplift (Faleide *et al.* 2008).

The Caledonian basement in the Barents Sea region is locally covered by more than 10-km-thick, semi-continuous sedimentary successions from the Late Palaeozoic to the Cenozoic (Faleide *et al.* 1984; Worsley 2008). The sedimentary successions record significant climate changes, varying tectonic processes and sea-level fluctuations.

*Geology of the Svalbard Archipelago*

Tertiary uplift of the northwestern corner of the Barents Sea exposed basement rocks and their overlying sedimentary successions (Fig. 2) that, to a large degree, are representative of the geological evolution of the Western Barents Sea region (Steel & Worsley 1984; Worsley & Aga 1986; Harland & Geddes 1997; Worsley 2008).

The Precambrian to Palaeogene stratigraphy of Svalbard can be subdivided into three major depositional phases separated by tectonic events (Faleide *et al.* 1984; Worsley & Aga 1986). The lowermost part of the tectonostratigraphy on Svalbard consists of metamorphic and igneous basement rocks of Mesoproterozoic to Silurian age (Worsley & Aga 1986; Myhre *et al.* 2009; Pettersson *et al.* 2009*a*, *b*).

From the Late Silurian to the Middle Devonian, erosion of the Caledonian Orogen led to the deposition of molasse sediments – the Old Red Sandstones – which were folded and faulted in the latest Devonian (Faleide *et al.* 1984; Worsley & Aga 1986; Worsley 2008). During the Carboniferous, the coal-bearing non-marine sands of the Billefjorden Group were deposited, followed by deposition of the clastic and shallow-marine carbonate deposits of the Gipsdalen Group. By the Middle Permian, carbonate-dominated sedimentation was replaced by fine clastics of the Tempelfjorden Group, reflecting regional transgression. In the Early Mesozoic, tectonic quiescence around the Barents Sea region led to the formation of a stable platform. With the continued northwards drift of Svalbard, the warm arid conditions changed to a more temperate, humid climate that would largely hold throughout the Mesozoic (Worsley & Aga 1986; Dallmann 1999).

The Permian–Triassic boundary is marked by a transgression and deposition of the Sassendalen Group shales. The Sassendalen Group includes the Vardebukta and Bravaiseberget formations (Figs 3 & 4), and contains silts and sands reflecting repeated coastal progradations throughout the Early Triassic (Buchan *et al.* 1965; Mørk *et al.* 1982, 1989, 1999; Worsley 2008). The overlying Kapp Toscana Group of Late Triassic to Middle Jurassic age (Buchan *et al.* 1965; Mørk *et al.* 1982, 1999) comprises shallow-marine, coastal and deltaic deposits in the form of shale-, silt- and sandstones. This group includes the De Geerdalen Formation, the Wilhelmøya Subgroup, the Knorringfjellet Formation and the Svenskøya Formation, among others (Figs 3 & 4). The overlying Adventdalen Group (Fig. 4) of Late Jurassic–Early Cretaceous age is dominated by shales with significant organic carbon content. In the Cretaceous, magmatic activity was widespread throughout Svalbard (Corfu *et al.* 2013). During the Early Cretaceous, clastic sedimentation decreased, and existing delta systems retreated with rising sea-level and basement subsidence (Worsley & Aga 1986; Worsley 2008). In Late Cretaceous times, the Svalbard Platform was uplifted, and a major hiatus in the sedimentary record on Svalbard suggests erosion and/or a period of non-deposition. The extent of this uplift was progressively greater towards the north (Worsley & Aga 1986; Worsley 2008). Thus, Paleocene strata were deposited directly on Early Cretaceous sediments. With the development of the Tertiary fold-and-thrust belt the basement and sedimentary successions were folded, faulted and tilted.

## Detrital zircon data

### Sample material

Samples were collected at different localities in the Svalbard Archipelago to determine the provenance signature at different times, and investigate potential lateral variations (Fig. 2). The easily accessible Festningen section (Fig. 2) on western Spitsbergen is the westernmost outcrop of Mesozoic sedimentary successions on Svalbard. Four samples were collected along the Festningen section, one sample from Wilhelmøya, one from Edgeøya and two from the island of Hopen (Figs 2 & 3). The samples are fine- to medium-grained sandstones from formations at different stratigraphic levels, with ages spanning from the Early Triassic to the Middle Jurassic (Fig. 4).

All the sampled units belong to the Early to Middle Triassic Sassendalen Group or the Carnian to Bathonian Kapp Toscana Group. The oldest sampled units, the Vardebukta and Bravaiseberget formations, were sampled along the Festningen section and are part of the Sassendalen Group (Fig. 2; Mørk *et al.* 1982). The depositional ages of the Vardebukta Formation and the Bravaiseberget Formation have been constrained by ammonoid and bivalve biostratigraphy to the Induan stage and the Middle Triassic, respectively (Tozer & Parker 1968; Mørk *et al.* 1999, and references therein). The remaining samples were collected in the Kapp Toscana Group at different geographical locations. The correlations of the formations and subgroups across the different sample localities are illustrated schematically in Figure 4. The oldest studied formation in the Kapp Toscana Group is the Carnian to early Norian De Geerdalen Formation (Mørk *et al.* 1982), which has been dated using ammonoids (Tozer & Parker 1968; Mørk *et al.* 1999 and references therein). The formation was sampled at three different geographical locations on Svalbard (Figs 2 & 3). The overlying unit is the Wilhelmøya

| Sample | Locality | Coordinates (Lat/Long) | Depositional age | Group | Formation | Depositional environment |
|---|---|---|---|---|---|---|
| W7-496 | Wilhelmøya | 79.05447 N 20.43763 E | Uppermost Triassic (Norian-Bathonian) | Kapp Toscana Wilhelmøya Subgroup | Wilhelmøya Subgroup | Deltaic to shallow marine |
| S9 | Spitsbergen Festningen | 78.098889 N 13.914167 E | Uppermost Triassic (Norian-Bathonian) | Kapp Toscana Wilhelmøya Subgroup | Knorringfjellet | Shallow marine |
| P-09-35 | Hopen Lyngefjellet (282 masl) | 76.41246 N 25.26208 E | Lower Jurassic (latest Triassic-Pliensbachian) | Kapp Toscana Wilhelmøya Subgroup | Svenskøya | Tidal flat, tidal channel and coastal plain |
| P-09-39 | Edgeøya Klink-hammeren | 78.021736 N 21.155092 E | Uppermost Triassic (Carnian-early Norian) | Kapp Toscana | De Geerdalen | Shallow shelf to deltaic |
| P-09-36 | Hopen Lyngefjellet (51 masl) | 76.687645 N 25.411741 E | Uppermost Triassic (Carnian-early Norian) | Kapp Toscana | De Geerdalen | Shallow shelf to deltaic |
| P-09-34 | Spitsbergen Festningen | 78.099444 N 13.894444 E | Uppermost Triassic (Carnian-early Norian) | Kapp Toscana | De Geerdalen | Shallow shelf to deltaic |
| S5 | Spitsbergen Festningen | 78.098056 N 13.876111 E | Middle Triassic (Anisian-Ladinian) | Sassendalen | Bravaiseberget | Deltaic influenced regressive unit |
| P-09-33 | Spitsbergen Festningen | 78.095 N 13.850278 E | Lower Triassic (Induan) | Sassendalen | Vardebukta | Coastal to shallow marine |

**Fig. 3.** Sample information. Depositional environments after Mørk *et al.* (1999).

Subgroup (Mørk *et al.* 1999), which has been sampled at its type locality on Wilhelmøya and analysed for this study. The depositional age ranges from Norian to Bathonian (Mørk *et al.* 1999). The Wilhelmøya Subgroup is the superior unit of the Knorringfjellet Formation, which is found in western and west-central Spitsbergen and has a depositional age of Norian to Bathonian with two major hiati. The Knorringfjellet Formation was sampled at the type locality on Spitsbergen along the Festningen section (Fig. 2). On Hopen, in the eastern part of the Svalbard Archipelago, the Knorringfjellet Formation corresponds to the Flatsalen, Svenskøya and Kongsøya formations. The Svenskøya Formation in the Wilhelmøya Subgroup has a depositional age of ?Latest Triassic–Early Jurassic, estimated from fossils such as foraminifera (Mørk *et al.* 1999, and references therein). Typical depositional environments of the sampled sands are deltaic to shallow-marine and coastal settings, such as delta lobes and offshore bars. The formations have been studied extensively, and are described and interpreted by Buchan *et al.* (1965), Worsley & Mørk (1978) and Dallmann (1999).

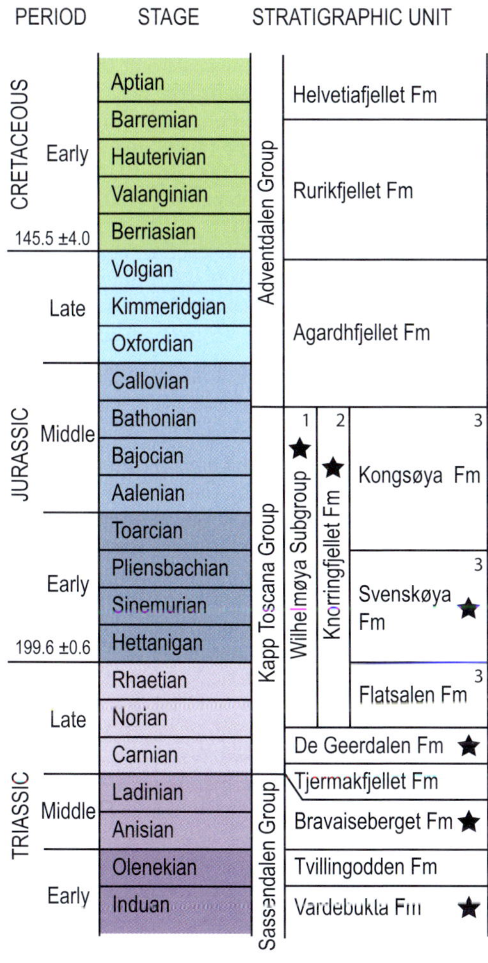



**Fig. 4.** Schematic stratigraphic column (simplified after Mørk *et al.* 1999) with sampled units indicated by stars. In the Late Triassic to Middle Jurassic the figure illustrates the relations between samples from the Festningen section (Knorringfjellet Formation), Wilhelmøya (Wilhelmøya Subgroup) and Hopen (Svenskøya Formation).

## Analytical methods

Zircons were separated using standard methods. After washing and drying the hand specimens, the samples were crushed in a jaw-crusher to a grain size of <1 cm, then further reduced in grain size to *c.* 0.5 mm using a Retch percussion mill. Zircons were separated from minerals of lower density using a Wilfley table and heavy liquid (sodium heteropolytungstate, $\rho = 2.80 \pm 0.02$ g ml$^{-1}$). The zircon grains were hand-picked under a binocular

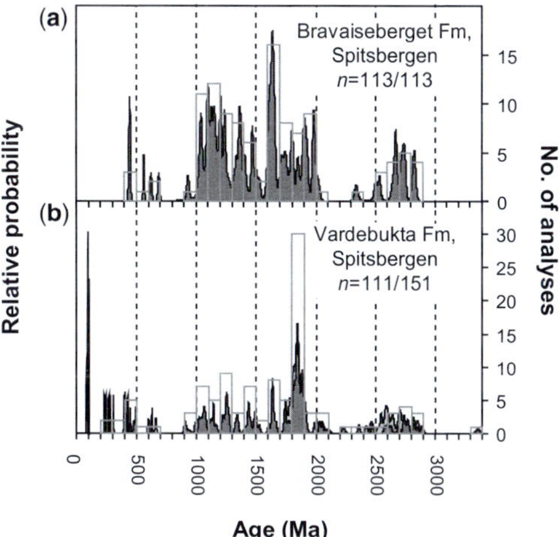



**Fig. 5.** (**a, b**) Zircon age data from analysed samples. Histograms display <10% discordant analyses. Probability density plots show <10% discordant ages and >10% discordant ages. The plotted data are $^{207}$Pb/$^{206}$Pb ages for zircon ages of >700 Ma and $^{206}$Pb–$^{238}$U ages for zircon ages of <700 Ma. *n* = number of <10% discordant analyses/total number of analyses. Samples (a) and (b) show a wide variety of ages up to and including Archaean ages.

microscope and mounted on double-adhesive tape. The picking was random, independent of colour, the presence of inclusions or the size of the zircons. The zircon grains were then mounted in epoxy, polished, carbon-coated and imaged using catho doluminescence on a JEOL-JSM6460LV scanning electron microscope at the Department of Geosciences, University of Oslo. After thorough cleaning of the samples, U–Pb isotope compositions were determined by LA-ICPMS using a Nu Plasma HR mass spectrometer and a NewWave LUV213 laser microprobe at the Department of Geosciences, University of Oslo. The analytical protocols described in detail by Rosa *et al.* (2009) and Andersen *et al.* (2009) were used for U–Pb geochronology. U–Pb ages were calculated using ISOPLOT version 3.71 (Ludwig 2003), with errors reported at $\pm 1\sigma$ confidence level. For each sample, between 122 and 221 zircons were analysed, giving a total of 1375 analyses. All analyses were evaluated with respect to background-corrected common Pb content in relation to total radiogenic Pb, and discordance. A total of 1129 analyses were used for probability density and Concordia plots (Figs 5–7 & Supplementary material). The plotted data are $^{207}$Pb/$^{206}$Pb ages for zircon ages >700 Ma and $^{206}$Pb–$^{238}$U ages for zircons ages <700 Ma. Analyses with more than 10% central discordance are shown separately in probability density plots and

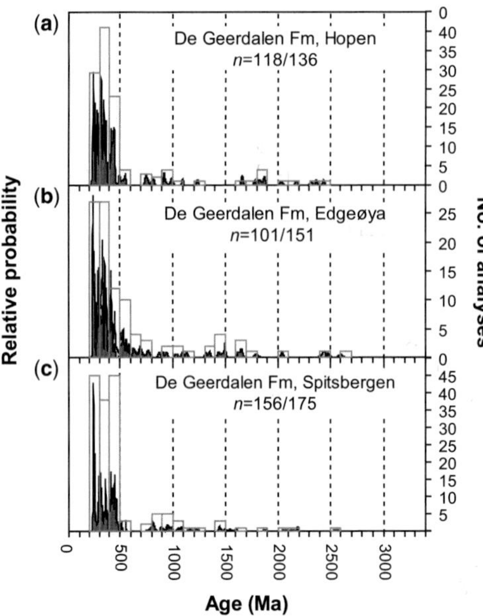

**Fig. 6.** (a–c) Zircon age data from analysed samples (for explanation see Fig. 5). Samples (a), (b) and (c) are dominated by Palaeozoic ages.

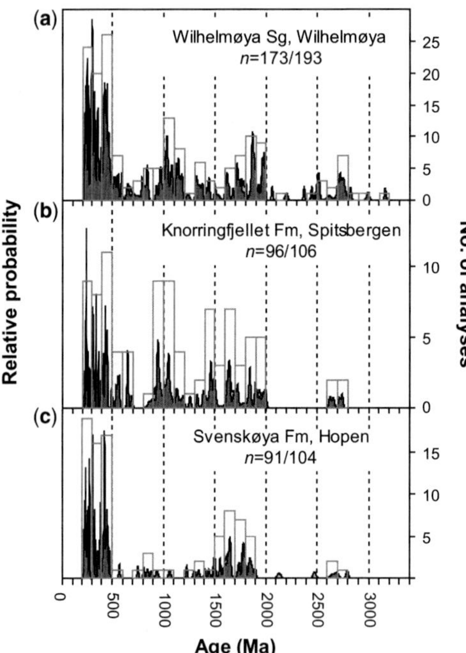

**Fig. 7.** (a–c) Zircon age data from analysed samples (for explanation see Fig. 5). Samples (a), (b) and (c) display characteristics from both of the two endmembers in Figures 5 and 6.

Concordia diagrams (Figs 5–7 & Supplementary material).

The age data were evaluated using the Kolmogorov–Smirnoff (K–S) statistical test. The K–S test compares detrital zircon age distributions from two samples to evaluate the probability (expressed by the *P*-value) that the two distributions were drawn from the same parent population (Berry *et al.* 2001; Guynn & Gehrels 2010). A *P*-value of 0.05 corresponds to a 95% confidence level that the samples were drawn from the same parent distribution. The results are shown in the Supplementary material.

## Results

Zircons were analysed from eight Triassic and Jurassic samples representing four different geographical areas on Svalbard (Figs 2–4); the results are shown in Figures 5–7 and the Supplementary material.

The main variations in the zircon age spectra reflect the stratigraphic positions of the samples, with no prominent lateral variations. Samples from the Early and Middle Triassic Vardebukta and Bravaisberget formations and from the Early Jurassic Wilhelmøya Subgroup contain *c.* 9–14% Archaean grains. In the Early Jurassic Svenskøya Formation and the Knorringfjellet Formation, *c.* 3–4% of the analysed grains yield Archaean ages, whereas the three Late Triassic samples from the De Geerdalen Formation contain very few Archaean grains.

The Palaeoproterozoic part of the age spectra varies from 19% to 47% of the analysed zircons in the samples, except for in the three Late Triassic samples from the De Geerdalen Formation, where few grains of this age are present. Neo- and Mesoproterozoic ages are present in all samples. However, the three Late Triassic samples of the De Geerdalen Formation and the Early Jurassic sample of the Svenskøya Formation have low proportions of these zircon ages (11–22%), whereas they account for more than 30% in the remaining samples.

Palaeozoic ages are represented in all the samples, but vary in relative importance. Zircon ages from 480 to 400 Ma roughly correspond to the age of the Caledonian orogeny and are present in all samples. However, these ages make up only 3–4% of the grains in the two Early and Middle Triassic samples, whereas this age span accounts for 10–27% in the remaining samples. Ages younger than 400 Ma are absent from the Bravaisberget Formation sample, whereas the De Geerdalen Formation samples are dominated by young, mid-Devonian–Triassic populations, representing 53–59% of the analysed grains. In the remaining samples these ages account for 4–38% of the analyses. Six of the samples display distinct peaks at *c.* 240 Ma; this peak is missing from samples of the Bravaisberget Formation and Vardebukta Formation, although the latter yields one grain of this age. Mesozoic ages represent 6–18% of the analyses in the Late Triassic and Jurassic samples, but these ages are absent in the Early and Middle Triassic samples.

## Discussions

### *Sample differentiation based on zircon age data*

To a first approximation the samples appear to form two endmember groups with distinctly different zircon age patterns. A third group displays mixed characteristics (Figs 5–8).

The first endmember group is represented by the two Early and Middle Triassic samples (Figs 5a, b & 8) from the Vardebukta and Bravaiseberget formations in outer Isfjorden (Fig. 2). These samples yield minor Neoproterozoic, Palaeozoic and Mesozoic zircon age populations, sparse Caledonian (500–390 Ma) zircons and three grains younger than 400 Ma. The dominant zircon age populations are composed of Palaeo- and Mesoproterozoic ages. The Archaean population makes up a considerable proportion of the analyses, with one grain >3.0 Ga.

The second endmember consists of early Late Triassic samples from the De Geerdalen Formation (Figs 6a–c & 8), sampled on Spitsbergen, Hopen and Edgeøya (Fig. 2). Archaean zircon ages are barely represented in two of the De Geerdalen Formation samples, and are missing entirely from the third. Proterozoic ages are well represented, and increasingly numerous towards the younger parts of the Proterozoic. The three samples yield a high proportion of mainly Ordovician–Silurian (Caledonian) grains. Palaeozoic and Mesozoic grains dominate the age spectra, with a well-defined peak at 240 Ma.

The intermediate group comprises three Late Triassic–Middle Jurassic samples (Figs 7a–c & 8) from localities on Wilhelmøya (Wilhelmøya Subgroup), Spitsbergen (Knorringfjellet Formation) and Hopen (Svenskøya Formation; Fig. 2). These samples display subdued Archaean populations, with one >3.0 Ga grain. Proterozoic ages make up the bulk of the zircon ages and the samples also display a high proportion of Palaeozoic grains, with a considerable portion of all analysed grains yielding Ordovician–Silurian (Caledonian) ages. Zircons with an age of <400 Ma make up, on average, just under one-third of the grains.

The provenance data thus suggest input from at least two different source areas. The different provenance signatures reflect the stratigraphic position of the samples, and do not change with geographical location. Early Late–Triassic sands from three different geographical locations yield similar zircon age signatures, as do Late Triassic to Middle Jurassic sands sampled in three different areas. The intermediate samples may represent a mix of sediments with endmember characteristics, potentially reflecting reworking of older sediments, or a third provenance region.

The K–S test results support the division of the samples into three groups based on their zircon age populations. The first group comprises samples from the Vardebukta and Bravaiseberget formations. Comparison of the Vardebukta and Bravaiseberget formations gives a *P*-value of 0.139, indicating a common parent population. Comparison of the Vardebukta and Bravaiseberget formations samples with the other samples in this study yields a *P*-value of 0.000, indicating different parent populations. The second group comprises samples from the De Geerdalen Formation collected at three different localities on Svalbard. The K–S test produces *P*-values ranging from 0.090 to 0.648, indicating a common parent population for these samples. Comparison of the De Geerdalen Formation with the remaining samples gives *P*-values ranging from 0.000 to 0.015, indicating different parent populations. The final group includes the Wilhelmøya Subgroup, Knorringfjellet Formation and the Svenskøya Formation. Comparing the former with the latter two yields *P*-values of 0.053 and 0.314, indicating that the samples are related. Interestingly, the *P*-value for the comparison between the Knorringfjellet Formation and the Svenskøya Formation is 0.001, suggestive of different parent populations for the zircons in these samples. However, given the similar ages present in the samples, and the sensitivity of the K–S test to varying proportions of similar ages in samples (Guynn & Gehrels 2010), a more likely interpretation is that these formations sample similar parent populations in different proportions.

### *Potential source areas*

The land areas surrounding the Barents Sea have a long and highly variable geological history. The zircon age spectra from the studied sediments, and the changing age signatures over time, are compatible with derivation from different continental domains and eroding orogens adjacent to the Barents Sea in northern Laurentia and in the northern East European Craton (Figs 1 & 9).

Both Laurentia and the East European Craton formed around Archaean nuclei and grew through

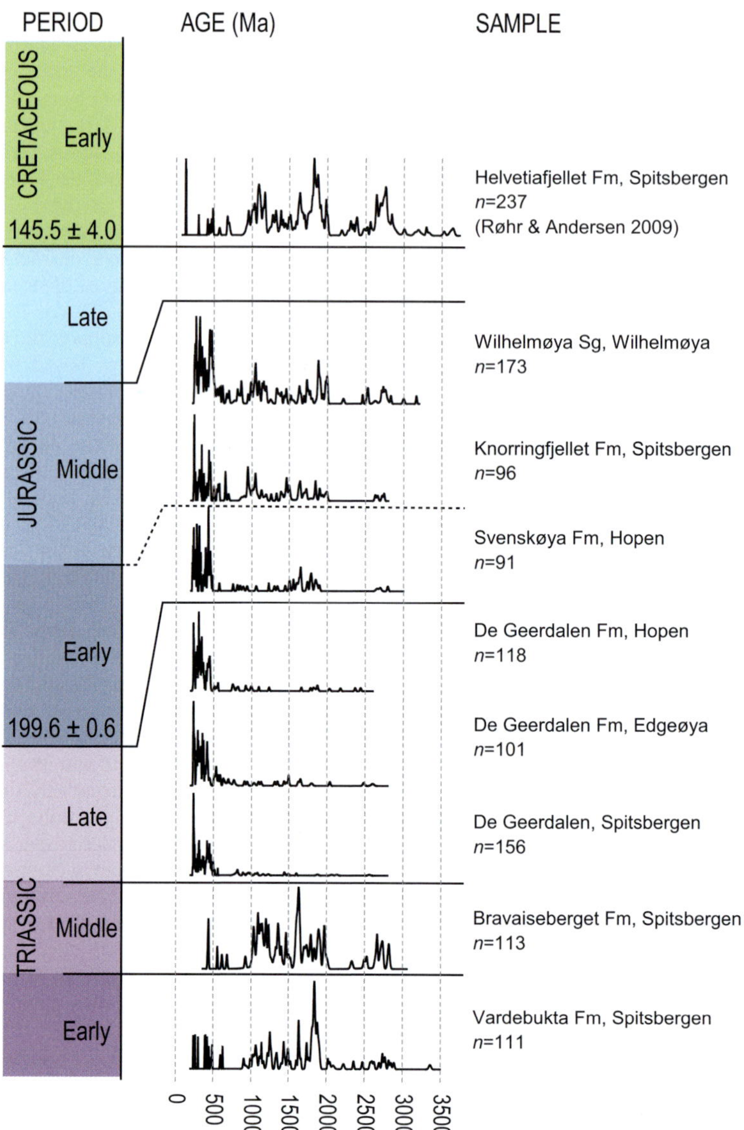



**Fig. 8.** Provenance zircon age signatures of Mesozoic sands from Svalbard including three Early Cretaceous samples from Røhr & Andersen (2009). Probability density plots show <10% discordant ages. The plotted data are $^{207}Pb/^{206}Pb$ ages for zircon ages of >700 Ma and $^{206}Pb-^{238}U$ ages for zircon ages of <700 Ma. $n$ = number of <10% discordant analyses/total number of analyses. The Early and Middle Triassic samples contain zircon age populations similar to that of the Early Cretaceous sands. Samples from the De Geerdalen Formation (Late Triassic) are clearly dominated by different zircon ages (mainly Palaeozoic) compared with the Vardebukta and Bravaiseberget formations. Samples of Late Triassic to Middle Jurassic sands display characteristics consistent with mixing of the two sample groups formed by the Vardebukta and Bravaiseberget formations, and the De Geerdalen Formation, respectively.

successive additions of juvenile crust and accretion of terrains. Eoarchaean (>3.6 Ga) rocks are known from Greenland (Whitehouse & Kamber 2005), but are practically absent from the northernmost East European Craton (Hölttä *et al.* 2008), although sparse *c.* 3.6 Ga zircons have been reported from

sediments on the Kola Peninsula (Myskova *et al.* 2005) and from magmatic Archaean complexes in Karelia (*c.* 3.6–3.7 Ga, Lauri *et al.* 2011). Zircon ages of 3.0 Ga or slightly older correspond to Archaean crystalline basement in Greenland, but similar ages in Baltica are less common (Steiger

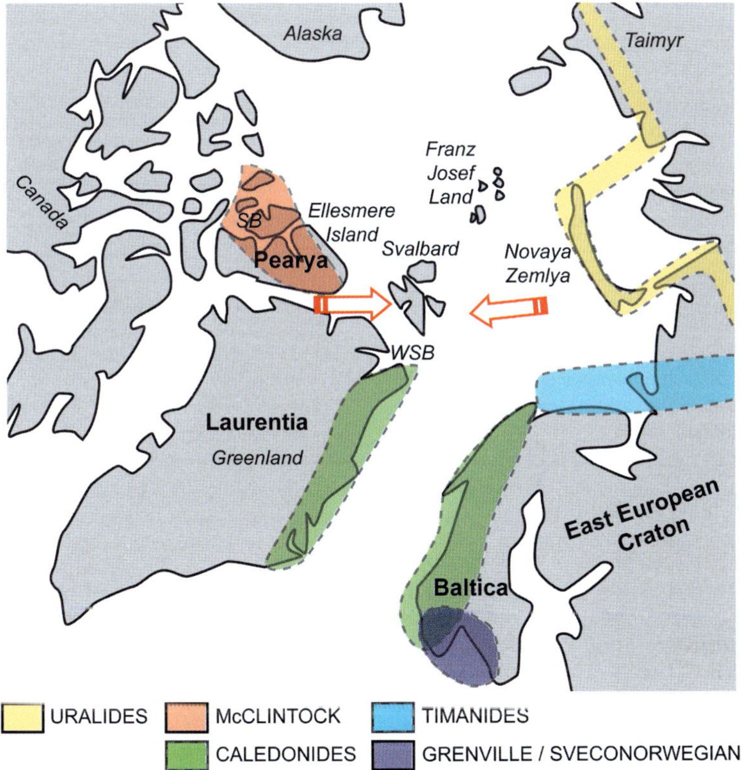



**Fig. 9.** Schematic palaeogeographical (pre-Paleocene) map showing the most important source areas during the Triassic and Jurassic (based on Worsley & Aga 1986; Omma *et al.* 2011). Red arrows indicate sediment transport directions, from the west in the Early and Middle Triassic (Vardebukta Formation and Bravaiseberget Formation) and from the east in the Late Triassic (De Geerdalen Formation). The Neoproterozoic to Mesozoic Timanide, McClintock, Caledonian and Uralian orogens are indicated. SB, Sverdrup Basin; WSB, Wandel Sea Basin.

*et al.* 1976; Mutanen & Huhma 2003; Bibikova *et al.* 2004; Hölttä *et al.* 2008). Rocks ranging from Eoarchaean to Neoarchaean are also present in the Canadian Shield (e.g. Hoffman 1989). Neoarchaean and Palaeoproterozoic ages are represented on both Greenland and Baltica (Henriksen *et al.* 2000; Bogdanova *et al.* 2008; Hölttä *et al.* 2008; Bingen & Solli 2009; Kuznetsov *et al.* 2010). Ages of 2.8–2.6 Ga are found in Greenland, the Canadian Shield (Henriksen *et al.* 2000; Kirkland *et al.* 2008, and references therein), as well as in Baltica (Bingen & Solli 2009; Lahtinen 2012). On Svalbard these ages are rare. However, an age of *c.* 2710 Ma has been reported for quartz-monzonite in the Eastern Terrane (Hellmann *et al.* 2001).

The northern East European Craton includes Palaeo- and Mesoproterozoic domains with ages spanning from 2.5 to 1.0 Ga (Patchett & Kuovo 1986; Åhäll & Larson 2000; Bogdanova *et al.* 2008; Lahtinen *et al.* 2008; Bingen & Solli 2009; Kuznetsov *et al.* 2010). Similar ages are also known from northern Laurentia (Hoffman 1988; Gower *et al.* 1990; Henriksen *et al.* 2000).

Igneous Grenvillian/Sveconorwegian (1.14–0.9 Ga, Andersson *et al.* 1999; Bogdanova *et al.* 2008) rocks are not common in the areas adjacent to the Barents Sea. However, *c.* 1.1–1.0 Ga granites are known from the Pearya Terrane (Trettin 1987), and *c.* 960–995 Ma gneiss is present on Svalbard (Myhre *et al.* 2009; Pettersson *et al.* 2009*a*, *b*). Kirkland *et al.* (2006) reported minor 981 Ma and 973 Ma granitic plutons in the Kalak Nappe, and Grenvillian/Sveconorwegian ages have also been reported from the Taimyr Peninsula (Pease *et al.* 2001) and from pre-Mesozoic sediments on Novaya Zemlya (Pease & Scott 2009). In the East Greenland Caledonides, 970–960 Ma granitoids intrude Neoproterozoic metasediments that contain zircons ranging in age from 1.8 to 1.1 Ga (Strachan *et al.* 1995; Kalsbeek *et al.* 2000; Watt & Thrane 2001), and Sveconorwegian/Grenvillian granites are widespread in southern Baltica (Andersen *et al.* 2002) and eastern Canada (Gower *et al.* 1990; Gower & Krogh 2002).

Neoproterozoic ages have been reported from Svalbard (Peucat *et al.* 1989; Gromet & Gee 1998; Majka *et al.* 2008; Pettersson *et al.* 2009*a*, *b*). In the Kalak Nappe, magmatic events have been dated to *c.* 876 Ma, 826 Ma and 710–680 Ma (Kirkland *et al.* 2006; Corfu *et al.* 2007). Rift-related *c.* 630–615 Ma rocks have been reported in Taimyr's Central Belt (Pease & Vernikovsky 2000), Vendian (650 Ma – Cambrian; Chumakov & Semikhatov 1981) and Neoproterozoic detrital zircons have been found in Taimyr's Northern Belt (Pease & Scott 2009), and Neoproterozoic detrital zircons have also been reported from Novaya Zemlya (Pease & Scott 2009), and these ages are also common in granitic rocks in the western part of the Siberian Craton (Vernikovsky *et al.* 2003*a*).

Neoproterozoic magmatism associated with the Timanian orogeny represents an important zircon-producing event in the Barents Sea region at *c.* 610–540 Ma (Beckholmen & Glodny 2004; Larionov *et al.* 2004; Pease 2011). Gabbroic and alkaline rocks of similar ages are also found in the Kalak Nappe (Pedersen *et al.* 1989; Roberts *et al.* 2006).

The Caledonian orogeny on East Greenland and in Scandinavia includes numerous magmatic and metamorphic events that ultimately provide detrital zircons ranging in age from *c.* 500 to 390 Ma (Kalsbeek *et al.* 2001; Bingen *et al.* 2002; Roberts 2003, and references therein; Kirkland *et al.* 2007; Rehnström 2010; Augland *et al.* 2012). Caledonian rocks ranging in age from *c.* 470 to 410 Ma are well represented on Svalbard, and similar ages (490–410 Ma) have also been reported from October Revolution Island, east of Svalbard (Lorenz *et al.* 2007; Myhre *et al.* 2009; Pettersson *et al.* 2010, and references therein). In the Pearya Terrane, granitic intrusions and volcanic rocks of the McClintock orogeny yield ages from 500 to 390 Ma (Trettin 1987; Gee & Teben'kov 2004).

The Palaeozoic Uralian orogenic cycle included important zircon producing magmatic events, such as granitic magmatism in subduction and intra-continental settings from 365 to 250 Ma (Bea *et al.* 2002; Scarrow *et al.* 2002). The northern Uralian Orogen includes the Polar Urals, Novaya Zemlya and the Taimyr fold belt (e.g. Hamilton 1970; Zonenshain *et al.* 1990; Doré 1991, and references therein; Brown *et al.* 2006; Petrov *et al.* 2008).

In the Triassic the formation of the Siberian traps was accompanied by syenitic magmatism between *c.* 249 Ma and 232 Ma. These rocks are known from the Taimyr Peninsula and from islands in the Kara Sea (Dobretsov & Vernikovsky 2001; Vernikovsky *et al.* 2003*b*; Dobretsov *et al.* 2008).

## Early and Middle Triassic Vardebukta and Bravaiseberget formations (endmember group 1)

The zircon age signature of the Early and Middle Triassic sands is dominated by Proterozoic, mainly Palaeo- and Mesoproterozoic zircons, accompanied by a considerable Archaean population (Fig. 5a, b).

The dominant age groups are 1.1–1.2, 1.35–1.45, 1.6 and 1.8–1.9 Ga. In the Bravaiseberget Formation, 2.6–2.8 Ga zircons are also present. In contrast to endmember group 2, there are no significant Palaeozoic or Mesozoic zircon age populations. This appears to exclude the Caledonides and the Uralides as important source areas. Furthermore, zircon ages between 600 and 700 Ma are sparsely represented and there are apparent gaps between 500 and 600 Ma and between 700 and *c.* 900 Ma, which makes the Timanides, Siberia and the enigmatic northern landmass of Crockerland unlikely source areas (Embry 1993, 2009; Anfinson *et al.* 2012 and references therein).

The large 1.1–1.2 Ga zircon population, which makes up a larger proportion of the analyses in the Bravaiseberget Formation than in the Vardebukta Formation, appears to have no important magmatic/metamorphic correlatives in the regions surrounding the Barents Sea. However, provenance data from Neoproterozoic–mid-Palaeozoic sediments in east Greenland, NW Svalbard and Arctic Canada display significant Grenvillian/Sveconorwegian zircon age populations (Kirkland *et al.* 2009; Røhr & Andersen 2009; Pettersson *et al.* 2010; Anfinson *et al.* 2012). These older sediments and the Grenvillian collisional front in eastern Canada, directly or indirectly, represent likely source areas of Grenvillian/Sveconorwegian zircons.

The Vardebukta and Bravaiseberget formations contain only sparse Caledonian zircons, four and three grains, respectively. The few Caledonian grains are compatible with source areas in Greenland west of the East Greenland Caledonides or in North Greenland, either in the Precambrian basement or the recycled Neoproterozoic to Palaeozoic sequences involved in the non-magmatic North Greenland fold belt.

Detrital zircon age data from Cretaceous sediments in the Wandel Sea Basin in northern Greenland, in the Sverdrup Basin in Arctic Canada, on Svalbard (Helvetiafjellet Formation; Røhr *et al.* 2008, 2010; Røhr & Andersen 2009; see Fig. 8), from Neoproterozoic to late Devonian deposits in the Franklinian Basin in Arctic Canada (Anfinson *et al.* 2012), from Cambrian sandstones in northern

Canada (Northwest Territories; Hadlari *et al.* 2012) and from Mesoproterozoic to early Cambrian siliciclastic units in Peary Land, North Greenland (Kirkland *et al.* 2009) display age signatures that are similar to those of the Vardebukta and Bravaiseberget formations (Fig. 8). This implies that the Early and Middle Triassic sediments on Svalbard (this study) and the Neoproterozoic–Late Devonian and Early Cretaceous sediments in NE Greenland, on Svalbard and in Arctic Canada largely shared the same provenance, namely Greenland and Canada. Because the Franklinian Basin deposits are older than those of the Vardebukta and Bravaiseberget formations, they could (partly) be the source of the Early and Middle Triassic sediments on Svalbard. For the Vardebukta and Bravaiseberget formations on Svalbard this means a westerly source area (Fig. 9).

## Late Triassic De Geerdalen Formation (endmember group 2)

The detrital zircon age signature in the De Geerdalen Formation is distinctly different from that of the Vardebukta and Bravaiseberget formations (Figs 5, 6 & 8). The Late Triassic sediments contain sparse Archaean grains, and zircons older than 3.0 Ga are not found. Palaeo- and Mesoproterozoic ages are minor in these samples, whereas Neoproterozoic ages are somewhat more common. The most distinct feature is the dominance of Palaeozoic ages. The main populations are of Ordovician and Silurian ages, which match both the age of the Caledonian Orogen and granitoids on Severnaya Zemlya (Lorenz *et al.* 2007). Even more significant is the presence of numerous Carboniferous age zircons, linking the De Geerdalen Formation sands to an easterly source area in the Uralides and Taimyr. The key characteristic of these sediments is the main population of Permo-Triassic ages that correspond to the Uralian orogeny (Pease *et al.* 2001).

The detrital zircon age data from the De Geerdalen Formation clearly point to a source area in the east (Fig. 9). This interpretation is consistent with the interpretation of seismic data on Triassic clinoforms in the Barents Sea, and results from sedimentological and sequence stratigraphic studies (Mørk 1999; Riis *et al.* 2008; Smelror *et al.* 2009; Glørstad-Clark *et al.* 2010, 2011).

## Late Triassic and Early (Middle) Jurassic Wilhelmøya Subgroup and Knorringfjellet and Svenskøya formations

The sediments of this group display characteristics from both endmember groups described above

(Figs 5–8). In terms of the presence of Archaean, Proterozoic and Palaeozoic zircons in the Late Triassic–Early Jurassic deposits (Wilhelmøya Subgroup & Knorringfjellet and Svenskøya formations) these samples appear to be a mix between the two endmember groups. The main zircon age peaks are Palaeozoic, with prominent Caledonian (Ordovician–Silurian) and Uralian (Carboniferous–Triassic) signatures. The provenance of this group of sediments could reflect either (1) a new and different source area or (2) mixing of sediments from the two source areas supplying the endmember groups. The Palaeozoic population indicates a generally eastern source area. The presence of Proterozoic and Archaean grains could be explained by additional input of material from Baltica to the south. Alternatively, the Precambrian zircons could reflect reworking of local sediments that originally had a westerly source, or renewed input of sediments from westerly sources.

The mature character of the sands in this group of samples (Dallmann 1999; Mørk 1999) compared to the immature De Geerdalen Formation sands, and the fact that the Late Triassic–Early Jurassic sedimentary successions on Svalbard include several hiati support the interpretation that the signature (partly) reflects reworking of older sediments at this time.

## Implications for regional Mesozoic sediment routing

Previous studies have suggested different source areas for the Mesozoic successions on Svalbard. One application of U–Pb zircon age data is to test and refine large-scale depositional models based on seismic, sequence stratigraphic and sedimentological datasets.

Early Triassic sediments have traditionally been attributed to a westerly source area (i.e. Greenland; e.g. Mørk 1999). This interpretation is supported by the zircon age data in this study. Grenvillian and Palaeozoic Sm–Nd provenance ages in the Late Triassic De Geerdalen Formation were interpreted by Mørk (1999) to reflect a source area to the north or NE of Svalbard, potentially including Nordaustlandet. Whereas the zircon age signature of the De Geerdalen Formation (this study) confirms a change in provenance compared to the older westerly derived sediments, the zircon age populations point to the Uralides and potentially Taimyr as the main source areas, together with minor input from Caledonian and possibly Timanian rocks. Mørk (1999) also showed that a notable change in sandstone maturity occurred in the latest Triassic–Early Jurassic, and attributed this change from immature to mature quartz-dominated sands to reworking of the sediments.

This finding supports the interpretation that the mixed provenance signature of latest Triassic–Early Jurassic sands in Svalbard (this study) reflects reworking of older sediments, as opposed to a change in provenance.

Based on a petrographic study of shallow stratigraphic cores and seismic data from the northern Barents Sea region, Riis *et al.* (2008) proposed that sediments sourced in the east and SE reached the Svalbard Archipelago in mid-Carnian, which is the depositional age of the De Geerdalen Formation (Figs 3, 4 & 8). The source of the De Geerdalen Formation on western Spitsbergen was regarded as uncertain. The present study confirms the arrival of sediments from the east in the mid-Carnian, and furthermore supports the idea that the De Geerdalen Formation across the Svalbard Archipelago was sourced from the east.

In a regional study of the Barents Sea based on seismic data combined with sequence stratigraphic analysis, Glørstad-Clark *et al.* (2010) tracked the filling of the Barents Sea in the Triassic from the SE. The main sediment supply was inferred to be derived from the southeasterly Uralian Orogen, with possible contributions from the Timanides and/or Baltica.

The data in the present study demonstrate that the sediments from the east yield a distinct and easily recognizable provenance signature. Where core samples are available, it should thus be possible to test individual sands for the presence or absence of an eastern signature. Following the filling of the East Barents Sea Basin in Late Permian–earliest Triassic, building out towards the west led to the arrival of Uralian sourced sediments on Svalbard at *c.* 230 Ma. The arrival of easterly derived sediments on Svalbard in the Late Triassic–Early Jurassic demonstrate that some 20 myr elapsed before the clastics that were shed from the Uralides finally arrived on Svalbard to form the De Geerdalen Formation.

To the south of Svalbard, Late Triassic prograding clinoforms have been identified that indicate a source area to the west, possibly Greenland (Glørstad-Clarck *et al.* 2010). This indicates that the intermediate group in this study, represented by the Late Triassic–Early Jurassic Wilhelmøya Subgroup and Knorringfjellet and Svenskøya formations, in addition to representing reworking of older sediments, also may represent renewed sediment input from the west.

Røhr & Andersen (2009) reported detrital zircon ages from Early Cretaceous sediments on Svalbard (Fig. 8) that they interpreted to be derived from a westerly source (i.e. Greenland). The Early Cretaceous zircon age signature resembles that of the Vardebukta and Bravaiseberget formations in this study (Figs 5a, b & 8). The appearance of this signature in the Early and Middle Triassic, in the Late Triassic–Early Jurassic and in the Early Cretaceous indicates that the same or similar provenance areas in the west repeatedly supplied sediments to Svalbard in the Mesozoic, although the Late Triassic–Early Jurassic sediments at least to some extent are interpreted to reflect reworking of older sediments derived from the west.

An enigmatic landmass, Crockerland (north of Svalbard and Ellesmere Island), has been proposed by several workers as a potential source area for sediments in the Sverdrup Basin and in north Greenland (Embry 1993, 2009; Miller *et al.* 2006; Omma *et al.* 2011; Anfinson *et al.* 2012). Pchelina (1988, referred to in Mørk 1999) suggested a land area to the NE of Franz Josef Land as a possible source of sediments in the Barents Sea region in part of the Triassic based on detritus in deltaic deposits. Riis *et al.* (2008) did not find evidence for a northerly source area and concluded that, compared to the voluminous easterly derived sediments, other sources of sediments had to be minor. Because the (zircon) ages (450–370 Ma and 700–500 Ma; Anfinson *et al.* 2012) assigned to Crockerland are found in other potential provenance regions adjacent to the Barents Sea (e.g. the Caledonides and Timanides), it is not possible to distinguish this hypothetical region from other sources based on the age signatures alone.

## Conclusions

Zircon U–Pb age data from eight samples collected on Svalbard demonstrate changing provenance signatures with time during the Mesozoic. The provenance signature of Early and Middle Triassic samples (Vardebukta and Bravaiseberget formations) on Svalbard is dominated by Proterozoic to Neoarchaean zircons and is interpreted to reflect derivation from westerly sources, most likely in northern Greenland and potentially Canada. The provenance signature of Late Triassic samples (De Geerdalen Formation) is dominated by Palaeozoic ages, and is interpreted to reflect derivation from eastern sources, most likely in the northern Uralides, with a minor input from the Timanides. Late Triassic to Early and Middle Jurassic samples (Knorringfjellet Formation, Svenskøya Formation and Wilhelmøya Subgroup) display a mix of the detrital zircon ages found in the Early to Late Triassic samples. These deposits are interpreted to mainly reflect reworking of older sediments, potentially together with renewed influx of sediments from the west (northern Greenland). The published provenance signature from Early Cretaceous sediments on Svalbard is similar to the Early and Middle Triassic signatures in this study, suggesting that sediments at both times were derived from the same or similar sources in the west.

The provenance data from Svalbard thus demonstrate the predominance of western sources in the Mesozoic, briefly interrupted in the Late Triassic by the influx of sediments from the Uralides in the east that reached the western edge of the Barents Sea some 20 million years after the formation of the Uralian Orogen.

The results of this study constrain and refine earlier interpretations of the origin of the Triassic and Jurassic sands on Svalbard and demonstrate the feasibility of using U–Pb zircon age data for testing models of Mesozoic sediment provenance and routing based on seismic, sedimentological, petrographic and heavy mineral investigations in the Barents Sea.

The project was funded by VISTA – a basic research programme funded by Statoil – and conducted in close collaboration with The Norwegian Academy of Science and Letters. The Norwegian Petroleum Directorate provided samples from Wilhelmøya, Hopen and Edgeøya from their 2007 and 2009 expeditions (special thanks to T. Brekke, R. S. Rød and I. Hynne). A. M. Lundmark provided one sample from Spitsbergen. R. Solbakken performed the LA-ICPMS analyses for c. 20% of the data. G. B. Fjeld, J. Lamminen and D. E. Førsund helped with mineral separation. S. L. Simonsen provided support for the LA-ICPMS work. A. M. Lundmark, T. Andersen, F. Corfu and J. I. Faleide are thanked for valuable comments. The manuscript was further improved by the helpful suggestions from reviewers V. Pease and A. V. Soloviev. This is publication no. 35 from the Isotope Geology Laboratory at the Department of Geosciences, University of Oslo.

# References

ÅHÄLL, K.-I. & LARSON, S. Å. 2000. Growth-related 1.85–1.55 Ga magmatism in the Baltic Shield: a renewed addressing the tectonic characteristics of Svecofennian, TIB 1-related and Gothian events. *GFF*, **122**, 193–206.

ANDERSEN, T., ANDRESEN, A. & SYLVESTER, A. G. 2002. Timing of late- to post-tectonic Sveconorwegian granitic magmatism in South Norway. *NGU Bulletin*, **440**, 5–18.

ANDERSEN, T., ANDERSSON, U. B., GRAHAM, S., ÅBERG, G. & SIMONSEN, S. L. 2009. Granitic magmatism by melting of juvenile continental crust: new constraints on the source of Palaeoproterozoic granitoids in Fennoscandia from Hf isotopes in zircon. *Journal of the Geological Society, London*, **166**, 233–248.

ANDERSSON, J., SODERLUND, U., CORNELL, D., JOHANSSON, L. & MOLLER, C. 1999. Sveconorwegian ( Grenvillian) deformation, metamorphism and leucosome formation in SW Sweden, SW Baltic Shield: constraints from a Mesoproterozoic granite intrusion. *Precambrian Research*, **98**, 151–171.

ANFINSON, O. A., LEIER, A. L., EMBRY, A. F. & DEWING, K. 2012. Detrital zircon geochronology and provenance of the Neoproterozoic to Late Devonian Franklinian Basin, Canadian Arctic Islands. *Geological Society of America Bulletin*, **124**, 415–430, http://dx.doi.org/10.1130/B30503.1

AUGLAND, L. E., ANDRESEN, A., CORFU, F. & DAVIKNES, H. K. 2012. Late Ordovician to Silurian ensialic magmatism in Liverpool Land, East Greenland: new evidence extending the northeastern branch of the continental Laurentian magmatic arc. *Geological Magazine*, **149**, 561–577.

BEA, F., FERSHTATER, G. B. & MONTERO, P. 2002. Granitoids of the Uralides: implications for the evolution of the orogen. *In*: BROWN, D., JUHLIN, C. & PUCHKOV, V. (eds) *Mountain Building in the Uralides: Pangea to the Present*. American Geophysical Union, Geophysical Monograph, Washington, DC, **132**, 211–233.

BECKHOLMEN, M. & GLODNY, J. 2004. Timanian blueschist-facies metamorphism in the Kvarkush metamorphic basement, Northern Urals, Russia. *In*: GEE, D. G. & PEASE, V. (eds) *The Neoproterozoic Timanide Orogen of Eastern Baltica*. Geological Society, London, Memoirs, **30**, 125–134.

BERRY, R. F., JENNER, G. A., MEFFRE, S. & TUBRETT, M. N. 2001. A North American provenance for Neoproterozoic to Cambrian sandstones in Tasmania? *Earth and Planetary Science Letters*, **192**, 207–222.

BIBIKOVA, E. V., BOGDANOVA, S. V., GLEBOVITSKY, V. A., CLAESSON, S. & SKIOLD, T. 2004. Evolution of the Belomorian Belt: NORDSIM U–Pb zircon dating of the Chupa Paragneisses, magmatism and metamorphic stages. *Petrology, Moscow*, **12**, 195–210.

BINGEN, B. & SOLLI, A. 2009. Geochronology of magmatism in the Caledonian and Sveconorwegian belts of Baltica: synopsis for detrital zircon provenance studies. *Norwegian Journal of Geology*, **89**, 267–290.

BINGEN, B., NORDGULEN, Ø. & SOLLI, A. 2002. U–Pb geochronology of Paleozoic events in the mid Scandinavian Caledonides. *In*: EIDE, E. A. (ed.) *BATLAS – Mid Norway Plate Reconstruction Atlas with Global and Atlantic Perspectives*. Geological Survey of Norway, 66–67.

BJØRLYKKE, K. 2010. *Petroleum Geoscience, from Sedimentary Environments to Rock Physics*. Springer-Verlag, Berlin.

BOGDANOVA, S. V., BINGEN, B., GORBATSCHEV, R., KHERASKOVA, T. N., KOZLOV, V. I., PUCHKOV, V. N. & VOLOZH, YU. A. 2008. The East European Craton (Baltica) before and during the assembly of Rodinia. *Precambrian Research*, **160**, 23–45.

BREIVIK, A. J., MJELDE, R., GROGAN, P., SHIMAMURA, H., MURAI, Y., NISHIMURA, Y. & KUWANO, A. 2002. A possible Caledonide arm through the Barents Sea imaged by OBS data. *Tectonophysics*, **355**, 67–97.

BROWN, D., PUCHKOV, V., ALVAREZ-MARRON, J., BEA, F. & PEREZ-ESTAUN, A. 2006. Tectonic processes in the Southern and Middle Urals: an overview. *In*: GEE, D. G. & STEPHENSON, R. A. (eds) *European Lithosphere Dynamics*. Geological Society, London, Memoirs, **32**, 407–419.

BUCHAN, S. H., CHALLINOR, A., HARLAND, W. B. & PARKER, J. R. 1965. *The Triassic Stratigraphy of Svalbard*. Skrifter, **135**. Norsk Polarinstitutt, Tromsø.

CHUMAKOV, N. M. & SEMIKHATOV, M. A. 1981. Riphean and Vendian of the USSR. *Precambrian Research*, **15**, 229–253.

CORFU, F., ROBERTS, R. J., TORSVIK, T. H., ASHWAL, L. D. & RAMSAY, D. M. 2007. Peri-Gondwanan elements in the Caledonian nappes of Finnmark, northern Norway: implications for the paleogeographic framework of the Scandinavian Caledonides. *American Journal of Science*, **307**, 434–458.

CORFU, F., POLTEAU, S., PLANKE, S., FALEIDE, J. I., SVENSEN, H., ZAYONCHECK, A. & STOLBOV, N. 2013. U–Pb geochronology of Cretaceous magmatism on Svalbard and Franz Josef Land, Barents Sea Large Igneous Province. *Geological Magazine*, First View, 1–9.

DALLAND, A., WORSLEY, D. & OFSTAD, K. 1988. *A Lithostratigraphic Scheme for The Mesozoic and Cenozoic Succession Offshore Mid- and Northern Norway*. Norwegian Petroleum Directorate Bulletin, **4**.

DALLMANN, W. K. 1999. Introduction. *In*: DALLMANN, W. K. (ed.) *Lithostratigraphic Lexicon of Svalbard. Review and Recommendations for Nomenclature Use. Upper Palaeozoic to Quaternary bedrock*. Norwegian Polar Institute, Tromsø, 11–21.

DALLMANN, W. K., OHTA, Y., ELVEVOLD, S. & BLOMEIER, D. (eds) 2002. *Bedrock Map of Svalbard and Jan Mayen*. Temakart No. 33. Norsk Polarinstitutt, Tromsø.

DOBRETSOV, N. L. & VERNIKOVSKY, V. A. 2001. Mantle plumes and their geologic manifestations. *International Geology Review*, **43**, 771–787.

DOBRETSOV, N. L., KIRDYASHKIN, A. A., KIRDYASHKIN, A. G., VERNIKOVSKY, V. A. & GLADKOV, I. N. 2008. Modelling of thermochemical plumes and implications for the origin of the Siberian traps. *Lithos*, **100**, 66–92.

DORÉ, A. G. 1991. The structural foundation and evolution of Mesozoic seaways between Europe and the Arctic. *Palaeogeography Palaeoclimatology Palaeoecology*, **87**, 441–492.

DRACHEV, S. & SAUNDERS, A. 2006. The early Cretaceous Arctic LIP: its geodynamic setting and implication for Canada Basin opening. *In*: SCOTT, R. A. & THURSTON, D. K. (eds) *Proceedings of the 4th International Conference on Arctic Margins, OCS Study MMS 2006–003*. US Department of the Interior, Anchorage, Alaska, 216–223.

EMBRY, A. F. 1993. Crockerland; the northwest source area for the Sverdrup Basin, Canadian Arctic Islands. *In*: VORREN, T. O., BERGSAGER, E., DAHL STAMNES, O. A., HOLTER, E., JOHANSEN, B., LIE, E. & LUND, T. B. (eds) *Arctic Geology and Petroleum Potential. Proceedings of the Norwegian Petroleum Society Conference*, Elsevier, New York, **2**, 205–216.

EMBRY, A. F. 2009. Crockerland – the source area for the Triassic to Middle Jurassic Strata of Northern Axel Heiberg Island, Canadian Arctic Islands. *Bulletin of Canadian Petroleum Geology*, **57**, 129–140.

FALEIDE, J. I., GUDLAUGSSON, S. T. & JACQUART, G. 1984. Evolution of the western Barents Sea. *Marine and Petroleum Geology*, **1**, 123–150.

FALEIDE, J. I., VÅGNES, E. & GUDLAUGSSON, S. T. 1993. Late Mesozoic–Cenozoic evolution of the southwestern Barents Sea in a regional riftshear tectonic setting. *Marine and Petroleum Geology*, **10**, 186–214.

FALEIDE, J. I., TSIKALAS, F. *ET AL.* 2008. Structure and evolution of the continental margin off Norway and Barents Sea. *Episodes*, **3**, 82–91.

GRADSTEIN, F. M. & OGG, J. G. 2009. The geologic time scale. *In*: HEDGES, S. B. & KUMAR, S. (eds) *The Timetree of Life*. Oxford University Press, Oxford, 26–34.

GEE, D. G. & TEBEN'KOV, A. M. 2004. Svalbard: a fragment of the Laurentian margin. *In*: GEE, D. G. & PEASE, V. (eds) *The Neoproterozoic Timanide Orogen of Eastern Baltica*. Geological Society, London, Memoirs, **30**, 191–206.

GEE, D. G., FOSSEN, H., HENRIKSEN, N. & HIGGINS, A. K. 2008. From the early Paleozoic platforms of Baltica and Laurentia to the Caledonide orogen of Scandinavia and Greenland. *Episodes*, **31**, 44–51.

GLØRSTAD-CLARK, E., FALEIDE, J. I., LUNDSCHIEN, B. A. & NYSTUEN, J. P. 2010. Triassic seismic sequence stratigraphy and paleogeography of the western Barents Sea area. *Marine and Petroleum Geology*, **27**, 1448–1475.

GLØRSTAD-CLARK, E., BIRKELAND, E. P., NYSTUEN, J. P., FALEIDE, J. I. & MIDTKANDAL, I. 2011. Triassic platform–margin deltas in the western Barents Sea. *Marine and Petroleum Geology*, **28**, 1294–1314.

GOWER, C. F. & KROGH, T. E. 2002. A U–Pb geochronological review of the Proterozoic history of the eastern Grenville Province. *Canadian Journal of Earth Sciences*, **39**, 795–829.

GOWER, C. F., RIVERS, T. & RYAN, A. B. 1990. Mid-Proterozoic Laurentia–Baltica: an overview of its geological evolution and a summary of the contributions made by this volume. *In*: GOWER, C. F., RIVERS, T. & RYAN, A. B. (eds) *Mid-Proterozoic Laurentia–Baltica*. Geological Association of Canada, Department of Earth Sciences, St. John's, NF, Canada, 1–22.

GROMET, L. P. & GEE, D. G. 1998. An evaluation of the age of high-grade metamorphism in the Caledonides of Biskayerhalvøya, NW Svalbard. *GFF*, **120**, 199–208.

GUYNN, J. & GEHRELS, G. 2010. *Comparision of Detrital Zircon Age Distributions using the K–S test*. University of Arizona, Department of Geosciences, Tuscon, AZ.

HADLARI, T., DAVIS, W. J. *ET AL.* 2012. Two detrital zircon signatures for the Cambrian passive margin of northern Laurentia highlighted by new U–Pb results from northern Canada. *Geological Society of America Bulletin*, **124**, 1155–1168, http://dx.doi.org/10.1130/B30530.1

HAMILTON, W. 1970. The Uralides and the motion of the Siberian and Russian platforms. *Geological Society of America Bulletin*, **81**, 2553–2576.

HARLAND, W. B. & GEDDES, I. 1997. Chapter 18 Triassic history. *In*: HARLAND, W. B. (ed.) *The Geology of Svalbard*. Geological Society, London, Memoirs, **17**, 340–362.

HELLMAN, F. J., GEE, D. G. & WITT-NILSSON, P. 2001. Late Archean basement in the Bangenhuken Complex of the Nordbreen Nappe, western Ny-Friesland, Svalbard. *Polar Research*, **20**, 49–59, http://dx.doi.org/10.1111/j.1751-8369.2001.tb00038.x

HENRIKSEN, N., HIGGINS, A. K., KALSBEEK, F. & PULVERTAFT, T. C. R. 2000. *Greenland from Archean to Quaternary*. Geological Survey of Denmark and Greenland Bulletin, **185**.

HENRIKSEN, E., RYSETH, A. E., LARSSEN, G. B., HEIDE, T., RØNNING, K., SOLLID, K. & STOUPAKOVA, A. V. 2011. Tectonostratigraphy of the greater Barents Sea: implications for petroleum systems. *In*: SPENCER, A. M., EMBRY, A. F., GAUTIER, D. L., STOUPAKOVA, A. V. & SØRENSEN, K. (eds) *Arctic Petroleum Geology*. Geological Society, London, Memoirs, **35**, 163–195.

HOFFMAN, P. F. 1988. United Plates of America, the birth of a craton: early Proterozoic assembly and growth of Laurentia. *Annual Review of the Earth and Planetary Sciences*, **16**, 543–603.

HOFFMAN, P. F. 1989. Precambrian geology and tectonic history of North America. *In*: BALLY, A. W. & PALMER, A. R. (eds) *The Geology of North America – An Overview*. Geological Society of America, Boulder, 447–512.

HÖLTTÄ, P., BALAGANSKY, V. *ET AL.* 2008. Archean of Greenland and Fennoscandia. *Episodes*, **31**, 13–19.

JACKSON, H. R. & GUNNARSSON, K. 1990. Reconstructions of the Arctic: Mesozoic to Present. *Tectonophysics*, **172**, 303–322.

JOHANSEN, S. E., OSTISTY, B. K. *ET AL.* 1993. Hydrocarbon potential in the Barents Sea region: play distribution and potential. *In*: VORREN, T., BERGSAGER, E., DAHL-STAMNES, Ø., HOLTER, E., JOHANSEN, B., LIE, E. & LUND, T. (eds) *Arctic Geology and Petroleum Potential*. Norwegian Petroleum Society (NPF), Special Publications, **2**, 273–320.

JOHANSSON, Å., GEE, D. G., LARIONOV, A. N., OHTA, Y. & TEBENKOV, A. M. 2005. Grenvillian and Caledonian evolution of eastern Svalbard – a tale of two orogenies. *Terra Nova*, **17**, 317–325.

KALSBEEK, F., THRANE, K., NUTMAN, A. P. & JEPSEN, H. F. 2000. Late Mesoproterozoic to early Neoproterozoic history of the East Greenland Caledonides: evidence for Grenvillian orogenesis? *Journal of the Geological Society, London*, **157**, 1215–1225.

KALSBEEK, F., JEPSEN, H. F. & NUTMAN, A. P. 2001. From source migmatites to plutons: tracking the orogin of ca. 435 Ma S-type granites in the East Greenland Caledonian Orogen. *Lithos*, **57**, 1–21.

KIRKLAND, C. L., DALY, J. S. & WHITEHOUSE, M. J. 2006. Granitic magmatism of Grenvillian and late Neoproterozoic age in Finnmark, Arctic Norway – constraining pre-Scandian deformation in the Kalak Nappe Complex. *Precambrian Research*, **145**, 24–52.

KIRKLAND, C. L., DALY, J. S., EIDE, E. A. & WHITEHOUSE, M. J. 2007. Tectonic evolution of the Arctic Norwegian Caledonides from a texturally- and structurally-constrained multi-isotopic (Ar/Ar, Rb–Sr, Sm–Nd, U–Pb) study. *American Journal of Science*, **307**, 459–526.

KIRKLAND, C. L., DALY, J. S. & WHITEHOUSE, M. J. 2008. Basement–cover relationships of the Kalak Nappe Complex, Arctic Norwegian Caledonides and constraints on Neoproterozoic terrane assembly in the North Atlantic region. *Precambrian Research*, **160**, 245–276.

KIRKLAND, C. L., PEASE, V., WHITEHOUSE, M. J. & INESON, J. R. 2009. Provenance record of Protero-zoic strata from Peary Land, North Greenland: implications for the ice-covered Greenland Shield and Laurentian palaeogeography. *Precambrian Research*,

170, 43–60, http://dx.doi.org/10.1016/j.precamres.2008.11.006

KUZNETSOV, N. B., NATAPOV, L. M., BELOUSOVA, E. A., O'REILLY, S. Y. & GRIFFIN, W. L. 2010. Geochronological, geochemical and isotopic study of detrial zircon suites from late Neoproterozoic clastic strata along the NE margin of the East European Craton: implications for the plate tectonic models. *Gondwana Research*, **17**, 583–601.

LAHTINEN, R. 2012. Evolution of the bedrock of Finland: an overview. *Lecture Notes in Earth Sciences*, **137**, 47–59.

LAHTINEN, R., GARDE, A. A. & MELEZHIK, V. A. 2008. Paleoproterozoic evolution of Fennoscandia and Greenland. *Episodes*, **31**, 20–28.

LARIONOV, A. N., ANDREICHEV, V. A. & GEE, D. G. 2004. The Vendian alkaline igneous suite of northern Timan: ion microprobe U–Pb zircon ages of gabbros and syenite. *In*: GEE, D. G. & PEASE, V. (eds) *The Neoproterozoic Timanide Orogen of Eastern Baltica*. Geological Society, London, Memoirs, **30**, 69–74.

LAURI, L. S., ANDERSEN, T., HÖLTTÄ, P., HUHMA, H. & GRAHAM, S. 2011. Evolution of the Archaean Karelian Province in the Fennoscandian Shield in the light of U–Pb zircon ages and Sm–Nd and Lu–Hf isotope systematics. *Journal of the Geological Society, London*, **168**, 201–218.

LOCK, B. E., PICKTON, C. A. G., SMITH, D. G., BATTEN, D. J. & HARLAND, W. B. 1978. The geology of Edgeøya and Barentsøya, Svalbard. *Norsk Polarinstitutt Skrifter, Oslo*, 168

LORENZ, H., GEE, D. G. & WHITEHOUSE, M. J. 2007. New geochronological data on Paleozoic igneous activity and deformation in the Severnaya Zemlya Archipelago, Russia, and implications for the development of the Eurasian Arctic margin. *Geological Magazine*, **144**, 105–125.

LUDWIG, K. R. 2003. *Users manual for Isoplot/Excel 3.00. A geochronological toolkit for Microsoft Excel*. Berkeley Geochronology Center, Special Publications, Berkeley, CA, USA, **4**, 59.

MAJKA, J., MAZUR, S., MANECKI, M., CZERNY, J. & HOLM, D. 2008. Late Neoproterozoic amphibolite facies metamorphism of a pre-Caledonian basement block in southwest Wedel Jarlsberg Land, Spitsbergen: new evidence from U–Th–Pb dating of monazite. *Geological Magazine*, **145**, 822–830.

MILLER, E. L., TORO, J. *ET AL.* 2006. New insights into Arctic paleogeography and tectonics from U–Pb detrital zircon geochronology. *Tectonics*, **25**, TC3013, http://dx.doi.org/10.1029/2005TC001830

MØRK, M. B. E. 1999. Compositional variations and provenance of Triassic sandstones from the Barents Shelf. *Journal of Sedimentary Research*, **69**, 690–710.

MØRK, A. & WORSLEY, D. 2006. Triassic of Svalbard and the Barents Shelf. *NGF, Abstracts and Proceedings*, **3**, 23–29.

MØRK, A., KNARUD, R. & WORSLEY, D. 1982. Depositional and diagenetic environments of the Triassic and Lower Jurassic succession of Svalbard. *In*: EMBRY, A. F. & BALKWILL, H. R. (eds) *Arctic Geology and Geophysics: Proceedings of the Third International Symposium on Arctic Geology*. Canadian Society of Petroleum Geologists, Calgary, 371–398.

MØRK, A., EMBRY, A. F. & WEITSCHAT, W. 1989. Triassic transgressive–regressive cycles in the Sverdrup Basin, Svalbard and the Barents Shelf. *In*: COLLINSON, J. D. (ed.) *Correlation in Hydrocarbon Exploration*. Norwegian Petroleum Society, Graham & Trotman, London, 113–130.

MØRK, A., DALLMANN, W. K. *ET AL.* 1999. Mesozoic lithostratigraphy. *In*: DALLMANN, W. K. (ed.) *Lithostratigraphic Lexicon of Svalbard. Review and Recommendations for Nomenclature Use. Upper Palaeozoic to Quaternary Bedrock*. Norwegian Polar Institute, Tromsø, 127–214.

MUTANEN, T. & HUHMA, H. 2003. The 3.5 Ga Siurua trondhjemite gneiss in the Archaean Pudasjärvi Granulite Belt, northern Finland. *Bulletin of the Geological Society of Finland*, **75**, 51–68.

MYHRE, P. I., CORFU, F. & ANDRESEN, A. 2009. Caledonian anatexis of Grenvillian crust: a U–Pb study of Albert I Land, NW Svalbard. *Norwegian Journal of Geology*, **89**, 173–191.

MYSKOVA, T. A., BEREZHNAYA, N. G. *ET AL.* 2005. Findings of the oldest (3600 Ma) Zircon in age from gneisses of the Kola Group, Central Kola Block, Baltic Shield: evidence from U–Pb (SHRIMR-II) data. *Doklady Earth Sciences*, **402**, 82–85.

NIKISHIN, A. M., ZIEGLER, P. A. *ET AL.* 1996. Late Precambrian to Triassic history of the East European Craton: dynamics of sedimentary basin evolution. *Tectonophysics*, **268**, 23–63.

OMMA, J. E., PEASE, V. & SCOTT, R. A. 2011. U–Pb SIMS zircon geochronology of Triassic and Jurassic sandstones on northwestern Axel Heiberg Island, northern Sverdrup Basin, Arctic Canada. *In*: SPENCER, A. M., EMBRY, A. F., GAUTIER, D. L., STOUPAKOVA, A. V. & SØRENSEN, K. (eds) *Arctic Petroleum Geology*. Geological Society, London, Memoirs, **35**, 559–566, http://dx.doi.org/10.1144/M35.37

OTTO, S. C. & BAILEY, R. J. 1995. Tectonic evolution of the northern Ural orogen. *Journal of the Geological Society, London*, **154**, 903–906, http://dx.doi.org/10.1144/GSL.JGS.1995.152.01.03

PATCHETT, J. & KUOVO, O. 1986. Origin of continental crust of 1.9–1.7 Ga age: Nd isotopes and U–Pb zircon ages in the Svecokarelian terrain of South Finland. *Contributions to Mineralogy and Petrology*, **92**, 1–12.

PCHELINA, T. M. 1988. The Mesozoic sedimentary complexes, Chapters 6.1 and 9.7. *In*: GRAMBERG, I. S. *ET AL.* (eds) *Barents Shelf Plate: VNII Okeangeologiya*. Institut Geologii I Mineral'nykh Resursov Mirovogo Okeana; All-Union Research Institute of the Geology and Mineral Resources of the World Ocean, Trudy, **196**, 263 [in Russian].

PEASE, V. 2011. Eurasian orogens and Arctic tectonics: an overview. *In*: SPENCER, A. M., EMBRY, A. F., GAUTIER, D. L., STOUPAKOVA, A. V. & SØRENSEN, K. (eds) *Arctic Petroleum Geology*. Geological Society, London, Memoirs, **35**, 311–324, http://dx.doi.org/10.1144/M35.20

PEASE, V. & SCOTT, R. 2009. Crustal affinities in the Arctic Uralides, northern Russia: significance of detrital zircon ages from Neoproterozoic and early Paleozoic sediments in Novaya Zemlya and Taimyr. *Journal of*

*the Geological Society, London*, **166**, 517–527, http://dx.doi.org/10.1144/0016-76492008-093

PEASE, V. & VERNIKOVSKY, V. 2000. *The Tectono–Magmatic Evolution of the Taimyr Peninsula: Further Constraints from New Ion-Microprobe Data*. Polarforschung, Alfred Wegener Institute for Polar and Marine Research & German Society of Polar Research, Bremerhaven, Germany, **68**, 171–178.

PEASE, V., GEE, D., VERNIKOVSKY, V., VERNIKOVSKAYA, A. & KIREEV, S. 2001. Geochronological evidence for late-Grenvillian magmatic and metamorphic events in central Taimyr, northern Siberia. *Terra Nova*, **13**, 270–280.

PEDERSEN, R. B., DUNNING, G. R. & ROBINS, B. 1989. U–Pb ages of nepheline syenite pegmatites from the Seiland Magmatic Province, N Norway. *In*: GAYER, R. A. (ed.) *The Caledonide Geology of Scandinavia*. Graham & Trotman, London, 3–8.

PETROV, G. A., RONKIN, YU. L., MASLOV, A. V., SVYAZHINA, I. A., RYBALKA, A. V. & LEPIKHINA, O. P. 2008. Timing of the onset of collision in the central and northern Urals. *Doklady Earth Sciences*, **422**, 1050–1055.

PETTERSSON, C. H., PEASE, V. & FREI, D. 2009a. U–Pb zircon provenance of metasedimentary basement of the Northwestern Terrane, Svalbard: implications for the Grenvilian–Sveconorwegian orogeny and the development of Rodinia. *Precambrian Research*, **175**, 206–220, http://dx.doi.org/10.1016/j.precamres.2009.09.010

PETTERSSON, C. H., TEBENKOV, A. M., LARIONOV, A. N., ANDRESEN, A. & PEASE, V. 2009b. Timing of migmatization and granite genesis of the Northwestern Terrane, Svalbard. *Journal of the Geological Society, London*, **166**, 147–158, http://dx.doi.org/10.1144/0016-76492008-023

PETTERSSON, C. H., PEASE, V. & FREI, D. 2010. Detrital zircon-Pb ages of Silurian–Devonian sediments from NW Svalbard: a fragment of Avalonia and Laurentia? *Journal of the Geological Society, London*, **167**, 1019–1032, http://dx.doi.org/10.1144/0016-76492010-062

PEUCAT, J. J., OHTA, Y., GEE, D. G. & BERNARD-GRIFFITHS, J. 1989. U–Pb, Sr and Nd evidence for Grenvillian and latest Proterozoic tectonothermal activity in the Spitsbergen Caledonides, Arctic Ocean. *Lithos*, **22**, 275–285.

POLTEAU, S., PLANKE, S., FALEIDE, J. I., SVENSEN, H. & MYKLEBUST, R. 2010. The Cretaceous high Arctic large igneous province. *Geophysical Research, Abstracts*, **12**, 13216.

PUCHKOV, V. N. 1997. Structure and geodynamics of the Uralian Orogen. *In*: BURG, J. P. & FORD, M. (eds) *Orogeny Through Time*. Geological Society, London, Special Publications, **121**, 201–236.

REHNSTRÖM, E. 2010. Prolonged Paleozoic magmatism in the East Greenland Caledonides: some constraints from U–Pb ages and Hf isotopes. *The Journal of Geology*, **118**, 447–465.

RIIS, F., LUNDSCHIEN, B. A., HØY, T., MØRK, A. & MØRK, M. B. E. 2008. Evolution of the Triassic shelf in the northern Barents region. *Polar Research*, **27**, 318–337.

RITZMANN, O. & FALEIDE, J. I. 2007. Caledonian basement of the western Barents Sea. *Tectonics*, **26**, 417–435.

ROBERTS, D. 2003. Tha Scandinavian Caledonides: event chronology, palaeogeographic settings and likely modern analogues. *Tectonophysics*, **365**, 283–299.

ROBERTS, D. & OLOVYANISHNIKOV, V. 2004. Structural and tectonic development of the Timanide orogen. *In*: GEE, D. G. & PEASE, V. (eds) *The Neoproterozoic Timanide Orogen of Eastern Baltica*. Geological Society, London, Memoirs, **30**, 47–57.

ROBERTS, R. J., CORFU, F., TORSVIK, T. H., ASHWAL, L. D. & RAMSAY, D. M. 2006. Short-lived mafic magmatism at 560–570 Ma in the northern Norwegian Caledonides: U–Pb zircon ages from the Seiland Igneous Province. *Geological Magazine*, **143**, 887–903.

ROSA, D. R. N., FINCH, A. A., ANDERSEN, T. & INVERNO, C. M. C. 2009. U–Pb geochronology and Hf isotope ratios of magmatic zircons from the Iberian Pyrite Belt. *Mineralogy and Petrology*, **95**, 47–69.

RØHR, T. S. & ANDERSEN, T. 2009. Detrital zircons from the high Arctic; evidence for extensive recycling of sediment from Devonian through Mesozoic times. *In*: RØHR, T. S. (ed.) *Sedimentary Provenance Analysis of Lower Cretaceous Sedimentary Successions in The Arctic; Constraints From Detrital Zircon data*. PhD thesis, Faculty of Mathematics and Natural Sciences, University of Oslo, 55–105.

RØHR, T. S., ANDERSEN, T. & HENNING, D. 2008. Provenance of Lower Cretaceous sediments in the Wandel Sea Basin, North Greenland. *Journal of the Geological Society, London*, **165**, 755–767.

RØHR, T. S., ANDERSEN, T., HENNING, D. & EMBRY, A. F. 2010. Detrital zircon characteristics of the Lower Cretaceous Isachsen Formation, Sverdrup Basin: source constraints from age and Hf isotope data. *Canadian Journal of Earth Sciences*, **47**, 255–271.

SCARROW, J. H., HETZEL, R. ET AL. 2002. Four decades of geochronological work in the Southern and Middle Urals: a review. *In*: BROWN, D., JUHLIN, C. & PUCHKOV, V. (eds) *Mountain Building in the Uralides: Pangea to the Present*. American Geophysical Union, Geophysical Monograph, Washington, DC, **132**, 233–255.

SMELROR, M., PETROV, O. V., LARSSEN, G. B. & WERNER, S. C. 2009. *Atlas: Geological History of the Barents Sea*. Geological Survey of Norway, Trondheim, **134**.

STEEL, R. J. & WORSLEY, D. 1984. Svalbard's post-Caledonian strata – an atlas of sedimentational patterns and palaeogeographic evolution. *In*: SPENCER, A. M. (ed.) *Petroleum Geology of the North European Margin*. Graham & Trotman, London, 109–135.

STEIGER, R. H., HARNIK-ŠOPTRAJANOVA, G., ZIMMERMAN, E. & HENRIKSEN, N. 1976. Isotopic age and metamorphic history of the banded gneiss at Danmarkshavn, East Greenland. *Contributions to Mineralogy and Petrology*, **57**, 1–24.

STRACHAN, R. A., NUTMAN, A. P. & FRIDERICHSEN, J. D. 1995. SHRIMP U–Pb geochronology and metamorphic history of the Smallefjord sequence, NE

Greenland Caledonides. *Journal of the Geological Society, London*, **152**, 779–784.

TEGNER, C., STOREY, M., HOLM, P. M., THORARINSSON, S. B., ZHAO, X., LO, C.-H. & KNUDSEN, M. F. 2011. Magmatism and Eurekan deformation in the High Arctic Large Igneous Province: $^{40}$Ar–$^{39}$Ar age of Kap Washington Group volcanics, North Greenland. *Earth and Planetary Science Letters*, **303**, 203–214.

TOZER, E. T. & PARKER, J. R. 1968. Notes on the Triassic biostratigraphy of Svalbard. *Geological Magazine*, **105**, 526–542.

TRETTIN, H. P. 1987. Pearya: a composite terrane with Caledonian affinities in northern Ellesmere Island. *Canadian Journal of Earth Sciences*, **24**, 224–245.

ULMISHEK, G. 1982. *Petroleum Geology and Resource Assessment of the Timan–Pechora Basin, USSR, and The Adjacent Barents–Northern Kara Shelf*. Argonne National Laboratory, Report **ANL/EES/-TM-199**.

ULMISHEK, G. 1985. *Geology and Petroleum Resources of The Barents-Northern Kara Shelf in Light of New Geologic Data*. Argonne National Laboratory, Report **ANL/ES-148**.

VERNIKOVSKY, V. A., VERNIKOVSKAYA, A. E., KOTOV, A. B., SAL'NIKOVA, E. B. & KOVACH, V. P. 2003*a*. Neoproterozoic accretion-collisional events on the western margin of the Siberian Craton: new geological and geochronological evidence from the Yenisey Ridge. *Tectonophysics*, **375**, 147–168.

VERNIKOVSKY, V. A., PEASE, V. L., VERNIKOVSKAYA, A. E., ROMANOV, A. P., GEE, D. G. & TRAVIN, A. V. 2003*b*. First report of early Triassic A-type granite and syenite intrusions from Taimyr: product of the northern Eurasian superplume? *Lithos*, **66**, 23–36.

WATT, G. R. & THRANE, P. D. 2001. Early Neoproterozoic events in East Greenland. *Precambrian Research*, **110**, 165–184.

WHITEHOUSE, M. J. & KAMBER, B. S. 2005. Assigning dates to thin gneissic veins in high-grade metamorphic terranes: a cautionary tale from Akilia, Southwest Greenland. *Journal of Petrology*, **46**, 291–318.

WORSLEY, D. 2008. The post-Caledonian development of Svalbard and the western Barents Sea. *Polar Research*, **27**, 298–317.

WORSLEY, D. & AGA, O. J. 1986. *The Geological History of Svalbard*. Den norske stats oljeselskap a.s., Stavanger.

WORSLEY, D. & MØRK, A. 1978. *The Triassic stratigraphy of Southern Spitsbergen*. Norsk Polarinstitutt, Årbok, **1977**, 43–60.

ZIEGLER, P. A. 1989. *Evolution of Laurussia: A Study in Late Palaeozoic Plate Tectonics*. Kluwer, Dordrecht.

ZIEGLER, P. A. 1990. *Geological Atlas of Western and Central Europe*. Shell International Petroleum, Maatschappij B.V.

ZONENSHAIN, L. P., KUZMIN, M. I. & NATAPOV, L. M. 1990. Uralian foldbelt and foldbelts of the northeast USSR, Taimyr and the Arctic. *In*: PAGE, B. M. (ed.) *Geology of the USSR: a Plate-Tectonic Synthesis*. American Geophysical Union, Geodynamics Series, Washington, DC, **21**, 27–54 121–146.

# Insights into crust formation and recycling in North Africa from combined U–Pb, Lu–Hf and O isotope data of detrital zircons from Devonian sandstone of southern Libya


GUIDO MEINHOLD[1,2]*, ANDREW C. MORTON[1,3], C. MARK FANNING[4], JAMES P. HOWARD[1], RICHARD J. PHILLIPS[1,5], DOMINIC STROGEN[1,6] & ANDREW G. WHITHAM[1]

[1]*CASP, University of Cambridge, West Building, 181A Huntingdon Road, Cambridge CB3 0DH, UK*

[2]*Geowissenschaftliches Zentrum der Universität Göttingen, Abt. Sedimentologie und Umweltgeologie, Goldschmidtstraße 3, 37077 Göttingen, Germany*

[3]*HM Research Associates, 2 Clive Road, Balsall Common, West Midlands CV7 7DW, UK*

[4]*Research School of Earth Sciences, Australian National University, Canberra ACT 0200, Australia*

[5]*School of Earth and Environment, University of Leeds, Leeds LS2 9JT, UK*

[6]*GNS Science, PO Box 30368, Lower Hutt 5010, New Zealand*

**Corresponding author (e-mail: guido.meinhold@geo.uni-goettingen.de)*



**Abstract:** Combined U–Pb, Lu–Hf and O isotope data of detrital zircons from the Devonian sandstone of southern Libya provide important new boundary parameters for reconstruction of palaeosource areas and sediment transport and may lead to novel approaches to test current plate tectonic models, with important implications for our understanding of the evolution of northern Gondwana (in present-day coordinates) during the Palaeozoic. Detrital zircon U–Pb ages from Devonian sandstone of the eastern margin of the Murzuq Basin show four main age populations: 2.7–2.5 Ga (13%), 2.1–1.9 Ga (10%), 1.1–0.9 Ga (25%) and 0.7–0.5 Ga (46%). The ubiquitous occurrence of *c.* 1.0 Ga detrital zircons is characteristic of the Saharan Metacraton sedimentary cover sequence and provides new insights into palaeogeographic reconstructions of Gondwana-derived terranes in the Eastern Mediterranean and SW Europe. The Lu–Hf isotope data suggest that zircons crystallized within a narrow time interval from magmas with heterogeneous Hf isotope compositions. These magmas were derived by melting of pre-existing rocks, rather than being juvenile. The calculated Hf model ages range from 3.7 Ga to 1.3 Ga, with a major population between 2.8 Ga and 1.9 Ga, indicating prominent recycling of Archaean and Palaeoproterozoic crust.


North Africa is made up of several intracratonic basins that contain a thick pile of Palaeozoic sediment. Some of the Palaeozoic sandstones host large hydrocarbon reserves and are therefore a target for hydrocarbon exploration and production (e.g. Macgregor 1996; Boote *et al.* 1998; Craig *et al.* 2008). The Murzuq Basin of southern Libya is one of these intracratonic basins and has large proven hydrocarbon reserves in Cambrian, Ordovician and Devonian sandstones (e.g. Boote *et al.* 1998; Echikh & Sola 2000; Hallett 2002; Craig *et al.* 2008). Sandstones with good porosity and permeability are one of the main targets for exploration. The porosity and permeability depend on the whole-rock composition of the sediment, which is determined by the original provenance of the detritus and the diagenetic history of the sediment (e.g. Ramm 2000; Ajdukiewicz & Lander 2010). Another important point to consider is the correlativity of sedimentary units within and between basins. For instance, this is of particular importance for the Palaeozoic succession of southern Libya, which consists of substantial thicknesses of clastic, often poorly dated marine and non-marine sediments (due to the lack of body fossils), the exception being fossiliferous clastic sediments of Carboniferous age (e.g. Klitzsch 1964; Klitzsch & Ziegert 2000). In marine units, macrofauna are sparse, and dating relies on palynomorphs and other microfossils. Although biostratigraphic studies are possible


*From*: SCOTT, R. A., SMYTH, H. R., MORTON, A. C. & RICHARDSON, N. (eds) 2014. *Sediment Provenance Studies in Hydrocarbon Exploration and Production*. Geological Society, London, Special Publications, **386**, 281–292. First published online June 11, 2013, http://dx.doi.org/10.1144/SP386.1

on fossil material recovered from boreholes (e.g. Paris *et al.* 2012), extreme surface weathering ensures that such studies are impossible on material from outcrop. The lack of biostratigraphic data means that other tools have to be sought to test the correlation of units within and between basins. Key provenance parameters may be identified and could be used for chemostratigraphy to correlate the sedimentary units within and between basins, as has been shown for the Clair Group west of the Shetland Islands (Morton *et al.* 2010). In general, understanding the origin of sedimentary units is important for reconstructions of palaeosource areas and sediment transport.

In this study, we present coupled U–Pb, Lu–Hf and O isotope data from detrital zircons of Devonian sandstone from the eastern margin of the Murzuq Basin, southern Libya. These data provide important new boundary parameters for the reconstruction of palaeosource areas and sediment transport and may lead to novel approaches to test current plate tectonic models, with important implications for our understanding of the evolution of northern Gondwana (in present-day coordinates) during the Palaeozoic.

## Geological setting

North Africa has experienced a complex and polyphase geological history. The assembly of North Africa was related to a number of 'Pan-African' orogenic cycles, which lasted from *c.* 900 Ma to 550 Ma (e.g. Kröner *et al.* 1987; Stern 1994; Abdelsalam *et al.* 2002; Kröner & Stern 2004; Collins & Pisarevsky 2005; Küster *et al.* 2008; Johnson *et al.* 2011). The central part of North Africa is referred to as the Saharan Metacraton (Abdelsalam *et al.* 2002, 2011) and is bordered by the Hoggar Shield to the west, the Congo Craton to the south and the Arabian–Nubian Shield to the east (Fig. 1), and is largely covered by Palaeozoic and younger sediments (e.g. Boote *et al.* 1998; Burke *et al.* 2003). Substantial outcrops of Palaeozoic rocks are found, for instance, along the margins of the Kufra and Murzuq basins of southern Libya (Fig. 1). The Murzuq Basin is located in the SW of Libya, and continues into northwestern Chad, northern Niger and eastern Algeria. Its present-day sedimentary fill reaches a maximum thickness of *c.* 4000 m in the basin centre (Davidson *et al.* 2000). The present-day eastern margin of the Murzuq Basin,

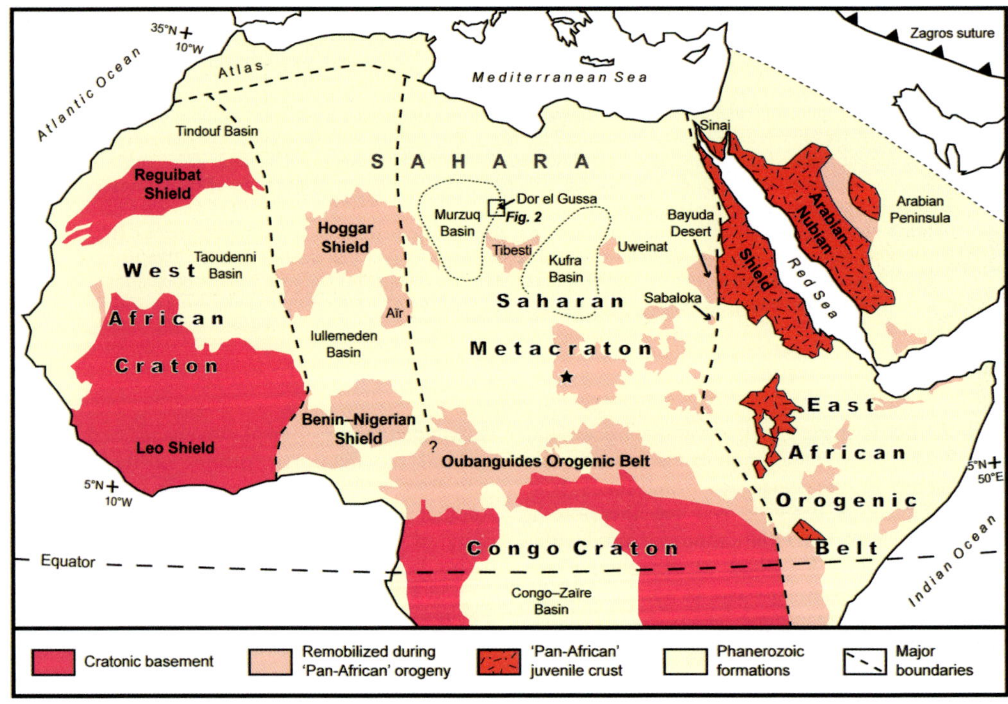


**Fig. 1.** Map of North Africa showing exposures of Precambrian rocks and major tectonic units (after Avigad *et al.* 2003; Kolodner *et al.* 2006; Meinhold *et al.* 2011). The black star denotes the region in western Dharfur (eastern Chad) from which de Wit *et al.* (2005) reported extensive *c.* 0.9 Ga and *c.* 0.6 Ga granitoids, predominantly derived from melting of 3.0–1.0 Ga continental lithosphere.

the Dor el Gussa region, is the focus of this study. In this area, the Precambrian basement comprises metasedimentary and igneous rocks. The latter are mainly granites, granodiorites and diorites, with minor amounts of gabbro and quartz monzonites (Pegram *et al.* 1976; Baumann *et al.* 1992) of a typical calcalkaline suite (Ghuma & Rogers 1978). Isotopic studies yielded ages of *c.* 586–533 Ma (Rb–Sr method; Pegram *et al.* 1976) for granites, pegmatites and gabbros south of Dor el Gussa and *c.* 554–551 Ma (Rb–Sr method; Baumann *et al.* 1992) for granites north of Dor el Gussa.

The stratigraphic framework of Dor el Gussa (Figs 2 & 3) is based predominantly on the work by Klitzsch (1964). The Ediacaran–Cambrian basement is unconformably overlain by a dominant clastic sedimentary sequence of Cambrian to Carboniferous age. A simplified stratigraphic column is shown in Figure 3. Details of the Dor el Gussa stratigraphy are provided in Klitzsch (1964) and Klitzsch & Ziegert (2000), and are summarized in Morton *et al.* (2011) and Le Heron *et al.* (2013).

Within Dor el Gussa, the Devonian Tadrart Formation (≤415 m thick) consists of massive sandstone, fine-grained to slightly conglomeratic, thick-bedded, cross-stratified, very ferruginous in some beds, but normally without matrix and very

friable, deposited in a fluvial environment (Klitzsch 1964). In eastern Dor el Gussa, the formation rests with an unconformity of 3–5° on the Silurian succession (Fig. 4) and is conformably overlain towards the west by the Ouan Kasa Formation (70–120 m thick), which consists of shale and silty shale alternating with fine-grained sandstone and calcareous siltstone (Klitzsch 1964; Klitzsch & Ziegert 2000).

For many years, the Tadrart Formation in Dor el Gussa has been suggested to be of Early Devonian age (Klitzsch 1964; Lejal-Nicol & Massa 1980; Klitzsch & Ziegert 2000). However, recent

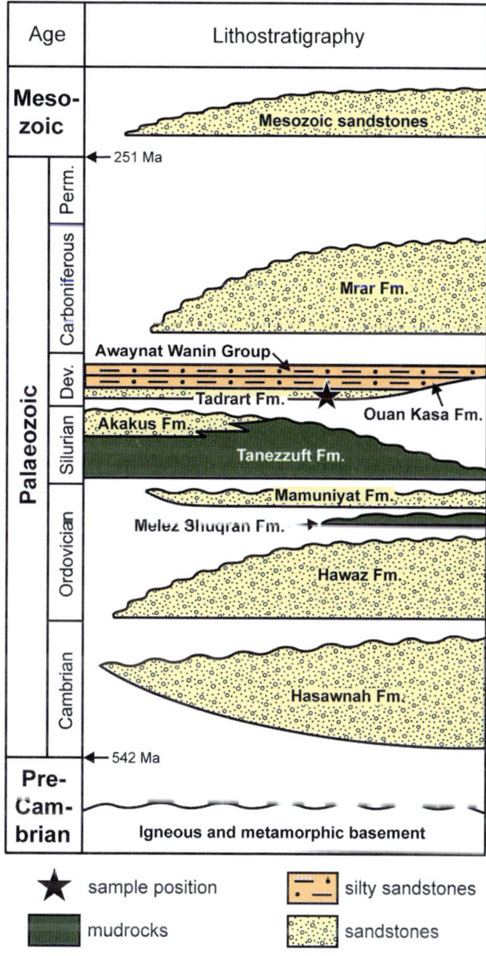


**Fig. 3.** Stratigraphic chart for the Dor el Gussa region, eastern Murzuq Basin (after Morton *et al.* 2011). Because of uncertainties over the stratigraphy the term 'Mesozoic sandstones' is used here for all Mesozoic-aged sediments. The black star indicates the stratigraphic position of the analysed Devonian sandstone sample H6032.

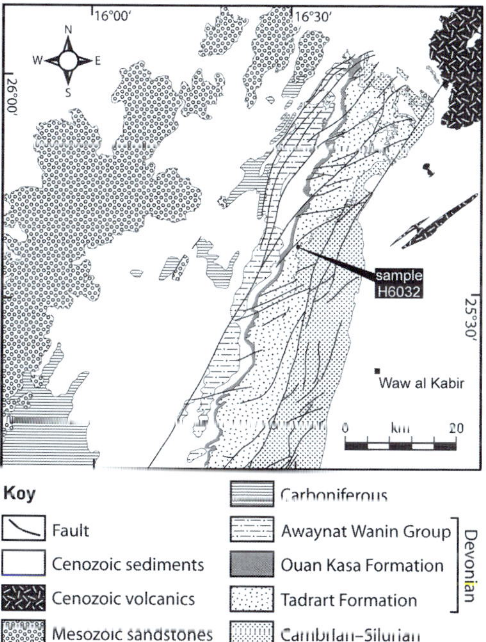


**Fig. 2.** Simplified geological map of the Dor el Gussa region, eastern Murzuq Basin (after Collomb 1962), showing the location of the analysed sandstone sample H6032 from the Devonian Tadrart Formation.

**Fig. 4.** Photograph showing the unconformity between the Silurian Tanezzuft Formation and the Devonian Tadrart Formation in central Dor el Gussa (geographic coordinates: 25°36′20.70″N, 16°30′43.10″E), *c.* 70 m to the east of sample locality H6032. Yellow field notebook (*c.* 20 cm in height) for scale.

observations made during CASP fieldwork and palaeobotanical studies (C. Berry in Meinhold *et al.* 2009) suggest that in central and northern Dor el Gussa the Tadrart and Ouan Kasa formation sediments may be of Middle to early Late Devonian age (*c.* 397–374 Ma).

## Sample description, preparation and analytical techniques

For this study we used a sandstone sample (H6032) collected at outcrop from central Dor el Gussa (geographic coordinates: 25°36′20.35″N, 16°30′40.67″E; Fig. 2), from the lower part of the Tadrart Formation, deposited in a braided river environment. Sample H6032 is classified as a quartz arenite, with quartz–feldspar–lithic fragments (Q–F–L) endmembers of 97–3–0, following the Gazzi–Dickinson method (Ingersoll *et al.* 1984). The grains are subangular to well-rounded. The fabric is grain-supported, with most clasts in long contact, suggesting moderate compaction. Zircon (86.5%), rutile (10%) and tourmaline (2.5%) are the main heavy minerals, with anatase present in minor amounts (1%) (Morton *et al.* 2011). The zircon grains are rounded to well-rounded, with aspect ratios varying between 1:1 and 1:4.

Lu–Hf and O isotope analyses were carried out on zircon grains that had previously been dated by SHRIMP. The U–Pb isotope data of Meinhold *et al.* (2011) are presented and are discussed together with the Lu–Hf and O isotope data from this study. Details about sample preparation and analytical techniques are given in the Appendix.

## Results

### Zircon U–Pb data

In total, 71 zircons from the Devonian Tadrart Formation sample H6032 were analysed, of which 61 have been used for interpretation. The corresponding U–Pb isotope data are given in Meinhold *et al.* (2011). Major zircon age populations are at 2.7–2.5 Ga (13%), 2.1–1.9 Ga (10%), 1.1–0.9 Ga (25%) and 0.7–0.5 Ma (46%) (Fig. 5). Two zircon grains have ages of $1.4 \pm 0.12$ Ga and $1.5 \pm 0.01$ Ga. The youngest zircon grains have ages of $506 \pm 6$ Ma and $516 \pm 6$ Ma. The oldest zircon grain of this sample is $2736 \pm 7$ Ma.

### Zircon Lu–Hf–O data

The $^{176}Hf/^{177}Hf$ isotope data for the detrital zircons of sample H6032 are plotted as a function of their crystallization age in Figure 6a. The Lu–Hf isotope zircon data yield mostly subchondritic to near-chondritic $\varepsilon_{Hf}$ values of −31.5 to −0.41, with initial $^{176}Hf/^{177}Hf$ values ranging from 0.280785 to 0.282446 (Table 1). Two zircons have positive $\varepsilon_{Hf}$ values of +2.12 and +4.53, with initial $^{176}Hf/^{177}Hf$ values ranging from 0.282392 to 0.282446. The calculated Hf model ages ($T_{DM}Hf$) range from 3.7 Ga to 1.3 Ga, with a major population between 2.8 Ga and 1.9 Ga (Fig. 7).

The $\delta^{18}O$ isotope data (Table 1) for the detrital zircons of sample H6032 are plotted as a function of their crystallization age in Figure 6b. The analysed $^{18}O/^{16}O$ values range from 0.002161 to 0.002176 ($\delta^{18}O = 2.3–9.2$).

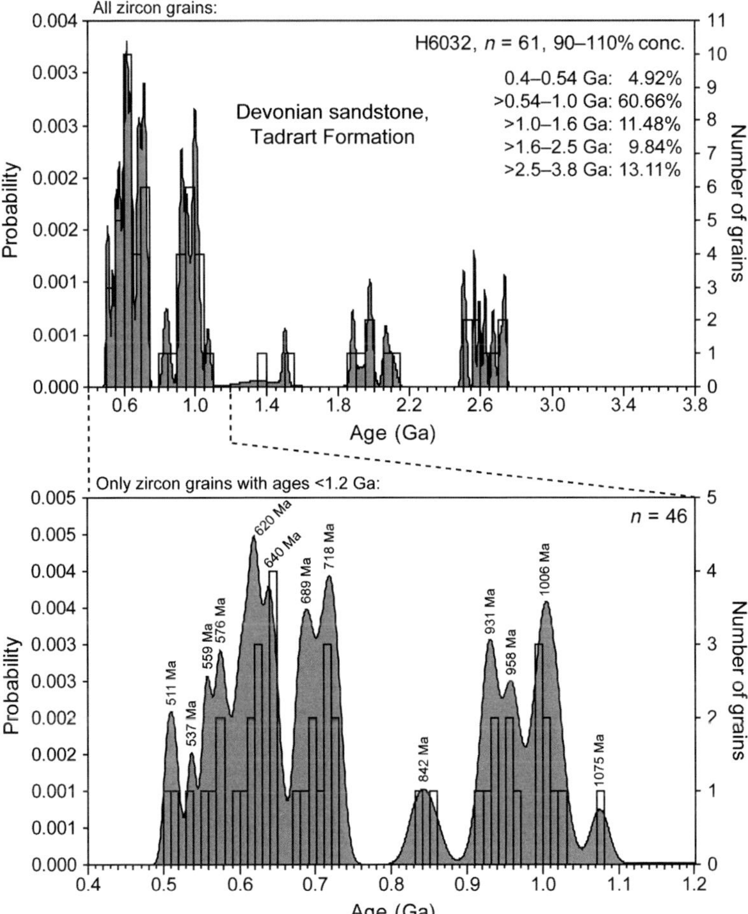


**Fig. 5.** Combined probability density distribution and frequency histogram plots for the set of U–Pb analytical zircon data from the Devonian sandstone sample H6032 of southern Libya. *n*, number of zircon grains analysed.

## Discussion and conclusions

### Zircon U–Pb data

The Cryogenian–Ediacaran age group represents orogenic events associated with the assembly of northern Gondwana (e.g. Kröner *et al.* 1987; Stern 1994; Abdelsalam *et al.* 2002; Kröner & Stern 2004; Collins & Pisarevsky 2005; Küster *et al.* 2008; Johnson *et al.* 2011). The Saharan Metacraton is considered to be the most likely source for the Neoproterozoic zircons because basement of such age is recognized within it (e.g. Pegram *et al.* 1976; Baumann *et al.* 1992; Abdelsalam *et al.* 2002, and references therein). The Neoarchaean–earliest Proterozoic (2.7–2.5 Ga) and Palaeoproterozoic (2.1–1.9 Ga) groups resemble the ages of the basement in the cratonic areas of the Saharan Metacraton (Abdelsalam *et al.* 2002). Basement rocks of

*c.* 2.7 Ga are also known from the Reguibat Shield of the West African Craton (Key *et al.* 2008). Furthermore, Neoarchaean–earliest Proterozoic (2.7–2.5 Ga) granites are known from the Congo craton (e.g. Begg *et al.* 2009, and references therein) and from affiliated crustal blocks in Uganda and Tanzania (e.g. Johnson *et al.* 2003; Begg *et al.* 2009; Link *et al.* 2010). These rocks may also account for the *c.* 2.7–2.5 Ga detrital zircons in the Saharan Metacraton cover sequence.

The presence of a large number of *c.* 1.0 Ga zircons is significant, because igneous or metamorphic rocks of similar age have not been reported previously in the zircon spectra of the Saharan Metacraton basement (Abdelsalam *et al.* 2002, fig. 3), yet they appear ubiquitous in the Palaeozoic succession of southern Libya (Meinhold *et al.* 2011). Interestingly, the Cambrian–Ordovician sandstones of Israel and Jordan contain similarly

G. MEINHOLD *ET AL.*

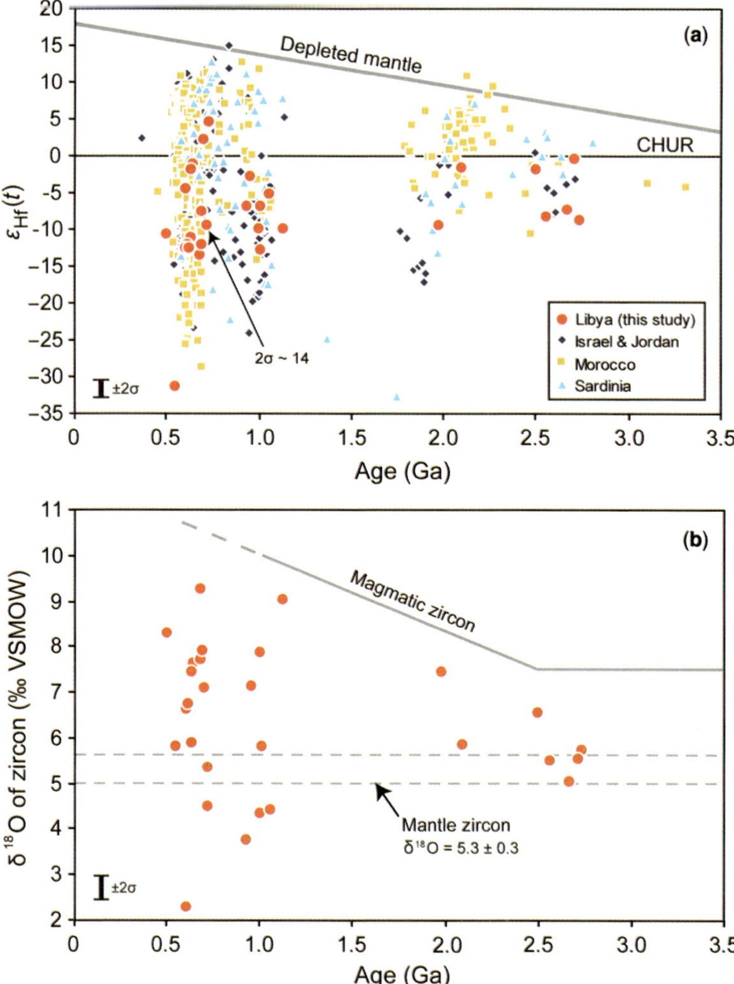

**Fig. 6.** (**a**) $\varepsilon_{Hf}(t)$ plotted v. inferred crystallization U–Pb age for detrital zircon grains of sample H6032. The standard epsilon Hf ($\varepsilon_{Hf}$) notation indicates deviation of the measured $^{176}Hf/^{177}Hf$ value from that of the chondritic uniform reservoir (CHUR) in parts per ten thousand (Patchett *et al.* 1981). Samples with higher than chondritic $^{176}Hf/^{177}Hf$ at time $t$ have positive $\varepsilon_{Hf}$ values; those with lower than chondritic $^{176}Hf/^{177}Hf$ have negative $\varepsilon_{Hf}$ values, and hence chondrites per definition have $\varepsilon_{Hf} = 0$. In general, the Lu–Hf isotope composition of the Earth's crust–mantle system varies widely between the depleted mantle (Lu/Hf > chondritic), the enriched crust (Lu/Hf < chondritic) and any remaining unfractionated source (Lu/Hf = chondritic). Different Lu/Hf ratios reflect the long-term evolution of these reservoirs (e.g. Choukroun *et al.* 2005). Hafnium preferentially enters the melt phase, thereby leaving a residuum with relatively higher Lu/Hf ratios for the subcontinental lithospheric mantle. However, metasomatism by melts derived from the asthenospheric mantle can introduce Hf and thus produces lithospheric mantle with relatively low Lu/Hf ratios (e.g. Choukroun *et al.* 2005). The black bar in the lower left corner of the plot approximates the external $2\sigma$ precision for the $\varepsilon_{Hf}(t)$ values. The single value with exceptional high error is indicated (see also Table 1). The depleted mantle evolution line is based on present-day values of $^{176}Lu/^{177}Hf = 0.0385$ and $^{176}Hf/^{177}Hf = 0.283225$ (Vervoort & Blichert-Toft 1999). Isotope data of detrital zircons from Cambrian–Ordovician sandstones, taken from literature (Morocco and Sardinia, Avigad *et al.* 2012; Israel and Jordan, Morag *et al.* 2011), are shown for comparison. (**b**) Oxygen isotope composition plotted v. inferred crystallization U–Pb age for detrital zircon grains of sample H6032. The black bar in the lower left corner of the plot approximates the external $2\sigma$ precision for the oxygen isotope values. Zircons in equilibrium with pristine mantle-derived melts have $\delta^{18}O$ values of $5.3 \pm 0.3‰$ (Valley *et al.* 2005), bracketed by the dashed horizontal lines. Higher $\delta^{18}O$ values reflect a component of $^{18}O$-enriched, supracrustal material in the magma from which the zircon precipitated, which could include sedimentary rock (10–40‰) or hydrothermally altered oceanic crust (10–20‰, Eiler 2001). The solid line brackets the upper limit for $\delta^{18}O$ values of magmatic zircon from the continental crust (Valley *et al.* 2005).

**Table 1.** *Lu–Hf and O isotope data for detrital zircon grains of sample H6032*

| Spot no. | $^{176}\text{Hf}/^{177}\text{Hf}$ | $\pm 2\sigma$ | $^{176}\text{Lu}/^{177}\text{Hf}$ | $\pm 2\sigma$ | U–Pb age (Ma) | $\varepsilon_{\text{Hf}}(0)$ | $^{176}\text{Lu}/^{177}\text{Hf}\,(t)$ | $\varepsilon_{\text{Hf}}(t)$ | $\pm 2\sigma$ | $T_{\text{DM}}(\text{Hf})$ | $^{18}\text{O}/^{16}\text{O}$ | $\pm 2\sigma$ | $\delta^{18}\text{O}$ | $\pm 2\sigma$ |
|---|---|---|---|---|---|---|---|---|---|---|---|---|---|---|
| 1 | 0.282139 | 18 | 0.001053 | 3 | 690 | −22.4 | 0.282125 | −7.7 | 1.3 | 2047 | 0.002176 | 0.5 | 9.25 | 1.10 |
| 2 | 0.282450 | 29 | 0.000311 | 5 | 728 | −11.4 | 0.282446 | 4.5 | 2.1 | 1303 | 0.002168 | 0.9 | 5.35 | 1.16 |
| 3 | 0.280897 | 19 | 0.000596 | 8 | 2670 | −66.3 | 0.280866 | −7.4 | 1.3 | 3556 | 0.002167 | 0.8 | 5.03 | 1.14 |
| 4 | 0.282344 | 41 | 0.000831 | 58 | 647 | −15.1 | 0.282334 | −1.2 | 2.9 | 1607 | 0.002173 | 0.4 | 7.61 | 1.09 |
| 5 | 0.281540 | 25 | 0.000300 | 20 | 557 | −43.6 | 0.281537 | −31.4 | 1.7 | 3422 | 0.002169 | 0.5 | 5.81 | 1.10 |
| 11 | 0.282063 | 195 | 0.000893 | 42 | 727 | −25.1 | 0.282051 | −9.5 | 13.8 | 2190 | 0.002166 | 0.6 | 4.49 | 1.11 |
| 15 | 0.280809 | 24 | 0.000456 | 41 | 2736 | −69.4 | 0.280785 | −8.8 | 1.7 | 3691 | 0.002168 | 0.5 | 5.74 | 1.10 |
| 18 | 0.281179 | 27 | 0.000937 | 19 | 2501 | −56.3 | 0.281134 | −1.8 | 1.9 | 3070 | 0.002170 | 0.8 | 6.55 | 1.14 |
| 19 | 0.282398 | 31 | 0.000467 | 7 | 706 | −13.2 | 0.282392 | 2.1 | 2.2 | 1439 | 0.002171 | 0.4 | 7.07 | 1.09 |
| 22 | 0.282062 | 22 | 0.000425 | 10 | 640 | −25.1 | 0.282057 | −11.2 | 1.6 | 2230 | 0.002172 | 0.6 | 7.42 | 1.11 |
| 25 | 0.282000 | 55 | 0.000362 | 45 | 936 | −27.3 | 0.281993 | −6.9 | 3.9 | 2186 | 0.002164 | 0.8 | 3.75 | 1.13 |
| 26 | 0.281798 | 41 | 0.001226 | 68 | 1018 | −34.4 | 0.281775 | −12.8 | 2.9 | 2622 | 0.002169 | 0.7 | 5.80 | 1.12 |
| 27 | 0.282157 | 25 | 0.000552 | 8 | 506 | −21.8 | 0.282151 | −10.8 | 1.8 | 2102 | 0.002174 | 0.3 | 8.30 | 1.08 |
| 29 | 0.281798 | 42 | 0.000938 | 97 | 1133 | −34.5 | 0.281778 | −10.1 | 2.9 | 2542 | 0.002176 | 0.5 | 9.03 | 1.10 |
| 30 | 0.281503 | 34 | 0.002721 | 58 | 2101 | −44.9 | 0.281395 | −1.8 | 2.4 | 2760 | 0.002169 | 0.4 | 5.83 | 1.09 |
| 31 | 0.282050 | 32 | 0.000810 | 16 | 621 | −25.5 | 0.282041 | −12.2 | 2.3 | 2278 | 0.002171 | 0.8 | 6.75 | 1.13 |
| 32 | 0.282038 | 19 | 0.000388 | 3 | 610 | −26.0 | 0.282033 | −12.7 | 1.4 | 2302 | 0.002170 | 0.9 | 6.62 | 1.16 |
| 33 | 0.282273 | 27 | 0.000556 | 8 | 610 | −17.6 | 0.282267 | −4.4 | 1.9 | 1781 | 0.002161 | 0.6 | 2.26 | 1.11 |
| 35 | 0.281877 | 30 | 0.000928 | 8 | 1006 | −31.7 | 0.281859 | −10.0 | 2.1 | 2441 | 0.002165 | 0.4 | 4.34 | 1.09 |
| 36 | 0.280916 | 21 | 0.000189 | 2 | 2565 | −65.6 | 0.280907 | −8.4 | 1.5 | 3539 | 0.002168 | 0.6 | 5.50 | 1.11 |
| 39 | 0.281956 | 29 | 0.000511 | 4 | 1011 | −28.8 | 0.281947 | −6.8 | 2.0 | 2243 | 0.002173 | 0.6 | 7.87 | 1.11 |
| 41 | 0.282111 | 32 | 0.001006 | 66 | 956 | −23.4 | 0.282093 | −2.9 | 2.3 | 1951 | 0.002171 | 0.7 | 7.13 | 1.12 |
| 48 | 0.281254 | 24 | 0.000258 | 7 | 1981 | −53.3 | 0.281254 | −9.5 | 1.7 | 3158 | 0.002172 | 0.4 | 7.45 | 1.09 |
| 56 | 0.281971 | 20 | 0.000748 | 6 | 686 | −28.3 | 0.281962 | −13.5 | 1.4 | 2413 | 0.002173 | 0.5 | 7.69 | 1.10 |
| 58 | 0.282003 | 25 | 0.000348 | 11 | 692 | −27.2 | 0.281998 | −12.1 | 1.8 | 2329 | 0.002173 | 0.5 | 7.91 | 1.10 |
| 61 | 0.282322 | 28 | 0.000478 | 21 | 642 | −15.9 | 0.282317 | −1.9 | 2.0 | 1649 | 0.002169 | 0.7 | 5.86 | 1.13 |
| 62 | 0.282029 | 24 | 0.000418 | 18 | 625 | −26.3 | 0.282024 | −12.7 | 1.7 | 2312 | 0.002171 | 0.4 | 6.73 | 1.09 |
| 63 | 0.281978 | 30 | 0.000741 | 16 | 1062 | −28.1 | 0.281963 | −5.1 | 2.1 | 2174 | 0.002166 | 1.0 | 4.39 | 1.17 |
| 66 | 0.281049 | 24 | 0.000304 | 1 | 2716 | −60.9 | 0.281034 | −0.4 | 1.7 | 3145 | 0.002168 | 0.6 | 5.51 | 1.11 |

We have used a $^{76}$Lu decay constant of $1.857 \times 10^{-11}\,\text{a}^{-1}$ (Söderlund *et al.* 2004), $^{176}\text{Lu}/^{177}\text{Hf}_{\text{crust,today}} = 0.015$ (Goodge & Vervoort 2006) and present-day CHUR values of $^{176}\text{Lu}/^{177}\text{Hf} = 0.0332$ and $^{176}\text{Hf}/^{177}\text{Hf} = 0.282772$ (Blichert-Toft & Albarède 1997) in all calculations. Note the $\delta^{18}\text{O}$ value for grain 33 at *c.* 2.3 is quite anomalous and so may be considered an analytical artefact/outlier. However, this grain may have crystallized in magma that resulted from melting of hydrothermally altered wall rock. See main text for discussion.

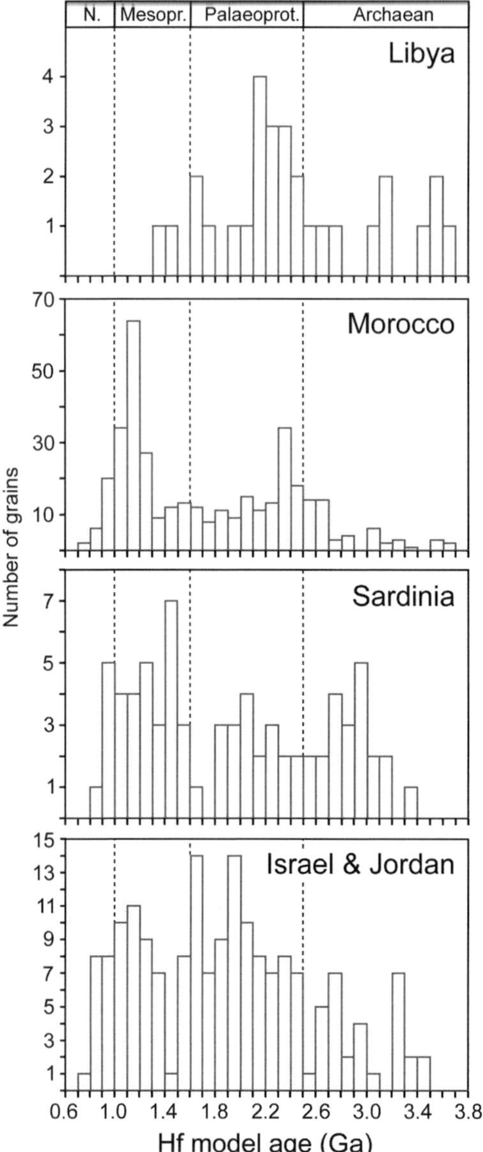



**Fig. 7.** Histograms of calculated Hf model ($T_{DM}Hf$) ages for detrital zircon grains of sample H6032 and for detrital zircon grains from Cambrian–Ordovician sandstones of Morocco, Sardinia, and Israel and Jordan for comparison (bin width = 100 Ma). See Figure 6 for references. N., Neoproterozoic; Mesopr., Mesoproterozoic; Palaeoprot., Palaeoproterozoic.

aged detrital zircon, with long-distance transport proposed for their distribution (e.g. Avigad *et al.* 2003; Kolodner *et al.* 2006; Morag *et al.* 2011).

The origin of *c.* 1.0 Ga detrital zircons in the Palaeozoic succession of southern Libya is controversial (Meinhold *et al.* 2011, 2013). Besides direct sourcing from *c.* 1.0 Ga igneous rocks in eastern Chad and *c.* 1.0 Ga igneous rocks along the southeastern margins of the Congo and Tanzania cratons, recycling of Neoproterozoic sediments is an alternative hypothesis to explain their presence in the Palaeozoic sedimentary sequence of central North Africa. The ubiquitous occurrence of such zircons in the Palaeozoic succession of southern Libya provides insights into the correlation and palaeotectonic arrangement of Gondwana-derived terranes, for example in the Eastern Mediterranean (e.g. Kröner & Şengör 1990; Keay & Lister 2002; Zulauf *et al.* 2007; Meinhold *et al.* 2008, 2010), Sardinia (Avigad *et al.* 2012) and NW Iberia (Fernández-Suárez *et al.* 2002*a*, *b*). Consequently, current palaeotectonic models of dextral terrane transport along the northern Gondwana margin during the early Palaeozoic (Fernández-Suárez *et al.* 2002*a*, *b*) may need to be revised (Bea *et al.* 2010; Meinhold *et al.* 2011, 2013).

### Zircon Lu–Hf–O data

Reworking of crustal material can be tested, for example by Lu–Hf isotope studies on detrital zircons (e.g. Munizaga *et al.* 2008; Morag *et al.* 2011; Avigad *et al.* 2012). Except for two grains, the initial $\varepsilon_{Hf}(t)$ values of detrital zircons of the Devonian sandstone sample H6032 are between −0.41 and −31.45, and initial $T_{DM}Hf$ between 3.7 Ga and 1.6 Ga. No zircons therefore have Hf isotope compositions that approach that of the depleted mantle at the time of crystallization (Fig. 6a; Table 1). The large range in initial $^{176}Hf/^{177}Hf$ values for the various zircon populations (e.g. at 730–550 Ma and *c.* 1 Ga) suggests that the zircons crystallized within a relative narrow time interval from magmas that had heterogeneous Hf isotope compositions. The magmas from which these zircons formed were derived by melting of pre-existing crustal rocks, rather than being juvenile. Calculated $T_{DM}Hf$ ages suggest the zircons are mainly derived from remelting of Archaean and Palaeoproterozoic crust (Fig. 7, Table 1). Focusing on the Archaean $T_{DM}Hf$ ages, reworking of the 3.7–3.5 Ga crust occurred between 2.7 Ga and 2.5 Ga, while reworking of the 3.2–2.7 Ga crust occurred between 2.7 Ga and 1.9 Ga. The *c.* 1.0 Ga zircons are derived from reworked late Neoarchaean and early to middle Palaeoproterozoic crust, as indicated by $T_{DM}Hf$ ages of *c.* 2.6–1.9 Ga. Detrital zircons with ages between *c.* 730 Ma and 500 Ma show a range of $T_{DM}Hf$ ages from 2.4 Ga to 1.3 Ga; the exception is a 557 Ma zircon grain with a $T_{DM}Hf$ age of 3.4 Ga.

According to the $\delta^{18}O$ values (Fig. 6b, Table 1), 5 of 29 grains from sample H6032 are below the lower limit for mantle-derived zircon ($\delta^{18}O$ <5‰). Low $\delta^{18}O$ values (as low as 0%) are known from Phanerozoic subvolcanic granites and rhyolites, the magmas of these rocks having resulted from melting of hydrothermally altered wall rock (Valley 2003; Valley *et al.* 2005). Thus, some of the zircon grains with $\delta^{18}O$ < 5‰ identified in this study probably crystallized in magma that resulted from melting of hydrothermally altered wall rock. All other grains seem to be, within uncertainty, of mantle zircon $\delta^{18}O$ (5.3 ± 0.3‰, Valley *et al.* 2005) or show elevated values (i.e. 6.5–9.5‰). The latter grains indicate supracrustal input and were likely formed from melts that contained a sedimentary component. This derives from the observation that O isotope ratios in igneous rocks are especially sensitive to sedimentary input, as revealed by the elevated $\delta^{18}O$ of S-type granites (O'Neil & Chappell 1977; Appleby *et al.* 2010).

Finally, it is important to point out that the detrital zircon Hf isotope data from the Devonian sandstone sample of Libya are similar to those from Cambrian–Ordovician sandstones of Israel, Jordan, Morocco and Sardinia (Fig. 6a). This suggests that detrital zircon grains from these Palaeozoic sandstones experienced a similar history of crust formation and recycling. The only exceptions seem to be *c.* 2.3–1.8 Ga and *c.* 1.0–0.9 Ga detrital zircons from early and middle Cambrian sandstones of Morocco. Zircons of such ages show a higher proportion of positive $\varepsilon_{Hf}(t)$ values (Fig. 6a). Furthermore, detrital zircons from the Cambrian sandstones of Morocco show a higher proportion of $T_{DM}Hf$ ages around 1 Ga that are absent in the Devonian sandstone sample of southern Libya (Fig. 7). A likely explanation is that the early Palaeozoic succession in Morocco was sourced from a different region of Gondwana than the cover sequences of the Saharan Metacraton and the Arabian–Nubian Shield.

The consortium of subscribing oil and gas companies is thanked for its financial support to the CASP Southern Basins of Libya Project. We thank Y. Abutarruma, F. Said, A. I. Asbali, R. Aburawi, B. Belgasem and B. Thusu for their scientific support and guidance during the project work in Libya. We are also grateful to the logistics team, provided by B. Grenat, for assistance in the field. The administrative and logistical support of staff at the Libyan Petroleum Institute is gratefully acknowledged. The authors would like to thank R. M. Key and J. von Raumer for their helpful comments on the original manuscript.

# Appendix

## *Sample preparation and analytical techniques*

The Devonian sandstone sample H6032 was prepared for heavy mineral separation using standard techniques, as described by Meinhold *et al.* (2011) and Morton *et al.* (2011). The resulting zircon-rich heavy mineral concentrate was then poured on to double-sided tape, mounted in epoxy together with chips of reference zircons (Temora and SL13), sectioned approximately in half, and polished. Reflected and transmitted light photomicrographs and cathodoluminescence (CL) scanning electron microscope (SEM) images were prepared for all zircon grains. The CL images were used to decipher the internal structures of the sectioned grains and to ensure that the analysis spot (*c.* 20 μm size) was wholly within a single age component (usually the youngest) within the sectioned grains.

U–Pb isotope analyses of individual zircon grains were performed using the sensitive high-resolution ion microprobe (SHRIMP RG) at the Research School of Earth Sciences (RSES), The Australian National University, Canberra, Australia. The SHRIMP analytical method follows Williams (1998, and references therein), as outlined by Meinhold *et al.* (2011).

Analyses with 90–110% concordance (calculated from 100 × ($^{206}Pb$–$^{238}U$ age)/($^{207}Pb$/$^{206}Pb$ age)) are considered to be concordant. Unless stated otherwise, $^{206}Pb$–$^{238}U$ ages are used for zircon grains <1.2 Ga, whereas older grains are quoted using their $^{207}Pb$/$^{206}Pb$ ages. This is because the $^{207}Pb$/$^{206}Pb$ ages become increasingly imprecise below <1.2 Ga due to small amounts of $^{207}Pb$. For data presentation, combined probability density distribution–histogram plots were produced using the AgeDisplay program (Sircombe 2004). Unless stated otherwise, ages reported in the text are given at the $1\sigma$ level. The geological timescale of Ogg *et al.* (2008) was used as stratigraphic reference for data interpretation.

Oxygen isotope analyses of individual zircon grains were made using the RSES-ANU SHRIMP II fitted with a Cs source and electron gun for charge compensation, following methods described by Ickert *et al.* (2008). The SHRIMP U–Pb analytical spots, craters *c.* 20 μm in diameter with a depth of 1–2 μm, were polished from the mount surface. The oxygen isotope analyses were then made on exactly the same location used for the U–Pb analyses. Oxygen isotope ratios were determined in multiple collector mode using an axial continuous electron multiplier (CEM) triplet collector, and two floating heads with interchangeable CEM – Faraday Cups. The Temora reference zircon was analysed to monitor and correct for isotope fractionation. The $\delta^{18}O$ ratios have been normalized relative to a Temora weighted mean value of +8.2‰. Reproducibility in the Temora reference zircon $\delta^{18}O$ ratios was ±1.08‰ ($2\sigma$ uncertainty) for the analytical session. The oxygen isotope ratios are expressed in the standard $\delta^{18}O$ notation, signifying deviation of the measured $^{18}O$/$^{16}O$ value from Vienna standard mean ocean water (VSMOW) in parts per thousand. For discussion, these values are plotted as a function of the detrital zircon crystallization ages.

Lu–Hf isotope measurements of individual zircon grains were conducted by laser ablation multicollector

Inductively coupled plasma mass spectroscopy (LA-MC-ICP-MS) using a RSES-ANU Neptune MC-ICP-MS coupled with a 193 nm ArF Excimer laser, similar to procedures described in Munizaga *et al.* (2008). Samples were photographed under a reflected light microscope, following the oxygen isotope analyses by SHRIMP, to reveal the location on each grain of the ion probe sputter pit. Laser ablation analyses were performed on the same locations. For all analyses of unknowns or secondary standards, the laser spot size on the sample had a diameter of 47 μm. The mass spectrometer was first tuned to optimal sensitivity using a large grain of zircon from the Monastery kimberlite. The masses 171, 173, 174, 175, 176, 177, 178, 179 and 181 were measured simultaneously in static collection mode. A gas blank was acquired at regular intervals throughout the analytical session (approximately every 10 analyses). The laser was fired with a typical repetition rate of 5–8 Hz and energy of 60 mJ. Data were acquired over 100 s, but in many cases only a selected interval from the total acquisition was used in data reduction. In each analytical session between gas blank measurements, several widely used reference zircons (91500, FC-1, Temora-2, Monastery and Mud Tank) were measured as a check on data quality, as well as several unknowns. Signal intensity was typically *c.* 5–6 V for total Hf at the beginning of ablation, and decreased over the acquisition time to 2 V or less. Isobaric interferences of $^{176}$Lu and $^{176}$Yb on the $^{176}$Hf signal were corrected by monitoring signal intensities of $^{175}$Lu and $^{173}$Yb, $^{172}$Yb and $^{171}$Yb. The calculation of signal intensity for $^{176}$Hf also involved independent mass bias corrections for Lu and Yb.

# References


ABDELSALAM, M. G., LIEGEOIS, J.-P. & STERN, R. J. 2002. The Saharan Metacraton. *Journal of African Earth Sciences*, **34**, 119–136.

ABDELSALAM, M. G., GAO, S. S. & LIEGEOIS, J.-P. 2011. Upper mantle structure of the Saharan Metacraton. *Journal of African Earth Sciences*, **60**, 328–336.

AJDUKIEWICZ, J. M. & LANDER, R. H. 2010. Sandstone reservoir quality prediction: the state of the art. *American Association of Petroleum Geologists Bulletin*, **94**, 1083–1091.

APPLEBY, S. K., GILLESPIE, M. R., GRAHAM, C. M., HINTON, R. W., OLIVER, G. J. H. & KELLY, N. M. 2010. Do S-type granites commonly sample infracrustal sources? New results from an integrated O, U–Pb and Hf isotope study of zircon. *Contributions of Mineralogy and Petrology*, **160**, 115–132.

AVIGAD, D., KOLODNER, K., MCWILLIAMS, M., PERSING, H. & WEISSBROD, T. 2003. Origin of northern Gondwana Cambrian sandstone revealed by detrital zircon SHRIMP dating. *Geology*, **31**, 227–230.

AVIGAD, D., GERDES, A., MORAG, N. & BECHSTÄDT, T. 2012. Coupled U–Pb–Hf of detrital zircons of Cambrian sandstones from Morocco and Sardinia: implications for provenance and Precambrian crustal evolution of North Africa. *Gondwana Research*, **21**, 690–703.

BAUMANN, A., EL CHAIR, M. & THIEDIG, F. 1992. Panafrican granites from deep wells of the Murzuk Basin (Fezzan), western Libya. *Neues Jahrbuch für Geologie und Paläontologie Monatshefte*, **1**, 1–14.

BEA, F., MONTERO, P., TALAVERA, C., ABU ANBAR, M., SCARROW, J. H., MOLINA, J. F. & MORENO, J. A. 2010. The palaeogeographic position of Central Iberia in Gondwana during the Ordovician: evidence from zircon chronology and Nd isotopes. *Terra Nova*, **22**, 341–346.

BEGG, G. C., GRIFFIN, W. L. *ET AL.* 2009. The lithospheric architecture of Africa: seismic tomography, mantle petrology, and tectonic evolution. *Geosphere*, **5**, 23–50.

BLICHERT-TOFT, J. & ALBARÈDE, F. 1997. The Lu–Hf isotope geochemistry of chondrites and the evolution of the mantle–crust system. *Earth and Planetary Science Letters*, **148**, 243–258.

BOOTE, D. R., CLARK-LOWES, D. D. & TRAUT, M. W. 1998. Palaeozoic petroleum systems of North Africa. *In*: MACGREGOR, D. S., MOODY, R. T. J. & CLARK-LOWES, D. D. (eds) *Petroleum Geology of North Africa*. Geological Society, London, Special Publications, **132**, 7–68.

BURKE, K., MACGREGOR, D. S. & CAMERON, N. R. 2003. Africa's petroleum systems: four tectonic 'Aces' in the past 600 million years. *In*: ARTHUR, T. J., MACGREGOR, D. S. & CAMERON, N. R. (eds) *Petroleum Geology of Africa: New Themes and Developing Technologies*. Geological Society, London, Special Publications, **207**, 21–60.

CHOUKROUN, M., O'REILLY, S. Y., GRIFFIN, W. L., PEARSON, N. J. & DAWSON, J. B. 2005. Hf isotopes of MARID (mica–amphibole–rutile–ilmenite–diopside) rutile trace metasomatic processes in the lithospheric mantle. *Geology*, **33**, 45–48.

COLLINS, A. S. & PISAREVSKY, S. A. 2005. Amalgamating eastern Gondwana: the evolution of the Circum-Indian Orogens. *Earth-Science Reviews*, **71**, 229–270.

COLLOMB, G. R. 1962. *Carte géologique du Fezzan Oriental, 1:500 000*. Compagnie Française des Pétroles, Paris.

CRAIG, J. L., RIZZI, C. *ET AL.* 2008. Structural styles and prospectivity in the Precambrian and Palaeozoic hydrocarbon systems of North Africa. *In*: SALEM, M. J., OUN, K. M. & ESSED, A. S. (eds) *The Geology of East Libya*. Gutenberg Press, Malta, **4**, 51–122.

DAVIDSON, L., BESWETHERICK, S. *ET AL.* 2000. The structure, stratigraphy and petroleum geology of the Murzuq Basin, southwest Libya. *In*: SOLA, M. A. & WORSLEY, D. (eds) *Geological Exploration in Murzuq Basin*. Elsevier, Amsterdam, 295–320.

DE WIT, M. J., BOWRING, S. A., DUDAS, F. & KAMGA, G. 2005. The great Neoproterozoic central Saharan Arc and the amalgamation of the North African Shield. *GAC–MAC–CSPG–CSSS Joint Meeting, Halifax, Nova Scotia, Abstracts*, **30**, 42–43.

ECHIKH, K. & SOLA, M. A. 2000. Geology and hydrocarbon occurences in the Murzuq Basin, SW Libya. *In*: SOLA, M. A. & WORSLEY, D. (eds) *Geological Exploration in Murzuq Basin*. Elsevier, Amsterdam, 175–222.

EILER, J. M. 2001. Oxygen isotope variations of basaltic lavas and upper mantle rocks. *In*: VALLEY, J. W. &

COLE, D. R. (eds) *Stable Isotope Geochemistry*. Reviews in Mineralogy and Geochemistry, Mineralogical Society of America, Washington, DC, **43**, 319–364.

FERNÁNDEZ-SUÁREZ, J., GUTIÉRREZ ALONSO, G., COX, R. & JENNER, G. A. 2002*a*. Assembly of the Armorican microplate: a strike–slip terrane delivery? Evidence from U–Pb ages of detrital zircons. *Journal of Geology*, **110**, 619–626.

FERNÁNDEZ-SUÁREZ, J., GUTIÉRREZ-ALONSO, G. & JEFFRIES, T. E. 2002*b*. The importance of along-margin terrane transport in northern Gondwana: insights from detrital zircon parentage in Neoproterozoic rocks from Iberia and Brittany. *Earth and Planetary Science Letters*, **204**, 75–88.

GHUMA, M. A. & ROGERS, J. J. W. 1978. Geology, geochemistry, and tectonic setting of the Ben Ghnema batholith, Tibesti massif, southern Libya. *Geological Society of America Bulletin*, **89**, 1351–1358.

GOODGE, J. & VERVOORT, J. D. 2006. Origin of Mesoproterozoic A-type granites in Laurentia: Hf isotope evidence. *Earth and Planetary Science Letters*, **243**, 711–731.

HALLETT, D. 2002. *Petroleum Geology of Libya*. 1st edn. Elsevier, Amsterdam, 1–508.

ICKERT, R. B., HIESS, J. *ET AL*. 2008. Determining high precision, *in situ*, oxygen isotope ratios with a SHRIMP II: analyses of MPI-DING silicate–glass reference materials and zircon from contrasting granites. *Chemical Geology*, **257**, 114–128.

INGERSOLL, R. V., BULLARD, T. F., FORD, R. L., GRIMM, J. P., PICKLE, J. P. & SARES, S. W. 1984. The effect of grain size on detrital modes: a test of the Gazzi–Dickinson Point Counting method. *Journal of Sedimentary Petrology*, **54**, 103–116.

JOHNSON, S. P., CUTTEN, H. N. C., MUHONGO, S. & DE WAELE, B. 2003. Neoarchaean magmatism and metamorphism of the western granulites in the central domain of the Mozambique Belt, Tanzania: U Pb shrimp geochronology and *PT* estimates. *Tectonophysics*, **375**, 125–145.

JOHNSON, P. R., ANDRESEN, A. *ET AL*. 2011. Late Cryogenian–Ediacaran history of the Arabian–Nubian Shield: a review of depositional, plutonic, structural, and tectonic events in the closing stages of the northern East African Orogen. *Journal of African Earth Sciences*, **61**, 167–232.

KEAY, S. & LISTER, G. S. 2002. African provenance for the metasediments and metaigneous rocks of the Cyclades, Aegean Sea. *Geology*, **30**, 235–238.

KEY, R. M., LOUGHLIN, S. C. *ET AL*. 2008. Two Mesoarchaean terranes in the Reguibat Shield of NW Mauritania. *In*: ENNIH, N. & LIEGEOIS, J.-P. (eds) *The Boundaries of the West African Craton*. Geological Society, London, Special Publications, **297**, 33–52.

KLITZSCH, E. 1964. Zur Geologie am Ostrand des Murzukbeckens (Provinz Fezzan, Libyen). *Oberrheinische Geologische Abhandlungen*, **13**, 51–73.

KLITZSCH, E. & ZIEGERT, H. 2000. *Short Notes and Guidebook on the Geology of the Dor el Gussa – Jabal Bin Ghanimah area*. Sedimentary Basins of Libya, Second Symposium, Geology of Northwest Libya. Gutenberg Press, Malta, 1–52.

KOLODNER, K., AVIGAD, D., MCWILLIAMS, M., WOODEN, J. L., WEISSBROD, T. & FEINSTEIN, S. 2006. Provenance of north Gondwana Cambrian–Ordovician sandstone: U–Pb SHRIMP dating of detrital zircons from Israel and Jordan. *Geological Magazine*, **143**, 367–391.

KRÖNER, A. & ŞENGÖR, A. M. C. 1990. Archean and Proterozoic ancestry in late Precambrian and early Paleozoic crustal elements in southern Turkey as revealed by single zircon dating. *Geology*, **18**, 1186–1190.

KRÖNER, A. & STERN, R. J. 2004. Africa: Pan-African Orogeny. *In*: SELLEY, R. C., COCKS, L. R. M. & PLIMER, I. R. (eds) *Encyclopedia of Geology*. Elsevier, Amsterdam, **1**, 1–12.

KRÖNER, A., STERN, R. J., DAWOUD, A. S., COMPSTON, W. & REISCHMANN, T. 1987. The Pan-African continental margin in northeastern Africa: evidence from a geochronological study of granulites at Sabaloka, Sudan. *Earth and Planetary Science Letters*, **85**, 91–104.

KÜSTER, D., LIÉGEOIS, J.-P., MATUKOV, D., SERGEEV, S. & LUCASSEN, F. 2008. Zircon geochronology and Sr, Nd, Pb isotope geochemistry of granitoids from Bayuda Desert and Sabaloka (Sudan): evidence for a Bayudian event (920–900 Ma) preceding the Pan-African orogenic cycle (860–590 Ma) at the eastern boundary of the Saharan Metacraton. *Precambrian Research*, **164**, 16–39.

LE HERON, D. P., MEINHOLD, G. & BERGIG, K. 2013. Neoproterozoic–Devonian stratigraphic evolution of the eastern Murzuq Basin, Libya: a tale of tilting in the central Sahara. *Basin Research*, **25**, 52–73, http://dx.doi.org/10.1111/j.1365-2117.2012.00555.x

LEJAL-NICOL, A. & MASSA, D. 1980. Sur des vegetaux du Devonien inferieur de la Libye. *Review of Palaeobotany and Palynology*, **29**, 221–239.

LINK, K., KOEHN, D., BARTH, M. G., TIBERINDWA, J. V., BARIFAIJO, E., AANYU, K. & FOLEY, S. F. 2010. Continuous cratonic crust between the Congo and Tanzania blocks in western Uganda. *International Journal of Earth Sciences*, **99**, 1559–1573.

MACGREGOR, D. S. 1996. The hydrocarbon systems of North Africa. *Marine and Petroleum Geology*, **13**, 329–340.

MEINHOLD, G., REISCHMANN, T., KOSTOPOULOS, D., LEHNERT, O., MATUKOV, D. & SERGEEV, S. 2008. Provenance of sediments during subduction of Palaeotethys: detrital zircon ages and olistolith analysis in Palaeozoic sediments from Chios Island, Greece. *Palaeogeography, Palaeoclimatology, Palaeoecology*, **263**, 71–91.

MEINHOLD, G., BERRY, C. *ET AL*. 2009. *Stratigraphic constraints of plant fossils on Palaeozoic strata of Dor el Gussa, southern Libya*. Murzuq Basin Project, Unpublished CASP Report **11**, Cambridge, 1–37.

MEINHOLD, G., KOSTOPOULOS, D., FREI, D., HIMMERKUS, F. & REISCHMANN, T. 2010. U–Pb LA–SF–ICP–MS zircon geochronology of the Serbo Macedonian Massif, Greece: palaeotectonic constraints for Gondwana-derived terranes in the Eastern Mediterranean. *International Journal of Earth Sciences*, **99**, 813–832.

MEINHOLD, G., MORTON, A. C. *ET AL*. 2011. Evidence from detrital zircons for recycling of Mesoproterozoic and Neoproterozoic crust recorded in Paleozoic and

Mesozoic sandstones of southern Libya. *Earth and Planetary Science Letters*, **312**, 164–175.

MEINHOLD, G., MORTON, A. C. & AVIGAD, D. 2013. New insights into peri-Gondwana paleogeography and the Gondwana super-fan system from detrital zircon U–Pb ages. *Gondwana Research*, **23**, 661–665, http://dx.doi.org/10.1016/j.gr.2012.05.003

MORAG, N., AVIGAD, D., GERDES, A., BELOUSOVA, E. & HARLAVAN, Y. 2011. Detrital zircon Hf isotopic composition indicates long-distance transport of North Gondwana Cambrian–Ordovician sandstones. *Geology*, **39**, 955–958.

MORTON, A. C., HALLSWORTH, C. R., KUNKA, J., LAWS, E., PAYNE, S. N. J. & WALDER, D. 2010. Heavy-mineral stratigraphy of the Clair Group (Devonian–Carboniferous) in the Clair Field, West of Shetland, U.K. *In*: RATCLIFFE, K. T. & ZAITLIN, B. A. (eds) *Application of Modern Stratigraphic Techniques: Theory and Case Histories*. SEPM Special Publications, Tulsa, **94**, 183–199.

MORTON, A. C., MEINHOLD, G. *ET AL.* 2011. A heavy mineral study of sandstones from the eastern Murzuq Basin, Libya: constraints on provenance and stratigraphic correlation. *Journal of African Earth Sciences*, **61**, 308–330.

MUNIZAGA, F., MAKSAEV, V., FANNING, C. M., GIGLIO, S., YAXLEY, G. & TASSINARI, C. C. G. 2008. Late Paleozoic–Early Triassic magmatism on the western margin of Gondwana: Collahuasi area, Northern Chile. *Gondwana Research*, **13**, 407–427.

OGG, J. G., OGG, G. & GRADSTEIN, F. M. 2008. *The Concise Geologic Time Scale*. Cambridge University Press, Cambridge.

O'NEIL, J. R. & CHAPPELL, B. W. 1977. Oxygen and hydrogen isotope relations in the Berridale Batholith. *Journal of the Geological Society, London*, **133**, 559–571.

PARIS, F., THUSU, B. *ET AL.* 2012. Palynological and palynofacies analysis of early Silurian shales from borehole CDEG-2a in Dor el Gussa, eastern Murzuq Basin, Libya. *Review of Palaeobotany and Palynology*, **174**, 1–26.

PATCHETT, P. J., KOUVO, O., HEDGE, C. E. & TATSUMOTO, M. 1981. Evolution of continental crust and mantle heterogeneity: evidence from Hf isotopes. *Contribution to Mineralogy and Petrology*, **78**, 279–297.

PEGRAM, B. J., REGISTER, J. K., FULLAGAR, P. D., GHUMA, M. A. & ROGERS, J. J. W. 1976. Pan-African ages from a Tibesti Massif batholith, southern Libya. *Earth and Planetary Science Letters*, **30**, 123–128.

RAMM, M. 2000. Reservoir quality and its relationship to facies and provenance in Middle to Upper Jurassic sequences, northeastern North Sea. *Clay Minerals*, **35**, 77–94.

SIRCOMBE, K. N. 2004. AgeDisplay: an EXCEL workbook to evaluate and display univariate geochronological data using binned frequency histograms and probability density distributions. *Computers and Geosciences*, **30**, 21–31.

SÖDERLUND, U., PATCHETT, P. J., VERVOORT, J. D. & ISACHSEN, C. E. 2004. The $^{176}$Lu decay constant determined by Lu–Hf and U–Pb isotope systematics of Precambrian mafic intrusions. *Earth and Planetary Science Letters*, **219**, 311–324.

STERN, R. J. 1994. Arc assembly and continental collision in the Neoproterozoic East African Orogen: implications for the assembly of Gondwanaland. *Annual Review of Earth and Planetary Sciences*, **22**, 319–351.

VALLEY, J. W. 2003. Oxygen isotopes in zircon. *In*: HANCHAR, J. M. & HOSKIN, P. W. O. (eds) *Zircon*. Reviews in Mineralogy and Geochemistry, Mineralogical Society of America, Washington, DC, **53**, 343–385.

VALLEY, J. W., LACKEY, J. S. *ET AL.* 2005. 4.4 billion years of crustal maturation: oxygen isotope ratios of magmatic zircon. *Contribution to Mineralogy and Petrology*, **150**, 561–580.

VERVOORT, J. & BLICHERT-TOFT, J. 1999. Evolution of the depleted mantle: Hf isotope evidence from juvenile rocks through time. *Geochimica et Cosmochimica Acta*, **63**, 533–556.

WILLIAMS, I. S. 1998. U–Th–Pb geochronology by ion microprobe. *In*: MCKIBBEN, M. A., SHANKS, W. C. III & RIDLEY, W. I. (eds) *Applications of Microanalytical Techniques to Understanding Mineralising Processes*. Reviews in Economic Geology, Littleton, Colorado, **7**, 1–35.

ZULAUF, G., ROMANO, S. S., DÖRR, W. & FIALA, J. 2007. Crete and the Minoan terranes: age constraints from U–Pb dating of detrital zircons. *In*: LINNEMANN, U., NANCE, R. D., KRAFT, P. & ZULAUF, G. (eds) *The Evolution of the Rheic Ocean: From Avalonian–Cadomian Active Margin to Alleghenian–Variscan Collision*. Geological Society of America Special Papers, **423**, 401–411.

# Detrital monazite geochronology, Upper Jurassic–Lower Cretaceous of the Scotian Basin: significance for tracking first-cycle sources

GEORGIA PE-PIPER[1]*, DAVID J. W. PIPER[2] & S. TRIANTAFYLLIDIS[1]

[1]*Department of Geology, Saint Mary's University, Halifax, Nova Scotia, B3H 3C3, Canada*

[2]*Geological Survey of Canada (Atlantic), Bedford Institute of Oceanography, P.O. Box 1006, Dartmouth, Nova Scotia, B2Y 4A2, Canada*

**Corresponding author (e-mail: gpiper@smu.ca)*

**Abstract:** Monazite geochronology was applied to an east–west transect of latest Jurassic and Lower Cretaceous deltaic sandstones of the Scotian Basin, to assess sediment sources and dispersal pathways. More than 200 detrital monazite grains yielded 694 electron microprobe age determinations with $1\sigma$ errors $<\pm20\%$. Based on age, external morphology, zoning, inclusions and major element chemistry (rare earth element [REE], Th, Y), monazite grains represent more than 20 discrete sources. Similar proportions of euhedral and subhedral compared with irregular and rounded monazite grains in most age classes, together with comparison with detrital muscovite and zircon geochronology, suggest that most monazite is first cycle. Six types of REE distribution are recognized (A–F). Many igneous monazites show chemical zoning, contain sparse euhedral inclusions, and have REE distributions of types A and E. Many metamorphic monazites contain inclusions, commonly aligned, are generally rounded–subhedral to rounded, and have REE distributions of types B, C and D. Monazite geochronology shows important supply to the Scotian Basin from the Labrador rift shoulder as early as Tithonian; from Avalonian sources in the Tithonian; from Ordovician sources in northern New Brunswick, apparently via the Chaswood River; and from the inner continental shelf, particularly in the Hauterivian–Barremian.

The Scotian Basin (Fig. 1) is a Mesozoic–Cenozoic passive-margin basin that is producing natural gas hosted by uppermost Jurassic and Lower Cretaceous deltaic sandstones (Wade & MacLean 1990). The basin is located near the northeastern limit of the central Atlantic Ocean that rifted in the Triassic and earliest Jurassic (Withjack *et al.* 2009). Break-up and the start of sea-floor spreading off the Scotian Basin was probably in the late Toarcian (Sibuet *et al.* 2012). This early segment of the Atlantic Ocean terminated against the Newfoundland–Gibraltar Transform along the SW Grand Banks transform margin. Until the latest Jurassic, the development of the continental margin was that of a typical sub-tropical passive margin, with abundant Jurassic shelf carbonate rocks (Wade & MacLean 1990). A striking change took place in the Late Jurassic–Early Cretaceous: sandy deltaic successions several kilometres in thickness were deposited on the shelf, implying a 3–4-fold increase in terrigenous sediment supply. This enhanced sand supply may have been related to a more humid climate (Valdes *et al.* 1996), as North America moved northwestward during the later Jurassic and early Cretaceous (Beck & Housen 2003). It is also likely that rift-related tectonics played a role in enhanced sediment supply: rifting of the Grand Banks from Iberia began in the Tithonian (Tucholke *et al.*

2007) and of Labrador from Greenland in the Valanginian (Dickie *et al.* 2011). Provenance studies using detrital petrology provide a method of discriminating between these various hypotheses.

Previous provenance studies of the Scotian Basin based on whole-rock geochemistry (Pe-Piper *et al.* 2008) and heavy minerals (Tsikouras *et al.* 2011) have shown that the western, central and eastern parts of the basin received sediment from different river basins. A large proportion of the sediment supply in the central part of the basin is interpreted to be polycyclic, reworked from Palaeozoic sandstones, but with less polycyclic reworking in the eastern part of the basin (Tsikouras *et al.* 2011). Geochronology of detrital muscovite (Reynolds *et al.* 2010, 2012) shows important supply of muscovite from the outboard Meguma Terrane close to the basin margin (Fig. 1a). In contrast, detrital zircon (Piper *et al.* 2012) indicates important to predominant supply of zircon from the Canadian Shield, although a high proportion of the zircon appears to be polycyclic.

In this study, we use geochronology of detrital monazite as a tool for resolving some of the apparently contradictory provenance information in the Scotian Basin that is a result of the large amount of polycyclic sediment supply. Because of its hardness, monazite can be transported long distances

*From*: SCOTT, R. A., SMYTH, H. R., MORTON, A. C. & RICHARDSON, N. (eds) 2014. *Sediment Provenance Studies in Hydrocarbon Exploration and Production.* Geological Society, London, Special Publications, **386**, 293–311.
First published online July 15, 2013, http://dx.doi.org/10.1144/SP386.13

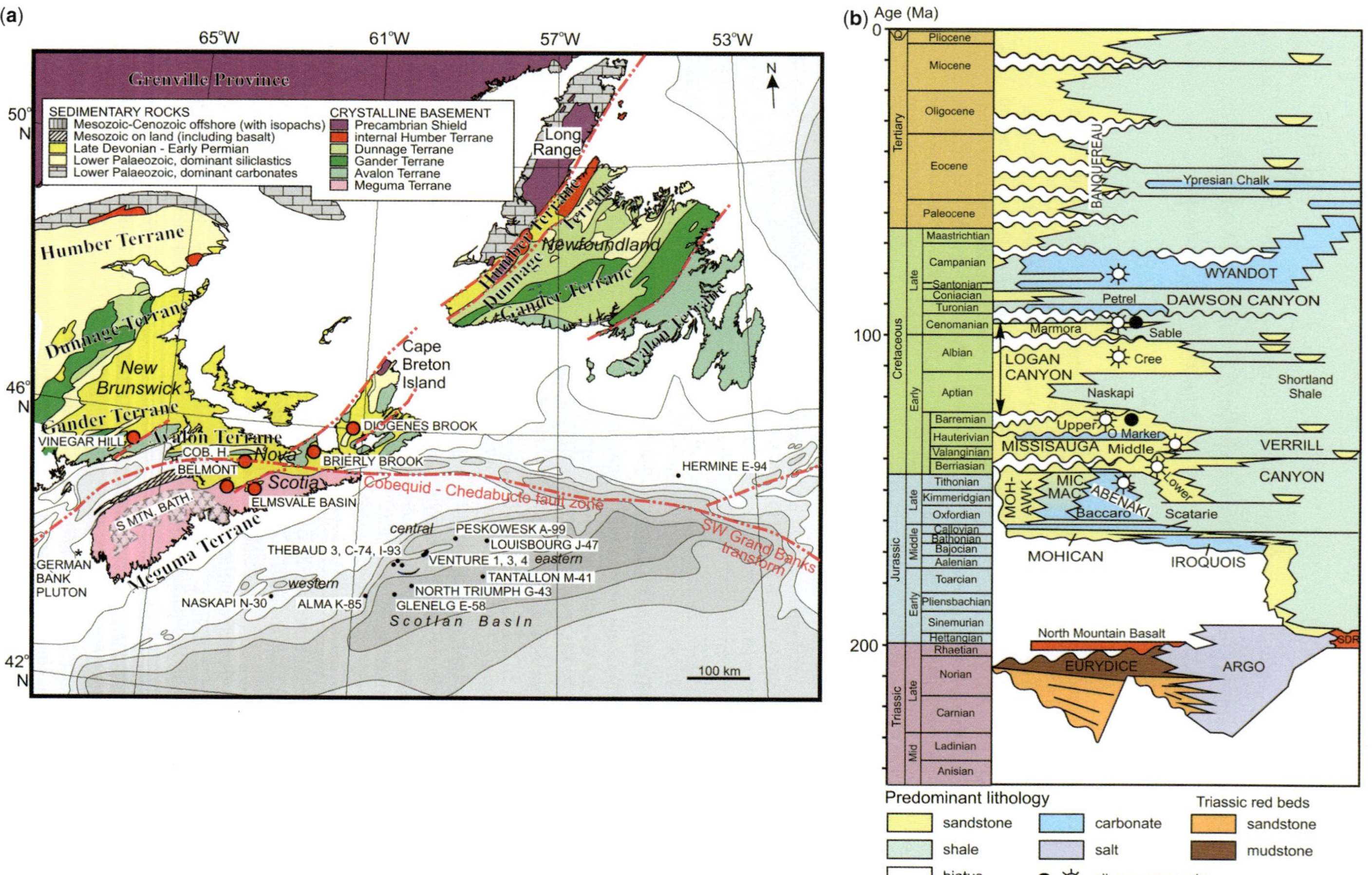


**Fig. 1.** (**a**) Location map showing isopachs of the Scotian Basin (in 4 km increments), sampled wells, distribution of Chaswood Formation on land (red dots), and regional bedrock geology (after Williams & Grant 1998). COB. H., Cobequid Highlands; S MTN. BATH, South Mountain Batholith. (**b**) Summary of stratigraphy of the Scotian Basin (modified from OETR 2011).

by rivers. On the other hand, it is susceptible to dissolution under acidic conditions (Polyakov *et al.* 2010), and thus second cycle monazite is rare (Ross *et al.* 2005). We use morphology of monazite grains and comparison of age spectra with muscovite and zircon to demonstrate that most dated monazite is first cycle. We then use the monazite geochronology to assess the source areas for the Scotian Basin and the role of climate change and tectonics in creating abundant sand supply.

## Geological setting

The Appalachian Orogen forms the basement to the Scotian Basin and on land is exposed in a 500 km-wide belt parallel to the length of the basin (Fig. 1). It comprises a series of distinct terranes that were deformed and assembled through the latest Neoproterozoic and Palaeozoic (van Staal 2005). Of these terranes, the Meguma Terrane is the most outboard; it is exposed in southern Nova Scotia and on the Scotian Shelf. It consists principally of Neoproterozoic–Ordovician metasedimentary rocks, metamorphosed in the early Devonian and intruded by mid-Devonian peraluminous granites. Successive inboard terranes are termed Avalon, Gander, Dunnage and Humber, comprising a variety of upper Neoproterozoic and Lower Palaeozoic metasedimentary and metavolcanic rocks, intruded by plutonic rocks that range in age from late Neoproterozoic to Devonian. The Humber Terrane was thrust over the cratonic margin of the Canadian Shield. The medium to high grade metamorphic and plutonic rocks of this part of the shield belong to the Mesoproterozoic Grenville Province. The Gulf of St Lawrence and adjacent land areas host the thick terrestrial sedimentary rocks of the Maritimes Basin, of late Devonian to early Permian age. Most of the contacts between major rock types in the Appalachians are steep, and apatite fission track data suggest no more than 1 km of denudation since the Jurassic (Hendriks *et al.* 1993; Grist & Zentilli 2003). Taken together, this implies that the modern distribution of rock types (Fig. 1) is a reasonable proxy for Early Cretaceous sources of detrital sediment. The principal difference is that Carboniferous cover rocks may have been more widespread in the Early Cretaceous (Li *et al.* 2012).

The uppermost Jurassic–Lower Cretaceous sedimentary rocks of the offshore Scotian Basin (Fig. 1b) comprise fluvial, deltaic and shelf sediments of the Missisauga and Logan Canyon formations (Wade & MacLean 1990). The sandstone-rich Missisauga Formation is of Kimmeridgian to Barremian age (Williams *et al.* 1990) and passes seaward into the shales of the Verrill Canyon Formation. The Missisauga Formation is divided into three members: Upper, Middle and Lower, with the Upper Jurassic Lower Member (for convenience hereafter, Lower Missisauga Formation) passing laterally into mixed carbonate and clastic rocks of the Mic Mac Formation. The Aptian to early Cenomanian Logan Canyon Formation is also predominantly deltaic, comprising two shale units (Naskapi and Sable members), separated by two sandier units (Cree and Marmora members). The equivalent Valanginian–Albian fluvial rocks on land are the Chaswood Formation (Stea & Pullan 2001; Falcon-Lang *et al.* 2007), known from boreholes and sparse outcrops in a few small outliers in Nova Scotia and New Brunswick (Fig. 1). Previous monazite geochronology from the Chaswood Formation (Pe-Piper & MacKay 2006) shows a predominant source from Lower Palaeozoic rocks of the inboard terranes of the northern Appalachians.

## Materials and methods

Sandstones were sampled from conventional core from 14 wells representing an east to west transect of the Scotian Basin and additionally from cuttings samples from the Hermine E-94 well (Figs 1a & 2). Collectively, these samples provide good stratigraphic coverage of the Logan Canyon and Missisauga Formations in the central and eastern parts of the basin. Monazite grains were identified from polished thin sections either of sandstone or of heavy mineral separates. As a result, systematic information on monazite concentrations is lacking. All thin sections were polished using loose diamond powders on cloth-covered aluminum laps, eliminating the chance of Pb contamination.

Chemical analyses for U–Pb age dating and chemical characterization of monazite grains was carried out at the Regional Electron Microprobe Centre, Dalhousie University, using a JEOL 8200 electron microprobe equipped with five wavelength spectrometers and a 131 eV Noran Energy Dispersive Detector. The Energy Dispersive Detector was used in the initial location and identification of the monazite grains. For analysis of the major elements (La, Ce, Pr, Nd, Sm, Gd, Dy, Y, Th, U, P, Si, Ca) to chemically characterize the monazite grains, the probe was operated with an acceleration voltage of 15 kV and probe current of 20 nA using a focused beam. A counting time of 20 s on the peaks was used with a background time of 10 s. For geochronological analysis, only the elements U, Pb, Th and Y were determined, operating the probe at 15 kV and 200 nA probe current. Peak counting time was 360 s and background counting time 180 s. Full analytical data

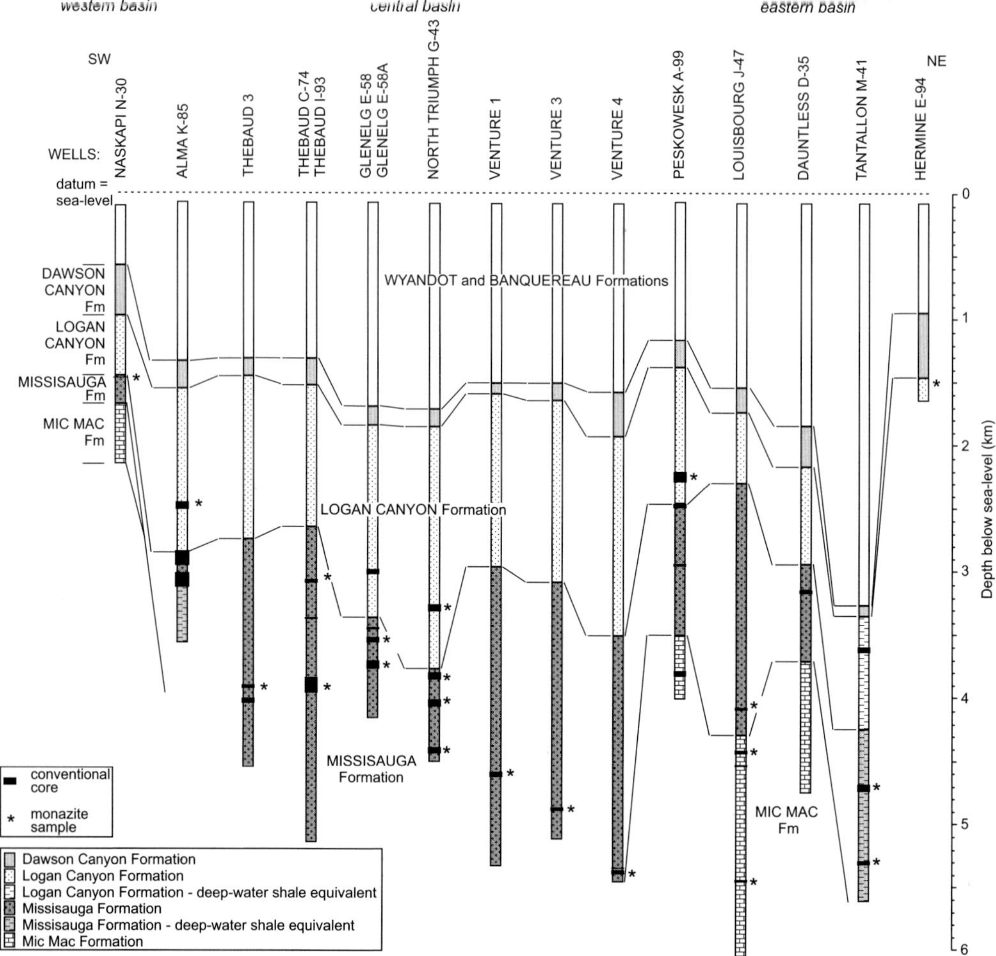



**Fig. 2.** Stratigraphic sections in a west–east transect showing location of samples. No horizontal depiction of scale. Stratigraphic picks from MacLean & Wade (1993).

are available in an open-file report (Triantafyllidis *et al.* 2010). A control monazite sample GSC-8153 was analysed prior to each batch of trace element analyses. Thirteen dates from this monazite yielded an average age of 506 Ma with a standard deviation of 17.13 Ma (3.39%) and a $1\sigma$ standard error of 37.9 Ma (7.51%).

Quantification of errors in monazite chemical analysis is based on counting statistics. This arises from both the nature of X-rays and the functioning of an electron microprobe (Pyle *et al.* 2005). Errors are principally due to analysis of U and Pb and if these elements are present in very low concentrations, the errors are much higher than those for the GSC-8153 control monazite, regardless of the calculated age. A total of 997 trace element analyses were performed on 324 grains. After

propagation of errors, analyses with a $1\sigma$ error larger than $\pm20\%$ were excluded. In addition, a few grains showing signs of dissolution that yielded unreasonably old dates (>3.5 Ga), were excluded. The final dataset used for geochronology studies is based on 694 spot analyses. Probability plots combined with histograms (Fig. 3) were created using Isoplot (Ludwig 1999). Precision and accuracy are difficult to determine with samples of unknown ages.

The average age of several spot analyses in each individual monazite grain was determined. All spot dates with $1\sigma$ errors better than $\pm20\%$ were accepted and the mean age determined, omitting any dates that appeared to be outliers. The outliers may represent inherited cores of monazite, or rims or patches that experienced late hydrothermal

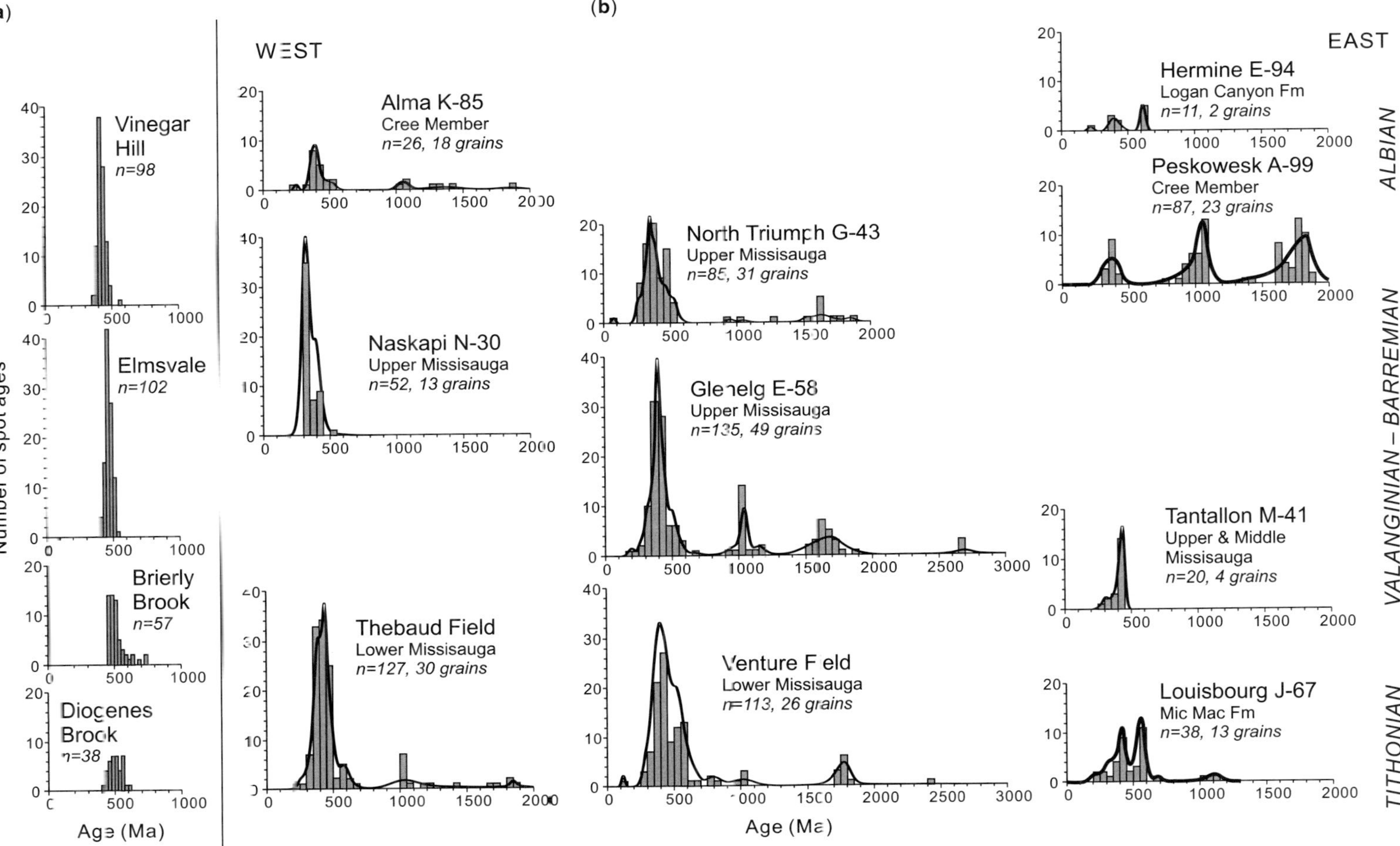


**Fig. 3.** Histograms and probability plots of all spot dates on monazite with 1σ error better than 20%. Plots are arranged schematically by stratigraphy and geography. (**a**) Chaswood Formation, from Pe-Piper & Mackay (2006) in 25 Ma bins and no probability information; (**b**) new data from the Scotian Basin in 50 Ma bins with probability density.

resetting or imprecise ages due to leaching of U. The mean grain ages are shown in histograms binned at 50 Ma (Fig. 4). Grains in which the average age is based on three or more age determinations make up 58% of the dataset, whereas 26% of the grains have only a single age determination better than $\pm 20\%$. No robust error analysis is possible, so that probabilities are not displayed.

Secondary electron (SE) or backscattered electron (BSE) images were available for most grains and were used to classify morphology, textures and chemical zoning and to identify the presence of inclusions and pits. Inclusions in selected grains were investigated by scanning electron microscope and semi-quantitative chemical analyses of inclusions were made by energy dispersive spectroscopy.

## Results: morphology, inclusions and chemistry

### *Morphology*

Many analysed grains are euhedral to subhedral, with some planar faces meeting at sharp corners (e.g. Fig. 5a, b). In other cases, such corners appear abraded, but the planar faces are preserved (e.g. Fig. 5c–f). All analysed grains were classified

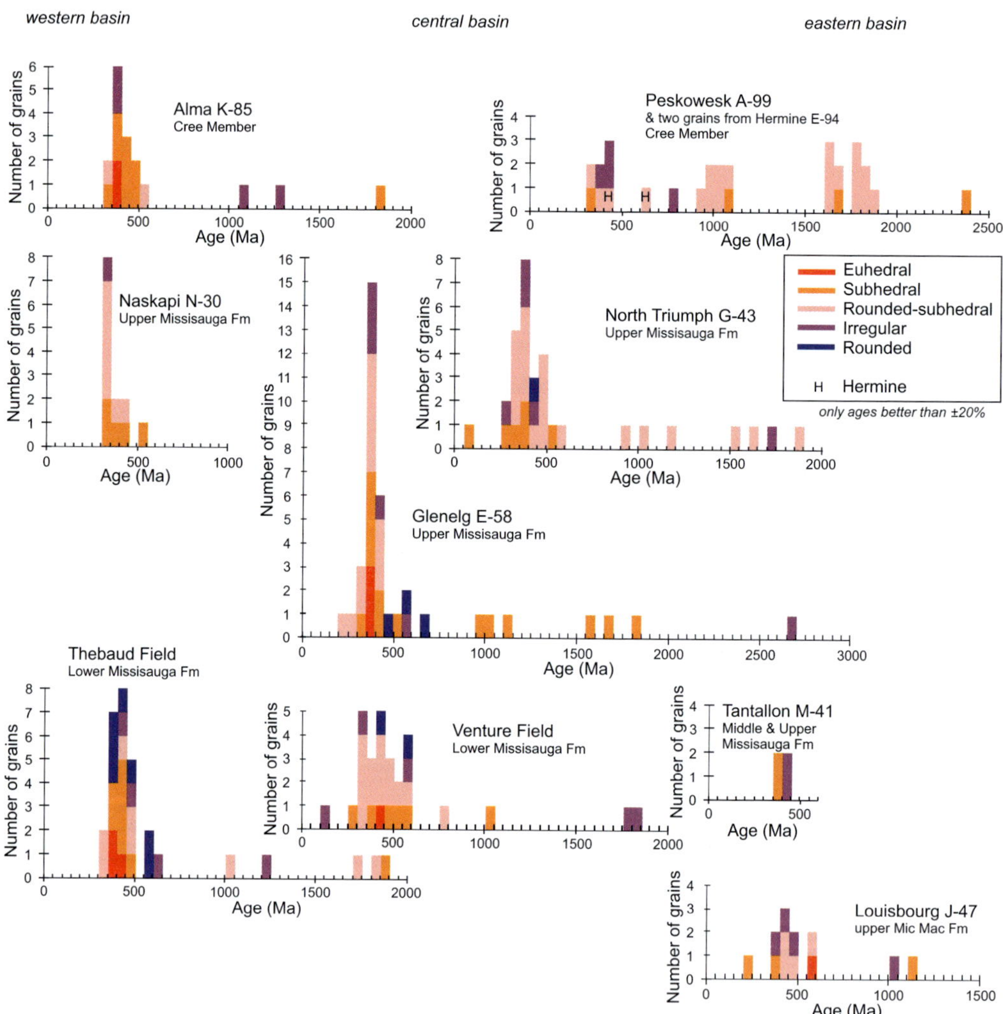



**Fig. 4.** Histograms of average age of individual monazite grains, showing external grain morphology. Plot differs from Figure 3 in showing individual grains rather than all spot ages and it is not possible to estimate dating errors rigorously. Plots are arranged schematically by stratigraphy and geography.

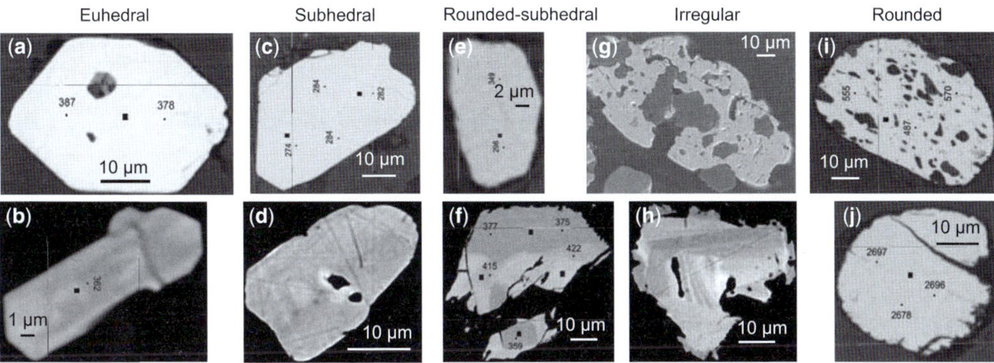



**Fig. 5.** Types of external morphology in monazite grains. Morphological types are defined as follows: *Euhedral*, with straight edges, fairly sharp corners, and good crystal form (e.g. pyramidal, prism). *Subhedral*, with somewhat straight edges, rather rounded corners, and crystal shape not always evident. *Rounded–subhedral*, similar to subhedral with one or more fairly straight edges, but most corners are rounded or the margins are corroded. *Irregular*, characterized by an irregular shape, with almost no straight edges. *Rounded*, lacking straight edges or sharp corners.

into five morphological types from euhedral to rounded (Fig. 5).

Except in the eastern part of the basin, grains with Precambrian age have a disproportionate number of irregular grains (Fig. 4), but some irregular and rounded grains are present in many age classes.

## Inclusions

Most monazite grains studied contain no inclusions. Some distinctive grains (Figs 5i & 6b, c) contain numerous inclusions, commonly aligned. These inclusions comprise quartz, muscovite, biotite and chlorite, together with kaolinite interpreted as a diagenetic alteration product of either feldspar or muscovite (cf. Karim *et al.* 2010). Some other grains contain only a few prismatic inclusions (Figs 5a & 6a) or more anhedral inclusions (Fig. 5d, h).

## Major element chemistry

Chondrite-normalized rare earth element (REE) patterns were plotted for each analysed grain and classified into six types (following Pe-Piper & Mackay 2006; Fig. 7). These types are:

- *Type A*: steady and steep fractionation from the light rare earth element (LREE) to the middle rare earth element (MREE).
- *Type B*: similar to type A, but has higher MREE contents.
- *Type C*: relatively enriched in the Nd and Sm. These patterns are typical for some monazites deriving from igneous rocks (Schandl & Gorton 2004).

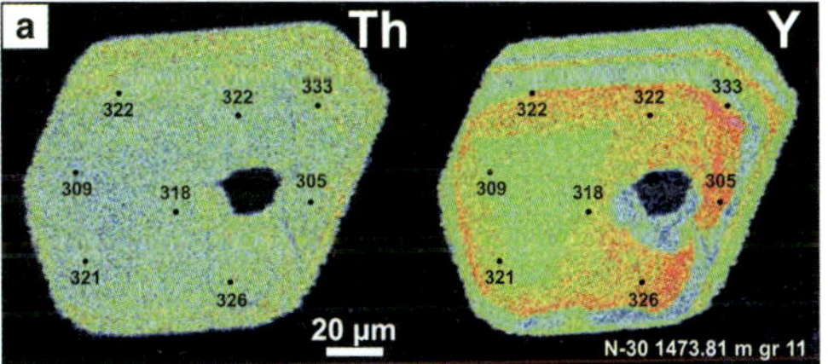



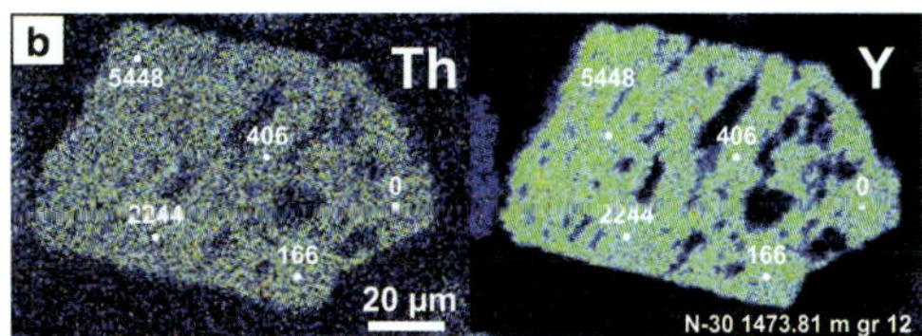



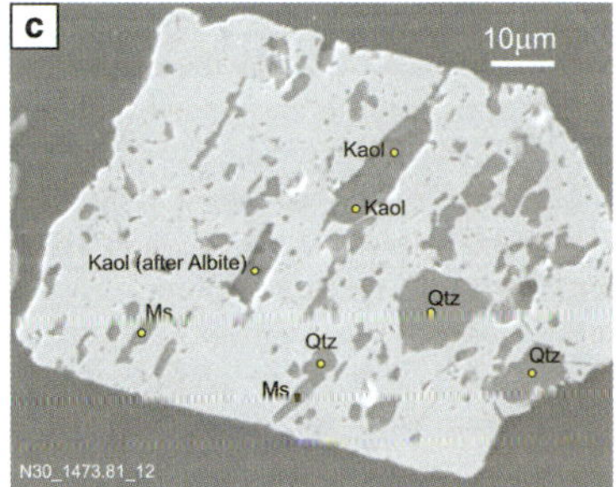



**Fig. 6.** Examples of zoning and inclusions in monazite grains from Naskapi N-30, showing location of spot ages. (**a**) Alleghanian plutonic monazite, zoned, with subhedral inclusion. (**b, c**) Metamorphic monazite, with subparallel inclusions of quartz, muscovite and kaolinite, low Th and Y, inconsistent spot dates with large error. (a) and (b) are X-ray maps of Th and Y distribution to show zoning; (c) is a backscattered electron image with spot EDS analyses of inclusions.

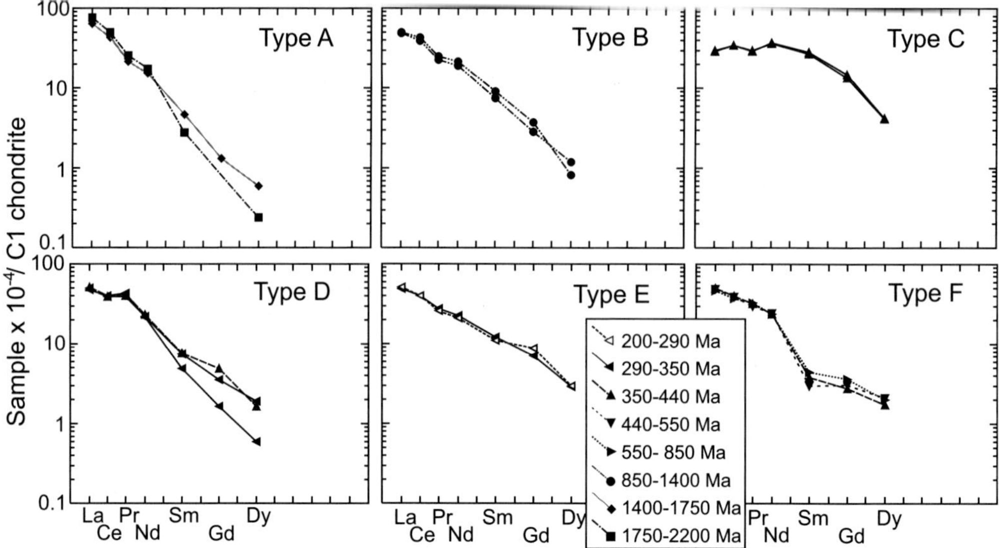



**Fig. 7.** Types of chondrite-normalized REE distributions recognized in monazite grains, plotted by age groups.

- *Type D*: shows almost no fractionation of the LREE, but strong fractionation of the MREE. These patterns are typical for some monazites derived from igneous rocks (Förster 1998).
- *Type E*: less strongly fractionated from LREE to MREE than any other type.
- *Type F*: LREE similar to type E, MREE from Sm are much less abundant.

There is significant geographical variation in predominant REE type (Fig. 8). Type B predominates in the western part of the basin at the Naskapi N-30 well, with lesser types D and A. This group of REE patterns is found in similar relative proportions in the central part of the basin and in the Mic Mac Formation at Louisbourg in the east, but is diluted by common type E grains, particularly abundant at the Alma K-85 well. The Lower Cretaceous samples from the eastern part of the basin (Peskowesk, Tantallon, Hermine) are dominated by types E, A and lesser F. Mesoproterozoic grains are predominantly types E and A with lesser B; otherwise no systematic variation with determined age was recognized.

Variation of Y v. Th (illustrated in Triantafyllidis *et al.* 2010, fig. 9) show that most grains with aligned inclusions, likely metamorphic, have low Y (<0.03 atoms per formula unit, afu). There are no clear patterns of variation of Y and Th with determined age. Grains with type E and to a lesser extent type B REE patterns commonly have high Y (>0.03 afu), whereas grains with type A REE patterns have low Y (<0.03 afu). Not unexpectedly, most grains with high errors on age have <0.01 afu Th.

## Stratigraphical and geographical variation

### Albian Cree Member of Logan Canyon Formation

*Alma K-85 well.* Samples from the Alma K-85 well, in the SW part of the central Scotian Basin, provided 26 spot dates from 18 monazite grains. Based on the probability plot (Fig. 3), the major age mode is Devonian (*c.* 400 Ma). Two Mesoproterozoic grains are rounded; one Palaeoproterozoic grain is subhedral. All Proterozoic grains are of REE type E.

*Peskowesk A-99 well.* Samples from the Peskowesk A-99 well, in the eastern Scotian Basin, resulted in 87 spot dates from 23 monazite grains (Fig. 3). There are modes with Mesoproterozoic (*c.* 1050 Ma) and Palaeoproterozoic (*c.* 1800 Ma) ages, with a small mode at *c.* 350 Ma (basal Carboniferous). Most grains are rounded–subhedral. Proterozoic grains are of REE type E, with lesser type A.

*Hermine E-94 well.* Only two monazite grains were found in cuttings from the Logan Canyon Formation in Hermine E-94, at the extreme eastern margin of the Scotian Basin. These grains had mean ages

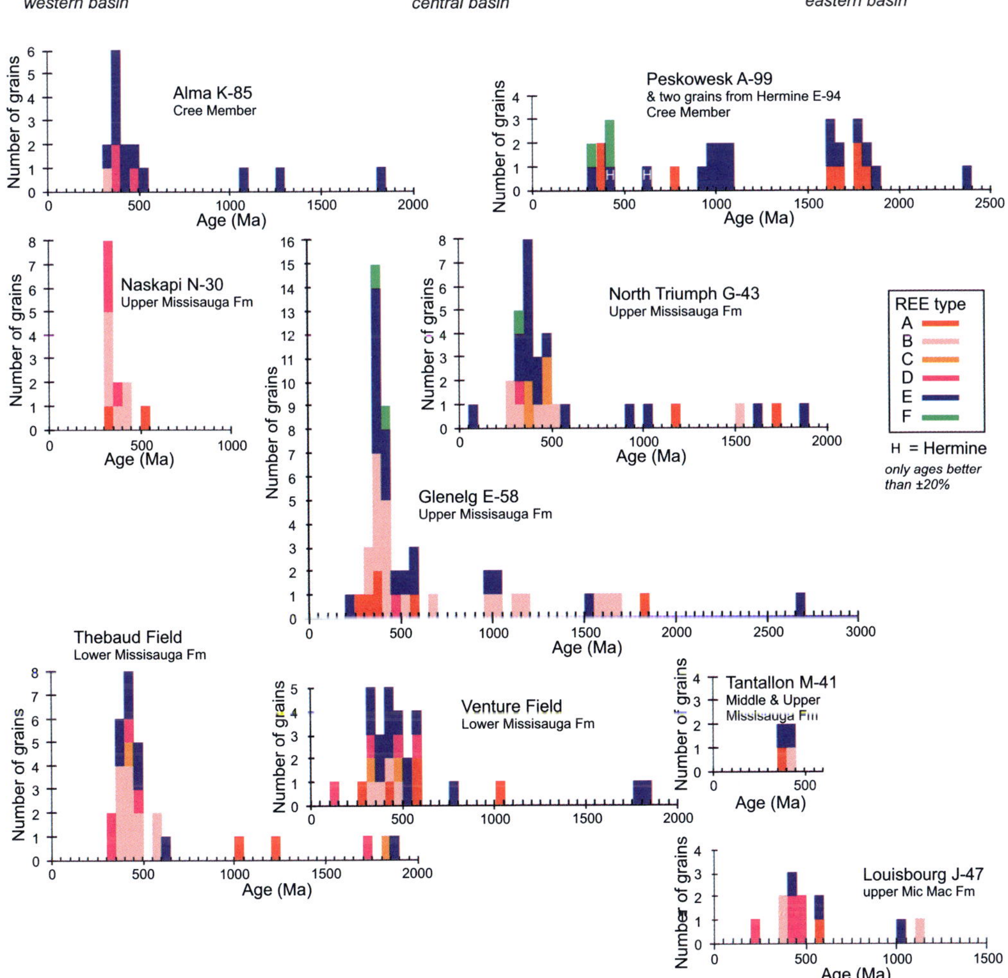



**Fig. 8.** Histograms of average age of individual monazite grains, showing REE types (cf. Fig. 7). Plots are arranged schematically by stratigraphy and geography.

of 403 Ma (early Devonian) and 613 Ma (latest Neoproterozoic) based on 11 spot analyses. Both are rounded–subhedral with type E REE.

## Hauterivian–Barremian Upper Missisauga Formation

*Naskapi N-30 well.* Samples from the only core in the Naskapi N-30 well, in the western part of the basin, yielded 52 spot dates from 13 monazite grains. Two multi-grain modes are recognized in the probability plot (Fig. 3): the larger clusters around 320 Ma (Carboniferous), whereas the smaller gives ages of c. 400 Ma (Devonian). One late Neoproterozoic grain is present. Grains are

subhedral or rounded–subhedral and many have abundant inclusions, commonly aligned (Fig. 6).

*Glenelg E-58 and E-58A wells.* Samples from the Glenelg Field in the central Scotian Basin yielded 135 spot dates from 49 monazite grains. The principal mode is at c. 400 Ma (Devonian), with secondary modes at c. 1050 Ma (Mesoproterozoic) and c. 500 Ma (Cambrian), the latter with a majority of rounded grains. Six Palaeoproterozoic and Mesoproterozoic grains for which images are available are all subhedral, and REE types are mostly E with lesser B.

*North Triumph G-43 well.* Samples from the North Triumph Field, SE of the Glenelg Field, provided

85 spot dates from 31 monazite grains, mostly sub-hedral or rounded–subhedral. The major mode is *c.* 400 Ma (Devonian) with lesser modes at *c.* 320 Ma (Carboniferous) and *c.* 450 Ma (Ordovician). Seven Mesoproterozoic and Palaeoproterozoic grains for which images are available are mostly rounded–subhedral.

*Tantallon M-41 well.* Samples from the Upper and Middle Members of the Missisauga Formation in the Tantallon M-41 well, in the eastern Scotian Basin, contained only four monazite grains, which provided 20 spot dates (Fig. 3). The major mode is at *c.* 440 Ma (Silurian), but in the Upper Member samples there is a small mode at *c.* 320 Ma (Carboniferous).

## Tithonian Mic Mac Formation and Lower Missisauga Formation

*Thebaud field.* Samples from several wells in the Thebaud field resulted in 127 spot dates from 30 monazite grains (four of which are from the Middle Missisauga Formation, grouped with the Lower Member for convenience). Major modes are of Silurian (*c.* 430 Ma) and Devonian age (*c.* 390 Ma). A small mode at *c.* 580 Ma (Neoproterozoic) comprises rounded or rounded–irregular grains. There are small modes at *c.* 1050 Ma (Mesoproterozoic) and *c.* 1850 Ma (Palaeoproterozoic), the latter comprising subhedral or subhedral–rounded grains.

*Venture field.* Samples from the Lower Missisauga Formation resulted in 113 spot dates from 26 monazite grains. The major mode is Devonian (*c.* 400 Ma), with a smaller mode at *c.* 530 Ma (Cambrian). The two Palaeoproterozoic grains for which images are available are rounded–irregular.

*Louisbourg J-47 well.* Twelve monazite grains were analysed from the MicMac Formation (coeval with the Lower Missisauga Formation) in Louisbourg J-47, together with one grain from the extreme base of the Middle Missisauga Formation, yielding a total of 38 spot ages. The three modes are *c.* 420 Ma (Silurian), *c.* 560 Ma (latest Neoproterozoic) and *c.* 1100 Ma (Mesoproterozoic).

## Discussion

### Alteration of monazite

Several lines of evidence suggest that some monazite grains have experienced early diagenetic alteration, resulting in ages that either have high errors or are indeterminate. Some grains have Pb below detection, despite high Th (plotted as + in Fig. 9), suggesting that Pb (and commonly U) have been leached out. Many grains with inclusions show alteration of muscovite and feldspar to kaolinite (Fig. 6c), characteristic of early diagenesis by meteoric water in the Chaswood Formation (Piper *et al.* 2009) and the Scotian Basin (Karim *et al.* 2010). Some of the grains with indeterminate ages appear highly corroded and have an irregular external morphology unlikely to have survived transportation, suggesting that it developed after deposition by corrosion (e.g. Fig. 5h).

## Stratigraphical and geographical variation in age modes

Based on the probability plots for spot dates with errors $< \pm 20\%$ (Fig. 3), a series of discrete age populations are identified that are reasonably consistent from one well or field to another (Table 1). In addition, there are many grains with an undeterminable or uncertain age, in which the age error is $> \pm 20\%$, or the grain yields a wide range of ages.

In the western part of the Scotian Basin, at the Naskapi N-30 well, most dates fall in the Carboniferous group, with a secondary mode in the Silurian–Devonian group (Fig. 9b). There are no data to evaluate stratigraphic variability.

In the central part of the Scotian Basin, from Alma to Venture, 40–50% of the age determinations fall in the Silurian–Devonian group, with a secondary mode of 13–33% in the Cambrian–Ordovician. The proportion of dates in this latter mode tends to be lowest in the Cree Member and highest in the Lower Missisauga Formation. All wells except Alma K-85 (where only 27 spot dates are available) have 7–31% of dates in the Carboniferous group (Fig. 9a). There is also a secondary Late Neoproterozoic mode (550–850 Ma) in the Lower Missisauga Formation at both Venture and Thebaud: this group is absent at higher stratigraphic levels except for a 2.3% presence in the Upper Missisauga Formation at Glenelg. Spot dates $> 850$ Ma make up only 12–14% of the Lower Missisauga dates, but 24–26% in the Cree Member, and are highly variable in the Upper Missisauga Formation.

In the eastern Scotian Basin, the abundance of monazite is low and large numbers of dates are available only from the Cree Member at Peskowesk A-99. There, 83% of spot dates are $> 850$ Ma, with a secondary mode in the Silurian–Devonian with 10% of the dates. The assemblage in the Upper and Middle Missisauga Formation at Tantallon is quite different, although only four grains are available, with no dates $> 550$ Ma and a strong mode in Silurian–Devonian with 55% of the dates. The small number of dates obtained from Louisbourg J-47

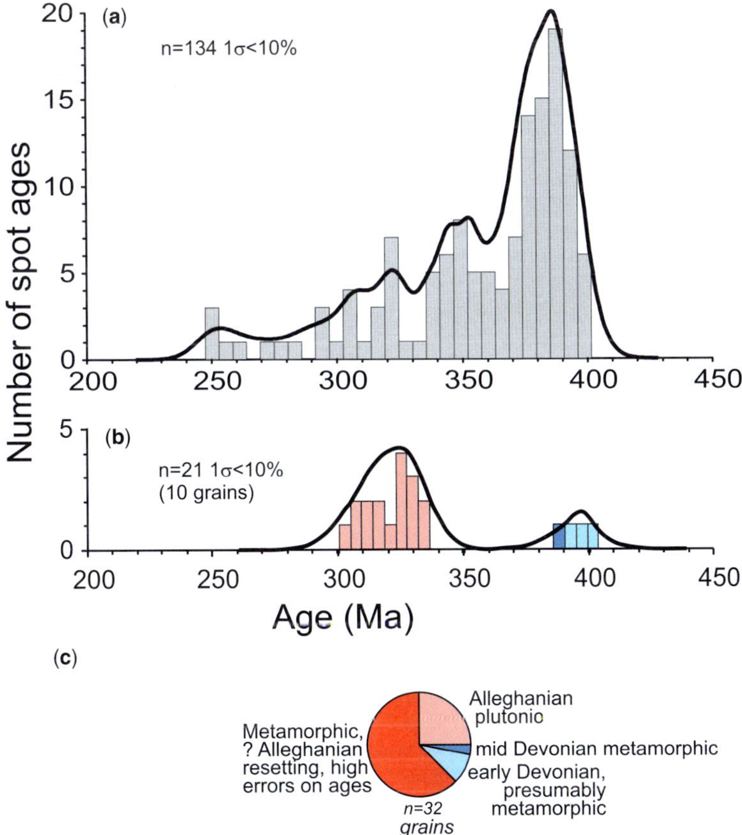



**Fig. 9.** Devonian–Carboniferous dates with $1\sigma < 10\%$ for (**a**) central Scotian Basin and (**b**) Naskapi N-30 well. (**c**) Interpreted types of monazite grain in the Naskapi N-30 well.

and Hermine E-94 are mostly in the range 350–650 Ma.

Thus, the Scotian Basin can be separated in at least three major domains based on the monazite geochronological data. In the western part, Carboniferous ages predominate, in the central part the largest monazite populations cluster around 400–440 Ma (Silurian–Devonian), whereas in the eastern part of the basin, Proterozoic dates are abundant relative to Palaeozoic dates.

The database is too small to evaluate stratigraphic variability except in the central part of the Scotian Basin. There, the Lower Missisauga Formation at Thebaud and Venture has an unusually low number of dates >850 Ma and an unusually high number of dates in group 5 (550–850 Ma) compared with any other stratigraphic level. No systematic variation was detected between the Upper Missisauga Formation and the Cree Member.

Comparison can be made with age determinations from the on-land Lower Cretaceous Chaswood Formation reported by Pe-Piper & MacKay (2006), which shows most ages between 380 and 550 Ma (Fig. 3). The westernmost outcrop, at Vinegar Hill, has an important *c.* 400 Ma mode, but borehole RR-97-23 in the Elmsvale Basin and boreholes at Brierly Brook and Diogenes Brook

**Table 1.** *Discrete age populations identified*

| Group | Age (Ma) |
| --- | --- |
| Permian–Triassic | 200–290 |
| Carboniferous | 290–350 |
| Silurian–Devonian | 350–440 |
| Cambrian–Ordovician | 440–550 |
| Late Neoproterozoic | 550–850 |
| Early Mesoproterozoic | 850–1400 |
| Late Mesoproterozoic | 400–1750 |
| Palaeoproterozoic | 1750–2200 |
| Early Palaeoproterozoic and Archaean | >2200 |

have a dominant mode in the Ordovician (440–500 Ma) with lesser Cambrian ages.

## Source-rock fertility and the provenance at the Naskapi N-30 well (western Scotian Basin)

The provenance of the sampled sandstone from the Naskapi well appears to have been restricted and simple. Monazite geochronology shows exclusively Devonian and Carboniferous ages (within error), except one grain yielding a single spot age of 515 Ma (Fig. 9). There is no evidence for supply of Lower Palaeozoic monazite characteristic of Chaswood Formation deposits in central Nova Scotia and southern New Brunswick (Fig. 3; Pe-Piper & MacKay 2006).

Age modes of monazite can be compared with the results of single-grain $^{40}Ar/^{39}Ar$ dating of muscovite (Reynolds *et al.* (2009). Muscovite is susceptible to physical abrasion and may be comminuted during river transport. Muscovite from the Naskapi N-30 well ranges in age from 372 to 350 Ma, whereas the monazite shows two modes centred on 330 Ma and 400 Ma. Muscovite grains of Carboniferous age (300–360 Ma) were evidently derived from basement metasedimentary rocks of the inner continental shelf that experienced resetting during the Alleghanian Orogen (Pe-Piper *et al.* 2010). Devonian (360–417 Ma) muscovite grains were derived from metasedimentary rocks and granite plutons of the Acadian Orogen on land. It seems reasonable to suggest that monazite grains with these modal ages have a similar origin. This would imply that the Naskapi well could have been supplied by sediment almost exclusively from the Meguma Terrane.

Muscovite geochemistry shows that the large detrital grains that were analysed for geochronology are predominantly of igneous origin (Reynolds *et al.* 2010), interpreted as being derived principally from the South Mountain Batholith (Fig. 1a), with some from inner shelf plutonic rocks hydrothermally modified during Alleghanian metamorphism (Reynolds *et al.* 2009). Monazite grains interpreted to be derived from granites show euhedral to subhedral morphology, commonly with some concentric zoning (Fig. 6a). Some such grains have mean ages as young as 315–320 Ma and type B REE patterns, and are interpreted to be derived from inner shelf plutons of similar age to the German Bank pluton (Fig. 1a), which yielded concentrically zoned monazite of similar modal age and chemistry (Pe-Piper *et al.* 2010). Grains with similar age, morphology and zoning also have types D and A REE patterns, and such grains have rare euhedral inclusions. Monazite grains interpreted as metamorphic, on the basis of subhedral form and abundant subparallel inclusions,

generally give high errors on ages, but one grain has a good mid-Devonian age with type D REE patterns. Another type of monazite has early Devonian dates, rounded–subhedral form, no inclusions, type B REE patterns and low Y. Some similar grains of the same age from the central Scotian Basin have aligned inclusions. The age of these monazites corresponds to the peak metamorphism of the Meguma Terrane at 410–395 Ma (Kontak *et al.* 1998; Hicks *et al.* 1999) prior to igneous intrusion in the Meguma Terrane and thus these grains are also interpreted as being of metamorphic origin. The most abundant type of monazite grain at Naskapi has abundant inclusions, generally aligned, but has no age determinations with errors better than 20%. Most grains have very low Th (<0.02 afu) and Y (<0.01 afu), the latter associated with type C REE patterns. Their textures appear metamorphic. Their inconsistent geochronology may result from the widespread Alleghanian thermal event that reset many metamorphic muscovites in the outboard part of the Meguma Terrane (Reynolds *et al.* 2009). No igneous monazites of age similar to the South Mountain Batholith were recognized at Naskapi, but euhedral or rounded–subhedral, zoned monazite of Late Devonian age makes up <2% of the monazite grains from the central Scotian Basin.

The relative abundances of different types of monazite in the Naskapi N-30 well illustrate the influence of source-rock fertility on the eventual monazite assemblage. No monazites are recognized from the South Mountain Batholith, the largest igneous body in the Appalachians, which was unroofed in the early Carboniferous (Martel *et al.* 1993). Most of the dated monazites come from cryptic Alleghanian plutons, which geophysical data show to be of limited size and distribution on the inner shelf, now largely covered by younger sedimentary rocks (Pe-Piper *et al.* 2010). The remaining dated monazites date the initial metamorphism of the Meguma Terrane. The predominant (62%) monazite type appears to be metamorphic, but yields incoherent spot ages with large errors.

## Provenance of monazite in the central Scotian Basin

The central Scotian Basin appears to have been supplied by a large 'Sable River' that transported sediment derived from the Appalachians and the Canadian Shield (Fig. 11), including large amounts of polycyclic sediment reworked from Palaeozoic sandstones and shales of the Maritimes Basin (Pe-Piper *et al.* 2008; Tsikouras *et al.* 2011). Monazite grains with a wide range of ages, morphology and chemistry have been distinguished, probably representing more than twenty discrete sources.

*Palaeozoic grains.* Most samples have small numbers of grains of Carboniferous–Permian age (Fig. 9a), similar to those at Naskapi, and likely from a similar source on the inner continental shelf that experienced Alleghanian resetting. Abundances are too low to interpret statigraphic or geographical variation. Almost all dated detrital muscovite from the central Scotian Basin (Reynolds *et al.* 2009) ranges in age from *c.* 420 to 240 Ma (Fig. 10). Devonian monazite grains are dominant

in the Upper Missisauga Formation of the central basin and (based on a smaller dataset) in the Cree Member (Fig. 4). Mid to late Devonian grains (8% of central basin grains) appear to be of igneous origin, are generally zoned, and correspond to widespread Devonian granites in the Appalachian Orogen, particularly in the Meguma Terrane. Early Devonian and Silurian grains (14%) lack zoning and many resemble the early Devonian grains at Naskapi, interpreted to be of metamorphic origin. Ordovician–Cambrian grains (9%) form a secondary mode in the Upper Missisauga Formation and Cree Member, but are subequal in abundance to the Silurian–Devonian in the Lower Missisauga Formation. Such grains predominate in all Chaswood Formation outcrops except Vinegar Hill (VH in Fig. 11; Pe-Piper & MacKay 2006), indicating that the Chaswood rivers in Nova Scotia were supplied principally from the Taconic Orogen of the Dunnage Terrane of Northern New Brunswick (Fig. 11). The course of the Chaswood River was likely deflected eastward along the Cobequid–Chedabucto fault zone (CCFZ in Fig. 11) to join the Sable River that flowed through Cabot Strait (Reynolds *et al.* 2012). Although metamorphic and igneous rocks of Ordovician age are also widespread in central Newfoundland, they are unlikely to have been a source, since monazite of this age is sparse in the eastern part of the basin supplied by the Banquereau River (Fig. 11).

*Late Neoproterozoic grains.* These make up 5% of the total grains and are all rounded–subhedral, irregular or rounded and all but one lack zoning. They correspond to igneous activity, deformation and metamorphism in the Avalon Terrane. They are surprisingly rare, given the geographical extent of the Avalon Terrane. Rare grains of this age are also found in the Chaswood Formation at Diogenes Brook and Brierly Brook (DB and BB, Fig. 11), but are absent in other Chaswood localities (Pe-Piper & MacKay 2006), despite evidence from pebbles of an Avalonian source for some sediment (Gobeil *et al.* 2006). Our previous studies of hundreds of thin sections from Avalonian rocks of the Cobequid Highlands (Pe-Piper & Piper 2003) did not reveal the presence of monazite. The subduction-related Avalonian rocks thus appear to have low monazite fertility.

*Late Mesoproterozoic grains.* Such grains are mostly subhedral or rounded–subhedral, some show igneous zoning, and most have type E REE distribution. They form a prominent mode in the Upper Missisauga Formation at Glenelg, but are sparse elsewhere in the central basin at all stratigraphic levels, overall making up 7% of total grains. The peak in the probability function is generally at

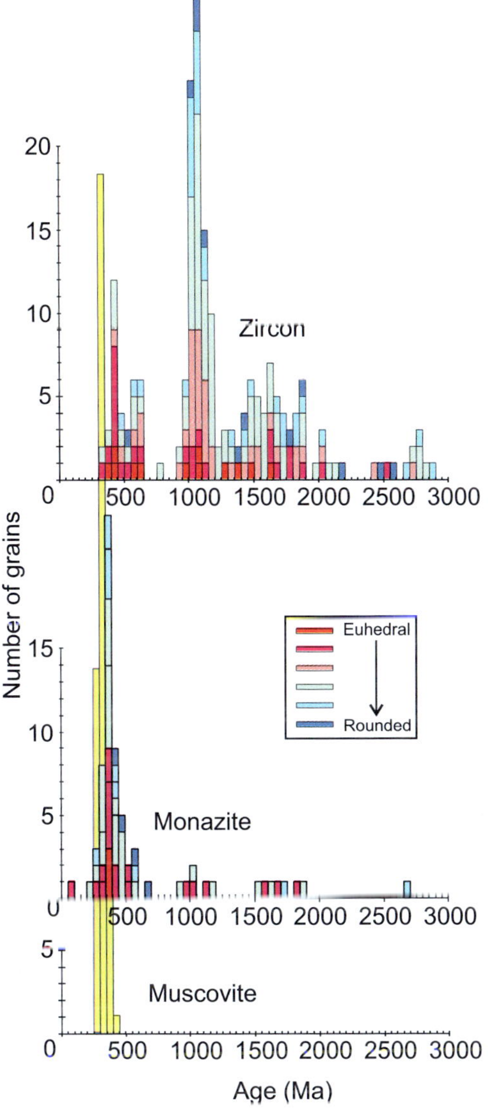



**Fig. 10.** Comparison of monazite with muscovite and zircon geochronology for wells in the central Scotian Basin, also showing morphology of monazite and zircon grains.

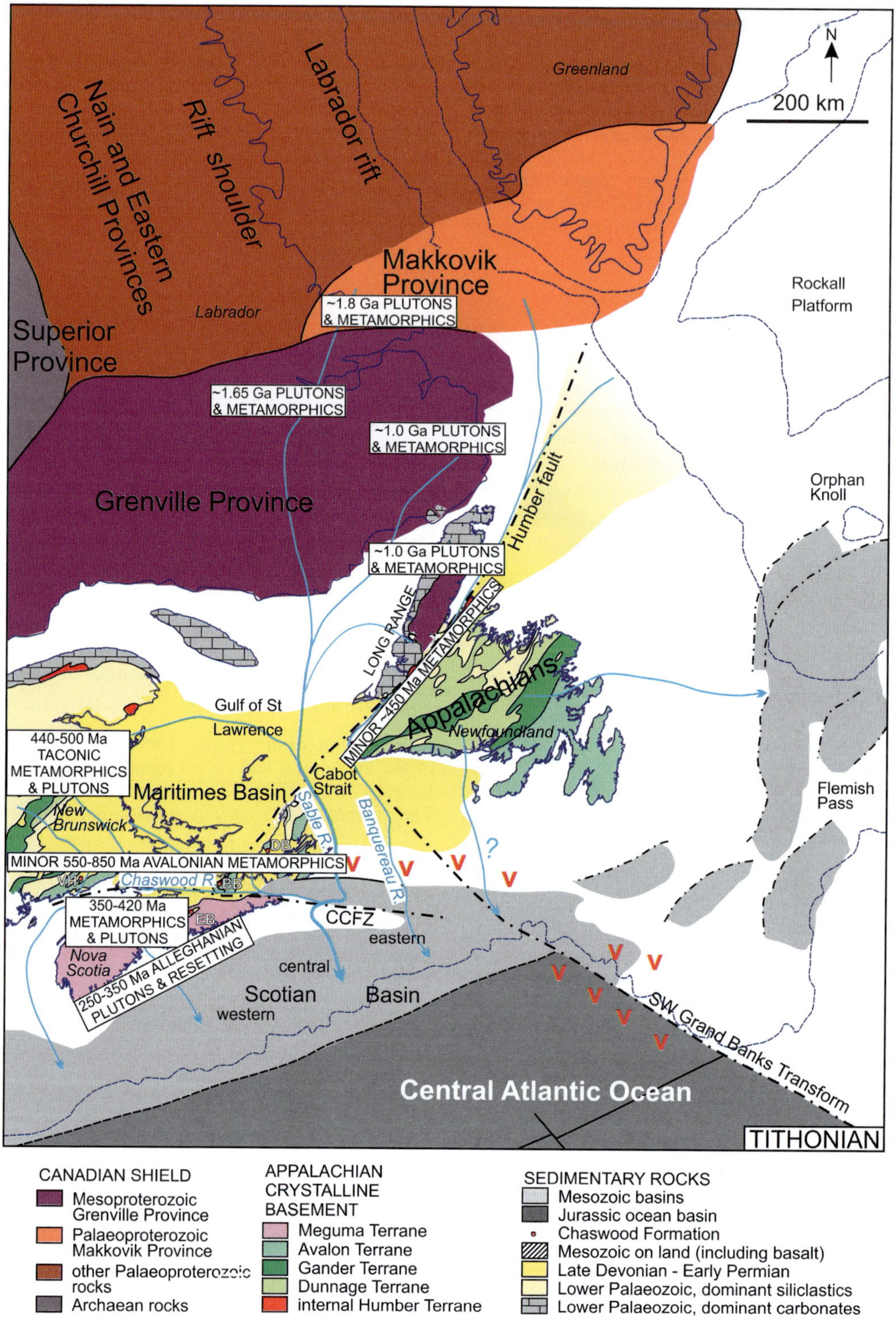

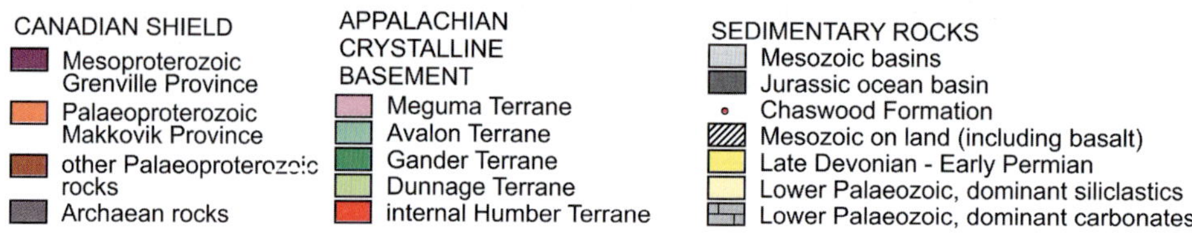
N
200 km
Greenland
Rockall Platform
Labrador rift
Rift shoulder
Labrador
Nain and Eastern Churchill Provinces
Makkovik Province
~1.8 Ga PLUTONS & METAMORPHICS
Superior Province
~1.65 Ga PLUTONS & METAMORPHICS
~1.0 Ga PLUTONS & METAMORPHICS
Grenville Province
Humber fault
Orphan Knoll
~1.0 Ga PLUTONS & METAMORPHICS
LONG RANGE
MINOR ~450 Ma METAMORPHICS
Appalachians
Newfoundland
Gulf of St Lawrence
440-500 Ma TACONIC METAMORPHICS & PLUTONS
Maritimes Basin
Cabot Strait
Sable R.
Banquereau R.
Flemish Pass
New Brunswick
MINOR 550-850 Ma AVALONIAN METAMORPHICS
DB
VHc
Chaswood R.
BB
?
350-420 Ma METAMORPHICS & PLUTONS
EB
Nova Scotia
250-350 Ma ALLEGHANIAN PLUTONS & RESETTING
CCFZ
eastern
central
Scotian
Basin
western
SW. Grand Banks Transform
Central Atlantic Ocean
TITHONIAN
CANADIAN SHIELD
Mesoproterozoic Grenville Province
Palaeoproterozoic Makkovik Province
other Palaeoproterozoic rocks
Archaean rocks
APPALACHIAN CRYSTALLINE BASEMENT
Meguma Terrane
Avalon Terrane
Gander Terrane
Dunnage Terrane
internal Humber Terrane
SEDIMENTARY ROCKS
Mesozoic basins
Jurassic ocean basin
Chaswood Formation
Mesozoic on land (including basalt)
Late Devonian - Early Permian
Lower Palaeozoic, dominant siliciclastics
Lower Palaeozoic, dominant carbonates

*c.* 1050 Ma (Fig. 3), similar to that observed for detrital zircons (Piper *et al.* 2012), suggesting a source in the *c.* 1060 Ma main Grenville plutons of western Newfoundland and coastal Labrador rather than the 960–980 Ma post-tectonic plutons of southern Labrador (Gower *et al.* 2008). The lack of both zircon and monazite of this age suggests that the principal sediment supply was from the Long Range of western Newfoundland and from coastal Labrador, which would have formed the uplifted rift shoulder prior to opening of the Labrador Sea.

*Older Mesoproterozoic grains.* These grains are mostly subhedral monazite and are sparse except at Glenelg, where there is a probability peak at *c.* 1650 Ma which corresponds to the Labradorian intrusions of the Grenville Province (Gower *et al.* 2008). Detrital zircons also show a probability peak at *c.* 1480 (Fig. 10; Piper *et al.* 2012), corresponding to the Pinwarian intrusions of the SE Labrador coast and the Long Range (Owen & Currie 1991; Gower *et al.* 2008), but this event is not represented by monazite.

*Late Palaeoproterozoic grains.* A small mode at *c.* 1850 Ma (3%) is present in the Lower Missisauga Formation at both Thebaud and Venture and single grains are present at higher stratigraphic levels. Most of the grains are euhedral to subhedral and thus apparently first cycle, some are zoned, and all have type A or E REE distribution. The age corresponds to plutonism and metamorphism of the Makkovik Province on the central Labrador coast (Ketchum *et al.* 2001); sparse euhedral zircon grains of similar age are also known from the Scotian Basin (Fig. 10; Piper *et al.* 2012).

*Inferred river patterns.* Thus, the Sable River appears to have supplied sediment throughout the Missisauga and Logan Canyon formations from as distant as the Makkovik Province of central Labrador, with evidence of sources in southern coastal Labrador, the Long Range inlier of western Newfoundland, and from all the terranes of the Appalachians. Everywhere, Devonian monazite predominates. Where Taconic (Ordovician) monazite is relatively important, Meso- and Palaeoproterozoic monazite is less important, and vice versa (as at Glenelg), but sampling density is insufficient to resolve stratigraphic variability in river supply. Evidence from the Chaswood Formation (Pe-Piper & MacKay 2006) suggests that Taconic

monazite was supplied from New Brunswick, whereas the Meso- and Palaeoproterozoic monazite can be linked to sources in Labrador and western Newfoundland (Fig. 11). Comparison with detrital zircon abundance suggests a lack of monazite fertility in Avalonian and Pinwarian rocks. In contrast, there is proportionally greater supply of late Devonian and Carboniferous monazite compared to zircon (Fig. 10), reflecting growth of monazite during the Alleghanian Orogen.

## Provenance of monazite in the eastern Scotian Basin

Previous studies have shown that the eastern Scotian Basin was supplied by a separate river, the Banquereau River, which appears to have occupied the Humber Fault valley in western Newfoundland (Pe-Piper & Piper 2012). Detrital monazite is less abundant in the eastern Scotian Basin than in the central basin, so that fewer grains have been analysed. Only in the Cree Member at Peskowesk have a statistically acceptable number of analyses been made. In general, the same types of monazite grains are recognized as in the central basin, but the relative abundances are different. For example, subhedral to rounded–subhedral late Palaeoproterozoic monazites from the Makkovik Province are relatively more important in the eastern basin (13%). Cambrian–Ordovician grains are lacking, with the possible exception of two grains from Louisbourg that appear to be reset Neoproterozoic grains with a wide range of spot ages. This suggests that Cambrian–Ordovician grains in the central basin are derived from New Brunswick via the Chaswood River, rather than from central Newfoundland. On the other hand, Silurian grains are more abundant in the east (9%) than in the central basin, probably reflecting the abundance of Silurian metamorphic rocks in southwestern Newfoundland (e.g. Waldron *et al.* 1998).

## What geological events are dated by the monazite?

It is difficult to determine with certainty whether monazite is derived from igneous or metasedimentary rocks. Systematic studies of monazite petrography in potential source rocks are lacking. In the case of the Naskapi well, where the potential source areas are limited, source rocks can be

---

**Fig. 11.** Palaeogeographical cartoon for the Tithonian, showing the river dispersal pattern inferred from monazite and other data. Base map modified from Lowe *et al.* (2011). CCFZ, Cobequid-Chedabucto Fault Zone. Chaswood Formation localities: VH, Vinegar Hill; EB, Elmsvale Basin; BB, Brierly Brook; DB, Diogenes Brook. V = early Cretaceous volcanic centres (Bowman *et al.* 2012).

linked to monazite morphology and texture. Monazite from metasedimentary rocks have aligned inclusions, are commonly rounded or irregular, and yield spot ages with high errors suggesting resetting and mobility of Pb and U. Monazite from igneous rocks may have well-developed concentric zoning and euhedral inclusions. REE pattern types A and E are most common in clearly igneous monazite and types B, C and D are most common in clearly metamorphic monazite, but are not sufficiently systematic to be of diagnostic value on their own. Different tectonothermal events have different monazite fertility. Igneous monazite is common in Alleghanian and some Devonian plutons (but not the South Mountain Batholith), the 1.05 Ga Grenvillian plutons, the 1.65 Ga Labradorian plutons and the 1.85 Ga Makkovik Province plutons. In contrast, the 1.5 Ga Pinwarian and 0.6 Ga Avalonian plutons that yield common zircons are not a significant source of monazite, and most if not all Avalonian monazite appears to be metamorphic.

Although monazite is relatively resistant to mechanical abrasion, it is readily broken down chemically under acid conditions. Monazite may thus have a polycyclic origin, reworked out of older sedimentary rocks, and such polycyclic monazite is likely to be represented by rounded or irregular grains. There is no systematic variation of morphology with age, except that there is a disproportionate number of euhedral grains of middle Palaeozoic age, characteristic of the outboard Appalachian terranes, and involving short transport distances. Nevertheless, some Palaeoproterozoic grains are also euhedral. We interpret this variation to mean that most monazite is likely to be of first cycle origin, but the presence of some polycyclic monazite grains cannot be excluded. Comparison with detrital zircon ages and morphology in the central Scotian Basin (Fig. 10) shows that, whereas about 50% of all dated zircons are rounded and interpreted as polycyclic (Piper *et al.* 2012), only 23% of the monazite is rounded or irregular and some (probably most) of such grains are either of metamorphic origin or have experienced diagenetic dissolution.

### Refining the palaeogeographical model

Previous papers based on heavy minerals (Tsikouras *et al.* 2011), detrital muscovite geochronology (Reynolds *et al.* 2012) and detrital zircon geochronology (Piper *et al.* 2012) have resulted in incremental improvements in our understanding of the palaeodistribution pattern of Late Jurassic–Early Cretaceous rivers. This monazite study has resulted in confirmation of some results that were previously supported by only limited data and has provided some new insights (Fig. 11). A distant first cycle source from the Makkovik Province and the Labradorian plutons first identified from a few euhedral zircons (Piper *et al.* 2012) is confirmed and suggests that uplift of the Labrador rift shoulder was important in supplying sediment to the Scotian Basin. The monazite data demonstrate that this Labrador supply dates back to the Tithonian.

The monazite data also provide a clearer distinction of Lower Palaeozoic sources than do the detrital zircons. Abundant, principally Ordovician monazite, was supplied to the Sable River mainly through the headwaters of the Chaswood River draining northern New Brunswick. Very small amounts of principally Silurian monazite were supplied to the Banquereau River from western or central Newfoundland.

Monazite data also confirm the importance of Alleghanian tectonothermal resetting on the Scotian Shelf and the significance of plutons with Alleghanian monazite in supplying monazite to the Scotian Basin. Such supply requires considerable unroofing of the Meguma Terrane, which was more important in the Upper than in the Lower Missisauga Formation. This confirms inferences made from muscovite geochronology (Reynolds *et al.* 2012).

### Petroleum significance

This provenance study is more than a regional investigation of a particular sedimentary basin. The Scotian Basin is a producing gas-prone basin with significant new deep-water exploration commitments at the time of writing. Geochronological studies of detrital mineral provenance have revealed details of how river supply to the basin has varied through time, with implications for the distribution of thick potential reservoir sandstones in parts of the basin not yet drilled (Pe-Piper & Piper 2012). Changes in the supply of detritus through time have a strong potential to influence reservoir character through different diagenetic processes. For example, the best development of chlorite rims on detrital quartz grains, leading to preservation of porosity (Gould *et al.* 2010), is the Lower Missisauga Formation of the central Scotian Basin – a stratigraphic level with distinctive sources based on detrital monazite geochronology.

## Conclusions

(1)   The combination of external morphology, texture, chemistry and geochronology of detrital monazite grains is a powerful tool for interpreting provenance.

(2) The occurrence of similar abundances of irregular or rounded monazite grains in most age classes suggests that most monazite is first cycle and irregular grains probably result from diagenetic processes. Such diagenetic processes also lead to leaching of Pb and U, resulting in indeterminate dates.

(3) Many igneous monazites show chemical zoning, contain sparse commonly euhedral inclusions, and have REE patterns of types A and E. Many metamorphic monazites contain inclusions that may be aligned, are more commonly rounded–subhedral to rounded, and have REE patterns of types B, C and D.

(4) The lack of petrographic studies of monazite in potential source rocks limits the precise identification of sources of monazite grains. However, this study has shown the value of multiple techniques in determining provenance. For example, comparison with detrital zircon geochronology shows that monazite fertility of source rocks is highly variable, with low fertility in Avalonian and Pinwarian sources. Finding evidence of a particular source by more than one technique, such as zircon and monazite from the Makkovik Province, improves confidence in palaeodrainage models.

(5) Detrital monazite confirms supply to the Scotian Basin from the Labrador rift shoulder and shows that it occurred already in the Tithonian, prior to the onset of volcanism in the Labrador rift. Monazite shows that the Sable River received significant supply from Ordovician sources in northern New Brunswick, apparently via the Chaswood River. Monazite also confirms the great importance of the inner continental shelf as a sediment source, particularly in the Hauterivian–Barremian. All this information is applicable to defining exploration targets in the Scotian Basin and understanding the observed variability in sandstone reservoir quality.

This work was supported by a grant from the Nova Scotia Offshore Energy Technical Research Association (OETR). Error calculation for monazite ages was based on an Excel workbook developed by S. Gagné; R. MacKay provided advice on monazite dating, and G. Strathdee carried out the morphological and textural analysis of the monazite grains.

# References

BECK, M. E. & HOUSEN, B. A. 2003. Absolute velocity of North America during the Mesozoic from paleomagnetic data. *Tectonophysics*, **377**, 33–54.

BOWMAN, S. J., PE-PIPER, G., PIPER, D. J. W., FENSOME, R. A. & KING, E. L. 2012. Early Cretaceous volcanism in the Scotian Basin. *Canadian Journal of Earth Sciences*, **49**, 1523–1539.

DICKIE, K., KEEN, C. E., WILLIAMS, G. L. & DEHLER, S. A. 2011. Tectonostratigraphic evolution of the Labrador margin, Atlantic Canada. *Marine and Petroleum Geology*, **28**, 1663–1675.

FALCON-LANG, H. J., FENSOME, R. A., GIBLING, M. R., MALCOLM, J., FLETCHER, K. R. & HOLLEMAN, M. 2007. Karst-related outliers of the Cretaceous Chaswood Formation of Maritime Canada. *Canadian Journal of Earth Sciences*, **44**, 619–642.

FÖRSTER, H. J. 1998. The chemical composition of REE-Y-Th-U-rich accessory minerals in peraluminous granites of the Erzgebirge-Fichtelgebirge region, Germany, Part I: The monazite-(Ce)-brabantite solid solution series. *American Mineralogist*, **83**, 259–272.

GOBEIL, J.-P., PE-PIPER, G. & PIPER, D. J. W. 2006. The Early Cretaceous Chaswood Formation in the West Indian Road pit, central Nova Scotia. *Canadian Journal of Earth Sciences*, **43**, 391–403.

GOULD, K., PE-PIPER, G. & PIPER, D. J. W. 2010. Relationship of diagenetic chlorite rims to depositional facies in Lower Cretaceous reservoir sandstones of the Scotian Basin. *Sedimentology*, **57**, 587–610.

GOULD, K. M., PIPER, D. J. W. & PE-PIPER, G. 2012. Lateral variation in sandstone lithofacies from conventional core, Scotian Basin: implications for reservoir quality and conductivity. *Canadian Journal of Earth Sciences*, **49**, 1478–1503.

GOWER, C. F., KAMO, S. L., KWOK, K. & KROGH, T. E. 2008. Proterozoic southward accretion and Grenvillian orogenesis in the interior Grenville Province in eastern Labrador: evidence from U–Pb geochronological investigations. *Precambrian Research*, **165**, 61–95.

GRIST, A. M. & ZENTILLI, M. 2003. Post-Paleocene cooling in the southern Canadian Atlantic region: evidence from apatite fission track models. *Canadian Journal of Earth Sciences*, **40**, 1279–1297.

HENDRIKS, M., JAMIESON, R. A., WILLETT, S. D. & ZENTILLI, M. 1993. Burial and exhumation of the Long Range Inlier and surroundings, western Newfoundland: results of an apatite fission track study. *Canadian Journal of Earth Sciences*, **30**, 1594–1606.

HICKS, R. J., JAMIESON, R. A. & REYNOLDS, P. H. 1999. Detrital and metamorphic $^{40}$Ar/$^{39}$Ar ages from muscovite and whole-rock samples, Meguma Supergroup, southern Nova Scotia. *Canadian Journal of Earth Sciences*, **36**, 23–32.

KARIM, A., PE-PIPER, G. & PIPER, D. J. W. 2010. Controls on diagenesis of Lower Cretaceous reservoir sandstones in the western Sable Subbasin, offshore Nova Scotia. *Sedimentary Geology*, **224**, 65–83.

KETCHUM, J. W. F., BARR, S. M., CULSHAW, N. G. & WHITE, C. E. 2001. U–Pb ages of granitoid rocks in the northwestern Makkovik Province, Labrador: evidence for 175 million years of episodic synorogenic and postorogenic plutonism. *Canadian Journal of Earth Sciences*, **38**, 359–372.

KONTAK, D. J., HORNE, R. J., SANDEMAN, H., ARCHIBALD, D. & LEE, J. K. W. 1998. $^{40}$Ar/$^{39}$Ar dating of ribbon-textured veins and wall-rock material from Meguma lode gold deposits, Nova Scotia: implications for timing and duration of vein formation in slate-belt hosted vein gold deposits. *Canadian Journal of Earth Sciences*, **35**, 746–761.

LI, G., PE-PIPER, G. & PIPER, D. J. W. 2012. The provenance of Middle Jurassic sandstones in the Scotian Basin: petrographic evidence of passive margin tectonics. *Canadian Journal of Earth Sciences*, **49**, 1465–1477.

LOWE, D. G., SYLVESTER, P. J. & ENACHESCU, M. E. 2011. Provenance and paleodrainage patterns of Upper Jurassic and Lower Cretaceous synrift sandstones in the Flemish Pass Basin, offshore Newfoundland, east coast of Canada. *American Association of Petroleum Geologists Bulletin*, **95**, 1295–1320.

LUDWIG, K. R. 1999. *User's Manual for Isoplot/Ex, version 2.06: A Geochronological Toolkit for Microsoft Excel*. Berkeley Geochronological Center, Berkeley, CA, Special Publications No. 1a.

MACLEAN, B. C. & WADE, J. A. 1993. *Seismic Markers and Stratigraphic Picks in Scotian Basin Wells*. Atlantic Geoscience Centre, Geological Survey of Canada, Ottawa.

MARTEL, A. T., MCGREGOR, D. C. & UTTING, J. 1993. Stratigraphic significance of Upper Devonian and Lower Carboniferous miospores from the type area of the Horton Group, Nova Scotia. *Canadian Journal of Earth Sciences*, **30**, 1091–1098.

OETR (OFFSHORE ENERGY TECHNICAL RESEARCH ASSOCIATION) 2011. *Play Fairway Analysis Atlas – Offshore Nova Scotia*. Nova Scotia Department of Energy Report 88-11-0004-01, http://www.novascotiaoffshore.com/analysis (accessed 4 June 2013).

OWEN, J. V. & CURRIE, K. L. 1991. The disappointment hill complex: Proterozoic granulites in southwestern Newfoundland. *Transactions – Royal Society of Edinburgh: Earth Sciences*, **82**, 55–63.

PE-PIPER, G. & MACKAY, R. M. 2006. Provenance of Lower Cretaceous sandstones onshore and offshore Nova Scotia from electron microprobe geochronology and chemical variation of detrital monazite. *Bulletin of Canadian Petroleum Geology*, **54**, 366–379.

PE-PIPER, G. & PIPER, D. J. W. 2003. A synopsis of the geology of the Cobequid Highlands, Nova Scotia. *Atlantic Geology*, **38**, 145–160.

PE-PIPER, G. & PIPER, D. J. W. 2012. Cretaceous re-activation of the passive-margin Scotian Basin. Chapter 13. *In*: BUSBY, C. & AZOR, A. (eds) *Tectonics of Sedimentary Basins: Recent Advances*. Blackwell Publishing, Oxford, 270–287.

PE-PIPER, G., TRIANTAFYLLIDIS, S. & PIPER, D. J. W. 2008. Geochemical identification of clastic sediment provenance from known sources of similar geology: the Cretaceous Scotian Basin, Canada. *Journal of Sedimentary Research*, **78**, 595–607.

PE-PIPER, G., KAMO, S. L. & MCCALL, C. 2010. The German Bank pluton, offshore SW Nova Scotia: petrology, age and regional significance for Alleghanian plutonism. *Geological Society of America Bulletin*, **122**, 690–700.

PIPER, D. J. W., HUNDERT, T., PE-PIPER, G. & OKWESE, A. C. 2009. The roles of pedogenesis and diagenesis in clay mineral assemblages: Lower Cretaceous fluvial mudrocks, Nova Scotia, Canada. *Sedimentary Geology*, **213**, 51–63.

PIPER, D. J. W., PE-PIPER, G., TUBRETT, M., TRIANTAFYLLIDIS, S. & STRATHDEE, G. 2012. Detrital zircon geochronology and polycyclic sediment sources, Upper Jurassic–Lower Cretaceous of the Scotian Basin, southeastern Canada. *Canadian Journal of Earth Sciences*, **49**, 1540–1557.

POLYAKOV, E., VOLKOV, I., SURIKOV, V., PERELYAEVA, L. & SHVEIKIN, G. 2010. Dissolution of monazite in humic solutions. *Radiochemistry*, **52**, 429–434.

PYLE, J. M., SPEAR, F. S., WARK, D. A., DANIEL, C. G. & STORM, L. C. 2005. Contributions to precision and accuracy of monazite microprobe ages. *American Mineralogist*, **90**, 547–577.

REYNOLDS, P. H., PE-PIPER, G., PIPER, D. J. W. & GRIST, A. M. 2009. Single-grain detrital muscovite ages from Lower Cretaceous sandstones, Scotian basin, and their implications for provenance. *Bulletin of Canadian Petroleum Geology*, **57**, 25–42.

REYNOLDS, P. H., PE-PIPER, G. & PIPER, D. J. W. 2010. Sediment sources and dispersion as revealed by single-grain $^{40}Ar/^{39}Ar$ ages of detrital muscovite from Carboniferous and Cretaceous rocks in mainland Nova Scotia. *Canadian Journal of Earth Sciences*, **47**, 957–970.

REYNOLDS, P. H., PE-PIPER, G. & PIPER, D. J. W. 2012. Detrital muscovite geochronology and the Cretaceous tectonics of the inner Scotian Shelf, southeastern Canada. *Canadian Journal of Earth Sciences*, **49**, 1558–1566.

ROSS, G. M., PATCHETT, P. J., HAMILTON, M., HEAMAN, L., DECELLES, P. G., ROSENBERG, E. & GIOVANNI, M. K. 2005. Evolution of the Cordilleran orogen (southwestern Alberta, Canada) inferred from detrital mineral geochronology, geochemistry, and Nd isotopes in the foreland basin. *Geological Society of America Bulletin*, **117**, 747–763.

SCHANDL, E. S. & GORTON, M. P. 2004. A textural and geochemical guide to the identification of hydrothermal monazite: criteria for selection of samples for dating epigenetic hydrothermal ore deposits. *Economic Geology*, **99**, 1027–1035.

SIBUET, J.-C., ROUZO, S. & SRIVASTAVA, S. 2012. Plate tectonic reconstructions and paleogeographic maps of the central and North Atlantic oceans. *Canadian Journal of Earth Sciences*, **49**, 1395–1415.

STEA, R. & PULLAN, S. 2001. Hidden Cretaceous basins in Nova Scotia. *Canadian Journal of Earth Sciences*, **38**, 1335–1354.

TRIANTAFYLLIDIS, S., PE-PIPER, G., MACKAY, R., PIPER, D. J. W. & STRATHDEE, G. 2010. *Monazite as a provenance indicator for the Lower Cretaceous reservoir sandstones, Scotian Basin*. Geological Survey of Canada, Open File, **6732**, 452, ftp://ftp2.cits.rncan.gc.ca/pub/geott/ess_pubs/287/287317/of_6732.pdf (accessed 14 June 2013).

TSIKOURAS, B., PE-PIPER, G., PIPER, D. J. W. & SCHAFFER, M. 2011. Varietal heavy mineral analysis of sediment provenance, Lower Cretaceous Scotian Basin, eastern Canada. *Sedimentary Geology*, **237**, 150–165.

TUCHOLKE, B. E., SAWYER, D. S. & SIBUET, J.-C. 2007. Breakup of the Newfoundland–Iberia rift. *In*: KARNER, G. D., MANATSCHAL, G. & PINHEIRO, L.

M. (eds) *Imaging, Mapping and Modelling Continental Lithosphere Extension and Breakup*. Geological Society, London, Special Publications, **282**, 9–46.

VALDES, P. J., SELLWOOD, B. W. & PRICE, G. D. 1996. Evaluating concepts of Cretaceous equability. *Palaeoclimates*, **2**, 139–158.

VAN STAAL, C. R. 2005. Northern appalachians. *In*: SELLEY, R. C., COCKS, R. L. M. & PLIMER, I. R. (eds) *Encyclopedia of Geology*. Elsevier, Amsterdam, **4**, 81–91.

WADE, J. A. & MACLEAN, B. C. 1990. Aspects of the geology of the Scotian Basin from recent seismic and well data. Chapter 5. *In*: KEEN, M. J. & WILLIAMS, G. L. (eds) *Geology of the Continental Margin Off Eastern Canada*. Geological Survey of Canada, Ottawa, Geology of Canada, no. 2, 190–238.

WALDRON, J. W. F., ANDERSON, S. D. *ET AL.* 1998. Evolution of the Appalachian Laurentian margin: lithoprobe results in western Newfoundland. *Canadian Journal of Earth Sciences*, **35**, 1271–1287.

WILLIAMS, H. & GRANT, A. C. 1998. *Tectonic assemblages, Atlantic region, Canada. [1:3m map]*. Geological Survey of Canada Open File, **3657**, ftp://ftp2.cits.rncan.gc.ca/pub/geott/ess_pubs/209/209977/gscof_3657_e_1998_mn01.pdf (accessed 20 June 2013).

WILLIAMS, G. L., ASCOLI, P., BARSS, M. S., BUJAK, J. P., DAVIES, E. H., FENSOME, R. A. & WILLIAMSON, M. A. 1990. Biostratigraphy and related studies. Chapter 3. *In*: KEEN, M. J. & WILLIAMS, G. L. (eds) *Geology of the Continental Margin off Eastern Canada*. Geological Survey of Canada, Ottawa, Geology of Canada, **2**, 87–137.

WITHJACK, M. O., SCHLISCHE, R. W. & BAUM, M. S. 2009. Extensional development of the Fundy rift basin, southeastern Canada. *Geological Journal*, **44**, 631–651.

# Tracing provenance and pathways of late Holocene fluvio-deltaic sediments by heavy-metal spatial distribution (Po Plain–Northern Apennines system, Italy)

ALESSANDRO AMOROSI* & IRENE SAMMARTINO

*Dipartimento di Scienze Biologiche, Geologiche e Ambientali, University of Bologna, Via Zamboni 67, 40127 Bologna, Italy*

*Corresponding author (e-mail: alessandro.amorosi@unibo.it)*



**Abstract:** The bulk geochemistry of 435 near-surface sediment samples from the southern Po Plain was used to identify the major sources of sediment delivered through distinct tracts of the routing system, from the Apenninic catchments to the Po Delta and the Adriatic coast. Sediment composition from the downstream reaches of the Po River and 23 Apenninic channel-levee river systems is fingerprinted by distinctive heavy metal (chromium and nickel) concentrations, which vary primarily as a function of the local ultramafic rock contribution. For any constant provenance domain, fine-grained (floodplain) sediments are invariably enriched in trace metals relative to their coarser-sized, channel-related counterparts, thus reflecting hydraulic sorting by crevasse and overbank processes. Once the geochemical signatures of fluvial endmembers are established, the relative contribution of the individual detrital sources to the downstream segments of the system can be assessed. Using an example from a multisourced supplied system, we outline the reconstruction of source-rock lithology and sediment pathways by combined sedimentological and geochemical studies as the basis for reliable estimates of sediment budgets in a source-to-sink context.


Quantitative provenance analysis (Weltje & von Eynatten 2004), that is, the possibility of reconstructing sediment transfer from the source area to the final sink, represents one of the most intriguing applications of sequence-stratigraphic concepts, with obvious high potential for hydrocarbon exploration and production. However, sediment-budget modelling from ancient successions is rarely constrained by accurate stratigraphic data, because of objective difficulties in unravelling sediment pathways and their evolution through time. In the case of late Holocene analogues, where high-resolution chronological control is available and sedimentological interpretation can be constrained on geomorphological grounds, the single factors that dictate stratigraphic architecture can be estimated independently, and processes of sediment transport and deposition and their evolution through time can be ascertained (Blum & Törnqvist 2000). This provides the basis for quantitative assessment of sediment storage and transfer across distinct segments of the routing system (Sømme *et al.* 2009; Martinsen *et al.* 2010; Paola & Leeder 2011).

Sandstone petrography is commonly used to detect changes in sediment flux within sedimentary successions. Because of the unsuitability of the finest fractions for petrographic analysis, however, although this technique can provide an important window into the reconstruction of sediment dispersal patterns, it cannot provide the full picture. In this article, we document the advantage of a geochemical approach, a method that is applicable to the whole spectrum of lithologies (McLennan *et al.* 1993; Pearce *et al.* 2010*b*).

Among the geochemical indicators that may help in the discrimination of distinct provenance signals, a large body of research has documented the high potential of Cr and Ni for fingerprinting ultramafic contributions in clastic sediments. This is because of the naturally elevated concentrations of Cr and Ni in ultramafic rocks, which can be orders of magnitude higher than values from normal continental crust rocks (Hiscott 1984; McLennan *et al.* 1993; Garver *et al.* 1996) In the Po Plain (northern Italy), one of the largest alluvial plains in Europe, the abundance of ultramafic detritus supplied by the extensive ophiolitic successions of the Western Alps and Northern Apennines (Fig. 1) has recently been interpreted as the cause of the 'anomalously' high Cr and Ni concentrations observed from middle Quaternary to modern sediments (Amorosi *et al.* 2002, 2007; Amorosi & Sammartino 2007). Such elevated Cr and Ni concentrations, which are recorded over wide areas (see the abandoned Po Delta lobe in Fig. 1), invariably exceed the Italian threshold limits designated for contaminated areas (150 mg/kg and


*From*: SCOTT, R. A., SMYTH, H. R., MORTON, A. C. & RICHARDSON, N. (eds) 2014. *Sediment Provenance Studies in Hydrocarbon Exploration and Production*. Geological Society, London, Special Publications, **386**, 313–325. First published online June 11, 2013, http://dx.doi.org/10.1144/SP386.6

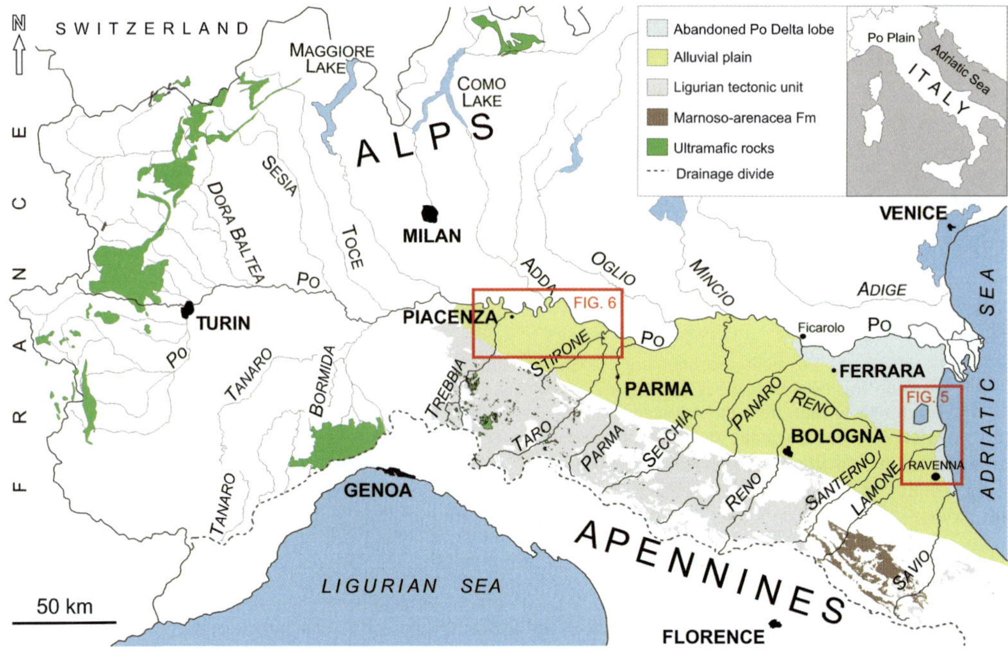



**Fig. 1.** Geological sketch map of southern Po Plain, and major structural units of the Northern Apennines (after *Regione Emilia–Romagna Geological Map, 1:250 000 scale*). Outcrops of ultramafic rocks supplying high Cr and Ni contents to the Po Plain via the Po River and its tributaries are shown in green.

120 mg/kg, respectively), with a consequent profound impact on environmental legislative measures (Amorosi & Sammartino 2011). Over the past two decades, assessment of heavy metal concentrations in the Po Plain area has focused primarily on stream sediments from the Apenninic catchments (Dinelli *et al.* 1999) and near-surface samples from the modern coastal plain and the delta (Curzi *et al.* 2006; Amorosi & Sammartino 2007; Amorosi *et al.* 2008; Amorosi 2012). Ironically, the wider alluvial plain between these two, non-adjacent segments of the sediment routing system has been mostly overlooked, with only a few exceptions (Bianchini *et al.* 2002; Sammartino *et al.* 2007; Amorosi & Sammartino 2011), and genetic relationships between metal distribution and putative source areas have been inferred, rather than demonstrated.

The primary objective of this work, which is part of a geochemical mapping project carried out in conjunction with the Geological Survey of Emilia-Romagna, is to fill this gap, using trace element evidence for sediment sources from the 20 000 km$^2$ wide alluvial plain south of the Po River (Fig. 1). Specific objectives are (1) to discuss the impact of transport mechanisms in the physical dispersal of heavy metals from a sedimentological (i.e. depositional facies) perspective, an

issue that has received little attention, with few exceptions (Miller 1997; Box & Wallis 2002; Garcia *et al.* 2004*a, b*; Myers & Thorbjornsen 2004; Jin *et al.* 2006; Amorosi & Sammartino 2007; Wright *et al.* 2010), and (2) to provide geochemical (heavy metal) characterization of crevasse and proximal (sandy) levee deposits. Within the fluvial system, comprising high-quality reservoir channel sandstones versus non-reservoir floodplain fines, this facies association represents a permeable conduit that ensures channel connectivity (Larue & Hovadik 2006) and may significantly contribute to the flow in terms of reservoir volumes. To this purpose, we present here a comprehensive heavy metal study of crevasse and overbank samples from the lower reaches of 23 roughly parallel, SW–NE trending, Apenninic rivers, 12 of which are tributaries of the Po River (Fig. 2). In total, the dataset comprises 435 samples from sand crevasse splays, heterolithic levee successions and distal floodplain settings.

## Geological setting

The Po Plain is the surface expression of a subsiding foreland basin of two collision orogens: the Alps and the Apennines (Fig. 1). These mountain chains represent thrust-and-fold belts with opposite structural vergence (Ricci Lucchi 1986; Doglioni 1993). The Po Plain is made up of a major trunk river, the Po (the largest Italian river by length), which flows from west to east into the Adriatic Sea, and by a series of transverse tributaries. Limited detrital production is associated with the Apenninic alluvial fans, in contrast to the higher detrital supply from the Alpine catchments (Ori 1993; Garzanti *et al.* 2011). Apenninic and especially Alpine catchments show strongly heterogeneous compositional signatures, revealing the complex geological history of these two orogens.

A variety of petrographic studies have been carried out north of the Po River (Stefani 2002; Garzanti *et al.* 2011). Reconstructing the sediment contribution from the Alpine tributaries is beyond the scope of this paper, which is focused instead on the southern part (Apennines supplied) of the plain. On the Apenninic side of the Po Plain, petrographic studies have focused on either modern stream sands, close to the basin margin (Di Giulio *et al.* 2003; Lugli *et al.* 2007), or coastal-plain and deltaic deposits (Gazzi & Zuffa 1970; Marchesini *et al.* 2000). One of the first studies to address the sediment composition of rivers draining the Northern Apennines, with specific reference to ophiolitic provenance, is the one by Garzanti *et al.* (1998).

A prominent feature of the Western Alpine chain is the presence of extensive ultramafic (ophiolitic) units of Mesozoic age (Fig. 1). Comparatively smaller volumes crop out in the Northern Apennines. These rocks, which are tectonically included at the base of the External Ligurian Units (see Ligurian tectonic unit in Fig. 1), are representative of a strongly thinned continent–ocean transition zone at the edge of the Tethyan ocean (Marroni *et al.* 1998).

The Po Plain south of the Po River consists of late Holocene (mostly post-Roman age) deposits that can be grouped into two main depositional systems (Fig. 1): (1) alluvial deposits, extending from the Apenninic foothills to the Po River (and Reno River, NE of Bologna) and (2) the abandoned Po Delta lobe, which was active SE of Ferrara before a major Po River avulsion in Ficarolo in AD 1152 (Ciabatti 1967; Bondesan *et al.* 1995). As a result of river avulsion, the Po Delta switched northward, towards the modern position, whereas the southern branch of the formerly active delta (*Po di Primaro*) was incorporated into the Apenninic fluvial network (modern Reno River). Late Holocene coastal-plain progradation is recorded east of Ravenna by a series of roughly parallel, north–south elongated and closely spaced beach ridges that reflect this complex depositional history.

The alluvial plain depositional system can be subdivided into two major facies associations (Fig. 2): channel-related and flood basin, respectively. Channel-related deposits include crevasse sand, with either fining-upward (crevasse channel) or coarsening-upward (crevasse splay) tendencies, and levee sandy silt to fine sand intercalations. Flood basin deposits, corresponding to overbank muds, consist of silt and clay, with rare, thin sand intercalations. Organic-rich layers are commonly encountered. These latter layers increase in frequency and thickness close to the tributary–trunk junctions, where lowland swamps were most abundant.

## Sampling and analytical procedures

A total of 435 samples were collected by hand drilling, using Eijkelkamp Agrisearch equipment. To avoid possible anthropogenic effects on metal determinations, alluvial sediment samples were collected at depths between 120 and 130 cm and stream sediments were not considered. The selection of sampling sites was not performed using regular grids, but was finalized to cover the entire spectrum of potential Apenninic sources (see river catchments in Fig. 2). Sampling was restricted to late Holocene deposits, for which weathering effects are negligible.

Sediments were analysed in bulk by X-ray fluorescence spectrometry (XRF) at Bologna University. Samples, pressed into tablets, were analysed for major and trace elements in a Philips PW1480 spectrometry with a Rh tube, using the matrix correction methods of Franzini *et al.* (1972; 1975), Leoni & Saitta (1976) and Leoni *et al.* (1982). The accuracy of determinations for trace elements is 5% (<3% for Cr, <2% for Ni, *c.* 2% for Cu, *c.* 3% for Zn), except for elements with concentrations of 10 ppm or lower, where the accuracy is 10–15%.

Although samples at depths greater than 1 m have been shown to be generally suitable for determining natural background values for potentially toxic metals, even in highly industrialized areas (Huisman *et al.* 1997), we ruled out a possible anthropogenic influence on measured metal concentrations by carrying out a comparison with geochemical analyses of 435 superficial samples, collected at the same sites, but at shallower depths (20–30 cm). The geochemical database for these soil samples is not presented here. Only samples showing either enrichment factor (EF) values equal to zero (Rubio *et al.* 2000) or negative values of geoaccumulation index $I_{geo}$ (Förstner & Müller 1981; Banat *et al.* 2005; Sainz & Ruiz 2006) were considered for this study.

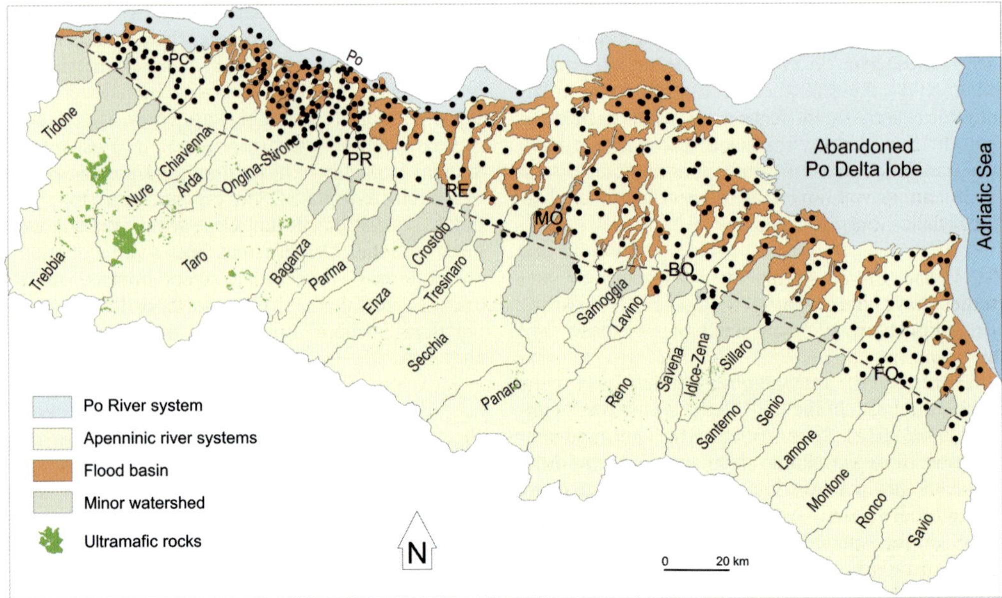


**Fig. 2.** Simplified geological map of the Po Plain south of the Po and Reno rivers (see Fig. 1), showing sampling localities (black dots). The dashed line marks the boundary between the Apenninic catchments and the Po Plain. River systems north of this line include undifferentiated fluvial channel, crevasse and levee deposits. Minor flood basins are not shown. PC, Piacenza; PR, Parma; RE, Reggio Emilia; MO, Modena; BO, Bologna; FO, Forlì.

## Heavy metals as tracers of sediment provenance in a source-to-sink context

Assessing the patterns of heavy metal distribution from the southern Po Plain provides a link between previous studies from the Apenninic catchments and the coastal area. The following sections explore previously published material and new data on metal distribution from the subaerial portion of the Po Plain–Adriatic system, with special emphasis on the major tracers of sediment provenance related to the ultramafic component.

### The Apenninic catchments

Ophiolitic rocks are widely exposed in the headwaters and upstream reaches of several Apenninic rivers draining the Ligurian tectonic unit, west of Sillaro River (Figs 1 & 2). The largest ultramafic-rock outcrops that contribute ophiolitic detritus to the Po Plain and to the adjacent Adriatic coastal system, via the Po River, are clustered at the NW tip of the Apennines, especially in the Trebbia, Nure and Taro river catchments (Fig. 2). At these locations, the ophiolitic successions include large volumes of serpentinite, gabbro, basalts and thick ophiolitic breccias. Small ultramafic-rock outcrops are located in the other Apenninic catchments, between Tidone and Sillaro rivers (Fig. 2).

A markedly different erosion pattern typifies the Apenninic catchments between Santerno and Savio river valleys (Fig. 2). The Ligurian tectonic unit is lacking in this part of the Apennines (Fig. 1), so no ultramafic detritus is delivered to the alluvial plain and to the Adriatic coastal area from this part of the orogen, which is virtually ophiolite-free. In this area, the Apenninic rivers supply essentially monogenic detritus from a thick sedimentary cover made up almost entirely of turbiditic sandstone–marlstones alternations (Miocene Marnoso-arenacea Formation; Fig. 1).

Few geochemical studies have documented the spatial metal distribution from this part of the Apenninic chain. Distinctly low Cr and Ni concentrations have been reported from the turbiditic Marnoso-arenacea Formation (Dinelli *et al.* 1999), and consistent values have been obtained from stream sediment samples (see Reno, Lamone and Savio rivers in Dinelli & Lucchini 1999). By contrast, high Cr and Ni values have been reported from stream sediments of the Po River (Dinelli & Lucchini 1999) and its tributaries in the western Alpine region (Madrid *et al.* 2006; Bonifacio *et al.* 2010).

### The alluvial plain

Among the mutually interacting variables that may influence sediment composition, sediment provenance and grain size have been observed to represent the major controlling factors of spatial metal distribution in sediments (Salminen & Tarvainen 1997; Ruiz 2001; Pearce *et al.* 2010*a*). The following sections document to what extent the spatial distribution of selected trace metals from crevasse, levee and floodplain sediments of the Po Plain can reveal information about the depositional facies and source areas.

*Zn, V and Cu as grain size indicators.* Several studies have documented the generally positive correlation between trace metal concentration and the proportion of fine-grained material in alluvial sediment (Wolfenden & Lewin 1978; Singh *et al.* 1999; Singh & Rajamani 2001; Whitmore *et al.* 2004; Wyżga & Ciszewski 2010). This tendency has been interpreted to reflect textural modifications that occur while fluvial sediment undergoes hydraulic selection and sorting with increasing distance from the channel (see intersample variability of Garzanti *et al.* 2009). In alluvial samples from the Po Plain, the distribution of selected heavy metals also appears to be markedly size-dependent, and significant lateral variations in metal concentrations occur as a function of the

depositional facies. In particular, the progressive increase in mud away from the channel axis is paralleled by systematically increasing metal contents (Fig. 3). Fine-grained overbank (flood basin) deposits display invariably higher Zn and V contents than their coarser-grained, channel-related (crevasse and levee) counterparts, irrespective of sediment provenance and composition (Fig. 3a), and the same holds for Cu and V, though with lower correlation coefficient (Fig. 3b).

The strongly positive correlation between metal contents and grain size suggests systematic variations in the mineralogy at the transition between finer bed load (crevasse channel sand), deep suspended load (levee silty sand) and shallow suspended load (flood basin mud; see Garzanti *et al.* 2012). These are controlled largely by the particular hydrodynamic behaviour associated with traction and traction-plus-fallout processes, which are active simultaneously in the different parts of the alluvial plain during floods. Lacking quantitative data on particle size distribution, we are not able to explore in detail the relation of stream sediment source area, grain size and composition to trace element chemistry (Horowitz & Elrick 1987).

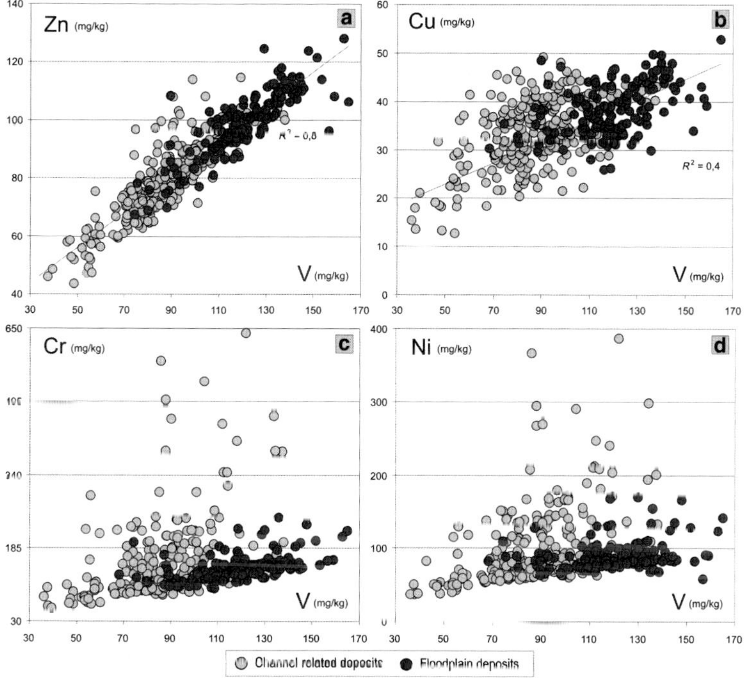



**Fig. 3.** Positive covariance of Zn/V (**a**), Cu/V (**b**), Cr/V (**c**) and Ni/V (**d**), invariably showing higher metal values in fine-grained (floodplain) deposits relative to their coarser-grained (channel-related) counterparts. Note the local Cr and Ni enrichments from highly weathered, alluvial-fan deposits (see text).

A generally strong positive correlation is also observed between Cr and V (Fig. 3c) and between Ni and V (Fig. 3d) for comparatively low Cr and Ni values (Fig. 3c, d), confirming the marked control of grain size on metal distribution. In the case of these two metals, however, a few tens of samples display distinctly higher Cr/V and Ni/V ratios, thus reflecting significant Cr and Ni enrichments. This is likely to reflect additional metal supply from a separate mineral form (see next section).

*Ni and Cr as tracers of sediment provenance.* In order to compensate for variability in grain size, a feature that may deeply affect sediment provenance reconstructions (Weltje & von Eynatten 2004; Lim *et al.* 2006; Garzanti *et al.* 2009; Liu *et al.* 2010), we used $Al_2O_3$ as a grain-size proxy to normalize the geochemical data (see Loring 1991; Daskalakis & O'Connor 1995; Liaghati *et al.* 2003). The $Cr/Al_2O_3$ ratio has been shown to be poorly influenced by grain size variations (Dinelli *et al.* 2007), and represents a powerful marker of sediment provenance in the Po coastal plain (Amorosi *et al.* 2002, 2008; Amorosi & Sammartino 2007), separating ophiolite-bearing detritus ($Cr/Al_2O_3 > 12$) from the ophiolite-free sediment contribution ($Cr/Al_2O_3 < 12$).

The application of $Cr/Al_2O_3$ and $Ni/Al_2O_3$ ratios to 258 channel-related (crevasse and proximal levee) sediment samples, for which provenance from the corresponding catchment is indubitable, provides clues to the composition of the feeding sources, reflecting erosion of different types (ultramafic v. non-ultramafic) rocks exposed in the corresponding drainage basins (Fig. 4). The spatial distribution trends of these two ratios highlight substantial differences in the proportions of Cr- and Ni-rich sediment being delivered to the Po Plain. On a basin scale, the threshold value of $Cr/Al_2O_3 = 12$ enables the prompt distinction of two major provenance-related components, based on the presence/lack of ultramafic detritus supplied downstream to the aggrading floodplains (compare Fig. 4 with Fig. 2). Specifically, sediment supplied by Po River and five Apenninic tributaries (Trebbia, Nure, Arda, Taro and Baganza rivers, from west to east) exhibits high Cr and Ni concentrations, with $Cr/Al_2O_3$ and $Ni/Al_2O_3$ values invariably higher than 12 and 8, respectively (Fig. 4). This is in agreement with the abundance of ultramafic rocks in the related catchments (see Figs 1 & 2), and recent estimates of a high ophiolitic contribution to the Po River (3.5–7.0% of total detritus) and to selected Apenninic rivers, such as Trebbia and Nure (up to 20% of bedload sand; Garzanti *et al.* 2012).

Among the Cr- and Ni-rich alluvial facies, sediment delivered to the Po Plain by the Trebbia and Nure rivers displays the highest $Cr/Al_2O_3$ and $Ni/Al_2O_3$ values. Two major observations may account for this particular metal distribution: (1) the Trebbia and Nure drainage basins exhibit the largest exposed volumes of ultramafic rocks in the study area (Fig. 2) and (2) owing to the proximity of the Apenninic foothills to the Po River in this area (Fig. 2), the sediment sampled by hand drilling from the Trebbia and Nure systems corresponds to a highly weathered, pre-Holocene (Last Glacial Maximum) alluvial-fan (coarse sand) deposit rather than a genuine (late Holocene) crevasse-levee (sandy-silt) facies. The markedly higher ophiolitic contribution to bed-load sand relative to suspended-load mud reported by Garzanti *et al.* (2012) for the Trebbia and Nure river systems is likely to account for such significantly higher $Cr/Al_2O_3$ and $Ni/Al_2O_3$ values.

The relatively high $Cr/Al_2O_3$ values, close to 12, recorded from Tidone, Enza and Idice-Zena river deposits (Fig. 4), as opposed to the comparatively small volumes of ultramafic rocks in the related catchments (Fig. 2), suggest that even trivial amounts of certain exotic components, such as Cr and Ni, and thus a markedly diluted ultramafic contribution, can substantially influence sediment composition, especially if transport is short and the catchment areas are small.

Finally, sediment delivered from all other catchments displays comparably low $Cr/Al_2O_3$ and $Ni/Al_2O_3$ ratios (Fig. 4). These values are consistent with catchments characterized by negligible ultramafic rock contribution (see Fig. 2).

## The coastal plain and the delta

Before being transferred to the downdip components of the sediment routing system, significant volumes of sediment from the Apenninic catchments and Po River are trapped or temporarily stored in coastal areas. Here, sediment contributed by the Po River through its delta merges with sediment supplied directly to the Adriatic Sea from the Apenninic chain (east of Panaro River, see Fig. 1), being redistributed in nearshore subenvironments by marginal marine (waves, longshore currents, etc.) processes.

A detailed geochemical investigation of the abandoned Po Delta lobe and of the adjacent strandplains (Fig. 1) has been carried out by previous workers (Amorosi & Sammartino 2007; Amorosi 2012) and will not be reiterated here in detail. We simply remark that, like for alluvial plain sediment, Cr and Ni contents in the delta are positively correlated to the relative proportion of mud, the highest values being recorded within the finest-grained, interdistributary (swamp, bay and lagoonal) deposits (Amorosi & Sammartino 2007).

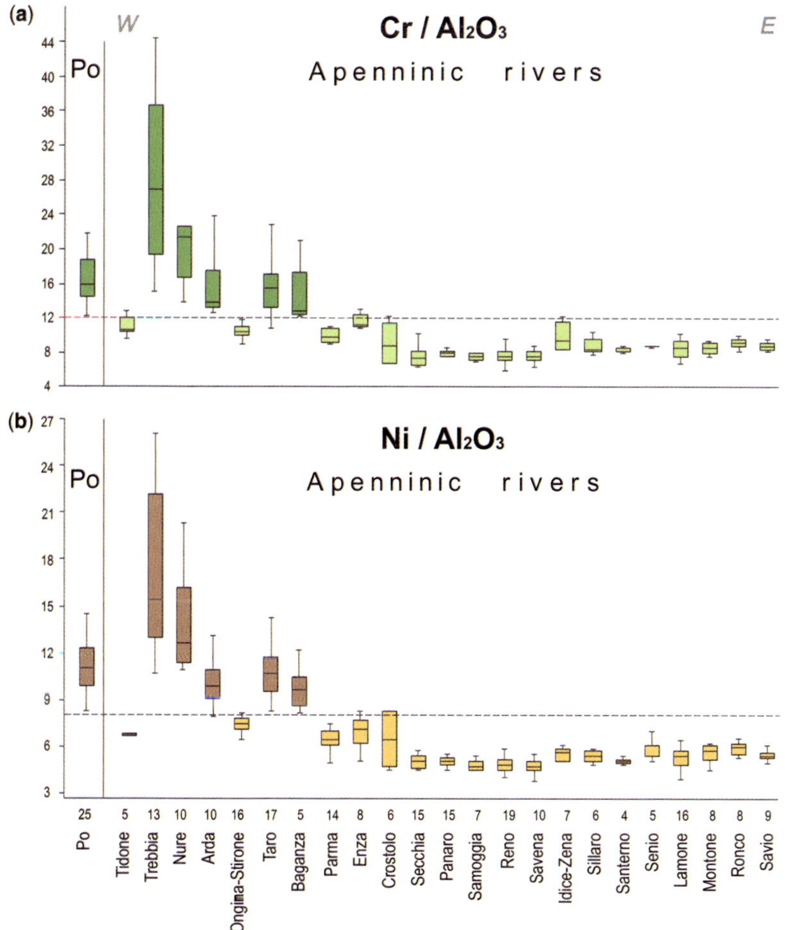



**Fig. 4.** Boxplots of Cr/Al₂O₃ (**a**) and Ni/Al₂O₃ (**b**) from the Po River and 23 Apenninic river systems, emphasizing the influence of sediment provenance on natural metal concentrations in the Po Plain (compare with the distribution of ultramafic rocks in the catchments, Fig. 2). Darker shades denote ultramafic contributions. Dashed lines mark the threshold values of Cr/Al₂O₃ = 12 and Ni/Al₂O₃ = 8 used for the discrimination between ophiolite-bearing detritus and ophiolite-free sediment contribution. The lower boundary of each box is the 25th percentile; the upper boundary is the 75th percentile; the bold line within the box corresponds to the median; the 'whiskers' define the minimum and maximum values. The number of analyses for each river system is indicated. River systems for which a minimum of four analyses were not available were not considered in this study.

As for the alluvial plain, the Cr/Al₂O₃ ratio also acts as a powerful provenance indicator for the delta and coastal system, allowing fingerprinting of markedly distinct sediment sources irrespective of textural features. Specifically, the same value (12) of Cr/Al₂O₃ ratio used for the alluvial plain deposits to differentiate ophiolite-bearing from ophiolite-free detritus can be applied readily to the modern coastal plain for the discrimination between Po-supplied material and Apenninic, ophiolite-free detritus (Fig. 5). For example, the modern Reno River drains an almost ophiolite-free catchment (Figs 2 & 4), delivering to the Adriatic coast and its modern beaches detritus with low (≪12) Cr/Al₂O₃ values (yellow beach ridges in Fig. 5). Before the historical Po River avulsion in Ficarolo (twelfth century AD), which shifted the Po Delta northward, the same fluvial channel was active as the southernmost branch (=distributary channel) of a now abandoned Po Delta lobe, thus feeding an ancient wave-dominated delta front and strandplain system (green beach ridges in Fig. 5). This results in significantly higher (≫12) Cr/Al₂O₃ ratios, which reflect direct supply from a Po River mouth. Similarly, alternating high and low Cr/Al₂O₃ values from crevasse sand of the *Po di*

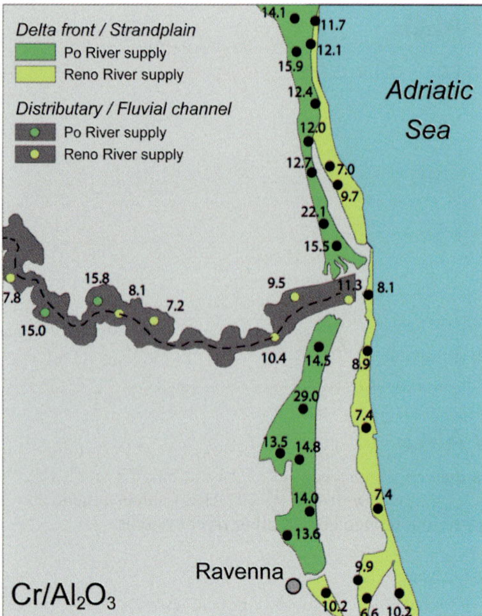


**Fig. 5.** Provenance and pathways of Po River v. Apenninic supplied detritus in the Po coastal plain (see Fig. 1 for location), as deduced from its heavy metal ($Cr/Al_2O_3$) signature (compare with Fig. 4). Contrasting metal values from lithologically homogeneous, sand (crevasse splay or beach ridge) lithosomes reflect reoccupation by the Reno River of a formerly active Po Delta distributary channel (modified after Amorosi & Sammartino 2007).

*Primaro*/Reno river system (Fig. 5) are interpreted to reflect crevasse lobes supplied by the Po distributary channel and Reno River, before and after the Ficarolo avulsion, respectively.

Trace metal geochemistry thus enables the discrimination of lithologically homogeneous crevasse and beach facies, resulting in a reliable reconstruction of sediment dispersal pathways and their evolution through time.

## Implications for source-to-sink studies and hydrocarbon exploration

The case study explored in this article represents a modern analogue for an intricate, older fluviodeltaic system, and provides an example of the complexity of a multisource supplied system that can be unravelled using trace element geochemistry. In a hydrocarbon exploration and production perspective, the combined use of selected heavy metals as both provenance and grain size indicators and its framing into a sequence stratigraphic context can be used pragmatically (1) to depict realistic scenarios of subsurface facies distribution, (2) to reduce the geological uncertainty about reservoir geometries, and (3) to help predict the spatial distribution of petrophysical properties through geochemical characterization of fining trends.

Geochemical signatures based on single indicators are in general poorly sensitive, that is, not distinct enough to warrant clear provenance assessments. For this reason, source discrimination is commonly based on a combination of geochemical parameters plus calibration with heavy mineral analysis (see Ratcliffe *et al.* 2004). The example shown here is different in that Ni and Cr act as powerful source discriminants for the Apenninic–Po Plain–Adriatic system, resulting in an accurate reconstruction of the sediment dispersal patterns from the river catchments to the coastal area. The interpretive framework shown in this article is based on a geomorphologically constrained system (Fig. 2), where channel–source area relationships are clearly observed rather than inferred.

Once geochemical signatures of source-rock composition are established in the primary endmembers (channel-related facies), the geochemical signal can be efficiently exported to areas where drainages are merging and mixing of source lithologies can affect discrimination of sediment provenance. To this purpose, we offer a real example of sediment mixing from a group of small flood basins, separated by the lower reaches of nine Apenninic rivers, east of Piacenza (Fig. 6). These flood basins, all located at the tributaries–trunk (Po River) junction (see Fig. 2), exhibit homogeneous silt–clay lithology, but highly contrasting geochemical signatures (Fig. 4). Cr-rich sediment is supplied by the Po River and by four Apenninic tributaries (Trebbia, Nure, Arda and Taro), whereas Cr-poor material is fed by Tidone, Chiavenna, Ongina-Stirone and Parma rivers (Fig. 6). Application of the $Cr/Al_2O_3$ provenance index to 28 floodplain samples reveals the coherent spatial distribution of the $Cr/Al_2O_3$ ratios as a function of the contributing river systems (compare with Fig. 4). This enables prompt attribution of each sample to either a specific endmember, or to mixing between the two (Fig. 6).

There are several valuable lessons that can be learned from the Holocene examples documented in this article, and which can be applied to ancient analogues to complement chemostratigraphic techniques (see Wright *et al.* 2010):

(1)  Unlike ancient successions, where individual causes of compositional variability are generally very difficult to unravel, recent (late Holocene) sedimentary successions provide an excellent opportunity to understand the

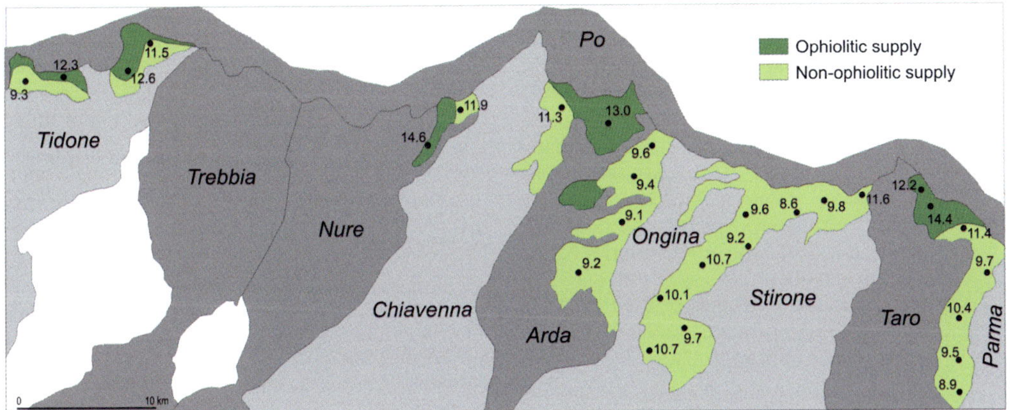


**Fig. 6.** Tracing sediment provenance from lithologically homogeneous, mud (flood basin) lithosomes (in colour) through their heavy metal ($Cr/Al_2O_3$) signature at the NW tip of the study area (see Fig. 1 for location). Tributaries that supply ophiolitic detritus to the Po River are shown in dark grey (compare with Fig. 4). $Cr/Al_2O_3$ values around 12 reflect sediment mixing of ophiolitic and non-ophiolitic detritus by the related contributing river systems.

relative impact on heavy metal distribution of source-rock composition and changes in grain size, helping to predict whether and to what extent these two variables might be combined or superposed in the rock record.

(2) Among the variety of facies associations that compose the sediment routing system, channel-related deposits, for which connection between river transport and hinterland sediment supply can reasonably be inferred, represent the building block for a comprehensive reconstruction of sediment transfer and storage. For trace metal characterization, levee and crevasse fine to medium sand appears more suitable to provenance analysis than its coarser-grained (bar sand) counterpart.

(3) Identification of endmembers for individual catchment–river systems with sufficient precision (see Garzanti *et al.* 2012) may help disentangle provenance mixing from lithologically homogeneous, alluvial (floodplain), deltaic (bay, lagoon, delta front) and nearshore (beach) facies associations, for which the relative contribution from individual detrital sources is virtually unknown

The technique shown in this study can also be applied successfully to diagenetically modified sediments. Several studies have shown that diagenetic processes are unlikely to significantly modify the distribution of Cr, owing to the very low postdepositional mobility of this metal (Mange & Morton 2007). In sedimentary successions undergoing extensive diagenetic alteration, Cr spinels are resistant to dissolution and generally left

as mineral residue (Hounslow 1996). As a consequence, the geochemical signature of Cr as provenance indicator will hardly be masked completely by diagenesis.

The three segments analysed in this article represent only the subaerial and transitional parts of the whole Po Plain–Adriatic sediment routing system. Recent studies conducted in the open-marine tract (the Adriatic Sea) suggest that the ophiolitic signature (i.e. high natural Cr and Ni concentrations) identified from the Apenninic catchments to the delta may serve as a reliable provenance indicator for other components of the system, such as the prodelta (Amorosi *et al.* 2007, 2008), the shelf (Picone *et al.* 2008) and the deep-marine environment (Lucchini *et al.* 2003). An additional Cr and Ni contribution to the Adriatic Sea (Prohic & Juracic 1989; Rivaro *et al.* 2004) is expected from the eastern Adriatic catchments (Albania, Croatia), which drain huge volumes of Cr- and Ni-rich ophiolitic material from the Dinaric chain (Lužar-Oberiter *et al.* 2009). Further studies, however, are needed to corroborate this hypothesis.

## Conclusions

Reliable estimates of sediment transport and storage from a multisourced system across a variety of depositional environments require accurate identification of sediment sources and their role in modulating sediment supply to the downdip components of the dispersal system. By using a modern example from the Po Plain, involving three segments (drainage area, alluvial plain and coastal plain/delta) of the system, we provide additional insight on the critical role of a combined sedimentological and geochemical approach, which can be highly reproducible in the ancient record.

In sedimentary basins characterized by multiple sediment sources, reconstructing sediment pathways is not straightforward. This study shows that, if vastly contrasting source rock compositions are present, trace element geochemistry of surface sediments can play a decisive role in the discrimination of sediment provenance.

The key points and outcomes derived from the spatial metal distribution in the southern Po Plain can be summarized as follows:

(1) Contrasting heavy metal signatures can be used to trace inheritance from source rocks and sediment pathways across the distinct segments of the Apennines–Po Plain–Adriatic routing system. Based on geochemical analysis of 435 alluvial samples from the southern Po Plain, this study outlines the presence of Cr and Ni in different proportions within alluvial and coastal plain sediment as an efficient tool to infer distinct ultramafic contribution from the catchment areas.

(2) Consistent variations of heavy metal contents across floodplain and channel-related (crevasse and levee) depositional facies provide substantial constraints on the effects of overbank processes. These tend to accumulate trace metals within the finer-grained (mud) fractions as a function of hydraulic sorting during sediment transport and settling.

(3) Calibration of geochemical data against sedimentological information can enhance the geochemical interpretation in a source-to-sink context. Once the relative importance of provenance and grain size control on sediment composition is assessed, heavy metal distribution across different components of the routing system can be established, and sediment mixing from alluvial to deep-marine deposits of unknown affinity can be disentangled.

(4) The use of Cr distribution for assessing sediment provenance may have great potential for hydrocarbon exploration and mineral prospecting in sedimentary rocks. Cr minerals, such as Cr-spinels, are highly stable and generally preserve their compositional signature after burial. Owing to its poor sensitivity to diagenetic control, Cr distribution is likely also to represent a reliable provenance indicator in diagenetically modified successions.

This study was financially supported by Regione Emilia–Romagna as part of a research project on the pedogeochemical mapping of southern Po Plain (scientific coordinator, Alessandro Amorosi). We are indebted to Raffaele Pignone and Marina Guermandi for permission to publish these data. Marina Guermandi and Nazaria Marchi are warmly thanked for providing soil data and interpretations. We are highly indebted to Daniel Garcia and Eduardo Garzanti for their constructive reviews, and to Robert Scott and Andy Morton for editorial suggestions.

## References

AMOROSI, A. 2012. Heavy metals as indicators of source-to-sink sediment transfer in a Holocene alluvial and coastal system (Po Plain, Italy). *Sedimentary Geology*, **280**, 260–269.

AMOROSI, A. & SAMMARTINO, I. 2007. Influence of sediment provenance on background values of potentially toxic metals from near-surface sediments of Po coastal plain (Italy). *International Journal of Earth Sciences*, **96**, 389–396.

AMOROSI, A. & SAMMARTINO, I. 2011. Assessing natural contents of hazardous metals by different analytical methods and its impact on environmental legislative measures. *International Journal of Environment and Pollution*, **46**, 164–177.

AMOROSI, A., CENTINEO, M. C., DINELLI, E., LUCCHINI, F. & TATEO, F. 2002. Geochemical and mineralogical variations as indicators of provenance changes in Late Quaternary deposits of SE Po Plain. *Sedimentary Geology*, **151**, 273–292.

AMOROSI, A., COLALONGO, M. L., DINELLI, E., LUCCHINI, F. & VAIANI, S. C. 2007. Cyclic variations in sediment provenance from late Pleistocene deposits of the eastern Po Plain, Italy. *In*: ARRIBAS, J., CRITELLI, S. & JOHNSSON, M. J. (eds) *Sedimentary Provenance and Petrogenesis: Perspectives from Petrography and Geochemistry*, Geological Society of America Special Papers, **420**, 13–24.

AMOROSI, A., DINELLI, E., ROSSI, V., VAIANI, S. C. & SACCHETTO, M. 2008. Late Quaternary palaeoenvironmental evolution of the Adriatic coastal plain and the onset of Po River Delta. *Palaeogeography, Palaeoclimatology, Palaeoecology*, **268**, 80–90.

BANAT, K. M., HOWARI, F. M. & AL-HAMAD, A. A. 2005. Heavy metals in urban soils of central Jordan: should we worry about their environmental risks? *Environmental Research*, **97**, 258–273.

BIANCHINI, G., LAVIANO, R., LOVO, S. & VACCARO, C. 2002. Chemical–mineralogical characterisation of clay sediments around Ferrara (Italy): a tool for environmental analysis. *Applied Clay Science*, **21**, 165–176.

BLUM, M. D. & TÖRNQVIST, T. E. 2000. Fluvial response to climate and sea-level change: a review and look forward. *Sedimentology*, **47**, 2–48.

BONDESAN, M., FAVERO, V. & VIÑALS, M. J. 1995. New evidence on the evolution of the Po-delta coastal plain during the Holocene. *Quaternary International*, **29/30**, 105–110.

BONIFACIO, E., FALSONE, G. & PIAZZA, S. 2010. Linking Ni and Cr concentrations to soil mineralogy: does it help to assess metal contamination when the natural background is high? *Journal of Soil and Sediments*, **10**, 1475–1486.

Box, S. & Wallis, J. C. 2002. *Surficial geology along the Spokane River, Washington and its relationship to the metal content of sediments*. US Geological Survey, Open-File Report **02–126**, 1–16.

Ciabatti, M. 1967. Ricerche sull'evoluzione del Delta Padano. *Giornale di Geologia*, **34**, 381–410.

Curzi, P. V., Dinelli, E., Ricci Lucchi, M. & Vaiani, S. C. 2006. Palaeoenvironmental control on sediment composition and provenance in the late Quaternary deltaic successions: a case study from the Po delta area (Northern Italy). *Geological Journal*, **41**, 591–612.

Daskalakis, K. D. & O'Connor, T. P. 1995. Normalization and elemental sediment contamination in the coastal United States. *Environmental Science and Technology*, **29**, 470–477.

Di Giulio, A., Ceriani, A., Ghia, E. & Zucca, F. 2003. Composition of modern stream sands derived from sedimentary source rocks in a temperate climate (Northern Apennines, Italy). *Sedimentary Geology*, **158**, 145–161.

Dinelli, E. & Lucchini, F. 1999. Sediment supply to the Adriatic Sea basin from the Italian rivers: geochemical features and environmental constraints. *Giornale di Geologia*, **61**, 121–132.

Dinelli, E., Lucchini, F., Mordenti, A. & Paganelli, L. 1999. Geochemistry of Oligocene–Miocene sandstones of the northern Apennines (Italy) and evolution of chemical features in relation to provenance changes. *Sedimentary Geology*, **127**, 193–207.

Dinelli, E., Tateo, F. & Summa, V. 2007. Geochemical and mineralogical proxies for grain size in mudstones and siltstones from the Pleistocene and Holocene of the Po River alluvial plain, Italy. *In*: Arribas, J., Critelli, S. & Johnsson, M. J. (eds) *Sedimentary Provenance and Petrogenesis: Perspectives from Petrography and Geochemistry*, Geological Society of America Special Papers, **420**, 25–36.

Doglioni, C. 1993. Some remarks on the origin of foredeeps. *Tectonophysics*, **228**, 1–20.

Förstner, U. & Müller, G. 1981. Concentrations of heavy metals and polycyclic aromatic hydrocarbons in river sediments: geochemical background, man's influence and environmental impact. *Geojournal*, **5**, 417–432.

Franzini, M., Leoni, L. & Saitta, M. 1972. A simple method to evaluate the matrix effects in X-ray fluorescence analysis. *X-Ray Spectrometry*, **1**, 151–154.

Franzini, M., Leoni, L. & Saitta, M. 1975. Revisione di una metodologia analitica per fluorescenza-X basata sulla correzione completa degli effetti di matrice. *Rendiconti della Società Italiana di Mineralogia e Petrologia*, **31**, 365–378.

Garcia, D., Joseph, P., Maréchal, B. & Moutte, J. 2004a. Patterns of geochemical variability in relation to turbidite facies in the Grès d'Annot Formation. *In*: Joseph, P. & Lomas, S. A. (eds) *Deep-Water Sedimentation in the Alpine Basin of SE France: New Perspectives on the Grès d'Annot and Related Systems*. Geological Society, London, Special Publications, **221**, 349–365.

Garcia, D., Ravenne, C., Maréchal, B. & Moutte, J. 2004b. Geochemical variability induced by entrainment sorting: quantified signals for provenance analysis. *Sedimentary Geology*, **171**, 113–128.

Garver, J. I., Royce, P. R. & Smick, T. A. 1996. Chromium and nickel in shale of the Taconic foreland: a case study for the provenance of fine-grained sediments with an ultramafic source. *Journal of Sedimentary Research*, **66**, 100–106.

Garzanti, E., Scutellà, M. & Vidimari, C. 1998. Provenance from ophiolites and oceanic allochthons: modern beach and river sands from Liguria and the Northern Apennines (Italy). *Ofioliti*, **23**, 65–82.

Garzanti, E., Andò, S. & Vezzoli, G. 2009. Grain-size dependance of sediment composition and environmental bias in provenance studies. *Earth and Planetary Science Letters*, **277**, 422–432.

Garzanti, E., Vezzoli, G. & Andò, S. 2011. Paleogeographic and paleodrainage changes during Pleistocene glaciations (Po Plain, Northern Italy). *Earth-Science Reviews*, **105**, 25–48.

Garzanti, E., Resentini, A., Vezzoli, G., Andò, S., Malusà, M. & Padoan, M. 2012. Forward compositional modelling of Alpine orogenic sediments. *Sedimentary Geology*, **280**, 149–164.

Gazzi, P. & Zuffa, G. G. 1970. Le arenarie paleogeniche dell'Appennino emiliano. *Mineralogica et Petrografica Acta*, **12**, 97–137.

Hiscott, R. N. 1984. Ophiolitic source rocks for Taconic-age flysch: trace-element evidence. *Geological Society of America Bulletin*, **95**, 1261–1267.

Horowitz, A. J. & Elrick, K. A. 1987. The relation of stream sediment surface area, grain size and composition to trace element geochemistry. *Applied Geochemistry*, **2**, 437–451.

Hounslow, M. W. 1996. Ferrimagnetic Cr and Mn spinels in sediments: residual magnetic minerals after diagenetic dissolution. *Geophysical Research Letters*, **23**, 2823–2826.

Huisman, D. J., Vermeulen, F. J. H., Baker, J., Veldkamp, A., Kroonenberg, S. B. & Klaver, G. Th. 1997. A geological interpretation of heavy metal concentrations in soils and sediments in the southern Netherlands. *Journal of Geochemical Exploration*, **59**, 163–174.

Jin, Z., Li, F., Cao, J., Wang, S. & Yu, J. 2006. Geochemistry of Daihai Lake sediments, Inner Mongolia, north China: implications for provenance, sedimentary sorting, and catchment weathering. *Geomorphology*, **80**, 147–163.

Larue, D. K. & Hovadik, J. 2006. Connectivity of channelized reservoirs: a modelling approach. *Petroleum Geoscience*, **12**, 291–308.

Leoni, L. & Saitta, M. 1976. X-ray fluorescence analysis of 29 trace elements in rock and mineral standard. *Rendiconti della Società Italiana di Mineralogia e Petrologia*, **32**, 497–510.

Leoni, L., Menichini, M. & Saitta, M. 1982. Determination of S, Cl and F in silicate rocks by X-ray fluorescence analysis. *X-Ray Spectrometry*, **11**, 156–158.

Liaghati, T., Preda, M. & Cox, M. 2003. Heavy metal distribution and controlling factors within coastal plain sediments, Bells Creek catchment, southeast Queensland, Australia. *Environment International*, **29**, 935–948.

LIM, D. I., JUNG, H. S., CHOI, J. Y., YANG, S. Y. & AHN, K. S. 2006. Geochemical compositions of river and shelf sediments in the Yellow Sea: grain-size normalization and sediment provenance. *Continental Shelf Research*, **26**, 15–24.

LIU, H., HE, Q., WANG, Z. B., WELTJE, G. J. & ZHANG, J. 2010. Dynamics and spatial variability of near-bottom sediment exchange in the Yangtze Estuary, China. *Estuarine, Coastal Shelf Science*, **86**, 322–330.

LORING, D. H. 1991. Normalization of heavy-metal data from estuarine and coastal sediments. *ICES Journal of Marine Science*, **48**, 101–115.

LUCCHINI, F., DINELLI, E. & MORDENTI, A. 2003. Geochemical records of palaeoenvironmental changes during Late Quaternary in Adriatic Sea sediments. *GeoActa*, **2**, 43–62.

LUGLI, S., MARCHETTI, D. S. & FONTANA, D. 2007. Alluvial sand composition as a tool to unravel late Quaternary sedimentation of the Modena Plain, Northern Italy. *In*: ARRIBAS, J., CRITELLI, S. & JOHNSSON, M. J. (eds) *Sedimentary Provenance and Petrogenesis: Perspectives from Petrography and Geochemistry*. Geological Society of America Special Papers, **420**, 57–72.

LUŽAR-OBERITER, B., MIKES, T., VON EYNATTEN, H. & BABIC, L. 2009. Ophiolitic detritus in Cretaceous clastic formations of the Dinarides (NW Croatia): evidence from Cr-spinel chemistry. *International Journal of Earth Sciences*, **98**, 1097–1108.

MADRID, L., DIAZ-BARRIENTOS, E. *ET AL.* 2006. Variability in concentrations of potentially toxic elements in urban parks from six European cities. *Journal of Environmental Monitoring*, **8**, 1158–1165.

MANGE, M. A. & MORTON, A. C. 2007. Geochemistry of heavy minerals. *In*: MANGE, M. A. & WRIGHT, D. T. (eds) *Heavy Minerals in Use*. Elsevier, Amsterdam, Developments in Sedimentology, **58**, 345–391.

MARCHESINI, L., AMOROSI, A., CIBIN, U., ZUFFA, G. G., SPADAFORA, E. & PRETI, D. 2000. Sand composition and sedimentary evolution of a Late Quaternary depositional sequence, northwestern Adriatic Coast, Italy. *Journal of Sedimentary Research*, **70**, 829–838.

MARRONI, M., MOLLI, G., MONTANINI, A. & TRIBUZIO, R. 1998. The association of continental crust rocks with ophiolites in the Northern Apennines (Italy): implications for the continent–ocean transition in the Western Tethys. *Tectonophysics*, **292**, 43–66.

MARTINSEN, O. J., SØMME, T. O., THURMOND, J. B., HELLAND-HANSEN, W. & LUNT, I. , 2010. Source-to-sink systems on passive margins: theory and practice with an example from the Norwegian continental margin. *In*: VINING, B. A. & PICKERING, S. C. (eds) *Petroleum Geology: From Mature Basins to New Frontiers. Proceedings of the 7th Petroleum Geology Conference*. Geological Society, London, 913–920.

MCLENNAN, S. M., HEMMING, S., MCDANIEL, D. K. & HANSON, G. N. 1993. Geochemical approaches to sedimentation, provenance, and tectonics. *In*: JOHNSSON, M. J. & BASU, A. (eds) *Processes Controlling the Composition of Clastic Sediments*. Geological Society of America Special Papers, **284**, 21–40.

MILLER, J. R. 1997. The role of fluvial geomorphic processes in the dispersal of heavy metals from mine sites. *Journal of Geochemical Exploration*, **58**, 101–118.

MYERS, J. & THORBJORNSEN, K. 2004. Identifying metals contamination in soil: a geochemical approach. *Soil and Sediment Contamination*, **13**, 1–16.

ORI, G. G. 1993. Continental depositional systems of the Quaternary of the Po Plain (northern Italy). *Sedimentary Geology*, **83**, 1–14.

PAOLA, C. & LEEDER, M. 2011. Environmental dynamics: simplicity versus complexity. *Nature*, **469**, 38–39.

PEARCE, T. J., MARTIN, J. H., COOPER, D. & WRAY, D. S. 2010a. Chemostratigraphy of Upper Carboniferous (Pennsylvanian) sequences from the southern North Sea (United Kingdom). *In*: RATCLIFFE, K. T. & ZAITLIN, B. A. (eds) *Application of Modern Stratigraphic Techniques: Theory and Case Histories*. SEPM, Special Publications, Tulsa, **94**, 109–127.

PEARCE, T. J., MCLEAN, D., MARTIN, J. H., RATCLIFFE, K. & WRAY, D. S. 2010b. A whole-rock geochemical approach to the recognition and correlation of 'marine bands'. *In*: RATCLIFFE, K. T. & ZAITLIN, B. A. (eds) *Application of Modern Stratigraphic Techniques: Theory and Case Histories*. SEPM Special Publications, Tulsa, **94**, 221–238.

PICONE, S., ALVISI, F., DINELLI, E., MORIGI, C., NEGRI, A., RAVAIOLI, M. & VACCARO, C. 2008. New insights on late Quaternary palaeogeographic setting in the Northern Adriatic Sea (Italy). *Journal of Quaternary Science*, **23**, 489–501.

PROHIC, E. & JURACIC, M. 1989. Heavy metals in sediments – problems concerning determination of the anthropogenic influence. Study in the Krka River estuary, eastern Adriatic coast, Yugoslavia. *Environmental Geology and Water Science*, **13**, 141–151.

RATCLIFFE, K. T., WRIGHT, A. M., HALLSWORTH, C., MORTON, A., ZAITLIN, B. A., POTOCKI, D. & WRAY, D. S. 2004. An example of alternative correlation techniques in a low-accommodation setting, nonmarine hydrocarbon system: the (Lower Cretaceous) Mannville Basal Quartz succession of southern Alberta. *American Association of Petroleum Geologists Bulletin*, **88**, 1419–1432.

RICCI LUCCHI, F. 1986. Oligocene to Recent foreland basins of northern Apennines. *In*: ALLEN, Ph. & HOMEWOOD, P. (eds) *Foreland Basins*. International Association of Sedimentologists, Special Publications, **8**, 105–139.

RIVARO, P., IANNI, C., MASSOLO, S., RUGGIERI, N. & FRACHE, R. 2004. Heavy metals in Albanian coastal sediments. *Toxicological & Environmental Chemistry*, **86**, 87–99.

RUBIO, B., NOMBELA, M. A. & VILAS, F. 2000. Geochemistry of major and trace elements in sediments of the Ria de Vigo (NW Spain): an assessment of metal pollution. *Marine Pollution Bulletin*, **40**, 968–980.

RUIZ, F. 2001. Trace metals in estuarine sediments from the Southwestern Spanish Coast. *Marine Pollution Bulletin*, **42**, 482–490.

SAINZ, A. & RUIZ, F. 2006. Influence of the very polluted inputs of the Tinto–Odiel system on the adjacent littoral sediments of southwestern Spain: a statistical approach. *Chemosphere*, **62**, 1612–1622.

SALMINEN, R. & TARVAINEN, T. 1997. The problem of defining geochemical baselines. A case study of selected elements and geological materials in Finland. *Journal of Geochemical Exploration*, **60**, 91–98.

SAMMARTINO, I., AMOROSI, A., GUERMANDI, M. & MARCHI, N. 2007. The pedogeochemical map of Parma alluvial plain: contribution of soil studies to geochemical mapping. *GeoActa*, **6**, 11–23.

SINGH, P. & RAJAMANI, V. 2001. Geochemistry of the floodplain sediments of the Kaveri River, southern India. *Journal of Sedimentary Research*, **71**, 50–60.

SINGH, A. K., HASNAIN, S. I. & BANERJEE, D. K. 1999. Grain size and geochemical partitioning of heavy metals in sediments of the Damodar River – a tributary of the lower Ganga, India. *Environmental Geology*, **39**, 90–98.

SØMME, T. O., HELLAND-HANSEN, W., MARTINSEN, O. J. & THURMOND, J. B. 2009. Relationships between morphological and sedimentological parameters in source-to-sink systems: a basis for predicting semi-quantitative characteristics in subsurface systems. *Basin Research*, **21**, 361–387.

STEFANI, C. 2002. Variation in terrigenous supplies in the Upper Pliocene to recent deposits of the Venice area. *Sedimentary Geology*, **153**, 43–55.

WELTJE, G. J. & VON EYNATTEN, H. 2004. Quantitative provenance analysis of sediments: review and outlook. *Sedimentary Geology*, **171**, 1–11.

WHITMORE, G. P., CROOK, K. A. W. & JOHNSON, D. P. 2004. Grain size control of mineralogy and geochemistry in modern river sediment, New Guinea collision, Papua New Guinea. *Sedimentary Geology*, **171**, 129–157.

WOLFENDEN, P. J. & LEWIN, J. 1978. Distribution of metal pollutants in active stream sediments. *Catena*, **5**, 67–78.

WRIGHT, A. M., RATCLIFFE, K. T., ZAITLIN, B. A. & WRAY, D. S. 2010. The application of chemostratigraphic techniques to distinguish compound incised valleys in low-accommodation incised-valley systems in a foreland-basin setting: an example from the Lower Cretaceous Mannville Group and Basal Colorado Sandstone (Colorado Group), Western Canadian sedimentary basin. *In*: RATCLIFFE, K. T. & ZAITLIN, B. A. (eds) *Application of Modern Stratigraphic Techniques: Theory and Case Histories*. SEPM Special Publications, Tulsa, **94**, 93–107.

WYŻGA, B. & CISZEWSKI, D. 2010. Hydraulic controls on the entrapment of heavy metal-polluted sediments on a floodplain of variable width, the upper Vistula River, southern Poland. *Geomorphology*, **117**, 272–286.

# Reconstruction of the evolution of the Niger River and implications for sediment supply to the Equatorial Atlantic margin of Africa during the Cretaceous and the Cenozoic


KATHELIJNE P. M. BONNE

*Getech, Kitson House, Elmete Hall, Elmete Lane, Leeds LS8 2LJ, UK*
*(e-mail: kathelijne_bonne@hotmail.com)*



**Abstract:** This paper presents a reconstruction of the palaeodrainage evolution of the Niger River in West Africa in order to contribute to the understanding of sediment supply to the Niger Delta. It has been covered extensively in literature that the Niger River has undergone changes along its course in the Holocene, as implied by the large bend it makes in Mali. However, other enigmatic bends further downstream are indicative of an older and more complicated history that has yet to be understood, and is the focus of this paper. Until now, sediment supply from the Niger River has been considered as being negligible compared to that of the Benue River. The results of this study imply that the contribution from the Niger River was more important than previously thought. The Niger River obtained its present-day geometry in three phases: a Bida Basin phase (Maastrichtian–Miocene); a Iullemmeden Basin phase (Miocene–Pleistocene); and a present-day Niger River phase (Holocene). In the Miocene, an important capture event occurred, increasing the incipient drainage basin by $10^6$ km$^2$, thereby changing the provenance of the sediment supplied to the Niger Delta from mainly crystalline basement to mixed lithologies including sandstone, shale, limestone and volcanic outcrops.


The geological history of large rivers in the world has been studied extensively because of both the scientific and commercial value of the large deltas and submarine fans fed by such systems. There is an increasing interest towards understanding the link between the evolution of large river systems supplying sediment from the hinterland, and the marine record; that is, the source to sink relationships. Reconstruction of palaeodrainage and the sediment routing systems contributes to the understanding of these source to sink relationships, thereby allowing the quantification of sediment supply parameters such as type, flux, volume and timing, which in petroleum exploration ultimately leads towards the assessment of reservoir quality. The overall Cenozoic histories of many of the world's largest rivers have been reconstructed, albeit with varying degrees of uncertainties within the proposed hypotheses, which are often the subject of debate. Many uncertainties are due to overprinting and the erosion of patterns in the drainage network, and the landscape that could have provided evidence, and to the complex interwoven nature of the processes that drive the dynamics leading to drainage changes.

The geological evolution of large rivers is closely linked to large-scale landscape evolution, primarily governed by tectonics and denudation that interact as an intricate coupled system. A considerable volume of literature is available on this topic, including Summerfield (1991), Burbank & Pinter (1999), Montgomery (2003), Pazzaglia (2003) and Allen (2008). The complex interplay of interdependent factors driving erosion, including climate, base level, vegetation, slope, relief, outcrop lithology and soil formation, all act upon the tectonically driven landscape by means of the river system, which ultimately forms the interface between the counteracting endogenic and exogenic forces. The consequent denudation of the landscape and the loading of the basins generate geodynamic feedback (i.e. isostatic and flexural response), which in turn rejuvenate the landscape. Climate is a fundamental component in the feedback system that affects the rate of erosion, and dictates sediment fluxes and volumes (Tucker & Slingerland 1997). Drainage changes, such as stream captures, caused by the evolving landscape, affect the size of the drainage basin or catchment of a river system, which eventually has an impact on sediment flux and type as the source areas have changed. Throughout the history of a river system, the dynamic events undergone are preserved at varying resolutions within the sedimentary successions beyond its mouth. Understanding the dynamics of landscape evolution affecting a given river system is therefore essential for the prediction of reservoir quality.

Many of the large-scale drainage rearrangements that have been detected in large rivers were triggered by regional tectonic events and directly affected distribution of petroleum reservoirs. The Amazon, for example, formed in the late Miocene


*From*: SCOTT, R. A., SMYTH, H. R., MORTON, A. C. & RICHARDSON, N. (eds) 2014. *Sediment Provenance Studies in Hydrocarbon Exploration and Production*. Geological Society, London, Special Publications, **386**, 327–349.
First published online April 9, 2014, http://dx.doi.org/10.1144/SP386.20
© The Geological Society of London 2014. Publishing disclaimer: www.geolsoc.org.uk/pub_ethics

or later due to uplift of the Andes, causing an eastward tilt of the Amazon Basin and an eastward shift in drainage (Figueiredo *et al.* 2009). The Yangtze River is at least 22 Ma old and formed during a regional drainage reorganization coeval with the formation of very high relief in Asia (Zheng *et al.* 2013). Several evolutionary hypotheses exist for the Nile River. Each of the hypotheses has in common that the sediment flux towards the Nile Delta is greatly influenced by the Miocene rift shoulder uplift of the Red Sea and by uplift of the Afar Dome (Macgregor 2011). It is accepted that the Congo River started flowing to its current outlet due to late Cenozoic capture near the Malebo Pool (Stankiewicz & de Wit 2006), but the location of the palaeo-outlet of the large Central African drainage is still debated. Rift-controlled drainage such as the Rufiji (Tanzania) (Stankiewicz & de Wit 2006) and Benue rivers (Nigeria) (Markwick & Valdes 2004) are possible candidates for the palaeo-Congo outlet.

The Niger Delta is one of the world's most important petroleum provinces, yet, surprisingly, only the Holocene history of the Niger River has been described in literature (Goudie 2005 and references therein). This recent history includes the formation of the large bend in Mali (Fig. 1), by the merging of NE- and SE-flowing sections of the Niger River, related to the formation of the Sahara Desert (Goudie 2005). The older history of this river remains enigmatic and referral to it in papers on African drainage is minimal (Summerfield 1991; Burke 1996; MacGregor *et al.* 2003; Goudie 2005; Gupta 2007). The Benue River is regarded as the only significant sediment supplier to the Niger Delta, and sediment contribution from the Niger River is assumed as negligible (e.g. Burke 1996). This perception is attributed to the very late capture of the Upper Niger and to the arid Sahelian climate presently dominating large parts of the drainage basin of the Niger River. However, even without the drainage of the Upper Niger, the catchment of the Middle and Lower Niger still covers an area of more than $10^6$ km$^2$ that should not be neglected as a possibly important provenance area for the Niger Delta. Furthermore, the present-day aridity is mainly a late Neogene phenomenon and climate conditions were more humid before (Micheels *et al.* 2009), suggesting that sediment fluxes from this region may have been higher than previously assumed. It is the aim of this paper to find out how the Niger River evolved through time as an integral part of the larger Niger–Benue system feeding the Niger Delta, and to assess how it could have contributed to deposition.

The Niger Delta is fed by two geologically distinct drainage systems: the Niger River draining a large part of the West African Craton and the Pan-African Mobile Belt (Dirks *et al.* 2009); and the Benue River, the downstream left-bank tributary of the Niger River that occupies the Benue Trough, a Cretaceous aulacogen related to the opening of the South Atlantic (Burke & Dewey 1973; Obaje 2009; Ukaegbu & Akpabio 2009; Opeloye 2012). This paper focuses mainly on the course of the actual Niger River. The Niger River is divided into three sections called the Upper, Middle and Lower Niger, as indicated in Figure 1. The main components of the Niger River and its drainage basin are shown in Figure 1, and the outcrop geology is shown in Figure 2.

The large bend of the Niger River, located in Mali (Fig. 1), is a well-known feature extensively covered in English and French literature. It is further referred to as the Large Bend to differentiate it from other bends discussed in this paper. As opposed to the Large Bend, the origin of two other prominent bends located in Nigeria, here called the Nigerian Bends (Fig. 1), remains largely uncovered. Anomalies in the stream networks, such as prominent bends, especially at large scales are often indicative of past changes along a river's course and can provide clues towards reconstructing the palaeodrainage evolution (Summerfield 1991; Twidale 2004). Following this line of thought, we assume that understanding the origin of the Nigerian Bends is pivotal to reconstruct the palaeodrainage of the pre-Holocene Niger.

## Physiography of the Holocene Niger River

The present-day Niger River is 4100 km long and has an annual suspended sediment load of 32 Mt (Gupta 2007). It is the largest river of West Africa and the third largest of Africa, after the Nile and the Congo. The present-day active catchment surface is $1.1 \times 10^6$ km$^2$ (Goudie 2005) but this surface can be expanded to $6.4 \times 10^6$ km$^2$ if it includes the entire area hydrologically contributing to the Niger Delta (Fig. 3). This includes the depressions of the Taoudenni and Chad basins. This full drainage capacity can potentially be reached if precipitation is sufficient to fill the latter endorheic basins to spill-point and to cause overflow. For the Chad Basin, overflow happened at a sill at Bongor (Bridges 1990; Goudie 2005), and, for the Taoudenni Basin, this would be north of the Large Bend (Fig. 3). The now nearly dry Chad Basin was filled to its maximum capacity, up to 320 m asl (metres above sea-level), forming Lake 'Mega-Chad' at 6500 BP, following several Pleistocene–Holocene dry–wet cycles (Thiemeyer 2000). These regions are now part of the Sahel region straddling the Sahara Desert, and are major dust pans (Varga 2012).

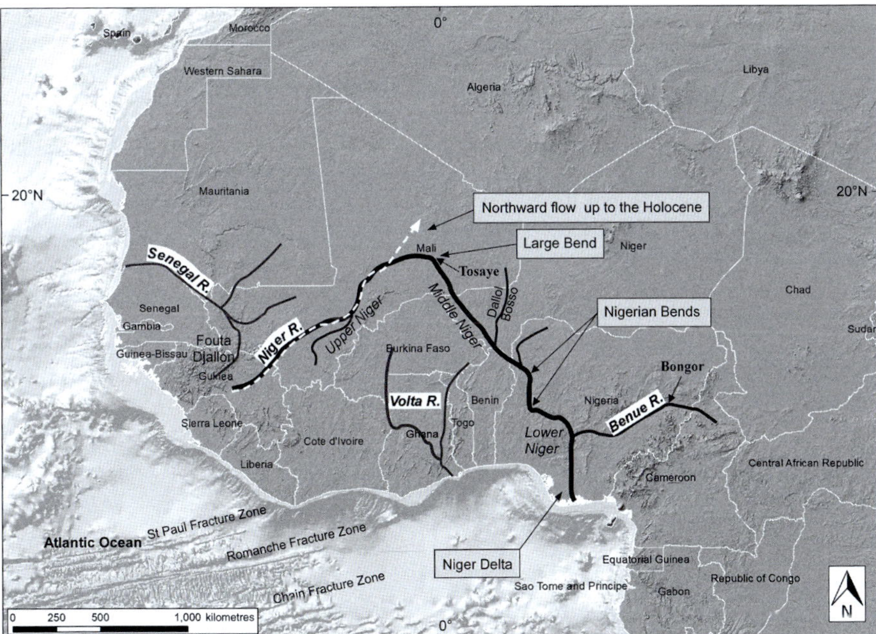


**Fig. 1.** Large rivers draining to the Equatorial Atlantic and details of the Niger River. The Large Bend and the Nigerian Bends are discussed in this paper.

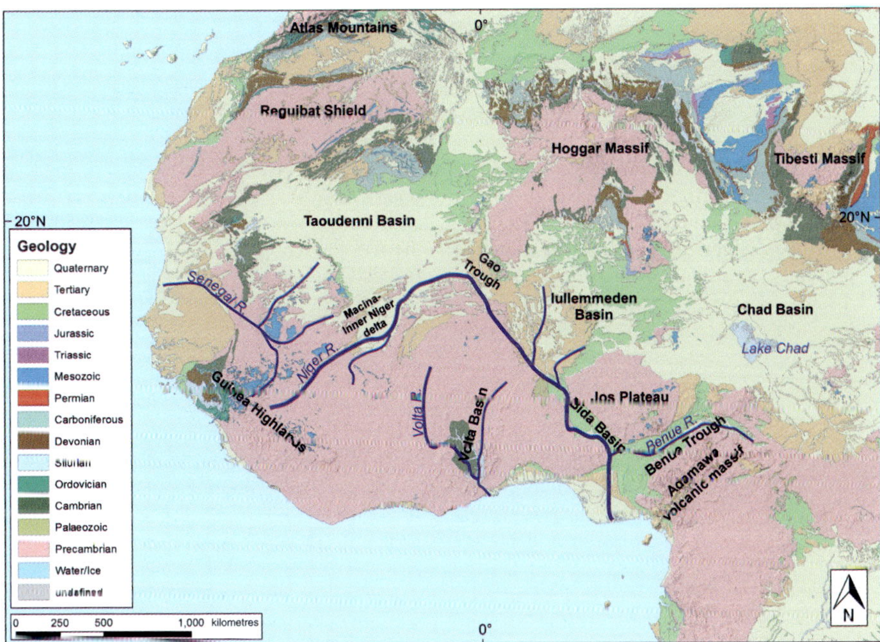


**Fig. 2.** Regional geology and large rivers of the Equatorial Atlantic. Geology from American Geoscience Institute (2014). The Niger River flows from its sources in the Guinea Highlands through the Taoudenni Basin, the Iullemmeden Basin and the Bida Basin. The Benue River joins the Niger River before it enters the Atlantic Ocean via the Niger Delta. The Benue Trough is shown in more detail in Figure 4.

The source of the Niger River is located in the Fouta Djallon Plateau in the Guinea Highlands, at an elevation of 800 m and at only 300 km from the Atlantic coast (Figs 1 & 2). The headwaters flow into the interior of West Africa, along a NE course into Mali, where the river's Large Bend is located. There the gradient decreases and the sediment load is deposited into the Inland Niger Delta (Fig. 2), a large area of marshes, swamps and pools, which is ecologically endangered by the encroaching Sahara Desert (Jacobberger 1981; Olivry & Boulègue 1995; Goudie 2005; Makaske et al. 2007). Owing to evaporation and infiltration, the flow rate decreases by two-thirds at the Inland Niger Delta (FAO 1997; Andersen et al. 2005). The Large Bend represents the location where two

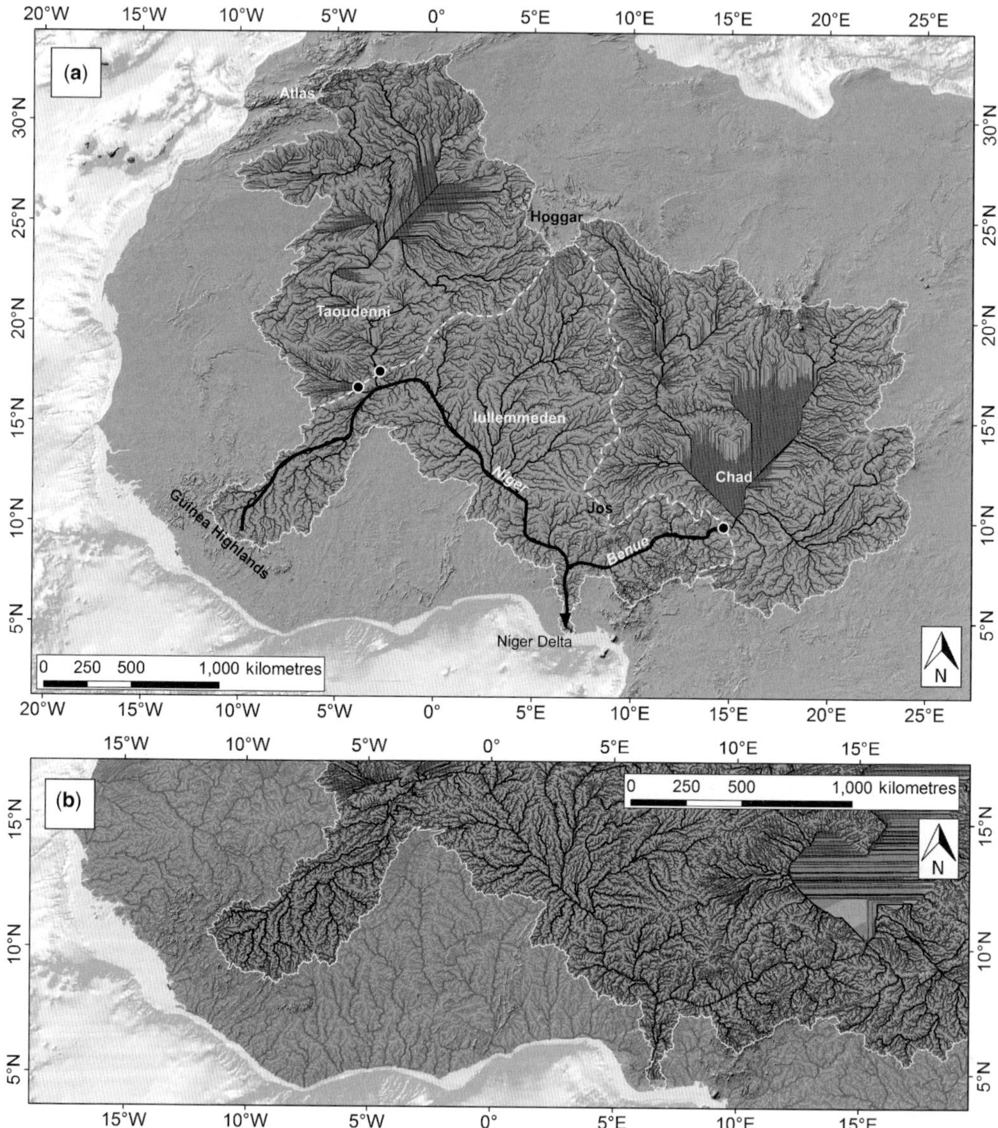


**Fig. 3.** (**a**) Drainage networks of the Niger River drainage basin generated from the SRTM30 DEM at a resolution of 100 km². The dashed white line represents the divide between the actively contributing area and the internally draining basins. The black dots show the locations of the spill-points between both. (**b**) Drainage networks of the Equatorial Atlantic margin of Africa generated from the SRTM3 DEM at a resolution of 10 km². Straight lines in the networks are GIS artefacts generated where topographical depressions are filled.

separate river systems merged in the Holocene, due to encroaching of the Sahara Desert (e.g. Goudie 2005), as discussed below.

East from the Inland Niger Delta, the Niger passes through the Large Bend and flows NE through the intracratonic Iullemmeden Basin in the Republic of Niger (Kogbe 1991; Obaje 2009) (Fig. 2). Several large tributaries, all of them located to the north, are sourced in the Hoggar Massif in the Sahara and are mostly dry, therefore the water flux remains low (Goudie 2005). An important tributary, now also dry, was the Dallol Bosso River (Fig. 1). Tributaries for the southern bank of the Niger River are short. After entering Nigeria, the Niger River makes two subsequent abrupt bends, the Nigerian Bends. The Lower Niger River is located within a subtropical environment, increasing the water flux (FAO 1997). The river flows through the Bida (Nupe) Basin (Fig. 2), a small Late Cretaceous extensional basin (Obaje 2009; Obaje et al. 2011). After the confluence with the Benue River, the Niger River reaches the Niger Delta. This delta is one of the world's largest deltas, spanning 19 135 km$^2$, starting progradation in the Oligocene (Gupta 2007; Reijers 2011).

The Large Bend of the Niger River has been known to people for millennia, as for many desert populations it was an important source of water and a meeting point for traders. Scientists recognized early that the Large Bend formed by the joining of two separate rivers, later recognized to be caused by processes related to climate oscillations, and the effect of those on the formation of the Sahara Desert and associated hydrological factors (e.g. Urvoy 1942; Palausi 1955; Monod 1964; Tricart 1965; Iloeje 1981; Jacobberger 1981; McIntosh 1983; Doust & Omatsola 1989; Bridges 1990; Burke 1996). According to Urvoy (1942), the Upper Niger used to flow NE into the Sahara, forming a large lake in the Azawagh (also Azawad), a region located north of the Large Bend covering a large part of NE Mali (Fig. 2). The Azawagh is geologically located in the Taoudenni Basin (Craig et al. 2009). Palausi (1955) discovered abandoned river channels within the same region, the presence of which was later confirmed by Jacobberger (1988) by the use of satellite images. Climate fluctuations in the Quaternary, responsible for alternations between lacustrine and aeolian phases in the Sahara (e.g. Gasse et al. 1990; Gasse 2000; Giresse 2008; Lespez et al. 2011), were most probably the cause of obstructing the previous NE flow by dune formation. In following wetter phases, a lake formed at the location of the Inland Niger Delta, with dunes effectively blocking throughflow. This lake, referred to as Lake Araouane (Iloeje 1981; Akaa et al. 2008), episodically spilled over a sill at Tosaye (Fig. 1).

Lake Araouane is suggested by Bridges (1990) to have risen to spill-point during the pluvial period between approximately 15 000 and 10 000 BP, and was possibly drained at around 5000 BP to form the Inland Niger Delta as a remnant of the ancient large lake. The actual connection between the Upper and Middle Niger was established by a combination of two interacting processes: overflow of Lake Araouane (e.g. Bridges 1990); and breaching of the sill due to headward erosion of the Middle Niger (e.g. Akaa et al. 2008). A possible chronology could be that excess overflow from Lake Araouane flowed into the Iullemmeden Basin (Kogbe 1991; Burke 1996), which was drained by the palaeo-Middle Niger, this extra discharge increasing the stream power of the latter. Consequently, headward erosion along the Middle Niger possibly occurred or increased, resulting in the upstream propagation of incision, eventually breaching the sill of Lake Araouane. As such, the water was drained from that lake, and the present-day Niger River and Inland Niger Delta may have formed.

The idea of the Upper Niger previously flowing NE into a lake in the Azawagh region in the interior of the African continent (Urvoy 1942) is consistent with the Cenozoic history of Africa during which a 'basin-and-swell' dynamic topography formed (Burke 1996). This caused many large rivers to flow into endorheic basins (e.g. into the Congo, Chad, Sudd, Taoudenni and Iullemmeden basins: Burke 1996). At several locations in Africa, large pluvial lakes existed but they have since dried up (e.g. Varga 2012).

## Geological setting

Most drainage changes are caused by the response of the landscape to tectonic events. Therefore, any palaeodrainage reconstruction should be based on the geological background. Whereas the drainage patterns at a given moment in time give clues of past events and changes, the timing and process of the change can only be understood through the knowledge of the geological history. An overview of this history is, hence, given in order to understand the setting of the landscape upon which the stream network of the Niger River has evolved.

In the catchment of the Niger River, the outcrop geology is dominated by Precambrian basement comprising gneisses, migmatites and granites (Obaje 2009). The Archaean Reguibat and Leo-Man shields are the cores of the Palaeoproterozoic West African Craton (Bertrand-Sarfati et al. 1991; Lompo 2009), surrounded by Pan-African deformation zones (e.g. the 3000 km-long Trans-Saharan Belt to the east: Kröner & Stern 2004) (Fig. 2).

**Table 1.** *Summary of Cretaceous events*

| Period | Epoch | Age | Regional events | Equatorial Atlantic | Benue Trough | Iullemmeden Basin | Anambra Basin | Bida Basin | Drainage events |
|---|---|---|---|---|---|---|---|---|---|
| Cretaceous | Late | Maastrichtian | Uplift affecting the Benue Trough and Anambra Basin | | Benue Trough became entirely emergent | Marine transgression from the Tethys Sea | End-Cretaceous uplift temporarily terminated marine sedimentation | Subsidence and deposition in the Bida Basin | Drainage from palaeo-Benue, which was a small river sourced in the Middle Benue Trough, and palaeo-Niger in the Bida Basin. Both deposited in the Anambra Basin |
| | | Campanian | | | | | Marine sedimentation | | Palaeo-Niger formed in the Bida Basin, draining a small catchment centred on the Nigerian Shield |
| | | Santonian | 'Santonian' event due to plate reorganization caused regional deformation and compression. Trans-Saharan Seaway migrated to the west of the Hoggar Massif | Transform motion ceased, establishment of passive margins | Compression and uplift, regression of seas. The Middle Benue Trough formed a drainage divide, with seas from the north (Tethys) still submerging the Upper Benue Trough. Westward shift in the depositional axis | | Subsidence of the Anambra Basin, which received sediment from the uplifted Benue Trough | | Disruption of integrated drainage through Upper Benue Trough |
| | | Coniacian | | | | Marine transgression from the Tethys Sea | | | |
| | | Turonian | Trans-Saharan Seaway connected Tethyan and Atlantic waters via the Chad Basin and the Benue Trough | | Sea levels rose to establish a marine connection between the Benue Trough and rifts in the Chad Basin | | | | |

| | | | | | |
|---|---|---|---|---|---|
| | Cenomanian | | | The Bima Formation formed a delta in the Upper Benue Trough. The rest of the Benue Trough remained submerged | Drainage from a possibly substantial river forming a delta in the Upper Benue Trough, draining rift shoulders of basins along the Central African Shear Zone. The palaeo-Congo may have been part of this system |
| Early | Albian | | Establishment of open marine conditions | Marine transgression from the Equatorial Atlantic | |
| | Aptian | Break-up in the South and Equatorial Atlantic | Seafloor spreading and establishment of a narrow seaway across very narrow, deep, rapidly subsiding basins undergoing transform motion | Main rifting phase | |
| | Barremian | Development of the Central African Rift System | Initial shearing, forming a series of dextral en echelon strike-slip basins | Initial shearing | |

The Dahomeyides (Fig. 2) are a part of the Trans-Saharan Belt located in Ghana, Togo and Benin, comprising suture zone nappes thrust westwards over the West African Craton (Castaing *et al.* 1993; Kröner & Stern 2004; Guiraud *et al.* 2005; Attoh & Brown 2008; Attoh & Nude 2008; Dos Santos *et al.* 2008; Obaje 2009). The Nigerian Shield and the Jos Plateau, both in Nigeria, are also part of the Trans-Saharan Belt, and consist of strongly varying lithologies that represent reworked Palaeoproterozoic basement and Neoproterozoic oceanic assemblages (Kröner & Stern 2004; Obaje 2009). At 200 Ma, magmatic rocks were intruded as, for example, ring complexes in many parts of West Africa, Cameroon and Nigeria during the Central Atlantic Magmatic Event (Wilson 1992; Torsvik *et al.* 2012).

The actual history of the Niger Delta begins with the creation of the accommodation space for its sedimentary successions; that is, the formation of the triple junction between the South and Equatorial Atlantic and the failed rift of the Benue Trough in Cretaceous times (Burke & Dewey 1973; Benkhelil 1989; Basile *et al.* 2005; Antobreh *et al.* 2009; Nemčok *et al.* 2012). A summary of the most noteworthy events of the Cretaceous Period is given in Table 1. The geological history can be summarized into three main events that significantly affected the palaeogeography and palaeotopography of the study area: (1) the Cretaceous rifting in Africa leading to the oblique opening of the Equatorial Atlantic (Wilson & Williams 1979; Guiraud *et al.* 1992; Wilson & Guiraud 1992; Genik 1992; Jones *et al.* 1995; Gasperini *et al.* 2001; Mohriak & Rosendahl 2003; Basile *et al.* 2005; Bigot-Cormier *et al.* 2005; Antobreh *et al.* 2009; Turner & Wilson 2009; Moulin *et al.* 2010); (2) the marine transgressions from the Tethys and the Trans-Saharan Seaway (Guiraud *et al.* 2005); and (3) the creation of the basin-and-swell topography since the Oligocene (e.g. Burke 1996).

## Cretaceous rifting

The Cretaceous rifting episode comprised the oblique dextral rifting in the Equatorial Atlantic (Basile *et al.* 2005) and sinistral rifting in the Benue Trough (Benkhelil 1989). Initial shearing started between the late Barremian and the Aptian, forming a series of dextral en echelon strike-slip basins (Basile *et al.* 2005). The continental break-up started in the Aptian, and a narrow seaway was established since that time (Jones *et al.* 1995). This seaway also transgressed the Benue Trough since the Albian. While oceanic accretion took place along the spreading axes, deformation occurred along the transform faults, forming very deep, narrow, rapidly subsiding basins, comparable

to that of the present Dead Sea (Basile & Allemand 2002). Open marine conditions appeared in the late Albian (Basile *et al.* 2005). In the Late Cretaceous, the newly formed basins were active transform basins until the Santonian, when they finally became passive margins (Basile *et al.* 2005).

The Benue Trough was connected to the north and east to other Cretaceous rift basins of, respectively, the West and Central African Rift System (Grant 1971; Fitton 1980; Adighije 1981; Ojo & Pinna 1982; Ofoegbu 1985; Benkhelil 1989; Doust 1990; Fairhead & Okereke 1990; Ofoegbu & Onuoha 1991; Genik 1992; Shemang *et al.* 2001; Basile *et al.* 2005; Obaje 2009; Ukaegbu & Akpabio 2009; Chukwuebuka *et al.* 2010; Akande *et al.* 2012). The Benue Trough contains over 6 km of sediments, is 150 km wide and 800 km long, and is arbitrarily subdivided from west to east in the Lower, Middle and Upper Benue (Obaje 2009) (Fig. 4).

The Benue Trough underwent rifting and subsidence in the Aptian coeval with the Equatorial Atlantic, and during Albian–Cenomanian times a delta (the Bima Formation) developed in the Upper Benue Trough (Benkhelil 1989). The Bima Delta was probably fed by sediments eroded from the rift shoulders of basins located along the Central African Shear Zone (Binks & Fairhead 1992; Genik 1992) and arguably also by the palaeo-Congo River (Markwick & Valdes 2004). Throughout the Cenomanian and Turonian, a marine connection with the Tethys formed across the Benue Trough (Petters 1978; Benkhelil 1989; Guiraud *et al.* 2005; Mathey *et al.* 2006; Obaje 2009). In the Santonian, the Middle Benue Trough became emergent as a result of regional uplift and deformation, creating a regional drainage divide between Tethyan and Atlantic waters (Ofoegbu 1985; Benkhelil 1989; Genik 1992). The depositional axis of the Benue Trough had since then shifted westwards to the Anambra Basin (Fig. 4) (Obaje 2009; Akande *et al.* 2011). Uplift at the end of the Cretaceous left the Benue Trough entirely emergent (Benkhelil 1989). Palaeogeographies of the Benue Trough (Benkhelil 1989, p. 258, fig. 4) imply that the palaeo-Benue River was a short river at the start of the Cenozoic, because large parts of the Middle Benue Trough were uplifted and were a sediment source. Deposition continued in the SW Anambra Basin in the Cenozoic, and from the Eocene onwards progradation of the Niger Delta started (Obaje 2009).

The Bida Basin (Fig. 2) is a NW–SE-trending fault-bounded basin that forms an embayment of the Anambra Basin and is structurally linked to the Benue Trough, possibly in a pull-apart setting (Benkhelil 1989; Kogbe *et al.* 1983). It underwent continental–marginal marine sedimentation

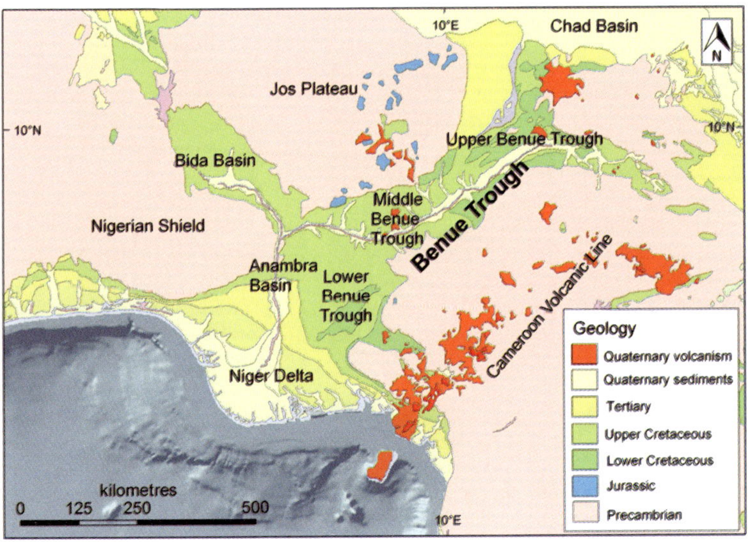



**Fig. 4.** Outcrop geology of the Benue Trough (from American Geoscience Institute 2014).

throughout the Campanian and the Maastrichtian. The Lower Niger River flows along its axis of the Bida Basin.

## Trans-Saharan Seaway

While tectonics affected the Equatorial Atlantic margin of Africa, the interior of Saharan Africa was submerged by major transgressions (Late Cretaceous–Eocene), due both to the high eustatic sea level and the overall low elevation of the continent (Petters 1978; Benkhelil 1989; Genik 1992; Jones *et al.* 1995; Guiraud *et al.* 2005; Burke & Gunnell 2008; Obaje 2009; Markwick *et al.* 2010). The Trans-Saharan Seaway connected the Tethys in the north with the incipient Atlantic Ocean (Fig. 5). The shape and timing of this seaway gives indications towards the palaeogeography throughout Cretaceous and Cenozoic times (Fig. 5), which enable recognition of the regional drainage divides through time. The Trans-Saharan Seaway was across the Benue Trough via the Chad Basin from Albian to Turonian times (Guiraud *et al.* 2005). Since the uplift of the Benue Trough in the Santonian, the marine connection was terminated (Guiraud *et al.* 2005; Obaje 2009). The Iullemmeden Basin was transgressed via the Taoudenni Basin through the narrow Gao Trough (Fig. 2) (Kogbe 1991; Obaje 2009). A short lived Paleocene connection between the Atlantic and the Tethys via the Iullemmeden Basin and the Bida Basin has not been proven (e.g. Guiraud *et al.* 2005) but is strongly suggested by Kogbe (1989) based on similarities between Tethyan and Atlantic fauna

during that time, a view supported by Obaje (2009). The marine conditions over North and West Africa continued into the Cenozoic, with Tethyan seas in the Iullemmeden Basin finally retreating in Bartonian times, as indicated by the depositional environment of sediments (Kogbe 1991). Then fluvial and, to a minor extent, lacustrine deposition took place in that basin (Kogbe 1991). Figure 5 shows the palaeogeographical evolution of the Trans-Saharan Seaway during each stage of the Cretaceous Period (Markwick *et al.* 2010). These palaeogeographies are based on plate tectonic models and interpretations from literature (Petters 1978; Genik 1992; Guiraud & Maurin 1992; Guiraud *et al.* 1992; Jones *et al.* 1995; Burke 1996; Benkhelil *et al.* 1998; Gonçalves & Ewert 1998; Basile *et al.* 2005; Guiraud *et al.* 2005; Antobreh *et al.* 2009; Obaje 2009 amongst others).

## Oligocene uplift

A final return to continental deposition took place in the Oligocene, at 30 Ma, when epeirogenic processes started to cause uplift throughout Africa creating a 'basin-and-swell' topography (Burke 1996; Burke & Gunnell 2008). The basin-and-swell topography was responsible for the creation of endorheic basins all over Africa, including the Taoudenni Basin into which the palaeo-Upper Niger drained. The uplifts have been sustained by upper mantle convection throughout the Neogene and represent Africa's main topographical features (Al-Hajri *et al.* 2009). The overall starting age of the magmatic events and volcanism is Oligocene but varies locally.

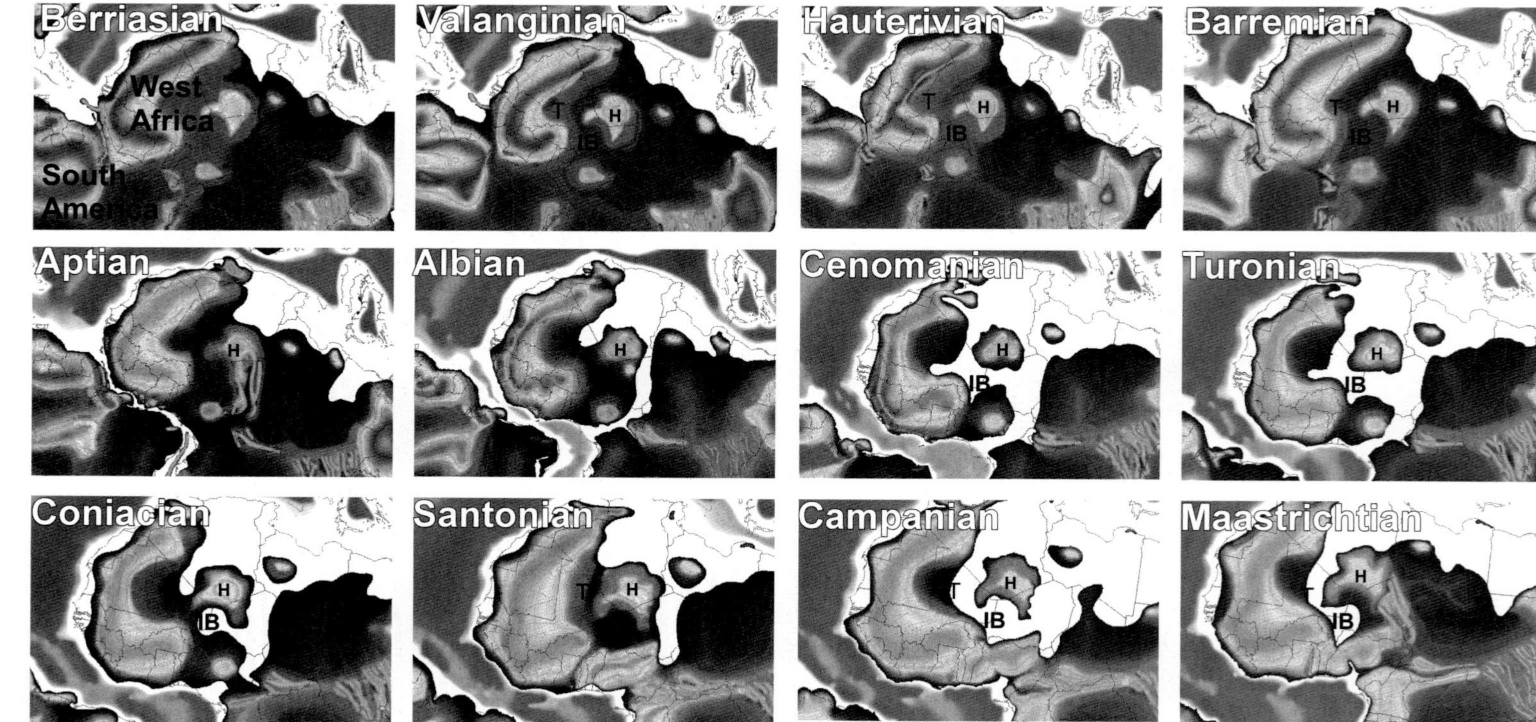



**Fig. 5.** Palaeogeography of the stages of the Cretaceous Period (Markwick *et al.* 2010). Top left of the figure shows the relative locations of Africa and South America. T, Taoudenni Basin; IB, Iullemmeden Basin; H, Hoggar Massif. Shallow seas are in white and lowlands are dark grey. The Aptian stage shows the onset of marine conditions between Africa and South America. The epicontinental seas submerge large parts of northern and equatorial Africa throughout the Cretaceous. The Iullemmeden Basin is flooded from Cenomanian to Coniacian times, and again from Campanian to Maastrichtian times. At the location of the Niger Delta and in the Benue Trough, a seaway has been present since Albian times. In the Campanian and Maastrichtian, rifting in the Bida Basin caused subsidence, forming a small marine embayment.

In the Hoggar Massif, the first volcanic rocks are upper Eocene (Liégeois 2006). In the Adamawa and Hoggar massifs, and on the Jos Plateau, volcanism and/or doming continued through the Neogene and the Quaternary (Stuart *et al.* 1985; Liégeois 2006; Obaje 2009). Important uplift of the Jos Plateau is recorded to have started in the Neogene (Obaje 2009). Erosional products from the swells accumulated in the basins (e.g. in the Iullemmeden Basin).

## The Niger Delta

The Niger Delta is located at the SW end of the Benue Trough and progradation has been taking place since the Eocene (Reijers 2011). Syntheses of the structure, stratigraphy and petroleum geology of the Niger Delta are given by Doust & Omatsola (1989), Doust (1990), Tuttle *et al.* (1999), Reijers (2011) and references therein. In the Eocene, the delta started to spread out over the continental–lithospheric transition zone, and since the Oligocene it has been prograding over oceanic crust of the Gulf of Guinea (Obaje 2009). The total distance of progradation exceeds 250 km and sediments reach a thickness of 12 km. It is one of the world's largest deltas and its three megasequences show an overall upward transition from marine shales (Akata Formation) through a sand–shale paralic interval (Agbada Formation) to continental sands (Benin Formation) (Obaje 2009).

The Niger Delta developed in pulses dictated by hinterland movements and sea-level oscillations. Reijers (2011) summarized the megasequences of the Niger Delta, and during the following time spans increased progradation rates occurred. At 21.8 Ma, sediment supply increased. In the Early Miocene, between 19.4 and 15.9 Ma, progradation occurred in pulses and reached in places 8–15 km Ma$^{-1}$. In the early Middle Miocene, between 14.4 and 14.0 Ma, there was an increase in progradation from 2 to 16–22 km Ma$^{-1}$. In the Middle Miocene, between 12.8 and 11.5 Ma, an increase in progradation to 16–22 km Ma$^{-1}$ was observed. Reijers (2011) attributes this specifically to the increased uplift of the hinterland. Between 9.5 and 5.0 Ma, progradation reached 13–17 km Ma$^{-1}$. Especially after 8.5 Ma, Reijers (2011) postulates that the hinterland shedded extensive amounts of clastic sediment. Generally during the Miocene, the average progradation rate was 1 km Ma$^{-1}$, with a rising hinterland increasing progradation through the Middle–Late Miocene.

## Methodology

The palaeodrainage of the Niger River is reconstructed using a multidisciplinary approach comprising geographical information system (GIS)-based geomorphological techniques. This approach aims at tying the drainage anomalies, and the chronology of events extracted from the analysis of the drainage network and the landscape, to the known geological history.

## Data and processing

Shuttle Radar Topography Mission (SRTM) digital elevation models (DEMs) were used for landscape and geomorphological interpretation. SRTM30 (30 arcsec) and SRTM3 (3 arcsec) DEMs were processed to generate stream networks and drainage basins at 100 and 10 km$^2$, respectively (Fig. 3). The resolution refers to the minimum area that each generated stream segment drains. Topographically generated stream networks were preferred over mapped river networks because the latter can be prone to different perceptions of scale, mapping resolution and mapping exploration at different locations. Since a large part of the study area is arid, dry riverbeds would be overlooked using mapped rivers. The Hoggar Massif, for example, was once drained by large rivers, leaving behind well-developed river valleys with an integrated network of tributaries. Almost none of those are fully displayed on regional maps. It is necessary to check whether the generated rivers match the mapped ones, which they most often do in non-arid or non-glacial regions. Where dunes occur in arid landscapes, such as the Sahara, the generated network appears artificial. The generated stream lines occupy the interdune valleys, creating parallel patterns. In such cases, the drainage network cannot be used as a starting point to interpret palaeodrainage.

The stream network is generated through an iterative process, including the filling of internal topographical depressions to force flow from each given point on the land to drain into the oceans (O'Callaghan & Mark 1984; Tarboton *et al.* 1991; Verdin 1997; Verdin & Verdin 1999; Tarboton & Ames 2001). Primarily, the filling of depressions is essential to reduce the effect of DEM artefacts such as pits (one pixel surrounded by higher pixels; hence not allowing outward flow) and lakes or dams along the river courses. On a larger scale, this also integrates the large naturally occurring depressions, such as the Chad or Taoudenni Basin, in the drainage basin of the Niger River via the lowest point along their drainage divides (i.e. via the spill-points: Fig. 3).

Orders of magnitude (stream numbers) were assigned to each stream segment of the network, allowing better visual interpretation and highlighting the higher order streams (Fig. 3). The ordering hierarchy used here is that of Strahler (1957), where first-order streams start at the sources and a higher order forms when two same-order streams join.

## Landscape and drainage patterns

The landscape and drainage analysis starts by identifying anomalous patterns that are indicative of changes in the stream network, as described extensively by Summerfield (1991) and Twidale (2004). Drainage patterns give clues towards past events and the chronology of events (Twidale 2004). If a stream network develops upon a flat and homogenous surface, a tree-like dendritic pattern develops. Such patterns are often observed on recently deposited alluvium in the downstream reaches of many rivers. However, at most scales, heterogeneous lithologies, underlying structure, uplift and subsidence, and the impact of climate, volcanism, glaciations and mass movements, and so on will cause the dendritic patterns to alter or not to develop at all. Patterns diverging from dendritic patterns can therefore be interpreted as being indicative of an underlying mechanism and/or structure. Amongst textbook examples of typical patterns occurring at varying scales are centripetal (in internal depressions), radial (around volcanoes and doming regions), parallel (due to tilting or uplift), trellis (in fold and thrust belts) and rectangular patterns (in limestone) (Summerfield 1991). In essence, all such patterns are determined by slope and structure (e.g. Twidale 2004).

Equally indicative of drainage change are anomalous bends (Summerfield 1991; Twidale 2004). Although somewhat arbitrary, we here define an anomalous bend as a bend that is abrupt, large and/or singular in map view. The Large Bend and the Nigerian Bends of the Niger River are examples. With abrupt, a sudden/sharp change in flow direction is referred to, very often due to underlying structure. Large bends are those along the trunk stream or along an important tributary, representing an obvious regional feature, and having an impact on the overall geometry and shape of the river network. Finally, with singular, we indicate that the bend in question is not part of a network of many smaller anomalous bends; for example, the many abrupt bends observed in trellis and rectangular drainage patterns (both caused by very local drainage readjustments influenced by the bedrock and structure) or along meandering streams. A main exception is when smaller grouped abrupt bends occur as barbed confluences. In that case the abrupt bends occur along several tributaries of a trunk stream and they indicate that the flow direction in this stream was opposite in the past, as illustrated in Figure 6. Anomalous bends can be elbows of captures that form when a more erosive stream beheads the head waters of another stream,

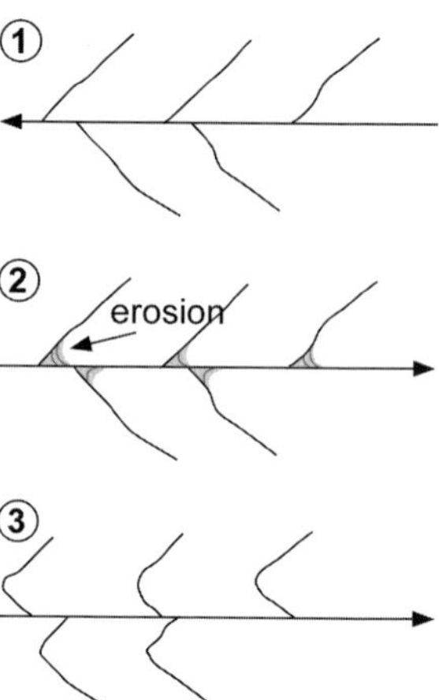



**Fig. 6.** Development of barbed confluences: (1) a river flows to the left with tributaries joining at sharp angles; (2) the river changes flow direction and the inner banks of the tributaries start being eroded; (3) in time, the barbed confluences migrate away from the main stream and become less sharp, as erosion on the inner banks continues. The patterns eventually disappear.

as described, for example, in the Yangtze River (Zheng *et al.* 2013). In many rivers, anomalous bends at varying scales are prominent features that are very often associated with important drainage reorganizations.

Whereas for recent and active geological events, the effects on the drainage network and the topography are evident, older patterns systematically are scarcer and more complex to interpret owing to overprinting, erosion or deposition. Over time, geomorphic evidence of more ancient events eventually becomes obliterated at rates dependent on, for example, lithology, uplift rate and climate. Therefore, understanding geological evolution and palaeogeography becomes increasingly important when taking palaeodrainage reconstructions back in time. Finally, for the more ancient reconstructions, only geological evidence and palaeogeographies and palaeotopographies can give clues of past drainage. For this reason, Cretaceous palaeogeographies of the study (Fig. 5) have been used to interpret the early palaeodrainage history of the Niger River.

Several geomorphological features in the landscape are major sources of information on uplift, and subsequent erosion, denudation and incision (Burbank & Pinter 1999). Such features include, but are not limited to, erosional scarps such as at palaeosurfaces and amphitheatre-shaped valleys. Palaeosurfaces stand as highs in the landscape, and are the relics of ancient erosion surfaces that formed due to denudation and peneplanation during an episode of tectonic stability (Twidale 1994; Burke & Gunnell 2008). Amphitheatre-shaped valleys are caused by headward erosion, and their presence indicates that an erosive river system is actively expanding its catchment at the expense of the area of the adjacent drainage basin. Understanding the above processes is useful because, if the age of the incised strata is known, a lower age for incision and, hence, a lowering in base level can be fixed and linked to a geological or eustatic event.

Longitudinal (long) profiles of rivers are used to identify knickpoints that give information on the propagation of erosion after an episode of uplift, or are caused by a drainage capture. The long profile of the Niger River is shown in Figure 7. The long profile consists of a nearly graded upstream section (between 450 and 270 m in elevation), which corresponds to the Upper Niger. The profile of the Upper Niger flattens out at 270 m elevation where it merges into the Inland Niger Delta. The maturity suggested by the concave-up shape of the profile supports the hypothesis of a long-lived drainage setting in which the palaeo-Upper Niger flowed into an endorheic basin in the Azawagh region in the Taoudenni Basin. The knickpoint (i.e. the convex-up segment between 250 and 200 m in elevation: Fig. 7) marks the location where Holocene capture took place.

## Results

In the Niger River the most obvious anomaly is the Large Bend, where the river turns 90° from a NE to SE course since the recent past. The Nigerian Bends, a northern and a southern one, show a more abrupt change in flow, even though the change in flow direction is only 45°, towards the south and then the SE, respectively. The northern Nigerian Bend is located at the southern end of the Iullemmeden Basin, whereas the southern Nigerian Bend is located at the SW side of the Bida Basin, 200 km further downstream. The Nigerian Bends suggest that a drainage reorganization may have occurred between the Iullemmeden and the Bida basins. This is in agreement with the palaeogeographical setting indicating that, in the Cretaceous, these basins were not connected hydrographically (Fig. 5).

The Lower Niger flows along the axis of the Bida Basin, which underwent fault-controlled subsidence in the Campanian and Maastrichtian (Obaje 2009). The palaeo-Lower Niger started to drain the newly exposed land since the sea had regressed from the Bida Basin in the latest Cretaceous. No major tectonic activity has later affected the Bida Basin, therefore it is assumed that the position of the Lower Niger has not significantly changed since then. Coevally, the Iullemmeden Basin was submerged by Tethyan seas that transgressed from the NW via the Taoudenni Basin (e.g. Guiraud et al. 2005). There was, hence, a sill or drainage divide between the Tethyan and Atlantic domains in the Iullemmeden and Bida basins, respectively, and this divide ran across the 200 km present-day stream segment between the Nigerian Bends (Fig. 8). This sill was later breached to form the integrated Niger River.

The drainage patterns near the Nigerian Bends show evidence of the above setting: (1) tributaries of the Niger diverge from a SW-trending ridge, interpreted to be the palaeodrainage divide (Fig. 9); and (2) barbed confluences in the SE Iullemmeden Basin, which are indicative of a change in flow direction (Fig. 9). Flow in that part of this basin was, hence, to the NW; that is, towards the centre of the Iullemmeden Basin. The palaeo-Middle Niger in the Iullemmeden Basin was either endorheic or drained to the Taoudenni Basin. The endorheic option is favoured because, after final retreat of the sea in Bartonian times (41–38 Ma), lacustrine deposition occurred, which is suggestive of the lack of an efficient outlet (Kogbe 1991). Furthermore, centripetal drainage is common in intra-cratonic basins (e.g. Summerfield 1991).

Other drainage patterns are also observed. (1) Along the Middle Niger in the Iullemmeden Basin, drainage is asymmetrical, with long tributaries to the north and only short ones to the south (Fig. 3). The asymmetry can be linked to southward tilting, which causes a shift to the south of the drainage in the Iullemmeden Basin. This tilting was caused by the epeirogenic uplift of the Hoggar Massif that started in the late Eocene and continued in pulses through the Neogene (Liégeois 2006). (2) Radial drainage at the Jos Plateau (Fig. 9), which is typical for doming and volcanism that has been continuous since the Pliocene up to very recent time (Obaje 2009).

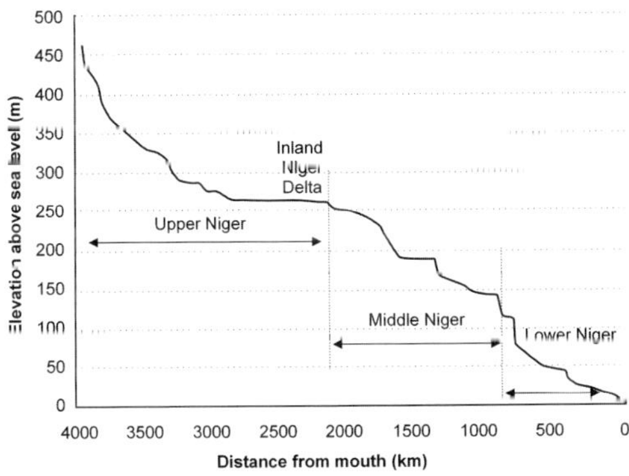


**Fig. 7.** Longitudinal profile of the Niger River.

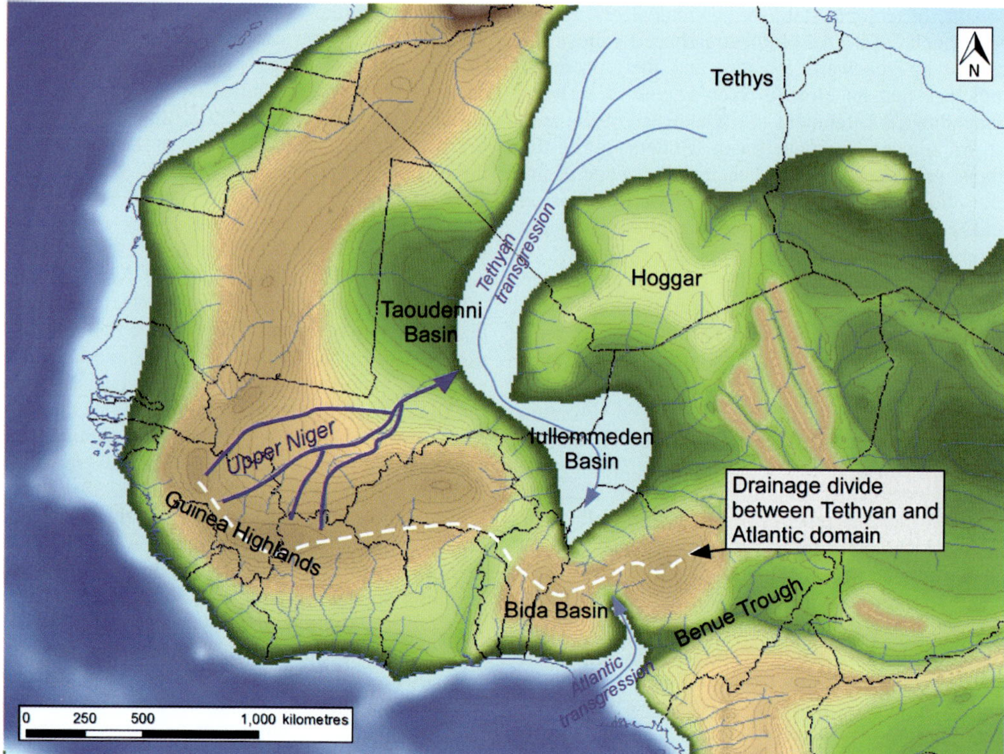



**Fig. 8.** Maastrichtian palaeogeography (Markwick *et al.* 2010). The distribution of land and sea shows that the Tethyan transgression submerged the Iullemmeden Basin. The last time this took place was in Bartonian times (Eocene). Until at least then, there was a continental drainage divide between the depositional Iullemmeden Basin as part of the Tethyan domain, and the Bida Basin, part of the Atlantic domain. The palaeo-Upper Niger flowed from the Guinea Highlands into the Taoudenni Basin, which was depositional throughout the Cenozoic.

The palaeodrainage divide between the Nigerian Bends could not have been breached earlier than the late Eocene (post-Bartonian) because until then there was a marine embayment in the Iullemmeden Basin, connected to the Tethys Sea. To better constrain the age of breaching, we analysed the geomorphology of the Iullemmeden Basin. This basin is recognized by a regional incised palaeosurface (Fig. 10), in the landscape appearing as flat-topped, steep-sided hills capped by ironstone (Obaje 2009). The strata on which the palaeosurface developed are intercalated massive white clays and sandstones of the fluvial/lacustrine Gwandu Formation (Kogbe 1991; Obaje 2009). The Gwandu Formation is tentatively of Eocene–Miocene age and outcrops extensively in the Iullemmeden Basin (Kogbe 1991). The observed incision can only have started after a base-level lowering that terminated the lacustrine/fluvial deposition. This could have occurred when a hydrological connection to the sea was made; that is, by breaching of the aforementioned drainage divide. Detritus from the Iullemmeden

Basin and its catchment has since then been transported to the Niger Delta. The upper age of the Gwandu Formation is the Miocene; hence the capture occurred in the Miocene or later.

A possible trigger for the capture event to happen was the combination of the onset of headward erosion along the NW edge of the Bida Basin, and the southward tilting and uplift in the Iullemmeden Basin, which facilitated the effective draining of the formerly endorheic basin. The epeirogenic uplift of the Hoggar Massif and the Jos Plateau played a pivotal role in initiating both processes.

The catchment of the palaeo-Lower Niger that has drained the Bida Basin since the Maastrichtian included the flanks of the Nigerian Shield in the south and the Jos Plateau in the north. The Jos Plateau started to undergo epeirogenic uplift in the Oligocene (Burke 1996), with the oldest volcanic deposits being pre-Neogene (Obaje 2009). This regional uplift may also have affected, to a lesser degree, the Nigerian Shield. The uplift caused a relative lowering in base level, triggering a pulse

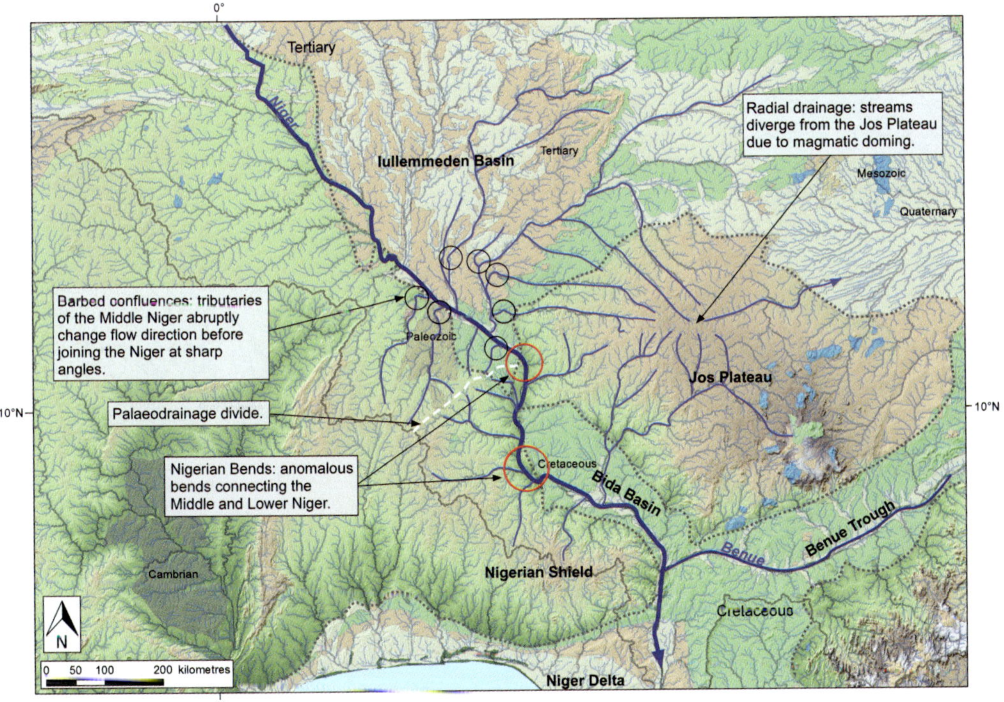


**Fig. 9.** Stream patterns of the Middle and Lower Niger. Red circles show the Nigerian Bends. Blue lines accentuate several streams showing evidence of drainage change. Dashed grey lines surround Cretaceous–Cenozoic basins. Barbed confluences (in small black circles) are shown along the Middle Niger in the Iullemmeden Basin. The white dashed line west of the Niger River shows the palaeodrainage divide between the Iullemmeden and Bida basins. Lower-order tributaries diverge from this ridge. East of the Niger River, overprinting by the radial drainage patterns of the Jos Plateau may have erased geomorphological evidence of the drainage divide.

of erosion into the Jos Plateau. Thereby significant volumes of sediment may have been removed since the Oligocene. Headward erosion was facilitated near the NW edge of the Bida Basin due to the presence of a NNE-trending Pan-African schist belt (Obaje 2009), detectable in the topography as lineaments (Fig. 11). Propagation of erosion along this structurally weaker zone eventually resulted in breaching of the palaeodrainage divide and capture of the drainage in the Iullemmeden Basin. Several features in the geomorphology and the drainage network are the result of the latter events: (1) the amphitheatre-shaped valleys (Fig. 11) observed in the landscape are indicative of headward erosion; (2) the radial drainage developed around the Jos Plateau testifies of the domal uplift origin of this plateau (Fig. 9); (3) the Maastrichtian marine sediments in the Bida Basin are incised, which is the result of lowering of the base level due to the uplift of the Jos Plateau; and (4) the Lower Niger River is shifted to the south with respect to the depositional axis of the Bida Basin, also interpreted to be due to the uplift.

The process of the capture may have been gradual rather than sudden. Erosion first took place in the SE of the Iullemmeden Basin and then propagated into the rest of the basin as a result of the increased potential energy of the stream network. An effective stream draining SE out of the basin became ultimately established. This was the newly formed Middle Niger, which until now has been recognized by barbed confluences along its course.

As a last phase of the evolution of the Niger River, the Upper Niger was captured in the Holocene, as explained at the start of this paper. Owing to deposition in the Inland Niger Delta, the sediment contribution of the Upper Niger River is interpreted as negligible for the Niger Delta.

At present, the water flux within the Iullemmeden Basin is greatly reduced, especially in the north where many streams are defunct owing to the formation of the Sahara Desert. As a result of the aridity, the sediment flux of the Niger River is lower than that of the Benue River, which drains an area with much a higher run-off due to the

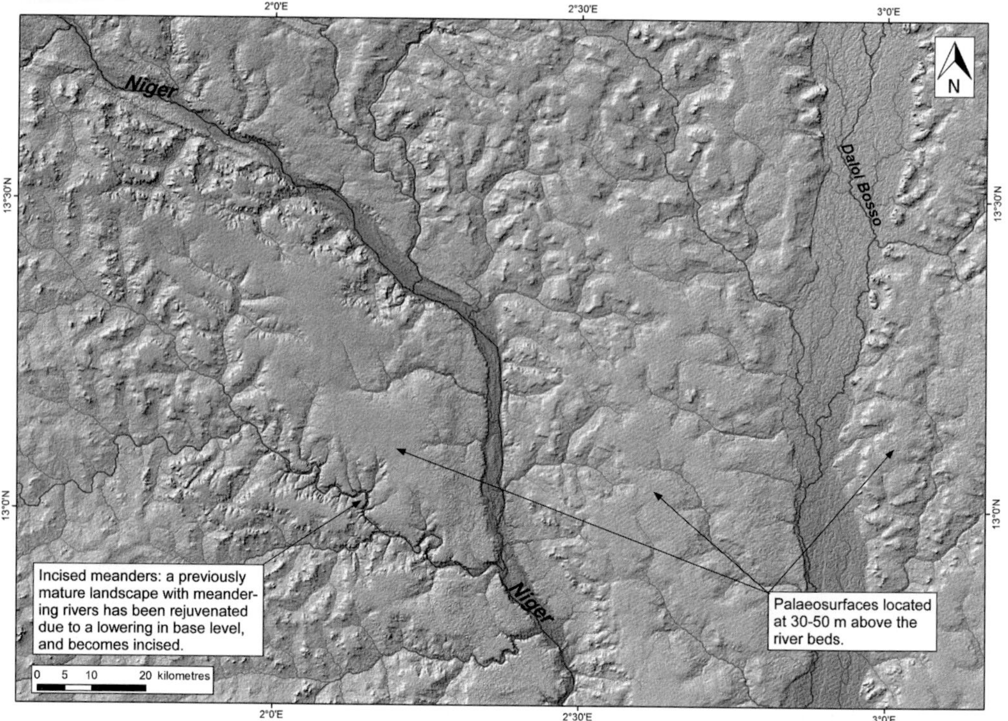



**Fig. 10.** Shaded relief of the Iullemmeden Basin showing palaeosurfaces incised by the Middle Niger and tributaries. The palaeosurfaces are the flat areas between the steep river valleys.

tropical climate. The climate in the Niger River drainage basin has not always been as arid as today. The climate was tropical until the end of the Miocene when open grasslands, replacing forests, started to appear (Micheels *et al.* 2009). Overall, the Sahara Desert started developing in the Pliocene (Micheels *et al.* 2009). In the Nigerian part of the Iullemmeden Basin, the region had a hot and humid climate until the Quaternary, as evidenced by the presence of laterites of Pliocene–Early Quaternary age (Obaje 2009). Therefore, past run-off contributed to higher sediment loads and considerably affected sedimentation in the Niger Delta.

## Discussion

We have discovered that the Niger River formed through three phases: the Bida Basin phase (Maastrichtian–Miocene); the Iullemmeden phase (Miocene–Holocene); and the present-day Niger River phase (Holocene). The capture events to form the Nigerian Bends and the Large Bend represent the transitions between the three phases, and hence their prominence in today's drainage patterns has been proved to be indicative of major

events within the palaeodrainage evolution of the Niger River. The chronological summary of the evolution of the Niger River is given together with the implications on sediment supply to the Niger Delta. Figure 12 illustrates the suggested evolution of the Niger River.

The Bida Basin phase was preceded by the development of the Cretaceous triple junction between the Benue Trough, and the South and Equatorial Atlantic. The Bida Basin formed in the Campanian as an embayment of the Benue Trough, and marine deposition continued into the Maastrichtian. After retreat of the sea in the embayment of the Bida Basin, the palaeo-Lower Niger started to drain the emerged basin. At the same time, the palaeo-Benue River drained the Lower Benue Trough, and has had sources along the flanks of the Benue Trough and the uplifted Middle Benue Trough since the Santonian. The palaeo-Lower Niger and palaeo-Benue Rivers were similar in length, and gradually filled in the Anambra Basin until the Niger Delta started to prograde. The catchment of the palaeo-Lower Niger remained limited to the Bida Basin and its immediate surroundings until the Miocene. Despite the small size of that basin, it might have had a considerable impact on the progradation of

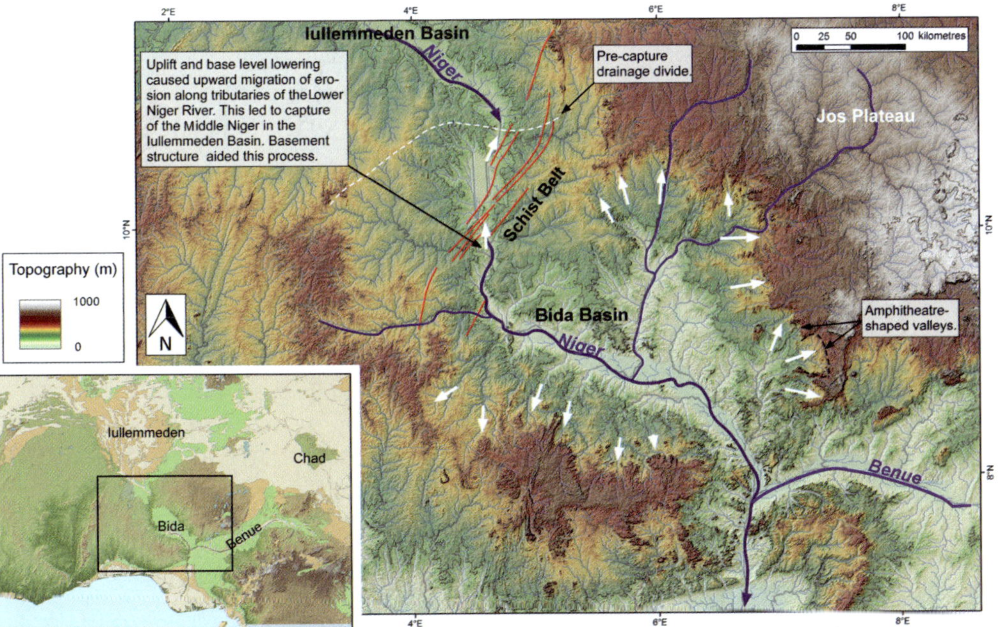



**Fig. 11.** Topography of the Bida Basin, with the Lower Niger and confluence of the Benue and Niger rivers. White arrows show the direction of headward erosion, most of them within amphitheatre-shaped valleys. The red lines show the structural fabric of the Pan-African belt. Eroding streams probably exploited those structures while eroding backwards into the Iullemmeden Basin and eventually capturing the drainage of that basin. The dashed white line is the interpreted palaeodrainage divide between the drainage systems of the Bida and Iullemmeden basins. The inset figure shows the location and the Meso- and Cenozoic geology, highlighting the distribution of the Cretaceous–Cenozoic basins.

the Niger Delta since the Oligocene, when uplift and volcanism started in the Jos Plateau. The onset of progradation of the Niger Delta coincides with the onset of regional magmatic uplifts of Burke (1996). Erosion both north and south of the Bida Basin was of Precambrian units of the Nigerian Basement Complex of Pan-African age (Obaje 2009). The Basement Complex consists mainly of the 'Migmatite–Gneiss–Quartzite Complex', and with secondary 'Schist Belts' made up of phylites, schists, pelites, quartzites, marbles and amphibolites (Obaje 2009). Even though the Schist Belts were less abundant in surface area, they were eroded more deeply due to their less resistant lithologies and foliated texture. Overall, the quartz content in the eroded rock is high, which has a positive impact on reservoir quality. Uplift of the Jos Plateau became more important through the Miocene (Obaje 2009), and hence this area remained an important sediment source for deposition in the Niger Delta. The several Miocene pulses in progradation in the Niger Delta are possibly related to pulses in uplift of the Jos Plateau, the latter supplying sediment via both the palaeo-Lower Niger and the palaeo-Benue.

The Iullemmeden phase started in the Miocene. A more precise timing cannot be determined. The palaeo-Lower Niger had breached through the palaeodrainage divide, eroding through the structurally controlled weaker lithologies of the Schist Belt (Obaje 2009) in the NW of the Bida Basin. The deposition of the Gwandu Formation stopped and, due to upstream progradation of erosion, the entire Iullemmeden Basin and its catchment, an area of roughly $10^6$ km$^2$, was incorporated to the provenance area of the Niger Delta. An increase in sediment supply to the Niger Delta is therefore expected to have occurred. At the end of the Miocene, the climate started to become more arid but savannah vegetation still covered large parts of North Africa, and Saharan conditions were not established until the Pliocene and Pleistocene (Micheels et al. 2009). This suggests that run-off and erosion in the catchment of the palaeo-Middle Niger was more important than at present. Magmatic uplift affecting the Hoggar Massif further increased erosion and the sediment load of the palaeo-Middle Niger. Not knowing exactly when the capture happened, the postulated increase in sediment load can be linked to either of the

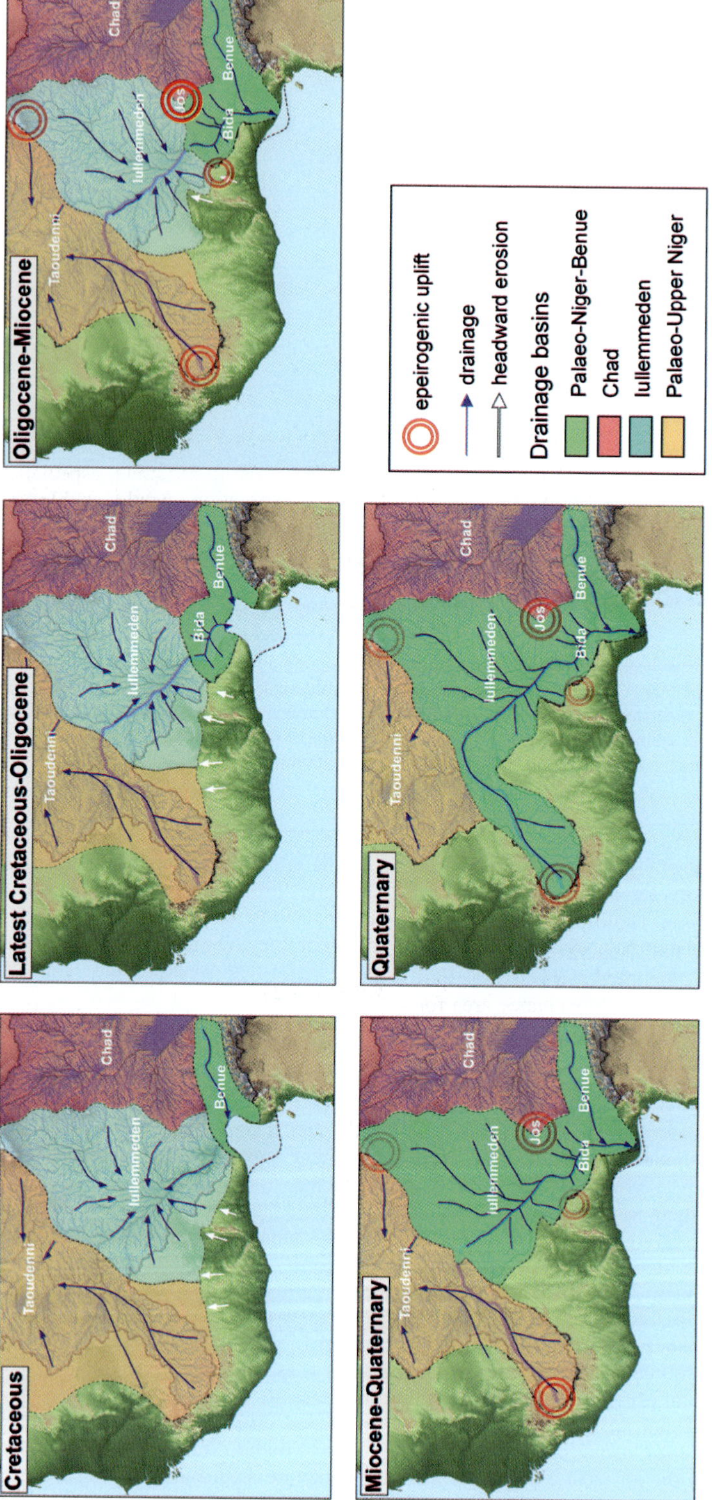


**Fig. 12.** Overview of the evolution of the Niger River drainage basin. The coloured areas are the previously separate drainage basins that were significant in the evolution of the Niger River. The green drainage basin is the one that evolves into the present-day Niger River drainage basin. Red circles highlight the locations of the magmatic uplifts that started in the Oligocene.

Miocene increases in progradation at 14.4–14.0, 12.8–11.5 or post-8.5 Ma. A change in sediment composition must have accompanied the increase in sediment supply. Cretaceous–Cenozoic sediments were eroded from the Iullemmeden Basin. The most widely exposed sediments are those of the Cenozoic, including the Gwandu Formation. They consist of clays and sands (Kogbe 1991). The Cretaceous, Paleocene and Eocene rocks are mainly marine limestones with some sandstones and mudstones (Kogbe 1991). Palaeozoic rocks, exposed on the SE side of the Hoggar Massifs contain conclomerates, sandstones (some with ferruginous oolites), arkoses, shales, calcareous and gypsiferous shales, and magmatic detritus (Kogbe 1991). In the Hoggar Massif and the Adrar des Iforas, Pan-African basement is exposed, comprising oceanic island arcs, cratonic cores, ophiolites and eclogites (Liégeois 2006). Volcanism occurred between 35 and 30 Ma, the voluminous phase between 20 and 12 Ma, and the last phase between 3 Ma and the Late Quaternary (Liégeois 2006). Depending on the timing of drainage capture, the volcanic phases could have negatively influenced sediment type and reservoir quality within the Niger Delta, even though the volcanic impact of the Hoggar would probably have been swamped by volcanic rocks transported by the Benue River. From the southern side of the Iullemmeden Basin, the West African Craton also contributed sediment to the Niger Delta, eroded mainly from gneisses, migmatites and granites.

The third phase of the present-day Niger River did not significantly affect sediment supply to the Niger Delta. Rather, the opposite happened. The climate started to become more arid throughout the Pliocene and into the Quaternary, decreasing the importance of the Iullemmeden Basin and especially its northern regions as a provenance area for the Niger Delta, and dune formation and deposition in the Inland Niger Delta have further hindered any significant transport of sediment from the Upper Niger.

## Conclusions

In this study, the pre-Holocene palaeodrainage evolution of the Niger River was reconstructed and it was shown that this river evolved through three phases, of which only the last one was previously described. Sediment supply of the Niger River to the incipient Niger Delta was initially from a small catchment near the Bida Basin, draining mainly crystalline basement including migmatites, gneisses and granites, and positively influencing reservoir quality. The second phase started in the Miocene when the drainage basin increased by

$10^6 \text{ km}^2$ due to capture of the drainage in the Iullemmeden Basin. This increased the sediment flux to the Niger Delta, and changed the sediment type to mixed lithologies eroded from large exposures of clay, sand and limestone, and smaller exposures of conglomerates, arkoze, shales, volcanics and ultramafics. This change in lithology could have negatively affected reservoir quality. The onset of aridification in the Pliocene reduced the run-off and, hence, sediment supply from the Iullemmeden area. Owing to the aridity and further development of the Sahara in the Holocene, the capture of the Upper Niger (third phase) did not affect sediment supply.

The drainage patterns and anomalies observed within the network of the Niger River match the geological history, and were used as a tool to reconstruct palaeodrainage. For future development, better dating of strata, especially that of the Gwandu Formation, would enable the proposed timing of stream capture between the Bida and Iullemmeden basins to be refined. This would also allow a better correlation with the episodes of increased progradation in the Niger Delta and, hence, to make interpretations about the sediment type at specific locations within the delta stratigraphy. A detailed palaeodrainage study of the Benue River would increase knowledge on the combined effect of both the Niger and Benue river systems on sediment supply to the Niger Delta.

This study was initiated at Getech (Leeds, UK), which is gratefully acknowledged for giving permission to publish. Thanks are extended to L. M. Wilson for processing morphometric and GIS data, and for discussions and support, and to P. J. Markwick for methodology support.

## References

ADIGHIJE, C. I. 1981. A gravity interpretation of the Benue trough, Nigeria. *Tectonophysics*, **79**, 109–128, http://dx.doi.org/10.1016/0040-1951(81)90235-3

AKAA, O. O., NICHOLSON, U., CLIFT, P. & MACDONALD, D. 2008. Niger or Benue Delta: A new insight. AAPG Search and Discovery Article 90078, presented at the AAPG Annual Convention, 20–23 April 2008, San Antonio, Texas, http://www.searchanddiscovery.com/abstracts/html/2008/annual/abstracts/409675.htm

AKANDE, S. O., OJO, O. J., ADEKEYE, O. A., EGENHOFF, S. O., OBAJE, N. G. & ERDTMANN, B. D. 2011. Stratigraphic evolution and petroleum potential of Middle Cretaceous sediments in the Lower and Middle Benue trough, Nigeria: insights from New Source Rock. *Petroleum Technology Development Journal: An International Journal*, **1**, 1–34.

AKANDE, S. O., EGENHOFF, S. O., OBAJE, N. G., OJO, O. J., ADEKEYE, O. A. & ERDTMANN, B. D. 2012. Hydrocarbon potential of Cretaceous sediments in the Lower and Middle Benue Trough, Nigeria: insights from new source rock facies evaluation. *Journal of African Earth Sciences*, **64**, 34–47, http://dx.doi.org/10.1016/j.jafrearsci.2011.11.008

AL-HAJRI, Y., WHITE, N. & FISHWICK, S. 2009. Scales of transient convective support beneath Africa. *Geology*, **37**, 883–886, http://dx.doi.org/10.1130/G25703A.1

ALLEN, P. A. 2008. From landscapes into geological history. *Nature*, **451**, 274–276, http://dx.doi.org/10.1038/nature06586

AMERICAN GEOSCIENCE INSTITUTE. 2014. Geological Age, http://www.agiweb.org/

ANDERSEN, I., DIONE, O., JAROSEWICH-HOLDER, M. & OLIVRY, J.-C. 2005. *The Niger Basin: A Vision for Sustainable Management*. Directions in Development, **34518**. The World Bank, Washington, DC.

ANTOBREH, A. A., FALEIDE, J. I., TSIKALAS, F. & PLANKE, S. 2009. Rift–shear architecture and tectonic development of the Ghana margin deduced from multichannel seismic reflection and potential field data. *Marine and Petroleum Geology*, **26**, 345–368, http://dx.doi.org/10.1016/j.marpetgeo.2008.04.005

ATTOH, K. & BROWN, L. 2008. Deep structure of the southeastern margin of the West African Craton from seismic reflection data, offshore Ghana. *In*: WIBBERLEY, C. A. J., KURZ, W., IMBER, J., HOLDSWORTH, R. E. & COLLETTINI, C. (eds) *The Internal Structure of Fault Zones: Implications for Mechanical and Fluid-Flow Properties*. Geological Society, London, Special Publications, **297**, 499–508, http://dx.doi.org/10.1144/SP297.24

ATTOH, K. & NUDE, P. M. 2008. Tectonic significance of carbonatite and ultrahigh-pressure rocks in the Pan-African Dahomeyide suture zone, southeastern Ghana. *In*: WIBBERLEY, C. A. J., KURZ, W., IMBER, J., HOLDSWORTH, R. E. & COLLETTINI, C. (eds) *The Internal Structure of Fault Zones: Implications for Mechanical and Fluid-Flow Properties*. Geological Society, London, Special Publications, **297**, 217–231, http://dx.doi.org/10.1144/SP297.10

BASILE, C. & ALLEMAND, P. 2002. Erosion and flexural uplift along transform faults. *Geophysical Journal International*, **151**, 646–653.

BASILE, C., MASCLE, J. & GUIRAUD, R. 2005. Phanerozoic geological evolution of the Equatorial Atlantic domain. *Journal of African Earth Sciences*, **43**, 275–282, http://dx.doi.org/10.1016/j.jafrearsci.2005.07.011

BENKHELIL, J. 1989. The origin and evolution of the Cretaceous Benue Trough (Nigeria). *Journal of African Earth Sciences (and the Middle East)*, **8**, 251–282, http://dx.doi.org/10.1016/S0899-5362(89)80028-4

BENKHELIL, J., MASCLE, J. & GUIRAUD, M. 1998. Sedimentary and structural characteristics of the Cretaceous along the Côte d'Ivoire-Ghana transform margin and in the Benue Trough: a comparison. *In*: MASCLE, J., LOHMANN, G. P. & MOULLADE, M. (eds) *Proceedings of the ODP, Scientific Results*, **159**. Ocean Drilling Program, College Station, TX, 93–99.

BERTRAND-SARFATI, J., MOUSSINE-POUCHKINE, A., AFFATON, P., TROMPETTE, R. & BELLION, Y. 1991. Chapter 4: Cover sequences of the West African Craton. *In*: DALLMEYER, R. D. & LÉCORCHÉ, J. P. (eds) *The West African Orogens and Circum-Atlantic Correlatives*. Springer, Berlin, 65–68, http://dx.doi.org/10.1007/978-3-642-84153-8_4

BIGOT-CORMIER, F., BASILE, C., POUPEAU, G., BOUILLIN, J.-P. & LABRIN, E. 2005. Denudation of the Côte d'Ivoire-Ghana transform continental margin from apatite fission tracks. *Terra Nova*, **17**, 189–195, http://dx.doi.org/10.1111/j.1365-3121.2005.00605.x

BINKS, R. M. & FAIRHEAD, J. D. 1992. A plate tectonic setting for Mesozoic rifts of West and Central Africa. *Tectonophysics*, **213**, 141–151, http://dx.doi.org/10.1016/0040-1951(92)90255-5

BRIDGES, E. M. 1990. *World Geomorphology*. Cambridge University Press, Cambridge.

BURBANK, D. W. & PINTER, N. 1999. Landscape evolution: the interactions of tectonics and surface processes. *Basin Research*, **11**, 1–6, http://dx.doi.org/10.1046/j.1365-2117.1999.00089.x

BURKE, K. 1996. The African Plate. *South African Journal of Geology*, **99**, 341–409.

BURKE, K. & DEWEY, J. F. 1973. Plume-generated triple junctions: key indicators in applying plate tectonics to old rocks. *Journal of Geology*, **81**, 406–433.

BURKE, K. & GUNNELL, Y. 2008. *The African Erosion Surface: A Continental-Scale Synthesis of Geomorphology, Tectonics and Environmental Change Over the Past 180 Million Years*. Geological Society of America, Memoirs, **201**.

CASTAING, C., TRIBOULET, C., FEYBESSE, J. L. & CHÈVREMONT, P. 1993. Tectonometamorphic evolution of Ghana, Togo and Benin in the light of the Pan-African/Brasiliano orogeny. *Tectonophysics*, **218**, 323–342, http://dx.doi.org/10.1016/0040-1951(93)90322-B

CHUKWUEBUKA, C. K., EGESI, N. & BEKA, F. T. 2010. Structural geology of the Lower Benue Trough: a case study of the Ishiagu Area of Ebonyi State, Nigeria. AAPG Search and Discovery Article 90115, presented at the AAPG Africa Region Annual Conference, 14–19 November 2010, Abuja, Nigeria, http://www.searchanddiscovery.com/abstracts/pdf/2010/african_regional/abstracts/ndx_chukwuebuka.pdf

CRAIG, J., THUROW, J., THUSU, B., WHITHAM, A. & ABUTARRUMA, Y. 2009. Global Neoproterozoic petroleum systems: the emerging potential in North Africa. *In*: CRAIG, J., THUROW, J., THUSU, B., WHITHAM, A. & ABUTARRUMA, Y. (eds) *Global Neoproterozoic Petroleum Systems: The Emerging Potential in North Africa*. Geological Society, London, Special Publications, **326**, 1–25.

DIRKS, P. H. G. M., BLENKINSOP, T. G. & JELSMA, H. A. 2009. The geological evolution of Africa. *In*: *Encyclopedia of Life Systems (EOLSS). Volume 4: Geology*. UNESCO Publishing, Paris.

DOS SANTOS, T. J. S., FETTER, A. H. & NETO, J. A. N. 2008. Comparisons between the northwestern Borborema Province, NE Brazil, and the southwestern Pharusian Dahomey Belt, SW Central Africa. *In*: PANKHURST, R. J., TROUW, R. A. J., BRITO NEVES, B. B. & DE WIT, M. J. (eds) *West Gondwana: Pre-Cenozoic Correlations Across the South Atlantic Region*. Geological Society, London, Special Publications, **294**, 101–120, http://dx.doi.org/10.1144/SP294.6

DOUST, H. 1990. Petroleum geology of the Niger Delta. *In*: BROOKS, J. (ed.) *Classic Petroleum Provinces*. Geological Society, London, Special Publications, **50**, 365–365, http://dx.doi.org/10.1144/GSL.SP.1990.050.01.21

DOUST, H. & OMATSOLA, E. 1989. Niger Delta. *In*: EDWARDS, J. D. & SANTOGROSSI, P. A. (eds) *Divergent/Passive Margin Basins*. American Association of Petroleum Geologists, Memoirs, **48**, 201–238.

FAIRHEAD, J. D. & OKEREKE, C. S. 1990. Crustal thinning and extension beneath the Benue Trough based on gravity studies. *Journal of African Earth Sciences (and the Middle East)*, **11**, 329–335, http://dx.doi.org/10.1016/0899-5362(90)90011-3

FAO. 1997. *Irrigation Potential in Africa: A Basin Approach*. FAO Land and Water Bulletin, Report **4**. Food and Agriculture Organization of the United Nations, Rome.

FIGUEIREDO, J., HOORN, C., VAN DER VEN, P. & SOARES, E. 2009. Late Miocene onset of the Amazon River and the Amazon deep-sea fan: evidence from the Foz do Amazonas Basin. *Geology*, **37**, 619–622, http://dx.doi.org/10.1130/G25567A.1

FITTON, J. G. 1980. The Benue Trough and Cameroon Line – a migrating rift system in West Africa. *Earth and Planetary Science Letters*, **51**, 132–138.

GASPERINI, L., BERNOULLI, D. ET AL. 2001. Lower Cretaceous to Eocene sedimentary transverse ridge at the Romanche Fracture Zone and the opening of the equatorial Atlantic. *Marine Geology*, **176**, 101–119, http://dx.doi.org/10.1016/S0025-3227(01)00146-3

GASSE, F. 2000. Hydrological changes in the African tropics since the Last Glacial Maximum. *Quaternary Science Reviews*, **19**, 189–211, http://dx.doi.org/10.1016/S0277-3791(99)00061-X

GASSE, F., TÉHET, R., DURANT, A., GIBERT, E. & FONTES, J. C. 1990. The arid-humid transition in the Sahara and the Sahel during the last deglaciation. *Nature*, **346**, 141–156.

GENIK, G. J. 1992. Regional framework, structural and petroleum aspects of rift basins in Niger, Chad and the Central African Republic (C.A.R.). *Tectonophysics*, **213**, 169–185, http://dx.doi.org/10.1016/0040-1951(92)90257-7

GIRESSE, P. 2008. *Tropical and Sub-Tropical West Africa – Marine and Continental Changes During the Late Quaternary*. Developments in Quaternary Science, **10**. Elsevier, Amsterdam.

GONÇALVES, C. A. & EWERT, L. 1998. Development of the Cote D'Ivoire–Ghana transform margin: evidence from the integration of core and wireline log data. *In*: HARVEY, P. K. & LOVELL, M. A. (eds) *Core–Log Integration*. Geological Society, London, Special Publications, **136**, 375–389, http://dx.doi.org/10.1144/GSL.SP.1998.136.01.30

GOUDIE, A. S. 2005. The drainage of Africa since the Cretaceous. *Geomorphology*, **67**, 437–456, http://dx.doi.org/10.1016/j.geomorph.2004.11.008

GRANT, N. K. 1971. South Atlantic, Benue Trough, and Gulf of Guinea Cretaceous Triple Junction. *Geological Society of America Bulletin*, **82**, 2295–2298, http://dx.doi.org/10.1130/0016-7606(1971)82[2295:SABTAG]2.0.CO;2

GUIRAUD, R. & MAURIN, J.-C. 1992. Early Cretaceous rifts of Western and Central Africa: an overview. *Tectonophysics*, **213**, 153–168, http://dx.doi.org/10.1016/0040-1951(92)90256-6

GUIRAUD, R., BINKS, R. M., FAIRHEAD, J. D. & WILSON, M. 1992. Chronology and geodynamic setting of Cretaceous-Cenozoic rifting in West and Central Africa. *Tectonophysics*, **213**, 227–234, http://dx.doi.org/10.1016/0040-1951(92)90260-D

GUIRAUD, R., BOSWORTH, W., THIERRY, J. & DELPLANQUE, A. 2005. Phanerozoic geological evolution of Northern and Central Africa: an overview. *Journal of African Earth Sciences*, **43**, 83–143, http://dx.doi.org/10.1016/j.jafrearsci.2005.07.017

GUPTA, A. 2007. *Large Rivers: Geomorphology and Management*. John Wiley, Chichester.

ILOEJE, N. P. 1981. *A New Geography of Nigeria*. Longman, London.

JACOBBERGER, P. 1981. Geomorphology of the upper Inland Niger Delta. *Journal of Arid Environments*, **13**, 95–112.

JONES, E. J. W., CANDE, S. C. & SPATHOPOULOS, F. 1995. Evolution of a major oceanographic pathway: the Equatorial Atlantic. *In*: SCRUTTON, R. A., STOKER, M. S., SHIMMIELD, G. B. & TUDHOPE, A. W. (eds) *The Tectonics, Sedimentation and Palaeoceanography of the North Atlantic Region*. Geological Society, London, Special Publications, **90**, 199–213, http://dx.doi.org/10.1144/GSL.SP.1995.090.01.12

KOGBE, C. A. 1989. *Geology of Nigeria*. Rockview (Nigeria), Jos, Nigeria.

KOGBE, C. A. 1991. Stratigraphy and tectonic history of the Iullemmeden Basin in West Africa. *South African Journal of Geology*, **94**, 19–32.

KOGBE, C. A., AJAKAIYE, D. E. & MATHEIS, G. 1983. Confirmation of rift structure along the Mid-Niger Valley. *Journal of African Earth Sciences*, **1**, 127–131.

KRÖNER, A. & STERN, R. J. 2004. Pan-African orogeny. *In*: SELLEY, R. C., COCKS, L. R. M. & PLIMER, I. R. (eds) *Encyclopedia of Geology, Volume 1*. Elsevier, Amsterdam, 1–12.

LESPEZ, L., LE DREZEN, Y. ET AL. 2011. High-resolution fluvial records of Holocene environmental changes in the Sahel: the Yamé River at Ounjougou (Mali, West Africa). *Quaternary Science Reviews*, **30**, 737–756, http://dx.doi.org/10.1016/j.quascirev.2010.12.021

LIÉGEOIS, J.-P. 2006. The Hoggar Swell and Volcanism, Tuareg Shield, Central Sahara: Intraplate Reactivation of Precambrian Structures as a Result of Alpine Convergence, http://www.mantleplumes.org/Hoggar.html

LOMPO, M. 2009. Geodynamic evolution of the 2.25–2.0 Ga Palaeoproterozoic magmatic rocks in the Man-Leo Shield of the West African Craton. A model of subsidence of an oceanic plateau. *In*: REDDY, S. M., MAZUMDER, R., EVANS, D. A. D. & COLLINS, A. S. (eds) *Palaeoproterozoic Supercontinents and Global Evolution*. Geological Society, London, Special Publications, **323**, 231–254, http://dx.doi.org/10.1144/SP323.11

MACGREGOR, D. 2011. Rift shoulder source to prodelta sink: The Cenozoic development of the Nile drainage system. AAPG Search and Discover Article 50506, presented at the AAPG International Conference and Exhibition, 23–26 October 2011, Milano, Italy, http://www.searchanddiscovery.com/documents/2011/50506macgregor/ndx_macgregor.pdf

MACGREGOR, D. S., ROBINSON, J. & SPEAR, G. 2003. Play fairways of the Gulf of Guinea transform margin. *In*: ARTHUR, T. J., MACGREGOR, D. S. & CAMERON, N. (eds) *Petroleum Geology of Africa: New Themes and Developing Technologies*. Geological Society, London, Special Publications, **207**, 131–150, http://dx.doi.org/10.1144/GSL.SP.2003.207.7

MAKASKE, B., VRIES, E. De, TAINTER, J. A. & MCINTOSH, R. J. 2007. Aeolian and fluviolacustrine landforms and prehistoric human occupation on a tectonically influenced floodplain margin, the Méma, central Mali. *Netherlands Journal of Geosciences – Geologie en Mijnbouw*, **86**, 241–256.

MARKWICK, P. J. & VALDES, P. J. 2004. Palaeo-digital elevation models for use as boundary conditions in coupled ocean–atmosphere GCM experiments: a Maastrichtian (late Cretaceous) example. *Palaeogeography, Palaeoclimatology, Palaeoecology*, **213**, 37–63, http://dx.doi.org/10.1016/j.palaeo.2004.06.015

MARKWICK, P. J., RADDADI, M. C. *ET AL.* 2010. The evolution of global source-to-sink relationships during the Cretaceous and Tertiary using stage level palaeogeographies and DEMs. AAPG Annual Convention and Exhibition, 11–14 April 2010, New Orleans, Louisiana, AAPG Search and Discover Article, 90104. http://www.searchanddiscovery.com/abstracts/pdf/2010/annual/abstracts/ndx_markwick.pdf

MATHEY, B., ALZOUMA, K., LANG, J., MEISTER, C., NERAUDEAU, D. & PASCAL, A. 2006. Marine deposits, faunal assemblages and palaeoenvironments of the Late Cenomanian – Early Turonian interval in the Iullemmeden Basin and the Northern Chad basin (Niger). *In: African Basins Biogeoscience and Environment: Part 2. Sedimentary Geosystems – South of the Sahara. Africa Geoscience Review*, **13**, 323–329.

MCINTOSH, R. J. 1983. Floodplain geomorphology and human occupation of the upper inland delta of the Niger. *The Geographical Journal*, **149**, 182–201.

MICHEELS, A., ERONEN, J. & MOSBRUGGER, V. 2009. The Late Miocene climate response to a modern Sahara desert. *Global and Planetary Change*, **67**, 193–204, http://dx.doi.org/10.1016/j.gloplacha.2009.02.005

MOHRIAK, W. U. & ROSENDAHL, B. R. 2003. Transform zones in the South Atlantic rifted continental margins. *In*: STORTI, F., HOLDSWORTH, R. E. & SALVINI, F. (eds) *Intraplate Strike-Slip Deformation Belts*. Geological Society, London, Special Publications, **210**, 211–228, http://dx.doi.org/10.1144/GSL.SP.2003.210.01.13

MONOD, T. 1964. The Late Tertiary and Pleistocene in the Sahara. *In*: HOWEL, F. C. & BOULIÈRE, F. (eds) *African Ecology and Human Evolution*. Methuen, London, 117–229.

MONTGOMERY, D. R. 2003. Predicting landscape-scale erosion rates using digital elevation models. *Comptes Rendus Geoscience*, **335**, 1121–1130, http://dx.doi.org/10.1016/j.crte.2003.10.005

MOULIN, M., ASLANIAN, D. & UNTERNEHR, P. 2010. A new starting point for the South and Equatorial Atlantic Ocean. *Earth-Science Reviews*, **98**, 1–37, http://dx.doi.org/10.1016/j.earscirev.2009.08.001

NEMČOK, M., HENK, A., ALLEN, R., SIKORA, P. J. & STUART, C. 2012. Continental break-up along strike-slip fault zones; observations from the Equatorial Atlantic. *In*: MOHRIAK, W. U., DANFORTH, A., POST, P. J., BROWN, D. E., TARI, G. C., NEMČOK, M. & SINHA, S. T. (eds) *Conjugate Divergent Margins*. Geological Society, London, Special Publications, **369**, 537–556, http://dx.doi.org/10.1144/SP369.14

O'CALLAGHAN, J. F. & MARK, D. M. 1984. The Extraction of Drainage Networks from Digital Elevation Data. *Computer Vision, Graphics and Image Processing*, **28**, 328–344.

OBAJE, N. G. 2009. *Geology and Mineral Resources of Nigeria*. Springer, Berlin.

OBAJE, N. G., MUSA, M. K., ODOMA, A. N. & HAMSA, H. 2011. The Bida Basin in north-central Nigeria: sedimentology and petroleum geology. *Journal of Petroleum and Gas Exploration Research*, **1**, 1–13.

OFOEGBU, C. O. 1985. A review of the geology of the Benue Trough, Nigeria. *Journal of African Earth Sciences*, **3**, 283–291, http://dx.doi.org/10.1016/0899-5362(85)90001-6

OFOEGBU, C. O. & ONUOHA, K. M. 1991. Analysis of magnetic data over the Abakaliki Anticlinorium of the Lower Benue Trough, Nigeria. *Marine and Petroleum Geology*, **8**, 174–183, http://dx.doi.org/10.1016/0264-8172(91)90005-L

OJO, O. M. & PINNA, P. 1982. Palaeogeographical and structural evolution of the Upper Benue trough of Nigeria. *Cretaceous Research*, **3**, 195–207, http://dx.doi.org/10.1016/0195-6671(82)90020-9

OLIVRY, J.-C. & BOULÈGUE, J. 1995. *Grands bassins fluviaux périatlantiques: Congo, Niger, Amazone.* ORSTOM editions, Paris.

OPELOYE, S. A. 2012. Lithofacies Association in the Bima Sandstone of the Upper Benue Trough, Nigeria. *Pacific Journal of Science and Technology*, **13**, 417–427.

PALAUSI, G. 1955. Au sujet du Niger fossile dans la région de Timbouctou. *Revue de Geomorphologie Dynamique*, **6**, 217–218.

PAZZAGLIA, F. J. 2003. Landscape evolution models. *Development in Quaternary Science*, **1**, 247–274, http://dx.doi.org/10.1016/S1571-0866(03)01012-1

PETTERS, S. W. 1978. Mid-Cretaceous paleoenvironments and biostratigraphy of the Benue Trough, Nigeria. *Geological Society of America Bulletin*, **89**, 151–154, http://dx.doi.org/10.1130/0016-7606(1978)89<151:MPABOT>2.0.CO;2

REIJERS, T. 2011. Stratigraphy and sedimentology of the Niger Delta. *Geologos*, **17**, 133–162, http://dx.doi.org/10.2478/v10118-011-0008-3

SHEMANG, E. M., AJAYI, C. O. & JACOBY, W. R. 2001. A magmatic failed rift beneath the Gongola arm of the upper Benue trough, Nigeria. *Journal of Geodynamics*, **32**, 355–371, http://dx.doi.org/10.1016/S0264-3707(01)00034-5

STANKIEWICZ, J. & DE WIT, M. J. 2006. A proposed drainage evolution model for Central Africa—Did the Congo flow east? *Journal of African Earth Sciences*,

**44**, 75–84, http://dx.doi.org/10.1016/j.jafrearsci.2005.11.008

STRAHLER, A. N. 1957. Quantitative analysis of watershed geomorphology. *Transactions of the American Geophysical Union*, **38**, 913–920.

STUART, G. W., FAIRHEAD, J. D., DORBATH, L. & DORBATH, C. 1985. Crustal structure of the Adamawa Plateau, Cameroon. *Revue Sciences et Techniques, Série Sciences de la Terre*, **1**, 25–35.

SUMMERFIELD, M. A. 1991. *Global Geomophology: An Introduction to the study of Landforms*. Longman, London.

TARBOTON, D. G., BRAS, R. L. & RODRIGUEZ-ITURBE, I. 1991. On the extraction of channel networks from digital elevation data. *Hydrologic Processes*, **5**, 81–100.

TARBOTON, D. G. & AMES, D. P. 2001. Advances in the mapping of flow networks from digital elevation data. Paper presented at the World Water and Environmental Resources Congress, 20–24 May, Orlando, Florida, USA.

THIEMEYER, H. 2000. From Megachad to Microchad - environmental changes during the Holocene. *Berichte des Sonderforschungsbereichs*, **268**, 11–19.

TORSVIK, T. H., VAN DER VOO, R. ET AL. 2012. Phanerozoic polar wander, palaeogeography and dynamics. *Earth-Science Reviews*, **114**, 325–368, http://dx.doi.org/10.1016/j.earscirev.2012.06.007

TRICART, J. 1965. *Rapport de la mission reconnaissance géomorphologique de la vallée moyenne du Niger*. Institut Fondamental d'Afrique Noire, Memoirs, **72**.

TUCKER, G. E. & SLINGERLAND, R. 1997. Drainage basin responses to climate change. *Water Resources Research*, **33**, 2031–2047, http://dx.doi.org/10.1029/97WR00409

TURNER, J. P. & WILSON, P. G. 2009. Structure and composition of the ocean–continent transition at an obliquely divergent transform margin, Gulf of Guinea, West Africa. *Petroleum Geoscience*, **15**, 305–311, http://dx.doi.org/10.1144/1354-079309-846

TUTTLE, M. L. W., CHARPENTIER, R. R. & BROWNFIELD, M. E. 1999. *The Niger Delta Petroleum System: Niger Delta Province, Nigeria, Cameroon, and Equatorial Guinea, Africa*. United States Geological Survey, Open File Report, **99-50-H**.

TWIDALE, C. R. 1994. Gondwanan (Late Jurassic and Cretaceous) palaeosurfaces of the Australian craton. *Palaeogeography, Palaeoclimatology, Palaeoecology*, **112**, 157–186, http://dx.doi.org/10.1016/0031-0182(94)90139-2

TWIDALE, C. R. 2004. River patterns and their meaning. *Earth-Science Reviews*, **67**, 159–218, http://dx.doi.org/10.1016/j.earscirev.2004.03.001

UKAEGBU, V. U. & AKPABIO, I. O. 2009. Geology and Stratigraphy of Middle Cretaceous Sequences Northeast of Afikpo Basin, Lower Benue Trough, Nigeria. *Pacific Journal of Science and Technology*, **10**, 518–527.

URVOY, Y. 1942. *Les basins du Niger*. Institut Fondamental d'Afrique Noire, Memoirs, **4**, 139.

VARGA, G. 2012. Spatio-temporal distribution of dust storms – a global coverage using NASA TOMS aerosol measurements. *Hungarian Geographical Bulletin*, **61**, 275–298.

VERDIN, K. L. 1997. A system for topologically coding global drainage basins and stream networks. Paper presented at the 17th Annual ESRI Users Conference, July 1997, San Diego, California.

VERDIN, K. L. & VERDIN, J. P. 1999. A topological system for delineation and codification of the Earth's river basins. *Journal of Hydrology*, **218**, 1–12.

WILSON, M. 1992. Magmatism and continental rifting during the opening of the South Atlantic Ocean: a consequence of Lower Cretaceous super-plume activity? *In*: STOREY, B. C., ALABASTER, T. & PANKHURST, R. J. (eds) *Magmatism and the Causes of Continental Break-up*. Geological Society, London, Special Publications, **68**, 241–255, http://dx.doi.org/10.1144/GSL.SP.1992.068.01.15

WILSON, M. & GUIRAUD, R. 1992. Magmatism and rifting in Western and Central Africa, from Late Jurassic to recent times. *Tectonophysics*, **213**, 203–225, http://dx.doi.org/10.1016/0040-1951(92)90259-9

WILSON, R. C. L. & WILLIAMS, C. A. 1979. Oceanic transform structures and the development of Atlantic continental margin sedimentary basins – a review. *Journal of the Geological Society, London*, **136**, 311–320, http://dx.doi.org/10.1144/gsjgs.136.3.0311

ZHENG, H., CLIFT, P. D., WANG, P., TADA, R., JIA, J., HE, M. & JOURDAN, F. 2013. Pre-Miocene birth of the Yangtze River. *Proceedings of the National Academy of Sciences*, **110**, 7556–7561, http://dx.doi.org/10.1073/pnas.1216241110

# Metamorphic grade of source rocks revealed by chemical fingerprints of detrital amphibole and garnet

SERGIO ANDÒ[1], ANDREW MORTON[2,3] & EDUARDO GARZANTI[1]*

[1]*Laboratory for Provenance Studies, Department of Earth and Environmental Sciences, Università di Milano-Bicocca, Piazza della Scienza 4, 20126 Milano, Italy*

[2]*HM Research Associates, 2 Clive Road, Balsall Common, West Midlands CV7 7DW, UK*

[3]*CASP, University of Cambridge, 181a, Huntingdon Road, Cambridge CB3 0DH, UK*

**Corresponding author (e-mail: eduardo.garzanti@unimib.it)*

**Abstract:** Amphibole and garnet are among the most widespread heavy minerals in orogenic sediments. Their chemical composition and optical properties vary markedly and systematically with temperature and pressure conditions during growth, and thus provide important information on the metamorphic evolution of source areas that is crucial in palaeotectonic and palaeogeodynamic reconstructions. This study investigates the chemical composition of detrital amphiboles and garnets derived from parent rocks of progressively increasing metamorphic grade through a well-studied composite section across the Central and Southern Alps, including the granulite-facies core of the Late Palaeozoic orogen exposed in the Ivrea–Verbano Zone and the amphibolite-facies core of the Cenozoic orogen exposed in the Lepontine Dome. We specifically focus on metamorphic grade because it represents the best proxy for tectono-stratigraphic crustal level, and hence degree of unroofing of source areas. In river sands collected between metamorphic isograds corresponding to crystallization temperatures ranging from *c.* 500 °C to *c.* 850 °C, $TiO_2$ gradually increases in detrital amphibole while its colour progressively changes from blue-green in the lower amphibolite-facies where actinolite, hornblende and tschermakite are most abundant, to brown in the granulite facies where pargasite is dominant. Detrital garnets display moderate gradual changes across the amphibolite-facies Lepontine Dome, where low-Mg 'type B' garnets predominate. Almandine-spessartine is spatially associated with abundance of pegmatites while entering the zone of anatexis (Southern Steep Belt), where grossular or grossular-andradite-spessartine are occasionally found. A sharp change occurs while reaching granulite-facies in the Ivrea–Verbano Zone, where high-Mn garnets disappear and 'type A' almandine-pyrope (from 'stronalite' metasediments) and 'type C' almandine-pyrope-grossular (from metagabbros of the Mafic Complex) predominate. Also redefined in this article are a series of numerical indices based on amphibole colour and relative abundances of diverse key minerals (chloritoid, staurolite, andalusite, kyanite, fibrolitic and prismatic sillimanite), useful to accurately assess the average metamorphic grade of meta-igneous and metasedimentary source rocks.

**Supplementary material:** Chemical composition of detrital amphiboles and garnets and full information on the location and mineralogical composition of studied samples is available at http://www.geolsoc.org.uk/SUP18618

By picking up a handful of sand, you have collected a half-million grains that come from all over a catchment. It is a very potent integrator of the information and obviously far more efficient than sampling hundreds of outcrops.(Doug Burbank, in Greensfelder 2002, p. 256)

Provenance analysis involves the use of a great variety of techniques, and allows us to efficiently extract palaeogeographic information from siliciclastic deposits accumulated in sedimentary basins (for up-to-date reviews see Sylvester 2012; von Eynatten & Dunkl 2012; von Eynatten *et al.* 2012). Such information is indispensable to unravelling the tectonic and erosional history of source areas in space and time, and to reconstruct geodynamic sceneries of the past (Dickinson 1988). It is particularly important to understand the operation of the largest factories of terrigenous sediments, including the Alpine–Himalayan orogenic belts formed by continental collision (Garzanti *et al.* 2007).

Although the structure of each orogen has its own peculiarities and stands as a case apart, collision orogens typically include an axial belt containing the remnants of stretched continental-margin lithosphere, which have undergone high-pressure metamorphism before being exhumed and re-equilibrated at greenschist-facies, amphibolite-facies or even granulite-facies conditions (Chopin

*From*: SCOTT, R. A., SMYTH, H. R., MORTON, A. C. & RICHARDSON, N. (eds) 2014. *Sediment Provenance Studies in Hydrocarbon Exploration and Production*. Geological Society, London, Special Publications, **386**, 351–371.
First published online June 20, 2013, http://dx.doi.org/10.1144/SP386.5

1984; Compagnoni *et al.* 1995; Hacker *et al.* 2006). Such an axial backbone of neometamorphic rocks, deformed at subcrustal depths, represents a fossil continental-subduction zone (Garzanti *et al.* 2010).

Unroofing of the axial belt, which accelerates markedly during the later 'morphogenic stage' associated with hard collision and rapid tectonic uplift (Gansser 1982; Garzanti & Malusà 2008), produces sediments with diagnostic mineralogical signatures. Greenschist-facies cover strata and basement rocks typically shed, respectively, quartzolithic and litho-quartzose detritus with moderately poor epidote-dominated and epidote-garnet heavy-mineral suites. Amphibolite-facies gneisses, instead, shed feldspatho-quartzose detritus with richer hornblende-dominated suites including garnet, epidote and generally minor staurolite, kyanite, andalusite or sillimanite. Amphibole and garnet, occurring in abundance in medium-grade meta-igneous and metasedimentary rocks, respectively, invariably account together for half of the transparent heavy minerals in sands of major Alpine and Himalayan rivers, including the Po, the Rhône, the Indus, the Ganga and the Brahmaputra (52 ± 10%; Garzanti & Andò 2007*a*, table 4).

The information needed to reconstruct the erosional evolution of a growing orogen is recorded in sedimentary sequences of the adjacent foreland to remnant-ocean basins (Ingersoll *et al.* 2003). For instance, the successive appearance and increase of index minerals such as chloritoid, staurolite, kyanite, fibrolitic sillimanite and finally prismatic sillimanite upward in the stratigraphic column, if not due to preservation in superficial layers less affected by diagenetic dissolution, can be and has long been used to reveal progressive deepening of erosion into higher-grade metasedimentary rocks (e.g. Szulc *et al.* 2006; Najman *et al.* 2009). These mineralogical trends can be usefully expressed by simple numerical parameters (e.g. the Metasedimentary Mineral Index (MMI) and Sillimanite

Index; Garzanti *et al.* 2006). Particularly useful for tracing metamorphic provenance is the Hornblende Colour Index (HCI), based on the relative abundance of blue-green, green, green-brown and brown hornblende grains (Table 1). Because it considers one single mineral, the HCI index is not severely biased by selective dissolution, and can be used to obtain unambiguous results in any sediment or sedimentary rock where amphibole is preserved. Although optical determination of hornblende colour is not entirely objective, it has been empirically demonstrated to provide valuable and consistent information in studies of both metamorphic rocks and metamorphiclastic sediments (Miyashiro 1972, p. 254; Raase 1974; Wenk *et al.* 1974; Garzanti *et al.* 2004, 2006). In order to verify the reliability and limitations of the HCI index, it is crucial to integrate the indications it provides with full quantitative chemical information. This is precisely the goal of the present article, which also investigates how the chemical composition of detrital garnet – which typically resists diagenetic dissolution better than amphibole and is thus more commonly preserved in ancient strata (Morton & Hallsworth 2007; Andò *et al.* 2012) – can be used in the accurate assessment of metamorphic provenance.

Excellent conditions for such a test are offered by the Alps, a thoroughly studied orogenic belt where every valley is accessible and patterns of metamorphic isograds have been carefully traced by numerous detailed studies throughout the last half century (Wenk 1962; Niggli & Niggli 1965; Trommsdorff 1966; Wenk & Keller 1969; Zingg 1983; Henk *et al.* 1997; Todd & Engi 1997; Frey *et al.* 1999; Nagel *et al.* 2002; Engi 2011). Amphibolite-facies rocks of Cenozoic age constitute the Lepontine heart of the axial Central Alps (Fig. 1), whereas exposed in the Southalpine external belt to the SW is a widely studied Upper Palaeozoic lower-crustal section including granulite-facies

**Table 1.** *Heavy-mineral indices useful to define average metamorphic grade and tectono-stratigraphic level of source rocks*

| Hornblende Colour Index | $(1/3$ green Hbl $+ 2/3$ green-brown Hbl $+$ brown Hbl$)$/total hornblende $\times$ 100 |
|---|---|
| Metasedimentary Mineral Index | $(St/2 + Ky/2 + And/2 + Sil)/(Cld + St + Ky + And + Sil) \times 100$ |
| Sillimanite Index | Prismatic sillimanite/total (fibrolitic $+$ prismatic) sillimanite $\times$ 100 |

| | Amphibolite facies | | | Granulite facies | |
|---|---|---|---|---|---|
| | Lower | Middle | Upper | Metasediments | Metagabbro |
| Hornblende Colour Index | 0–10 | 10–30 | 30–60 | 60–100 | 80–100 |
| Metasedimentary Mineral Index | 50–70 | 70–90 | 90–100 | 100 | – |
| Sillimanite Index | – | 0 | 0–30 | 75–100 | – |

*Source*: Modified after Garzanti *et al.* (2004, 2006).
All three indices vary from 0 to 100. Hbl, hornblende; Cld, chloritoid; St, staurolite; Ky, kyanite; And, andalusite; Sil, sillimanite.

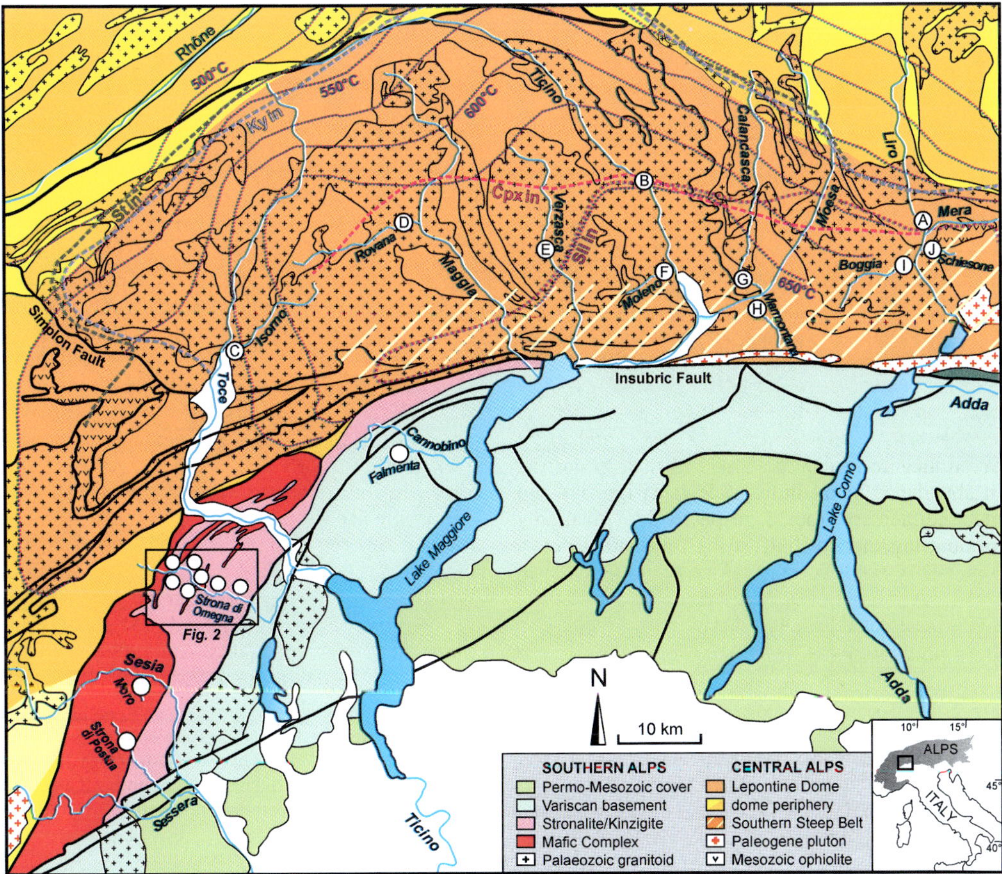



**Fig. 1.** Geological sketch map of the study area (redrawn after Frey *et al.* 1999 and Engi *et al.* 2004 and other sources cited in the text). Isograds and locations of studied samples are shown in the Lepontine Dome (A, Liro; B, Ticino; C, Isorno; D, Rovana; E, Verzasca; F, Moleno; G, Calancasca; H, Marmontana; I, Boggia; J, Schiesone). St, staurolite; Ky, kyanite; Cpx, clinopyroxene; Sil, sillimanite.

metasediments and metagabbros (Fig. 2). We collected river sand in small monolithological catchments ('first-order sampling scale' of Ingersoll 1990) along this composite transect, and thus obtained representative samples of detritus produced by diverse parent lithologies spanning the complete range of metamorphic temperatures from *c.* 500 °C to *c.* 850 °C. Taking advantage of such a suitable natural laboratory, we will thus test here if and to what extent the chemical composition of detrital amphibole and garnet faithfully reflects the metamorphic grade of source rocks. In particular, we will show that the chemistry of amphibole grains crystallized at progressively higher temperatures correlates remarkably well with their colour. By a simple optical assessment of amphibole colour in heavy-mineral slides (a standard for comparison is provided in Fig. 3), we can thus estimate the metamorphic temperatures of the source rocks with remarkable accuracy.

## The Alps as a source of metamorphic detritus

The Alps are a doubly-vergent collision orogen, formed during convergence between Europe and Africa–Adria from the Late Cretaceous (Polino *et al.* 1990; Schmid *et al.* 2008). Closure of the Piemont–Ligurian Ocean and consequent subduction of oceanic and continental slivers beneath the Adriatic palaeomargin took place in the Eocene, and was followed by continental indentation from the late Oligocene (Malusà *et al.* 2011).

The Alpine orogen includes an axial stack of nappes, flanked by opposite-verging, thick-skinned

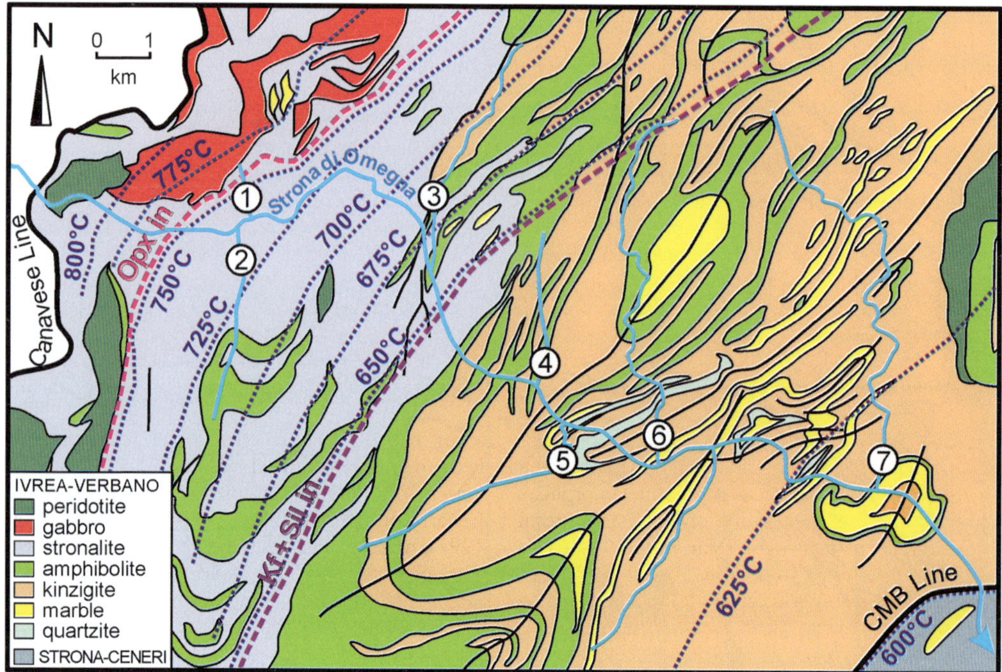



**Fig. 2.** The Southern Alps lower crustal section in the Strona di Omegna Valley. The Ivrea–Verbano Zone includes granulite-facies metagabbros (Mafic Complex) and metasediments (stronalites) in the west, passing upward to upper-amphibolite-facies metasediments (kinzigites) in the east (geological map and isograds based on Zingg 1983; Henk *et al.* 1997; Rutter *et al.* 2007). Locations of studied samples are shown (1, Piana di Forno; 2, Nagarone; 3, Forno; 4, Ovasca; 5, Sambughetto; 6, Luzzogno; 7, Prelo). Opx, orthopyroxene; Kf, K-feldspar; Sil, sillimanite.

**Fig. 3.** The colour of detrital amphiboles and garnets reveals compositional changes largely associated with changing temperature and pressure conditions during metamorphism of the parent rocks. Careful recording of amphibole colour allows determination of the HCI index and assessment of average metamorphic grade (and hence average tectono-stratigraphic level and degree of unroofing) of source areas.

external thrust belts on both sides (Schlunegger & Willett 1999). The axial nappe-pile, delimited by the Penninic Front to the north and by the Insubric Fault to the south, includes remnants of strongly thinned outer continental margins (continental Penninic and Austroalpine nappes) and of the intervening oceanic domain (Piemontese Zone), which have undergone high-pressure metamorphism during subduction at subcrustal depth (Dal Piaz & Lombardo 1986; Rubatto & Hermann 2001).

The Southalpine external thrust belt began to form in the Late Cretaceous along the Austroalpine margin during southward subduction of the Alpine Tethys (Zanchetta *et al.* 2012), and continued to grow outward in the Miocene (Schumacher *et al.* 1997), when the inner part of the European margin (Dauphinois-Helvetic Domain) was also involved in deformation (Pfiffner *et al.* 2000). The Southalpine belt did not undergo Cenozoic metamorphism and still largely preserves the original structure of the Late Palaeozoic orogen (Vai *et al.* 1984). This can be followed from lower-crustal Ivrea–Verbano granulites in the west to low-grade basement units and anchimetamorphic Palaeozoic sediments exposed in the Palaeocarnic Range to the east, stratigraphically overlain by largely carbonate continental-margin successions of Upper Carboniferous to Eocene age.

## *The Southern Alps crustal section*

The Ivrea-Verbano Zone, representing the deepest tectono-stratigraphic crustal level exposed in the Southern Alps (Fig. 1), consists of an up to 10-km-thick gabbroic body (Mafic Complex), intruded at a depth of 18–25 km into high-grade metasediments during Early Permian times (Quick *et al.* 1994). It is delimited to the NW by the Canavese Line (a greenschist mylonitic belt representing the westernmost segment of the Insubric system) and to the SE by the Cossato–Mergozzo–Brissago Line (CMB Line; Boriani *et al.* 1990*a*).

The Mafic Complex includes ultramafic cumulates, layered gabbros, norites and anorthosites ('Layered Series'), followed by 5- to 7-km-thick gabbronorite ('Main Gabbro'), grading upward into monzogabbros and diorites (Rivalenti *et al.* 1981). Primary magmatic phases and textures are locally preserved, but sub-solidus re-equilibration under static conditions gave rise to anhydrous assemblages including clinopyroxene, orthopyroxene and garnet. Retrograde hydrous associations developed subsequently, with growth of brown amphibole eventually overgrown by rims of green hornblende coexisting with biotite, oligoclase and titanite. Orthopyroxene changed into cummingtonite and tremolite/actinolite in the presence of fluids. Final retrogression, confined in shear bands

within the gabbros, is pervasive in the diorites (Colombo & Tunesi 1999).

Ivrea-Verbano metapelites, metapsammites, marbles and amphibolites display a progressive southeastward decrease in metamorphism from granulite facies to amphibolite facies (Fig. 2; Sills & Tarney 1984). Granulite-facies metapelites ('stronalites') are intermingled with leucosomes of granitic/granodioritic composition and contain garnet, prismatic sillimanite and minor biotite (Schmid & Wood 1976; Sills 1984; Bea & Montero 1999). Upper-amphibolite-facies schists and gneisses ('kinzigites') include biotite associated with fibrolitic sillimanite. Muscovite occurs only in middle amphibolite facies metapelites at the top of the unit.

The CMB Line marks the abrupt tectonic transition to kyanite-bearing metasediments, coarsergrained gneisses with calcsilicate nodules, and large lenses of Ordovician metagranitoids (Strona–Ceneri Zone; Boriani *et al.* 1990*b*; Zurbriggen *et al.* 1997), representing lower-grade mid-crustal levels. The main metamorphic event took place at lower amphibolite-facies conditions during the Variscan Orogeny (320–340 Ma). Subsequent retrogression to greenschist facies is documented locally.

## *Garnet in Ivrea–Verbano source rocks*

In Ivrea–Verbano metapelites, garnet compositions are reported to vary from almandine-spessartine in the middle amphibolite facies ($Al_{67}Py_9Sp_{20}Gr_4$; *c.* 4 kbar, 560 °C) to almandine-pyrope in the granulite facies ($Al_{57}Py_{37}Sp_1Gr_5$; *c.* 8kbar, 800 °C; Bea *et al.* 1997). In the Mafic Complex, they range from pyrope-almandine-grossular to almandine-pyrope-grossular, mostly depending on bulk-rock composition ($Al_{27}Py_{55}Sp_1Gr_{17}$ to $Al_{56}Py_{32}Sp_2Gr_{21}$; Mazzucchelli *et al.* 1992). Marked changes are shown by chondrite-normalized REE patterns, which display increasing uptake of elements with larger ions at increasing pressure. In the kinzigites, garnets have steep REE patterns with negligible Eu anomalies. In the stronalites, REE patterns progressively become nearly flat from Sm to Lu, with increasingly strong negative Eu anomalies. Garnets from mafic granulites display remarkably similar patterns to those of metapelitic garnets equilibrated at the same pressure, except that they have negligible or even positive Eu anomalies (Bea & Montero 1999).

## *Detritus from the Southern Alps crustal section*

The Ivrea Mafic Complex sheds hypersthene rich feldspathic detritus with plagioclase, amphibolite, gabbro/metagabbro, pyroxenite, dunite and diorite grains, together with mostly brown hornblende, opaques, garnet and clinopyroxene. Largely metasedimentary granulite-facies rocks ('stronalites') shed garnet-rich litho-quartzo-feldspathic detritus with plagioclase, garnet-sillimanite metapelite/metapsammite rock fragments, mainly brown hornblende, hypersthene, clinopyroxene, minor biotite and dominantly prismatic sillimanite. Largely metasedimentary upper-amphibolite-facies rocks of the Ivrea–Verbano Zone ('kinzigites') shed heavy-mineral-rich litho-feldspatho-quartzose detritus with garnet-sillimanite metapelite/metapsammite rock fragments, biotite, blue-green to green-brown hornblende, clinopyroxene, garnet and dominantly fibrolitic sillimanite. Diopside, tremolite/actinolite and epidote are abundant where impure meta-carbonates are exposed; epidote is also possibly flushed from retrogressed shear zones. Metasedimentary to meta-igneous lower-amphibolite-facies rocks of the Strona–Ceneri Zone supply litho-feldspatho-quartzose detritus with high-rank felsic metamorphic rock fragments, biotite, muscovite, mainly blue-green hornblende, epidote, garnet and kyanite.

*The Lepontine Dome*

Exposed in the Lepontine Alps, between the Simplon Fault in the west and the Austroalpine lid in the east, are several kilometre-thick sheets of granitoid gneisses interleaved with Mesozoic calcschists (Trümpy 1980; Berger *et al.* 2005). Gneiss units, deformed by major post-nappe isoclinal folds with axial-plane foliations formed under amphibolite-facies conditions (Steck & Hunziker 1994), were eventually intruded by plutonic rocks in the Oligocene (Rosenberg 2004). Metamorphic isograds cross-cut the nappe pile, and outline a broad culmination classically referred to as the Lepontine Dome (Fig. 1). Polyphase foliation defines two distinct antiformal structures (Toce and Ticino subdomes; Merle *et al.* 1989), separated by a submeridian steep zone located along the Maggia Valley and bounded to the south by the Southern Steep Belt (Milnes 1974; Berger *et al.* 2008). The latter is characterized by the widespread occurrence of migmatites, formed during Alpine partial melting at $700 \pm 50\,°C$ (Burri *et al.* 2005). Metamorphic grade ranges up to middle amphibolite-facies in the Toce subdome and to upper amphibolite-facies in the Ticino subdome, where peak temperatures of $675–700\,°C$ were reached at 33–32 Ma toward the Insubric Fault in the south (Frey & Ferreiro Mählmann 1999; Engi *et al.* 2004). Granulite-facies orthogneisses, paragneisses and charnockites, formed at $920–940\,°C$ and 8.5–9.5 kbar in the Early–Middle Permian (282–260 Ma) and later overprinted and partly

re-equilibrated during Alpine migmatization at $720–740\,°C$ and 7–7.5 kbar in the Early Oligocene (34–29 Ma), are exposed east of the Mera Valley in the southeastern corner of the Lepontine Dome, structurally below the Bregaglia pluton (Gruf Complex; Galli *et al.* 2011, 2012).

*Amphibole in Lepontine source rocks*

Amphibolite-facies rocks of the Lepontine Dome are reported to contain common hornblende, rarely associated with actinolite and displaying limited chemical variation (e.g. $Al^{tot}$ mostly ranging from 1.75 to 2.4; atoms per formula units). MgO is seen to increase in amphiboles at increasing temperature, with interdependence between the Mg/Fe ratios of hornblende, biotite and source rocks, whereas the mean values of $Fe_2O_3$, $Na_2O$ and $H_2O$ appear to decrease, but with large variance (Wenk *et al.* 1974).

*Detritus from the Lepontine Dome*

Granitoid gneisses of the Lepontine Dome supply mainly feldspatho-quartzose detritus, including very high rank gneissic rock fragments. Metabasite rock fragments are common, and metacarbonate grains locally significant (Ticino River). Micas are abundant, with biotite slightly more frequent than white mica. Moderately rich to very rich heavy-mineral suites are dominated by amphiboles, progressively changing in colour and increasing in abundance from the periphery of the dome (where predominantly blue-green hornblende represents <50% of transparent heavy minerals) to the core of the dome (where blue-green to brown hornblende represents up to 78% of transparent heavy minerals). Chloritoid occurs only in sand of the Liro River, draining partly outside the dome. Epidote-group minerals decrease markedly from the periphery to the core of the dome, as well as tourmaline and Ti oxides. Clinozoisite and zoisite persist in sand of the Calancasca River, draining high-pressure rocks of the Adula Nappe (Dale & Holland 2003). Titanite, garnet, staurolite and kyanite also tend to decrease relative to the other minerals, whereas fibrolitic sillimanite increases and is associated with a few prismatic sillimanite grains in sands from the Southern Steep Belt and Gruf Complex (Metasedimentary Mineral Index 100 and Sillimanite Index 14 in Schiesone sand). Diopsidic clinopyroxene tends to increase towards the core of the dome, and is associated with common olivine, enstatite and minor hypersthene and cummingtonite in sand of the Moleno River draining the lherzolites of the Cima–Lunga Unit (Evans & Trommsdorff 1978; Nimis & Trommsdorff 2001). Diopside, enstatite, olivine, hypersthene and cummingtonite also occur in sand of the Schiesone River, draining the Gruf Complex granulites (Galli *et al.* 2011) and marginally the Mesozoic metaophiolites of the Chiavenna Unit (Talerico 2000).

## Rationale and methods

In this study we follow a strictly empirical approach, and largely exploit descriptions of the chemical composition and colour of amphiboles and garnets dating back to the 1960s and 1970s. Huge advances have been made in metamorphic petrology since then, including thermo-barometric studies based on physical–chemical laws and reactions that typically involve mineral pairs in equilibrium. Sedimentary petrologists cannot follow the same path to any great extent, because mineral grains of any kind can freely coexist in sediments without obeying such laws. On the other hand, detrital systems efficiently provide us with a weighted average of the various endmember signatures of rocks exposed and eroded in the drainage basin. Diverse numerical indices statistically representative of their metamorphic grade can thus be determined by analysing a sufficient number of mineral grains.

Here, we specifically focus on the metamorphic grade because it represents the best proxy for tectono-stratigraphic level, and hence degree of unroofing. The composition of detrital garnets and amphiboles, however, is not controlled only by the metamorphic grade of the parent rocks, but also depends on the chemical composition of the protoliths and on pressure conditions. Consequently, we do not expect to (and did not) find precisely identical results in different first-order catchments. However, considering that sediments are effectively mixed and homogenized at the scale of larger basins (Ingersoll 1990), we do expect to find the same average patterns and trends in other regions worldwide, as confirmed by numerous studies carried out with the same approach in the last decade, not only in diverse orogenic belts (including the Himalayas) but also in different geodynamic settings (Garzanti *et al.* 2007). The results presented in this article, therefore, are believed to be fully applicable to those ancient sandstones where detrital amphiboles and garnets have not undergone severe diagenetic dissolution. This may and generally does represent a serious limitation in provenance analysis of ancient sedimentary rocks (Garzanti & Andò 2007*b*; Garzanti *et al.* 2012), particularly for amphiboles that are widely documented to be less stable than garnets and already rapidly dissolved at burial depths exceeding 1–2 km (Morton & Hallsworth 2007; Andò *et al.* 2012).

## Mineralogical analyses

To determine their chemical composition, we carried out electron-microprobe analyses of 1139 amphibole grains and 1142 garnet grains from 20 selected samples of river sands, 10 from the Lepontine Dome and 10 from the Southern Alps. Lepontine samples, collected along a west–east transverse from the Ossola Valley to Lake Como (Fig. 1), provide a representative coverage of diverse areas of the dome that have undergone Early Oligocene metamorphism under conditions ranging from lower amphibolite-facies at its periphery to upper amphibolite facies in the Southern Steep Belt. Southalpine samples were collected from bottom to top of an ideally complete lower-crustal section, including two sands from the granulite-facies Mafic Complex, 7 sands from granulite-facies stronalites and upper-amphibolite-facies kinzigites exposed continuously along the Strona di Omegna Valley, and one sand from the lower amphibolite-facies Strona–Ceneri Zone (Fig. 2).

Samples were dry-sieved and heavy minerals separated by centrifuging in sodium metatungstate (density of 2.90 $g/cm^3$) from the very fine sand and fine sand classes (63–250 $\mu$m), which were treated with oxalic and acetic acids. High-resolution bulk-petrography and heavy-mineral analyses were carried out according to traditional methods (Ingersoll *et al.* 1984; Mange & Maurer 1992; Garzanti & Vezzoli 2003). The identities of dubious heavy minerals were checked with Raman spectroscopy (Andò & Garzanti 2013). In the case of zoned, composite or partly retrogressed grains (e.g. Tiepolo *et al.* 2012), as occasionally revealed in our samples by actinolite rims on detrital hornblende or pargasite, amphibole colour was assigned according to the predominant area of the grain (in most cases the core).

Amphibole and garnet grains were picked from the heavy-mineral separates, cut, mounted in epoxy, polished, and analysed at the British Geological Survey, Keyworth, UK, using a Cameca SX50 electron microprobe fitted with three wavelength-dispersive spectrometers. Throughout the procedure, an accelerating voltage of 15 kV, a sample current of 20 nA and a beam focused to *c.* 1 $\mu$m were used. A range of natural minerals and synthetic materials was used for calibration, and raw data were processed using the Quantiview software provided by Cameca. OH and F contents were calculated following the method described by Deer *et al.* (1992). To avoid grain-to grain bias and treat all the samples equally, each analysis was carried out close to the core of the detrital mineral, taking care to avoid inclusions and structural imperfections.

Amphibole grains were classified according to the scheme of Leake *et al.* (1997) by using Program AMPH13 developed by Jeremy Preston at Aberdeen University. The molar proportion of garnet endmembers was calculated from chemical analyses with the Microsoft Excel spreadsheet program devised by Locock (2008), which considers 15 different species and 14 hypothetical endmembers. Amphibole and garnet formulae were calculated on the basis of 23 O,OH and 24 O,OH, respectively. $Fe^{2+}/Fe^{3+}$ was determined by standard charge-balancing.

Amphibole grains represent $55 \pm 15\%$ of transparent heavy minerals in sands shed by amphibolite-facies units (with increasing temperatures epidote-group minerals decrease and sillimanite increases) and $20 \pm 10\%$ in sands shed by granulite-facies units. Garnet grains represent $15 \pm 10\%$ of transparent heavy minerals in sands shed by amphibolite-facies units, increase to $50 \pm 15\%$ in sands from Ivrea–Verbano stronalites (where pyroxenes increase), and drop to $15 \pm 10\%$ in sands from the Mafic Complex (where pyroxenes and opaques become dominant).

The complete dataset on the chemical composition of detrital amphiboles and garnets is provided in the Supplementary material, together with detailed information about the studied samples including their precise location and full petrographic and mineralogical composition.

# Detrital amphibole as a metamorphic marker

Amphibole is formed under a wide range of pressures, temperatures and chemical environments, and its composition reflects the crystallization conditions. Leake (1965, 1971) recognized that hornblendes in magmatic and contact-metamorphic rocks generally have lower $Al^{IV}$ and Si content than hornblendes in regional metamorphic rocks, and suggested that $Al^{IV}$ is pressure-dependent (Schmidt 1992). Changes in the chemical composition of Ca-amphiboles associated with increasing metamorphic grade have been investigated by Shido & Miyashiro (1959), Engel & Engel (1962), Binns (1965) and Bard (1970). They have documented increases in $Mg/(Mg + Fe)$ and Ti, Al, Na and K contents, and commensurate decreases in Si and total $Fe + Mg + Mn \pm Ca$, reflecting increases in the tschermakitic, pargasitic and edenitic substitutions as a consequence of net-transfer reactions and exchange equilibria (Ernst & Liu 1998). These authors recognized that the pleochroic scheme of hornblende varies with increasing crystallization temperatures, the colour parallel $n\gamma$ changing from blue-green to green, to green-brown and finally brown as Ti increases relative to Fe in the hornblende lattice (Raase 1972, 1974).

The present article focuses on amphiboles of metamorphic origin, which in sediments are commonly mixed in variable proportions with (and are generally optically indistinguishable from) amphiboles of magmatic origin. However, high-temperature plutonic amphiboles also have high Ti content (e.g. kaersutite), and consequently a brown colour similar to granulite-facies amphiboles. Including them in the determination of the HCI will therefore not introduce a serious bias in the assessment of the tectono-stratigraphic level of parent rocks, provided that volcanic provenance is duly recognized (e.g. oxy-hornblende). For such a correct interpretation of provenance, it is always fundamental to obtain accurate information on the volcanic, plutonic or metamorphic nature of source rocks by framework-petrography and heavy-mineral analyses of the bulk sediment, and in particular by careful observation of rock fragments in thin section (Garzanti & Vezzoli 2003).

## Detrital amphiboles from the Lepontine Dome

In Liro, upper Ticino, Isorno, Rovana and Verzasca sands derived from the outer part of the Lepontine Dome below the 650 °C isograd, hornblende grains are dominantly blue-green. In the Moleno and Calanzasca sands derived from the central part of the dome including the 650 °C isograd, hornblende grains are still mainly blue-green but also frequently green and green-brown. In the Marmontana, Boggia and Schiesone sands derived from gneissic rocks of the Southern Steep Belt well above the 650 °C isograd, hornblende grains are largely green-brown to brown (Fig. 4). Hornblende colour thus varies steadily across the Lepontine Dome, reflecting progressive compositional changes with increasing metamorphic grade (Fig. 3).

From the periphery to the core of the dome, the average composition of detrital amphiboles displays a progressive enrichment in $TiO_2$ (Table 2), associated with a decrease in $SiO_2$ (Fig. 5). Other elements do not display equally regular trends. Amphibole grains with $Al_2O_3 > 16\%$ are found only in sands of the Ticino, Rovana, Verzasca and Moleno rivers, draining high-pressure rocks in the central part of the Lepontine Dome (Engi 2011).

Most amphibole grains in Lepontine river sands are classified as hornblende and tschermakite (Fig. 4). Amphibole variety is greatest in sand of the Liro River draining the periphery of the dome, which includes tremolite-actinolite, barroisite and edenite. Tremolite-actinolite decreases and finally disappears towards the core of the dome, where cummingtonite was detected in several samples

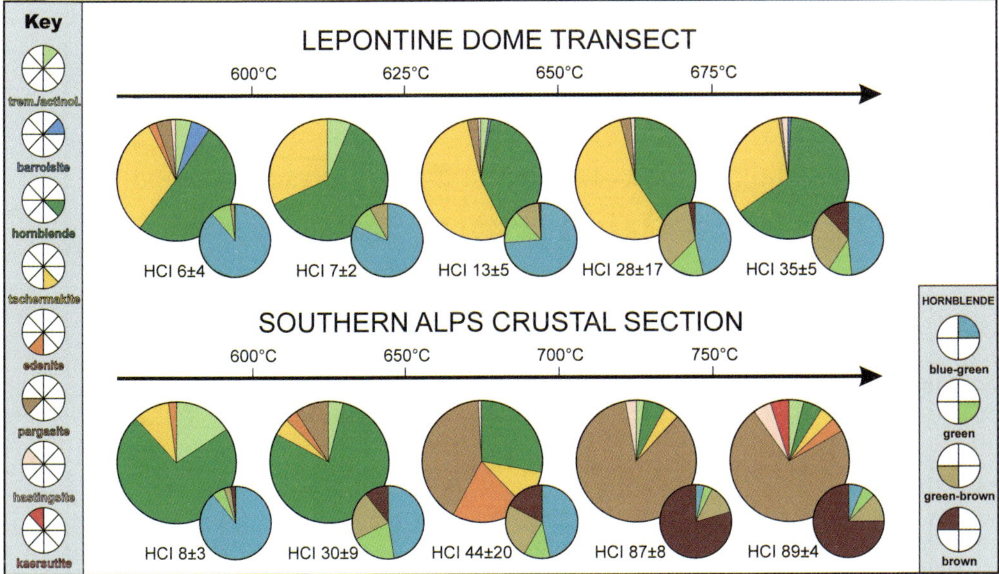



**Fig. 4.** Amphibole chemistry changes progressively with increasing metamorphic grade of source rocks. In spite of inevitable operator inconsistencies and empirical definition of colour, the increase in metamorphic grade can be readily revealed by the simple observation of amphibole colour under the microscope (HCI index; Fig. 3).

with the help of Raman spectroscopy. A few pargasite or hastingsite grains generally occur.

## Detrital amphiboles from the Southern Alps crustal section

Hornblende is dominantly blue-green in sand derived from gneisses of the Strona–Ceneri Zone, ranges from blue-green to brown in sands from the Ivrea–Verbano kinzigites, and is dominantly brown in sands from the Ivrea–Verbano stronalites and Mafic Complex (Fig. 4). Such a change in amphibole colour observed in river sands from east to west across the Southern Alps crustal section reflects the progressive compositional change associated with increasing metamorphic grade.

The average composition of detrital amphiboles displays a marked enrichment in $TiO_2$ from amphibolite-facies upper crustal levels to granulite-facies lower-crustal levels (Fig. 5). $Al_2O_3$, $Na_2O$ and $K_2O$ also increase progressively, whereas $SiO_2$ and MnO decrease (Table 2). Other elements (e.g. CaO, FeO, MgO and $Cr_2O_3$) are roughly constant throughout the section.

Amphibole grains are classified as mostly hornblende in sands derived from the Strona–Ceneri Zone and Ivrea–Verbano kinzigites, and as pargasite in sands derived from granulite-facies metasediments and metagabbros (Fig. 4; Table 3).

Tremolite-actinolite, most abundant in sand from the Strona–Ceneri Zone, decreases progressively in sands from higher-grade rocks but does not disappear even in granulite-derived sands, reflecting the invariable local occurrence of retrogressed source rocks. A few tschermakite grains are observed in all samples, whereas kaersutite is only found in sands from the Mafic Complex where hastingsite is also significant. Edenite is particularly abundant in kinzigite-derived Ovasca sand.

## Amphibole chemistry, amphibole colour and metamorphic grade

If detrital assemblages constitute a representative sample of rock units exposed in each catchment, then the studied samples characterize a complete composite section through the middle (Lepontine Dome, Strona Ceneri Zone) and lower crust (Ivrea Verbano Zone). The trends in chemical composition displayed by detrital minerals derived from source rocks metamorphosed at temperatures increasing progressively from *c.* 500 °C to *c.* 850 °C can thus be determined and illustrated in full detail. Throughout the composite Lepontine/Ivrea–Verbano section, $TiO_2$ steadily increases and $SiO_2$ decreases in the lattice of detrital amphiboles with increasing metamorphic grade of source rocks (Fig. 5). Among other elements, $Na_2O$ and

**Table 2.** *Chemistry of amphibole and garnet grains derived from the Lepontine Dome and Southern Alps crustal section*

| Sample | River | Site | Drained isograds (°C) | Analysed amphiboles | Amphiboles SiO$_2$ | TiO$_2$ | Al$_2$O$_3$ | Cr$_2$O$_3$ | MgO | CaO | MnO | FeO | Na$_2$O | K$_2$O | H$_2$O | F | Total | Analysed garnets | Garnets SiO$_2$ | TiO$_2$ | Al$_2$O$_3$ | Cr$_2$O$_3$ | Fe$_2$O$_3$ | MgO | CaO | MnO | FeO | Total |
|---|---|---|---|---|---|---|---|---|---|---|---|---|---|---|---|---|---|---|---|---|---|---|---|---|---|---|---|---|
| **Lepontine Dome** | | | | | | | | | | | | | | | | | | | | | | | | | | | | |
| S1210 | Liro | Mese | 500–600 | 46 | **48.5** | **0.4** | **8.5** | **0.04** | **13.2** | **11.2** | **0.27** | **14.2** | **1.2** | **0.4** | **2.0** | **0.2** | 100.2 | 56 | **37.2** | **0.06** | **21.0** | **0.03** | **0.5** | **2.3** | **4.6** | **2.1** | **33.4** | 101.2 |
| | | | | | *4.3* | *0.3* | *3.5* | *0.04* | *3.5* | *0.9* | *0.11* | *4.4* | *0.6* | *0.3* | *0.1* | *0.1* | *0.1* | | | *0.4* | *0.05* | *0.7* | *0.04* | *0.2* | *1.1* | *3.2* | *1.6* | *3.3* |
| S1894 | Ticino | Biasca | 525–640 | 59 | **44.9** | **0.4** | **12.4** | **0.04** | **11.2** | **11.2** | **0.24** | **15.3** | **1.4** | **0.4** | **1.9** | **0.2** | 99.7 | 56 | **37.4** | **0.06** | **21.0** | **0.02** | **0.6** | **2.6** | **5.6** | **1.9** | **32.0** | 101.2 |
| | | | | | *2.3* | *0.1* | *2.7* | *0.04* | *1.8* | *0.6* | *0.08* | *2.1* | *0.3* | *0.2* | *0.0* | *0.1* | *0.1* | | | *0.3* | *0.04* | *0.2* | *0.03* | *0.2* | *0.9* | *2.4* | *2.4* | *3.4* |
| S1877 | Isorno | Montecrestese | 600–630 | 67 | **47.7** | **0.4** | **10.5** | **0.08** | **14.8** | **11.8** | **0.19** | **10.8** | **1.3** | **0.2** | **2.0** | **0.2** | 99.9 | 50 | **37.2** | **0.08** | **21.0** | **0.02** | **0.5** | **3.4** | **2.8** | **1.7** | **34.8** | 101.4 |
| | | | | | *3.1* | *0.2* | *3.1* | *0.10* | *2.6* | *0.6* | *0.06* | *2.8* | *0.4* | *0.2* | *0.1* | *0.1* | *0.1* | | | *0.4* | *0.27* | *0.2* | *0.03* | *0.2* | *0.6* | *2.0* | *0.9* | *2.2* |
| S1888 | Rovana | Cevio | 630–635 | 68 | **47.3** | **0.4** | **10.9** | **0.07** | **14.7** | **11.7** | **0.17** | **10.8** | **1.4** | **0.2** | **2.0** | **0.2** | 99.8 | 51 | **37.1** | **0.06** | **21.0** | **0.03** | **0.5** | **3.0** | **3.6** | **2.7** | **33.3** | 101.1 |
| | | | | | *3.8* | *0.2* | *3.9* | *0.08* | *3.2* | *0.6* | *0.08* | *3.3* | *0.4* | *0.2* | *0.1* | *0.1* | *0.1* | | | *0.4* | *0.19* | *0.3* | *0.03* | *0.4* | *0.9* | *2.6* | *3.4* | *4.7* |
| S1890 | Verzasca | Frasco | 625–645 | 64 | **45.4** | **0.5** | **11.9** | **0.07** | **12.0** | **11.6** | **0.22** | **14.2** | **1.2** | **0.6** | **1.9** | **0.2** | 99.8 | 52 | **37.4** | **0.06** | **21.1** | **0.03** | **0.6** | **3.0** | **5.4** | **2.1** | **31.7** | 101.4 |
| | | | | | *3.5* | *0.2* | *3.2* | *0.06* | *2.9* | *0.5* | *0.09* | *3.2* | *0.3* | *0.4* | *0.1* | *0.1* | *0.1* | | | *0.4* | *0.06* | *0.2* | *0.06* | *0.2* | *0.8* | *2.7* | *1.3* | *3.0* |
| S1896 | Moleno | Moleno | 650–660 | 71 | **45.3** | **0.6** | **11.9** | **0.07** | **12.6** | **11.5** | **0.23** | **13.6** | **1.3** | **0.5** | **1.9** | **0.2** | 99.8 | 53 | **37.8** | **0.03** | **21.5** | **0.03** | **0.4** | **5.0** | **4.3** | **2.6** | **29.7** | 101.4 |
| | | | | | *3.8* | *0.2* | *3.2* | *0.07* | *3.1* | *0.5* | *0.09* | *3.3* | *0.4* | *0.3* | *0.1* | *0.1* | *0.1* | | | *0.9* | *0.06* | *0.4* | *0.04* | *0.3* | *2.1* | *4.0* | *2.6* | *5.5* |
| S1898 | Calancasca | Grono | 550–660 | 47 | **44.4** | **0.7** | **12.5** | **0.06** | **11.3** | **11.4** | **0.22** | **14.9** | **1.4** | **0.6** | **1.9** | **0.2** | 99.7 | 56 | **37.4** | **0.14** | **21.2** | **0.02** | **0.4** | **3.5** | **4.5** | **1.9** | **32.3** | 101.3 |
| | | | | | *2.2* | *0.2* | *1.8* | *0.07* | *2.1* | *0.4* | *0.08* | *2.6* | *0.3* | *0.3* | *0.1* | *0.1* | *0.1* | | | *0.6* | *0.69* | *0.3* | *0.03* | *0.2* | *1.3* | *2.7* | *1.3* | *2.5* |
| S1884 | Boggia | Gordona | 650–700 | 61 | **45.2** | **0.8** | **12.0** | **0.08** | **11.6** | **11.4** | **0.26** | **14.5** | **1.2** | **0.6** | **1.9** | **0.2** | 99.9 | 59 | **37.1** | **0.01** | **21.0** | **0.03** | **0.5** | **2.6** | **2.9** | **5.6** | **31.6** | 101.4 |
| | | | | | *2.5* | *0.2* | *2.0* | *0.07* | *2.3* | *0.5* | *0.10* | *2.6* | *0.2* | *0.3* | *0.1* | *0.1* | *0.1* | | | *0.7* | *0.02* | *0.3* | *0.03* | *0.2* | *1.3* | *1.9* | *3.8* | *3.0* |
| S1902 | Marmontana | Roveredo | 650–700 | 57 | **45.3** | **1.0** | **11.5** | **0.08** | **12.6** | **11.6** | **0.24** | **13.4** | **1.3** | **0.7** | **1.9** | **0.2** | 99.8 | 57 | **37.5** | **0.02** | **21.2** | **0.02** | **0.5** | **4.3** | **3.3** | **2.6** | **32.0** | 101.4 |
| | | | | | *2.1* | *0.3* | *1.9* | *0.09* | *2.1* | *0.4* | *0.08* | *2.6* | *0.3* | *0.3* | *0.1* | *0.1* | *0.1* | | | *0.5* | *0.03* | *0.2* | *0.03* | *0.2* | *1.2* | *3.1* | *1.6* | *3.8* |
| S1864 | Schiesone | Pr. Camportaccio | 650–750 | 60 | **46.7** | **1.1** | **9.8** | **0.07** | **14.0** | **11.4** | **0.20** | **12.7** | **1.5** | **0.3** | **2.0** | **0.2** | 99.8 | 51 | **37.3** | **0.02** | **21.2** | **0.02** | **0.5** | **4.4** | **2.1** | **2.2** | **33.7** | 101.4 |
| | | | | | *1.8* | *0.3* | *1.7* | *0.06* | *1.5* | *0.8* | *0.06* | *1.9* | *0.4* | *0.2* | *0.1* | *0.1* | *0.1* | | | *0.6* | *0.02* | *0.3* | *0.03* | *0.2* | *1.7* | *4.9* | *1.5* | *4.8* |
| **Strona–Ceneri Zone** | | | | | | | | | | | | | | | | | | | | | | | | | | | | |
| S2616 | Trib. Canobino | Falmenta | 550–600 | 51 | **48.3** | **0.4** | **8.7** | **0.08** | **13.2** | **12.0** | **0.27** | **12.5** | **0.9** | **0.5** | **1.9** | **0.2** | 98.9 | 59 | **37.2** | **0.02** | **20.6** | **0.02** | **0.5** | **2.8** | **1.9** | **4.7** | **31.9** | 99.7 |
| | | | | | *3.6* | *0.2* | *3.4* | *0.07* | *2.6* | *0.4* | *0.10* | *2.7* | *0.4* | *0.3* | *0.1* | *0.1* | *0.1* | | | *0.4* | *0.02* | *0.2* | *0.02* | *0.2* | *0.9* | *1.1* | *2.0* | *2.2* |
| **Ivrea–Verbano Metasediments** | | | | | | | | | | | | | | | | | | | | | | | | | | | | |
| S2609 | Bagnone | Prelo | 620–640 | 55 | **46.9** | **0.7** | **9.9** | **0.07** | **12.6** | **11.5** | **0.18** | **13.1** | **1.2** | **0.4** | **1.9** | **0.2** | 98.7 | 63 | **37.5** | **0.02** | **20.7** | **0.02** | **0.6** | **3.1** | **2.3** | **3.5** | **32.4** | 100.1 |
| | | | | | *3.2* | *0.3* | *3.1* | *0.05* | *2.2* | *0.6* | *0.08* | *2.5* | *0.5* | *0.3* | *0.1* | *0.1* | *0.1* | | | *0.6* | *0.02* | *0.3* | *0.03* | *0.2* | *1.7* | *2.3* | *2.1* | *3.1* |
| S2606 | Crosa | Luzzogno | 630–650 | 61 | **44.8** | **1.3** | **10.5** | **0.06** | **10.8** | **11.5** | **0.20** | **15.3** | **1.5** | **0.8** | **1.9** | **0.2** | 98.8 | 59 | **37.4** | **0.02** | **20.8** | **0.03** | **0.4** | **3.7** | **2.4** | **2.2** | **32.7** | 99.7 |
| | | | | | *2.0* | *0.3* | *1.5* | *0.04* | *1.3* | *0.3* | *0.06* | *1.5* | *0.3* | *0.3* | *0.1* | *0.1* | *0.1* | | | *0.4* | *0.02* | *0.3* | *0.03* | *0.2* | *1.1* | *2.0* | *1.0* | *2.6* |
| S3286 | Sambughetto | Sambughetto | 630–640 | 49 | **47.2** | **0.7** | **10.0** | **0.08** | **12.7** | **11.3** | **0.18** | **13.1** | **1.1** | **0.3** | **1.9** | **0.3** | 98.9 | 63 | **37.3** | **0.02** | **20.6** | **0.02** | **0.6** | **3.0** | **1.8** | **3.3** | **33.1** | 99.9 |
| | | | | | *2.0* | *0.2* | *2.2* | *0.06* | *1.4* | *0.4* | *0.09* | *1.6* | *0.4* | *0.1* | *0.1* | *0.1* | *0.1* | | | *0.5* | *0.03* | *0.2* | *0.03* | *0.3* | *0.8* | *4.4* | *1.3* | *4.6* |
| S3285 | Ovasca | Ovasca | 640–650 | 55 | **44.0** | **1.4** | **10.8** | **0.08** | **11.1** | **11.5** | **0.20** | **15.0** | **1.9** | **0.9** | **1.9** | **0.2** | 98.9 | 60 | **37.7** | **0.02** | **20.8** | **0.04** | **0.5** | **4.0** | **2.8** | **2.1** | **31.7** | 99.7 |
| | | | | | *2.3* | *0.3* | *1.7* | *0.06* | *2.3* | *0.3* | *0.06* | *2.7* | *0.5* | *0.4* | *0.1* | *0.1* | *0.1* | | | *0.5* | *0.03* | *0.2* | *0.03* | *0.2* | *1.2* | *2.4* | *0.8* | *2.6* |
| S2605 | Trib. Strona | Forno | 690–720 | 50 | **42.3** | **2.0** | **12.0** | **0.07** | **10.3** | **11.3** | **0.20** | **15.7** | **1.9** | **1.0** | **1.9** | **0.2** | 98.8 | 57 | **38.2** | **0.02** | **21.4** | **0.04** | **0.4** | **7.1** | **2.6** | **0.9** | **28.8** | 99.4 |
| | | | | | *2.2* | *0.6* | *1.5* | *0.05* | *1.5* | *0.4* | *0.06* | *1.9* | *0.6* | *0.1* | *0.1* | *0.1* | *0.1* | | | *0.6* | *0.02* | *0.3* | *0.03* | *0.3* | *1.8* | *1.8* | *0.5* | *2.2* |
| S3281 | Nagarone | Piana di Forno | 720–730 | 50 | **42.5** | **2.5** | **12.1** | **0.12** | **11.0** | **11.4** | **0.13** | **13.9** | **2.0** | **1.1** | **1.8** | **0.3** | 98.9 | 59 | **38.6** | **0.03** | **21.6** | **0.05** | **0.4** | **8.8** | **2.5** | **0.7** | **26.8** | 99.4 |
| | | | | | *2.2* | *0.6* | *1.8* | *0.09* | *1.3* | *0.3* | *0.07* | *1.8* | *0.4* | *0.5* | *0.1* | *0.1* | *0.1* | | | *0.6* | *0.03* | *0.4* | *0.05* | *0.3* | *1.9* | *2.0* | *1.0* | *1.7* |
| S3172 | Trib. Strona | Piana di Forno | 750–770 | 62 | **42.5** | **2.5** | **13.0** | **0.11** | **12.2** | **11.4** | **0.07** | **11.7** | **2.3** | **0.9** | **1.8** | **0.5** | 98.9 | 59 | **39.1** | **0.04** | **21.7** | **0.06** | **0.5** | **9.6** | **2.9** | **0.5** | **25.3** | 99.8 |
| | | | | | *1.7* | *0.8* | *1.8* | *0.15* | *1.6* | *0.4* | *0.07* | *2.6* | *0.5* | *0.5* | *0.2* | *0.5* | | | | *0.5* | *0.03* | *0.3* | *0.05* | *0.3* | *1.8* | *1.8* | *0.2* | *1.7* |
| **Ivrea Mafic Complex** | | | | | | | | | | | | | | | | | | | | | | | | | | | | |
| S2301 | Moro | Morca | 800–900 | 55 | **42.7** | **2.5** | **12.1** | **0.10** | **11.7** | **11.5** | **0.11** | **13.0** | **2.0** | **1.2** | **1.9** | **0.2** | 99.0 | 62 | **38.8** | **0.09** | **21.1** | **0.02** | **1.1** | **6.4** | **6.6** | **1.2** | **24.8** | 100.0 |
| | | | | | *3.3* | *1.0* | *2.8* | *0.12* | *2.2* | *0.3* | *0.06* | *2.5* | *0.6* | *0.5* | *0.1* | *0.1* | *0.1* | | | *0.7* | *0.05* | *0.4* | *0.03* | *0.3* | *1.8* | *0.7* | *0.5* | *2.0* |
| S2944 | Strona di Postua | Cravoso | 800–900 | 51 | **41.1** | **3.0** | **13.2** | **0.07** | **10.9** | **11.4** | **0.08** | **13.4** | **2.1** | **1.4** | **1.9** | **0.3** | 98.8 | 60 | **38.0** | **0.06** | **21.0** | **0.03** | **0.7** | **5.7** | **4.3** | **1.0** | **28.9** | 99.7 |
| | | | | | *2.8* | *1.1* | *2.4* | *0.08* | *1.7* | *0.3* | *0.05* | *2.5* | *0.6* | *0.4* | *0.1* | *0.1* | *0.1* | | | *0.4* | *0.05* | *0.3* | *0.03* | *0.3* | *1.1* | *2.4* | *0.5* | *2.0* |

Bold indicates mean; italics indicates standard deviation.

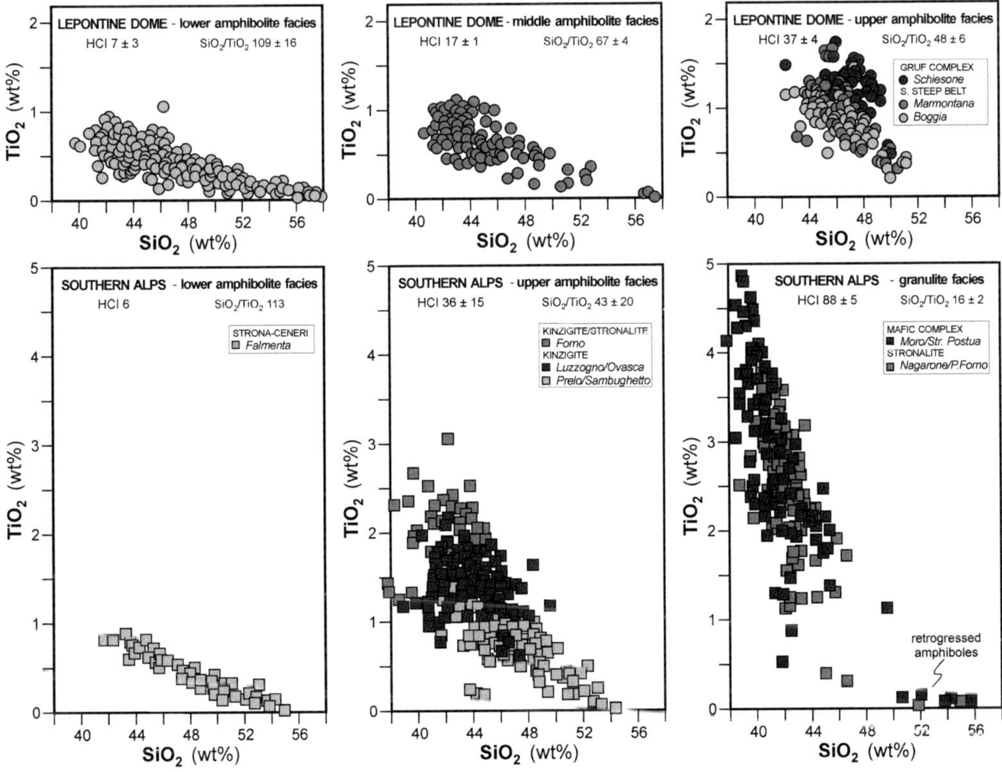



**Fig. 5.** Steady increase in $TiO_2$ and corresponding decrease in $SiO_2$ with increasing crystallization temperatures of detrital amphiboles. $TiO_2$ increases rapidly in the granulite facies, where biotite becomes unstable. Note the close correspondence of both the $SiO_2/TiO_2$ ratio and HCl with metamorphic grade. The few low $TiO_2$/high $SiO_2$ actinolite and Mg-hornblende grains found in granulite-derived sands most probably represent amphiboles retrogressed within shear zones. Means and standard deviations are given for each group of samples.

$K_2O$ tend to increase, whereas MnO tends to decrease.

At the transition between the amphibolite facies and the granulite facies, chiefly because of dehydration melting of biotite during prograde metamorphism (Bea & Montero 1999), the amphibole lattice becomes rapidly enriched in $TiO_2$ due to the high compatibility of Ti in the amphibole (Tiepolo *et al.* 2007). Blue green to green hornblende is thus replaced by brown pargasite, titanian pargasite and kaersutite. This general trend may be reversed at later stages, as a result of continuing magmatic crystallization (Tiepolo *et al.* 2011) or metamorphic retrogression. However, in spite of the complexities invariably involved in natural processes, and of inevitable inconsistencies among operators due to the individual empirical definition of colour, the correlations among metamorphic temperatures as inferred from detailed regional petrological studies and Ti content, HCl and the relative abundance of pargasite obtained in this case study are all remarkably good (correlation coefficient $\geq 0.90$, significance level 0.1%; Fig. 6). This proves the value of detrital amphibole as a provenance discriminator.

## Detrital garnet as a metamorphic marker

The common naturally occurring garnets are traditionally classified into two groups: pyralspites and ugrandites (Winchell 1933; Meagher 1980). However, interpreting crystal-chemical and solid solution properties in terms of such classification can lead to erroneous conclusions (Geiger 2008). In fact, solid solutions between garnets of both groups and containing a number of endmember components have been synthesized in the laboratory and can be found in nature as a function of the bulk chemistry of the system and of the $P-T$ crystallization conditions. Examples include grossular-pyrope solid solutions (called 'grospydite' by Sobolev *et al.*

**Table 3.** *Most typical amphibole and garnet types (Mange & Morton 2007) in detrital assemblages derived from source rocks of increasing metamorphic grade*

| | | Amphibolite facies | | | Granulite facies | |
|---|---|---|---|---|---|---|
| | | Lower | Middle | Upper | Metasediments | Metagabbro |
| Amphibole composition | Dominant | Mg-hornblende | Mg-hornblende | Mg-hornblende | Pargasite | Pargasite |
| | Major | Tschermakite | Tschermakite | Pargasite | – | – |
| | Minor | Trem.-actinol. | Pargasite | Tschermakite | Mg-hornblende | Mg-hornblende |
| | $TiO_2$ wt% | 0.4–0.5 | 0.6–0.7 | 0.8–1.5 | 2.5 | 2.5–3.0 |
| Hornblende colour | Dominant | Blue-green | Blue-green | Blue-green | Brown | Brown |
| | Major | – | Green | Green-brown | Green-brown | – |
| | Minor | Green | Green-brown | Green | Green | Green-brown |
| | HCI range | 0–10 | 10–30 | 30–60 | 60–100 | 80–100 |
| Garnet composition | Dominant | alm-prp | alm-prp | alm-prp-sps | alm-prp | alm-prp-grs |
| | Major | alm-grs-prp | alm-sps-prp | alm-prp | alm-prp-grs | alm-prp |
| | Minor | alm-sps-prp | alm-grs-prp | alm-sps-prp | | alm-grs-prp |
| Garnet type | Dominant | B | B | B | A | Ci |
| | Minor | | A, Ci | A, D | | |

*Abbreviations*: HCI, Hornblende Colour Index; alm, almandine; grs, grossular; prp, pyrope; sps, spessartine.

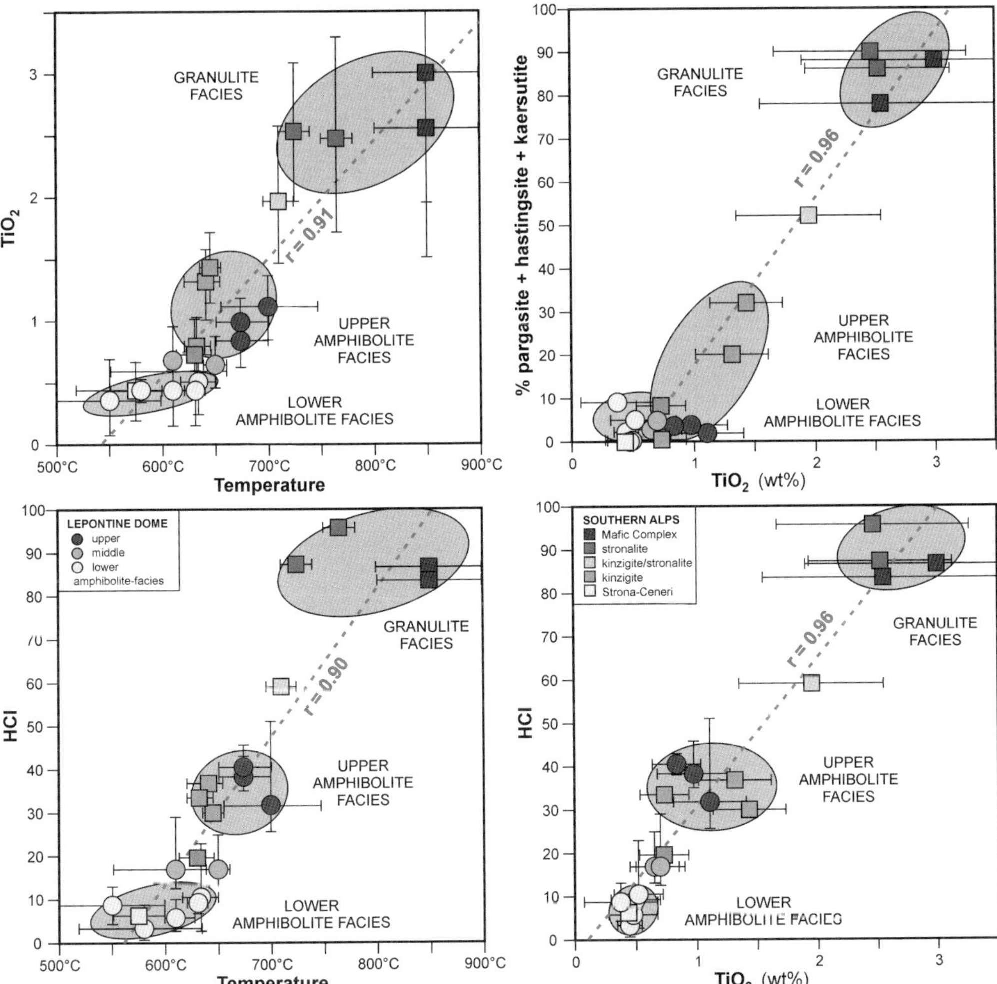


**Fig. 6.** Detrital amphibole as a metamorphic marker. Note the close relationships among amphibole crystallization temperatures, $TiO_2$ content, progressively changing colour as expressed by the HCI, and abundance of pargasite (associated with subordinate hastingsite and kaersutite in detritus from the Mafic Complex) relative to tremolite/actinolite, hornblende and tschermakite. Linear regressions are shown and correlation coefficients provided. Error bars are $1\sigma$ for $TiO_2$, or represent the range of isograds drained in each river catchment as shown in Figure 1 and Table 2 (temperature) and the range of values obtained by repeated analyses of Lepontine samples (HCI).

1968) and high-pressure metamorphic pyrope-grossular garnets (Sobolev *et al.* 2006). Taking into account such observations, we developed a flexible nomenclature scheme, and classified our Alpine detrital garnets according to the first three components listed in order of abundance, provided they exceed 5% (e.g. almandine-pyrope-grossular if almandine > pyrope > grossular > 5%).

The correspondence between garnet species and diverse igneous and metamorphic source rocks was first investigated by Wright (1938), and documented further by subsequent studies (e.g. Tröger

1959; Sobolev 1964). Although some overlap exists among different parageneses, garnet chemistry can be correlated with the physical–chemical conditions under which the host rocks were formed. With increasing metamorphic grade, garnets become reluctant to accommodate ions of larger ionic radii, and pyralspite garnets show progressive substitution from $Mn^{2+}$, $Fe^{2+}$ to $Mg^{2+}$, with their $Ca^{2+}$ content also corresponding to different pressure conditions (Miyashiro 1953; Nandi 1967; Deer *et al.* 1982). Such insights represent the basis for provenance diagnosis of garnet-rich suites (Mange

& Morton 2007; Pe-Piper *et al.* 2009; Suggate & Hall 2013), generally based on the classical pyrope-(almandine + spessartine)-grossular plot (Preston *et al.* 2002; Morton *et al.* 2003, 2004). The pyrope-spessartine-grossular plot, which does not consider almandine (commonly the dominant component in garnet grains), is an additional convenient way to illustrate the different origins of garnets (Teraoka *et al.* 1997, 1998; Win *et al.* 2007). Both diagrams will be used in this article to test the correspondence between garnet composition and metamorphic grade of source rocks. An additional set of plots, according to Suggate & Hall (2013) and kindly provided by Robert Hall, is included in the Supplementary material.

*Detrital garnets from the Lepontine Dome and Southern Alps crustal section*

No significant differences in the colour of garnet grains are observed across the Lepontine Dome, but pink varieties generally prevail over colourless varieties, with sporadic occurrence of yellow-green grains. Detrital garnet is largely colourless to pink in sands derived from amphibolite-facies to granulite-facies metasediments of the Ivrea–Verbano Zone, whereas dark-pink varieties characterize sands from the Mafic Complex (Fig. 3).

The average chemical composition of garnet grains does not change much throughout the Lepontine Dome, and also remains rather homogeneous in sands from the Strona–Ceneri Zone and Ivrea–Verbano kinzigites. However, detrital garnets in granulite-derived sands are distinctly richer in $SiO_2$ and $MgO$, and poorer in $FeO$ and $MnO$. Highest values are reached in garnets derived from stronalites for $Al_2O_3$ and $Cr_2O_3$, and from the Mafic Complex for $CaO$, $Fe_2O_3$ and $TiO_2$ (Table 2).

In sands derived from the Lepontine Dome, the diversity of garnet types tends to decrease with increasing metamorphic grade. Different varieties of almandine-grossular and almandine-pyrope characterize sands derived from the periphery of the dome (Fig. 7). Almandine-grossular decreases in sands from higher-grade rocks, whereas almandine-spessartine and almandine-spessartine-pyrope become more common, particularly in correspondence with pegmatite outcrops at the core of the dome (Hänny *et al.* 1975). Almandine-pyrope with variable amounts of Ca and Mn is invariably common.

Garnet is mostly Mn-rich in sands from the Strona–Ceneri Zone (almandine-spessartine-pyrope, almandine-pyrope-spessartine), whereas almandine-pyrope becomes dominant in the

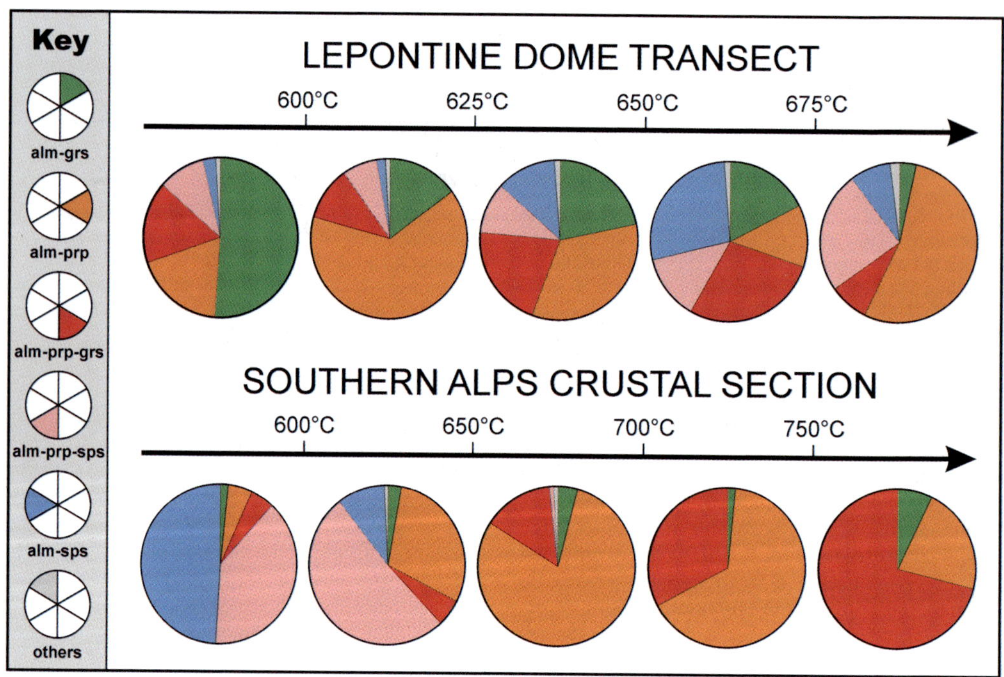



**Fig. 7.** Garnet chemistry changes progressively with increasing metamorphic grade of source rocks.

Ivrea–Verbano Zone (Table 3). Associated with almandine-spessartine-pyrope and rare almandine-grossular-spessartine in kinzigite sands, almandine-pyrope garnets become progressively richer in Ca and associated with rare almandine-grossular-pyrope across the kinzigite/stronalite transition, where Mn-rich garnets disappear. Stronalite sands contain almandine-pyrope garnets, with subordinate almandine-pyrope-grossular. In sands from the Mafic Complex, garnet is dominantly almandine-pyrope-grossular, with subordinate almandine-pyrope possibly derived from metasedimentary septa within the gabbros (Barboza *et al.* 1999; Quick *et al.* 2003).

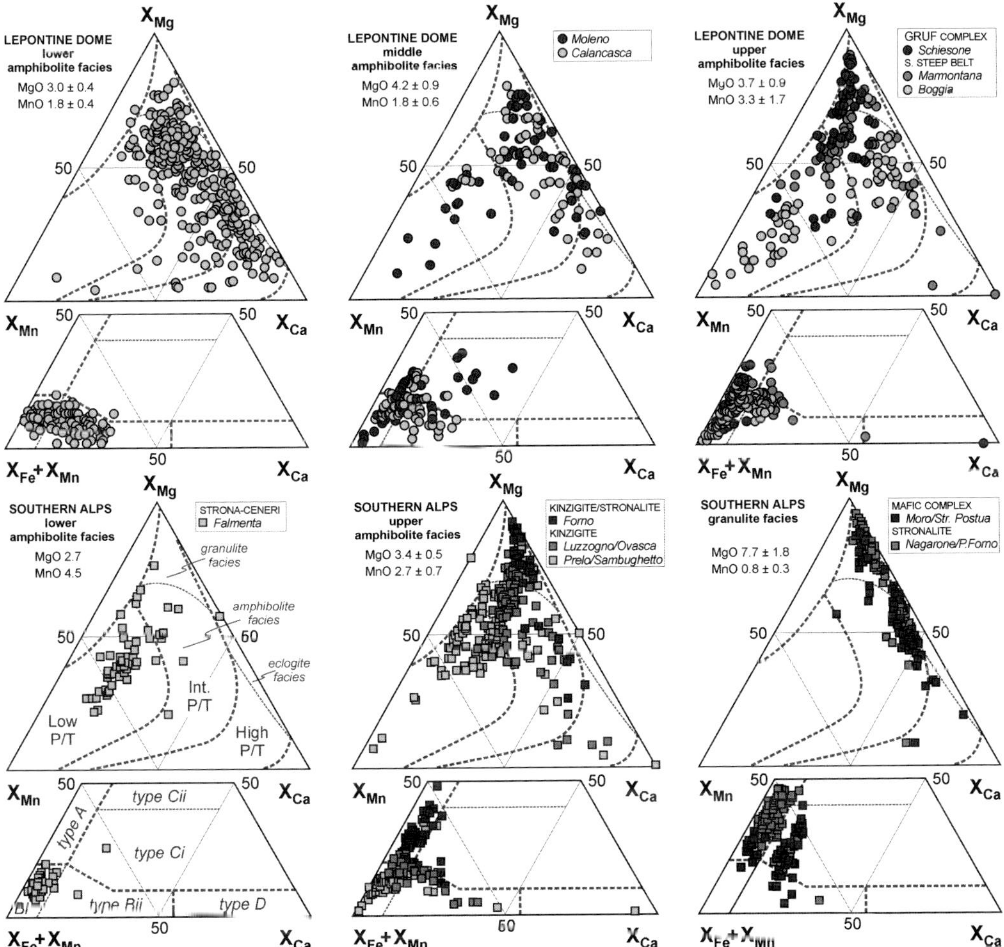

**Fig. 8.** Detrital garnet as a metamorphic marker. Lower amphibolite-facies units of the Strona–Ceneri Zone and Lepontine Dome periphery shed type B garnets (type Bi in the former, pointing to meta-igneous protoliths, and both types Bi and Bii in the latter, indicating both metasedimentary and meta-igneous protoliths). Granulite-facies metasediments (Ivrea–Verbano stronalites) and metagabbros (Ivrea Mafic Complex) dominantly shed type A and type C garnets, respectively. Bimodal distribution with clear compositional gap suggests that type A garnets in gabbro-derived sands may be derived from metasedimentary septa, and type C garnets in stronalite-derived sands from locally interlayered gabbro. Type A garnets in Schiesone sand are derived from Gruf granulites, and type C garnets in Moleno sand from Cima–Lunga garnet peridotites (Nimis & Trommsdorff 2001). Mn is most abundant in garnet from the Southern Steep Belt, reflecting the abundance of pegmatites, particularly in the Boggia catchment (Hänny *et al.* 1975), and decreases drastically in granulite-derived garnets where Mg increases. Fields in Fe + Mn–Mg–Ca and Mn Mg–Ca ternary plots are adapted after Mange & Morton (2007) and Win *et al.* (2007), respectively. $X_{Fe}$, $X_{Mg}$, $X_{Ca}$, $X_{Mn}$ = molecular proportions of $Fe^{2+}$, Mg, Ca and Mn, calculated on the basis of 24 oxygens and normalized to $Fe^{2+}$ + Mg + Ca + Mn.

### *Garnet chemistry and metamorphic grade*

Because garnet is generally abundant in sediments derived from collision orogens (Garzanti & Andò 2007*a*), resists diagenetic dissolution better than amphiboles, pyroxenes and epidote-group minerals (Morton & Hallsworth 2007), and displays a wide range of major-element compositions, garnet chemistry is a particularly valuable tool for determining the provenance of sedimentary rocks.

Detailed studies of both modern sands and ancient sandstones have shown that garnet composition varies systematically with both metamorphic grade and lithology of source rocks (Sabeen *et al.* 2002; Morton *et al.* 2003, 2004). Different types of detrital garnets have been empirically distinguished according to their provenance (Mange & Morton 2007). Type A garnets (high Mg, low Ca) are mainly derived from granulite-facies metasediments, type B garnets (low Mg, variable Ca) from amphibolite-facies metasediments, type C garnets (high Mg, high Ca) from high-grade metabasic rocks, and type D garnets (low Mg, very high Ca) from skarns, very low-grade metabasites or calc-silicate granulites. More specifically, type Bi ($X_{Mg} < 20\%$, $X_{Ca} < 10\%$) and type Cii ($X_{Mg} > 40\%$, $X_{Ca} > 10\%$) garnets point to provenance from intermediate-felsic igneous rocks and from pyroxenites and peridotites, respectively.

The value of such categorization is confirmed and further refined by the present study. Our results (Fig. 8) in fact document virtually exclusive type B garnets in sands derived from lower amphibolite-facies rocks of the Lepontine Dome and Strona–Ceneri Zone. Type A (almandine-pyrope) garnets appear in sands derived from middle to upper amphibolite-facies rocks and become dominant in sands derived from granulite-facies metasediments (Ivrea–Verbano stronalites). Type C (almandine-pyrope-grossular) garnets characterize sands from the Ivrea Mafic Complex, whereas only three type D garnet grains were detected overall (grossular in Schiesone and Sambughetto sands, grossular-andradite-spessartine in Marmontana sand). Types Bi and Bii garnets are both abundant in sands from the Lepontine Dome, reflecting the occurrence of both meta-igneous and metasedimentary protoliths, whereas type Bi garnets are dominant in Falmenta sand from the Strona–Ceneri Zone, suggesting prevailing meta-igneous protoliths in the catchment. A few garnet grains derived from both granulite-facies Ivrea–Verbano metasediments and metagabbros reach but do not cross the boundary between type Ci and type Cii garnets, the latter being virtually absent in our dataset. Several type Ci garnets also occur in Moleno sand, pointing to provenance from garnet peridotites of the Cima–Lunga Unit (Nimis & Trommsdorff 2001).

Considering that in pyralspite garnets $Mg^{2+}$ progressively substitutes for $Mn^{2+}$ and $Fe^{2+}$ with increasing metamorphic temperatures, whereas $Ca^{2+}$ increases at increasing pressures, Teraoka *et al.* (1997, 1998) used the Mn–Mg–Ca diagram to illustrate the different origins of detrital garnets. In such a diagram (Fig. 8), almandine-spessartine garnets derived from felsic protoliths including granitic pegmatites, aplites and thermal aureoles (Deer *et al.* 1982), particularly common in river sands from the Southern Steep Belt and Strona–Ceneri Zone, plot in the low $P/T$ field (Win *et al.* 2007). At the other end of the diagram, almandine with high Mg and Ca and lacking Mn, held to be typical of the eclogite facies, occurs sporadically in our samples and most frequently in sands of the Moleno and Calancasca rivers draining the eclogite-bearing Adula–Cima Lunga Nappe (Nimis & Trommsdorff 2001).

## Conclusion

The composition of detrital amphiboles and garnets contained in river sediments closely reflects the distribution of isograds in metamorphic source terranes, and thus represents a very sensitive provenance tool to trace erosion patterns in mountain belts. In this article we have analysed over a thousand sand-sized amphibole and garnet grains in selected monolithologic catchments across the Central and Southern Alps. These catchments ideally represent a complete section through middle and lower continental crust formed during Palaeozoic to Cenozoic orogenic events, at metamorphic temperatures from *c.* 500 °C to *c.* 850 °C.

Diverse amphibole and garnet types are found where source rocks barely reach lower-amphibolite-facies conditions, at temperatures not high enough to erase the inheritance of parent lithology. With increasing metamorphic grade, $TiO_2$ markedly increases and $SiO_2$ decreases in the lattice of detrital amphiboles, until blue-green/green hornblende and tschermakite are replaced by brown pargasite in sands derived from granulite-facies rocks (Fig. 4). High-Mn garnets, common in sands from the Southern Steep Belt of the Lepontine Dome where pegmatites are abundant and in the lower-amphibolite-facies Strona–Ceneri Zone (Fig. 7), decrease in sands from upper-amphibolite-facies metasediments of the Ivrea–Verbano Zone and finally disappear in sands from granulite-facies metasediments and metagabbros, respectively characterized by almandine-pyrope and almandine-pyrope-grossular ('type A' and 'type C' garnets of Mange & Morton 2007).

Some chemical properties of detrital minerals are directly revealed by optical features that are quickly identified and recorded during grain counting (e.g. hornblende colour; Garzanti & Andò 2007*a*). They represent accurate provenance tracers that can be profitably used in provenance studies of modern sands and ancient sandstones to identify the metamorphic grade and tectono-stratigraphic level of source rocks, and consequently reconstruct the palaeogeodynamic and erosional evolution of collision orogens.

The article benefited from excellent constructive reviews by H. von Eynatten and R. Hall. We are deeply grateful to B. Lombardo, who first provided the idea of using amphibole colour and garnet chemistry as detrital markers of metamorphic grade, to Grenville Turner, who undertook the microprobe analysis, and to A. Tunesi and M. Tiepolo, who offered precious advice on the mineralogy of Alpine units and amphibole chemistry. G. Vezzoli and A. Resentini carried out framework-petrography analyses. Danilo Bersani, Giuditta Fellin, Paolo Gentile, Marco Malusà and Pietro Vignola provided generous help at various stages of the work.

# References


ANDÒ, S. & GARZANTI, E. 2013. Raman spectroscopy in heavy-mineral studies. *In*: SCOTT, R., SMYTH, H., MORTON, A. & RICHARDSON, N. (eds) *Sediment Provenance Studies in Hydrocarbon Exploration and Production*. Geological Society, London, Special Publications, **386**. First published online June 20, 2013, http://dx.doi.org/10.1144/SP386.2.

ANDÒ, S., GARZANTI, E., PADOAN, M. & LIMONTA, M. 2012. Corrosion of heavy minerals during weathering and diagenesis: a catalogue for optical analysis. *Sedimentary Geology*, **280**, 165–178.

BARBOZA, S. A., BERGANTZ, G. W. & BROWN, M. 1999. Regional granulite facies metamorphism in the Ivrea zone: is the Mafic Complex the smoking gun or a red herring? *Geology*, **27**, 447–450.

BARD, J. P. 1970. Composition of hornblendes formed during the Hercynian progressive metamorphism of the Aracena metamorphic belt (SW Spain). *Contributions to Mineralogy and Petrology*, **28**, 117–134.

BEA, F. & MONTERO, P. 1999. Behaviour of accessory phases and redistribution of Zr, REE, Y, Th, and U during metamorphism and partial melting of metapelites in the lower crust: an example from the Kinzigite Formation of Ivrea–Verbano, NW Italy. *Geochimica et Cosmochimica Acta*, **63**, 1133–1153.

BEA, F., MONTERO, P., GARUTI, G. & ZACHARINI, F. 1997. Pressure-dependence of Rare Earth Element distribution in amphibolite- and granulite-grade garnets. A LA ICP–MS study. *Geostandards Newsletters*, **21**, 253–270.

BERGER, A., MERCOLLI, I. & ENGI, M. 2005. The central Lepontine Alps: notes accompanying the tectonic and petrographic map sheet Sopra Ceneri (1:100 000).

*Schweizerische Mineralogische und Petrographische Mitteilungen*, **85**, 109–146.

BERGER, A., BURRI, T., ALT-EPPING, P. & ENGI, M. 2008. Tectonically controlled fluid flow and water-assisted melting in the middle crust: an example from the Central Alps. *Lithos*, **102**, 598–615.

BINNS, R. A. 1965. The mineralogy of metamorphosed basic rocks from the Willyama complex, Broken Hill district, New South Wales. *Part I, Hornblendes*. *Mineralogical Magazine*, **35**, 306–326.

BORIANI, A., BURLINI, L. & SACCHI, R. 1990*a*. The Cossato–Mergozzo–Brissago line and the Pogallo line (Southern Alps, Northern Italy) and their relationships with the late Hercynian magmatic and metamorphic events. *Tectonophysics*, **182**, 91–102.

BORIANI, A., GIOBBI ORIGONI, E., BORGHI, A. & CAIRONI, V. 1990*b*. The evolution of the 'Serie dei Laghi' (Strona Ceneri and Scisti dei Laghi): the upper component of the Ivrea–Verbano crustal section; Southern Alps, North Italy and Ticino, Switzerland. *Tectonophysics*, **182**, 103–118.

BURRI, T., BERGER, A. & ENGI, M. 2005. Tertiary migmatites in the Central Alps: regional distribution, field relations, conditions of formation and tectonic implications. *Schweizerische Mineralogische Petrographische Mitteilungen*, **85**, 215–232.

CHOPIN, C. 1984. Coesite and pure pyrope in high-grade blueschists of the Western Alps: a first record and some consequences. *Contributions to Mineralogy and Petrology*, **86**, 107–118.

COLOMBO, A. & TUNESI, A. 1999. Alpine metamorphism of the Southern Alps west of the Giudicarie Line. *Schweizerische Mineralogische und Petrographische Mitteilungen*, **79**, 183–189.

COMPAGNONI, R., HIRAJIMA, T. & CHOPIN, C. 1995. Ultrahigh pressure metamorphic rocks in western Alps. *In*: COLEMAN, R. G. & WANG, X. (eds) *Ultrahigh-Pressure Metamorphism*. Cambridge University Press, New York, 206–243.

DALE, J. & HOLLAND, T. J. B. 2003. Geothermobarometry, *P–T* paths and metamorphic field gradients of high-pressure rocks from the Adula Nappe, Central Alps. *Journal of Metamorphic Geology*, **21**, 813–829.

DAL PIAZ, G. V. & LOMBARDO, B. 1986. Early Alpine eclogite metamorphism in the Penninic Monte Rosa–Gran Paradiso basement nappes of the northwestern Alps. *In*: EVANS, B. W. & BROWN, E. H. (eds) *Blueschists and Eclogites*. Geological Society of America Memoirs, **164**, 249–265.

DEER, W., HOWIE, R. A. & ZUSSMAN, J. 1982. *Rock-Forming Minerals. Vol. 1A, Orthosilicates*. Longman, London

DEER, W., HOWIE, R. A. & ZUSSMAN, J. 1992. *An Introduction to the Rock-Forming Minerals*. Longman, London.

DICKINSON, W. R. 1988. Provenance and sediment dispersal in relation to paleotectonics and paleogeography of sedimentary basins. *In*: KLEINSPEHN, K. L. & PAOLA, C. (eds) *New Perspectives in Basin Analysis*. Springer, Berlin, 3–25.

ENGEL, A. E. J. & ENGEL, C. G. 1962. Hornblendes formed during progressive metamorphism of amphibolites. Northwest Adirondack Mountains, New York. *Bulletin of the Geological Society of America*, **73**, 1499–1514.

ENGI, M 2011. Structure et évolution métamorphique des Alpes centrales. *Géochronique*, **117**, 22–26.

ENGI, M., BOUSQUET, R. & BERGER, A. 2004. Explanatory notes to the map: metamorphic structure of the Alps. Central Alps. *Mitteilungen der Österreichischen Geologischen Gesellschaft*, **149**, 157–173.

ERNST, W. G. & LIU, J. 1998. Experimental phase-equilibrium study of Al- and Ti-contents of calcic amphibole in MORB – a semiquantitative thermobarometer. *American Mineralogist*, **83**, 952–969.

EVANS, B. W. & TROMMSDORFF, V. 1978. Petrogenesis of garnet lherzolite, Cima di Gagnone, Lepontine Alps. *Earth and Planetary Science Letters*, **40**, 333–348.

FREY, M. & FERREIRO MÄHLMANN, R. 1999. Alpine metamorphism of the Central Alps. *Schweizerische Mineralogische und Petrographische Mitteilungen*, **79**, 135–154.

FREY, M., DESMONS, J. & NEUBAUER, F. 1999. The new metamorphic map of the Alps. *Schweizerische Mineralogische und Petrographische Mitteilungen*, **79**, 1–209.

GALLI, A., LE BAYON, B., SCHMIDT, M. W., BURG, J. P., CADDICK, M. J. & REUSSER, E. 2011. Granulites and charnockites of the Gruf Complex: evidence for Permian ultra-high temperature metamorphism in the Central Alps. *Lithos*, **124**, 17–45.

GALLI, A., LE BAYON, B., SCHMIDT, M. W., BURG, J. P., REUSSER, E., SERGEEV, S. A. & LARIONOV, A. 2012. U–Pb zircon dating of the Gruf Complex: disclosing the late Variscan granulitic lower crust of Europe stranded in the Central Alps. *Contributions to Mineralogy and Petrology*, **163**, 353–378.

GANSSER, A. 1982. The morphogenic phase of mountain building. *In*: HSÜ, K. J. (ed.) *Mountain Building Processes*. Academic Press, London, 221–228.

GARZANTI, E. & ANDÒ, S. 2007a. Plate tectonics and heavy-mineral suites of modern sands. *In*: MANGE, M. A. & WRIGHT, D. T. (eds) *Heavy Minerals in Use*. Elsevier, Amsterdam, Developments in Sedimentology, **58**, 741–763.

GARZANTI, E. & ANDÒ, S. 2007b. Heavy-mineral concentration in modern sands: implications for provenance interpretation. *In*: MANGE, M. A. & WRIGHT, D. T. (eds) *Heavy Minerals in Use*. Elsevier, Amsterdam, Developments in Sedimentology, **58**, 517–545.

GARZANTI, E. & MALUSÀ, M. G. 2008. The Oligocene Alps: domal unroofing and drainage development during early orogenic growth. *Earth and Planetary Science Letters*, **268**, 487–500.

GARZANTI, E. & VEZZOLI, G. 2003. A classification of metamorphic grains in sands based on their composition and grade. *Journal of Sedimentary Research*, **73**, 830–837.

GARZANTI, E., VEZZOLI, G., LOMBARDO, B., ANDÒ, S., MAURI, E., MONGUZZI, S. & RUSSO, M. 2004. Collision–orogen provenance (Western Alps): detrital signatures and unroofing trends. *The Journal of Geology*, **112**, 145–164.

GARZANTI, E., ANDÒ, S. & VEZZOLI, G. 2006. The Continental crust as source of sand (Southern Alps cross-section, Northern Italy). *The Journal of Geology*, **114**, 533–554.

GARZANTI, E., DOGLIONI, C., VEZZOLI, G. & ANDÒ, S. 2007. Orogenic belts and orogenic sediment provenances. *The Journal of Geology*, **115**, 315–334.

GARZANTI, E., RESENTINI, A., VEZZOLI, G., ANDÒ, S., MALUSÀ, M. G., PADOAN, M. & PAPARELLA, P. 2010. Detrital fingerprints of fossil continental–subduction zones (Axial Belt Provenance, European Alps). *The Journal of Geology*, **118**, 341–362.

GARZANTI, E., RESENTINI, A., VEZZOLI, G., ANDÒ, S., MALUSÀ, M. & PADOAN, M. 2012. Forward compositional modelling of Alpine orogenic sediments. *Sedimentary Geology*, **280**, 149–164.

GEIGER, C. A. 2008. Silicate garnet: a micro to macroscopic (re)view. *American Mineralogist*, **93**, 360–372.

GREENSFELDER, L. 2002. Subtleties of sand reveal how mountains crumble. *Science*, **295**, 256–258.

HACKER, B. R., MCCLELLAND, W. C. & LIHOU, J. C. 2006. *Ultrahigh-Pressure Metamorphism: Deep Continental Subduction*. Geological Society of America Special Papers, Boulder, **403**.

HÄNNY, R., GRAUERT, B. & SOPTRAJANOVA, G. 1975. Paleozoic migmatites affected by high-grade Tertiary metamorphism in the central Alps (Valle Bodengo, Italy). *Contributions to Mineralogy and Petrology*, **51**, 173–196.

HENK, A., FRANZ, L., TEUFEL, S. & ONCKEN, O. 1997. Magmatic underplating, extension, and crustal reequilibration: insights from a cross-section through the Ivrea Zone and Strona–Ceneri Zone, Northern Italy. *The Journal of Geology*, **105**, 367–377.

INGERSOLL, R. V. 1990. Actualistic sandstone petrofacies: discriminating modern and ancient source rocks. *Geology*, **18**, 733–736.

INGERSOLL, R. V., BULLARD, T. F., FORD, R. L., GRIMM, J. P., PICKLE, J. D. & SARES, S. W. 1984. The effect of grain size on detrital modes: a test of the Gazzi–Dickinson point-counting method. *Journal of Sedimentary Petrology*, **54**, 103–116.

INGERSOLL, R. V., DICKINSON, W. R. & GRAHAM, S. A. 2003. Remnant-ocean submarine fans: largest sedimentary systems on Earth. *In*: CHAN, M. A. & ARCHER, A. W. (eds) *Extreme Depositional Environments: Mega End Members in Geologic Time*. Geological Society of America Special Papers, Boulder, **370**, 191–208.

LEAKE, B. E. 1965. The relationship between tetrahedral aluminium and the maximum possible octahedral aluminium in natural calciferous and subcalciferous amphiboles. *American Mineralogist*, **50**, 843–851.

LEAKE, B. E. 1971. On aluminous and edenitic hornblendes. *Mineralogical Magazine*, **38**, 389–407.

LEAKE, B. E., WOOLEY, A. ET AL. 1997. Nomenclature of amphiboles. Report of the Subcommitee on Amphiboles of the International Mineralogical Association Commission on New Minerals and Mineral Names. *European Journal of Mineralogy*, **9**, 623–651.

LOCOCK, A. J. 2008. An Excel spreadsheet to recast analyses of garnet into end-member components, and a synopsis of the crystal chemistry of natural silicate garnets. *Computer & Geosciences*, **34**, 1769–1780.

MALUSÀ, M. G., FACCENNA, C., GARZANTI, E. & POLINO, R. 2011. Divergence in subduction zones and exhumation of high pressure rocks (Eocene Western Alps). *Earth and Planetary Science Letters*, **310**, 21–32.

MANGE, A. & MAURER, H. F. W. 1992. *Heavy Minerals in Colour*. Chapman and Hall, London.

MANGE, M. A. & MORTON, A. C. 2007. Geochemistry of heavy minerals. *In*: MANGE, M. A. & WRIGHT, D. T. (eds) *Heavy Minerals in Use*. Elsevier, Amsterdam, Developments in Sedimentology, **58**, 345–391.

MAZZUCCHELLI, M., RIVALENTI, G. *ET AL.* 1992. Trace element distribution between clinopyroxene and garnet in gabbroic rocks of the deep crust: an ion microprobe study. *Geochimica et Cosmochimica Acta*, **56**, 2371–2385.

MEAGHER, E. P. 1980. Silicate garnets. *Reviews in Mineralogy and Geochemistry*, **5**, 25–66.

MERLE, O., COBBOLD, P. R. & SCHMID, S. 1989. Tertiary kinematics in the Lepontine dome. *In*: COWARD, M. P., DIETRICH, D. & PARK, R. G. (eds) *Alpine Tectonics*. Geological Society, London, Special Publications, **45**, 113–134.

MILNES, A. G. 1974. Structure of the Pennine Zone (Central Alps): a new working hypothesis. *Geological Society of America Bulletin*, **85**, 1727–1732.

MIYASHIRO, A. 1953. Calcium-poor garnet in relation to metamorphism. *Geochimica et Cosmochimica Acta*, **4**, 179–208.

MIYASHIRO, A. 1972. *Metamorphism and metamorphic belts*. John Wiley, New York.

MORTON, A. & HALLSWORTH, C. 2007. Stability of detrital heavy minerals during burial diagenesis. *In*: MANGE, M. A. & WRIGHT, D. T. (eds) *Heavy Minerals in Use*. Elsevier, Amsterdam, Developments in Sedimentology, **58**, 215–245.

MORTON, A., ALLEN, M. *ET AL.* 2003. Provenance patterns in a neotectonic basin: Pliocene and Quaternary sediment supply to the South Caspian. *Basin Research*, **15**, 321–337.

MORTON, A., HALLSWORTH, C. & CHALTON, B. 2004. Garnet compositions in Scottish and Norwegian basement terrains: a framework for interpretation of North Sea sandstone provenance. *Marine and Petroleum Geology*, **21**, 393–410.

NAGEL, T., DE CAPITANI, C. & FREY, M. 2002. Isograds and *P–T* evolution in the eastern Lepontine Alps (Graubünden, Switzerland). *Journal of Metamorphic Geology*, **20**, 309–324.

NAJMAN, Y., BICKLE, M. *ET AL.* 2009. Reconstructing the exhumation history of the Lesser Himalaya, NW India, from a multi-technique provenance study of the foreland basin Siwalik Formation. *Tectonics*, **28**, TC5018, http://dx.doi.org/10.1029/2009TC002506.

NANDI, K. 1967. Garnets as indices of progressive regional metamorphism. *Mineralogical Magazine*, **36**, 89–93.

NIGGLI, E. & NIGGLI, C. 1965. Karten der Verbreitung einiger Mineralien der Alpidischen Metamorphose in den Schweizer Alpen (Stilpnomelan, Alkali-Amphibol, Chloritoid, Staurolith, Disthen, Sillimanit). *Eclogae Geologicae Helvetiae*, **58**, 335–368.

NIMIS, P. & TROMMSDORFF, V. 2001. Revised thermobarometry of Alpe Arami and other garnet peridotites from the central Alps. *Journal of Petrology*, **42**, 103–115.

PE-PIPER, G., TSIKOURAS, B., PIPER, D. J. W. & TRIANTAPHYLLIDIS, S. 2009. *Chemical Fingerprinting of Detrital Minerals in the Upper Jurassic Lower Cretaceous Sandstones, Scotian Basin*. Geological Survey of Canada, Ottawa, Open File **6288**.

PFIFFNER, O. A., ELLIS, S. & BEAUMONT, C. 2000. Collision tectonics in the Swiss Alps: insight from geodynamic modeling. *Tectonics*, **19**, 1065–1094.

POLINO, R., DAL PIAZ, G. V. & GOSSO, G. 1990. Tectonic erosion at the Adria margin and accretionary processes for the Cretaceous orogeny of the Alps. *Mémoires de la Société Géologique de France*, **156**, 345–367.

PRESTON, J., HARTLEY, A., MANGE-RAJETZKY, M., HOLE, M., MAY, G., BUCK, S. & VAUGHAN, L. 2002. The provenance of Triassic continental sandstones from the Beryl Field, northern North Sea: mineralogical, geochemical, and sedimentological constraints. *Journal of Sedimentary Research*, **72**, 18–29.

QUICK, J. E., SINIGOI, S. & MAYER, A. 1994. Emplacement dynamics of a large mafic intrusion in the lower crust, Ivrea–Verbano Zone, northern Italy. *Journal of Geophysical Research*, **99**, 21599–21573.

QUICK, J. E., SINIGOI, S., SNOKE, A. W., KALAKAY, T. J., MAYER, A. & PERESSINI, G. 2003. *Geologic Map of the Southern Ivrea–Verbano Zone northwestern Italy, scale 1:25 000*. US Geological Survey, Geologic Investigations Series Map **I-2776**.

RAASE, P. 1972. Über die Färbung von Hornblenden und die mineralogische Kennzeichnung von Barroisit und Karinthin. *Karinthin*, **66**, 273–280.

RAASE, P. 1974. Al and Ti contents of hornblende, indicators of pressure and temperature of regional metamorphism. *Contributions to Mineralogy and Petrology*, **45**, 231–236.

RIVALENTI, G., GARUTI, G., ROSSI, A., SIENA, F. & SINIGOI, S. 1981. Existence of different peridotite types and of a layered igneous complex in the Ivrea Zone of the Western Alps. *Journal of Petrology*, **22**, 127–153.

ROSENBERG, C. L. 2004. Shear zones and magma ascent: a model based on a review of the Tertiary magmatism in the Alps. *Tectonics*, **23**, 1–21.

RUBATTO, D. & HERMANN, J. 2001. Exhumation as fast as subduction? *Geology*, **29**, 3–6.

RUTTER, E., BRODIE, K., JAMES, T. & BURLINI, L. 2007. Large-scale folding in the upper part of the Ivrea–Verbano zone, NW Italy. *Journal of Structural Geology*, **29**, 1–17.

SABEEN, H. M., RAMANUJAM, N. & MORTON, A. C. 2002. The provenance of garnet: constraints provided by studies of coastal sediments from Southern India. *Sedimentary Geology*, **152**, 279–287.

SCHLUNEGGER, F. & WILLETT, S. 1999. Spatial and temporal variations in exhumation of the central Swiss Alps and implications for exhumation mechanisms. *In*: RING, U., BRANDON, M. T., LISTER, G. S. & WILLETT, S. (eds) *Exhumation Processes: Normal Faulting, Ductile Flow and Erosion*. Geological Society, London, Special Publications, **154**, 157–179.

SCHMID, R. & WOOD, B. J. 1976. Phase relationships in granulitic metapelites from the Ivrea–Verbano Zone (Northern Italy). *Contributions to Mineralogy and Petrology*, **54**, 255–279.

SCHMID, S., BERNOULLI, D. *ET AL.* 2008. The Alpine Carpathian Dinaridic orogenic system: correlation and evolution of tectonic units. *Swiss Journal of Geosciences*, **101**, 139–183.

SCHMIDT, M. W. 1992. Amphibole composition in tonalite as a function of pressure: an experimental calibration of the Al-in-hornblende barometer. *Contributions to Mineralogy and Petrology*, **110**, 304–310.

SCHUMACHER, M. E., SCHÖNBORN, G., BERNOULLI, D. & LAUBSCHER, H. P. 1997. Rifting and collision in the Southern Alps. *In*: PFIFFNER, O. A., HEITZMANN, P. & FREI, W. (eds) *Results of NRP 20*. Birkhäuser, Basel, 186–204.

SHIDO, F. & MIYASHIRO, A. 1959. Hornblende of basic metamorphic rocks. *Journal of the Faculty of Science Tokyo University*, *II*, **12**, 85–102.

SILLS, J. 1984. Granulite-facies metamorphism in the Ivrea Zone, N.W. Italy. *Schweizerische Mineralogische und Petrographische Mitteilungen*, **64**, 169–191.

SILLS, J. & TARNEY, J. 1984. Petrogenesis and tectonic significance of amphibolites interlayered with metasedimentary gneisses in the Ivrea Zone, Southern Alps, Northwest Italy. *Tectonophysics*, **107**, 187–206.

SOBOLEV, N. V. 1964. Classification of rock-forming garnets. *Doklady Akademii Nauk SSSR*, **157**, 353–356.

SOBOLEV, N. V., KUZNETSOVA, I. K. & ZYUZIN, N. I. 1968. The petrology of grospydite xenoliths from the Zagadochnaya Kimberlite Pipe in Yakutia. *Journal of Petrology*, **9**, 253–280.

SOBOLEV, N. V., SCHERTL, H.-P. & NEUSER, R. D. 2006. Composition and paragenesis of garnets from ultrahigh-pressure calc-silicate metamorphic rocks of the Kokchetav Massif. *Russian Geology and Geophysics*, **47**, 521–531.

STECK, A. & HUNZIKER, J. 1994. The Tertiary structural and thermal evolution of the Central Alps – compressional and extensional structures in an orogenic belt. *Tectonophysics*, **238**, 229–254.

SUGGATE, S. M. & HALL, R. 2013. Using detrital garnet compositions to determine provenance: a new compositional database and procedure. *In*: SCOTT, R., SMYTH, H., MORTON, A. & RICHARDSON, N. (eds) *Sediment Provenance Studies in Hydrocarbon Exploration and Production*. Geological Society, London, Special Publications, **386**. First published online July 18, 2013, http://dx.doi.org/10.1144/SP386.8.

SYLVESTER, P. (ed.) 2012. *Quantitative Mineralogy and Microanalysis of Sediments and Sedimentary Rocks*. Mineralogical Association of Canada, Short Course Series 42, Quebec.

SZULC, A. G., NAJMAN, Y. *ET AL.* 2006. Tectonic evolution of the Himalaya constrained by detrital $^{40}Ar$–$^{39}Ar$, Sm–Nd and petrographic data from the Siwalik Foreland Basin succession, SW Nepal. *Basin Research*, **18**, 375–391.

TALERICO, C. 2000. *Petrological and chemical investigation of a metamorphosed oceanic crust–mantle section (Chiavenna, Bergell Alps)*. PhD thesis, ETH Zürich.

TERAOKA, Y., SUZUKI, M., HAYASHI, T. & KAWAKAMI, K. 1997. Detrital garnets from Paleozoic and Mesozoic sandstones in the Onogawa Area, East Kyushu, Southwest Japan. *Bulletin of the Faculty of Education Hiroshima University*, **19**, 87–101 (in Japanese, English abstract).

TERAOKA, Y., SUZUKI, M. & KAWAKAMI, K. 1998. Provenance of Cretaceous and Paleogene sediments in the Median Zone of Southwest Japan. *Bulletin of the Geological Survey of Japan*, **49**, 395–411 [in Japanese, English abstract].

TIEPOLO, M., OBERTI, R. & ZANETTI, A. 2007. Trace-element partitioning between amphibole and silicate melt. *Reviews in Mineralogy & Geochemistry*, **67**, 417–452.

TIEPOLO, M., TRIBUZIO, R. & LANGONE, A. 2011. High-Mg andesite petrogenesis by amphibole crystallization and ultramafic crust assimilation: evidence from Adamello hornblendites (Central Alps, Italy). *Journal of Petrology*, **52**, 1011–1045.

TIEPOLO, M., LANGONE, A., MORISHITA, T. & YUHARA, M. 2012. On the recycling of amphibole-rich ultramafic intrusive rocks in the arc crust: evidence from Shikanoshima Island (Kyushu, Japan). *Journal of Petrology*, **53**, 1255–1285.

TODD, C. S. & ENGI, M. 1997. Metamorphic field gradients in the Central Alps. *Journal of Metamorphic Geology*, **15**, 513–530.

TRÖGER, E. 1959. Die Granatgruppe: Beziehungen zwischen Mineralchemismus und Gesteinsart. *Neues Jahrbuch für Mineralogie Abhandlungen*, **93**, 1–44.

TROMMSDORFF, V. 1966. Progressive Metamorphose kieseliger Karbonatgesteine in den Zentralalpen zwischen Bernina und Simplon. *Schweizerische Mineralogische und Petrographische Mitteilungen*, **46**, 431–460.

TRÜMPY, R. 1980. *Geology of Switzerland, a Guidebook. Part A. An Outline of the Geology of Switzerland*. Wepf, Basel.

VAI, G. B., BORIANI, A., RIVALENTI, G. & SASSI, P. 1984. Catena Ercinica e Paleozoico nelle Alpi Meridionali. *In*: CASTELLARIN, A. (ed.) *Cento anni di geologia italiana*, 1° Centenario della Società Geologica Italiana, Bologna, 133–154.

VON EYNATTEN, H. & DUNKL, I. 2012. Assessing the sediment factory: the role of single grain analysis. *Earth-Science Reviews*, **115**, 97–120.

VON EYNATTEN, H., CRITELLI, S., INGERSOLL, R. V. & WELTJE, G. J. 2012. Actualistic models of sediment generation. *Sedimentary Geology Special Issue*, **280**, 1–269.

WENK, E. 1962. Plagioklas als Indexmineral in den Zentralalpen. *Schweizerische Mineralogische und Petrographische Mitteilungen*, **42**, 139–152.

WENK, E. & KELLER, F. 1969. Isograde in Amphibolitserien der Zentralalpen. *Schweizerische Mineralogische und Petrographische Mitteilungen*, **49**, 157–198.

WENK, E., SCHWANDER, H. & STERN, W. 1974. On calcic amphiboles and amphibolites from the Lepontine Alps. *Schweizerische Mineralogische und Petrographische Mitteilungen*, **54**, 97–149.

WIN, K. S., TAKEUCHI, M. & TOKIWA, T. 2007. Changes in detrital garnet assemblages related to transpressive uplifting associated with strike–slip faulting: an example from the Cretaceous System in Kii Peninsula, southwest Japan. *Sedimentary Geology*, **201**, 412–431.

WINCHELL, A. N. 1933. *Elements of Optical Mineralogy, Part II*, 3rd edn. Wiley, New York.

WRIGHT, W. I. 1938. The composition and occurrence of garnets. *American Mineralogist*, **23**, 436–449.

ZANCHETTA, S., GARZANTI, E., DOGLIONI, C. & ZANCHI, A. 2012. The Alps in the Cretaceous: a doubly-vergent pre-collisional orogen. *Terra Nova*, **24**, 351–356.

ZINGG, A. 1983. The Ivrea and Strona–Ceneri Zones (Southern Alps, Ticino and N. Italy). A review. *Schweizerische Mineralogische und Petrographische Mitteilungen*, **63**, 361–392.

ZURBRIGGEN, R., FRANZ, L. & HANDY, M. R. 1997. Pre-Variscan deformation, metamorphism and magmatism in the Strona–Ceneri Zone (southern Alps of northern Italy and southwern Switzerland). *Schweizerische Mineralogische und Petrographische Mitteilungen*, **77**, 361–380.

# Using detrital garnet compositions to determine provenance: a new compositional database and procedure

SIMON M. SUGGATE* & ROBERT HALL

*SE Asia Research Group, Department of Earth Sciences, Royal Holloway University of London, Egham, Surrey TW20 0EX, UK*

**Corresponding author (e-mail: s.suggate@es.rhul.ac.uk)*

**Abstract:** Detrital garnet compositions can be used to help determine the provenance of sedimentary rocks. In this study a garnet compositional database consisting of more than 2500 wet chemical and electron microprobe analyses was compiled from the literature. For the garnets in the database the six principal garnet end-member compositions (pyrope, almandine, spessartine, uvarovite, grossular and andradite) were calculated. A multi-stage methodology was devised to match garnet compositions to source rocks, and a series of garnet provenance fields on ternary plots were identified. The method was tested using compositional data from detrital garnet studies in several areas where provenance has already been identified, with good results. The methodology was then used to assess the provenance of detrital garnets from Neogene sandstones of northern Sabah, Borneo for which provenance is unknown and in a region where there are few garnet analyses for comparison. Comparison of garnet compositions from possible sources on Palawan, Philippines and in Borneo combined with the ternary plots excludes some known garnet-bearing rocks as potential sources and suggests derivation of metamorphic and igneous garnets from Palawan during the Early Miocene.

**Supplementary material:** The compositional database and garnet data plotting spreadsheets are available at www.geolsoc.org.uk/SUP18649

Naturally occurring garnet is potentially a useful provenance indicator. It has a wide compositional variation that may be specific to certain lithologies and, therefore, source areas, it is mechanically resistant during transport, and it is resistant to chemical modification during transport, diagenesis and low-grade metamorphism (e.g. Mange & Morton 2007). Garnets are particularly common in metamorphic rocks, but are also found in acid volcanic rocks, granites and pegmatites, peridotites and kimberlites, and are widespread as detrital grains in sediments.

Chemical analyses of garnets are often expressed as percentages of real or hypothetical molecules – the 'garnet end-members' (Deer *et al.* 1962, 1966, 1997; Rickwood 1968). The end-member molecules provide a qualitative summary of garnet composition that is useful in provenance studies as the data can be displayed on ternary plots so that comparisons between sediment and potential sources can be made. Garnets have been used as a provenance indicator in a number of studies (Morton 1985; Haughton & Farrow 1989; Tebbens *et al.* 1995; Hutchison & Oliver 1998; Oliver 2001; Sabeen *et al.* 2002; Morton *et al.* 2004; Copjakova *et al.* 2005; Win *et al.* 2007; Hallsworth & Chisholm 2008). Many of these studies have successfully matched garnets to local sources. However, the range of garnet compositions is very wide, and it

is not known if all garnets can be matched to source rocks. Furthermore, in regions where the sources are unknown, or have been removed by erosion, or where there are no compositional data from garnet-bearing rocks, it is not clear if compositional data from other regions of the world can be reliably used to infer protoliths and provenance. This paper discusses a large garnet compositional database compiled from the literature which was used to examine how useful such data could be in identifying the provenance of detrital garnets. We find that although not all garnets can be matched uniquely to particular source rocks, many garnets are useful for provenance interpretation. It is necessary to calculate the common garnet end-members (almandine, andradite/schorlomite, grossular, pyrope, spessartine and uvarovite), then follow a series of steps before plotting compositions on ternary diagrams. The new procedure has been applied successfully to determine the provenance of detrital garnets from Neogene sandstones of northern Borneo.

## Garnet mineralogy

The garnet group can be considered as a number of common molecules (Table 1) which represent the

*From*: Scott, R. A., Smyth, H. R., Morton, A. C. & Richardson, N. (eds) 2014. *Sediment Provenance Studies in Hydrocarbon Exploration and Production*. Geological Society, London, Special Publications, **386**, 373–393. First published online July 18, 2013, http://dx.doi.org/10.1144/SP386.8

**Table 1.** *Garnet end-members of the isomorphous series*

| | Specific gravity (D) | |
|---|---|---|
| Pyrope | 3.58 | $Mg_3Al_2Si_3O_{12}$ |
| Almandine | 4.32 | $Fe_3Al_2Si_3O_{12}$ |
| Spessartine | 4.19 | $Mn_3Al_2Si_3O_{12}$ |
| Uvarovite | 3.90 | $Ca_3Cr_2Si_3O_{12}$ |
| Grossular | 3.59 | $Ca_3Al_2Si_3O_{12}$ |
| Andradite | 3.86 | $Ca_3Fe_2Si_3O_{12}$ |

end-members of isomorphous series (Deer *et al.* 1962, 1997). These species form two solid solution series: the Pyralspite group (pyrope, almandine and spessartine) and Ugrandite group (uvarovite, grossular, and andradite). It is unusual for any garnet to be a pure end-member composition, and most garnets are solid solutions of several end-members. The summary of garnet mineralogy and paragenesis below is based on Deer *et al.* (1962, 1966, 1997).

## Pyralspite group

There are three end-members for the Pyralspite garnets: pyrope, almandine and spessartine.

*Pyrope.* $Mg_3Al_2Si_3O_{12}$-rich garnets occur in certain ultrabasic rocks, including peridotites, kimberlites and xenocrysts in basalt pipes.

*Almandine.* $Fe_3Al_2Si_3O_{12}$-rich garnets are typical of garnetiferous schists resulting from the regional metamorphism of argillaceous sediments and basic igneous rocks. Almandine also occurs in some thermal or contact aureoles, in contaminated granites, and as xenocrysts in some volcanic rocks. The garnet in granulite facies rocks is commonly almandine or more rarely almandine–pyrope. In eclogite facies rocks the garnet is typically almandine–pyrope, often with almandine percentages greater than pyrope percentages.

*Spessartine.* $Mn_3Al_2Si_3O_{12}$-rich garnets are commonly found in granites, granitic pegmatites, often as spessartine–almandine. Spessartine also occurs in some skarn deposits, in Mn-rich assemblages of metasomatic origin, within veins in metamorphosed greywackes, and is also known from Mn-rich cherts in blueschist facies rocks.

## Ugrandite group

There are three end-members for Ugrandite garnets: uvarovite, grossular and andradite.

*Uvarovite.* $Ca_3Cr_2Si_3O_{12}$-rich garnets are the least common of the anhydrous garnet species. Garnets with uvarovite as the dominant molecule are found mainly in serpentinites and chlorite schists, often in association with chromite, and in metamorphosed limestones and skarn ore-bodies.

*Grossular.* $Ca_3Al_2Si_3O_{12}$-rich garnet is characteristic of thermally and regionally metamorphosed impure calcareous rocks. Hydrogrossular garnets are found in metamorphosed marls and are common in rodingites and altered rocks associated with serpentinites.

*Andradite.* $Ca_3Fe_2Si_3O_{12}$-rich garnet commonly occurs in contact or thermally metamorphosed impure calcareous sediments, as well as in metasomatic skarn deposits that are often associated with thermal metamorphism. Deer *et al.* (1962) considered melanite and schorlomite as Ti-bearing varieties of andradite but Rickwood (1968) suggested that *schorlomite* ($Ca_3Ti_2Fe_2TiO_{12}$) should be considered a separate molecule. The Ti-rich andradite garnets are found mainly in alkaline igneous rocks and skarns.

## Garnet end-members

Chemical analyses of garnets are often expressed as percentages of real or hypothetical molecules of extreme composition–the 'garnet end-member' molecules (Deer *et al.* 1962, 1997; Rickwood 1968). The end-member molecules provide a qualitative summary of garnet composition, as the data can be plotted quickly on ternary plots. As outlined above, Deer *et al.* (1962) considered six important end-members. Since then a number of authors (Rickwood 1968; Knowles 1987; Muhling & Griffin 1991; Locock 2008) have suggested that additional end-members should be considered. Deer *et al.* (1997) included several other end-members in their compilations of garnet compositions in addition to the six main species. Rickwood (1968) identified 19 end-members and Locock (2008) calculated 29. They developed procedures using stand-alone computer programs or spreadsheets to allow users to calculate the molecular proportions of garnet end-members from chemical analyses. The procedure suggested by Locock (2008) uses an Excel spreadsheet to calculate 29 end-members (15 naturally occurring and 14 hypothetical end-members) for each analysis. However, this procedure is not very useful in provenance studies because 14 of the end-members calculated are not naturally occurring and cannot be assigned to a protolith for provenance interpretations, and of the 15 naturally occurring garnets, a number are unusually rich in elements that are exceptionally rare in most geological environments and, in fact, have mainly not been reported from detrital garnets (Mange & Morton 2007). The large number of end-members make the data difficult to display on diagrams and, because of these limitations, a new approach has been devised.

We found that because of the rarity of extreme compositions it is sufficient to screen the data to identify those that are unusual, for example, in having high contents of Zr, Y, V and Ti. If such garnets were to be found in detrital assemblages they could be matched to source rocks very easily. We then identified a series of steps, described below, for examining the data, and finally the end-member compositions are plotted on ternary diagrams.

We used Pascal computer programs to input the data and to calculate garnet formulae on the basis of 24 oxygens in order to estimate ferric iron, and then calculated the common garnet end-member compositions: pyrope, almandine, spessartine, uvarovite, grossular and andradite. Schorlomite was combined with andradite for garnets with less than $2\,wt\%$ $TiO_2$. The programs are available from the authors on request. We compared the results of the calculations with garnet end-members listed by Deer *et al.* (1966) and Locock (2008). Molecular percentages of end-member components for six typical garnets (Table 2) of various compositions were recalculated from the oxide weight percentages listed by Deer *et al.* (1966) using the Locock (2008) method and the method of this study, and were compared with end-members given by Deer *et al.* (1966). The results (Table 3) are in good agreement.

## Garnet compositional database

The garnet compositional database (Supplementary material) consists of more than 2500 wet chemical and electron microprobe analyses that were compiled from more than 150 published data sources. Data are listed in a single table containing oxide wt% compositions ($SiO_2$, $TiO_2$, $Al_2O_3$, $Cr_2O_3$, $Fe_2O_3$, FeO, MnO, MgO, CaO, $Na_2O$ and rare oxides $V_2O_3$, $Y_2O_3$, $ZrO_2$ if present) and calculated end-member compositions, together with the published garnet source rock/protolith, grade, pressure, temperature and the literature source. A number of ternary plots of garnet compositions from the database are presented in an Excel spreadsheet (Supplementary material).

It can be problematic to determine provenance of detrital garnets using plots in which end-member compositions are combined (e.g. almandine + spessartine, grossular + andradite + uvarovite) in order to produce a single ternary diagram. An example of this is shown in Figure 1. The ternary plot shows that when end-member compositions are combined it may be very difficult to identify specific protoliths, as many of the data points plot as clusters in similar or overlapping areas. Different authors have chosen different ways in which to combine garnet end-members, commonly so that they can use just one triangle to display garnet compositions. We found a single triangle was too restrictive and often caused overlap of otherwise distinctive protoliths. We tried several different combinations and found that two triangular plots were most effective in discriminating between different protolith compositions and metamorphic grades. The apices of the triangles are almandine, pyrope, spessartine and grossular + andradite + schorlomite.

Using the database a multi-stage methodology was devised which identifies the protoliths and/or metamorphic grade of garnets therein. Stage 1 removed garnets with unusual contents of $Y_2O_3$, $V_2O_3$ and $ZrO_2$. Stage 2 excluded garnets with unusually high contents of $TiO_2$. The garnets identified in Stages 1 and 2 can be matched to unusual protoliths (e.g. ores, skarns, mafic pyroclastic rocks and nepheline syenites). After the unusual garnet compositions were excluded the next stages matched the remaining garnet compositions to specific protoliths and/or metamorphic grade. Stages 3–5 identified garnets with high uvarovite and pyrope contents which are mostly typical of

**Table 2.** *Analyses of garnets from Deer* et al. *(1966)*

|  |  | $SiO_2$ | $TiO_2$ | $Al_2O_3$ | $Cr_2O_3$ | $Fe_2O_3$ | FeO | MnO | MgO | CaO |
|---|---|---|---|---|---|---|---|---|---|---|
| Quartz–biotite gneiss | DHZ 1 | 38.03 | 0.00 | 22.05 | 0.00 | 0.88 | 29.17 | 1.57 | 6.49 | 1.80 |
| Metamorphosed andesite | DHZ 2 | 37.03 | 0.04 | 8.92 | 0.00 | 18.34 | 2.25 | 1.09 | 0.83 | 30.26 |
| Anorthite–clinozoisite–<br>corundum–garnet gneiss | DHZ 3 | 38.69 | 0.55 | 18.17 | 0.00 | 5.70 | 3.78 | 0.64 | 0.76 | 31.76 |
| Eclogite | DHZ 4 | 41.52 | 0.01 | 23.01 | 0.22 | 1.22 | 12.86 | 0.33 | 16.64 | 4.71 |
| Calcsilicate hornfels | DHZ 5 | 35.84 | 0.03 | 20.83 | 0.00 | 0.65 | 1.78 | 33.37 | 2.48 | 5.00 |
| Uvarovite–tremolite–<br>tawmawite–pyrrhotite rock | DHZ 6 | 35.88 | 0.00 | 1.13 | 27.04 | 2.46 | 0.00 | 0.03 | 0.04 | 33.31 |

**Table 3.** *Comparison of molecular percent end-members calculated for garnet compositions of Table 2 by Deer et al. (1966), Locock (2008) and this study*

| | This study | | | | | | Deer *et al.* (1966) | | | | | | Locock (2008) | | | | | |
|---|---|---|---|---|---|---|---|---|---|---|---|---|---|---|---|---|---|---|
| | DHZ 1 | DHZ 2 | DHZ 3 | DHZ 4 | DHZ 5 | DHZ 6 | DHZ 1 | DHZ 2 | DHZ 3 | DHZ 4 | DHZ 5 | DHZ 6 | DHZ 1 | DHZ 2 | DHZ 3 | DHZ 4 | DHZ 5 | DHZ 6 |
| Almandine | 63.7 | 6.7 | 8.7 | 26.2 | – | 0.3 | 65.4 | 5.2 | 8.1 | 26.1 | 4.0 | – | 63.8 | 5.1 | 8.1 | 25.9 | 0.2 | – |
| Andradite + Schorlomite | – | 58.9 | 16.2 | – | 7.6 | 8.2 | 2.7 | 56.9 | 18.2 | 3.3 | 2.2 | 7.7 | – | 56.4 | 16.6 | – | 2.0 | 5.6 |
| Grossular | 5.4 | 28.6 | 70.9 | 11.7 | 6.8 | 1.9 | 2.5 | 32.0 | 69.4 | 8.9 | 11.2 | 2.4 | 5.0 | 32.0 | 69.5 | 11.5 | 6.6 | 4.5 |
| Pyrope | 27.2 | 3.3 | 2.9 | 60.7 | 9.9 | 0.2 | 25.8 | 3.4 | 2.9 | 60.3 | 7.9 | 0.2 | 25.3 | 3.4 | 2.9 | 59.8 | 9.9 | 0.2 |
| Spessartine | 3.7 | 2.5 | 1.4 | 0.7 | 75.7 | 0.1 | 3.6 | 2.5 | 1.4 | 0.7 | 74.7 | 0.1 | 3.5 | 2.5 | 1.4 | 0.7 | 75.6 | 0.1 |
| Uvarovite | – | – | – | 0.6 | – | 89.4 | – | – | | 0.7 | – | 89.6 | – | – | – | 0.6 | – | 88.9 |
| Total | 100.0 | 100.0 | 100.1 | 99.9 | 100.0 | 100.1 | 100.0 | 100.0 | 100.0 | 100.0 | 100.0 | 100.0 | 97.6 | 99.4 | 98.4 | 98.6 | 94.2 | 99.2 |

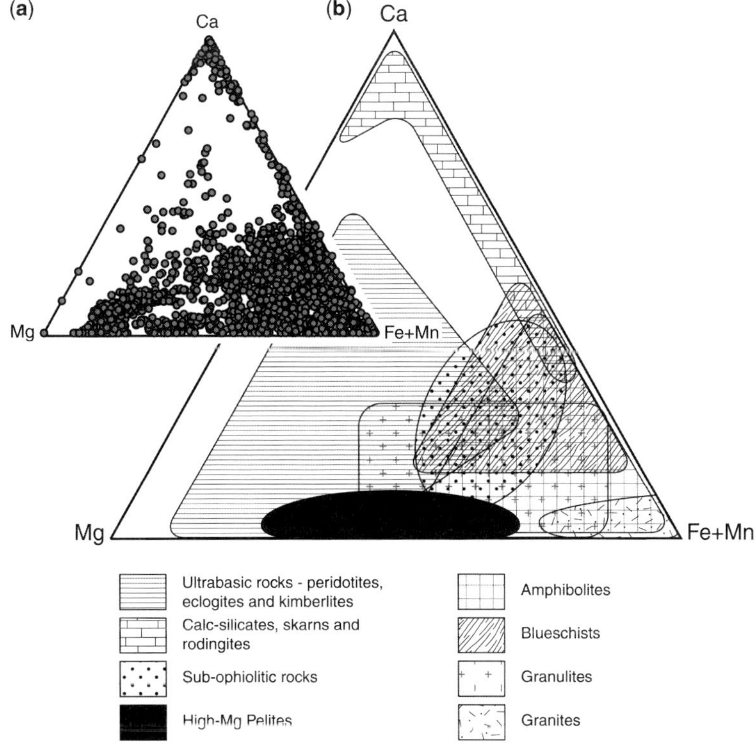



**Fig. 1.** Ternary plots showing all garnet data. (**a**) All garnets from the database ($n = 2400$) excluding those with unusual rare oxide contents ($V_2O_3$, $Y_2O_3$, $ZrO_2$) and $TiO_2$ >2 wt%, as discussed in the text. (**b**) All garnet compositions grouped by protolith. Ca, Mg, Fe and Mn are ionic contents of those elements calculated on the basis of 24 oxygens normalized to total $Ca + Mg + Fe^2 + Mn$. When end-member compositions are combined in a single ternary plot it is difficult to distinguish many protoliths because of overlaps of compositional areas for different types.

ultrabasic protoliths. Then different groups of garnets were plotted on the ternary diagrams to establish the compositional areas. The sequence and rationale is summarized below. Observations made from each stage are also detailed below. The areas for garnets from different protoliths and different metamorphic grades are plotted on a series of ternary plots in Figures 2 and 3.

## Stage 1: high rare oxide contents

This stage identified all garnets with extreme $Y_2O_3$, $V_2O_3$ and $ZrO_2$ contents and removed them from the database. There is a very small number of such garnets. Y-rich garnets ($n = 24$) range up to 3.4 wt% $Y_2O_3$. They are found in Y-rich gneisses, granite pegmatites and one hornfels. V-rich garnets ($n = 24$) range from 0.02 to 22.1 wt% $V_2O_3$. The V-rich garnets are found mainly in metaironstones, a few metasediments, unusual mafic pyroclastic rocks and ore deposits. Several of the V-rich garnets also have high Cr contents. There are

only nine Zr-rich garnets (up to 29.9 wt% $ZrO_2$) which are from a skarn/rodingites, a carbonatite and a calc-silicate. It is also notable that such unusual garnets are commonly enriched in more than one rare oxide; for example, scandium-rich garnets from Russia (Galuskina *et al.* 2005) are also rich in Hf, Y, V and Zr. It is unusual for any of these elements to be reported, which probably implies unusual compositions or environments of garnet formation when they are determined. We arbitrarily excluded all garnets that have greater than 0.1 wt% of these oxides. In detrital studies this value could probably be set higher, for example, at 1 wt%, since garnet compositions are typically determined for garnet cores, and the lower values in our dataset are commonly rims to garnet with cores that have larger enrichments than 1 wt%.

## Stage 2: high $TiO_2$ contents

This stage identified all garnets with high $TiO_2$ contents, and we chose an arbitrary cut-off value

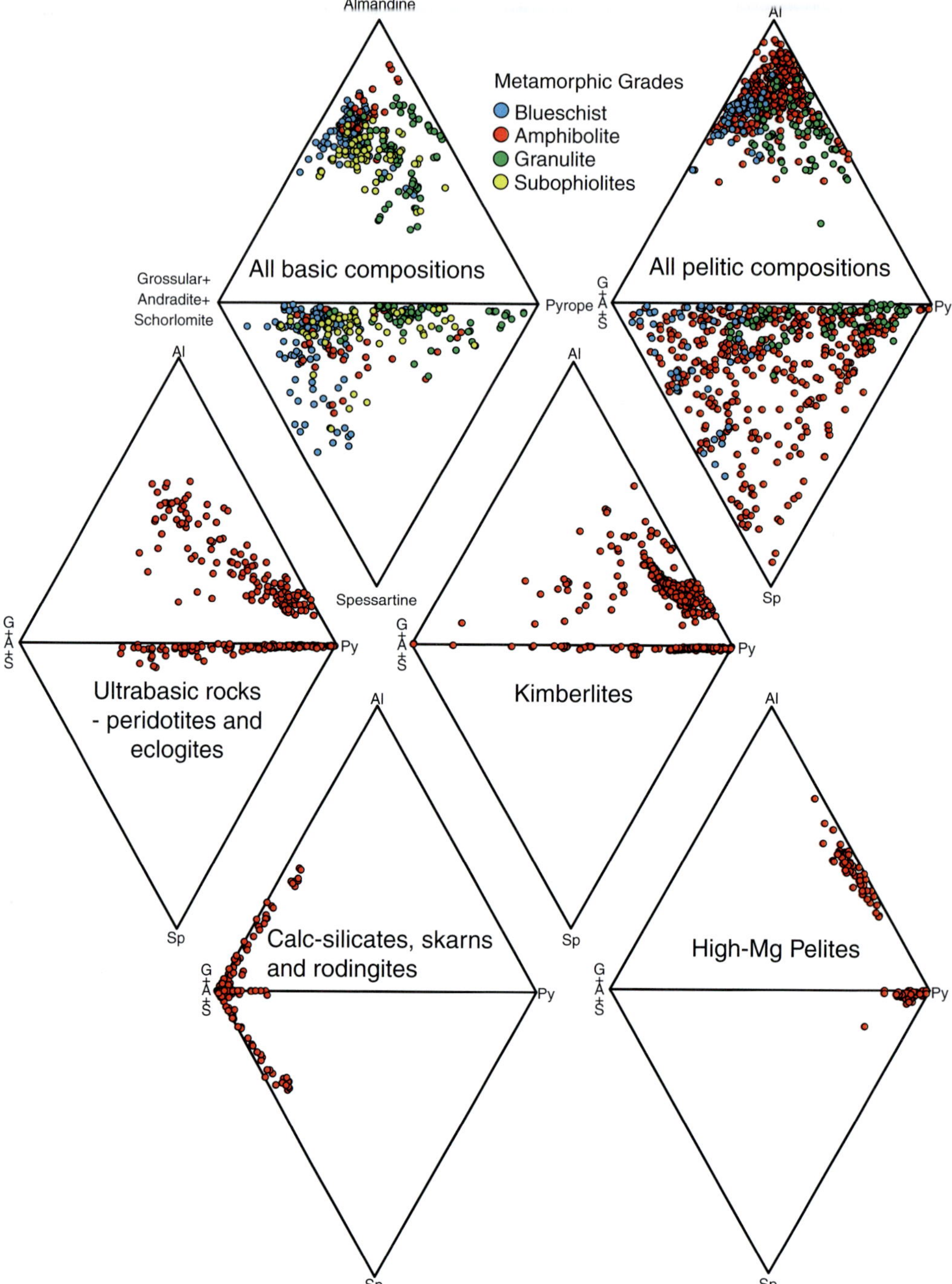


**Fig. 2.** Ternary plots using end-members grossular + andradite + schorlomite, almandine, pyrope and spessartine for various protolith compositions. Garnets were selected for different protoliths after excluding those with unusual rare oxide contents ($V_2O_3$, $Y_2O_3$, $ZrO_2$) and $TiO_2$ >2 wt% as discussed in the text.

of 2%, which excluded a relatively small number ($n = 44$) of garnets with up to 19.6 wt% $TiO_2$. Garnets excluded at this stage are almost all skarns and nepheline syenites, with a few unusual calc-silicates.

## Stage 3: high pyrope and high uvarovite contents

This stage identified garnets with pyrope contents >55% and uvarovite contents >1%. Based on trial and error we found that these values identified only garnets with ultrabasic protoliths, including various peridotites, kimberlites and some eclogites (types A and B of Coleman *et al.* 1965). Because these garnets all have very low spessartine contents they plot on the almandine, pyrope, grossular + andradite + schorlomite triangle in the area with pyrope >55%.

## Stage 4: high uvarovite contents

This stage identified all remaining garnets with high uvarovite content. There is a very small number of garnets ($n = 29$) that have uvarovite >1% and do not have pyrope contents >55%. These are mainly from ultrabasic rocks, ores, skarns or kimberlites.

## Stage 5: high pyrope contents

This stage identified all remaining garnets with high pyrope contents. There is a small number ($n = 118$) of garnets that have pyrope contents >55% and do not have uvarovite >1%. The majority of these have ultrabasic protoliths, mainly various peridotites and kimberlites. The remainder consists mainly of Mg-rich granulites. Thus after stage 3, garnets from ultrabasic rocks can be identified with confidence and, if the uvarovite content is ignored, 95% of the garnets that plot on the almandine, pyrope, grossular + andradite + schorlomite triangle in the area with pyrope >55% can be identified as having an ultrabasic protolith. Most of the remaining garnets that fall in this area are unusual high-Mg pelites of granulite facies grade.

## Stage 6: remaining garnets

After a number of trials testing various criteria for selecting values we found that simply plotting garnet compositions on the two triangles with apices of almandine, pyrope, spessartine and grossular + andradite + schorlomite was an effective way to identify garnet from many rock compositions, or that correspond to different metamorphic grades. Figure 4 shows the sub-areas characteristic of

garnets with different protoliths. The sub-areas we have identified are similar to those identified by Mange & Morton (2007) which they named as garnet types A, Bi, Bii, Ci and Cii. Hutchison & Oliver (1998) and Oliver (2001) used similar end-members. We were not able to replot on our triangles the garnets they used, as their data are not available, but we did plot our data on their $X_{Mg}$–$X_{Ca}$–$X_{Fe+Mn}$ triangle. This broadly confirmed that different protoliths can be identified using their single or our double triangle. However, we consider that a better separation of some garnet types can be made using the two triangles rather than the single $X_{Mg}$–$X_{Ca}$–$X_{Fe+Mn}$ triangle they used. Furthermore, some garnet types are missing or not shown on the earlier plots. For example, Mange & Morton (2007) have no blueschist or sub-ophiolite garnets, while Hutchison & Oliver (1998) and Oliver (2001) used only part of the triangle, so excluding many calc-silicate and ultrabasic garnets. In addition, we now have a much larger database of garnet compositions. This means that some fields are larger than those identified by Hutchison & Oliver (1998) and Oliver (2001); for example, their granulite field included a small number of Lewisian garnets, whereas we now have many more garnets from granulite rocks world-wide. Their plots were effective because it was certain that garnets had a Scottish source but their compositional areas may not be so useful when comparing garnets from other parts of the world. The larger database with more protolith types gives more confidence that garnets can be matched to sources, noting that it will still be valuable to continue enlarging the database, which we intend to do. We summarize below the main features of the database garnets and the ternary plots.

## Protolith compositions

Figure 4 shows the main sub-areas that we have identified using the methods explained above. Most garnets from ultrabasic rocks–peridotites, eclogites and kimberlites–can be identified by their pyrope contents >55% (Fig. 4a). Skarns and calc-silicates can also be identified confidently by their high grossular(+andradite + schorlomite) and very low pyrope contents. Granites (which include large intrusive bodies, leucosomes and migmatites) typically have very high almandine contents and can usually be distinguished from other almandine-rich garnets by their high spessartine contents. Blueschist, amphibolite, granulite facies and sub-ophiolite garnets overlap (Fig. 4b, c), but can often be separated, as explained below. Many basic eclogite garnets plot in the same areas, but can usually be distinguished by their very low spessartine contents.

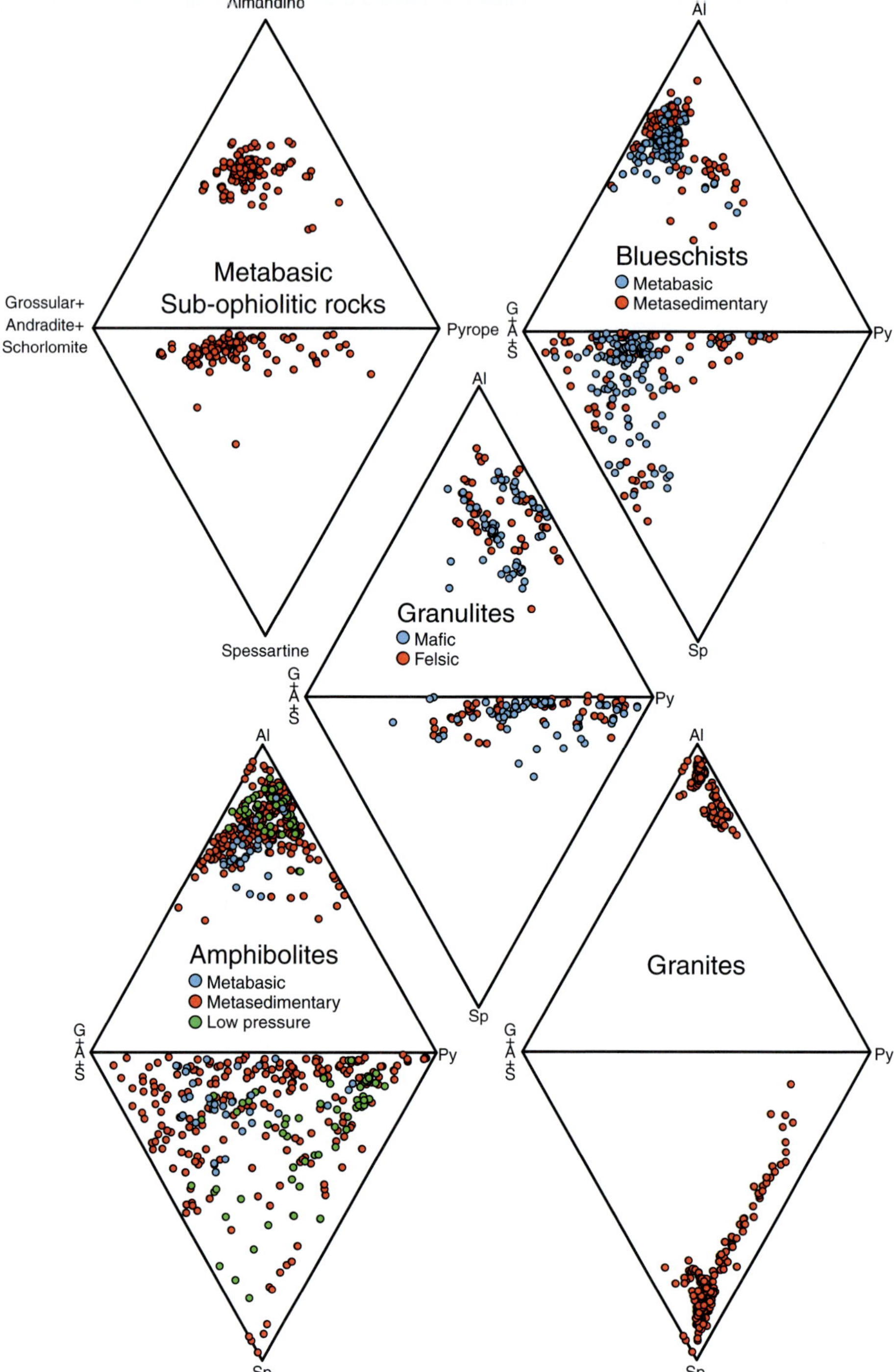


**Fig. 3.** Ternary plots using end-members grossular + andradite + schorlomite, almandine, pyrope and spessartine for protolith compositions and different metamorphic grades, and granites. Granites include large intrusive bodies, leucosomes and migmatites. Garnets were selected for the different categories after excluding those with unusual rare oxide contents ($V_2O_3$, $Y_2O_3$, $ZrO_2$) and $TiO_2$ >2 wt%, as discussed in the text.

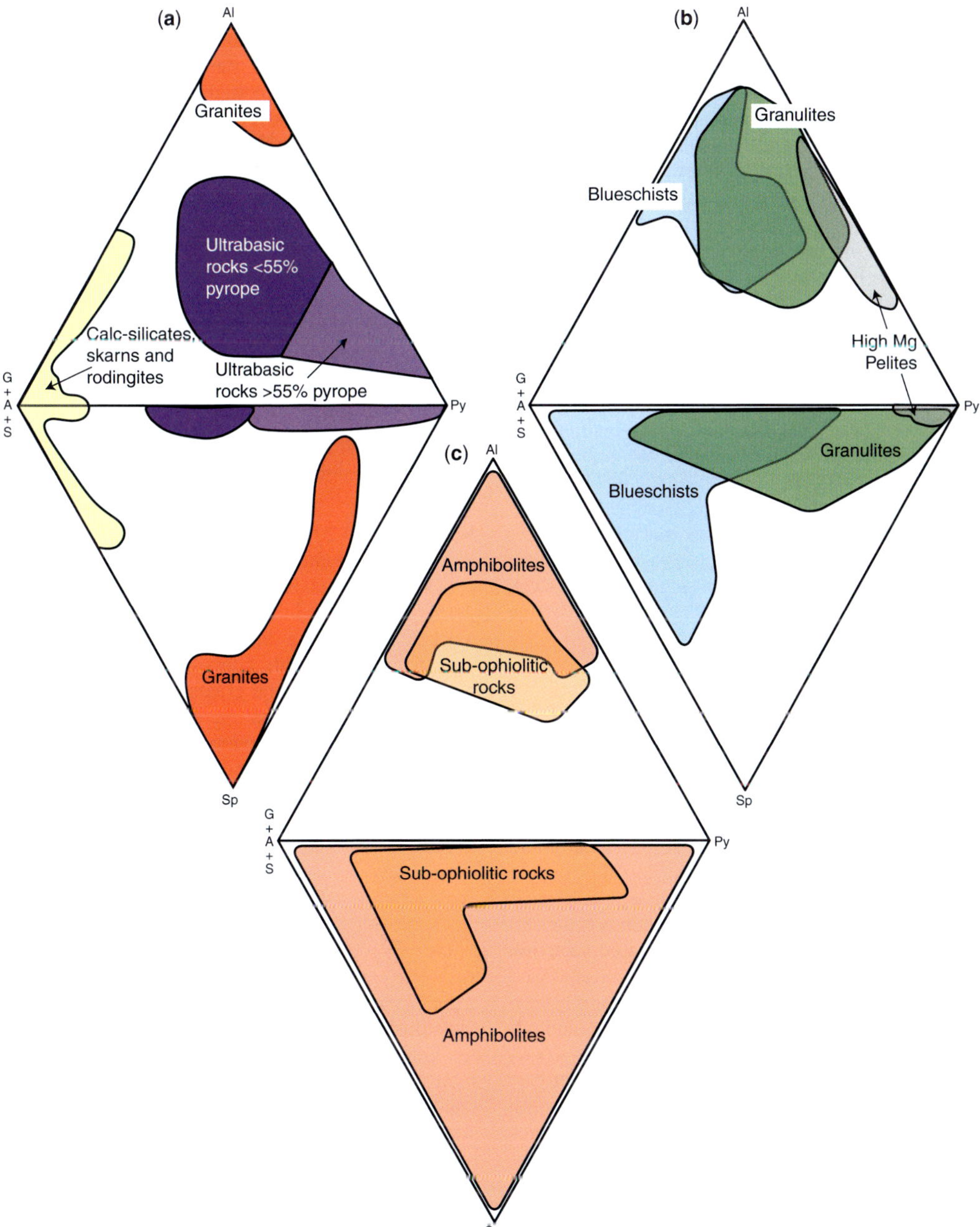



**Fig. 4.** Ternary plots using end-members grossular + andradite + schorlomite, almandine, pyrope and spessartine showing sub-areas characteristic of garnets with different protoliths. (**a**) Ultrabasic rocks (peridotites, eclogites and kimberlites); granites; and calc-silicates, skarns and rodingites. 95% of all ultrabasic garnets have pyrope >55%. (**b**) Granulites, granulite facies high-Mg pelites, and blueschists. (**c**) Amphibolites and metabasic sub-ophiolitic rocks.

## Metamorphic grade

After ultrabasic rocks, calc-silicates and granites (Fig. 4a) have been identified, there is a large area where many garnets plot. These garnets are mainly from blueschist, amphibolite and granulite facies rocks and sub-ophiolite metabasites. Figure 4b shows garnets of blueschists and granulite facies rocks from two common rock compositions: pelites and metabasites. Many blueschist and granulite facies garnets can be separated on the triangles, although there is some overlap. Metabasite sub-ophiolite garnets overlap with blueschist, amphibolite and granulite facies garnets (Fig. 4c). Compositions of amphibolite facies garnets from basic rocks are not abundant in the literature, presumably because metabasite mineral assemblages are not very useful in determining PT conditions. In contrast, compositions of pelitic garnets from amphibolite facies rocks are abundant and they tend to be more almandine-rich than either blueschist or granulite garnets. The plots show that, although many of these garnets cannot be matched to protoliths, a reasonable number can be. Pyrope-poor grossular-almandine garnets are likely to be from blueschists (Fig. 4b), and grossular-poor pyrope-almandine garnets are likely to be from granulites (Fig. 4b). Extremely almandine-rich pyrope-grossular-poor garnets are either amphibolite facies garnets from pelites (Fig. 4c) or from granites (Fig. 4a), but granite garnets can usually be identified by their higher spessartine contents.

## Application to detrital garnets

The methodology above was developed using the large database to evaluate how useful garnet compositions could be in provenance analysis and to identify specific garnet compositions that could be matched to protoliths. Here we test the method using earlier studies.

The Excel spreadsheet (Supplementary material) that contains the database allows a rapid selection of different garnets based on compositional and metamorphic grade criteria. We first examined the dataset garnets using plots used by authors including Mange & Morton (2007), Win *et al.* (2007) and Méres *et al.* (2012) who have shown that garnets can be subdivided into different protolith types and metamorphic grades. Using the triangles they suggested confirms that garnets from the database do fall into the fields they identified. However, in all cases we found the triangles chosen were too restrictive and resulted in overlaps of some garnet types. As explained above, we found the most effective separation of different garnet types was achieved using triangles with apices of almandine, pyrope, spessartine, and grossular + andradite + schorlomite. The papers of Mange & Morton (2007) and Win *et al.* (2007) do not include their large datasets of garnet analyses so we are unable to replot their garnets on our diagrams for comparison. Because they considered garnets from potential sources in the same region they

were able to make quite precise matches to protoliths. Although we cannot identify all their subtypes, we can confirm the main protoliths they identify. We have replotted the dataset of Méres *et al.* (2012), and this is discussed below.

Below we test the method by using a number of recent studies that published chemical analyses of detrital garnets and interpreted their provenance, and we compare the conclusions of the earlier works with our interpretations based on the ternary plots. In each case the earlier works compared garnets to those from protoliths in the studied region (Fig. 5). We then discuss the application of the method to garnets in Borneo sandstones where the provenance is unknown and there are few analyses of garnets from potential source rocks.

### Algeria

Kahoui *et al.* (2012) analysed detrital garnets from Pliocene to Quaternary sands and Lower Cretaceous sandstones in the El Kseibat area, where detrital diamonds have also been found. The primary sources of the diamonds and garnets are not known, but sediments containing numerous kimberlite indicator mineral grains such as pyrope garnet, chrome spinel and picroilmenite are known from an area up to 1000 km wide within which is the Djebel Aberraz diamond placer deposit. The compositions of the detrital garnets are plotted in Figure 5a. They fall unambiguously in the area of ultrabasic protoliths found in mantle xenoliths and kimberlites.

### Eastern Europe

Aubrecht *et al.* (2009) analysed heavy minerals from Cretaceous marly limestones of the Czorsztyn Unit in the western Carpathians. They interpreted the majority of detrital garnets to be derived from high pressure/ultra-high pressure (HP/UHP) parental rocks which recrystallized under granulite and amphibolite facies conditions, probably originally derived from magmatic and metamorphic rocks of the Oravic basement.

Based on the ternary plots (Fig. 5b1) it is possible to exclude protoliths such as ultrabasic rocks, kimberlites, eclogites and calc-silicates. On the ternary triangles the garnets plot in areas that include basic and pelitic rocks from a wide range of metamorphic conditions including blueschists, amphibolites, granulites and sub-ophiolite soles. Some of the garnets cannot be from blueschists or sub-ophiolite soles. It is possible to say that most of the garnets, notably those which are grossular-poor, could only have come from amphibolite and granulite facies rocks of basic and pelitic

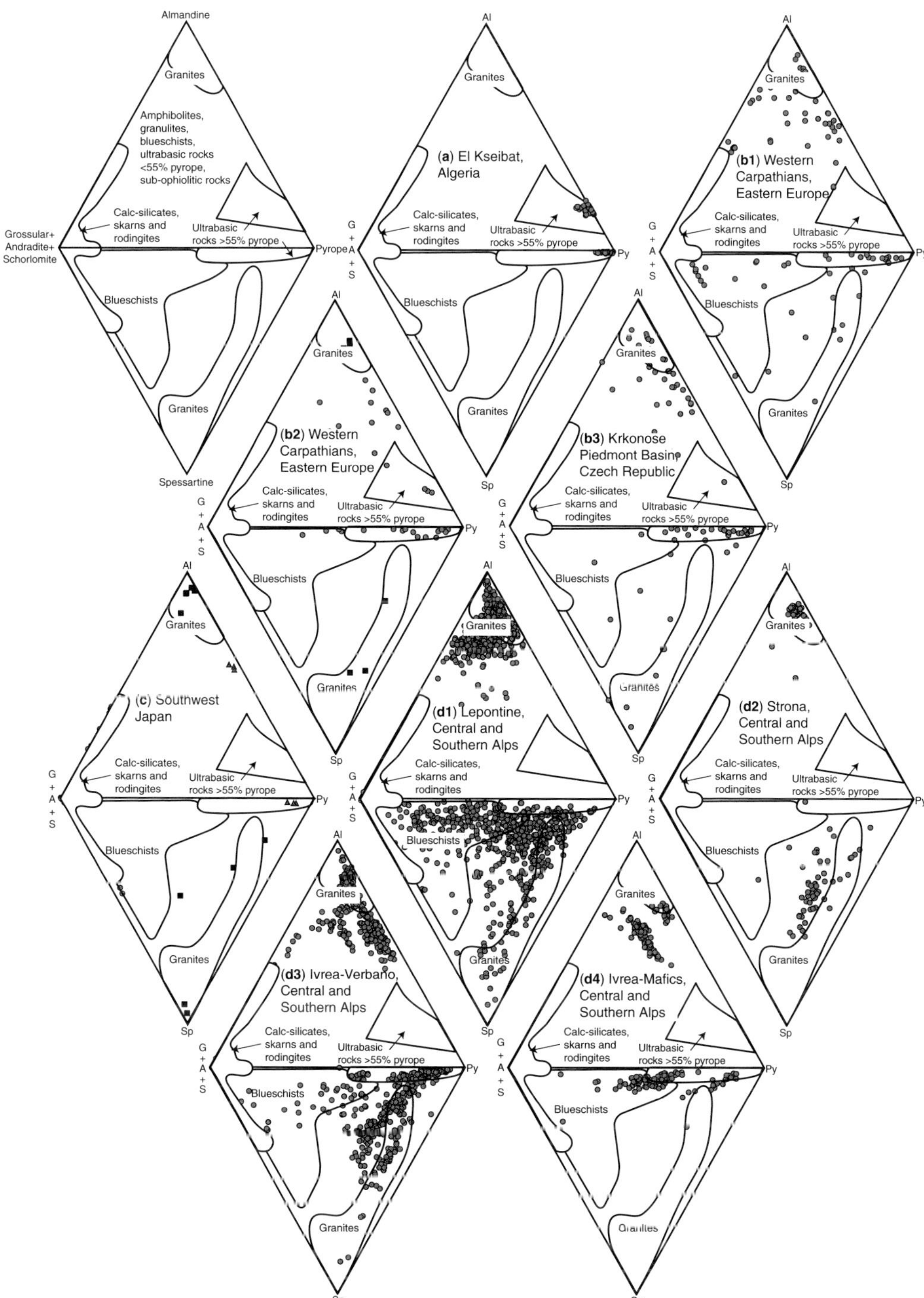


**Fig. 5.** Detrital garnet compositions from provenance studies in Algeria (Kahoui *et al.* 2012), Western Carpathians, Eastern Europe (Aubrecht *et al.* 2009; Méres *et al.* 2012), Czech Republic (Martínek & Štolfová 2009), Japan (Takeuchi 1994; Takeuchi *et al.* 2008), Central and Southern Alps, Italy–Switzerland (Andò *et al.* 2013).

compositions, and all of the garnet could be from amphibolite facies rocks. A few garnets could have granitic protoliths which were migmatites.

Méres *et al.* (2012) described detrital garnets from Jurassic sandstones which they interpreted to be derived from HP/UHP metamorphic rocks including garnet peridotites, eclogites and granulites. Their conclusions are supported by the presence of kyanite and omphacite inclusions in pyrope-rich garnets. The ternary plots (Fig. 5b2) show it is possible to exclude calc-silicate and blueschist facies protoliths, and they are unlikely to have amphibolite facies protoliths of any composition. A small number of the garnets have granitic protoliths. Some definitely have ultrabasic protoliths. The low spessartine contents of the majority most closely resemble garnets from granulites and deep crustal eclogites.

Martínek & Štolfová (2009) reported heavy mineral assemblages and garnet compositions from Permian terrestrial sediments of the Krkonoše Piedmont Basin in the Czech Republic. They interpreted detrital garnet compositions to indicate sources to include Moldanubian granulites, garnet clinopyroxenites, leucogranites and pegmatites. The ternary plots (Fig. 5b3) show that some garnets probably have granitic protoliths. One definitely has an ultrabasic protolith. Calc-silicate and blueschist facies protoliths can be ruled out. The majority most closely resembles garnets from granulite facies rocks, but amphibolite facies rocks cannot be excluded.

## Japan

Takeuchi (1994) and Takeuchi *et al.* (2008) reported analyses of detrital garnets from Permian–Jurassic sandstones of two different terranes in SW Japan. Although many garnets were analysed in each study only 10 analyses are reported in each paper so the two subsets have been combined here. The garnet compositions are quite different from those from Algeria and East Europe. They were interpreted to be derived from intrusive rocks, low-grade or contact rocks such as skarns or hornfels, and high grade metamorphic rocks including granulites. The ternary plots (Fig. 5c) show that blueschist facies and ultrabasic protoliths can be excluded. The majority of garnets in the two areas have calc-silicate protoliths. Several have granite protoliths. The remaining garnets most closely resemble garnets from granulite facies rocks.

## Central and southern Alps, Italy–Switzerland

Andò *et al.* (2013) compiled a very large number of detrital garnet analyses as part of a heavy mineral study of river sands derived from source rocks of high-grade metamorphic rocks in a section across the Central and Southern Alps, including the Cenozoic amphibolite facies core of the Lepontine Dome and Palaeozoic granulite facies rocks of the Ivrea–Verbano Zone. They were able to match the detrital garnet compositions to local source rocks and showed that they included amphibolite facies low-Mg garnets, high-Mn garnets from granite pegmatites, almandine–pyrope-rich garnets from granulite facies metasediments and almandine–pyrope–grossular garnets from metagabbros. They suggested an ultrabasic protolith for a few Lepontine garnets.

The ternary plots (Fig. 5d1–4) show garnets from the four sub-areas. The plots support the conclusions of Andò *et al.* (2013) and show that even in the absence of information about source rocks and garnet compositions, the garnets can provide valuable information about protoliths. Blueschist facies rocks can be ruled out for the Strona area (Fig. 5d2) and the Ivrea mafic rocks (Fig. 5d4), and probably for the Ivrea–Verbano area (Fig. 5d3). Blueschist facies rocks cannot be ruled out for some of the Lepontine garnets (Fig. 5d1), but most cannot be from blueschist facies protoliths and are typical of amphibolite facies rocks and granites (probably migmatites). There may be rare calc-silicate garnets in the Lepontine and Ivrea–Verbano areas, but not in the Strona or Ivrea Mafic areas. The majority of Strona garnets have a granitic protolith. The majority of Ivrea–Verbano garnets are likely to be from granulite facies rocks but the most almandine-rich garnets do not resemble granulites and can be matched in our database either by amphibolite facies or granitic garnets. The Ivrea Mafic area garnets are from granulite facies rocks. We differ only from Andò *et al.* (2013) in the suggestion of ultrabasic garnets. None of the alpine garnets resembles those with ultrabasic protoliths in our database, and they are all much less Mg-rich than those reported from potential Alpine peridotite sources (e.g. Evans & Trommsdorff 1978; Nimis & Trommsdorff 2001).

## Northern Sabah, Borneo

Neogene sandstones from northern Borneo contain varying amounts of garnet. There are a number of possible source areas for these garnets in Borneo itself, but there are no analyses of garnets from any of these areas. Heavy mineral studies (van Hattum 2005; Suggate 2011; van Hattum *et al.* 2013) suggested that the sandstones could have a Philippines source, and there are a small number of analyses of garnets from potential source rocks in Palawan.

## Sabah geology and stratigraphy

The sedimentary rocks of Sabah, northern Borneo (Fig. 6) were deposited in two distinctly different phases. The first phase predates the Top Crocker Unconformity (TCU) of Early Miocene age (van Hattum *et al.* 2006), and is characterized by deep-marine sedimentation. These sedimentary rocks include the Upper Cretaceous to Eocene Sapulut Formation (Rajang Group) and Eocene to Lower Miocene Trusmadi and Crocker formations (Crocker Group). The second phase of sedimentation post-dates the TCU and occurred after arc–continent collision in the Early Miocene uplifted and deformed the strata of the Rajang Group and Crocker Group. Subsidence resumed and sedimentation continued in a fluvio-deltaic to shallow-marine setting, depositing the thick sequences of Neogene clastic sediments in northern and central–eastern Sabah. In northern Sabah (Fig. 6), the Neogene rocks comprise the Lower to Middle Miocene Kudat Group which consists of the Lower to Middle Miocene Kudat Formation and the Upper Miocene Bongaya Formation.

The Kudat Formation (Fig. 6) is divided into the clastic Tajau Sandstone Member and Sikuati Member, which are overlain by a series of mud-stone and limestone members (Stephens 1956). The Kudat Formation is composed of interbedded shallow-marine sandstones and mudstones. The Tajau Sandstone Member consists predominantly of gently folded thick coarse sandy proximal debris flows. The Sikuati Member sits stratigraphi-cally above the Tajau Sandstone Member and con-sists of well-laminated tilted sandstones, siltstones and mudstones, with few sedimentary structures. Biostratigraphical re-evaluation of the foraminiferal assemblages in the Tajau Sandstone Member by BouDagher-Fadel (pers. comm. 2011) confirms initial suggestions by Clement (1958) and van Hattum (2005) that the Eocene foraminifera orig-inally used by Stephens (1956) as age determin-ing are reworked. BouDagher-Fadel (pers. comm. 2011) confirms assemblages present are of Late Oligocene–Early Miocene age, and the presence of *Miogypsinella ubaghsi* indicates a lowermost Early Miocene age of *c.* 21–23 Ma (Early Aquitanian–Early Te5 Letter Stage; BouDagher-Fadel, 2008) for the Tajau Sandstone Member.

## Provenance of northern Sabah Lower Miocene sandstones

A provenance study of the Lower Miocene Tajau Sandstone Member of northern Sabah (Suggate 2011) shows that based on light minerals the sand-stones are compositionally and texturally imma-ture. The sandstones are predominantly arkoses, feldspathic litharenites and litharenites. Monocrys-talline quartz (<60%) and potassium feldspar (*c.* 20–30%) are dominant. Chert is present, although always in small amounts, and is probably derived from the local basement. The heavy mineral assem-blage of the Tajau Sandstone Member is dominated by zircon (max. 45% and mean 29%) and garnet (max. 50% and mean 24%). Apatite is present in significant amounts (*c.* 10–15%). The zircons are typically a mixture of euhedral (21%), subhedral (42%) and subrounded (15%) grains. Medium- to high-grade metamorphic minerals present include kyanite, sillimanite, andalusite, epidote and stauro-lite. This is in marked contrast to the overlying Sikuati Member, in which heavy mineral assem-blages are dominated by zircon, lack kyanite, silli-manite, andalusite, epidote and staurolite, and which are interpreted to have been derived from the Crocker and Rajang Groups of northern Borneo (Suggate 2011).

U–Pb dating studies of detrital zircons from the Tajau Sandstone Member indicate that the dominant age population is Jurassic and Early Cre-taceous. Late Cretaceous, Permian–Triassic, Pala-eozoic and Proterozoic zircons are present but do not form significant populations. Cenozoic (Pale-ocene to Eocene, 56 Ma to 41 Ma) zircons are present. The presence of unabraded Jurassic and Cretaceous zircons, apatite, detrital garnet, and the presence of rare metamorphic minerals such as kyanite and andalusite suggest that the Tajau Sand-stone Member has an unusual metamorphic prove-nance. The main zircon age populations observed in the Sikuati Member are different to those of the Tajau Sandstone Member. The significant age groups observed are abraded Cretaceous, Permian–Triassic and Proterozoic (Neoproterozoic and Pala-eoproterozoic) zircons and are broadly the same as those seen in the other Neogene sandstones in northern Borneo (Suggate 2011).

One potential source is the Palaeogene Crocker Group sandstones of northern Borneo which is characterized by typical continent-derived quartz-rich sediments with heavy minerals such as zircon, tourmaline and apatite. The Crocker Group sand-stones were derived from the Schwaner Mountain granites and nearby Sunda Shelf and Malay–Thai Tin Belt granites (van Hattum 2005, van Hattum *et al.* 2013). The dominant zircon age populations are Cretaceous, Permian–Triassic and Palaeozoic. However, the Crocker Group Sandstones cannot be the only source for the Tajau Sandstone Mem-ber. Kyanite is unknown, and Jurassic zircons are largely absent in the Crocker Group sandstones. Palawan was suggested by van Hattum (2005) to be a possible source since kyanite is reported from high pressure metamorphic rocks interpreted to be related to subduction (Encarnación *et al.* 1995).

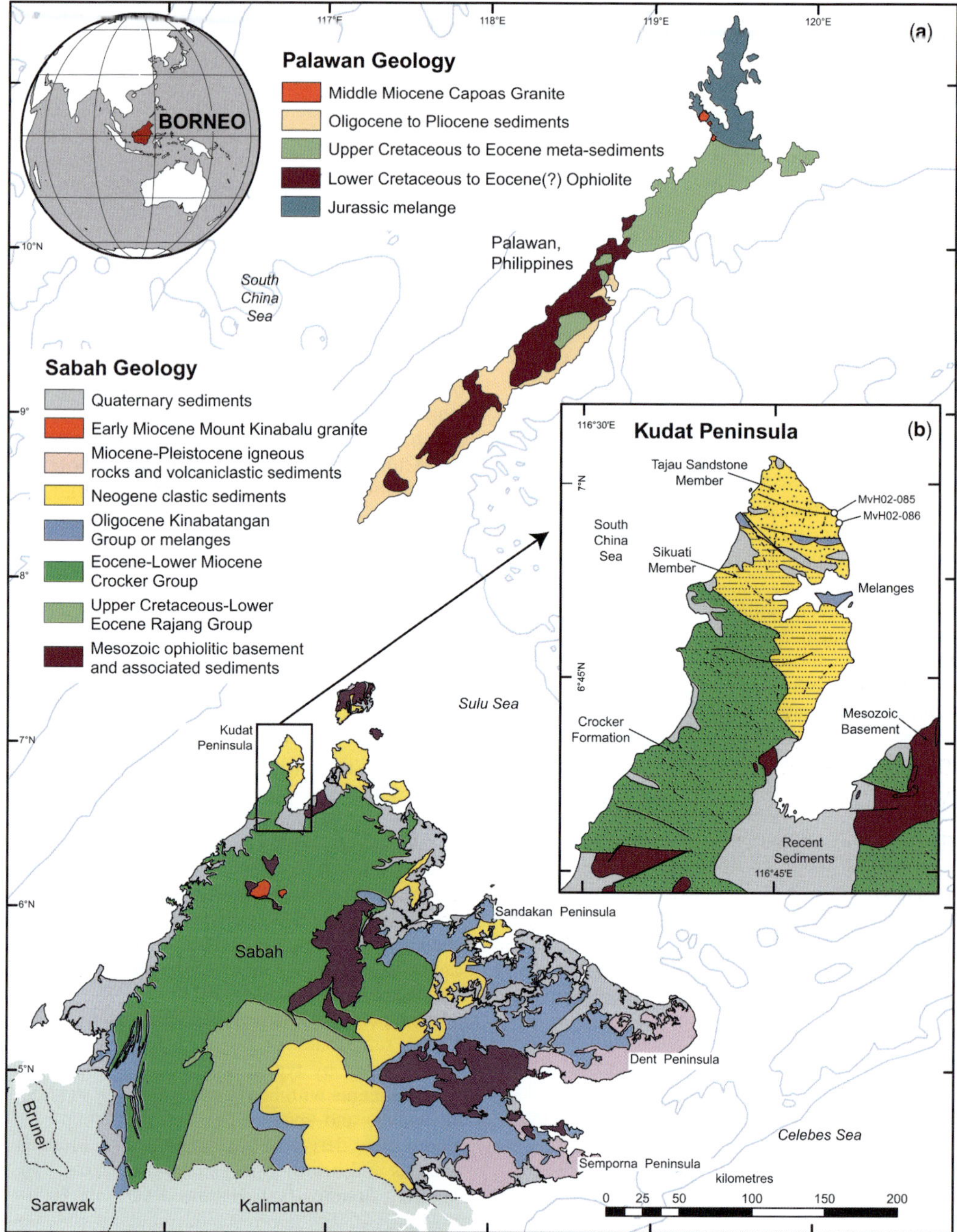



**Fig. 6.** (**a**) Geological map of northern Borneo and Palawan, Philippines, showing the main tectonostratigraphic provinces on land, modified from Lim & Heng (1985), Almasco *et al.* (2000), Mines & Geoscience Bureau, Philippines (2011) and Suggate (2011). (**b**) Inset shows geological map of the Kudat Peninsula, northern Sabah with locations of garnet-rich sandstone samples (van Hattum 2005) from the Tajau Sandstone Member, modified from Tongkul (1994) and Suggate (2011).

## Detrital garnets

Detrital garnets are a significant component (*c.* 45%) of the heavy mineral assemblage from the Tajau Sandstone Member. Detrital garnets in this abundance are not known in other sandstones from Sabah. To help establish their provenance the garnet compositional database was used. The detrital garnets were analysed for major elements and their end-member compositions calculated. These compositions were plotted on ternary plots to identify possible protoliths using the methods described above, and compared with garnets from Palawan.

## Sample locations and analytical techniques

A total of 43 detrital garnet grains hand picked from two samples (Fig. 6) from the Tajau Sandstone Member (MvH02-085 and MvH02-086) were embedded in epoxy resin and polished. Garnets from the Tajau Sandstone and a Kinabalu garnet peridotite were analysed for major elements ($SiO_2$, $TiO_2$, $Al_2O_3$, $Cr_2O_3$, $Fe_2O_3$, FeO, MnO, MgO, CaO and $Na_2O$) on a Jeol8100 Superprobe with an Oxford Instrument INCA micro-analytical system (EDS) at the Department of Earth and Planetary Sciences, Birkbeck College, University of London.

## Detrital garnet compositions of the Neogene sandstones

Electron microprobe results and garnet end-member compositions for the detrital garnet grains analysed from samples MvH02-085 and MvH02-086 of the Tajau Sandstone Member are in Tables 4 and 5. Stages 1 and 2 were skipped as there were no garnets with rare oxides and high $TiO_2$. The garnets from the Tajau Sandstone Member are plotted on a ternary plot with provenance fields and fall into three sub-areas (Fig. 7).

## Possible source areas of detrital garnets

One possible source for the Neogene sandstone detrital garnets are kyanite–garnet amphibolites exposed near the base of the pre-middle Eocene ophiolite described by Encarnación *et al.* (1995) on central Palawan, Philippines, north of the Kudat Peninsula. A second possible source is the garnet peridotites near Mount Kinabalu that form part of the Mesozoic crystalline basement of Sabah (Imai & Ozawa 1991).

Garnet end-member compositions for both possible sources were recalculated from major element analyses of eight garnets recorded by Encarnación *et al.* (1995), four garnets recorded by

Imai & Ozawa (1991), and 14 new microprobe analyses from a garnet peridotite from Kinabalu made during this study. The garnet compositions are plotted on ternary plots with the provenance fields (Fig. 7a–d).

The garnet compositions in the kyanite–garnet amphibolites of central Palawan (Encarnación *et al.* 1995) are dominated by almandine–pyrope garnets (Alm49–59% and Prp12–36%). The majority of garnets plot within the field of amphibolites, granulites or sub-ophiolite soles, apart from one grain which plots within the blueschist field (Fig. 7a).

The average garnet compositions of the basement garnet peridotites of the Mount Kinabalu are dominated by a group of pyrope-rich (Prp60–76%) garnets with a small compositional range (Fig. 7b). This population is represented by Prp60–76%Alm13–26%Gro3–15%Sps0–1%. These garnets plot within the ultrabasic field (Fig. 7b).

## Comparison between detrital garnets and possible sources

The garnets from the kyanite–garnet amphibolites on Palawan plot in an area that is typical of sub-ophiolite metabasites (Fig. 7c). If these were detrital garnets it would be possible to identify them as derived from amphibolites, granulites or sub-ophiolite soles; blueschists can be ruled out since several of the garnets are too enriched in pyrope.

The garnets from the basement peridotites of Mount Kinabalu area plot clearly within the peridotite field (Fig. 7b), demonstrating that the method correctly identifies them. There are no comparable detrital garnets from the Tajau Sandstone Member, indicating that they do not have a peridotite source. None of the detrital garnets from the Kudat Tajau Sandstone Member have a calc-silicate source.

On the ternary triangles the Kudat garnets fall into three groups (Fig. 7a). One group (Group A) is clearly derived from a granitic source (Fig. 7b). A second group (Group B) is almandine-rich. These garnets could be derived from amphibolites, granulites or sub-ophiolite soles (Fig. 7c), but not blueschists or granites. They plot close to the Palawan garnets. The third group (Group C) is almandine-rich and pyrope-poor. They could be derived from blueschists or amphibolites (Fig. 7d), but not granulites, sub-ophiolite soles, or granites.

Based on the ternary plots and the comparison to garnet compositions from Palawan and Kinabalu it is possible to rule out a Kinabalu source for any of the garnets, and imply a Palawan source for some of the garnets. The granitic garnets could come from Borneo or the Tin Belt, since granitic

**Table 4.** *Electron microprobe analyses of detrital garnets for sample MvH02-085, Tajau Sandstone Member, Kudat Formation*

| Wt% oxide | 1 | 2 | 3 | 4 | 5 | 6 | 7 | 8 | 9 | 10 | 11 | 12 | 13 | 14 | 15 | 16 | 17 | 18 | 19 | 20 | 21 |
|---|---|---|---|---|---|---|---|---|---|---|---|---|---|---|---|---|---|---|---|---|---|
| $SiO_2$ | 37.57 | 37.41 | 38.47 | 38.33 | 37.13 | 36.87 | 37.67 | 37.39 | 37.46 | 37.98 | 37.22 | 37.24 | 36.30 | 36.65 | 37.13 | 37.35 | 37.08 | 36.87 | 36.87 | 35.52 | 35.78 |
| $Al_2O_3$ | 19.65 | 20.51 | 20.90 | 20.86 | 20.65 | 20.96 | 21.52 | 21.18 | 20.96 | 20.91 | 20.60 | 20.73 | 19.46 | 20.55 | 20.60 | 21.36 | 20.83 | 20.98 | 19.86 | 18.76 | 19.08 |
| $TiO_2$ | 0.12 | 0.14 | 0.06 | 0.09 | 0.02 | 0.00 | 0.02 | 0.02 | 0.05 | 0.04 | 0.02 | 0.03 | 0.00 | 0.07 | 0.09 | 0.07 | 0.11 | 0.01 | 0.01 | 0.17 | 0.16 |
| $Cr_2O_3$ | 0.00 | 0.01 | 0.00 | 0.00 | 0.01 | 0.00 | 0.01 | 0.01 | 0.01 | 0.01 | 0.01 | 0.00 | 0.02 | 0.00 | 0.01 | 0.05 | 0.01 | 0.02 | 0.00 | 0.00 | 0.00 |
| $Fe_2O_3$ | 0.27 | 0.46 | 0.00 | 0.01 | 0.13 | 0.33 | 0.00 | 0.03 | 0.70 | 0.17 | 0.20 | 0.00 | 0.47 | 0.57 | 0.65 | 0.25 | 0.63 | 0.79 | 1.02 | 1.47 | 1.25 |
| $FeO$ | 23.55 | 23.87 | 27.17 | 27.02 | 34.76 | 34.45 | 30.34 | 30.36 | 28.37 | 28.85 | 34.92 | 34.93 | 32.35 | 32.94 | 29.28 | 29.82 | 27.77 | 24.74 | 24.27 | 19.31 | 19.56 |
| $MnO$ | 6.21 | 6.26 | 2.68 | 2.72 | 2.94 | 2.85 | 6.47 | 6.38 | 2.44 | 2.49 | 1.66 | 1.66 | 4.39 | 4.48 | 2.17 | 2.08 | 1.64 | 5.92 | 6.04 | 20.82 | 20.31 |
| $MgO$ | 1.29 | 1.28 | 5.19 | 5.32 | 2.85 | 2.84 | 3.71 | 3.61 | 4.68 | 4.57 | 3.16 | 3.26 | 2.99 | 3.04 | 3.90 | 3.79 | 0.69 | 1.37 | 1.34 | 0.76 | 0.75 |
| $CaO$ | 10.03 | 9.58 | 5.10 | 5.16 | 1.21 | 1.28 | 1.03 | 1.04 | 4.38 | 4.63 | 1.68 | 1.64 | 1.00 | 0.74 | 4.72 | 4.64 | 10.68 | 8.41 | 8.73 | 0.57 | 0.59 |
| $Na_2O$ | 0.01 | 0.02 | 0.02 | 0.01 | 0.01 | 0.01 | 0.03 | 0.03 | 0.01 | 0.00 | 0.03 | 0.00 | 0.00 | 0.01 | 0.00 | 0.02 | 0.02 | 0.03 | 0.03 | 0.03 | 0.02 |
| Total | 98.70 | 99.54 | 99.59 | 99.52 | 99.71 | 99.59 | 100.80 | 100.04 | 99.06 | 99.65 | 99.50 | 99.49 | 96.98 | 99.05 | 98.56 | 99.44 | 99.46 | 99.14 | 98.16 | 97.42 | 98.11 |

*Formulae calculated on the basis of 24 oxygens. $Fe_3$ calculated for 16 cations*

| | 1 | 2 | 3 | 4 | 5 | 6 | 7 | 8 | 9 | 10 | 11 | 12 | 13 | 14 | 15 | 16 | 17 | 18 | 19 | 20 | 21 |
|---|---|---|---|---|---|---|---|---|---|---|---|---|---|---|---|---|---|---|---|---|---|
| Si | 6.092 | 6.014 | 6.066 | 6.049 | 6.018 | 5.978 | 5.989 | 5.998 | 5.980 | 6.028 | 6.024 | 6.022 | 6.056 | 5.981 | 5.990 | 5.965 | 5.973 | 5.957 | 6.027 | 6.016 | 6.014 |
| Al | 3.756 | 3.887 | 3.885 | 3.881 | 3.946 | 4.007 | 4.034 | 4.005 | 3.945 | 3.913 | 3.931 | 3.952 | 3.827 | 3.954 | 3.918 | 4.022 | 3.956 | 3.996 | 3.827 | 3.746 | 3.781 |
| Ti | 0.015 | 0.017 | 0.007 | 0.011 | 0.002 | 0.000 | 0.002 | 0.002 | 0.006 | 0.005 | 0.002 | 0.004 | 0.000 | 0.009 | 0.011 | 0.008 | 0.013 | 0.001 | 0.001 | 0.022 | 0.020 |
| Cr | 0.000 | 0.001 | 0.000 | 0.000 | 0.001 | 0.000 | 0.001 | 0.001 | 0.001 | 0.001 | 0.001 | 0.000 | 0.003 | 0.000 | 0.001 | 0.006 | 0.001 | 0.003 | 0.000 | 0.000 | 0.000 |
| $Fe_3$ | 0.033 | 0.055 | 0.000 | 0.001 | 0.016 | 0.040 | 0.000 | 0.003 | 0.085 | 0.020 | 0.025 | 0.000 | 0.059 | 0.070 | 0.079 | 0.030 | 0.077 | 0.095 | 0.125 | 0.188 | 0.158 |
| $Fe_2$ | 3.193 | 3.209 | 3.583 | 3.566 | 4.712 | 4.672 | 4.034 | 4.072 | 3.787 | 3.830 | 4.726 | 4.724 | 4.513 | 4.495 | 3.951 | 3.984 | 3.741 | 3.343 | 3.317 | 2.736 | 2.764 |
| Mn | 0.853 | 0.852 | 0.358 | 0.364 | 0.404 | 0.391 | 0.871 | 0.867 | 0.330 | 0.335 | 0.228 | 0.227 | 0.620 | 0.619 | 0.297 | 0.281 | 0.224 | 0.810 | 0.836 | 2.987 | 2.963 |
| Mg | 0.312 | 0.307 | 1.220 | 1.251 | 0.688 | 0.686 | 0.879 | 0.863 | 1.113 | 1.081 | 0.762 | 0.786 | 0.743 | 0.739 | 0.938 | 0.902 | 0.166 | 0.330 | 0.326 | 0.192 | 0.188 |
| Ca | 1.743 | 1.650 | 0.862 | 0.873 | 0.210 | 0.222 | 0.175 | 0.179 | 0.749 | 0.787 | 0.291 | 0.284 | 0.179 | 0.129 | 0.816 | 0.794 | 1.843 | 1.456 | 1.529 | 0.103 | 0.106 |
| Na | 0.003 | 0.006 | 0.006 | 0.003 | 0.003 | 0.003 | 0.009 | 0.009 | 0.003 | 0.000 | 0.009 | 0.000 | 0.000 | 0.003 | 0.000 | 0.006 | 0.006 | 0.009 | 0.010 | 0.010 | 0.007 |

*End-members*

| | 1 | 2 | 3 | 4 | 5 | 6 | 7 | 8 | 9 | 10 | 11 | 12 | 13 | 14 | 15 | 16 | 17 | 18 | 19 | 20 | 21 |
|---|---|---|---|---|---|---|---|---|---|---|---|---|---|---|---|---|---|---|---|---|---|
| Almandine | 52.3 | 53.3 | 59.5 | 58.9 | 78.3 | 78.2 | 67.7 | 68.1 | 63.3 | 63.5 | 78.7 | 78.5 | 74.5 | 75.1 | 65.8 | 66.8 | 62.6 | 56.3 | 55.2 | 43.9 | 44.7 |
| Andradite + Schorlomite | 1.2 | 1.8 | 0.2 | 0.3 | 0.5 | 1.0 | 0.1 | 0.1 | 2.3 | 0.6 | 0.7 | 0.1 | 1.5 | 2.0 | 2.3 | 1.0 | 2.3 | 2.4 | 3.2 | 5.0 | 4.3 |
| Grossulur | 27.4 | 25.6 | 14.1 | 14.1 | 3.0 | 2.7 | 2.9 | 2.8 | 10.2 | 12.4 | 4.1 | 4.6 | 1.4 | 0.2 | 11.3 | 12.2 | 28.6 | 22.0 | 22.3 | 0.0 | 0.0 |
| Pyrope | 5.1 | 5.1 | 20.3 | 20.7 | 11.4 | 11.5 | 14.7 | 14.4 | 18.6 | 17.9 | 12.7 | 13.0 | 12.3 | 12.4 | 15.6 | 15.1 | 2.8 | 5.6 | 5.4 | 3.1 | 3.0 |
| Spessartine | 14.0 | 14.2 | 5.9 | 6.0 | 6.7 | 6.6 | 14.6 | 14.5 | 5.5 | 5.5 | 3.8 | 3.8 | 10.2 | 10.4 | 4.9 | 4.7 | 3.7 | 13.6 | 13.9 | 48.0 | 47.9 |
| Uvarovite | 0.0 | 0.0 | 0.0 | 0.0 | 0.0 | 0.0 | 0.0 | 0.0 | 0.0 | 0.0 | 0.0 | 0.0 | 0.1 | 0.0 | 0.0 | 0.2 | 0.0 | 0.1 | 0.0 | 0.0 | 0.0 |

**Table 5.** *Electron microprobe analyses of detrital garnets for sample MvH02-086, Tajau Sandstone Member, Kudat Formation*

| Wt% oxide | 1 | 2 | 3 | 4 | 5 | 6 | 7 | 8 | 9 | 10 | 11 | 12 | 13 | 14 | 15 | 16 | 17 | 18 | 19 | 20 | 21 | 22 |
|---|---|---|---|---|---|---|---|---|---|---|---|---|---|---|---|---|---|---|---|---|---|---|
| $SiO_2$ | 38.81 | 38.46 | 38.54 | 96.72 | 37.68 | 37.79 | 37.76 | 37.75 | 38.08 | 38.21 | 34.01 | 34.30 | 37.96 | 37.49 | 32.96 | 34.29 | 38.65 | 38.99 | 37.31 | 36.76 | 37.93 | 38.34 |
| $Al_2O_3$ | 21.65 | 21.11 | 21.28 | 0.02 | 21.13 | 21.28 | 21.19 | 20.92 | 21.38 | 21.51 | 0.01 | 0.00 | 20.81 | 20.81 | 0.02 | 0.00 | 21.84 | 21.74 | 20.18 | 20.28 | 21.05 | 21.48 |
| $TiO_2$ | 0.05 | 0.01 | 0.06 | 0.02 | 0.02 | 0.00 | 0.02 | 0.04 | 0.00 | 0.01 | 0.03 | 0.05 | 0.00 | 0.01 | 0.00 | 0.00 | 0.03 | 0.03 | 0.09 | 0.08 | 0.10 | 0.10 |
| $Cr_2O_3$ | 0.01 | 0.00 | 0.00 | 0.00 | 0.01 | 0.02 | 0.00 | 0.01 | 0.00 | 0.03 | 0.00 | 0.00 | 0.02 | 0.02 | 0.00 | 0.00 | 0.01 | 0.02 | 0.00 | 0.01 | 0.00 | 0.02 |
| FeO | 29.17 | 28.89 | 29.02 | 0.65 | 33.13 | 33.06 | 32.79 | 32.71 | 34.18 | 34.97 | 0.00 | 0.00 | 34.11 | 34.11 | 0.02 | 0.00 | 0.00 | 0.00 | 0.00 | 0.11 | 0.00 | 0.00 |
| MnO | 3.11 | 3.10 | 3.14 | 0.07 | 4.56 | 4.67 | 5.18 | 5.31 | 2.70 | 2.45 | 0.01 | 0.00 | 2.32 | 2.31 | 0.00 | 0.00 | 28.56 | 28.48 | 20.18 | 20.02 | 26.00 | 26.68 |
| MgO | 4.96 | 5.06 | 4.85 | 0.00 | 2.24 | 2.22 | 2.42 | 2.39 | 2.99 | 3.00 | 0.00 | 0.00 | 3.15 | 3.16 | 0.01 | 0.01 | 2.74 | 2.63 | 18.80 | 18.74 | 4.83 | 3.15 |
| CaO | 3.09 | 3.65 | 3.53 | 0.30 | 1.35 | 1.24 | 0.95 | 1.03 | 1.18 | 1.07 | 0.00 | 0.03 | 1.43 | 1.43 | 0.09 | 0.02 | 4.50 | 4.56 | 1.25 | 1.26 | 0.77 | 0.80 |
| $Na_2O$ | 0.00 | 0.01 | 0.00 | 0.02 | 0.02 | 0.04 | 0.04 | 0.04 | 0.03 | 0.02 | 0.02 | 0.02 | 0.02 | 0.01 | 0.06 | 0.01 | 4.60 | 4.72 | 1.75 | 1.93 | 9.54 | 10.31 |
| $K_2O$ | 0.01 | 0.00 | 0.00 | 0.01 | 0.00 | 0.00 | 0.00 | 0.00 | 0.01 | 0.00 | 0.01 | 0.00 | 0.00 | 0.00 | 0.04 | 0.01 | 0.01 | 0.03 | 0.06 | 0.06 | 0.01 | 0.02 |
| Total | 100.86 | 100.29 | 100.42 | 97.81 | 100.14 | 100.32 | 100.35 | 100.20 | 100.55 | 101.27 | 101.27 | 101.27 | 99.82 | 99.34 | 33.20 | 33.20 | 100.94 | 101.20 | 99.63 | 99.26 | 100.24 | 100.90 |

*Formulae calculated on the basis of 24 oxygens. $Fe_3$ calculated for 16 cations*

| | 1 | 2 | 3 | 4 | 5 | 6 | 7 | 8 | 9 | 10 | 11 | 12 | 13 | 14 | 15 | 16 | 17 | 18 | 19 | 20 | 21 | 22 |
|---|---|---|---|---|---|---|---|---|---|---|---|---|---|---|---|---|---|---|---|---|---|---|
| Si | 6.059 | 6.052 | 6.056 | 11.937 | 6.063 | 6.066 | 6.062 | 6.075 | 6.069 | 6.054 | 6.054 | 6.054 | 6.095 | 6.057 | 11.955 | 11.955 | 6.030 | 6.060 | 6.090 | 6.032 | 6.050 | 6.051 |
| Al | 3.985 | 3.916 | 3.942 | 0.003 | 4.009 | 4.027 | 4.011 | 3.969 | 4.017 | 4.018 | 4.018 | 4.018 | 3.939 | 3.964 | 0.009 | 0.009 | 4.017 | 3.983 | 3.883 | 3.923 | 3.959 | 3.996 |
| Ti | 0.006 | 0.001 | 0.007 | 0.002 | 0.002 | 0.000 | 0.002 | 0.005 | 0.000 | 0.001 | 0.001 | 0.001 | 0.000 | 0.001 | 0.000 | 0.000 | 0.004 | 0.004 | 0.011 | 0.010 | 0.012 | 0.012 |
| Cr | 0.001 | 0.000 | 0.000 | 0.000 | 0.001 | 0.003 | 0.000 | 0.001 | 0.000 | 0.000 | 0.004 | 0.004 | 0.004 | 0.003 | 0.001 | 0.000 | 0.001 | 0.002 | 0.000 | 0.001 | 0.000 | 0.002 |
| $Fe_2$ | 3.809 | 3.802 | 3.814 | 0.067 | 4.459 | 4.438 | 4.402 | 4.403 | 4.556 | 4.634 | 4.634 | 4.634 | 4.580 | 4.609 | 0.006 | 0.006 | 0.000 | 0.000 | 0.000 | 0.014 | 0.000 | 0.000 |
| Mn | 0.411 | 0.413 | 0.418 | 0.007 | 0.622 | 0.635 | 0.704 | 0.724 | 0.364 | 0.329 | 0.329 | 0.329 | 0.316 | 0.316 | 0.000 | 0.000 | 3.727 | 3.702 | 2.755 | 2.747 | 3.469 | 3.521 |
| Mg | 1.154 | 1.187 | 1.136 | 0.000 | 0.537 | 0.531 | 0.579 | 0.573 | 0.710 | 0.708 | 0.708 | 0.708 | 0.754 | 0.761 | 0.005 | 0.005 | 0.362 | 0.346 | 2.599 | 2.605 | 0.653 | 0.421 |
| Ca | 0.517 | 0.615 | 0.594 | 0.040 | 0.233 | 0.213 | 0.163 | 0.178 | 0.201 | 0.182 | 0.182 | 0.182 | 0.246 | 0.248 | 0.035 | 0.035 | 1.046 | 1.056 | 0.304 | 0.308 | 0.183 | 0.188 |
| Na | 0.000 | 0.003 | 0.000 | 0.005 | 0.006 | 0.012 | 0.012 | 0.012 | 0.009 | 0.006 | 0.005 | 0.006 | 0.006 | 0.003 | 0.042 | 0.042 | 0.769 | 0.786 | 0.306 | 0.339 | 1.631 | 1.743 |
| K | 0.002 | 0.000 | 0.000 | 0.002 | 0.000 | 0.000 | 0.000 | 0.000 | 0.002 | 0.000 | 0.000 | 0.000 | 0.000 | 0.000 | 0.019 | 0.019 | 0.003 | 0.009 | 0.019 | 0.019 | 0.003 | 0.006 |

*End-members*

| | 1 | 2 | 3 | 4 | 5 | 6 | 7 | 8 | 9 | 10 | 11 | 12 | 13 | 14 | 15 | 16 | 17 | 18 | 19 | 20 | 21 | 22 |
|---|---|---|---|---|---|---|---|---|---|---|---|---|---|---|---|---|---|---|---|---|---|---|
| Almandine | 64.7 | 63.2 | 64.0 | 58.8 | 76.2 | 76.3 | 75.3 | 74.9 | 78.1 | 79.2 | 79.2 | 79.2 | 77.7 | 77.7 | 13.1 | 13.1 | 63.1 | 62.8 | 46.2 | 45.8 | 58.4 | 59.9 |
| Andradite + Schorlomite | 0.1 | 0.0 | 0.2 | 2.4 | 0.1 | 0.0 | 0.1 | 0.1 | 0.0 | 0.0 | 0.0 | 0.0 | 0.0 | 0.0 | 0.0 | 0.0 | 0.1 | 0.1 | 0.3 | 0.6 | 0.3 | 0.3 |
| Grossular | 8.6 | 10.2 | 9.8 | 32.3 | 3.9 | 3.6 | 2.7 | 2.9 | 3.5 | 3.0 | 3.0 | 3.0 | 4.1 | 4.1 | 75.3 | 75.3 | 12.9 | 13.2 | 4.9 | 5.0 | 27.2 | 29.3 |
| Pyrope | 19.6 | 19.7 | 19.1 | 0.0 | 9.2 | 9.1 | 9.9 | 9.8 | 12.2 | 12.1 | 12.1 | 12.1 | 12.8 | 12.8 | 11.6 | 11.6 | 17.7 | 17.9 | 5.1 | 5.1 | 3.1 | 3.2 |
| Spessartine | 7.0 | 6.9 | 7.0 | 6.4 | 10.6 | 10.9 | 12.0 | 12.3 | 6.2 | 5.6 | 5.6 | 5.6 | 5.4 | 5.3 | 0.0 | 0.0 | 6.1 | 5.9 | 43.6 | 43.4 | 11.0 | 7.2 |
| Uvarovite | 0.0 | 0.0 | 0.0 | 0.0 | 0.0 | 0.1 | 0.0 | 0.0 | 0.0 | 0.1 | 0.1 | 0.1 | 0.1 | 0.0 | 0.0 | 0.0 | 0.0 | 0.1 | 0.0 | 0.0 | 0.0 | 0.1 |

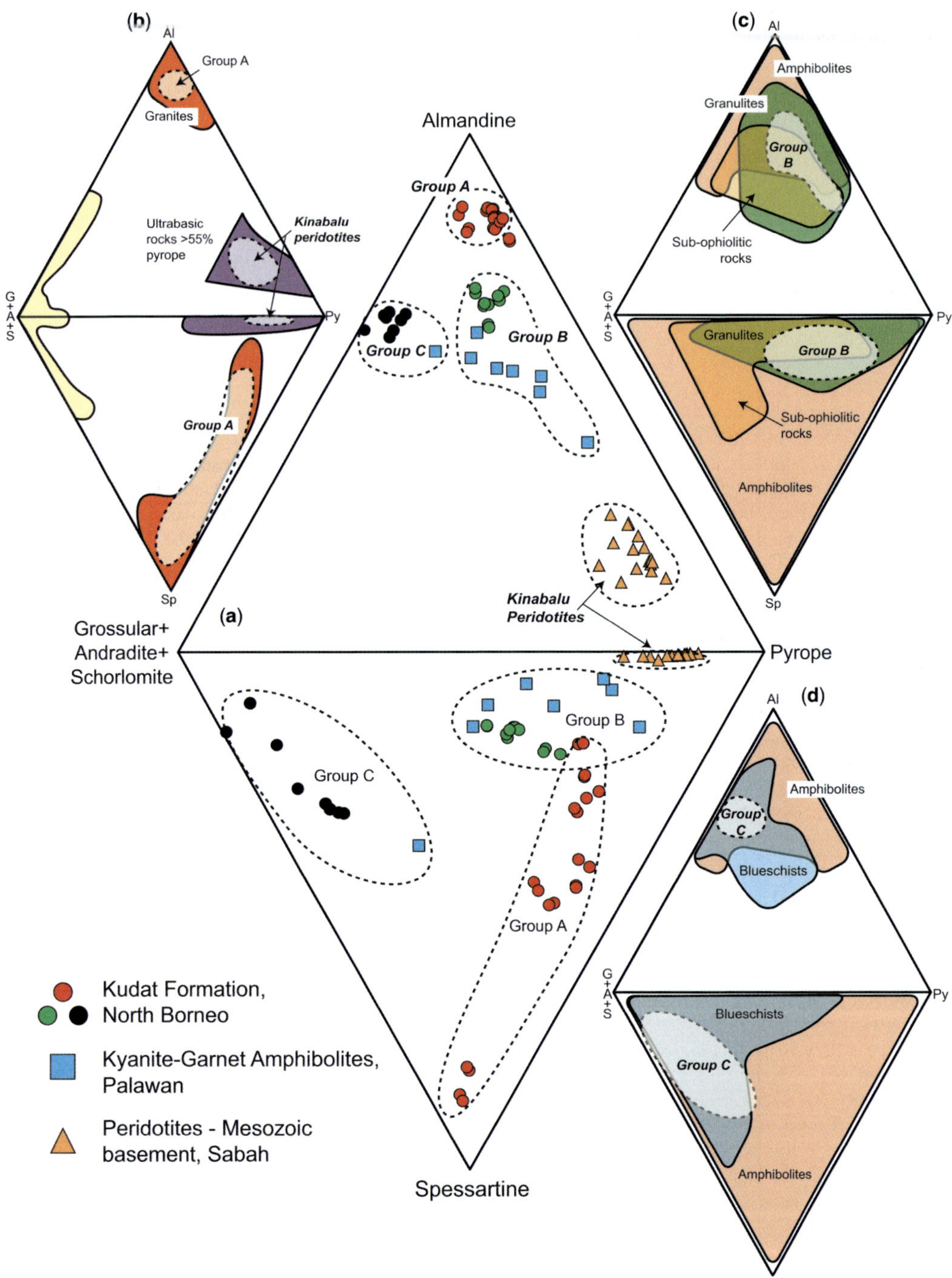


**Fig. 7.** Ternary plots using end-members grossular + andradite + schorlomite, almandine, pyrope and spessartine. (**a**) Individual grain compositions and provenance groups for detrital garnets from the Tajau Sandstone Member, Kudat Formation, Sabah compared to garnets from kyanite-garnet amphibolites on Palawan (Encarnación *et al.* 1995), and garnets from peridotites of the Sabah basement (Imai & Ozawa 1991; this study). (**b–d**) Ternary plots showing sub-areas characteristic of garnets with different protoliths, from Figure 4, with inferred protoliths of Kudat detrital garnets.

rocks are known to be the source of Crocker Group sediments which have been recycled into other Neogene sandstones. However, garnets are not abundant in the Crocker Group heavy mineral assemblages (van Hattum 2005; van Hattum *et al.* 2013). We suggest it is likely that the granitic garnets are derived from either Cretaceous and Jurassic granites that rifted away from the South China margin during the opening of the South China Sea in the Early Oligocene (Taylor & Hayes 1983; Hall 2002; Franke *et al.* 2011) and are now part of the North Palawan Continental Terrane (e.g. Encarnación *et al.* 1995; Encarnación & Mukasa 1997; Knittel *et al.* 2010), or they were derived from Middle Eocene South China Sea rift-related granites that are exposed in Central Palawan.

In addition to the unusual abundance of garnets, the presence of other metamorphic minerals such as kyanite supports a Palawan source. Kyanite is not known from Palaeogene Crocker Group sediments on Borneo, but is known from kyanite–garnet amphibolites in central Palawan (Encarnación *et al.* 1995). Thus a Palawan source for the Kudat detrital garnets seems more likely. The source of the third group of garnets is not known. Their compositions indicate a blueschist facies or amphibolite facies origin. Rare blueschists have been reported from Sabah (Leong 1978), but little is known of them. Subduction-related rocks are described from Palawan (Encarnación *et al.* 1995), so blueschists could be expected there. Amphibolite facies rocks are known from the Schwaner region of Kalimantan. Nothing is currently known of garnet compositions from any of these areas.

## *Early Miocene palaeogeography of Northern Borneo*

During the Early Miocene two river systems deposited sediment in offshore northern Borneo in a shallow-marine setting (Tajau Sandstone Member). Rivers flowed northwards draining the Crocker Mountains contributing recycled quartz-rich material from the Crocker Group sandstones, originally derived from granitic rocks in the Schwaner Mountain of SE Kalimantan, nearby Sunda Shelf and Malay–Thai Tin Belt (van Hattum *et al.* 2006, 2013). At the same time, southward-flowing rivers drained the granitic and metamorphic rocks on Palawan, depositing sediments containing garnets derived from granites, amphibolites, granulites and sub-ophiolite soles.

## Conclusions

There have been many studies that have related garnet compositions to sources. This study confirms that detrital garnet compositions are very useful in provenance studies, and can be used to determine protoliths. However, many common garnet compositions are not completely diagnostic of protoliths, although if the garnets have local sources, quite accurate matches can be made. This study shows that compositional data from a large global database can be reliably used to infer protoliths and provenance, which is particularly useful for regions where the sources are unknown, have been removed by erosion, or there are no compositional data from local garnet-bearing rocks. By using a specific sequence of steps and ternary plots, garnets can be matched to sources.

Testing the method with sandstones from northern Borneo shows that the garnet-rich heavy mineral assemblages of the Kudat Formation Tajau Sandstone Member probably came from granitic, blueschist and amphibolite sources. Such rocks are known from Palawan, and some garnet compositions match well. The discovery of the unusual heavy mineral assemblages of the Tajau Sandstone Member was unexpected and confirms the value of garnet as a provenance indicator in certain circumstances.

The project was funded by the SE Asia Research Group at Royal Holloway University of London, supported by an oil company consortium. We thank the Economic Planning Unit Malaysia for permission to conduct field work and Allagu Balaguru for field support. A. Carter and A. Beard at University College London and Birkbeck College, University of London are thanked for their help with analytical work. L. Davies and P. Hall are thanked for help in compiling the garnet compositional data from the published literature. We are grateful to J. Wakabayashi and A. Okay for information on high pressure garnets. We thank E. Garzanti for permission to use the Alpine garnet data and for invaluable discussion of garnet provenance.

## References

ALMASCO, J. N., RODOLFO, K., FULLER, M. & FROST, G. 2000. Paleomagnetism of Palawan, Philippines. *Journal of Asian Earth Sciences*, **18**, 369–389.

ANDÒ, S., MORTON, A. C. & GARZANTI, E. 2013. Metamorphic grade of source rocks revealed by chemical fingerprints of detrital amphibole and garnet. *In:* SCOTT, R. A., SMYTH, H. R., MORTON, A. C. & RICHARDSON, N. (eds) *Sediment Provenance Studies in Hydrocarbon Exploration and Production.* Geological Society, London, Special Publications, 386. First published online June 20, 2013, http://dx.doi.org/10.1144/SP386.5

AUBRECHT, R., MÉRES, Š., SÝKORA, M. & MIKUŠ, T. 2009. Provenance of the detrital garnets and spinels from the Albian sediments of the Czorsztyn Unit (Pieniny Klippen Belt, Western Carpathians, Slovakia). *Geologica Carpathica*, **60**, 463–483.

BouDagher-Fadel, M. 2008. *Evolution and Geological Significance of Larger Benthic Foraminifera*. Developments in Palaeontology and Stratigraphy, Elsevier, Amsterdam, **21**.

Clement, J. F. & Keij, J. 1958. Geology of the Kudat Peninsula, North Borneo (Compilation). GR783, unpublished reports of the Royal Dutch Shell Group of Companies in British Borneo.

Coleman, R. G., Lee, D. E., Beatty, L. B. & Brannock, W. W. 1965. Eclogites and eclogites: their differences and similarities. *Geological Society of America Bulletin*, **76**, 483–508.

Copjakova, R., Sulovsky, P. & Paterson, B. A. 2005. Major and trace elements in pyrope-almandine garnets as sediment provenance indicators of the Lower Carboniferous Culm sediments, Drahany Uplands, Bohemian Massif. *Lithos*, **82**, 51–70.

Deer, W. A., Howie, R. A. & Zussman, J. 1962. *Rock-Forming Minerals, Volume 1 (Ortho and Ring Silicates)*. Longmans, London.

Deer, W. A., Howie, R. A. & Zussman, J. 1966. *An Introduction to the Rock Forming Minerals*. Longmans, London.

Deer, W. A., Howie, R. A. & Zussman, J. (eds) 1997. *Rock-Forming Minerals, Volume 1A, Orthosilicates*. 2nd edn. Geological Society, London.

Encarnación, J. & Mukasa, S. B. 1997. Age and geochemistry of an 'anorogenic' crustal melt and implications for I-type granite petrogenesis. *Lithos*, **42**, 1–13.

Encarnación, J. P., Essene, E. J., Mukasa, S. B. & Hall, C. H. 1995. High pressure and temperature subophiolitic kyanite-garnet amphibolites generated during initiation of mid Tertiary subduction, Palawan, Philippines. *Journal of Petrology*, **36**, 1481–1503.

Evans, B. W. & Trommsdorff, V. 1978. Petrogenesis of garnet lherzolite, Cima di Gagnone, Lepontine Alps. *Earth and Planetary Science Letters*, **40**, 333–348.

Franke, D., Barckhausen, U. et al. 2011. The continent–ocean transition at the southeastern margin of the South China Sea. *Marine and Petroleum Geology*, **28**, 1187–1204.

Galuskina, I. O., Galuskin, E. V., Dzierzanowski, P., Armbruster, T. & Kozanecki, M. 2005. A natural scandian garnet. *American Mineralogist*, **90**, 1688–1692.

Hall, R. 2002. Cenozoic geological and plate tectonic evolution of SE Asia and the SW Pacific: computer-based reconstructions, model and animations. *Journal of Asian Earth Sciences*, **20**, 353–434.

Hallsworth, C. R. & Chisholm, J. I. 2008. Provenance of late Carboniferous sandstones in the Pennine Basin (UK) from combined heavy mineral, garnet geochemistry and palaeocurrent studies. *Sedimentary Geology*, **203**, 196–212.

Haughton, P. D. W. & Farrow, C. M. 1989. Compositional variation in Lower Old Red Sandstone detrital garnets from the Midland Valley of Scotland and the Anglo-Welsh Basin. *Geological Magazine*, **126**, 373–396.

Hutchison, A. R. & Oliver, G. J. H. 1998. Garnet provenance studies, juxtaposition of Laurentian marginal terranes and timing of the Grampian orogeny in Scotland. *Journal of the Geological Society, London*, **155**, 541–550.

Imai, A. & Ozawa, K. 1991. Tectonic implications of the hydrated garnet peridotites near Mt Kinabalu, Sabah, East Malaysia. *Journal of Southeast Asian Earth Sciences*, **6**, 431–446.

Kahoui, M., Kaminsky, F. V., Griffin, W. L., Belousova, E., Mahdjoub, Y. & Chabane, M. 2012. Detrital pyrope garnets from the El Kseibat area, Algeria: a glimpse into the lithospheric mantle beneath the northeastern edge of the West African Craton. *Journal of African Earth Sciences*, **63**, 1–11.

Knittel, U., Hung, C.-H., Yang, T. F. & Iizuka, Y. 2010. Permian arc magmatism in Mindoro, the Philippines: an early Indosinian event in the Palawan Continental Terrane. *Tectonophysics*, **493**, 113–117.

Knowles, C. R. 1987. A BASIC program to recast garnet end-members. *Computers & Geosciences*, **13**, 655–658.

Leong, K. M. 1978. The 'Sabah Blueschist Belt' – a preliminary note. *Warta Geologi, Geological Society of Malaysia Newsletter*, **4**, 45–51.

Lim, P. S. & Heng, Y. E. 1985. Geological Map of Sabah 1:500 000. *In*: *Geological Survey of Malaysia*, Ministry of Lands and Mines, Malaysia.

Locock, A. J. 2008. An Excel spreadsheet to recast analyses of garnet into end-member components, and a synopsis of the crystal chemistry of natural silicate garnets. *Computers & Geosciences*, **34**, 1769–1780.

Mange, M. A. & Morton, A. C. 2007. Geochemistry of heavy minerals. *In*: Mange, M. A. & Wright, D. T. (eds) *Heavy Minerals in Use*. Developments in Sedimentology, Elsevier, Amsterdam, **58**, 345–391.

Martínek, K. & Štolfová, K. 2009. Provenance study of Permian non-marine sandstones and conglomerates of the Krkonoše Piedmont Basin (Czech Republic): exotic marine limestone pebbles, heavy minerals and garnet composition. *Bulletin of Geosciences*, **84**, 555–568.

Méres, Š., Aubrecht, R., Gradziński, M. & Sýkora, M. 2012. High (ultrahigh) pressure metamorphic terrane rocks as the source of the detrital garnets from the Middle Jurassic sands and sandstones of the Cracow Region (Cracow-Wieluń Upland, Poland). *Acta Geologica Polonica*, **2**, 231–246.

MINES AND GEOSCIENCE BUREAU, PHILIPPINES 2011. *Geological map of Northern Palawan*. Mines and Geosciences Bureau, Quezon City, Phillipines.

Morton, A. C. 1985. A new approach to provenance studies–electron-microprobe analysis of detrital garnets from Middle Jurassic sandstones of the northern North Sea. *Sedimentology*, **32**, 553–566.

Morton, A., Hallsworth, C. & Chalton, B. 2004. Garnet compositions in Scottish and Norwegian basement terrains: a framework for interpretation of North Sea sandstone provenance. *Marine and Petroleum Geology*, **21**, 393–410.

Muhling, J. R. & Griffin, B. J. 1991. On recasting garnet analyses into end-member molecules–revisited short note. *Computers & Geosciences*, **17**, 161–170.

Nimis, P. & Trommsdorff, V. 2001. Revised thermobarometry of Alpe Arami and other garnet peridotites from the Central Alps. *Journal of Petrology*, **42**, 103–115.

OLIVER, G. J. H. 2001. Reconstruction of the Grampian episode in Scotland: its place in the Caledonian Orogeny. *Tectonophysics*, **332**, 23–49.

RICKWOOD, P. C. 1968. On recasting analyses of garnet into end-member molecules. *Contributions to Mineralogy and Petrology*, **18**, 175–198.

SABEEN, H. M., RAMANUJAM, N. & MORTON, A. C. 2002. The provenance of garnet: constraints provided by studies of coastal sediments from southern India. *Sedimentary Geology*, **152**, 279–287.

STEPHENS, E. A. 1956. The geology and mineral resources of the Kota Belud and Kudat area, North Borneo. *Malaysia Geological Survey Borneo Region, Memoir*, **5**, 137.

SUGGATE, S. 2011. *Provenance of Neogene sandstones of Subah, northern Borneo*. PhD thesis, Royal Holloway University of London.

TAKEUCHI, M. 1994. Changes in garnet chemistry show a progressive denudation of the source areas for Permian-Jurassic sandstones, Southern Kitakami Terrane, Japan. *Sedimentary Geology*, **93**, 85–105.

TAKEUCHI, M., KAWAI, M. & MATSUZAWA, N. 2008. Detrital garnet and chromian spinel chemistry of Permian clastics in the Renge area, central Japan: implications for the paleogeography of the East Asian continental margin. *Sedimentary Geology*, **212**, 25–39.

TAYLOR, B. & HAYES, D. E. 1983. Origin and history of the South China Sea Basin. *In*: HAYES, D. E. (ed.) *The Tectonic and Geologic Evolution of Southeast Asian Seas and Islands, Part 2. 27*, American Geophysical Union, Washington, DC, Geophysical Monographs Series, 23–56.

TEBBENS, L. A., KROONENBERG, S. B. & VAN DEN BERG, M. W. 1995. Compositional variation of detrital garnets in Quaternary Rhine, Meuse and Baltic River sediments in the Netherlands. *Geologie en Mijnbouw*, **74**, 213–224.

TONGKUL, F. 1994. The geology of northern Sabah, Malaysia – its relationship to the opening of the South China Sea Basin. *Tectonophysics*, **235**, 131–147.

VAN HATTUM, M. 2005. *Provenance of Cenozoic Sedimentary Rocks of Northern Borneo*. PhD thesis, University of London.

VAN HATTUM, M., HALL, R., PICKARD, A. L. & NICHOLS, G. J. 2006. Southeast Asian sediments not from Asia: provenance and geochronology of north Borneo sandstones. *Geology*, **34**, 589–592.

VAN HATTUM, M., HALL, R., PICKARD, A. L. & NICHOLS, G. J. 2013. Provenance and geochronology of Cenozoic sandstones of northern Borneo. *Journal of Asian Earth Sciences*, in press.

WIN, K. S., TAKEUCHI, M., IWAKIRI, S. & TOKIWA, T. 2007. Provenance of detrital garnets from the Yukawa Formation, Yanase district, Shimanto belt, Kii Peninsula, Southwest Japan. *Journal of the Geological Society of Japan*, **113**, 133–145.

# Raman spectroscopy in heavy-mineral studies

SERGIO ANDÒ & EDUARDO GARZANTI*

*Laboratory for Provenance Studies, Università di Milano-Bicocca,
Piazza della Scienza 4, 20126 Milano, Italy*

*Corresponding author (e-mail: eduardo.garzanti@unimib.it)*

**Abstract:** Raman spectroscopy is an innovative tool with tremendous potential, serving as a fundamental complement to a variety of provenance methods including heavy-mineral analysis and detrital geochronology. Because of its accuracy, efficiency and versatility, the results of the Raman technique are indispensable for fully reliable identification of heavy minerals in grain mounts or thin sections. Thorny long-standing problems that cannot be solved confidently with a polarizing microscope alone, such as the determination of opaque and altered heavy minerals, of detrital grains as small as a few microns, or of colourless crystals with uncertain orientation and rounded morphology, can finally be addressed. Although the method can be highly automatized, the full ability and experience of the operator is required to combine Raman data with the optical information obtained under the microscope on the same grains, which is essential for the efficient application of the method in provenance studies. This article provides exemplary Raman spectra useful for the comparison and determination of over 70 different opaque and transparent heavy-mineral species commonly found in sediments, conveying specific information on the genesis of their source rocks, and thus is particularly useful in provenance diagnoses and palaeotectonic reconstructions.

**Supplementary material:** Detailed information on the lasers used and the origin of the analysed minerals is available at http://www.geolsoc.org.uk/SUP18615

It is the scattering of light by atoms and molecules that gives us the light of the sky, the blue colour of the deep sea and the delicate opalescence of large masses of clear ice. (Raman 1928, pp. 368–369)

Raman spectroscopy is a user-friendly, efficient and innovative technique that is the ideal complementary tool to the heavy-mineral analysis carried out under a polarizing microscope. The method is non-destructive, does not require specific preparation, and can be used directly on both heavy-mineral slides and thin sections (Griffith 1969; McMillan 1989; Hope *et al.* 2001; Nasdala *et al.* 2004). It can be used to investigate the very same grains that have been counted under the optical microscope, to identify undetermined grains, including opaque or weathered turbid accessory minerals, and to facilitate the elimination of operator error. Moreover, additional information can be obtained on polymorphs, hydrated minerals, carbonaceous materials, weathering products and solid, liquid and gaseous inclusions within single grains (Mernagh & Liu 1991; Beyssac *et al.* 2002; Stefaniak *et al.* 2006; Bersani *et al.* 2009; Frezzotti *et al.* 2011). The Raman technique can also be used profitably in fission-track analysis

and detrital geochronology in general, to calibrate the crystallographic structure of apatite (Zattin *et al.* 2007) or to assess the degree of metamictization in zircon (Nasdala *et al.* 1996, 2001; Balan *et al.* 2001).

Finally, but perhaps most importantly, Raman spectroscopy allows us to reliably and routinely determine heavy minerals down to a few microns in size (Garzanti *et al.* 2011), and even to identify minerals in clay-rich muds (Villanueva *et al.* 2008), which is not feasible with standard optical techniques. Atmospheric particles can also be identified (Godoi *et al.* 2006; Potgieter-Vermaak *et al.* 2011). It is therefore unrivalled as a tool for the quantitative mineralogical analysis of fine-grained sediments – including suspended load in rivers, distal turbidites and wind-laid loess deposits – but has seldom been used so far because of operational difficulties (Blatt 1985; Totten & Hanan 2007). Raman spectroscopy can be readily and effectively applied in the mineralogical analysis of modern muds as well as ancient mudrocks (Andò *et al.* 2011). A whole frontier is thus opening up in provenance studies, given that mud represents a huge fraction of sediment transport, and mudrocks account for most of the stratigraphic

*From*: SCOTT, R. A., SMYTH, H. R., MORTON, A. C. & RICHARDSON, N. (eds) 2014. *Sediment Provenance Studies in Hydrocarbon Exploration and Production*. Geological Society, London, Special Publications, **386**, 395–412.
First published online June 20, 2013, http://dx.doi.org/10.1144/SP386.2

record (Blatt & Jones 1975) and are less affected by diagenetic dissolution than interlayered sandstones (Blatt & Sutherland 1969).

## Raman mineral analysis (RaMAn)

Raman spectroscopy is an inelastic light-scattering technique widely used to study the vibrational properties of solids, liquids and gases (Nasdala *et al.* 2004; Frezzotti *et al.* 2012). The technique is routinely used in the study of economic ore deposits (Potgieter-Vermak 2007; Wells & Ramanaidou 2012) and in gemology (e.g. to identify whether diamonds have been treated artificially to change their colour and hence value; Bersani & Lottici 2010), but applications are wide ranging, including art and archaeology (Sendova *et al.* 2005; Vandenabeele *et al.* 2007; Aliatis *et al.* 2009), forensic investigations (Palenik 2007) and analysis of commercial products (pharmaceuticals, polymers, thin films, semiconductors, nanomaterials; Burgio & Clark 2001).

When monochromatic radiation strikes the molecules of a sample, the predominant mode of scattering is elastic (photons are scattered with unchanged energy and frequency). Occasionally, the molecule takes up energy from, or gives up energy to, the photons, which are scattered at a different wavelength (Stokes and anti-Stokes Raman scattering). It is this change in wavelength of the scattered photon (only $1 \times 10^{-7}$ of the scattered light is Raman) that provides chemical and structural information.

The Raman technique can be applied to any mineral group, and is based on the identification of diagnostic peaks and comparison with reference spectra. The position of a Raman peak for a single crystal is constant, depending only on crystallographic structure and mineral chemistry. Detrital grains can thus be identified independent of their orientation. Peak intensity, however, is strongly dependent on crystal orientation relative to the polarization direction of the exciting and scattered light. Peak width yields images of crystallinity (e.g. metamictic zircon; Fig. 1). The ratio between the intensities of different peaks may be helpful in the identification of isomorphous series of isotropic minerals (e.g. garnets; Fig. 2).

The position of the peaks is influenced by grain size, as observed experimentally for crystals less than 10 μm in size, where the contribution of surface modes becomes significant, resulting in considerable lowering and increasing width of the Raman bands (Palmeri *et al.* 2009). In quartz, we have observed that such a shift is barely perceptible for grains coarser than *c.* 10 μm, and is limited to $1 \text{ cm}^{-1}$ in the 5–10 μm range. The effect is

therefore negligible for solid crystalline phases of sand to coarse-silt size, but becomes relevant when powders are analysed. The change in the Raman signal for powders is mainly due to a decrease in thermal conductivity; small grains are less able to dissipate the heat induced by the laser, and temperature therefore increases locally (Foucher *et al.* 2012). This problem can be circumvented by reducing the laser power and increasing the time for acquisition of the spectrum.

### The Raman microscope

Raman analyses are performed with a standard optical microscope using either reflected or transmitted light, coupled with a laser source, an optical system to focus the incident beam, and a detector to analyse the scattered photons. The resolution of the spot size is 1 μm with the ×100 objective, and single crystals down to a few microns in size can therefore be determined. Commonly used lasers range from the ultraviolet to the near-infrared, and can be interchanged automatically. Lasers in the 514–785 nm range are best used to avoid problems caused by fluorescence effects. Solid-state lasers require minimum maintenance, and the cost of the spectrometer is affordable. Operations are handled by a computer with suitable software for data storage and processing. Modern instruments automatically find particles and focus on the centre of each one to collect Raman spectra. Accurate mineralogical maps along predefined grids and automatic point-counting can be achieved with imaging functions and a software-governed motorized $x-y$ stage (Haskin *et al.* 1997; Nasdala *et al.* 2012). In this way, survey scans of representative areas in thin sections and heavy-mineral mounts can be obtained efficiently.

### Raman spectra acquisition and mineral identification

The collection of a single spectrum requires an acquisition time ranging from just a second for minerals with good Raman response (quartz, carbonates, zircon) to about a minute for minerals with poor Raman signal (e.g. phyllosilicates). The poor signal of Fe–Ti–Cr oxides can be improved by mounting opaque grains directly on pin stubs to avoid interference of the bonding resin (Fig. 3). Dark and strongly pleochroic minerals (e.g. amphibole) may have a marked fluorescence emission that hides Raman features. Fluorescence effects can be avoided by selecting a suitable laser excitation wavelength.

Because shifting of Raman peaks can be produced by heating of metallic parts during prolonged

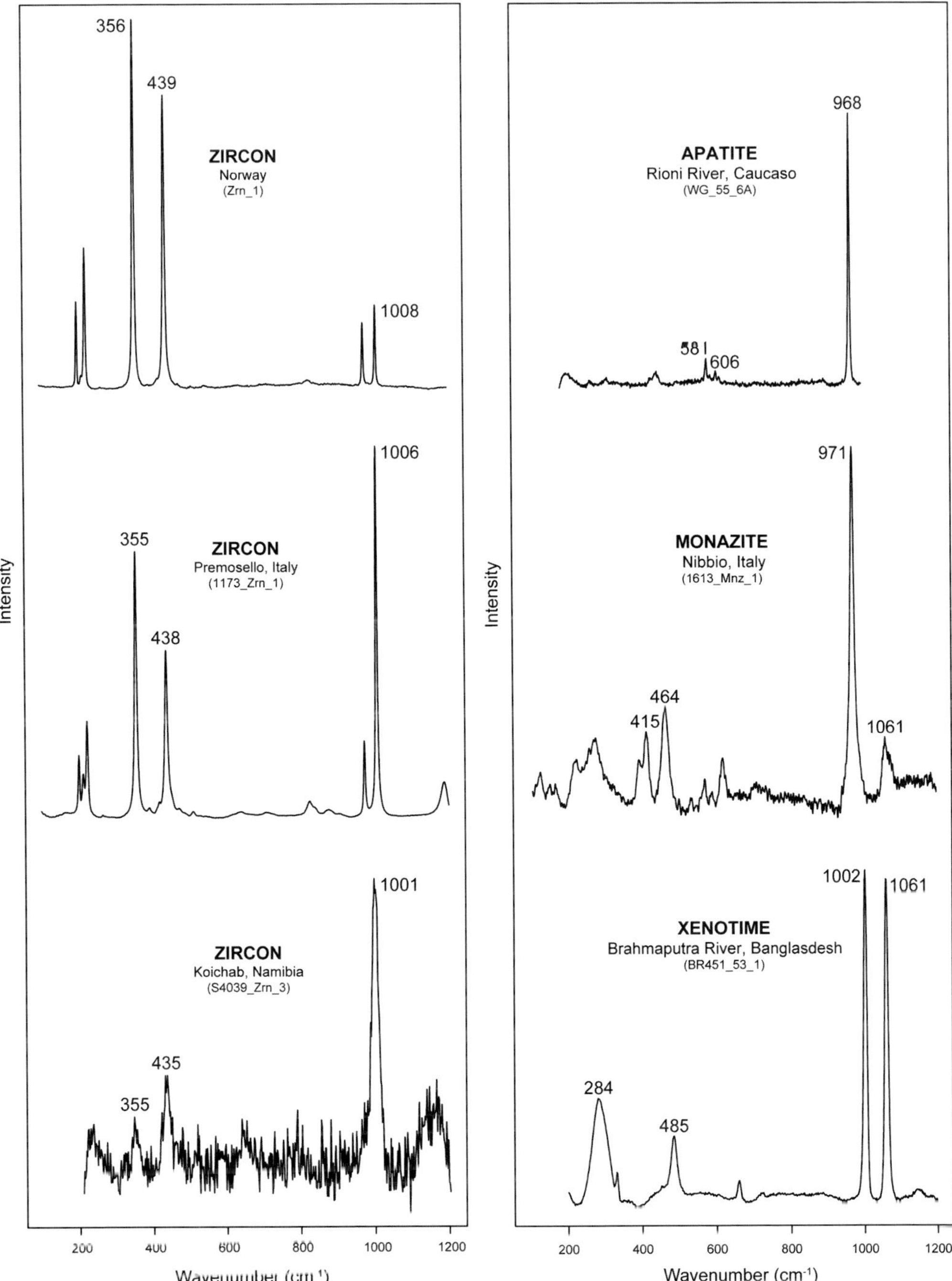


**Fig. 1.** Raman assessment of the degree of metamictization in zircon and discrimination of detrital phosphates. Well-ordered zircons show distinct, narrow internal and external vibrational modes between 200 and 1010 cm$^{-1}$. With increasing radiation damage, all main Raman bands decrease in intensity and become increasingly broader, and show a notable shift towards lower wavenumbers (Nasdala *et al.* 2001). Phosphates are strong Raman scatterers. For all figures, detailed information on the lasers used and the origin of the analysed minerals are provided in the Supplementary material.

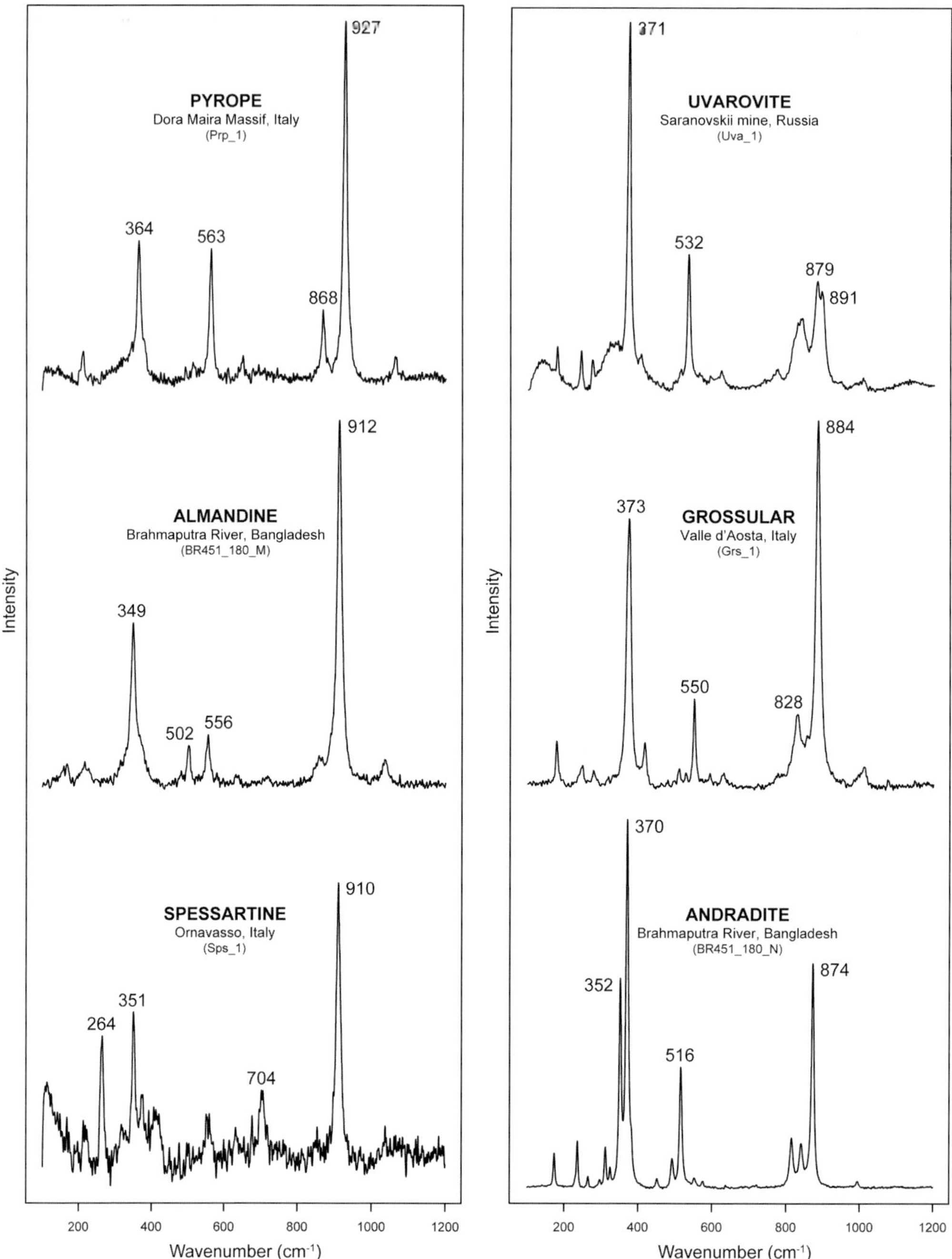



**Fig. 2.** Discriminating within the isomorphous series of garnets. Pyralspite and ugrandite garnets can be distinguished by the position of peaks found at high frequencies and caused by Si–O stretching modes (873–880 cm$^{-1}$ in ugrandites, 907–926 cm$^{-1}$ in pyralspites; Bersani *et al.* 2009).

use, spectra have to be calibrated frequently (e.g. hourly by using, as a reference, the signal from a silicon wafer at 520.7 cm$^{-1}$, or 'plasma lines' in opaque minerals with metallic lustre, or the position of characteristic lines for each laser). Because diagnostic phonon modes are found in the low Raman-

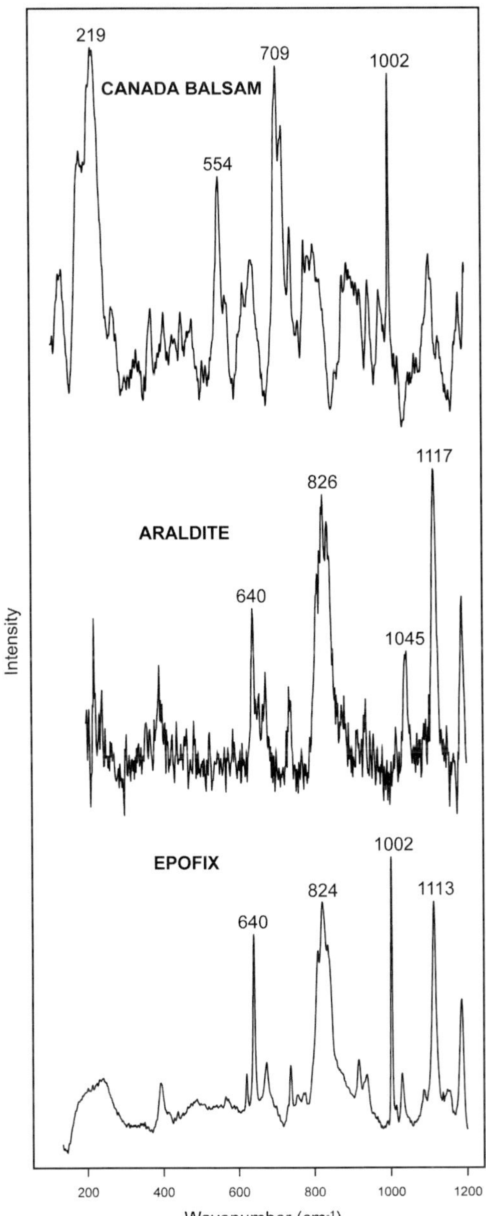



**Fig. 3.** Raman signal of common bonding resins. The spectrum of the bonding resin needs to be subtracted before interpreting the Raman spectrum obtained, because the most intense peaks may interfere with the diagnostic peaks of heavy minerals, particularly in the spectral region above 600 cm⁻¹.

shift region (e.g. 180–520 cm⁻¹ for quartz and feldspars) as well as in the high Raman-shift region (>800 cm⁻¹ for carbonates, phosphates, garnet,

zircon), calibration with a pure calcite spar, end-member garnet or non-metamict zircon is also advised. Modern acquisition systems have a self-calibrating device that operates before each measure and takes the whole spectral region into account. It is advisable to adopt a configuration that allows the collection of signals in low-frequency regions between 50 and 150 cm⁻¹, where diagnostic vibrational modes allowing more accurate identification of mineral species (e.g. feldspars, carbonates) may be present.

Detrital minerals are determined by comparison of the obtained Raman peaks with reference spectra reported in the literature (e.g. Wang *et al.* 2004 for Fe–Ti–Cr oxides; Kolesov & Geiger 1998 and Geiger 2008 for garnets; Freeman *et al.* 2008 for feldspars) and mineralogical databases available online (e.g. www.irug.org, www.rruff.info, Downs 2006; Parma University, www.fis.unipr.it/phevix/ramandb.php; Lyon University, www.ens-lyon.fr/LST/Raman/index.php; Romanian Database, http://rdrs.uaic.ro/). Additional in-house spectra obtained from minerals of precisely known specific chemical composition are commonly and profitably used.

In order to allow the comparison of Raman spectra with optical properties observed under both a stereomicroscope and a transmitted-light microscope, single grains within the counted area of a heavy-mineral slide or thin section must be encircled with a marker pen or identified on a photograph and numbered. Combining optical observations under the microscope and Raman analyses allows us to determine how peak intensities vary with crystal orientation relative to the laser vibrational direction, and to independently identify any anisotropic mineral from the crystal face under scrutiny. In general, prismatic faces elongated parallel to the *c*-axis are most commonly observed in grain mounts, whereas sections randomly cut oblique to the *c*-axis are typically observed in thin sections.

## Operational conditions of our experiments

Non-polarized micro-Raman spectra were obtained in nearly backscattered geometry with two different instruments and four different laser wavelengths at CNR ICVBC Milano, University of Parma and University of Milano-Bicocca.

The 632.8 nm line of a He–Ne laser and the 488 line of an Ar⁺ laser were used for excitation with a Jobin-Yvon Horiba LabRam apparatus, equipped with an Olympus microscope with ×10, ×50 and ×100 objectives and a motorized *x–y* stage. The system was calibrated using the 520.7 cm⁻¹ Raman silicon band before each experimental session. Spectra were generally collected with counting times ranging between 60 and 180 s.

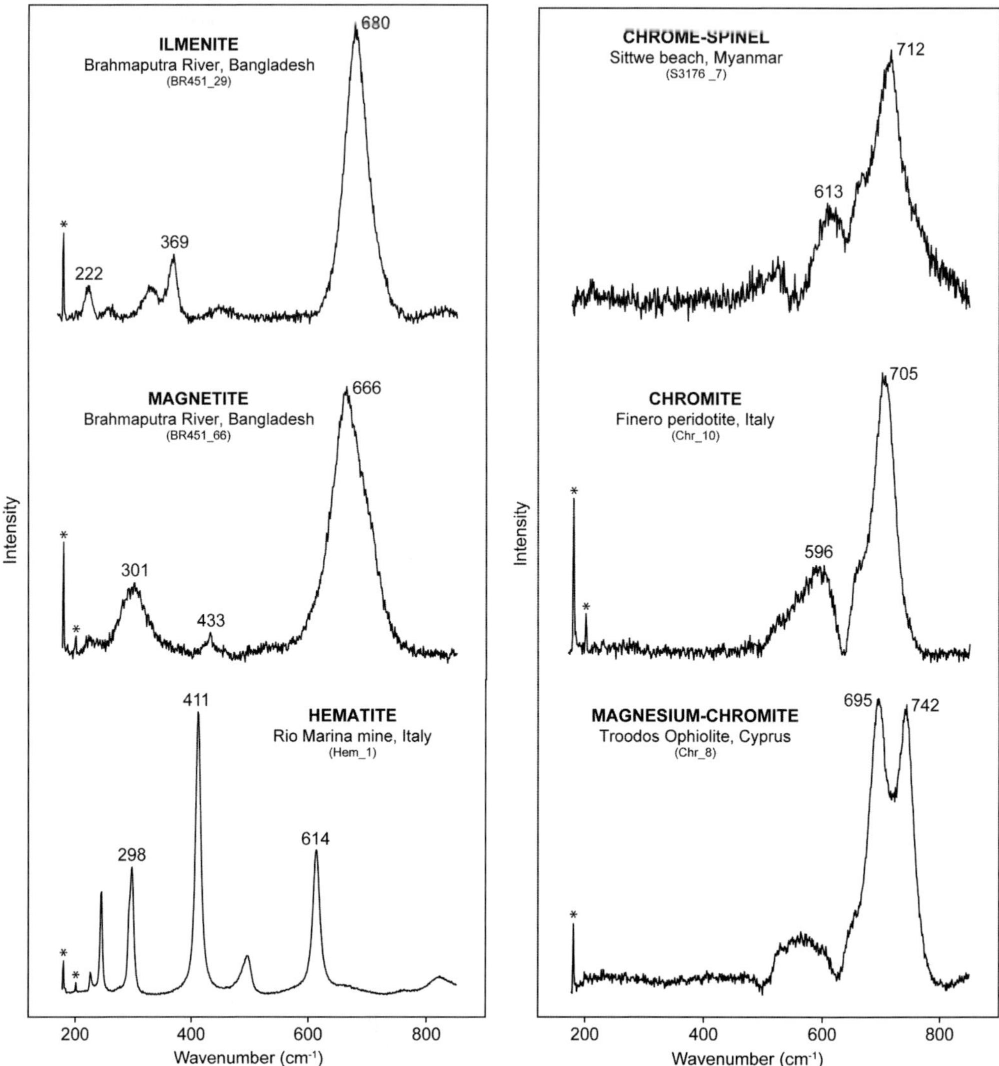


**Fig. 4.** Raman discrimination of opaque and non-opaque Fe–Ti–Cr oxides. The position and shape of the strongest peak in the 660–680 cm$^{-1}$ spectral region ($A_{1g}$ mode) is most useful for assessing Fe$^{3+}$–Ti–Cr–Al substitutions in the magnetite-ulvöspinel, ulvöspinel-chromite and chromite-spinel series. Minor peaks in the range 300–600 cm$^{-1}$ also assist in identification (Wang *et al.* 2004). *He–Ne, 633 nm; plasma lines, 180, 202 cm$^{-1}$.

The 785 nm line of a NIR laser and the 532 nm line of a solid-state laser were used for excitation with a Bruker Senterra dispersive spectrometer, equipped with an Olympus microscope with ×20, ×50 and ×100 objectives and a motorized $x$–$y$ stage. The system was automatically calibrated using the Raman frequencies of an internal neon lamp before each measurement. Spectra were collected with counting times ranging between 90 and 180 s.

For both instruments, the minimum lateral and depth resolution was set to c. 2 μm with a confocal hole. Laser power was controlled by means of a series of density filters in order to avoid heating effects. The wavenumbers of the Raman bands were then determined by fitting with Voigt functions using the LabSpec software, after polynomial background removal. The uncertainty on the measured wavenumbers was estimated at less than 1 cm$^{-1}$, which allowed us to confidently distinguish

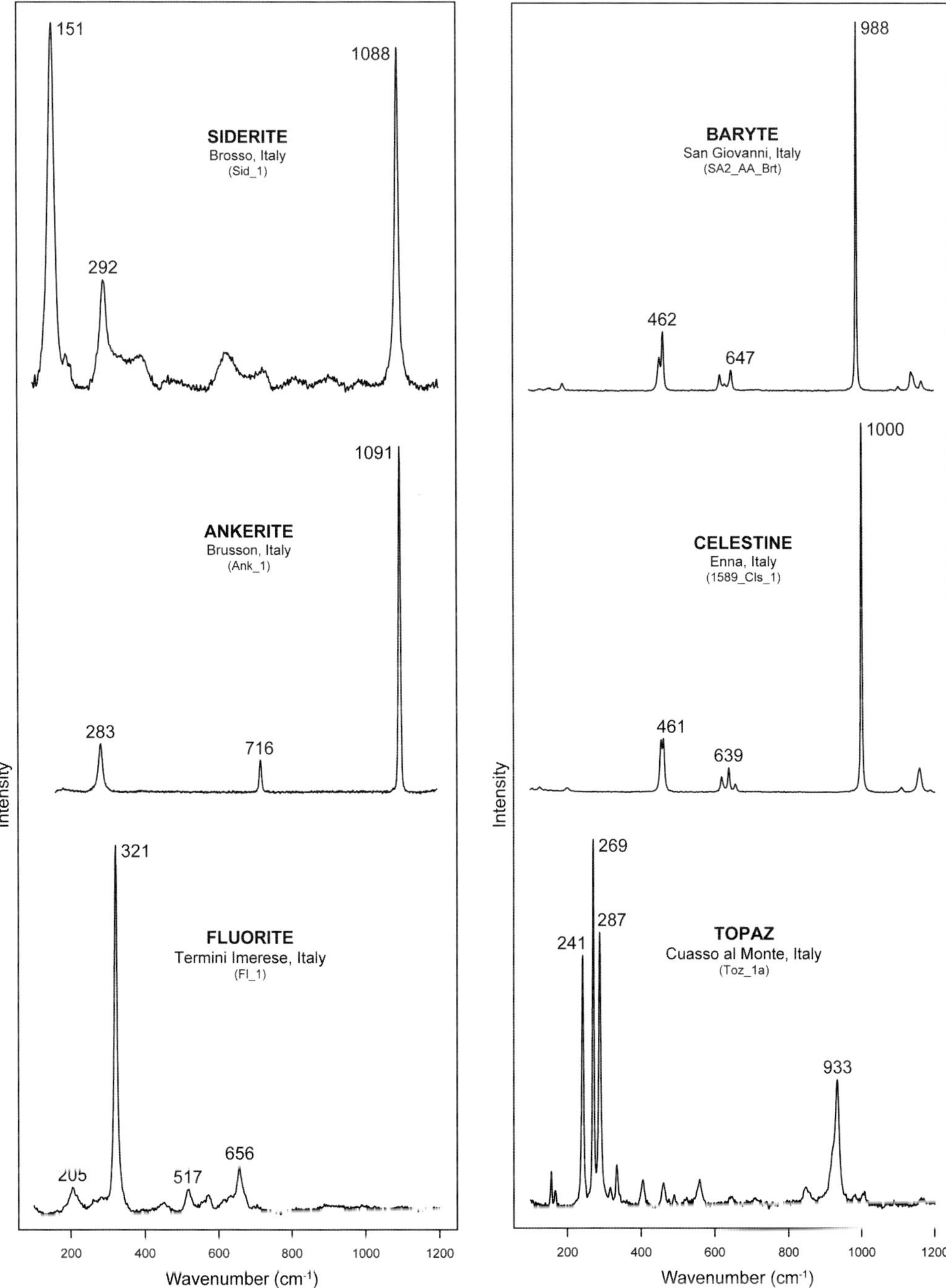



**Fig. 5.** Raman discrimination of heavy carbonates, sulphates and F-bearing minerals. Carbonates and sulphates are strong Raman scatterers. Carbonates display Raman modes at $c$. 1100 cm$^{-1}$ due to the symmetric stretching vibration ($v_1$) of the $CO_3$ group. The positions of their lattice modes over the range 100–350 cm$^{-1}$ are also diagnostic. Sulphates display Raman modes at $c$. 1000 cm$^{-1}$ due to the symmetric stretching vibration ($v_1$) of the $SO_4$ group.

S. ANDÒ & E. GARZANTI

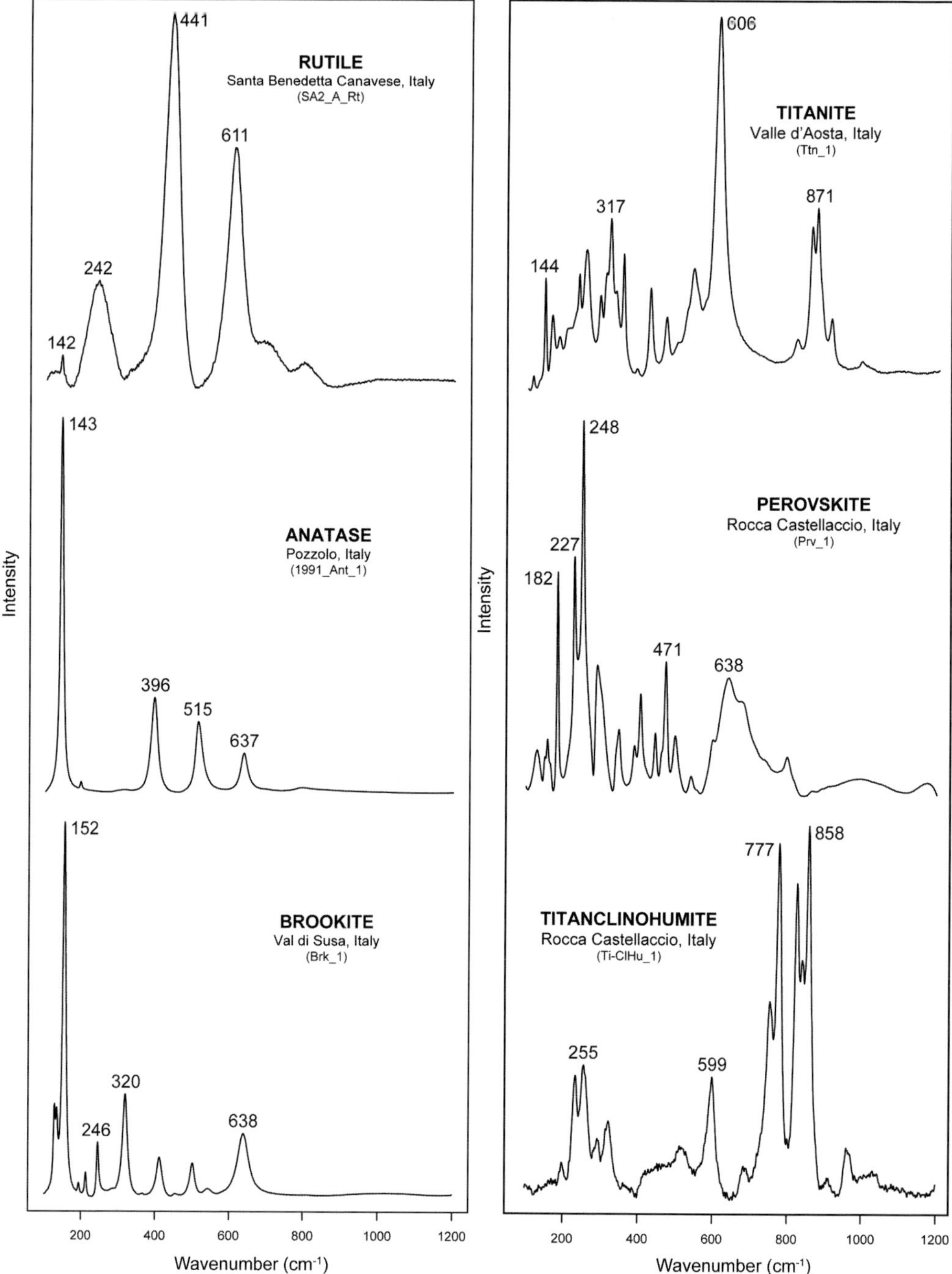


**Fig. 6.** Raman discrimination of Ti-oxide polymorphs and Ti-bearing minerals. Note the marked differences in rutile, anatase and brookite spectra.

mineralogical species even within isomorphous series (e.g. garnets, Bersani *et al.* 2009). Grains were individually selected for analysis, and time and focus were set by the operator. At least three different spots were analysed for each grain, and the best spectrum obtained was selected for identification.

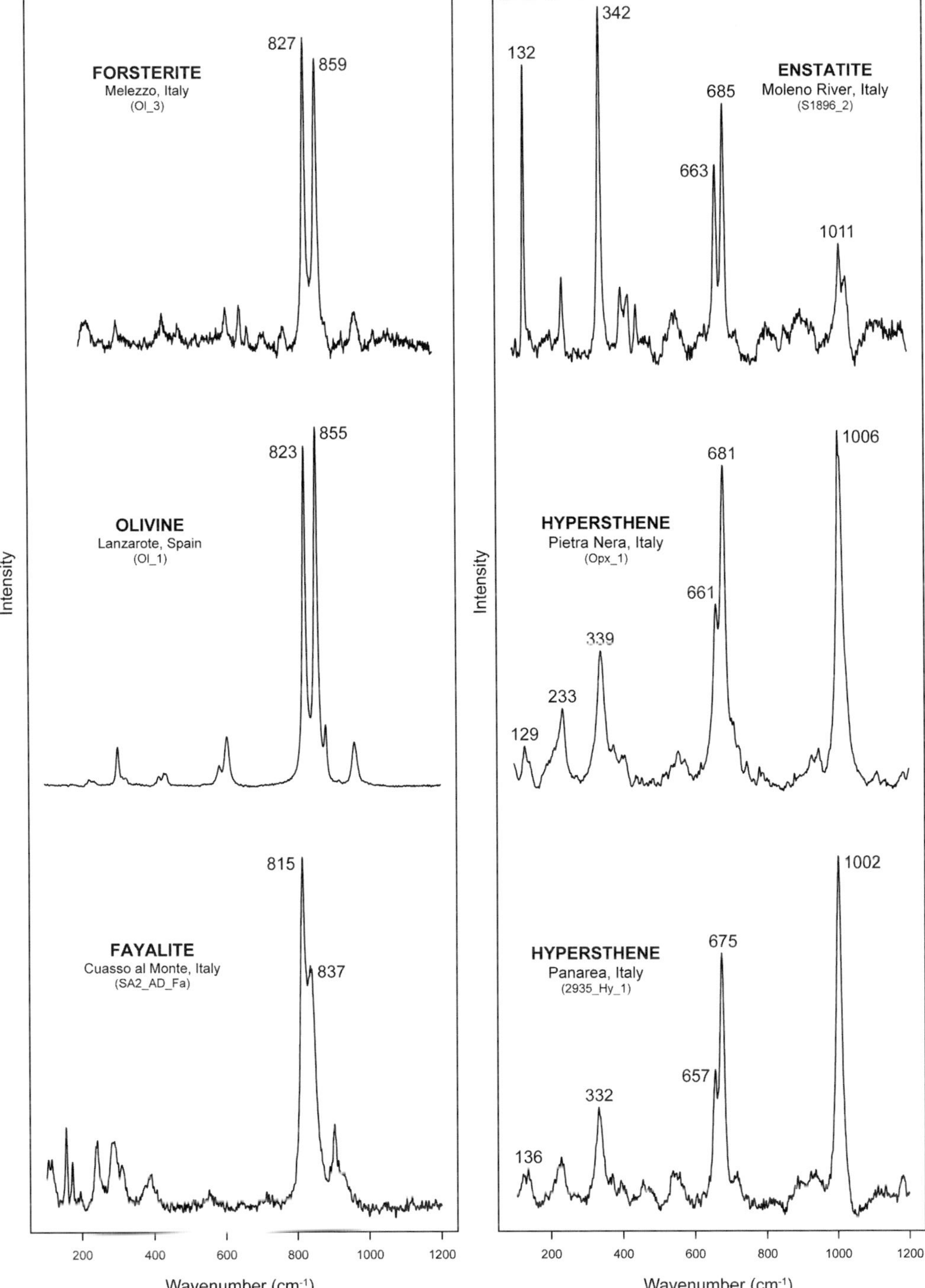



**Fig. 7.** Raman discrimination of olivines and orthopyroxenes. The very good correlation between cation substitution (Mg/Fe + Mg) and the wavenumbers of the $SiO_4$ main bands allows accurate determination of the chemical composition of olivines (Guyot *et al.* 1986; Kubler *et al.* 2006).

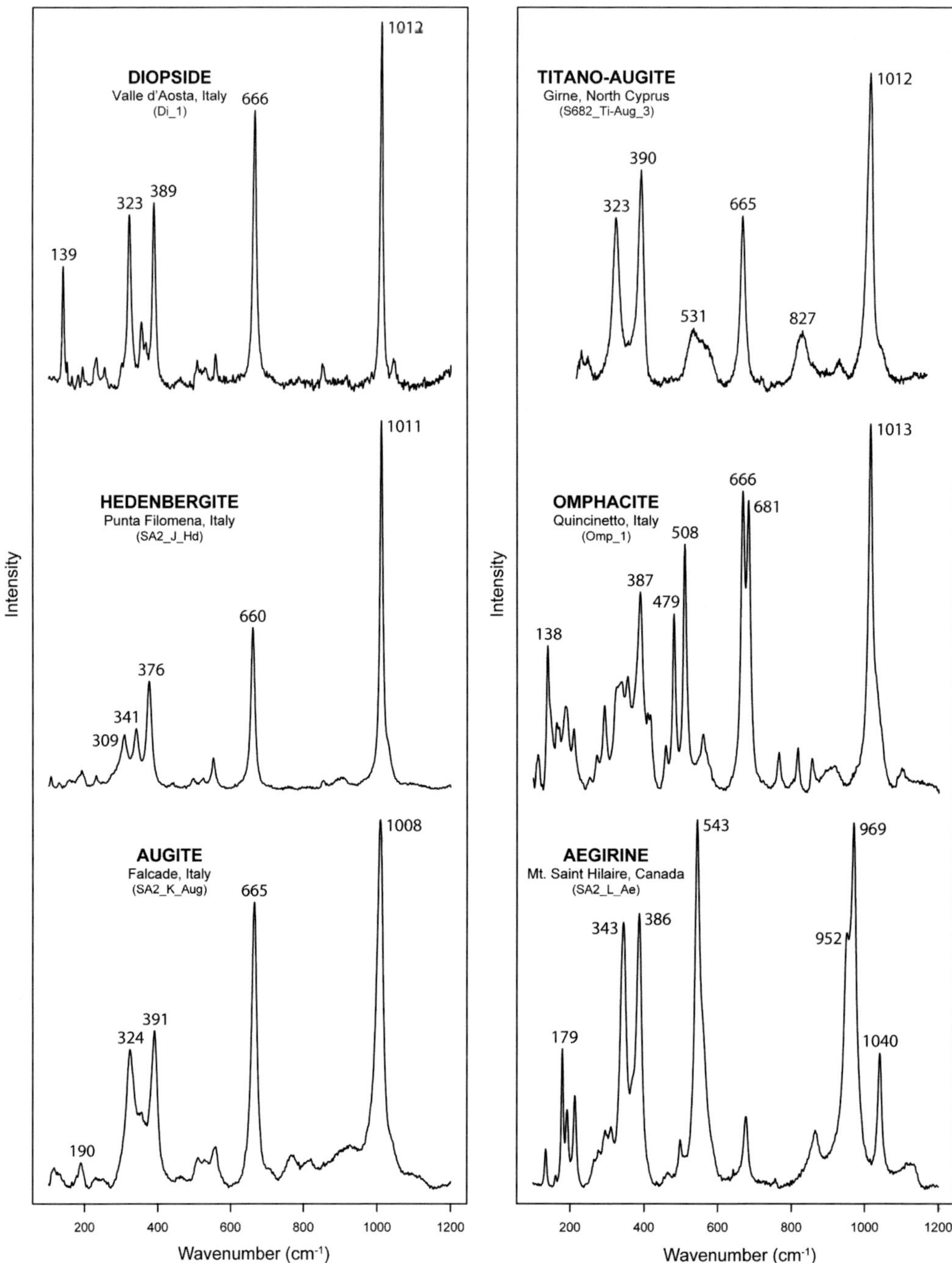



**Fig. 8.** Raman discrimination of clinopyroxenes. Pyroxenes can be distinguished by the number of bands in the $600-700$ cm$^{-1}$ region. Clinopyroxenes have one single band, whereas orthopyroxenes have two bands (Fig. 7).

## RaMAn in provenance studies

Raman mineral analysis allows us to confidently identify dubious detrital minerals independently of their size and orientation. During counting of heavy-mineral mounts or thin sections, grain types that cannot be specifically determined are usually assigned to generic groups (i.e. opaques, turbid

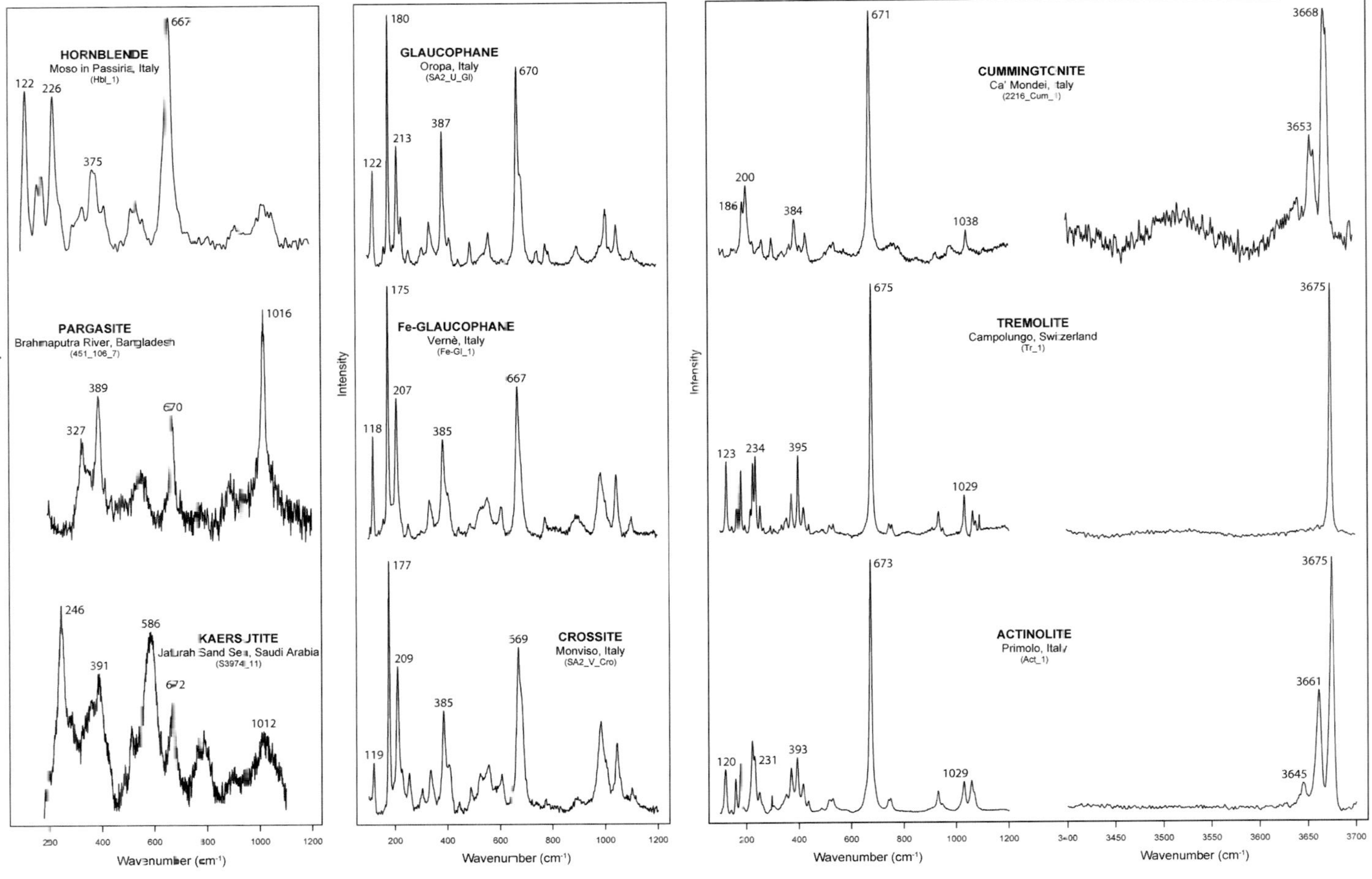



**Fig. 9.** Raman discrimination of amphiboles. Because they are very weak Raman scatterers due to both their dark colour and the low polarizability of their Si–O bonds, identification of amphiboles may be difficult. The diagnostic position and shape of the more intense OH stretching bands (frequencies between 3600 and 3700 $cm^{-1}$) is particularly helpful (as shown for cummingtonite, tremolite and actinolite).

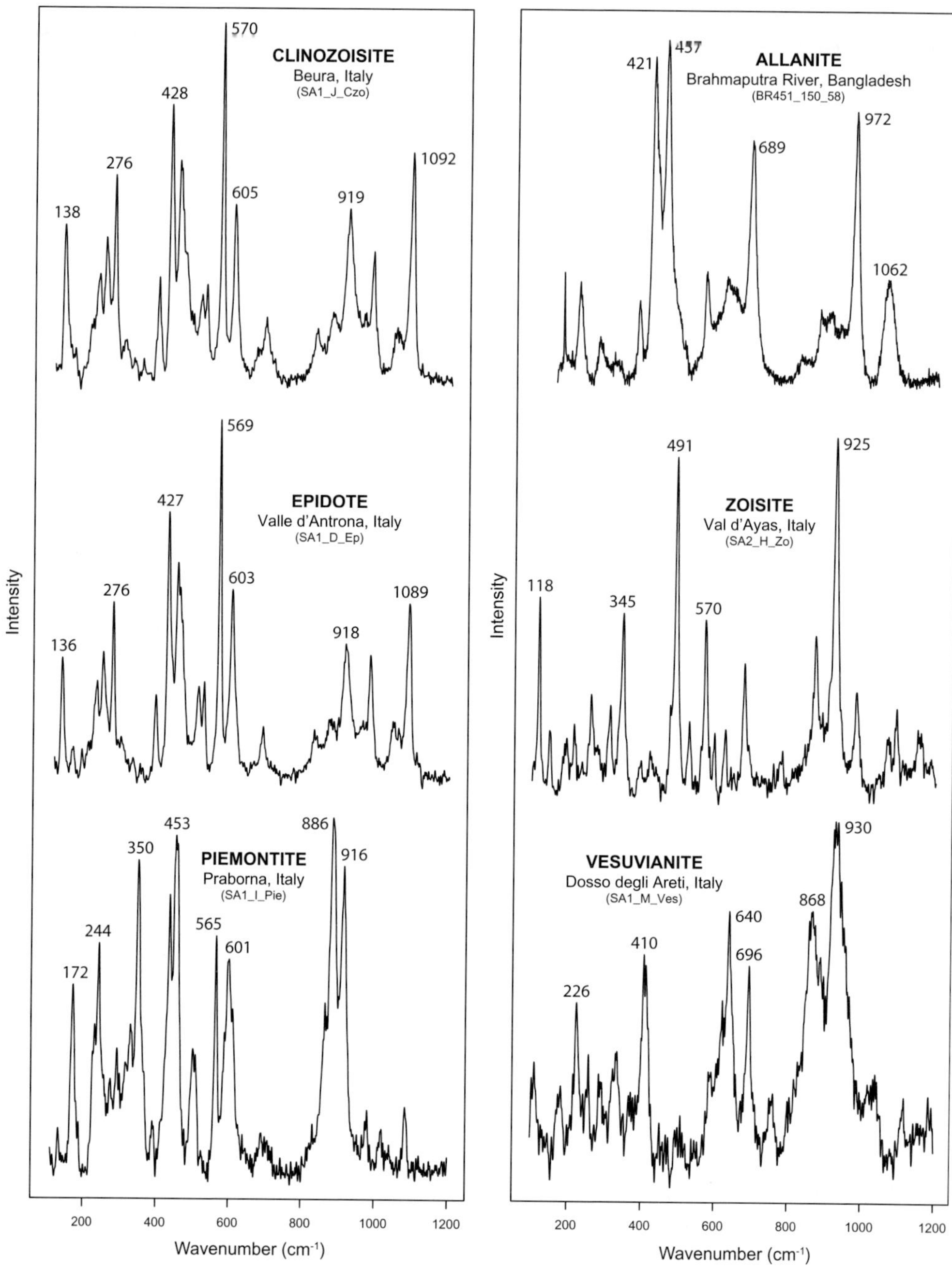



**Fig. 10.** Raman discrimination of epidote-group minerals and vesuvianite. The characteristic abundance of vibrational modes in sorosilicates facilitates distinction from other silicates (e.g. pyroxenes).

grains, alterites, undetermined rock fragments) and generally neglected in subsequent provenance considerations, which may result in a partial and seriously biased description and interpretation of the detrital assemblage.

The RaMAn technique, instead, allows us to identify various types of opaque and transparent accessory minerals as well as altered grains of uncertain origin, and even to quantitatively or semi-quantitatively analyse the chemistry of solid

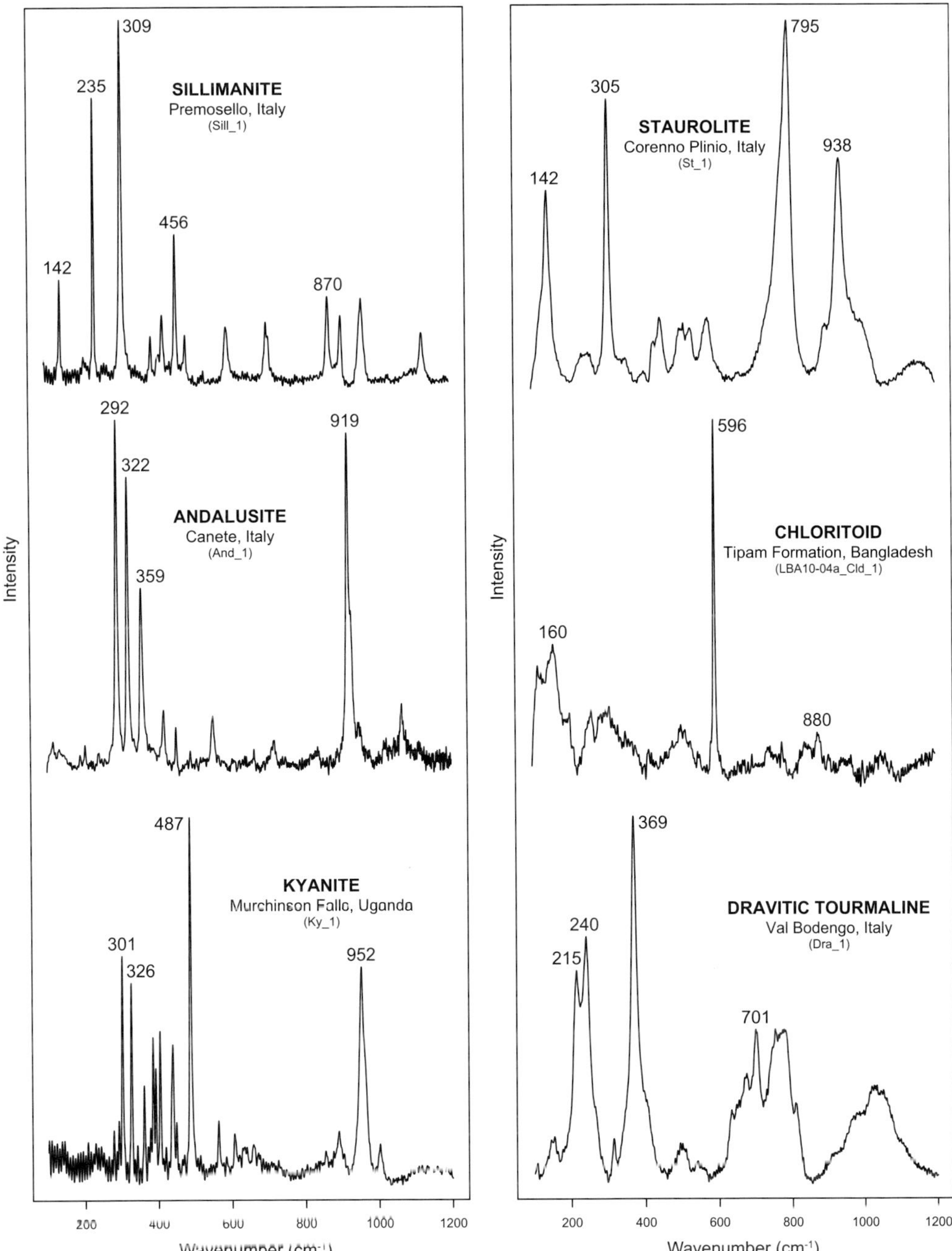


**Fig. 11.** Raman discrimination of minerals from metasedimentary rocks.

solutions and isomorphic series (amphiboles, pyroxenes, garnets, feldspars, carbonates). Among opaque accessory minerals, which may be strongly concentrated in placer deposits of economic interest (Worobiec *et al.* 2007) and are particularly relevant for both provenance (Basu & Molinaroli 1989) and settling-equivalence studies (Garzanti *et al.* 2008), magnetite, ilmenite or chromite can

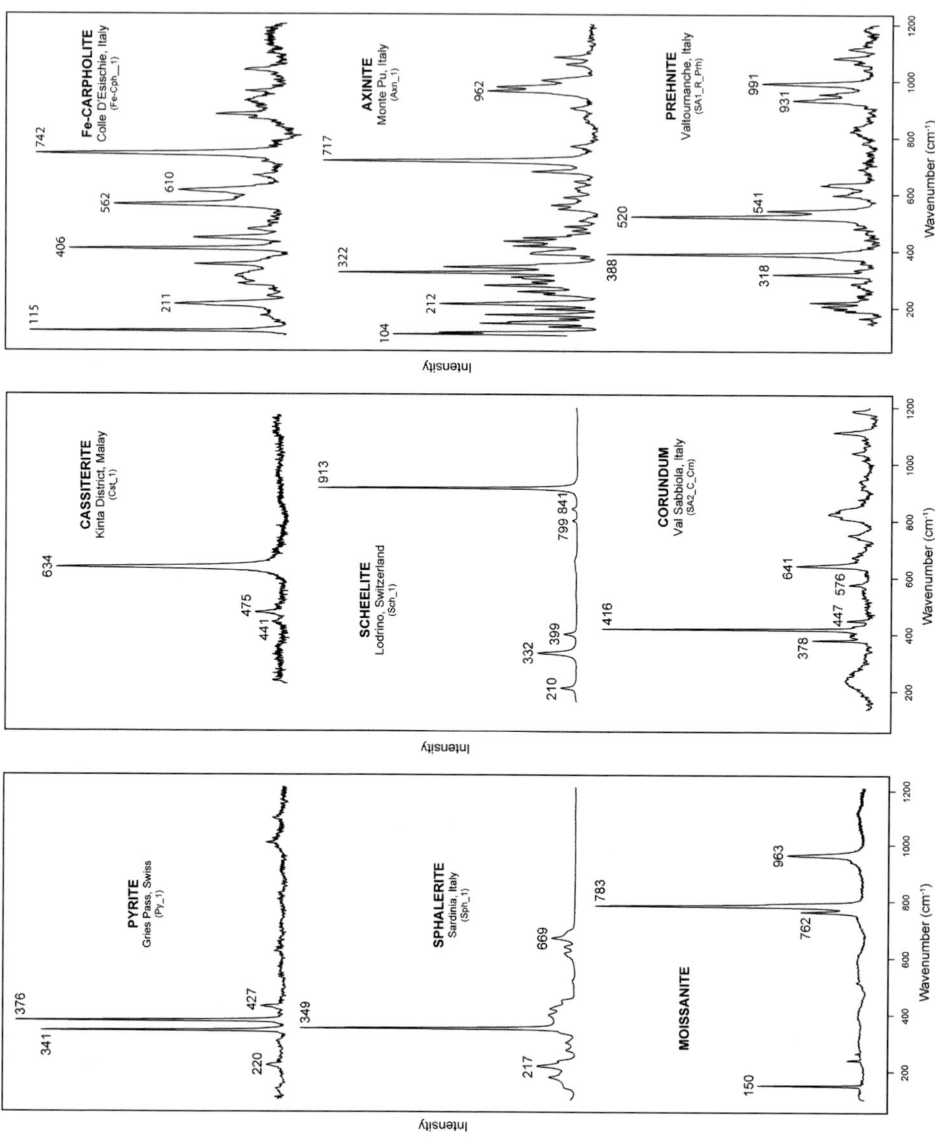



**Fig. 12.** Raman discrimination of other heavy minerals. The rare rock-forming minerals corundum and moissanite (e.g. Di Pierro *et al.* 2003) are commonly used as abrasive paste, and may thus occur in heavy-mineral slides due to anthropic contamination.

be discriminated (Fig. 4; Shebanova & Lazor 2003; Garzanti *et al.* 2010). Among rock fragments of uncertain origin, felsic volcanic grains containing anhedral feldspars can be reliably distinguished from impure chert and cherty mudstones, and limestone from dolostone. Of the detrital feldspars, which if untwinned and <100 μm in size are not invariably easy to identify under a polarizing microscope, we can readily detect Na-plagioclase from Ca-plagioclase, for example, and detrital orthoclase or microcline cores from adularia syntaxial overgrowths. Untwinned feldspars can be easily discriminated from quartz in silt-sized sediments, and sanidine from anorthoclase and plagioclase in volcanic rock fragments. Various types of phyllosilicates useful for provenance (e.g. phlogopite) or environmental diagnosis (e.g. clay minerals, glauconite; Ospitali *et al.* 2008) can be determined, as well as authigenic tectosilicates, phyllosilicates, carbonates or sulphates (Wang *et al.* 2006), which are particularly useful in the reconstruction of post-depositional evolution during burial, lithification and final exhumation (Fig. 5).

## RaMAn in heavy-mineral studies

Raman mineral analysis allows us to efficiently solve a number of problems found routinely in traditional heavy-mineral studies, but that cannot be confidently dealt with under the polarizing microscope, even by an experienced operator. Moreover, with Raman spectroscopy we can readily distinguish single phases and solid solutions of opaque Fe–Ti–Cr oxide grains, found in variable amounts in virtually all heavy mineral separates (Fig. 4). We can distinguish among Ti-oxide polymorphs (Fig. 6), identify altered minerals (particularly common in sediments derived from very low-grade metasedimentary source rocks), and analyse authigenic heavy minerals (which may occur in abundance in both ancient sandstones and in polycyclic sediments derived from them).

Even the identification of unweathered transparent heavy minerals is not invariably straightforward under the optical microscope. This is particularly true for colourless grains that cannot be easily oriented because they lack cleavage and evident crystal faces, for rounded high-relief and high-birefringence grains (e.g. REE phosphates; Silva *et al.* 2006) or for fibrous grains displaying incomplete extinction. Orientation problems are most serious in the study of aeolian sands, where detrital minerals are commonly transformed into quasi-perfect spheres.

With Raman spectroscopy we can rapidly solve the thorny identification problems routinely encountered in optical heavy-mineral analysis, and

reliably distinguish among colourless magnesian minerals (forsterite, enstatite, diopside, tremolite, Mg-cummingtonite, Mg-anthophyllite; Figs 7–9), for instance, or among pale-green weakly ferriferous minerals with intermediate birefringence (e.g. olivine, pigeonite, Fe-diopside, omphacite, actinolite, Fe-anthophyllite, grunerite, Fe-poor epidote; Figs 7–10). RaMAn can thus efficiently solve all long-standing problems in heavy-mineral analysis, including the following (Mange & Maurer 1992):

(1) olivine v. enstatite or prismatic sillimanite (Figs 7 & 11);
(2) anthophyllite v. fibrolitic sillimanite (Figs 9 & 11);
(3) prismatic sillimanite v. apatite or topaz (Figs 1, 5 & 11);
(4) baryte and celestite v. kyanite (Figs 5 & 11);
(5) pale-green augite v. epidote or clinozoisite (Figs 8 & 10);
(6) brown hornblende v. allanite (Figs 9 & 10);
(7) colourless zircon v. monazite, xenotime or titanite (Figs 1 & 6);
(8) metamictic zircon v. rutile, cassiterite or sphalerite (Figs 1, 6 & 12).

In varietal studies (Morton 1985; Mange & Wright 2007), Raman spectroscopy represents a particularly powerful, complementary time-saving technique (Andò *et al.* 2009), because precious chemical information can be obtained rapidly, on many grains, and directly on heavy-mineral mounts without further preparation. We can thus quantitatively or semi-quantitatively assess the chemistry of isomorphic series for detrital spinels, olivines or garnets (Guyot *et al.* 1986; Wang *et al.* 2004; Kübler *et al.* 2006; Bersani *et al.* 2009; Figs 2, 4 & 7), and readily identify silicate varieties such as sodic clinopyroxenes or amphiboles (Figs 8 & 9). Raman spectroscopy also offers an opportunity to:

(1) identify rare grains occasionally encountered in heavy-mineral slides or thin sections;
(2) investigate gaseous, liquid and solid inclusions in single detrital minerals (e.g. apatite, zircon, micas);
(3) calibrate the apatite chemistry and assess the degree of metamictization in zircon;
(4) assess varieties in REE-bearing minerals (e.g. Ce allanite v. La-allanite or Y-allanite),
(5) investigate anisotropic crystal behaviour for physical or chemical processes (e.g. weathering, diagenetic dissolution);
(6) investigate natural photoluminescence in garnet, tourmaline and monazite (Bersani *et al.* 2012; Lenz *et al.* 2012);
(7) distinguish among minerals containing OH⁻ groups (e.g. amphiboles, phyllosilicates; Wang *et al.* 1988; Frezzotti *et al.* 2012).

Moreover, RaMAn can be used to investigate the nature of anisotropic domains (e.g. in garnets), strain effects or the metamictic effect on the lattice structure produced by radiogenic α-decay (Nasdala *et al.* 2012).

## Conclusions

Traditional heavy-mineral analysis under the microscope provides us with a wealth of detailed information on the geology of source areas as well as on physical and chemical processes that took place subsequently during the sedimentary cycle. However, its full efficacy is hampered by limits in the identification of detrital grains that are barely overcome even by operators with ability and experience. Opaque minerals or turbid altered grains are generally impossible to determine under the transmitted-light petrographic microscope, and consequently the information they potentially convey is overlooked and lost. Furthermore, in the case of unweathered, colourless minerals of uncertain orientation, only a tentative determination may be obtained, particularly if grains are well-rounded or silt-sized. The best tool to overcome all of these routinely encountered obstacles is the Raman spectroscope, which, in a few seconds, allows us to obtain, even on the very same grain mounts or thin sections and without any specific additional preparation, a spectrum whose frequencies are diagnostic not only of the different mineralogical species, but also of subtle variations in the chemical composition and crystallographic structure of each detrital mineral. In several cases, Raman spectroscopy can provide virtually the same information on the chemistry of detrital minerals as obtained by more expensive and less user-friendly instruments such as the electron microprobe (Smith 2005; Bersani *et al.* 2009), and in some cases even information that cannot be easily obtained otherwise (e.g. on the nature of solid or fluid inclusions, the degree of metamictization or the occurrence of $OH^-$ groups). The Raman technique is also quite helpful in the identification of minute features of minerals most widely used in detrital geochronology (e.g. zircon, apatite, rutile), allowing us to discriminate among different apatite or zircon types that respond differently to chemical attack or geochronological analysis (Balan *et al.* 2001; Nasdala *et al.* 2001; Zattin *et al.* 2007; Malusà *et al.* 2013).

Last, but not least, Raman spectroscopy allows us to reliably determine detrital minerals as fine-grained or fine silt. Quantitative heavy-mineral investigation of suspended load in rivers, distal turbidite deposits and wind-laid loess becomes possible, with the same precision level reached in provenance analysis of sand-sized sediments (Andò *et al.* 2011). This powerful technique thus provides us with a new user-friendly key to unlock the sedimentary record preserved in ancient clastic wedges, a conspicuous part of which is represented by silty mudrocks. By considering not only the coarse fraction, which is easier to handle and analyse in the laboratory, but the entire sediment flux conveyed through fluvial–deltaic–turbiditic systems, we can better constrain how and to what extent sediment composition is modified physically and chemically during transfer from detrital sources to depositional sinks, and thus calculate sediment budgets more accurately and achieve a deeper understanding of sedimentary processes in general. Borrowing our conclusive words from the seminal article by Raman (1928, p. 376), 'we are obviously only at the fringe of a fascinating new region of experimental research, which promises to throw light on diverse problems. … It all remains to be worked out'!

We heartily thank D. Bersani (University of Parma), I. Aliatis (CNR-ICVBC Milano), A. Paleari (University of Milano-Bicocca) and P. Censi (University of Palermo) for their great help and advice through the years. M. Mange, P. Vignola, P. Gentile, A. Castiglioni, G. Castiglioni, M. Marzini, R. Sala and A. Resentini provided minerals from their collections and technical support. Discussions with, and insightful comments by, L. Nasdala, O. Beyssac and C. Lenz at EGU 2012 are gratefully acknowledged.

## References

ALIATIS, I., BERSANI, D. *ET AL.* 2009. Green pigments of the Pompeian artists' palette. *Spectrochimica Acta, Part A*, **73**, 532–538.

ANDÒ, S., BERSANI, D., VIGNOLA, P. & GARZANTI, E. 2009. Raman spectroscopy as an effective tool for high-resolution heavy-mineral analysis: examples from major Himalayan and Alpine fluvio-deltaic systems. *Spectrochimica Acta, Part A*, **73**, 450–455.

ANDÒ, S., VIGNOLA, P. & GARZANTI, E. 2011. Raman counting: a new method to determine provenance of silt. *Rendiconti Lincei*, **22**, 327–347.

BALAN, E., NEUVILLE, D. R., TROCELLIER, P., FRITSCH, E., MULLER, J. P. & CALAS, G. 2001. Metamictization and chemical durability of detrital zircon. *American Mineralogist*, **86**, 1025–1033.

BASU, A. & MOLINAROLI, E. 1989. Provenance characteristics of detrital opaque Fe–Ti oxide minerals. *Journal of Sedimentary Research*, **59**, 922–934.

BERSANI, D. & LOTTICI, P. P. 2010. Applications of Raman spectroscopy to gemology. *Analytical and Bioanalytical Chemistry*, **397**, 2631–2646.

BERSANI, D., ANDÒ, S., VIGNOLA, P., MOLTIFIORI, G., MARINO, I. G., LOTTICI, P. P. & DIELLA, V. 2009. Micro-Raman spectroscopy as a routine tool for garnet analysis. *Spectrochimica Acta, Part A*, **73**, 484–491.

BERSANI, D., PETRIGLIERI, J. R., ANDÒ, S. & LOTTICI, P. P. 2012. Identification of impurities in gemological materials by means of photoluminescence using micro-Raman apparatus. In: *GeoRaman Xth Meeting*, June 2012, Nancy, France, 11–13.

BEYSSAC, O., ROUZAUD, J. N., GOFFÉ, B., BRUNET, F. & CHOPIN, C. 2002. Graphitization in a high-pressure, low-temperature metamorphic gradient: a Raman microspectroscopy and HRTEM study. *Contributions to Mineralogy and Petrology*, **143**, 19–31.

BLATT, H. 1985. Provenance studies and mudrocks. *Journal of Sedimentary Research*, **55**, 69–75.

BLATT, H. & JONES, R. L. 1975. Proportions of exposed igneous, metamorphic, and sedimentary rocks. *Geological Society of America Bulletin*, **86**, 1085–1088.

BLATT, H. & SUTHERLAND, B. 1969. Intrastratal solution and non-opaque heavy minerals in mudrocks. *Journal of Sedimentary Petrology*, **39**, 591–600.

BURGIO, L. & CLARK, R. J. H. 2001. Library of FT-Raman spectra of pigments, minerals, pigment media and varnishes, and supplement to existing library of Raman spectra of pigments with visible excitation. *Spectrochimica Acta, Part A*, **57**, 1491–1521.

DI PIERRO, S., GNOS, E., GROBETY, B. H., ARMBRUSTER, T., BERNASCONI, S. M. & ULMER, P. 2003. Rock-forming moissanite (natural α-silicon carbide). *American Mineralogist*, **88**, 1817–1821.

DOWNS, R. T. 2006. The RUFF Project: an integrated study of the chemistry, crystallography, Raman and infrared spectroscopy of minerals. *19th General Meeting of the International Mineralogical Association*, 23–28 July 2006, Kobe, Japan, Program and Abstracts, O03–13.

FOUCHER, F., LÓPEZ REYES, G., BOST, N., RULL, F., RÜBMANN, P. & WESTALL, F. 2012. Effect of the crushing process on Raman analyses: consequences for the Mars 2018 mission. *Geophysical Research Abstracts*, **14**, EGU2012–4318.

FREEMAN, J. J., WANG, A., KÜBLER, K. E., JOLLIFF, B. L. & HASKIN, L. A. 2008. Characterization of natural feldspars by Raman spectroscopy for future planetary exploration. *Canadian Mineralogist*, **46**, 1477–1500.

FREZZOTTI, M. L., SELVERSTONE, J., SHARP, Z. D. & COMPAGNONI, R. 2011. Carbonate dissolution during subduction revealed by diamond-bearing rocks from the Alps. *Nature Geoscience*, **4**, 703–706.

FREZZOTTI, M. L., TECCE, F. & CASAGLI, A. 2012. Raman spectroscopy for fluid inclusion analysis. *Journal of Geochemical Exploration*, **112**, 1–20.

GARZANTI, E., ANDÒ, S. & VEZZOLI, G. 2008. Size-density sorting of detrital minerals and grain-size dependence of sediment composition. *Earth Planetary Science Letters*, **273**, 138–151.

GARZANTI, E., ANDÒ, S., FRANCE-LANORD, C., VEZZOLI, G., GALY, V. & NAJMAN, Y. 2010. Mineralogical and chemical variability of fluvial sediments. 1. Bedload sand (Ganga-Brahmaputra, Bangladesh). *Earth Planetary Science Letters*, **299**, 368–381.

GARZANTI, E., ANDÒ, S., FRANCE-LANORD, C., CENSI, P., VIGNOLA, P., GALY, V. & LUPKER, M. 2011. Mineralogical and chemical variability of fluvial sediments 2. Suspended-load silt (Ganga-Brahmaputra, Bangladesh). *Earth Planetary Science Letters*, **302**, 107–120.

GEIGER, C. A. 2008. Silicate garnet: a micro to macroscopic (re)view. *American Mineralogist*, **93**, 360–372.

GODOI, R. H. M., POTGIETER-VERMAAK, S., DE HOOG, J., KAEGI, R. & VAN GRIEKEN, R. 2006. Substrate selection for optimum qualitative and quantitative single atmospheric particles analysis using nano-manipulation, sequential thin-window electron probe X-ray microanalysis and micro Raman spectrometry. *Spectrochimica Acta, Part B*, **61**, 375–388.

GRIFFITH, W. P. 1969. Raman spectroscopy of minerals. *Nature*, **224**, 264–266.

GUYOT, F., BOYER, H., MADON, M., VELDE, B. & POIRIER, J. P. 1986. Comparison of the Raman microprobe spectra of $(Mg,Fe)_2SiO_4$ and $Mg_2GeO_4$ with olivine and spinel structures. *Physics and Chemistry of Minerals*, **13**, 91–95.

HASKIN, L. A., WANG, A., ROCKOW, K. M., JOLLIFF, B. L., KOROTEV, R. L. & VISKUPIC, K. M. 1997. Raman spectroscopy for mineral identification and quantification for *in situ* planetary surface analysis: a point count method. *Journal of Geophysical Research*, **102**, 19 293–19 306.

HOPE, G. A., WOODS, R. & MUNCE, C. G. 2001. Raman microprobe mineral identification. *Minerals Engineering*, **14**, 1565–1577.

KOLESOV, B. A. & GEIGER, C. A. 1998. Raman spectra of silicate garnets. *Physics and Chemistry of Minerals*, **25**, 142–151.

KÜBLER, K. E., JOLLIFF, B. L., WANG, A. & HASKIN, L. A. 2006. Extracting olivine (Fo–Fa) compositions from Raman spectral peak positions. *Geochimica et Cosmochimica Acta*, **70**, 6201–6222.

LENZ, C., PETAUTSCHNIG, C., AKHMADALIEV, S., HANF, D., TALLA, D. & NASDALA, L. 2012. Combined Raman and photoluminescence spectroscopic investigation of He-irradiation effects in monazite. *Geophysical Research Abstracts*, **14**, EGU2012–11138.

MALUSÀ, M. G., CARTER, A., LIMONCELLI, M., GARZANTI, E. & VILLA, I. M. 2013. Bias in detrital zircon geochronology and thermochronometry. *Chemical Geology*, in press.

MANGE, M. A. & MAURER, H. F. W. 1992. *Heavy Minerals in Colour*. Chapman and Hall, London.

MANGE, M. A. & WRIGHT, D. T. 2007. High-resolution heavy mineral analysis (HRHMA): a brief summary. *In*: MANGE, M. A. & WRIGHT, D. T. (eds) *Heavy Minerals in Use*. Elsevier, Amsterdam, Developments in Sedimentology, **58**, 433–436.

MCMILLAN, P. F. 1989. Raman spectroscopy in mineralogy and geochemistry. *Annual Review of Earth and Planetary Sciences*, **17**, 255–283.

MERNAGH, T. P. & LIU, L. 1991. Raman spectra from the $Al_2SiO_5$ polymorphs at high pressure and room temperature. *Physics and Chemistry of Minerals*, **18**, 126–130.

MORTON, A. C. 1985. Heavy minerals in provenance studies. *In*: ZUFFA, G. G. (ed.) *Provenance of Arenites*. Reidel, Dordrecht, NATO ASI Series, **148**, 249–277.

NASDALA, L., PIDGEON, R. T. & WOLF, D. 1996. Heterogeneous metamictization of zircon on microscale. *Geochimica et Cosmochimica Acta*, **60**, 1091–1097.

NASDALA, L., WENZEL, M., VAVRA, G., IRMER, G., WENZEL, T. & KOBER, B. 2001. Metamictisation of natural zircon: accumulation versus thermal annealing of radioactivity-induced damage. *Contributions to Mineralogy and Petrology*, **141**, 125–144.

NASDALA, L., SMITH, D. C., KAINDL, R. & ZIEMANN, M. 2004. Raman spectroscopy: analytical perspectives in mineralogical research. *In*: BERAN, A. & LIBOWITZKY, E. (eds) *Spectroscopic Methods in Mineralogy*. Eötvös University Press, Budapest, European Mineralogical Union Notes in Mineralogy, **6**, 281–343.

NASDALA, L., BEYSSAC, O., SCHOPF, J. W. & BLEISTEINER, B. 2012. Application of Raman-based images in the Earth Sciences. *In*: ZOUBIR, A. (ed.) *Raman Imaging*. Springer Series in Optical Sciences, Berlin, **168**, 145–187.

OSPITALI, F., BERSANI, D., DI LONARDO, G. & LOTTICI, P. P. 2008. 'Green earths': vibrational and elemental characterization of glauconites, celadonites and historical pigments. *Journal of Raman Spectroscopy*, **39**, 1066–1073.

PALENIK, S. 2007. Heavy minerals in forensic science. *In*: MANGE, M. A. & WRIGHT, D. T. (eds) *Heavy Minerals in Use*. Elsevier, Amsterdam, Developments in Sedimentology, **58**, 937–961.

PALMERI, R., FREZZOTTI, M. L., GODARD, G. & DAVIES, R. J. 2009. Pressure-induced incipient amorphization of α-quartz and transition to coesite in an eclogite from Antarctica: a first record and some consequences. *Journal of Metamorphic Geology*, **27**, 685–705.

POTGIETER-VERMAAK, S. 2007. Surface characterization of a heavy mineral sand with micro-Raman spectometry. *6th International Heavy Minerals Conference 'Back to Basics'*. The Southern African Institute of Mining and Metallurgy, Johannesburg, 49–53.

POTGIETER-VERMAAK, S., WOROBIEC, A., DARCHUK, L. & VAN GRIEKEN, R. 2011. Micro-Raman spectroscopy for the analysis of environmental particles. *In*: SIGNORELL, R. & REID, J. P. (eds) *Fundamentals and Applications in Aerosol Spectroscopy*. Taylor & Francis, Boca Raton, **8**, 193–208.

RAMAN, C. V. 1928. A new radiation. *Indian Journal of Physics*, **2**, 387–398.

SENDOVA, M., ZHELYASKOV, V., SCALERA, M. & RAMSEY, M. 2005. Micro-Raman spectroscopic study of pottery fragments from the Lapatsa Tomb, Cyprus, ca 2500 BC. *Journal of Raman Spectroscopy*, **36**, 829–833.

SHEBANOVA, O. N. & LAZOR, P. 2003. Raman spectroscopic study of magnetite ($FeFe_2O_4$): a new assignment for the vibrational spectrum. *Journal of Solid State Chemistry*, **174**, 424–430.

SILVA, E. N., AYALA, A. P., GUEDES, I., PASCHOAL, C. W. A., MOREIRA, R. L., LOONG, C. K. & BOATNER, L. A. 2006. Vibrational spectra of monazite-type rare-earth orthophosphates. *Optical Materials*, **29**, 224–230.

SMITH, D. C. 2005. The RAMANITA method for non-destructive and *in situ* semi-quantitative chemical analysis of mineral solid-solutions by multidimensional calibration of Raman wavenumber shifts. *Spectrochimica Acta, Part A*, **61**, 2299–2314.

STEFANIAK, E. A., WOROBIEC, A., POTGIETER-VERMAAK, S., ALSECZ, A., TÖRÖK, S. & VAN GRIEKEN, R. 2006. Molecular and elemental characterisation of mineral particles by means of parallel micro-Raman spectrometry and scanning electron microscopy/energy dispersive X-ray analysis. *Spectrochimica Acta, Part B*, **61**, 824–830.

TOTTEN, M. W. & HANAN, M. A. 2007. Heavy minerals in shales. *In*: MANGE, M. A. & WRIGHT, D. T. (eds) *Heavy Minerals in Use*. Elsevier, Amsterdam, Developments in Sedimentology Series, **58**, 323–341.

VANDENABEELE, P., EDWARDS, H. G. M. & MOENS, L. 2007. A decade of Raman spectroscopy in art and archaeology. *Chemical Reviews*, **107**, 675–686.

VILLANUEVA, U., RAPOSO, J. C., CASTRO, K., DE DIEGO, A., ARANA, G. & MADARIAGA, J. M. 2008. Raman spectroscopy speciation of natural and anthropogenic solid phases in river and estuarine sediments with appreciable amount of clay and organic matter. *Journal of Raman Spectroscopy*, **39**, 1195–1203.

WANG, A., DHAMELINCOURT, P. & TURRELL, G. 1988. Raman microspectroscopic study of the cation distribution in amphiboles. *Applied Spectroscopy*, **42**, 1441–1450.

WANG, A., KÜBLER, K. E., JOLLIFF, B. L. & HASKIN, L. A. 2004. Raman spectroscopy of Fe–Ti–Cr-oxides, case study: martian meteorite EETA79001. *American Mineralogist*, **89**, 665–680.

WANG, A., FREEMAN, J. J., JOLLIFF, B. L. & MING-CHOU, I. 2006. Sulfates on Mars: a systematic Raman spectroscopic study of hydration states of magnesium sulfates. *Geochimica et Cosmochimica Acta*, **70**, 6118–6135.

WELLS, M. A. & RAMANAIDOU, E. R. 2012. Raman spectroscopic characterisation of Australian banded iron formation and iron ore. *Geophysical Research Abstracts*, **14**, EGU2012–6847.

WOROBIEC, A., STEFANIAK, E. A., POTGIETER-VERMAAK, S., SAWLOWICZ, Z., SPOLNIK, Z. & VAN GRIEKEN, R. 2007. Characterisation of concentrates of heavy mineral sands by micro-Raman spectrometry and CC-SEM/EDX with HCA. *Applied Geochemistry*, **22**, 2078–2085.

ZATTIN, M., BERSANI, D. & CARTER, A. 2007. Raman microspectroscopy: a non-destructive tool for routine calibration of apatite crystallographic structure for fission-track analyses. *Chemical Geology*, **240**, 197–204.

# Index

Page numbers in *italic* denote figures. Page numbers in **bold** denote tables.